T0137020

Lecture Notes in Computer Science   11540

Commenced Publication in 1973
Founding and Former Series Editors:
Gerhard Goos, Juris Hartmanis, and Jan van Leeuwen

More information about this series at http://www.springer.com/series/7407

João M. F. Rodrigues · Pedro J. S. Cardoso ·
Jânio Monteiro · Roberto Lam ·
Valeria V. Krzhizhanovskaya ·
Michael H. Lees · Jack J. Dongarra ·
Peter M. A. Sloot (Eds.)

Computational Science – ICCS 2019

19th International Conference
Faro, Portugal, June 12–14, 2019
Proceedings, Part V

 Springer

Editors

João M. F. Rodrigues 🆔
University of Algarve
Faro, Portugal

Pedro J. S. Cardoso 🆔
University of Algarve
Faro, Portugal

Jânio Monteiro 🆔
University of Algarve
Faro, Portugal

Roberto Lam 🆔
University of Algarve
Faro, Portugal

Valeria V. Krzhizhanovskaya 🆔
University of Amsterdam
Amsterdam, The Netherlands

Michael H. Lees
University of Amsterdam
Amsterdam, The Netherlands

Jack J. Dongarra 🆔
University of Tennessee at Knoxville
Knoxville, TN, USA

Peter M. A. Sloot 🆔
University of Amsterdam
Amsterdam, The Netherlands

ISSN 0302-9743   ISSN 1611-3349   (electronic)
Lecture Notes in Computer Science
ISBN 978-3-030-22749-4   ISBN 978-3-030-22750-0   (eBook)
https://doi.org/10.1007/978-3-030-22750-0

LNCS Sublibrary: SL1 – Theoretical Computer Science and General Issues

This Springer imprint is published by the registered company Springer Nature Switzerland AG
The registered company address is: Gewerbestrasse 11, 6330 Cham, Switzerland

Preface

Welcome to the 19th Annual International Conference on Computational Science (ICCS - https://www.iccs-meeting.org/iccs2019/), held during June 12–14, 2019, in Faro, Algarve, Portugal. Located at the southern end of Portugal, Algarve is a well-known touristic haven. Besides some of the best and most beautiful beaches in the entire world, with fine sand and crystal-clear water, Algarve also offers amazing natural landscapes, a rich folk heritage, and a healthy gastronomy that can be enjoyed throughout the whole year, attracting millions of foreign and national tourists. ICCS 2019 was jointly organized by the University of Algarve, the University of Amsterdam, NTU Singapore, and the University of Tennessee.

The International Conference on Computational Science is an annual conference that brings together researchers and scientists from mathematics and computer science as basic computing disciplines, as well as researchers from various application areas who are pioneering computational methods in sciences such as physics, chemistry, life sciences, engineering, arts, and humanitarian fields, to discuss problems and solutions in the area, to identify new issues, and to shape future directions for research.

Since its inception in 2001, ICCS has attracted an increasingly higher quality and numbers of attendees and papers, and this year was no exception, with over 350 participants. The proceedings series have become a major intellectual resource for computational science researchers, defining and advancing the state of the art in this field.

ICCS 2019 in Faro was the 19th in this series of highly successful conferences. For the previous 18 meetings, see: http://www.iccs-meeting.org/iccs2019/previous-iccs/.

The theme for ICCS 2019 was "Computational Science in the Interconnected World," to highlight the role of computational science in an increasingly interconnected world. This conference was a unique event focusing on recent developments in: scalable scientific algorithms; advanced software tools; computational grids; advanced numerical methods; and novel application areas. These innovative novel models, algorithms, and tools drive new science through efficient application in areas such as physical systems, computational and systems biology, environmental systems, finance, and others.

ICCS is well known for its excellent line-up of keynote speakers. The keynotes for 2019 were:

- Tiziana Di Matteo, King's College London, UK
- Teresa Galvão, University of Porto/INESC TEC, Portugal
- Douglas Kothe, Exascale Computing Project, USA
- James Moore, Imperial College London, UK
- Robert Panoff, The Shodor Education Foundation, USA
- Xiaoxiang Zhu, Technical University of Munich, Germany

This year we had 573 submissions (228 submissions to the main track and 345 to the workshops). In the main track, 65 full papers were accepted (28%); in the workshops, 168 full papers (49%). The high acceptance rate in the workshops is explained by the nature of these thematic sessions, where many experts in a particular field are personally invited by workshop organizers to participate in their sessions.

ICCS relies strongly on the vital contributions of our workshop organizers to attract high-quality papers in many subject areas. We would like to thank all committee members for the main track and workshops for their contribution to ensure a high standard for the accepted papers. We would also like to thank Springer, Elsevier, and Intellegibilis for their support. Finally, we very much appreciate all the local Organizing Committee members for their hard work to prepare this conference.

We are proud to note that ICCS is an A-rank conference in the CORE classification.

June 2019

João M. F. Rodrigues
Pedro J. S. Cardoso
Jânio Monteiro
Roberto Lam
Valeria V. Krzhizhanovskaya
Michael Lees
Jack J. Dongarra
Peter M. A. Sloot

Organization

Workshops and Organizers

Advanced Modelling Techniques for Environmental Sciences – AMES

Jens Weismüller
Dieter Kranzlmüller
Maximilian Hoeb
Jan Schmidt

Advances in High-Performance Computational Earth Sciences: Applications and Frameworks – IHPCES

Takashi Shimokawabe
Kohei Fujita
Dominik Bartuschat

Agent-Based Simulations, Adaptive Algorithms, and Solvers – ABS-AAS

Maciej Paszynski
Quanling Deng
David Pardo
Robert Schaefer
Victor Calo

Applications of Matrix Methods in Artificial Intelligence and Machine Learning – AMAIML

Kourosh Modarresi

Architecture, Languages, Compilation, and Hardware Support for Emerging and Heterogeneous Systems – ALCHEMY

Stéphane Louise
Löic Cudennec
Camille Coti
Vianney Lapotre
José Flich Cardo
Henri-Pierre Charles

Biomedical and Bioinformatics Challenges for Computer Science – BBC

Mario Cannataro
Giuseppe Agapito
Mauro Castelli

Riccardo Dondi
Rodrigo Weber dos Santos
Italo Zoppis

Classifier Learning from Difficult Data – CLD^2

Michał Woźniak
Bartosz Krawczyk
Paweł Ksieniewicz

Computational Finance and Business Intelligence – CFBI

Yong Shi
Yingjie Tian

Computational Methods in Smart Agriculture – CMSA

Andrew Lewis

Computational Optimization, Modelling, and Simulation – COMS

Xin-She Yang
Slawomir Koziel
Leifur Leifsson

Computational Science in IoT and Smart Systems – IoTSS

Vaidy Sunderam

Data-Driven Computational Sciences – DDCS

Craig Douglas

Machine Learning and Data Assimilation for Dynamical Systems – MLDADS

Rossella Arcucci
Boumediene Hamzi
Yi-Ke Guo

Marine Computing in the Interconnected World for the Benefit of Society – MarineComp

Flávio Martins
Ioana Popescu
João Janeiro
Ramiro Neves
Marcos Mateus

Multiscale Modelling and Simulation – MMS

Derek Groen
Lin Gan
Stefano Casarin
Alfons Hoekstra
Bartosz Bosak

Simulations of Flow and Transport: Modeling, Algorithms, and Computation – SOFTMAC

Shuyu Sun
Jingfa Li
James Liu

Smart Systems: Bringing Together Computer Vision, Sensor Networks, and Machine Learning – SmartSys

João M. F. Rodrigues
Pedro J. S. Cardoso
Jânio Monteiro
Roberto Lam

Solving Problems with Uncertainties – SPU

Vassil Alexandrov

Teaching Computational Science – WTCS

Angela Shiflet
Evguenia Alexandrova
Alfredo Tirado-Ramos

Tools for Program Development and Analysis in Computational Science – TOOLS

Andreas Knüpfer
Karl Fürlinger

Programme Committee and Reviewers

Ahmad Abdelfattah
Eyad Abed
Markus Abel
Laith Abualigah
Giuseppe Agapito
Giovanni Agosta
Ram Akella

Elisabete Alberdi
Marco Aldinucci
Luis Alexandre
Vassil Alexandrov
Evguenia Alexandrova
Victor Allombert
Saad Alowayyed

Stanislaw
  Ambroszkiewicz
Ioannis Anagnostou
Philipp Andelfinger
Michael Antolovich
Hartwig Anzt
Hideo Aochi

Rossella Arcucci
Tomasz Arodz
Kamesh Arumugam
Luiz Assad
Victor Azizi Tarksalooyeh
Bartosz Balis
Krzysztof Banas
João Barroso
Dominik Bartuschat
Daniel Becker
Jörn Behrens
Adrian Bekasiewicz
Gebrail Bekdas
Stefano Beretta
Daniel Berrar
John Betts
Sanjukta Bhowmick
Bartosz Bosak
Isabel Sofia Brito
Kris Bubendorfer
Jérémy Buisson
Aleksander Byrski
Cristiano Cabrita
Xing Cai
Barbara Calabrese
Carlos Calafate
Carlos Cambra
Mario Cannataro
Alberto Cano
Paul M. Carpenter
Stefano Casarin
Manuel Castañón-Puga
Mauro Castelli
Jeronimo Castrillon
Eduardo Cesar
Patrikakis Charalampos
Henri-Pierre Charles
Zhensong Chen
Siew Ann Cheong
Andrei Chernykh
Lock-Yue Chew
Su Fong Chien
Sung-Bae Cho
Bastien Chopard
Stephane Chretien
Svetlana Chuprina

Florina M. Ciorba
Noelia Correia
Adriano Cortes
Ana Cortes
Jose Alfredo F. Costa
Enrique
    Costa-Montenegro
David Coster
Camille Coti
Carlos Cotta
Helene Coullon
Daan Crommelin
Attila Csikasz-Nagy
Loïc Cudennec
Javier Cuenca
Yifeng Cui
António Cunha
Ben Czaja
Pawel Czarnul
Bhaskar Dasgupta
Susumu Date
Quanling Deng
Nilanjan Dey
Ergin Dinc
Minh Ngoc Dinh
Sam Dobbs
Riccardo Dondi
Ruggero Donida Labati
Goncalo dos-Reis
Craig Douglas
Aleksandar Dragojevic
Rafal Drezewski
Niels Drost
Hans du Buf
Vitor Duarte
Richard Duro
Pritha Dutta
Sean Elliot
Nahid Emad
Christian Engelmann
Qinwei Fan
Fangxin Fang
Antonino Fiannaca
Christos
    Filelis-Papadopoulos
José Flich Cardo

Yves Fomekong Nanfack
Vincent Fortuin
Ruy Freitas Reis
Karl Frinkle
Karl Fuerlinger
Kohei Fujita
Wlodzimierz Funika
Takashi Furumura
Mohamed Medhat Gaber
Jan Gairing
David Gal
Marco Gallieri
Teresa Galvão
Lin Gan
Luis Garcia-Castillo
Delia Garijo
Frédéric Gava
Don Gaydon
Zong-Woo Geem
Alex Gerbessiotis
Konstantinos
    Giannoutakis
Judit Gimenez
Domingo Gimenez
Guy Gogniat
Ivo Gonçalves
Yuriy Gorbachev
Pawel Gorecki
Michael Gowanlock
Manuel Graña
George Gravvanis
Marilaure Gregoire
Derek Groen
Lutz Gross
Sophia
    Grundner-Culemann
Pedro Guerreiro
Kun Guo
Xiaohu Guo
Piotr Gurgul
Pietro Hiram Guzzi
Panagiotis Hadjidoukas
Mohamed Hamada
Boumediene Hamzi
Masatoshi Hanai
Quillon Harpham

Daniela Piccioni
Juan C. Pichel
Anna
  Pietrenko-Dabrowska
Laércio L. Pilla
Armando Pinto
Tomasz Piontek
Erwan Piriou
Yuri Pirola
Nadia Pisanti
Antoniu Pop
Ioana Popescu
Mario Porrmann
Cristina Portales
Roland Potthast
Ela Pustulka-Hunt
Vladimir Puzyrev
Alexander Pyayt
Zhiquan Qi
Rick Quax
Barbara Quintela
Waldemar Rachowicz
Célia Ramos
Marcus Randall
Lukasz Rauch
Andrea Reimuth
Alistair Rendell
Pedro Ribeiro
Bernardete Ribeiro
Robin Richardson
Jason Riedy
Celine Robardet
Sophie Robert
João M. F. Rodrigues
Daniel Rodriguez
Albert Romkes
Debraj Roy
Philip Rutten
Katarzyna Rycerz
Augusto S. Neves
Apaar Sadhwani
Alberto Sanchez
Gabriele Santin

Robert Schaefer
Olaf Schenk
Ulf Schiller
Bertil Schmidt
Jan Schmidt
Martin Schreiber
Martin Schulz
Marinella Sciortino
Johanna Sepulveda
Ovidiu Serban
Vivek Sheraton
Yong Shi
Angela Shiflet
Takashi Shimokawabe
Tan Singyee
Robert Sinkovits
Vishnu Sivadasan
Peter Sloot
Renata Slota
Grażyna Ślusarczyk
Sucha Smanchat
Maciej Smołka
Bartlomiej Sniezynski
Sumit Sourabh
Hoda Soussa
Steve Stevenson
Achim Streit
Barbara Strug
E. Dante Suarez
Bongwon Suh
Shuyu Sun
Vaidy Sunderam
James Suter
Martin Swain
Grzegorz Swisrcz
Ryszard Tadeusiewicz
Lotfi Tadj
Daniele Tafani
Daisuke Takahashi
Jingjing Tang
Osamu Tatebe
Cedric Tedeschi
Kasim Tersic

Yonatan Afework
  Tesfahunegn
Jannis Teunissen
Andrew Thelen
Yingjie Tian
Nestor Tiglao
Francis Ting
Alfredo Tirado-Ramos
Arkadiusz Tomczyk
Stanimire Tomov
Marko Tosic
Jan Treibig
Leonardo Trujillo
Benjamin Uekermann
Pierangelo Veltri
Raja Velu
Alexander von Ramm
David Walker
Peng Wang
Lizhe Wang
Jianwu Wang
Gregory Watson
Rodrigo Weber dos
  Santos
Kevin Webster
Josef Weidendorfer
Josef Weinbub
Tobias Weinzierl
Jens Weismüller
Lars Wienbrandt
Mark Wijzenbroek
Roland Wismüller
Eric Wolanski
Michał Woźniak
Maciej Woźniak
Qing Wu
Bo Wu
Guoqiang Wu
Dunhui Xiao
Huilin Xing
Miguel Xochicale
Wei Xue
Xin-She Yang

Contents – Part V

Track of Smart Systems: Computer Vision, Sensor Networks and Machine Learning

Effective Self Attention Modeling for Aspect Based Sentiment Analysis   3
  Ningning Cai, Can Ma, Weiping Wang, and Dan Meng

Vision and Crowdsensing Technology for an Optimal Response
in Physical-Security. 15
  Fernando Enríquez, Luis Miguel Soria, Juan Antonio Álvarez-García,
  Fernando Sancho Caparrini, Francisco Velasco, Oscar Deniz,
  and Noelia Vallez

New Intelligent Tools to Adapt NL Interface to Corporate Environments 27
  Svetlana Chuprina and Igor Postanogov

Asymmetric Deep Cross-modal Hashing 41
  Jingzi Gu, JinChao Zhang, Zheng Lin, Bo Li, Weiping Wang,
  and Dan Meng

Applying NSGA-II to a Multiple Objective Dial a Ride Problem 55
  Pedro M. M. Guerreiro, Pedro J. S. Cardoso,
  and Hortênsio C. L. Fernandes

Smart Campus Parking – Parking Made Easy. 70
  Amanda Vieira, Iolanda Rosa, Ivo Santos, Tiago Paulo, Nuno Costa,
  Marisa Maximiano, and Catarina I. Reis

The Network Topology of Connecting Things: Defence of IoT Graph
in the Smart City 84
  Marta Chinnici, Vincenzo Fioriti, and Andrea Arbore

SILKNOWViz: Spatio-Temporal Data Ontology Viewer 97
  Javier Sevilla, Cristina Portalés, Jesús Gimeno, and Jorge Sebastián

Ontology-Driven Automation of IoT-Based Human-Machine
Interfaces Development 110
  Konstantin Ryabinin, Svetlana Chuprina, and Konstantin Belousov

Towards Parameter-Optimized Vessel Re-identification Based on IORnet. . . . 125
  Amir Ghahremani, Yitian Kong, Egor Bondarev,
  and Peter H. N. de With

Towards Low-Cost Indoor Localisation Using a Multi-camera System   137
  Oualid Araar, Saadi Bouhired, Sami Moussiou, and Ali Laggoune

A New Shape Descriptor and Segmentation Algorithm for Automated
Classifying of Multiple-morphological Filamentous Algae   149
  *Saowanee Iamsiri, Nuttha Sanevas, Chakrit Watcharopas,
  and Pakaket Wattuya*

Application of Hierarchical Clustering for Object Tracking with a Dynamic
Vision Sensor.   164
  Tobias Bolten, Regina Pohle-Fröhlich, and Klaus D. Tönnies

Binarization of Degraded Document Images with Generalized
Gaussian Distribution   177
  Robert Krupiński, Piotr Lech, Mateusz Tecław, and Krzysztof Okarma

Nonlinear Dimensionality Reduction in Texture Classification:
Is Manifold Learning Better Than PCA?   191
  Cédrick Bamba Nsimba and Alexandre L. M. Levada

Event-Oriented Keyphrase Extraction Based on Bi-clustering Model   207
  Lin Zhao, Liangjun Zang, Longtao Huang, Jizhong Han, and Songlin Hu

Track of Solving Problems with Uncertainties

Path-Finding with a Full-Vectorized GPU Implementation of Evolutionary
Algorithms in an Online Crowd Model Simulation Framework   223
  Anton Aguilar-Rivera

Analysing the Trade-Off Between Computational Performance
and Representation Richness in Ontology-Based Systems   237
  *Salvatore F. Pileggi, Fabian C. Peña, Maria Del Pilar Villamil,
  and Ghassan Beydoun*

A Framework for Distributed Approximation of Moments
with Higher-Order Derivatives Through Automatic Differentiation   251
  Michel Schanen, Daniel Adrian Maldonado, and Mihai Anitescu

IPIES for Uncertainly Defined Shape of Boundary, Boundary Conditions
and Other Parameters in Elasticity Problems.   261
  Marta Kapturczak and Eugeniusz Zieniuk

Enabling UQ for Complex Modelling Workflows   269
  *Małgorzata J. Zimoń, Samuel Antão, Robert Sawko, Alex Skillen,
  and Vadim Elisseev*

Ternary-Decimal Exclusion Algorithm for Multiattribute Utility Functions . . .   282
    Yerkin G. Abdildin

Sums of Key Functions Generating Cryptosystems   293
    Nataliya Kalashnykova, Viktor V. Avramenko,
    and Viacheslav Kalashnikov

Consistent Conjectures in Globalization Problems   303
    Nataliya Kalashnykova, Mariel A. Leal-Coronado,
    Arturo García-Martínez, and Viacheslav Kalashnikov

Verification on the Ensemble of Independent Numerical Solutions   315
    A. K. Alekseev, A. E. Bondarev, and A. E. Kuvshinnikov

On the Estimation of the Accuracy of Numerical Solutions
in CFD Problems   325
    A. E. Bondarev

"Why Did You Do That?": Explaining Black Box Models
with Inductive Synthesis   334
    Görkem Paçacı, David Johnson, Steve McKeever, and Andreas Hamfelt

Predictive Analytics with Factor Variance Association   346
    Raul Ramirez-Velarde, Laura Hervert-Escobar,
    and Neil Hernandez-Gress

Track of Teaching Computational Science

Redesigning Interactive Educational Modules for Combinatorial
Scientific Computing   363
    M. Ali Rostami and H. Martin Bücker

A Learner-Centered Approach to Teaching Computational Modeling,
Data Analysis, and Programming   374
    Devin Silvia, Brian O'Shea, and Brian Danielak

Enabling Interdisciplinary Instruction in Computer Science and Humanities:
An Innovative Teaching and Learning Model Customized for Small Liberal
Arts Colleges   389
    William B. Crum Jr., Aaron Angello, Xinlian Liu, and Corey Campion

A Project-Based Course on Software Development
for (Engineering) Research   401
    Kyle E. Niemeyer

Programming Paradigms for Computational Science:
Three Fundamental Models   408
    Miguel-Angel Sicilia, Elena García-Barriocanal,
    Salvador Sánchez-Alonso, and Marçal Mora-Cantallops

Numerical Analysis Project in ODEs for Undergraduate Students   421
    Sigurdur Hafstein

Poster Track

Mixed Finite Element Solution for the Natural-Gas
Dual-Mechanism Model.   437
    Mohamed F. El-Amin, Jisheng Kou, Shuyu Sun, and Jingfa Li

On the Feasibility of Distributed Process Mining in Healthcare.   445
    Roberto Gatta, Mauro Vallati, Jacopo Lenkowicz, Carlotta Masciocchi,
    Francesco Cellini, Luca Boldrini, Carlos Fernandez Llatas,
    Vincenzo Valentini, and Andrea Damiani

How to Plan Roadworks in Urban Regions? A Principled Approach
Based on AI Planning   453
    Mauro Vallati, Lukáš Chrpa, and Diane Kitchin

Big Data Approach to Fluid Dynamics Visualization Problem   461
    Vyacheslav Reshetnikov, Egor Golubchikov, Andrey Pyatlin,
    Alexey Kuzin, Vladislav Kiev, Nikolay Shabrov, Alexey Zhuravlev,
    and Ekaterina Guseva

Dolphin Kick Swimmer Using the Unstructured Moving Mesh Method   468
    Masashi Yamakawa, Norihito Mizuno, and Yongmann M. Chung

The Performance Prediction and Improvement of SPH
with the Interaction-List-Sharing Method on PEZY-SCs.   476
    Natsuki Hosono and Mikito Furuichi

Influence of Architectural Features of the SNC-4 Mode of the Intel Xeon
Phi KNL on Matrix Multiplication   483
    Ruben Laso, Francisco F. Rivera, and José Carlos Cabaleiro

Improving Planning Performance in PDDL+ Domains via Automated
Predicate Reformulation.   491
    Santiago Franco, Mauro Vallati, Alan Lindsay,
    and Thomas Lee McCluskey

The Case of iOS and Android: Applying System Dynamics to Digital
Business Platforms   499
    Ektor Arzoglou, Tommi Elo, and Pekka Nikander

Sockpuppet Detection in Social Network via Propagation Tree   507
  Jiacheng Li, Wei Zhou, Jizhong Han, and Songlin Hu

Exploring the Performance of Fine-Grained Synchronization and Data
Exchange Across Process Boundaries on Modern Multi-core Architectures. . .   514
  Jiri Dokulil and Siegfried Benkner

Accelerating Wild Fire Simulator Using GPU   521
  C. Carrillo, T. Margalef, A. Espinosa, and A. Cortés

Augmented Reality for Real-Time Navigation Assistance to Wheelchair
Users with Obstacles' Management   528
  Sawssen Ben Abdallah, Faiza Ajmi, Sarah Ben Othman,
  Sébastien Vermandel, and Slim Hammadi

p3Enum: A New Parameterizable and Shared-Memory Parallelized Shortest
Vector Problem Solver.   535
  Michael Burger, Christian Bischof, and Juliane Krämer

Rendering Non-Euclidean Space in Real-Time Using Spherical
and Hyperbolic Trigonometry   543
  Daniil Osudin, Chris Child, and Yang-Hui He

Improving Academic Homepage Identification from the Web
Using Neural Networks   551
  Jiapeng Zhao, Tingwen Liu, and Jinqiao Shi

Combining Fuzzy Logic and CEP Technology to Improve Air Quality
in Cities.   559
  Hermenegilda Macià, Gregorio Díaz, Juan Boubeta-Puig,
  Edelmira Valero, and Valentín Valero

Parallel Parametric Linear Programming Solving, and Application
to Polyhedral Computations   566
  Camille Coti, David Monniaux, and Hang Yu

Automating the Generation of Comparison Weights for Enhancing the AHP
Decision-Making Process   573
  Karim Zarour, Djamel Benmerzoug, Nawal Guermouche,
  and Khalil Drira

Parallel Algorithm Based on Singular Value Decomposition for High
Performance Training of Neural Networks   581
  Gabriele Maria Lozito, Valentina Lucaferri, Mauro Parodi,
  Martina Radicioni, Francesco Riganti Fulginei, and Alessandro Salvini

In-Situ Visualization with Membrane Layer
for Movie-Based Visualization   588
    Kohei Yamamoto and Akira Kageyama

Genetic Algorithm based EV Scheduling for On-Demand Public
Transit System   595
    Thilina Perera, Alok Prakash, and Thambipillai Srikanthan

Short-Term Irradiance Forecasting on the Basis of Spatially
Distributed Measurements   604
    Antonino Laudani, Gabriele Maria Lozito, Valentina Lucaferri,
    and Martina Radicioni

Multi-GPU Acceleration of the iPIC3D Implicit Particle-in-Cell Code   612
    Chaitanya Prasad Sishtla, Steven W. D. Chien, Vyacheslav Olshevsky,
    Erwin Laure, and Stefano Markidis

Reducing Symbol Search Overhead on Stream-Based Lossless
Data Compression.   619
    Shinichi Yamagiwa, Ryuta Morita, and Koichi Marumo

Stabilized Variational Formulation for Solving Cell Response to Applied
Electric Field   627
    Cesar Augusto Conopoima, Bernardo Martins Rocha, Iury Igreja,
    Rodrigo Weber Dos Santos, and Abimael Fernando Dourado Loula

Data-Driven Partial Derivative Equations Discovery
with Evolutionary Approach.   635
    Mikhail Maslyaev, Alexander Hvatov, and Anna Kalyuzhnaya

Predicting Cervical Cancer with Metaheuristic Optimizers
for Training LSTM   642
    Andre Quintiliano Bezerra Silva

Top k 2-Clubs in a Network: A Genetic Algorithm   656
    Mauro Castelli, Riccardo Dondi, Sara Manzoni, Giancarlo Mauri,
    and Italo Zoppis

CA-RPT: Context-Aware Road Passage Time Estimation
for Urban Traffic   664
    Ying Liu, Zhenyu Cui, Tianlin Zhang, Jiaxu Leng, Weihong Xie,
    and Liang Zhang

Modelling and Analysis of Complex Patient-Treatment Process Using
GraphMiner Toolbox.   674
    Oleg Metsker, Sergey Kesarev, Ekaterina Bolgova, Kirill Golubev,
    Andrey Karsakov, Alexey Yakovlev, and Sergey Kovalchuk

Combining Algorithmic Rethinking and AVX-512 Intrinsics for Efficient
Simulation of Subcellular Calcium Signaling   681
    Chad Jarvis, Glenn Terje Lines, Johannes Langguth, Kengo Nakajima,
    and Xing Cai

Ocean Circulation Hindcast at the Brazilian Equatorial Margin   688
    Luiz Paulo de Freitas Assad, Raquel Toste, Carina Stefoni Böck,
    Dyellen Soares Queiroz, Anne Goni Guedes, Maria Eduarda Pessoa,
    and Luiz Landau

A Matrix-Free Eigenvalue Solver for the Multigroup Neutron
Diffusion Equation   702
    Amanda Carreño, Antoni Vidal-Ferràndiz, Damian Ginestar,
    and Gumersindo Verdú

Path-Dependent Interest Rate Option Pricing with Jumps
and Stochastic Intensities   710
    Allan Jonathan da Silva, Jack Baczynski,
    and João Felipe da Silva Bragança

Composite Data Types in Dynamic Dataflow Languages as Copyless
Memory Sharing Mechanism   717
    Aurelien Bloch, Endri Bezati, and Marco Mattavelli

A Coupled Food Security and Refugee Movement Model for the South
Sudan Conflict   725
    Christian Vanhille Campos, Diana Suleimenova, and Derek Groen

A Proposal to Model Ancient Silk Weaving Techniques
and Extracting Information from Digital Imagery - Ongoing Results
of the SILKNOW Project.   733
    Cristina Portalés, Javier Sevilla, Manolo Pérez, and Arabella León

A Comparison of Selected Variable Ordering Methods for NFA Induction . . .   741
    Tomasz Jastrząb

Traffic3D: A Rich 3D-Traffic Environment to Train Intelligent Agents   749
    Deepeka Garg, Maria Chli, and George Vogiatzis

Energy Efficiency Evaluation of Distributed Systems.   756
    James Phung, Young Choon Lee, and Albert Y. Zomaya

Support for High-Level Quantum Bayesian Inference   764
    Marcin Przewięźlikowski, Michał Grabowski, Dariusz Kurzyk,
    and Katarzyna Rycerz

Financial Time Series: Motif Discovery and Analysis Using VALMOD.   771
  Eoin Cartwright, Martin Crane, and Heather J. Ruskin

Profiling of Household Residents' Electricity Consumption Behavior Using
Clustering Analysis.   779
  *Christian Nordahl, Veselka Boeva, Håkan Grahn,
  and Marie Persson Netz*

DNAS-STriDE Framework for Human Behavior Modeling
in Dynamic Environments   787
  Muhammad Arslan, Christophe Cruz, and Dominique Ginhac

OPENCoastS: An Open-Access App for Sharing Coastal Prediction
Information for Management and Recreation   794
  *Anabela Oliveira, Marta Rodrigues, João Rogeiro, André B. Fortunato,
  Joana Teixeira, Alberto Azevedo, and Pedro Lopes*

Author Index   809

Track of Smart Systems: Computer Vision, Sensor Networks and Machine Learning

Effective Self Attention Modeling
for Aspect Based Sentiment Analysis

Ningning Cai[1,2] ⓘ, Can Ma[1(✉)], Weiping Wang[1], and Dan Meng[1]

[1] Institute of Information Engineering, Chinese Academy of Sciences, Beijing, China
{cainingning,macan,wangweiping,mengdan}@iie.ac.cn
[2] University of Chinese Academy of Sciences, Beijing, China

Abstract. Aspect Based Sentiment Analysis is a type of fine-grained sentiment analysis. It is popular in both industry and academic communities, since it provides more detailed information on the user generated text in product reviews or social network. Therefore, we propose a novel framework based on neural network to determine the polarity of a review given a specific target. Not only the words close to the target but also the words far from the target determine the polarity of the review given a certain target, so we use self attention to solve the problem of long distance dependence. Briefly, we do multiple linear mapping on the review, do multiple attention and combine them to attend to the information from different representation sub-spaces. Besides, we use domain embedding to get close to the real word embedding in a certain domain, since the meaning of the same word may be different in different situation. Moreover, we use position embedding to underline the target and pay more attention to the words that are close to the target to get better performance on the task. We validate our model on four benchmarks, they are SemEval 2014 restaurant dataset, SemEval 2014 laptop dataset, SemEval 2015 restaurant dataset and SemEval 2016 restaurant dataset. The final results show that our model is effective and strong, which brings a 0.74% boost averagely based on the previous state-of-the-art work.

Keywords: Aspect Based Sentiment Analysis ·
Long Short-Term Memory (LSTM) · Attention

1 Introduction

Aspect Based Sentiment Analysis (ABSA) is a subtask of sentiment analysis. Instead of predicting the polarity of the overall sentence, it's proposed to predict the sentence polarity towards a given target. There are two subtasks [27], namely Aspect Category Sentiment Analysis (ACSA) and Aspect Term Sentiment Analysis (ATSA). The goal of ACSA is to predict the polarity with regard to a given target, which is one of some prepared and specific categories. And

This work is supported by the National Key Research and Development Program (Grant No. 2016YFB1000604).

the ATSA is to predict the polarity towards the given target, which is a sub sequence of the sentence. For example, given a sentence *"I bought a new camera. The picture quality is amazing but the battery life is too short"*, it's ACSA if the target is *"price"* and it's ATSA if the target is *"picture quality"*. Here, we mainly deal with the second task. As for the ATSA, if the target is *"picture quality"*, the expected sentiment polarity is positive as the sentence expresses positive emotion towards the target, but if the target is *"battery life"*, the true prediction should be negative. In other words, the polarity of a sentence may be opposite towards different targets. So the main challenge of ABSA is to find the words that actually determine the polarity towards a given target.

Now, we'll introduce some core technique we used in this paper. LSTM has the remarkable capacity of modeling the sequence, so there are some previous works based on LSTM. [21] uses two LSTM to model the left and right sequence of the target. However, the key information could disappear if the key words are far from the target. Attention mechanism has been proven effective in many Natural Process Language task, such as machine translation [1]. Therefore, many great works that base on attention and LSTM make progress in dealing with the ABSA task. [25] builds up a variational attention layer on the top of LSTM, [19] stacks multiple attention layers and the experimental results show it is resultful. [2] does multiple attention operation and combines them with a non-linear method.

The self attention mechanism plays an important role in many tasks, such as [11,17,22]. In this paper, we propose a novel model with self attention which builds up a self attention layer on the top of the bi-LSTM layer. Specifically, we do multiple linear mapping on the input sentence, and do multiple attention operation on each of them, finally, we concatenate them.

Besides, we come up with an original multiple word embedding. As we all know, the same word may have different meaning in diverse situation, so either the word embedding. For example, "hot" in "hot dog" is totally different from "hot" in "Today is so hot" or "The girl is hot". So apart from the general embedding that is trained from large corpus [12], we introduce the domain embedding, which is trained from a certain domain corpus. For example, if the ATSA task is about restaurant, then the domain embedding is trained from large restaurant corpus. Moreover, we introduce another novel word embedding, the position embedding. The position information is so important that it has been used with different methods in previous works [7]. In our paper, we use one-dimensional vector to represent it. The target is 0 and others are the distance from the given target. This way not only highlights the target phrase but also emphasizes the words close to the target.

We evaluate our model on four benchmarks, SemEval 2014 [15], containing the reviews of restaurant domain and laptop domain, SemEval 2015 restaurant dataset [14] and SemEval 2016 restaurant dataset [13]. The results prove that our model perform better than other baselines for all of the benchmarks, it gets competitive or even state-of-the-art results.

In general, our contributions are as follows: (i) introduce the domain embedding and firstly use position embedding in the embedding layer as far as we know; (ii) to our knowledge. we firstly use self attention in this area and come up with a novel framework; (iii) get the state-of-the-art results on four benchmarks.

The remainder of the paper is organized as follows. Section 2 introduces other related excellent work in this area and the differences among us. Section 3 introduces our model with detailed information. Section 4 is the details of the experiments. Finally, Sect. 5 is a further analysis of our model.

2   Related Work

There are many abundant and excellent works in the area of ABSA, which in literature is a fine-grained classification task [16]. The previous works are basically rule based or statistic based. [28] incorporates target dependent features and employs Support Vector Machine (SVM) to get comparable results. [3] employs probabilistic soft logic model to solve the problem. They [5,9,24] usually need expensive artificial features, such as n-grams, part-of-speech tags, lexicon dictionaries, dependency parser information and so on.

Since the neural network has the ability to capture features automatically through multiple hidden layers, there are more and more outstanding models based on neural network in this area. [23] extracts a rich set of automatic features through multiple embedding and multiple neural pooling function. [4] uses the dependency parsing results, regards the target word as the tree root and propagates the sentiment of the words from the tree bottom to the tree root node. However, the use of the dependency parser makes it not effective enough if the data is noisy like twitter data. [27] comes up with a model Gated Convolutional network with Aspect Embedding (GCAE), which is a pure Convolution Neural Network and uses gating mechanism to assign different weights to the words. [19] uses two LSTM to model the sequence from the beginning and the tail to the target word. And it has to be noted that if the decisive words are far from the target, the model may fail.

Furthermore, attention based LSTM has gained a lot of attention due to their ability to capture the importance of the words. [21] stacks multiple attention layer and gets competitive results. [25] comes up with a variant of LSTM with attention, they add the target embedding to each of the hidden units. [2] also adopts multiple attention layers and combines the outputs with a Recurrent Neural Network (RNN) model. [7] incorporates syntactic information into the attention mechanism. We also use self attention based on bi-LSTM. The self attention does multiple linear mapping on the input sentence, does multiple attention and combines them. Besides the self attention, we also use domain embedding and position embedding. The former has been proved effective in extraction task [26], the latter is usually used in the attention layer and is computed by dependency parser [7], here we use it in the embedding layer in a simple but effective way.

3   Model

The architecture of our model is shown in Fig. 1, which consists of four modules, word embedding module, bi-LSTM module, self attention module and softmax output module. ATSA aims to determine the sentiment polarity of a sentence s towards a given target word or phrase a, a sub-sequence of s.

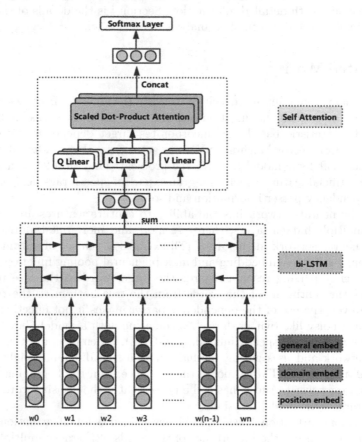

Fig. 1. The architecture of our model.

3.1   Word Embedding

The input is a sentence $s = (w_0, w_1, w_2..., w_n)$, which consists of a given target $a = (a_0, a_1, ..., a_m)$. Each of the above word w_i is presented as a continuous and dense numeric vector e_{wi} from a look up table named word-embedding-matrix $E \in \mathbb{R}^{V \times d}$, where V is the vocabulary size and d is the word embedding dimension. The word embedding concatenate three different components, which are general embedding $E^g \in \mathbb{R}^{V \times d_g}$, domain embedding $E^d \in \mathbb{R}^{V \times d_d}$ and position embedding $E^p \in \mathbb{R}^{V \times d_p}$. Usually, other related works use general embedding only, but we introduce the other two to improve the performance.

General Embedding. The general embedding matrix E^g is pre-trained from a large corpus irrelevant to the specific task, such as glove.840B.300d [12].

Domain Embedding. The domain embedding matrix E^d is pretrained from a corpus relevant to the specific task. For example, if the ABSA task is about restaurant, then we pretrain word embedding from large restaurant corpus like Yelp Dataset [20]. And the reason we introduce it is vectors trained from out-of-domain corpus can't express theirs true meaning properly. For instance, "hot" in "hot dog" would be close to warm or something about weather and "dog" would be close to something about animals. However, this kind of expression is far from its true meaning that unexpectedly it turns out to be a kind of food.

Position Embedding. Intuitively, not all words are equal important to classify the polarity of the sentence given a target, usually words appear near the target and words have translation relation need more attention. We use a one-dimensional vector to represent each word w_i. The number is the distance from the target. Suppose there is a sentence that "I love [the hot dog]$_{target}$ very much", then in our paper its position embedding is [2 1 0 0 0 1 2]. We mark the target with 0 to distinguish the target from others and other distances characterize the different importance to the classification task.

3.2 bi-LSTM Layer

Long Short-Term Memory (LSTM) [6] is a type of varietal RNN in order to overcome the vanishing gradient problem, so it's a powerful tool to model the long sequence. bi-LSTM can capture more information than LSTM, since both forward information and backward information can be used to infer. We use bi-LSTM to process the input sentence in opposite direction and sum the last hidden vectors as the output. The $h_t = LSTM(h_{t-1}, e_{wt})$ is calculated as follows, where W is the weight matrix and b is the bias:

$$f_t = \sigma(W_f \times [h_{t-1}, e_{wt}] + b_f) \tag{1}$$

$$i_t = \sigma(W_i \times [h_{t-1}, e_{wt}] + b_i) \tag{2}$$

$$o_t = \sigma(W_o \times [h_{t-1}, e_{wt}] + b_o) \tag{3}$$

$$\tilde{c}_t = tanh(W_c \times [h_{t-1}, e_{wt}] + b_c) \tag{4}$$

$$c_t = f_t \times c_{t-1} + i_t \times \tilde{c}_t \tag{5}$$

$$h_t = o_t \times tanh(c_t) \tag{6}$$

The bi-LSTM academic description is just like this:

$$\overrightarrow{h_t} = LSTM(\overrightarrow{h}_{t-1}, e_{wt}) \tag{7}$$

$$\overleftarrow{h_t} = LSTM(\overleftarrow{h}_{t-1}, e_{wt}) \tag{8}$$

$$output = \overrightarrow{h_t} + \overleftarrow{h_t} \tag{9}$$

where e_{wt} is the embedding vector of the word w_t which is the tth of the input sentence s and h_t is the corresponding hidden state.

3.3 Self Attention

Self-attention is a special attention mechanism to compute the representation of a sentence. It has been proved effective in many Natural Language Processing (NLP) tasks, such as Semantic Role Labeling [18], Machining Translation [22] and other tasks [11,17]. In this section, firstly, we introduce the self-attention, then we discuss its advantages.

Scaled Dot-Product Attention. Given a query matrix $Q \in \mathbb{R}^{n \times d}$, key matrix $K \in \mathbb{R}^{n \times d}$ and value matrix $V \in \mathbb{R}^{n \times d}$, we calculate the scaled dot-product attention *head* as follows. Here, n means we pack n queries, keys or values together into matrix Q, K and V, and d is the dimension of them:

$$head(Q, K, V) = softmax(\frac{QK}{\sqrt{d}}V) \qquad (10)$$

The divisor $\sqrt{d}$ is pushing the softmax function into regions where it has extremely small gradients [22].

Multi-head Attention. The mechanism firstly does linear mapping on the input matrices Q, K and V, repeats h times and then concatenates the results as output m. The h parallel operations allow the model to jointly attend to information from different representation sub-spaces.

$$m = concat(head_1, head_2, ..., head_h)W^m$$
$$where\,head_i = head(QW^q, KW^k, VW^v) \qquad (11)$$

In our paper, the input Q, K, V are all the output of the bi-LSTM layer. The self attention can capture dependencies even if the distance of the words are too far. The distance of each two words are 1 while it can be n (the sequence length) in RNN architecture. Also, it's highly parallel while RNN is not. At the same time, Features it captures are more abundant than CNN since CNN uses the fixed window size.

3.4 Softmax Layer

The ABSA is a three classification task whose label is positive, negative or neural. The self attention layer's output m is the representation of the given sentence, and we feed it into a softmax layer to predict the probability distribution p over sentiment label, where W_o is the weight matrix and b_o is the bias:

$$p = softmax(W_o\,m + b_o) \qquad (12)$$

The training object is minimizing cross-entropy function:

$$loss = \sum_{i \in C} log\,p_i(t_i) \qquad (13)$$

where C is the training corpus, p_i is the prediction label while t_i is the real label.

4   Experiments

4.1   Datasets and Preparations

We validate our model on four benchmarks, they are SemEval2014 [15], containing two datasets, SemEval2015 [14] and SemEval2016 [13]. The statistics of them is shown in Table 1. Following the previous work [8], we also remove the data with *conflicting* label.

Table 1. The positive, negative and neural examples statistics of Semval Datasets

Datasets	Train			Test		
	Pos	Neg	Neu	Pos	Neg	Neu
SemEval 2014_restaurant	2164	805	633	728	196	196
SemEval 2014_laptop	987	866	460	341	128	169
SemEval 2015_restaurant	1178	382	50	439	328	35
SemEval 2016 restaurant	1620	709	88	597	190	38

4.2   Evaluation Metric

We use the accuracy metric **acc** to evaluate our model. The method is:

$$acc = \frac{TP}{TP + FP} \tag{14}$$

where TP is true positive and FP is false positive.

4.3   Hyper-parameters Settings

In all of our experiments, 300-dimension E^g is initialed by Glove [12], 100-dimension E^d is trained by fasttext[1] with yelp [20] corpus and the Amazon Electronics dataset [10]. We randomly pick up 20% of training data as development data to keep the best parameters. The optimizer is Root Mean Square Prop (RMSProp) with initial learning rate 0.001. The dimension of the bi-LSTM is 400. The epoch is 25 and the mini batch is 32. We use dropout with 0.5 and early-stopping to prevent from overfitting. The number of multi-head h is 16.

4.4   Model Comparison

We compare some traditional models with our model, they are as follows.

SVM with labor features [9] is a typical statistic model. The SVM is trained with a lot of labor features, including n-grams, POS labels and large-scale lexicon dictionaries. We compare with the reported results on SemEval2014.

---

[1] ref: https://github.com/facebookresearch/fastText.

Table 2. Average accuracies over 3 runs with random initialization. The best results are in bold.

Model	SemEval 2014_res	SemEval 2014_lt	SemEval 2015_res	SemEval 2016_res
SVM	80.16	70.49	NA	NA
LSTM	75.27	66.55	74.06	82.39
LSTM-ATT	75.74	67.55	76.14	83.48
TD-LSTM	75.37	68.25	76.38	82.16
ATAE-LSTM	78.60	68.88	78.48	83.77
RAM	78.48	72.08	79.98	83.88
PRET+MULT	79.11	71.15	81.30	85.58
SA-LSTM (OURS)	**80.15**	**72.42**	**81.92**	**85.62**

LSTM [6] We build up a LSTM layer on the word embedding layer, and the output is the average of the hidden states.

LSTM + attention (ATT) Based on the above LSTM, we add an attention layer on the top of the LSTM layer. Briefly, we calculate the weight α of each hidden state h and multiply them as the sentence representation. The weight α is described with the following equation:

$$target = \frac{1}{m} \sum_{i=1}^{m} e_{ai} \tag{15}$$

$$d_i = tanh(h_i, target) \tag{16}$$

$$\alpha_i = \frac{exp(d_i)}{\sum_{j=1}^{n} exp(d_j)} \tag{17}$$

Target-dependent LSTM (TD-LSTM) [21] They use one LSTM to model the sequence from the beginning to the target and another LSTM to model the sequence from the end to the target, then they combine the results as the sentence representation.

Attention-based LSTM with Aspect Embedding (ATAE-LSTM) [25] is a variant of LSTM+ATT, they add target embedding vetor to each of the LSTM hidden states.

Recurrent Attention Network on Memory (RAM) [2] uses LSTM and multiple attention. Briefly, they use multiple attention operation and combines them with a RNN as the sentence representation.

Pre-train + Multi-task learning (PRET+MULT) [8] use pre-train and multi-task learning to get better performance. They use another document level sentiment analysis task as the auxiliary task.

The results are shown in Table 2, and it's the average value over three times with random initialization. The results indicate that our model is effective and strong in four different benchmarks. More abundant analyses will be in next section.

4.5 Analysis

Table 2 indicates that we can gain a lot from multi-embedding and self attention. Our model bring a 0.74% boost averagely based on the previous state-of-the-art work. We can find that the improvement of th previous two dataset are better than the SemEval2015_res and SemEval2016_res dataset, and we think it's because the problem of the label imbalance is less serious on the previous two dataset. For further verification, we do more experiments whose results are shown in Fig. 2.

To validate the effectiveness of the word embedding layer, (1) we remove the domain embedding from the model, the experiment result decreases by 0.60%–1.62%, the average is 1.05%. (2) we remove the position embedding from the model, the experiment result decreases by 1.74%–2.74%, the average is 2.18%. On the whole, the position embedding plays more important role than the domain embedding in the ATSA task. Intuitively, the phenomenon is reasonable because the position embedding not only stress the target information but also pay more attention to the words close to the target.

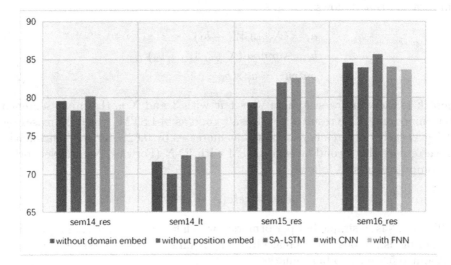

Fig. 2. More experiments to valid the effectiveness of the model. The five setting from the left to the right is: (1) without domain embedding in the embedding layer, (2) without position embedding in the embedding layer, (3) our model SA-LSTM, (4) replace the bi-LSTM with CNN in the bi-LSTM layer, (5) replace the bi-LSTM with FNN in the bi-LSTM layer. The y axis is the accuracy on the four datasets.

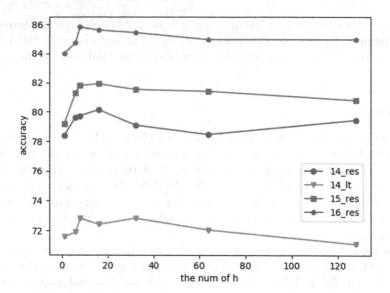

Fig. 3. The influence of the multi-head nums in the self attention layer. The y axis is the accuracy on the four datasets.

To approve the potential of the bi-LSTM layer, (1) we replace the second layer bi-LSTM with Convolutional Neural Networks (CNN) inspired by [27], the computation is as follows:

$$a_i = (X_{i,i+k} W_1 + b_1) \tag{18}$$

$$b_i = sigmoid(X_{i,i+k} W_2 + b_2) \tag{19}$$

$$output_i = a_i \times b_i \tag{20}$$

where k is the window size, here we set it with 3 and X is the input sentence after embedding. The result shows it isn't good as bi-LSTM, which decreases by 1.2% averagely on three benchmarks but increases by 0.62% on one benchmark. (2) we replace the second layer bi-LSTM with FNN (Forward Neural Network), the computation is as follows:

$$output = relu(X W + b_1) \tag{21}$$

The FNN is so simple but perform well, which follows Occam's razor principle that simple is the best. It decrease by over 2% on two benchmarks but increases by about 0.5% on two benchmarks.

Additionally, In order to get the influence of the factor multi-head number h, we draw the Fig. 3. The figure pinpoints that it's not the more the better, most benchmarks get theirs best performance when h is 16. However, SemEval2014_lt dataset gets its best performance when h is 32.

5 Conclusion

To our knowledge, our work is the first attempt to use the domain and position embedding in the embedding layer and the first attempt to use self attention in the ABSA area. We have validated the effectiveness of our model and we get competitive or even the state-of-the-art results on four benchmarks. In the future, we'll attempt to model sentence and target separately with self attention to get better performance and focus on the problem of label imbalance. Besides, we may also try other position embedding strategies to give the important words more attention.

References

1. Bahdanau, D., Cho, K., Bengio, Y.: Neural machine translation by jointly learning to align and translate. Computer Science (2014)
2. Chen, P., Sun, Z., Bing, L., Yang, W.: Recurrent attention network on memory for aspect sentiment analysis. In: Proceedings of the 2017 Conference on Empirical Methods in Natural Language Processing, pp. 452–461 (2017)
3. Deng, L., Wiebe, J.: Joint prediction for entity/event-level sentiment analysis using probabilistic soft logic models. In: Proceedings of the 2015 Conference on Empirical Methods in Natural Language Processing, pp. 179–189 (2015)
4. Dong, L., Wei, F., Tan, C., Tang, D., Zhou, M., Xu, K.: Adaptive recursive neural network for target-dependent twitter sentiment classification. In: Proceedings of the 52nd Annual Meeting of the Association for Computational Linguistics (Volume 2: Short Papers), vol. 2, pp. 49–54 (2014)
5. Ganapathibhotla, M., Liu, B.: Mining opinions in comparative sentences. In: Proceedings of the 22nd International Conference on Computational Linguistics, vol. 1, pp. 241–248. Association for Computational Linguistics (2008)
6. Gers, F.A., Schmidhuber, J., Cummins, F.: Learning to forget: continual prediction with LSTM (1999)
7. He, R., Lee, W.S., Ng, H.T., Dahlmeier, D.: Effective attention modeling for aspect-level sentiment classification. In: Proceedings of the 27th International Conference on Computational Linguistics, pp. 1121–1131 (2018)
8. He, R., Lee, W.S., Ng, H.T., Dahlmeier, D.: Exploiting document knowledge for aspect-level sentiment classification. arXiv preprint arXiv:1806.04346 (2018)
9. Kiritchenko, S., Zhu, X., Cherry, C., Mohammad, S.: NRC-Canada-2014: detecting aspects and sentiment in customer reviews. In: International Workshop on Semantic Evaluation, pp. 437–442 (2014)
10. Mcauley, J., Targett, C., Shi, Q., Hengel, A.V.D.: Image-based recommendations on styles and substitutes (2015)
11. Paulus, R., Xiong, C., Socher, R.: A deep reinforced model for abstractive summarization. arXiv preprint arXiv:1705.04304 (2017)
12. Pennington, J., Socher, R., Manning, C.: Glove: global vectors for word representation. In: Proceedings of the 2014 Conference on Empirical Methods in Natural Language Processing (EMNLP), pp. 1532–1543 (2014)
13. Pontiki, M., et al.: Semeval-2016 task 5: aspect based sentiment analysis. In: Proceedings of the 10th International Workshop on Semantic Evaluation (SemEval-2016), pp. 19–30 (2016)

14. Pontiki, M., Galanis, D., Papageorgiou, H., Manandhar, S., Androutsopoulos, I.: Semeval-2015 task 12: aspect based sentiment analysis. In: Proceedings of the 9th International Workshop on Semantic Evaluation (SemEval 2015), pp. 486–495 (2015)
15. Pontiki, M., Galanis, D., Pavlopoulos, J., Papageorgiou, H., Androutsopoulos, I., Manandhar, S.: Semeval-2014 task 4: aspect based sentiment analysis. In: Proceedings of International Workshop on Semantic Evaluation, pp. 27–35 (2014)
16. Rojas-Barahona, L.M.: Deep learning for sentiment analysis. Lang. Linguist. Compass **10**(12), 701–719 (2016)
17. Shen, T., Zhou, T., Long, G., Jiang, J., Pan, S., Zhang, C.: Disan: directional self-attention network for RNN/CNN-free language understanding. arXiv preprint arXiv:1709.04696 (2017)
18. Tan, Z., Wang, M., Xie, J., Chen, Y., Shi, X.: Deep semantic role labeling with self-attention. arXiv preprint arXiv:1712.01586 (2017)
19. Tang, D., Qin, B., Feng, X., Liu, T.: Effective LSTMs for target-dependent sentiment classification. arXiv preprint arXiv:1512.01100 (2015)
20. Tang, D., Qin, B., Liu, T.: Document modeling with gated recurrent neural network for sentiment classification. In: Proceedings of the 2015 Conference on Empirical Methods in Natural Language Processing, pp. 1422–1432 (2015)
21. Tang, D., Qin, B., Liu, T.: Aspect level sentiment classification with deep memory network. arXiv preprint arXiv:1605.08900 (2016)
22. Vaswani, A., et al.: Attention is all you need. In: Advances in Neural Information Processing Systems, pp. 5998–6008 (2017)
23. Vo, D.T., Zhang, Y.: Target-dependent twitter sentiment classification with rich automatic features. In: International Conference on Artificial Intelligence, pp. 1347–1353 (2015)
24. Wagner, J., et al.: DCU: aspect-based polarity classification for semeval task 4. In: Proceedings of the 8th International Workshop on Semantic Evaluation (SemEval 2014), pp. 223–229 (2014)
25. Wang, Y., Huang, M., Zhao, L., et al.: Attention-based LSTM for aspect-level sentiment classification. In: Proceedings of the 2016 Conference on Empirical Methods in Natural Language Processing, pp. 606–615 (2016)
26. Xu, H., Liu, B., Shu, L., Yu, P.S.: Double embeddings and CNN-based sequence labeling for aspect extraction. arXiv preprint arXiv:1805.04601 (2018)
27. Xue, W., Li, T.: Aspect based sentiment analysis with gated convolutional networks. arXiv preprint arXiv:1805.07043 (2018)
28. Zhou, M.: Target-dependent twitter sentiment classification. In: Proceedings of Annual Meeting of the Association for Computational Linguistics Human Language Technologies, vol. 1, pp. 151–160 (2011)

Vision and Crowdsensing Technology for an Optimal Response in Physical-Security

Fernando Enríquez[1], Luis Miguel Soria[1], Juan Antonio Álvarez-García[1(✉)], Fernando Sancho Caparrini[1], Francisco Velasco[1], Oscar Deniz[2], and Noelia Vallez[2]

[1] Universidad de Sevilla, Seville, Spain
{fenros,lsoria,jaalvarez,fsancho,velasco}@us.es
[2] VISILAB, E.T.S.I.I, University of Castilla-La Mancha, 13071 Ciudad Real, Spain
{oscar.deniz,noelia.vallez}@uclm.es

Abstract. Law enforcement agencies and private security companies work to prevent, detect and counteract any threat with the resources they have, including alarms and video surveillance. Even so, there are still terrorist attacks or shootings in schools in which armed people move around a venue exercising violence and generating victims, showing the limitations of current systems. For example, they force security agents to monitor continuously all the images coming from the installed cameras, and potential victims nearby are not aware of the danger until someone triggers a general alarm, which also does not give them information on what to do to protect themselves. In this article we present a project that is being developed to apply the latest technologies in early threat detection and optimal response. The system is based on the automatic processing of video surveillance images to detect weapons and a mobile app that serves both for detection through the analysis of mobile device sensors, and to send users personalised and dynamic indications. The objective is to react in the shortest possible time and minimise the damage suffered.

Keywords: Computer vision · Weapon detection · Crowdsensing

1 Introduction

Every city in the world suffers events of diverse nature that endanger the life of its citizens. Security specialists protect public and private institutions with professionalism, but the great diversity of possible events and the size and structure of the area to monitor make it very difficult to prevent them, and above all, to plan an optimal response for each threat. Unfortunately, the current global alert situation due to the proliferation of terrorist acts, directly oriented towards citizenship, has only emphasised the need to evolve current security systems to deal with threats in the best possible way.

The project presented in this paper, called VICTORY, aims to provide a next-generation security system, more intelligent, agile and effective, that significantly

© Springer Nature Switzerland AG 2019
J. M. F. Rodrigues et al. (Eds.): ICCS 2019, LNCS 11540, pp. 15–26, 2019.
https://doi.org/10.1007/978-3-030-22750-0_2

improves the reaction time and the response to a threat in a predefined area (for example a building) which we will call 'security zone'. Therefore, the main objectives of the system are twofold:

1. Global solution to detect, analyse and classify security threats using new technologies and generating progress in the state of the art, complementing and improving current solutions.
2. Provide quick and personalised information to potential victims within the security zone, as well as to security personnel to facilitate an optimised management of the threat.

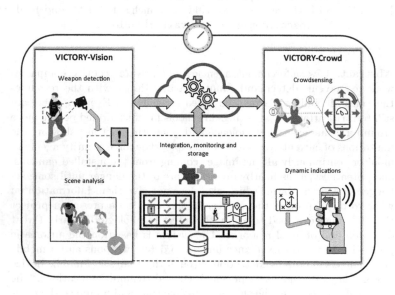

Fig. 1. System components

For this purpose, the design of the system has been divided into several components as can be seen in Fig. 1:

– A computer vision subsystem that automatically analyses the images captured by the security cameras installed in the security zone using Deep Learning techniques. The analysis contemplates two different and complementary approaches that will provide robustness to the system. On the one hand, the generation of detectors specific to relevant patterns (such as different types of weapons), and on the other hand the use of autoencoders that alert when anomalous patterns deviate from the usual scenes captured by the cameras.
– An efficient crowdsensing subsystem with indoor positioning for the recognition of falls and/or stampedes among other relevant physical activities of the occupants of the security zone using the inertial units of smartphones and machine learning techniques. These events allow detecting anomalous situations and generate specific indications to the potential victims.

– An integration framework that analyses the indications that are received from
the vision and crowdsensing systems, and processes them to establish levels
of priority, probability percentages and actions to be carried out. The results
will be shown to the users according to their role:

  • Security agents: a central console shall be provided with all information
    relating to the indications received, giving the possibility of confirming
    the threat or discarding it and offering a list of possible actions to be
    taken in response. The main objective in this case will be to optimise
    the precision in all the components involved so that the system is of the
    maximum utility for the security personnel.

  • All other users in the security zone: they will receive indications to follow
    through a mobile application (for example an escape route or a recom-
    mendation to lock the door and hide) based on their position relative
    to the threat, the nature of the threat and the user profile (e.g. reduced
    mobility). This functionality flows from the integration framework to the
    crowdsensing subsystem communicating with individual devices and the
    biggest challenge will be to achieve a method capable of adapting and
    reacting quickly in a changing scenario considering a large number of
    parameters.

The place where the system is being tested is the School of Computer Engineering at the University of Seville, Spain. The building takes up more than
$9500\,\mathrm{m}^2$ and can be seen in Fig. 2. It has a closed circuit television (CCTV)
composed by more than 50 cameras where five of them have been shared with
the research team. It also has a infrastructure of WiFi with more than 400 access
points.

Fig. 2. School of Computer Engineering building

In the next sections we will show the approaches that are being studied to
implement these components, specifically the vision subsystem in Sect. 2 and the
crowdsensing subsystem in Sect. 3, on which work is already underway. Finally,
we draw our conclusions in Sect. 4.

2 Computer Vision-Based Weapon Detection

2.1 Object Detection

From the International Joint Conference on Neural Networks in 2011 where IDSIA team [6] won the German traffic sign recognition benchmark and ImageNet Large Scale Visual Recognition Challenge 2012 where AlexNet won [14], object detection is evolved by leaps and bounds. Competitions such as Imagenet [21] or COCO [17] have promoted these advances.

In video surveillance, the detection of dangerous objects as weapons has been studied with the Deep Learning methodology, but only very recently. In [19], Faster-RCNN was shown as the best detector in this task using a dataset generated from violent movies, also giving a very low response time, 5 frames per second.

Unfortunately, these images, normally foreground images of pistols and several kind of guns, differ from fixed CCTV cameras that obtain a wide shot, transforming the problem into a hardest one: small object detection.

2.2 Autoencoder

The development of a detection system is normally driven to achieve good detection and false positive rates on a certain dataset. Ideally, training data would contain representative instances from all possible application scenarios. In practice, obtaining such a huge amount of data is not feasible in terms of time and resources. This problem forces data scientists to be cautious about overfitting and poor generalization when training new models [27]. Some techniques such as dataset partitioning, L1 and L2 regularizations or early stopping are applied to alleviate them [22]. However, misclassification of samples in new scenarios must be addressed. Thus, it is conceivable that a weapon detector could be trained using a dataset containing instances from all possible weapons that provides accurate detections and a small number of false positives. Then, when put into a surveillance system in a real scenario the result is generally an unbearable rate of false alarms [23]. This means that the system will almost certainly be switched off, specially in cases where the incidence of the event of interest is very low. In this context we propose to add an additional step that models and filters the typical false alarms of the new scenario while maintaining the ability to detect the objects it was trained for (Fig. 3).

In the first step of the process, the detector runs in the new scenario over a period of time, saving all the detections. Those detections, that with high probability will be false positives, are then used to train an autoencoder. Autoencoder networks learn how to compress input data into a short code and then reconstruct that code into something as close as possible to its original input and are commonly used for anomaly detection [10]. In this case the autoencoder is trained to model one class: the typical false positives of the new scenario. Finally, it is applied to reconstruct images from a test dataset that contains also instances from the searched objects. If the reconstruction error is compared between both

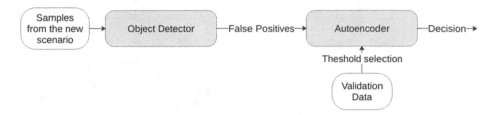

Fig. 3. Autoencoder training phase

classes, it is lower for the class used to train the autoencoder and thus, a threshold could be established to separate them. Figure 4 illustrates this difference by computing the reconstruction error as the mean squared error (MSE).

We have applied the proposed methodology with a handgun detector. A previously available detector has been used for this purpose [1]. For the dataset we downloaded 4 videos from YouTube where people appear testing handguns in the countryside. In this scenario most false positives of the detector are caused by the trees in the background (Fig. 5). Three of the videos were used to train and adjust the threshold of the autoencoder, following a 3-fold cross-validation approach. The remaining video was used only for testing both the detector and the detector+autoencoder configurations.

The experimental results show an improvement in the false positive rate at roughly the same detection rates (Fig. 6). To obtain the detector+autoencoder ROC curve the detector operational point is fixed (at the optimal threshold selected during training) and the autoencoder threshold is made to vary from the lowest to the highest value.

3 Crowdsensing and Action Policy

Mobile crowdsensing consists in the extraction and sharing of data coming from mobile device sensors carried by a group of users. The information can be analysed from all sources and draw conclusions about a common interest. Crowdsensing has been used in various security-related scenarios. We can highlight the systems for detecting incidents related to traffic [20], as well as natural disasters such as fires [18] or earthquakes [8]. Generally, an analysis is made of messages published in various social networks that allows monitoring of a specific emergency event [26], thereby making use of "social media" as a crowdsourcing mechanism [25] for the contribution and dissemination of information on the effects of the emergency on the population. There are also studies based on crowdsensing, using sensors from mobile devices such as GPS [5] to develop organizational strategies that avoid crowding and the risk of incidents. Indoors, localisation from WiFi footprint [11] or Bluetooth [4] is also used. For the detection of physical activities of individuals [15], inertial sensors are used, where the accelerometer is the most used device because of its low energy consumption. In this context, thanks to the use of data from inertial sensors and through the

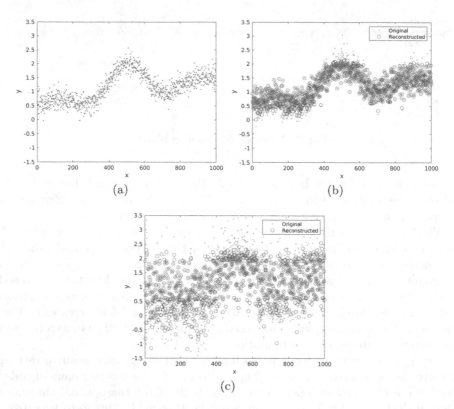

Fig. 4. Reconstruction error comparison. The more the data differs from the training set the higher the error is. (a) Autoencoder training data from the function $y = 1 + 0.05x + sin(x)/x + 0.2 * r$ where r is a normally distributed random number. (b) Reconstructed data from $y = 1 + 0.05x + sin(x)/x + 0.3 * r$ with $MSE = 0.533$. (c) Reconstructed data from $y = 1 + 0.05x + sin(x)/x + 0.8 * r$ with $MSE = 1.099$.

Fig. 5. Sample frame from one of the videos used and a typical false positive in this scenario (enlarged).

application of different classification algorithms, it is possible to recognize the physical activity that the user is doing at each moment. In the field of security, there are systems [2] capable of recognizing the steps and length of the user's stride to design an efficient evacuation system. Falls and other daily live activities [9] are other events that have been studied in depth based on accelerometers [7]. However, the vast majority of these works focus on the detection of falls in a context of Ambient Assisted Living [3], and there are no systems evaluated for the detection of falls in emergency contexts.

3.1   Indoor Positioning and Activity Recognition

As previously commented, one of the objectives of this work is to obtain the location of the different users of the system inside the building. This functionality allows to determine, in case of a threat, the exact position of the users and to propose, if so required, an escape route adapted according to its location and the characteristics of the route.

There exist several commercial solutions that require a high initial investment in infrastructure but due to the more than 400 WiFi access points (APs) allocated within the building a fingerprint method based on [13] has been developed. A fingerprint is composed of several access points, identified by its MAC address and the received signal strength (RSS) value observed by a mobile phone.

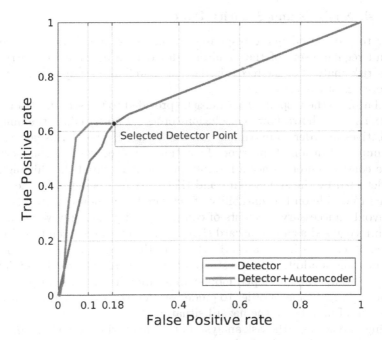

Fig. 6. ROC curves. The autoencoder shows a reduction of 8% in the number of false positives while maintaining the detection rate of the selected detector operational point.

To train the system hundreds of fingerprints have been collected in the middle of corridors every 2 m and in several parts of each classroom.

To test the indoor positioning system an Android app has been developed, as it can be seen in Fig. 7. The app obtains fingerprints every 30 s and compares them with the training set using overlapping of APs and the correlation with their RSS. Results show that the mean average error is 9.1 m. The accuracy can be improved using other methods such as dead reckoning or including deep learning but the system must be used by hundreds of students and lecturers in the building so a trade-off between accuracy and battery has been obtained simplifying the positioning technique and including some improvements: the accelerometer is used to detect steps, when no steps are detected during 3 min, the app stops the fingerprint gathering up to the next step, minimizing the battery draining.

Another important feature from the app is the use of the accelerometer to detect falls and run activities (stampedes). A simple approach is used to detect them using acceleration thresholds. Although false positives occur, crowdsensing here is primary to filter them: when two or more users are experimenting one of these activities in close locations, an emergency event is transmitted to the security personnel and they can check the CCTV and the location of all the application users to confirm or rebut the alarm. Isolated events can normally be avoided if there are people close to the event but nothing changes in their behaviour.

3.2   Risk Analysis and Security Policies

In order to provide adequate responses to the users (members of the security team and regular users of the monitored area) it is necessary to carry out a previous risk analysis of the recognized event as well as a study of the structural characteristics of the area.

In relation to the response that must be provided to the security team, previous sections have shown how the solutions designed must reduce the number of false positives in order to maximize user confidence in the alarms generated. A correct implementation of this type of solutions involves working collaboratively with the control centers, since a filtering process is required that highlights the events detected by the sub-systems and that must be evaluated by the security personnel to decide on the suitability of activate the response protocols.

To avoid unnecessary moments of confusion and panic, only when an alarm has been recognised as a real hazard should some action be taken to inform and guide the other users in the area. To this end, the project proposes channels of communication in addition to the standards (loudspeaker systems and light signals). Specifically, the developed mobile application allows to send personalized information in real time in order to guide users on the safest actions during the valuable initial minutes of the hazard situation.

Taking into account the risk analysis of the area (prior to the implementation of the system and that only needs to be updated when structural changes occur in the topology of the area), and the features of the detected threat, the platform will inform the user about the main actions to be carried out considering their

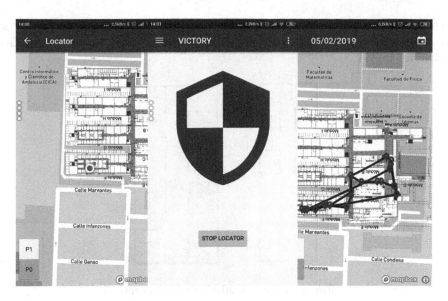

Fig. 7. User interface for Victory App

current position within the area and personal traits. Thus, the main actions that the platform will suggest, depending on the classification of the threat, are:

- Evacuation Route: on a map of the area, indications to follow a safe route to a nearby exit.
- Assurance Tips: actions that increase the user security during the first few minutes of the threat (for example, which accessible area is safe and what procedures to perform to maintain and improve the security).
- Collective Behaviour Tips: basic rules to increase the safety environment of other users in the area.

The main problem we encounter both in the risk assessment process and the generation of security protocols is the high dependence of these on the particular parameters of the area under control and the complexity of modelling and predicting human behaviour under panic situations. Although in the literature we can find some work about similar topics, most of the studies focus on the interactions between humans and their surroundings during evacuation in general emergencies [12], but there are very few studies about other threats (such as violent assaults, including terrorist attacks with planning) [24]. If it is normally hard to model the reaction of a group under a panic situation when the focus of danger is clear and inert, it is even harder when the focus requires additional modelling due to human causes [16].

As a first approach to this problem, we have decided to use multi-agent modelling that allows to parameterize in an open and independent way the possible dangerous conditions (fire, attack of mobile agents with weapons, bombs, attack in group, etc.) and a collection of diverse users that have to react to the detected threat. We aim to provide an experimental platform on which to test various response strategies and measure their effectiveness under a controlled environment (Fig. 8).

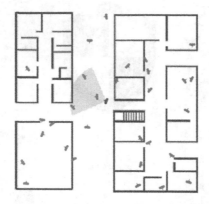

Fig. 8. Hazard Simulation environment.

4   Conclusions

We have introduced the VICTORY security system, which seeks to improve current security systems by providing better reaction time and the ability to generate an optimal response through automation and the use of mobile devices to help potential victims.

The system is mainly divided into a computer vision component and a crowd-sensing component, joined by an integration framework that aggregates all the information and generates the response. The vision component uses deep learning techniques to detect threats such as firearms or knives automatically. The crowd-sensing component analyses the data from the sensors of the mobile devices of the users who are in the security zone, detecting falls, stampedes or other signs of violence. These devices also provide users with personalised and dynamic guidance to help keep them safe.

The VICTORY system is still an ongoing project in which there is still a long way to go, but we are confident that the application of new technological advances to security systems can in the near future significantly reduce the damage caused by violent events that occur periodically worldwide.

References

1. Gun   detector.   https://github.com/swatz10/Weapon-Detection-Final-Year-Project. Accessed 09 Feb 2019
2. Ahn, J., Han, R.: An indoor augmented-reality evacuation system for the smartphone using personalized pedometry. Hum. Centric Comput. Inf. Sci. **2**(1), 18 (2012)

3. Álvarez-García, J.A., Barsocchi, P., Chessa, S., Salvi, D.: Evaluation of localization and activity recognition systems for ambient assisted living: the experience of the 2012 evaal competition. J. Ambient Intell. Smart Environ. **5**(1), 119–132 (2013)
4. Basalamah, A.: Sensing the crowds using bluetooth low energy tags. IEEE Access **4**, 4225–4233 (2016)
5. Blanke, U., Troster, G., Franke, T., Lukowicz, P.: Capturing crowd dynamics at large scale events using participatory GPS-localization. In: 2014 IEEE Ninth International Conference on Intelligent Sensors, Sensor Networks and Information Processing (ISSNIP), pp. 1–7. IEEE (2014)
6. Cireşan, D., Meier, U., Masci, J., Schmidhuber, J.: Multi-column deep neural network for traffic sign classification. Neural Netw. **32**, 333– 338 (2012). https://doi.org/10.1016/j.neunet.2012.02.023, http://www.sciencedirect.com/science/article/pii/S0893608012000524. Selected Papers from IJCNN 2011
7. de la Concepción, M.Á.Á., Morillo, L.M.S., García, J.A.Á., González-Abril, L.: Mobile activity recognition and fall detection system for elderly people using ameva algorithm. Pervasive Mob. Comput. **34**, 3–13 (2017)
8. Crooks, A., Croitoru, A., Stefanidis, A., Radzikowski, J.: # earthquake: Twitter as a distributed sensor system. Trans. GIS **17**(1), 124–147 (2013)
9. Gjoreski, H., et al.: Competitive live evaluations of activity-recognition systems. IEEE Pervasive Comput. **14**(1), 70–77 (2015)
10. Gutoski, M., Ribeiro, M., Aquino, N.M.R., Lazzaretti, A.E., Lopes, H.S.: A clustering-based deep autoencoder for one-class image classification. In: 2017 IEEE Latin American Conference on Computational Intelligence (LA-CCI), pp. 1–6 (2017)
11. He, S., Chan, S.H.G.: Wi-fi fingerprint-based indoor positioning: recent advances and comparisons. IEEE Commun. Surv. Tutorials **18**(1), 466–490 (2016)
12. Helbing, D., Farkas, I., Vicsek, T.: Simulating dynamical features of escape panic. Nature **407**, 487–490 (2000). https://doi.org/10.1038/35035023
13. Jiang, Y., et al.: Ariel: automatic wi-fi based room fingerprinting for indoor localization. In: Proceedings of the 2012 ACM Conference on Ubiquitous Computing, pp. 441–450. ACM (2012)
14. Krizhevsky, A., Sutskever, I., Hinton, G.E.: Imagenet classification with deep convolutional neural networks. In: Advances in Neural Information Processing Systems, pp. 1097–1105 (2012)
15. Lara, O.D., Labrador, M.A., et al.: A survey on human activity recognition using wearable sensors. IEEE Commun. Surv. Tutorials **15**(3), 1192–1209 (2013)
16. Li, S., Zhuang, J., Shen, S.: A three-stage evacuation decision-making and behavior model for the onset of an attack. Transp. Res. Part C: Emerg. Technol. **79**, 119–135 (2017). http://www.sciencedirect.com/science/article/pii/S0968090X17300840
17. Lin, T., et al.: Microsoft COCO: common objects in context. CoRR abs/1405.0312 (2014). http://arxiv.org/abs/1405.0312
18. Nunavath, V., Prinz, A.: Liferescue: a web based application for emergency responders during fire emergency response. In: 2016 3rd International Conference on Information and Communication Technologies for Disaster Management (ICT-DM), pp. 1–8. IEEE (2016)
19. Olmos, R., Tabik, S., Herrera, F.: Automatic handgun detection alarm in videos using deep learning. Neurocomputing **275**, 66–72 (2018)
20. Pan, B., Zheng, Y., Wilkie, D., Shahabi, C.: Crowd sensing of traffic anomalies based on human mobility and social media. In: Proceedings of the 21st ACM SIGSPATIAL International Conference on Advances in Geographic Information Systems, pp. 344–353. ACM (2013)

21. Russakovsky, O., et al.: ImageNet large scale visual recognition challenge. Int. J. Comput. Vis. (IJCV) **115**(3), 211–252 (2015). https://doi.org/10.1007/s11263-015-0816-y
22. Srivastava, N., Hinton, G., Krizhevsky, A., Sutskever, I., Salakhutdinov, R.: Dropout: a simple way to prevent neural networks from overfitting. J. Mach. Learn. Res. **15**(1), 1929–1958 (2014)
23. Vállez, N., Bueno, G., Déniz, O.: False positive reduction in detector implantation. In: Peek, N., Marín Morales, R., Peleg, M. (eds.) AIME 2013. LNCS (LNAI), vol. 7885, pp. 181–185. Springer, Heidelberg (2013). https://doi.org/10.1007/978-3-642-38326-7_28
24. Wang, H., Mostafizi, A., Cramer, L.A., Cox, D., Park, H.: An agent-based model of a multimodal near-field tsunami evacuation: decision-making and life safety. Transp. Res. Part C: Emerg. Technol. **64**, 86–100 (2016). http://www.sciencedirect.com/science/article/pii/S0968090X15004106
25. Xu, Z., et al.: Crowdsourcing based description of urban emergency events using social media big data. IEEE Trans. Cloud Comput. **10**, 1109 (2016)
26. Xu, Z., Liu, Y., Zhang, H., Luo, X., Mei, L., Hu, C.: Building the multi-modal storytelling of urban emergency events based on crowdsensing of social media analytics. Mob. Netw. Appl. **22**(2), 218–227 (2017)
27. Zhang, C., Bengio, S., Hardt, M., Recht, B., Vinyals, O.: Understanding deep learning requires rethinking generalization (2016)

New Intelligent Tools to Adapt NL Interface to Corporate Environments

Svetlana Chuprina$^{(\boxtimes)}$ and Igor Postanogov

Perm State University, Bukireva Str. 15, 614990 Perm, Russia
chuprinas@inbox.ru, ipostanogov@outlook.com

Abstract. This paper is devoted to new aspects of Natural Language Interface to Relational Database (NLIDB) integration into third-party corporate environments related to control data access. Because there is no schema information in the input NL query and the different relational database management system (RDBMS) requires different metadata types and rules to control data access, developers meet a problem addressed to automatic data access control in the case of NL interface implementation to relational databases. In the paper, we suggest a comprehensive approach which takes into account permissions throughout the pipeline of transforming NL query into SQL query with an intermediate SPARQL representation. Our integration solutions based on well-known Ontology Based Data Access (OBDA) approach, which gives us the opportunity to adapt the proposed solutions to the specifics of access control facilities of various RDBMS. Suggested approach has been implemented within intelligent service, named Reply.

Keywords: Natural Language Interface ·
Natural Language Query to Relational Database ·
Database Access Control · Ontology Based Data Access ·
Intelligent Information System

1 Introduction

Nowadays there is a trend for providing natural language interfaces (NL interfaces, NLI) for human-machine communication. According to the Gartner's hype cycle report for emerging technologies [11], in 2017 Virtual Assistants with NL interfaces, such as Siri (iOS), Google Assistant and Yandex.Alisa (Android), Cortana (Windows) were at the peak of the cycle. Now the advances in this area (such as NL understanding) have a major impact on the corporate environment. Because a lot of time of employees is still spent on formulating of queries for retrieving data (see, for example, [6]), it is beneficial for companies to use high-level tools to simplify this process by means of NLIDBs. NLI lowers the entry threshold for newcomers and gives an ability to formulate ad hoc queries.

NLIDBs can be classified as domain-dependent and domain-independent. The domain-dependent usually demonstrates better results but are highly restricted

© Springer Nature Switzerland AG 2019
J. M. F. Rodrigues et al. (Eds.): ICCS 2019, LNCS 11540, pp. 27–40, 2019.
https://doi.org/10.1007/978-3-030-22750-0_3

to the domain specifics. In [3] we have presented our domain-independent system Reply for transforming traditional information systems (IS) into intelligent ones with NL querying. End users are provided with new NLI to their existing IS by means of Reply without source code modification. It is also possible to start developing a new IS with NLI as a ready-made component to its database. To provide NLI to relational database (RDB) we have used an ontology-based data access approach that helps us automatically transform input NL query into SPARQL query and then into SQL query taking into account the specifics of concrete RDBMS, database schema, and its content.

Towards creating a platform which can be used as a self-service solution, we have taken a number preexisted stand-alone third-party modules and complemented them with our own modules to build a complete solution. Although many NLIDBs have been developed since 60s, there are only a few reports about real-world installation (e.g., [16]). Bearing in mind that "NLIDBs will not be widely used in businesses until their configuration is so easy that any computer professional could be able to perform it" [12], we have developed a number of tools, which automate the configuration throughout the whole pipeline. Thanks to that, our solution favorably distinguishes from others.

While bringing our system into real-world enterprise infrastructure we have encountered the problem that neither our, nor any third-party transformation module used in our pipeline was not taking into account user's access privileges for the data sources. E.g., SPARQL to SQL module expects that used configuration maps elements of the ontology to available elements of the database. Direct module integration led to permission violation exceptions at execution stage accompanied with error messages that are incomprehensible to the end user.

To the best of our knowledge, there is no published approach for automated handling data access policies in ontology-based NLIDBs, which use black box NL to SPARQL and SPARQL to SQL modules that are unaware of policy restrictions.

The main contributions of our research presented in the paper are the following:

1. adaptive ontology-driven approach to enforce practical value of NLIDBs taking into account the issues of restricted data source access;
2. algorithm for handling access policy in NLIDBs based OBDA solutions, where SPARQL is used as an intermediate representation language;
3. demonstration of the proposed approach with an example.

The paper is divided into six sections. Section 2 contains the description of the suggested approach. Section 3 includes a demo example. Section 4 presents a preliminary evaluation of the approach. Section 5 describes the related work. Section 6 contains a conclusion and future perspectives.

2 NLIDB Obstacles and OBDA Solutions

In this paper, we will focus on access policies for 2 types of database objects such as tables and columns. Access privileges for rows (row-level security) are usually implemented with the help of data views. The user cannot violate access privileges for rows and if some data row shouldn't be returned to the user according to the access control requirements, it should have already been filtered from the result. For cell-level privileges it is common to mask the not accessible actual value and inform the user that the value was masked. Before discussing the NLIDB obstacles, let's say a few words about data access control in traditional information system UIs.

For a traditional information systems UI it is very easy to report current access restrictions. If there are some UI controls that lead to the execution of queries that cannot be executed, they can be disabled or hidden. Queries that cannot be executed are queries that access restricted tables (for table-level restrictions) or restricted columns in table references and search conditions clauses (for column-level restrictions). Asterisk signs in select lists are explicitly expanded to the available columns. In the case of whole database search (e.g., by substring) the search should be done only in the available tables and columns.

User interface for an ad hoc query usually has controls for specifying blocks in the resulting SQL query, e.g., there are controls for choosing output fields, conditions, sorting order etc. For the specification stage, all restricted tables and columns should be disabled or hidden. If UI suggests possible values for columns that are affected by row or cell-level security, the suggestions should contain only visible values. For the building stage, where the resulting SQL query is formulated, the building engine should use only those paths and join conditions for connecting required tables, which avoid restricted objects. If this is not possible, advanced ad hoc query UIs suggest which fields or tables should be excluded to make query building successful.

For NLIDB it is usually difficult to inform the user what information is available via NLI. The easiest way to do this is to describe the available information in the documentation. More advanced user interfaces suggest autocompletion tools. Another problem is a lack of adequate reporting if the query fails, which can occur for several reasons. In this paper, we focus on the case when the query has been successfully interpreted and the execution has failed due to permission violation.

This problem can be solved in two main ways:

1. Interpreting the query without considering permissions, executing the query and catching the exception, interpreting the exception and reporting the reason why this exception has happened.
2. Considering permissions on the interpretation stage and trying to avoid permission violation.

The challenge of the first way is that the exception could have happened due to a number of reasons, so it is difficult to take into account all the related issues. That is why we chose the second way.

Considering permissions on the interpretation stage is specific to the NLIDB type. Different NLIDB types may be classified by architecture into four groups [1]:

1. pattern-matching;
2. syntax-based systems;
3. semantic grammar systems;
4. intermediate representation languages systems.

For the intermediate representation language system, one could use SPARQL. Therefore, an input NL query is transformed into SPARQL query which is later transformed into SQL query by an OBDA framework. Reply has this type of architecture with SPARQL as the intermediate representation language.

OBDA systems built on top RDBs can be implemented in two different ways: materialized and virtual. According to the materialized approach, all the data from RDBs should be transformed into ontology representation and stored in RDF triplestores. In the virtual approach, the data is stored only in RDBs, and querying the virtual RDF graph using SPARQL leads to execution the corresponding SQL query over the original source thus avoiding the cost of materialization [2].

From the security point of view, another benefit of the virtual approach is that data is not duplicated in RDF triplestore. Although access control can be specified at the RDF level [7], not so many IT-specialists who knows not only how to specify user access rights in the RDBMS, but also how to specify user access rights in RDF triplestore. With that in mind, many companies would reject such OBDA systems because the possibility of data leaks due to incorrect configuration and may overweight the potential benefit of the intelligent system applying. If the virtual approach is used, end user can only access the data which she/he can access using traditional ad hoc query interface.

In OBDA solutions for NLIDBs there are 4 components where permissions could be considered:

1. OBDA framework;
2. NL to SPARQL translator;
3. ontology;
4. mappings between elements of an RDB schema and elements of an ontology.

For this paper, we treat OBDA framework as a black box with ontology, mappings and SPARQL query as input and an SQL query as an output. With this assumption, we show how OBDA solutions for NLIDBs could be enhanced.

The original architecture is presented at Fig. 1. For the sake of simplicity do not depict the NL to SPARQL module errors on the diagram.

OBDA mapping specification consists of 2 blocks: *Source* and *Target* (in the Ontop mappings notation). *Source* is an SQL SELECT query that would be executed by RDB to raise triplets defined in *Target*.

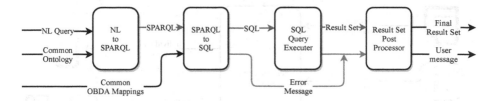

Fig. 1. Original Reply's execution stage.

On the assumption that *Target* contains only one template for an RDF triple, this triple has one of the following types:

1. `<instance_uri> a <class_uri>`
2. `<instance1_uri> <object_property_uri> <instance2_uri>`
3. `<instance_uri> <data_property_uri> <value>`

SQL SELECT query has the following structure: `SELECT <select_list> FROM <from_item> WHERE <condition>`

When considering the permission aspect, *select_list* and *conditions* from the SQL SELECT query could use fields, which could be forbidden. In addition, the *from_item* can refer to the forbidden tables or fields. Although SQL looks different from the NL query point of view, their building blocks are the same. NL query can vary output fields, requested entities types, and conditions. Like in SQL, output fields can be specified implicitly.

The easiest way to prevent execution failure of the translated query is to remove elements of the ontology that are mapped to the forbidden elements of the RDB. In this case, the queries that contain removed elements of ontology could be misinterpreted. Firstly, the closest element in the edited ontology maybe not related to the concept/relation which the user has asked for. Secondly, if the closest concept/relation could not be found, the related part of the query would be completely ignored, and the NLIDB has not enough data to inform the user about the situation and give some recommendations about the query reformulation. Worse, in this case, the user can conclude that the system cannot handle complex queries. Finally, the system would not know why the user was not satisfied with the query result and would not give any recommendations about query reformulation.

In order to tackle this problem, we suggest the following approach. For each user (or for each role if multiple users have the same privileges) automatically a new instance of ontology and a set of mappings are created by means of our personalization module. The individual user's instance of ontology is created by copying the common ontology and annotating the elements for which mappings have been removed and no other mappings left. The individual user's set of mappings is created by removing from the common mapping specification only those ones, *Sources* of which suppose access to restricted tables or fields. This process is presented at Fig. 2.

To provide the best user experience translators from NL to SPARQL should handle permission-based restrictions. Unfortunately, none of the tools that we

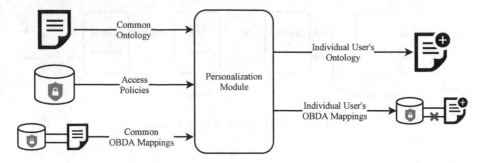

Fig. 2. Creating individual user's instance of ontology and set of mappings.

have analyzed [4,15,18] support this. Here we suggest an approach how the existing NL to SPARQL translators can be minimally modified to support permission-based restrictions.

One of the main requirements is that each output data property, that wasn't explicitly specified in the user's query, should be wrapped into OPTIONAL block of the related SPARQL query. If there are no mappings for the data property that is wrapped into OPTIONAL block [17], OBDA framework would return the corresponding SQL query where the not found column would have NULL values.

To make Reply more adaptable to the user preferences the default values for output attributes of the queried concepts are supported. E.g., if some user types the query "list clients", a predefined subset of fields would be returned per client. Additionally, the system supports application-wide defaults, and the user can also choose the preferred attributes. This approach is implemented by excluding OPTIONAL blocks with output data properties that are not in the corresponding preferred list. An alternative connection between the related classes via object properties can be specified by means of UNION.

Algorithm 1 describes steps of our approach in case of permission change. Algorithm 2 can be used for modifying intermediate SPARQL query for user's NL query. The complete transformation process is presented at Fig. 3. The new modules are marked by the blue color.

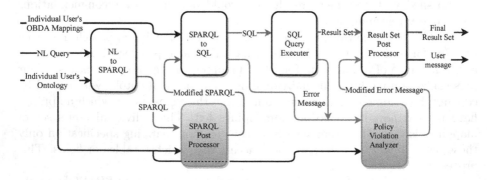

Fig. 3. Modified Reply's execution stage. (Color figure online)

foreach *user U in tracked users* do

> **retrieve** permissions from the RDB for U;
> **if** *permissions changed* **then**
>
>> **create** a copy of common mappings B for U;
>> **foreach** *table T used in the Sources of mappings* **do**
>>
>>> **if** *access to T is not allowed* **then**
>>>
>>>> | **remove** all mappings, that reference T;
>>>
>>> **end if**
>>
>> **end foreach**
>> **foreach** *column C used in the Sources of mappings* **do**
>>
>>> **if** *access to C is not allowed* **then**
>>>
>>>> | **remove** all mappings, that reference C;
>>>
>>> **end if**
>>
>> **end foreach**
>> **foreach** *ontology element E that had mappings in B* **do**
>>
>>> **if** *there are no spawning mappings left* **then**
>>>
>>>> | **mark** E as unavailable due to the policy restrictions;
>>>
>>> **end if**
>>
>> **end foreach**
>
> **end if**

end foreach

Algorithm 1: Generating users' mappings on access policy change.

transform NL query Q into SPARQL query SPQ;
init empty list of errors $ERRS$;
foreach *non-optional ontology element E in SPQ* **do**

> **if** *E is marked as unsupported* **then**
>
>> | **add** into $ERRS$ an error that E is unsupported;
>
> **else if** *E is marked as unavailable due to the policy restrictions*
> **then**
>
>> | **add** into $ERRS$ an error that E is unavailable;
>
> **end if**

end foreach
if *ERRS is not empty* **then**

> | **return** $ERRS$;

end if
foreach *optional ontology data property P in SPQ* **do**

> **if** *P is not in the preferred set of attributes* **then**
>
>> | **remove** P from SPQ;
>
> **end if**

end foreach
transform SPQ into SQL query SQ;
return execute SQ;

Algorithm 2: Executing NL query taking into account access policy.

3   Example

To demonstrate the suggested approach, we present an example of our demo database about apartment agreements. Each house contains a number of apartments. Some of the apartments are rented, and some of them have been sold. For each rented apartment there is a rental agreement, and for each sold apartment there is a service agreement.

A person who has a running rental agreement is called tenant, and a person who owns an apartment (has a running service agreement from the landlord's point of view) — owner. The fragment of this database schema is presented at Fig. 4.

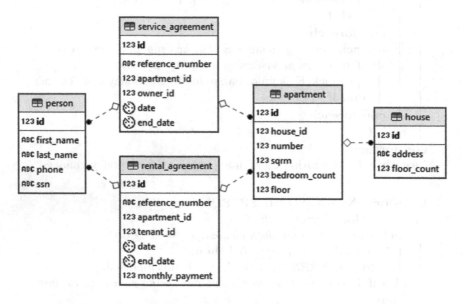

Fig. 4. Fragment of database schema about apartment agreements.

The restrictions on this database are the following: technical support specialist have no access to *service_agreement* table (therefore have no access to the information who are the owners in the list of people), and have no access to a *social security number* of any person. If the database has row-level access policies, the forbidden rows would be hidden by the database management system itself so there is no need in a special treatment on the OBDA layer. If there is a cell-level access policy, restricted sells would be marked as hidden (either by NULL or predefined value) and would not need the special treatment.

As it was suggested above, the individual user's set of mappings was created by removing from the source mapping specification only those ones, *Sources* of which suppose access to restricted tables or fields.

Figure 5 presents a result of comparison between source mapping rules file and the corresponding individual user's file. The latter lacks mappings related to *service_agreement* table and *SSN* column.

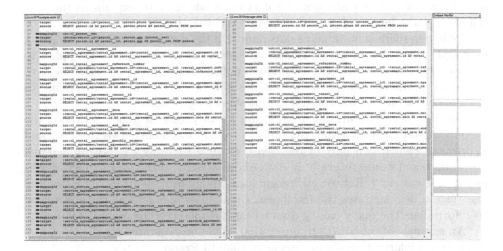

Fig. 5. Comparison between source and individual mapping rule files.

For NL query "Return all information about the tenants" the traditional NL to SPARQL engine would detect that person has an *id*, *first_name*, *last_name*, *phone*, and *SSN* and would formulate a SPARQL query returning all these columns. Running this query would lead to an access denied exception due to the *SSN*'s restriction. To solve this problem one could supply the NL to SPARQL engine information about the restricted fields or post-process the generated query. The easiest solution is to put the expected fields in OPTIONAL blocks. In that case, because the mapping for the forbidden fields has been removed, the OBDA framework would just skip them without any exception.

The same solution is adequate for the cases when all the output fields are available and explicitly specified (e.g., NL query "Return first and last name of tenants"). For NL query "Return first name, last name and SSN of tenants" OBDA framework would fail to generate the SQL query because of the lack of mappings for *SSN*. To report the user that the field *SSN* was detected in the query but was omitted because of the policy restrictions, our system parses the generated SPARQL query to find the data properties for which there are no mappings. This approach is error-prone, that's why we suggest for NL to SPARQL query engines to explicitly output all detected classes, objects and data properties as well as the role they play in the query.

All constraints, retrieved from the query should be specified in the SPARQL query as usual. If the user constrains the available field, the query would be successfully executed (e.g., "Return all information about tenants with last name Doe"). If the user constrains the restricted field ("Return all information about a person with SSN 078-05-1120"), the OBDA framework would fail to generate the corresponding SQL query due to the lack of mappings for *SSN* attribute. There is an option to remove the constraint automatically and to answer the query without this constraint, but we believe that most of the time it would be not the result the user expects and it is more adequate to explain to the user why this query has no opportunity to be executed.

Although there is no table named *tenant* in the database, we specify tenants as people who have a running rental agreement. The same approach applies to the owners. The user with the permission policy specified above would not have access to service agreements so she/he not able to get the expected response to the query "Return all information about service agreements".

For NL query "Return all information about people at 1650 El Prado" the system will establish that the only object property available between *person* and *apartment* (which is a part of the *house* at 1650 El Prado) is *having a rental agreement*, and would return only data about *tenants*. If the user explicitly asks about the *owners* ("Return all information about owners at 1650 El Prado") the system demonstrates the intelligent capabilities and returns the suggestion to reformulate the query asking about *people* (superclass) or *tenants* (sibling).

An example of the SPARQL query that would be generated for the NL query "Return all information about people at 1650 El Prado":

SELECT DISTINCT ?person_fn ?person_ln ?person_phone ?person_ssn {
    {
        ?agreement :service_agreement_person_id_fkey ?person.
        ?agreement :service_agreement_apartment_id_fkey ?apartment.
    } **UNION** {
        ?agreement :rental_agreement_tenant_id_fkey ?person.
        ?agreement :rental_agreement_apartment_id_fkey ?apartment.
    }
    OPTIONAL { ?person :person.first_name ?person_fn. }
    OPTIONAL { ?person :person.last_name ?person_ln. }
    OPTIONAL { ?person :person.phone ?person_phone. }
    OPTIONAL { ?person :person.ssn ?person_ssn. }
    ?apartment :apartment_house_id_fkey ?house.
    ?house :house.address "1650 El Prado".
}

The first block of the UNION clause and the SSN's OPTIONAL clause would be omitted by an OBDA framework due to the lack of mappings which have been removed according to our approach.

4   Implementation and Testing Specifics

The suggested approach was implemented in our system Reply which transforms traditional information systems into intelligent ones with a natural language query interface. In the previous version of Reply administrator had an ability to create multiple independent versions of intelligent systems by specifying multiple pairs of ontology and mappings. In the current version, in addition to that, a base pair of ontology and mappings could be specified and a permission tracking option could be enabled. In that case, our system automatically retrieves access policies from RDB metadata and generate multiple configurations per selected users according to the suggested approach. This process can be rerun from the system administrator's GUI, by an API call or automatically by our system. The automatic permission tracking is available for a subset of supported databases.

For example, to automatically track changes in user permissions for PostgreSQL we generate a code for creating an *event trigger* that is fired in case of GRANT or REVOKE DDL commands. The generated code should be executed by a database superuser. In the body of the trigger, we raise a *notification* to a channel that our system *listens* to. PostgreSQL's metadata about user permissions is retrieved from the *information_schema.column_privileges* view.

To parse SQL queries in the *Sources* of the mappings we use JSqlParser, and for manipulating generated SPARQL query — Apache Jena.

Because we haven't found a NL to SPARQL translator that outputs implicitly requested attributes in a consistent way to our approach, testing the whole pipeline starting with NL query would not provide more information than starting with SPARQL query. As it was said above, our approach is independent not only from the language used by the user but also from the specifics of the query, from the specifics of the implementations of NL to SPARQL and SPARQL to SQL translators.

To test our system, we use the database from the example above (see Fig. 4). The database schema contains one-to-many and many-to-many relations, as well as multiple paths between tables. We have generated SPARQL queries, which use combinations of available and forbidden tables and columns. Our system has successfully reported permission violation and suggested what elements of the NL query should be avoided (even though there was no one). In the real-world example, the reported elements might not be explicitly included in the user's NL query, and so to make the report even clearer, tighter integration with NL to SPARQL translator is needed. We have also tested whether our system correctly reacts to permission updates.

5   Related Work

Modern corporate NLIDBs have a number of additional requirements that are not usually considered by academia:

1. taking into account user's access privileges for the data sources;
2. providing out-of-the-box integration with corporate chat platforms;
3. reusing existing corporate knowledge resources for automation the configuration stage.

At present, there are no platforms, which provide a comprehensive OBDA solutions for NLIDB and at the same time cover all of its pipeline stages. Nevertheless, there are a number of tools, which implement the individual stages of the pipeline to transform NL query into SQL query with the intermediate SPARQL representation [2,4,15,18]. Almost all of them expect that either all instances in ontology or the whole database are available. This is not the case for corporate environment where the users usually have only limited access to data. The papers [10,12] present sufficiently complete overview of the NLIDB state of the art, including benefits and limitations of different solutions.

Another key factor for successful installation NLIDB into corporate environment is bringing a tight integration with existing NL tools. One of the most widely used types of them is corporate chat platforms that support chat bots. Amongst other things, chat bots can report a status of the running systems, run tests and even deploy code to the production environment. From the earliest versions, Reply had an API that can be called from user's NL query interface to retrieve the query result. While bringing Reply into real-world installations we have developed out of the box wrappers for the most popular corporate chat platforms. Reply can be integrated into third-party web-based information systems as a separate web-page. Due to the fact that we provide HTTP REST API, Reply can also be used by desktop applications.

Previously, to automate Reply configuring we have successfully implemented a number of our [14] and other's [5] state of art techniques. E.g., we automatically convert RDB's schema to ontology representation and enrich it with synonyms, parent and child concepts using our or third-party ontology or linguistic resources [8,13]. In corporate environment, we can also use corporate knowledge resources and document storages as a source for ontology learning. Nowadays, many of knowledge resources use wiki or wiki-based markup language. Our approach for constructing ontologies from knowledge bases being developed is similar to the approach used for building DBpedia [9].

To improve the comprehensiveness of NLIDB's and make the NL-interface enable users to get really practical results we suggest new intelligent tools to adapt NL-interface to corporate environments on self-service principles.

6   Conclusion

In the paper, we presented the new approach to intelligent handling database access policies within OBDA. It helps to tackle the problems of adaptation of NLIDBs to corporate environments. Our approach does not depend on the methods used to implement the transformation of NL query to SPARQL as well as SPARQL to SQL so it is applicable to a wide range of OBDA solutions for NLIDBs.

According to the algorithm, the system automatically modifies the generated SPARQL query taking into account the access policies of target DBMS and user's preferences. The algorithm is demonstrated on the apartment agreements database example.

In future, we plan to apply developed tools and services not only for querying RDBMS using natural language but also for inserting data into them. We believe that it is possible to use our intelligent tools to improve the human-computer interaction within traditional corporate environments and also in other environments, such as human-centric Internet of Things.

References

1. Androutsopoulos, I., Ritchie, G., Thanisch, P.: Natural language interfaces to databases - an introduction. Nat. Lang. Eng. **1**, 29–81 (1995). https://doi.org/10.1017/S135132490000005X
2. Calvanese, D., et al.: Ontop: answering SPARQL queries over relational databases. Semant. Web J. **8**(3), 471–487 (2017). https://doi.org/10.3233/SW-160217
3. Chuprina, S., Postanogov, I., Nasraoui, O.: Ontology Based Data Access Methods to Teach Students to Transform Traditional Information Systems and Simplify Decision Making Process. Procedia Comput. Sci. **80**, 1801–1811 (2016). https://doi.org/10.1016/j.procs.2016.05.458
4. Dubey, M., Dasgupta, S., Sharma, A., Höffner, K., Lehmann, J.: AskNow: a framework for natural language query formalization in SPARQL. In: Sack, H., Blomqvist, E., d'Aquin, M., Ghidini, C., Ponzetto, S.P., Lange, C. (eds.) ESWC 2016. LNCS, vol. 9678, pp. 300–316. Springer, Cham (2016). https://doi.org/10.1007/978-3-319-34129-3_19
5. Jiménez-Ruiz, E., et al.: BOOTOX: practical mapping of RDBs to OWL 2. In: Arenas, M., et al. (eds.) ISWC 2015. LNCS, vol. 9367, pp. 113–132. Springer, Cham (2015). https://doi.org/10.1007/978-3-319-25010-6_7
6. Kharlamov, E., et al.: How semantic technologies can enhance data access at siemens energy. In: Mika, P., Tudorache, T., Bernstein, A., Welty, C., Knoblock, C., Vrandečić, D., Groth, P., Noy, N., Janowicz, K., Goble, C. (eds.) ISWC 2014. LNCS, vol. 8796, pp. 601–619. Springer, Cham (2014). https://doi.org/10.1007/978-3-319-11964-9_38
7. Kirrane, S., Mileo, A., Decker, S.: Access control and the resource description framework: a survey. Semant. Web (2017). https://doi.org/10.3233/SW-160236
8. Kostareva, T., Chuprina, S., Nam, A.: Using ontology-driven methods to develop frameworks for tackling NLP problems. In: Supplementary Proceedings of the Fifth International Conference on Analysis of Images, Social Networks and Texts (AIST 2016), Yekaterinburg, Russia, 6–8 April 2016, pp. 102–113 (2016). http://ceur-ws.org/Vol-1710/paper11.pdf
9. Lehmann, J., et al.: DBpedia - a large-scale, multilingual knowledge base extracted from Wikipedia. Semant. Web **6**(2), 167–195 (2015). https://doi.org/10.3233/SW-140134
10. Nihalani, N., Silakari, S., Motwani, M.: Natural language interface for database: a brief review. Int. J. Comput. Sci. Issues (IJCSI) **8**(2), 600–608 (2011)
11. Panetta, K.: Top Trends in the Gartner Hype Cycle for Emerging Technologies (2017). https://www.gartner.com/smarterwithgartner/top-trends-in-the-gartner-hype-cycle-for-emerging-technologies-2017/. Accessed 05 Jan 2019
12. Pazos, R.R.A., González, B.J.J., Aguirre, L.M.A., Martínez, F.J.A., Fraire, H.J.: Natural language interfaces to databases: an analysis of the state of the art. In: Castillo, O., Melin, P., Kacprzyk, J. (eds.) Recent Advances on Hybrid Intelligent Systems. SCI, pp. 463–480. Springer, Heidelberg (2013). https://doi.org/10.1007/978-3-642-33021-6_36

13. Postanogov, I., Jastrzab, T.: Ontology reuse as a means for fast prototyping of new concepts. In: Kozielski, S., Mrozek, D., Kasprowski, P., Małysiak-Mrozek, B., Kostrzewa, D. (eds.) BDAS 2017. CCIS, vol. 716, pp. 273–287. Springer, Cham (2017). https://doi.org/10.1007/978-3-319-58274-0_23
14. Postanogov, I.: Towards automating the creation of OBDA systems. Procedia Comput. Sci. **150**, 511–517 (2019). https://doi.org/10.1016/j.procs.2019.02.086. Proceedings of the 13th International Symposium "Intelligent Systems 2018" (INTELS 2018), 22–24 October, 2018, St. Petersburg, Russia
15. Shekarpour, S., Marx, E., Ngonga Ngomo, A.C., Auer, S.: SINA: semantic interpretation of user queries for question answering on interlinked data. J. Web Semant. (2015). https://doi.org/10.1016/j.websem.2014.06.002
16. Waltinger, U., Tecuci, D., Olteanu, M., Mocanu, V., Sullivan, S.: Natural Language Access to Enterprise Data. AI Mag. **35**, 38 (2014). https://doi.org/10.1609/aimag.v35i1.2502
17. Xiao, G., Kontchakov, R., Cogrel, B., Calvanese, D., Botoeva, E.: Efficient handling of SPARQL OPTIONAL for OBDA. In: Vrandečić, D., et al. (eds.) ISWC 2018. LNCS, vol. 11136, pp. 354–373. Springer, Cham (2018). https://doi.org/10.1007/978-3-030-00671-6_21
18. Xu, K., Feng, Y., Zhao, D.: Xser@QALD-4: answering natural language questions via phrasal semantic parsing. In: CEUR Workshop Proceedings (2014). https://doi.org/10.1007/978-3-662-45924-9_30

Asymmetric Deep Cross-modal Hashing

Jingzi Gu[1,2], JinChao Zhang[1(✉)], Zheng Lin[1], Bo Li[1], Weiping Wang[1],
and Dan Meng[1]

[1] Institute of Information Engineering, Chinese Academy of Sciences, Beijing, China
{gujingzi,zhangjinchao,linzheng,libo,wangweiping,mengdan}@iie.ac.cn
[2] School of Cyber Security, University of Chinese Academy of Sciences,
Beijing, China

Abstract. Cross-modal retrieval has attracted increasing attention in recent years. Deep supervised hashing methods have been widely used for cross-modal similarity retrieval on large-scale datasets, because the deep architectures can generate more discriminative feature representations. Traditional hash methods adopt a symmetric way to learn the hash function for both query points and database points. However, those methods take an immense amount of work and time for model training, which is inefficient with the explosive growth of data volume. To solve this issue, an Asymmetric Deep Cross-modal Hashing (ADCH) method is proposed to perform more effective hash learning by simultaneously preserving the semantic similarity and the underlying data structures. More specifically, ADCH treats the query points and database points in an asymmetric way. Furthermore, to provide more similarity information, a detailed definition for cross-modal similarity matrix is also proposed. The training of ADCH takes less time than traditional symmetric deep supervised hashing methods. Extensive experiments on two widely used datasets show that the proposed approach achieves the state-of-the-art performance in cross-modal retrieval.

Keywords: Asymmetric hashing · Cross-modal · Retrieval

1 Introduction

A tremendous amount of data in heterogeneous modalities are being generated every day on the Internet, including image, text, audio, etc. Multimedia retrieval has been an essential technique in many applications. However, essential retrieval methods mainly focus on single-modal scenarios [1,2]. For example, image retrieval or text retrieval are homogeneous-modal, in which the query and result are from the same modality. These methods cannot directly measure the similarity between different modalities. Thus, effective retrieval of such massive amounts of media data from heterogeneous sources poses a great challenge.

Retrieval across multimedia data [3] is a relatively new paradigm. Recently, deep hashing methods [5] have been used in cross-modal retrieval. Hashing methods map high-dimensional representations in the original space to short binary

© Springer Nature Switzerland AG 2019
J. M. F. Rodrigues et al. (Eds.): ICCS 2019, LNCS 11540, pp. 41–54, 2019.
https://doi.org/10.1007/978-3-030-22750-0_4

codes in the Hamming space, which can bridge the "heterogeneity gap" between multimedia. For example, deep cross-modal hashing (DCMH) [7] integrates feature learning and hash-code learning into the same framework. Besides, collective deep quantization for efficient cross-modal retrieval (CDQ) [8] introduces quantization in end-to-end deep architecture for cross-modal retrieval.

However, the related previous methods adopt a symmetric strategy to learn deep hash function for both query points and database points. On one hand, the training of these symmetric deep hashing methods are typically time-consuming. The storage and computation of these data cost even more time. To make the training feasible, most deep hashing methods choose simple small datasets or subsets of a large dataset, which make it difficult to utilize the information adequately. On the other hand, these traditional supervised deep hashing methods measure the similarity of image-text using the semantic-level labels and define the similarity in a certain way. However, such similarity definition cannot reflect the similarity in detail on multiple labels datasets.

To solve these problems, an Asymmetric Deep Cross-modal Hashing (ADCH) method is proposed which treats the query points and database points in an asymmetric way. The binary hash codes of query points can be obtained from a deep hash function which is learned in this method, while the binary hash codes of database points are directly learned. The training of ADCH takes less time than traditional symmetric deep supervised hashing methods, because the training points are only query points that are much fewer than all of the databases points. Hence, the whole set of database points can be used for training even if the database is large. Furthermore, a detailed definition is proposed that the similarities between points are quantified into a percentage, which can provide more label information. In this paper, the main contributions of ADCH are outlined as follows:

(1) A novel method ADCH is proposed which learns cross-modal hash codes in an asymmetric way. ADCH generates hash codes for database points directly, while hash functions are only for query points. Therefore, it takes less training time than symmetric deep supervised hash methods.
(2) A detailed definition is proposed for cross-modal similarity matrix to make ADCH use more similarity information. The detailed definition of similarity matrix quantifies the similarity into a percentage with the normalized semantic labels, which provide more fine-grained information of labels in the loss function. Thus, it can improve the retrieval accuracy.
(3) ADCH takes less time and achieves high accuracy than the traditional symmetric deep supervised hashing methods. Experiments on two large-scale datasets show that ADCH can achieve the state-of-the-art performance.

2   Related Work

There are many traditional hashing methods [9,11] in cross-modal retrieval, which do not use deep networks. However, deep learning has shown its strength in modeling nonlinear correlation, and has achieved state-of-the-art performance in single-modal scenarios. Therefore, cross-modal hashing with deep methods has been proposed to meet retrieval demands in large-scale cross-modal databases.

Deep cross-modal hashing methods can be further divided into the following two categories: unsupervised and supervised hashing. Unsupervised cross-modal hashing methods map unlabeled input data into hash codes by learning a hash function. Zhang et al. make full use of Generative Adversarial Networks for unsupervised representation learning to exploit the underlying manifold structure of cross-modal data [4]. Even though unsupervised methods can get good performance, they still cannot satisfy the demanded accuracy of image retrieval. Therefore, lots of supervised methods were proposed to improve retrieval accuracy. Specifically, supervised cross-modal deep hashing methods [5,23] learn the hash function with supervised information.

The deep supervised architectures mainly include two ways in cross-modal retrieval. The first way is that inputs of different modal types pass through the same shared layer [12,13] to extract a unified representation that fuses modalities together, while the second way is that each modal passes through a sub-network and the output of these sub-networks are coupled by correlation constraints at the code layers [14,15]. For example, Cao et al. propose deep visual semantic hashing (DSVH) [6] model that generates compact hash codes of images and sentences in an end-to-end deep learning architecture, which captures the intrinsic cross-modal correspondences between visual data and a natural language. Wang [5] proposed an online learning method to learn the similarity function between heterogeneous modalities by preserving the relative similarity in the training data, which is modeled as a set of bidirectional hinge loss constraints on the cross-modal training triplets.

However, the related previous methods adopt a symmetric strategy to learn deep hash functions for both query points and database points. It is time-consuming to train data using these methods. Thus, most deep hashing methods choose simple small datasets or subsets of a large dataset. Unlike previous work, ADCH method is proposed which treats the query points and database points in an asymmetric way. It takes less time and achieves a better accuracy than the traditional symmetric deep supervised hashing methods.

3 Asymmetric Deep Cross-modal Hashing

3.1 Problem Definition

Although ADCH can be used in more than two modalities, we only focus on image and text for simplicity. Each point in the dataset has two feature modalities. Assume there is a query set image-text data-pair $\mathbf{P} = \{(x_i, y_i)\}_{i=1}^{m}$, in which m is the number of query data points, and a database set $\mathbf{D} = \{(x_j, y_j)\}_{j=1}^{n}$, in which n is the number of database points. In addition, we give a cross-modal similarity matrix $\mathbf{S}$ and it would be written as $\mathbf{S} = [S_{ij}]$ next. If $S_{ij} = 1$, $\{(x_i, y_i)\} \in P$ and $\{(x_j, y_j)\} \in D$ are similar; otherwise normally $S_{ij} = 0$, they are not similar. The matrix $\mathbf{S}$ is defined by semantic information such as class labels. For example, if image x_i and text y_j have the same label, then they are similar. Otherwise, they are dissimilar.

The goal of cross-modal hashing is to learn hash codes for both image and text. The hash codes preserve the semantic information of image and text. We learn two hash functions for query points: $f(x_i)$ for the image modality and $g(y_i)$ for the text modality. Additionally, $\mathbf{U} = \{(u_i^x, u_i^y)\}_{i=1}^m \in \{-1, +1\}^{m \times c}$ denotes the binary hash codes of query points, generating from $f(x_i)$ and $g(y_i)$. m is the number of query set, and c is the binary hash codes length. $\mathbf{V} = \{(v_j^x, v_j^y)\}_{j=1}^n \in \{-1, +1\}^{n \times c}$ denotes the directly learned binary hash codes of database. n is the number of database points. $\mathbf{U}$ and $\mathbf{V}$ are used to compute the similarity between the query points and the database points. The Hamming distance should be small if $S_{ij} = 1$, while the Hamming distance should be large if $S_{ij} = 0$.

In practice, there may be only database points $\mathbf{D} = \{(x_j, y_j)\}_{j=1}^n$ without query points. Hence, we randomly sample m points from the database as the query set. Set $\mathbf{P} = (\mathbf{D})^\Omega$, where $(\mathbf{D})^\Omega$ denotes the database points indexed by Ω. Here the indices of all the database points are denoted by $\Gamma = \{1, 2, ..., n\}$ and the indices of the sampled query points are denoted by $\Omega = \{i_1, i_2, ..., i_m\} \subseteq \Gamma$. Since $\Omega \subseteq \Gamma$, if a pair of points (x_i, y_i) belongs to Ω, then the points also belong to Γ. Accordingly, setting $\mathbf{S} = S^\Omega$, the supervised information (similarity) between pairs of all database points are denoted by $\mathbf{S} \in \{0, 1\}^{n \times n}$. $S^\Omega \in \{0, 1\}^{m \times n}$ denotes the submatrix formed by the rows of $\mathbf{S}$ indexed by Ω. The goal of asymmetric cross-modal hashing is that query set $\mathbf{P}$ is learned from the image and text modalities. And database set $\mathbf{D}$ is learned directly by loss function.

3.2   Model Formulation

The whole ADCH model is shown in Fig. 1, which contains the following three components: the image feature learning part, the text feature learning part and the asymmetric function part. In the image feature learning part, an appropriate image feature representation for image binary hash codes learning is extracted by using a deep convolutional neural network. Analogously, the text feature representation is extracted for the text binary hash codes learning in the text feature learning part. In the asymmetric loss part, an asymmetric loss function is proposed to train the end-to-end model.

Unlike previous methods, the query points and database points are treated in an asymmetric way in ADCH. Feature learning is only used to learn hash function for query points and not for database points, then the hash function of ADCH generates hash codes for query points; database points learn binary hash codes directly which is as a variable in the asymmetric loss. Because the training points are only query points, ADCH takes less time than traditional methods.

3.3   Feature Learning

In the ADCH model, feature learning contains the following two parts: image feature learning and text feature learning.

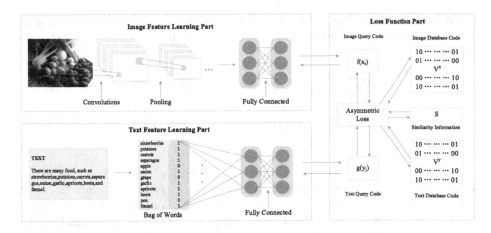

Fig. 1. The architecture of ADCH. The image feature learning part extracts image feature representations and the text feature part extracts text feature representations. The loss function preserves the similarity between query points and database points.

In the image feature learning part, the deep neural network is a convolutional neural network (CNN) model from CNN-F [16] model. In this model, there are eight layers in total, five of which are fully connected convolutional layers. The last layer of the CNN-F model is used to project the output of the first seven layers into R^c space, which represents the learned image features. The details of CNN-F can be found in [16].

In the text feature learning part, the text alternates into bag-of-words (BOW) representation. And the bag-of-words vectors are the input of a deep neural network with two fully-connected layers. The activation function of first layer is a ReLU and the identity function is used for second layers. Then the text representation can be obtained from the last layer of the deep neural network.

3.4 Asymmetric Loss

The hash codes of query points are generated from hash function, while the hash codes of database points are learned directly. Since the hash codes of query points and database points are generated in different manners, they are asymmetric. The basic idea of asymmetric cross-modal hash codes learning is to make the query point and the database point with the same labels as close as possible. In other words, the image-text pair of query points and database points are similar.

Firstly, to obtain the hashing functions $u_i^x = f(x_i)$, $u_i^y = g(y_i)$ and directly learned hashing codes v_j^x, v_j^y, we propose the J_1 loss. There are four kinds of relative similarity, i.e., the relative similarity query image to a database text, query text to a database text, query image to a database image, and query text to a database image. The object is to minimize the loss between the similarity and inner product of query-database binary codes pairs. It can be written as follows:

$$J_1 = \sum_{i=1}^{m}\sum_{j=1}^{n}[((u_i^x)^T v_j^x - cS_{ij})^2 + ((u_i^x)^T v_j^y - cS_{ij})^2$$
$$+ ((u_i^y)^T v_j^x - cS_{ij})^2 + ((u_i^y)^T v_j^y - cS_{ij})^2]$$
$$s.t.\ x \in I,\ y \in T \tag{1}$$

I is an image set and T is a text set. u_i^x and u_i^y are difficult to learn, because they are discrete data. Thus it sets $u_i^x = f(x_i) = tanh(F(x_i, \Theta_x))$ and $u_i^y = g(y_i) = tanh(G(y_i, \Theta_y))$, which are used to optimize the formulation (1). $F(x_i, \Theta_x) \in R^c$ denotes the learned image feature for point i, which is the output of the CNN-F model. Furthermore, $G(y_i, \Theta_y) \in R^c$ denotes the learned text feature for point i, which is the output of the deep neural network for text modality. Θ_x is the parameter of CNN-F model for image datasets, and Θ_y is the parameter of the deep neural network for text datasets.

Secondly, because the query point and database point are treated as asymmetrically, the hash codes of query points and database points in the same data-pair should be same. Thus, we propose J_2. There are two kinds of relative similarity, i.e., the relative similarity query image to a database image, and query text to a database text. Then J_2 can be defined as:

$$J_2 = \gamma \sum_{i \in \Omega} [v_i^x - u_i^x]^2 + \gamma \sum_{i \in \Omega} [v_i^y - u_i^y]^2 \tag{2}$$

where γ is a hyper-parameter.

Finally, the hash codes of image and text are from different neural networks, but they hold the same semantic information. So the hash codes should be correspondent. Thus, J_3 which can be defined as:

$$J_3 = \eta \sum_{i \in \Omega} [u_i^x - u_i^y]^2 \tag{3}$$

where η is a hyper-parameter.

By combining the above, we can get an asymmetric loss to perform deep hashing from similarity data by jointly preserving similarity between query points and database points. The asymmetric loss is designed as follows:

$$\min_{\Theta, \mathbf{V}} J(\Theta, \mathbf{V}) = J_1 + J_2 + J_3$$
$$= \sum_{i \in \Omega}\sum_{j \in \Gamma} ((u_i^x)^T v_j^x - cS_{ij})^2 + ((u_i^x)^T v_j^y - cS_{ij})^2$$
$$+ ((u_i^y)^T v_j^x - cS_{ij})^2 + ((u_i^y)^T v_j^y - cS_{ij})^2$$
$$+ \gamma \sum_{i \in \Omega} [v_i^x - (u_i^x)]^2 + \gamma \sum_{i \in \Omega} [v_i^y - (u_i^y)]^2$$

$$+ \eta \sum_{i \in \Omega} [(u_i^x)) - (u_i^y)]^2$$
$$s.t. \ x \in I, \ y \in T, \mathbf{V} = \{v_j^x, v_j^y\}_{j=1}^n \in \{-1, +1\}^{n \times c} \quad (4)$$

3.5 Detailed Definition of Similarity Matrix

The similarity matrix $\mathbf{S}$ is usually defined as $S_{ij} = 1$ if $\mathbf{P}_i$ and $\mathbf{D}_j$ have the same semantic label, and $S_{ij} = 0$ if $\mathbf{P}_i$ and $\mathbf{D}_j$ do not share any semantic label. This definition does not take the multi-label information into account, and leads to the loss of semantic similarity information.

Therefore, to relax the constraint of S_{ij}, we propose a detailed definition for cross-modal similarity matrix S_{ij}. The detailed definition of similarity matrix quantifies the similarity into a percentage with the normalized semantic labels, which provide more fine-grained information of labels in the loss function. The similarity of text and image can be passed into the following three cases: completely similar, partially similar, and dissimilar. Completely similar is that all the labels are the same in the compared points, partially similar is that some of the labels are same and the similarity is in percentages, and dissimilar is that all the labels are different. Thus, the detailed definition of similarity matrix can improve the retrieval accuracy. So the similarity matrix $\mathbf{S}$ is defined as:

$$S_{ij} = \begin{cases} pc_{ij} \ if \ \mathbf{P}_i \ and \ \mathbf{D}_j \ share \ some \ class \ labels \\ 0 \ \ otherwise \end{cases}$$

where pc_{ij} is the cosine distance of pairwise label vectors, which is formulated as:

$$pc_{ij} = \frac{< l_i{}^T, l_j >}{||l_i|| ||l_j||}$$

where l_i and l_j denote the semantic label vector of text and image $\mathbf{P}_i$ and $\mathbf{D}_j$, respectively, and $< l_i{}^T, l_j >$ calculates the inner product.

In the query procedure, we can use the functions $f(x_i) = sign(F(x_i, \Theta_x))$ and $g(y_i) = sign(G(y_i, \Theta_y))$ to generate binary hash codes. Then we can use hamming distance to compute the distance between query point and database point.

3.6 Learning Algorithm

We adopt an alternating learning strategy to learn $\Theta_x, \Theta_y, \mathbf{V}^x$ and $\mathbf{V}^y$. That is, in each iteration, one parameter is learned, and other parameters are fixed. Then, after several iterations, the training procedure finishes. In this paper, we repeat the learning for several times, and each time, we can sample a query set from Ω. The detailed derivation will be introduced in the following content of this subsection.

Learn the Parameter of Image Modality. When Θ_y, $\mathbf{V}^x$ and $\mathbf{V}^y$ are fixed, we learn the CNN-F parameter Θ_x of the image modality by using a back-propagation (BP) algorithm. More specifically, we sample a mini-batch of the query points and update the parameter Θ_x. To simplify the formula, $z_i^x = F(x_i, \Theta_x)$, $z_i^y = G(y_i, \Theta_y)$, and $\odot$ denotes the dot product. We compute the derivative of the cost function with respect to z_i^x.

$$\frac{\partial J}{\partial z_i^x} = \left\{ \begin{array}{l} 2 \sum_{j \in \Gamma} \left[\left((u_i^x)^T v_j^y - cS_{ij} \right) v_j^y \right] \\ + 2 \sum_{j \in \Gamma} \left[\left((u_i^x)^T v_j^x - cS_{ij} \right) v_j^x \right] \\ + 2\gamma \left(u_i^x - v_i^x \right) + 2\eta \left(u_i^x - u_i^y \right) \end{array} \right\} \odot \left(1 - u_i^{x2} \right) \tag{5}$$

Then we can compute $\frac{\partial J}{\partial \Theta_x}$ with $\frac{\partial J}{\partial z_i^x}$ by using the chain rule, based on BP algorithm to update the parameter Θ_x.

Learn the Parameter of Text Modality. When Θ_x, $\mathbf{V}^x$ and $\mathbf{V}^y$ are fixed, we also learn the parameter Θ_y of the text modality with a back-propagation (BP) algorithm. As simplified above, we compute the gradient of z_i^y. We can compute $\frac{\partial J}{\partial \Theta_y}$ with $\frac{\partial J}{\partial z_i^y}$ by using the chain rule. We can get the parameter Θ_y using the same method as calculating Θ_x.

Learn the Binary Codes of Image Database. When Θ_x, Θ_y and $\mathbf{V}^y$ are fixed, we rewrite the problem (4). $\mathbf{U}^x = [u_{i1}^x, u_{i2}^x, \ldots, u_{im}^x]^T \in [-1, +1]^{m \times c}$, $\mathbf{U}^y = [u_{i1}^y, u_{i2}^y, \ldots, u_{im}^y]^T \in [-1, +1]^{m \times c}$, and $(\mathbf{V}^x)^{\Omega} = [v_{i1}^x, v_{i2}^x, \ldots, v_{im}^x]^T$. $(\mathbf{V}^x)^{\Omega}$ represents the binary codes for the database which is in set Ω.

$$\begin{aligned} \min_{\mathbf{V}^x} J(\mathbf{V}^x) &= \parallel \mathbf{U}^x (\mathbf{V}^x)^T - c\mathbf{S} \parallel_F^2 + \parallel \mathbf{U}^y (\mathbf{V}^x)^T - c\mathbf{S} \parallel_F^2 \\ &\quad + \gamma \parallel (\mathbf{V}^x)^{\Omega} - \mathbf{U}^x \parallel_F^2 + L \\ &= \parallel \mathbf{U}^x (\mathbf{V}^x)^T \parallel_F^2 + \parallel \mathbf{U}^y (\mathbf{V}^x)^T \parallel_F^2 - 2c\mathrm{tr}((\mathbf{V}^x)^T \mathbf{S}^T \mathbf{U}^x) \\ &\quad - 2c\mathrm{tr}((\mathbf{V}^x)^T \mathbf{S}^T \mathbf{U}^y) - 2\gamma \mathrm{tr}((\mathbf{V}^x)^{\Omega} (\mathbf{U}^x)^T) + L \end{aligned} \tag{6}$$

where L is a constant independent of $\mathbf{V}^x$. We define $\bar{\mathbf{U}}^x = \{\bar{u}_j^x\}_{j=1}^n$, where the definition is as follows:

$$\bar{u}_j^x = \left\{ \begin{array}{l} u_j^x \ if \ j \in \Omega \\ 0 \ otherwise \end{array} \right.$$

Then we can rewrite the problem (6) as follows:

$$\begin{aligned} \min_{\mathbf{V}^x} J(\mathbf{V}^x) &= \parallel \mathbf{V}^x (\mathbf{U}^x)^T \parallel_F^2 + \parallel \mathbf{V}^x (\mathbf{U}^y)^T \parallel_F^2 + L \\ &\quad - 2\mathrm{tr}(\mathbf{V}^x [c(\mathbf{U}^x)^T \mathbf{S} + c(\mathbf{U}^y)^T \mathbf{S} + \gamma (\bar{\mathbf{U}}^x)^T]) \\ &= \parallel \mathbf{V}^x (\mathbf{U}^x)^T \parallel_F^2 + \parallel \mathbf{V}^x (\mathbf{U}^y)^T \parallel_F^2 + \mathrm{tr}(\mathbf{V}^x (\mathbf{Q}^x)^T) + L \end{aligned} \tag{7}$$

We set $\mathbf{Q}^x = -2(c\mathbf{S}^T\mathbf{U}^x + c\mathbf{S}^T\mathbf{U}^y + \gamma(\bar{\mathbf{U}}^x))$. Every time we update one column of $\mathbf{V}^x$ when other columns are fixed. $\mathbf{V}^x_{*k}$ denotes the k-th column of $\mathbf{V}^x$, $\hat{\mathbf{V}}^x_k$ denotes the matrix of $\mathbf{V}^x$ excluding $\mathbf{V}^x_{*k}$. We set $\mathbf{Q}^x_{*k}$ to denote the k-th column of $\mathbf{Q}^x$. Let $\mathbf{U}^x_{*k}$ denotes the k-th column of $\mathbf{U}^x$ and $\hat{\mathbf{U}}^x_k$ denotes the matrix of $\mathbf{U}^x$ excluding $\mathbf{U}^x_{*k}$. $\mathbf{V}^x_{*k} \in \{-1,+1\}^c$. $\mathbf{U}^y_{*k}$ denotes the k-th column of $\mathbf{U}^y$, and $\hat{\mathbf{U}}^y_k$ denotes the matrix of $\mathbf{U}^y$ excluding the k-th column. To optimize $\mathbf{V}^x_{*k}$, we can obtain the problem as follows:

$$J(\mathbf{V}^x_{*k}) = \| \mathbf{V}^x(\mathbf{U}^x)^T \|^2_F + \| \mathbf{V}^x(\mathbf{U}^y)^T \|^2_F + tr(\mathbf{V}^x(\mathbf{Q}^x)^T) + L$$
$$= tr(\mathbf{V}^x_{*k}[2(\mathbf{U}^x_{*k})^T\hat{\mathbf{U}}^x_k(\hat{\mathbf{V}}^x_k)^T + 2(\mathbf{U}^y_{*k})^T\hat{\mathbf{U}}^y_k(\hat{\mathbf{V}}^x_k)^T + (\mathbf{Q}^x_{*k})^T]) + L \quad (8)$$

Then, we must minimize $J(\mathbf{V}^x_{*k})$, so we obtain the problem (9) as follows:

$$\min_{\mathbf{V}^x_{*k}} J(\mathbf{V}^x_{*k}) = tr(\mathbf{V}^x_{*k}[2(\mathbf{U}^x_{*k})^T\hat{\mathbf{U}}^x_k(\hat{\mathbf{V}}^x_k)^T + 2(\mathbf{U}^y_{*k})^T\hat{\mathbf{U}}^y_k(\hat{\mathbf{V}}^x_k)^T + (\mathbf{Q}_{*k}{}^x)^T]) + L$$
$$(9)$$

Then, an optimal solution of problem (10) can be get as follows:

$$J(\mathbf{V}^x_{*k}) = -sign(2(\mathbf{U}^x_{*k})^T\hat{\mathbf{U}}^x_k(\hat{\mathbf{V}}^x_k)^T + 2(\mathbf{U}^y_{*k})^T\hat{\mathbf{U}}^y_k(\hat{\mathbf{V}}^x_k)^T + (\mathbf{Q}_{*k}{}^x)^T) \quad (10)$$

Learn the Binary Codes of Text Database. As the same as $J(\mathbf{V}^x_{*k})$, we can get $J(\mathbf{V}^y_{*k})$ as follows:

$$J(\mathbf{V}^y_{*k}) = -sign(2(\mathbf{U}^x_{*k})^T\hat{\mathbf{U}}^x_k(\hat{\mathbf{V}}^y_k)^T + 2(\mathbf{U}^y_{*k})^T\hat{\mathbf{U}}^y_k(\hat{\mathbf{V}}^y_k)^T + (\mathbf{Q}_{*k}{}^y)^T) \quad (11)$$

$\mathbf{V}^y_{*k}$ denotes the k-th column of $\mathbf{V}^y$, and $\hat{\mathbf{V}}^y_k$ denotes the matrix of $\mathbf{V}^y$ excluding $\mathbf{V}^y_{*k}$, $\mathbf{Q}^y_{*k}$ denotes the k-th column of $\mathbf{Q}^y$, $\mathbf{U}^y_{*k}$ denotes the k-th column of $\mathbf{U}^y$ and $\hat{\mathbf{U}}^y_k$ denotes the matrix of $\mathbf{U}^y$ excluding $\mathbf{U}^y_{*k}$.

Using the above method, we can get the image and text hash functions and database hash codes. The hash functions are used to generate the hash codes of query points, and database hash codes are used to index by query points.

4 Experiment

4.1 Datasets

In the experiments, we carry out cross-modal hashing on two widely-used datasets: MIRFLICKR-25K [17] and NUS-WIDE [18]. The MIRFLICKR-25K dataset consists of 25,000 images. Following the prior work (DCMH [11]), we only select the image-text pairs that have at least 20 textual tags. We randomly select 2000 points as test set and the remaining points as database set. The NUS-WIDE is a public web image dataset which contains 260648 images. There are 81 concept labels in the dataset and one or multiple labels in each point. We select 195834 image-text pairs that belong to the 21 most frequent concepts. We randomly select 2100 points as test set and the remaining points as database set.

4.2   Evaluation Protocol and Baselines

To evaluate ADCH, we choose three metric methods. Firstly, the mean average precision (MAP) is a widely used metric to measure the accuracy of the Hamming ranking protocol. Secondly, we can compute the precision and recall for the returned points given any Hamming radius. Finally, in this paper, we also compare the training time between hashing methods.

We compare our ADCH with eight state-of-the-art methods, including several shallow-structure-based methods (CCA [9], CMFH [22], STMH [20], SCM [21], SePH [19]) and several deep-structure-based methods (DCMH [9], CHN [10], SSAH [22]). These methods are all symmetric models.

For ADCH, in dataset MIRFLICKR-25K we set $\gamma = 200$, $T_{out} = 50$, $T_{in} = 3$ and $\Omega = 2000$ and in dataset NUS-WIDE we set $\gamma = 200$, $T_{out} = 50$, $T_{in} = 3$, $\Omega = 2100$. T_{in} is the iteration number of randomly generating index set Ω from Γ, and T_{out} is the total iteration number. For ADCH, the number of query points m will be much fewer than n. We use the CNN-F network pretrained on the ImageNet dataset to initialize the image modality.

Table 1. The MAP results of baselines and ADCH on MIRFLICKE-25K and NUS-WIDE datasets.

Task	Method	MIRFLICKE-25K			NUS-WIDE		
		16 bits	24 bits	32 bits	16 bits	24 bits	32 bits
I → T	CCA†	0.5719	0.5693	0.5672	0.3604	0.3485	0.3390
	CMFH†	0.6377	0.6418	0.6451	0.4900	0.5053	0.5095
	SCM†	0.6851	0.6921	0.7003	0.5409	0.5485	0.5553
	STMH†	0.6132	0.6219	0.6274	0.4710	0.4864	0.4942
	SePH†	0.7123	0.7194	0.7232	0.6037	0.6136	0.6211
	DCMH	0.7510	0.7425	0.7471	0.5922	0.6125	0.6108
	CHN	0.7531	0.7673	0.7721	0.6045	0.6187	0.6321
	SSAH	0.7732	0.7894	0.8045	0.6494	0.6312	0.6488
	ADCH	**0.8878**	**0.9011**	**0.9041**	**0.8197**	**0.8192**	**0.8196**
T → I	CCA†	0.5742	0.5713	0.5691	0.3614	0.3494	0.3395
	CMFH†	0.6365	0.6399	0.6429	0.5031	0.5187	0.5225
	SCM†	0.6939	0.7012	0.7060	0.5344	0.5412	0.5484
	STMH†	0.6074	0.6153	0.6217	0.4471	0.4677	0.4780
	SePH†	0.7216	0.7261	0.7319	0.5983	0.6025	0.6109
	DCMH	0.7727	0.7800	0.7832	0.6139	0.6611	0.6671
	CHN	0.7739	0.7847	0.7956	0.6182	0.6479	0.6455
	SSAH	0.7921	0.7942	0.8067	0.6790	0.6655	0.679
	ADCH	**0.8796**	**0.8864**	**0.8864**	**0.8034**	**0.8022**	**0.8036**

Table 2. The MAP Comparison of ADCH Variants on MIRFLICKE-25K and NUS-WIDE datasets.

Task	Method	MIRFLICKE-25K			NUS-WIDE		
		16 bits	24 bits	32 bits	16 bits	24 bits	32 bits
I → T	ADCH-b	0.8710	0.8965	0.8885	0.8103	0.8131	0.8193
	ADCH	**0.8878**	**0.9011**	**0.9041**	**0.8197**	**0.8192**	**0.8196**
T → I	ADCH-b	0.8610	0.8765	0.8785	0.7903	0.7931	0.8003
	ADCH	**0.8796**	**0.8864**	**0.8864**	**0.8034**	**0.8022**	**0.8036**

4.3 Accuracy

This experiment is to investigate the performance of the ADCH method on cross-modal retrieval on given datasets. The MAP results are presented in Table 1. We compare our ADCH with eight cross-modal methods with the output dimensions of 16 bits, 24 bits and 32 bits. We can find that ADCH significantly outperforms all the other baselines, including deep hashing baselines. DCMH is a well-known method, having a good performance in cross-modal retrieval. From the experimental results, the MAP of ADCH is higher than that of DCMH by approximately fifteen percent in MIRFLICKR-25K and twenty-five percent in NUS-WIDE. I → T denotes that the query is image and the database is text, and T → I denotes that the query is text and the database is image. † denotes the result cited from [7]. The best results for MAP are shown in bold.

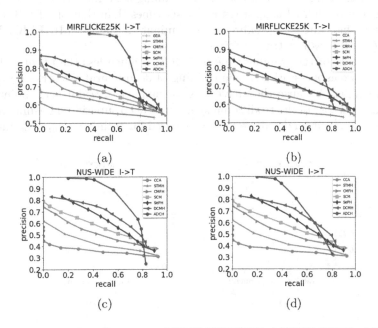

Fig. 2. The Precision-Recall curves on MIRFLICKR-25K and NUS-WIDE with 16 Bits Hash Code.

To go deeper with the effectiveness of ADCH, we design a variant of the proposed approach: ADCH-b is the ADCH variant that utilizes the cross-modal similarity matrix **S** with value 0 or 1. In Table 2, by introducing adaptive similarity matrix **S** into loss the detailed similarity is quantified into percentages, ADCH outperforms ADCH-b. The results show that the detailed definition of similarity matrix S_{ij} can provide more information of semantic similarity.

We report precision-recall curve in Fig. 2. According to the increasing of recall, the precision of ADCH is higher than other baselines. Our ADCH can also achieve the best performance on other cases with different values of codes length, such as 32 bits and 64 bits. $Precision = \frac{true\ positive\ data}{retrieval}$ data, and $Recall = \frac{true\ positive\ data}{data\ in\ database}$. With the same recall, precision of ADCH is higher than other methods.

When the precision is high, there are a large number of true positive data retrieved, and a small number of true positive data left in the database. To increase recall, we need more true positive data, so more retrieval data should be obtained. The number of retrieval data increase dramatically, but the true positive data show a slow growth. So the ADCH precision drops quickly.

4.4   Time Complexity

We compare ADCH with DCMH on the dataset MIRFLICKE-25K in Fig. 3. Y-axis is MAP of $I \rightarrow T$, and X-axis is training time. The length of hash codes is 16 bits, 24 bits, and 32 bits. We can see ADCH outperforms DCMH. With the increase of the length of hash codes, the training time increases gradually. The training time of DCMH is approximately six times than that of ADCH, and the ADCH training time is approximately one minute in dataset MIRFLICKE-25K. Hence, ADCH uses less training time.

The reasons of the training time of ADCH is much less than that of DCMH come into two aspects: Firstly, the training of deep neural networks is typically time-consuming, because they have to scan number of n database points. The computational cost of the traditional symmetric deep supervised hashing method is at least $O(n^2)$, if all database points are used for training. T_{out}, T_{in} and c are

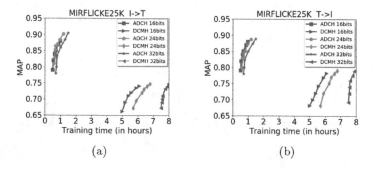

(a)   (b)

Fig. 3. Training time on MIRFLICKE-25K dataset with output dimensions 16 bits, 24 bits and 32 bits.

slightly relative to the number of database points n. Hence, the computational cost of ADCH is $O(n)$. Secondly, ADCH uses not only a direct hash code learning approach but also an asymmetric approach with the query and database points. The training time of ADCH is only the training time for the query points. Thus, the training time of ADCH is much less than that of DCMH.

In this paper, we only compare ADCH with DCMH on training time. Because the training time of directly learned hash codes methods are efficient than the methods of generating hash codes from hash function [2]. Thus, we do not compare ADCH with CHN and SSAH on training time.

4.5 Sensitivity to Parameters

As shown in Fig. 4a, the number of query points of ADCH effects the retrieval accuracy. With an increase in the number of query points on MIRFLICKE25K, the retrieval accuracy improves. When m > 2000, the MAP of I → T becomes gradually stable.

Figure 4b and c show the effect of the hyper-parameters γ and η on the MAP I → T result, and the binary codes lengths include 16 bits, 24 bits and 32 bits. We can see that the change in the hyper-parameters γ and η is not sensitive when $100 < \gamma < 500$ or $100 < \eta < 500$. In our experiments, we choose $m = 2000$, $\eta = \gamma = 200$, to get a stable and high MAP.

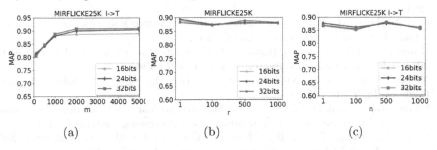

(a)   (b)   (c)

Fig. 4. The accuracy on MIRFLICKE25K of query number m, γ and η.

5 Conclusion

Effective retrieval of massive amounts of media data from heterogeneous sources is a great challenge. In this paper, we propose ADCH as a novel cross-modal deep supervised hashing method for cross-modal retrieval. ADCH uses two deep neural networks to extract image and text feature representations, and learns hash codes of query points and database points in an asymmetric way. Further, we propose a detailed definition of similarity matrix S_{ij} to improve the performance. Through experiments on real-word datasets, the results show that ADCH takes less time and achieves more accurate than traditional symmetric deep supervised hashing methods.

Acknowledgements. This work was supported by the Strategic Priority Research Program of the Chinese Academy of Sciences (XDC02050200).

References

1. Yiqun, H., Cheng, X., Chia, L.-T.: Coherent phrase model for efficient image near-duplicate retrieval. IEEE Trans. **11**, 1434–1445 (2007)
2. Jiang, Q.Y., Li, W.J.: Asymmetric deep supervised hashing. In: Proceedings of the 32nd AAAI Conference on Artificial Intelligence (2018)
3. Wang, W., Yang, X.: Effective deep learning-based multi-modal retrieval. VLDB J. **25**, 79–101 (2016)
4. Zhang, J., Peng, Y., Yuan, M.: Unsupervised Generative Adversarial Cross-modal Hashing. In: Proceedings of the 32nd AAAI Conference on Artificial Intelligence (2018)
5. Wang, Y., Wang, S., Huang, Q.: Online asymmetric similarity learning for cross-modal retrieval. In: 30th IEEE Conference on Computer Vision and Pattern Recognition (2017)
6. Cao, Y., Long, M., Wang, J.: Deep visual-semantic hashing for cross-modal retrieval. In: ACM SIGKDD Conference on Knowledge Discovery and Data Mining (2016)
7. Jiang, Q., Li, W.: Deep cross-modal hashing. In: Proceedings of the IEEE Conference on Computer Vision and Pattern Recognition (2017)
8. Cao, Y., Long, M., Wang, J., Liu, S.: Collective deep quantization for efficient cross-modal retrieval. In: AAAI Conference on Artificial Intelligence (2017)
9. Hardoon, D.R., Szedmak, S., Shawe-Taylor, J.: Canonical correlation analysis: an overview with application to learning methods. Neural Comput. **12**, 2639–2664 (2017)
10. Cao, Y., Long, M., Wang, J.: Correlation hashing network for efficient cross-modal retrieval. Neural Computation, BMVC (2017)
11. Ranjan, V., Rasiwasia, N., Jawahar, C.V.: Multi-label cross-modal retrieval. In: 2015 IEEE International Conference on Computer Vision (2015)
12. Ngiam, J., Khosla, A., Kim, M., Nam, J., Lee, H., Ng, A.Y.: Multimodal deep learning. In: ICML (2011)
13. Srivastava, N., Salakhutdinov, R.R.: Multimodal learning with deep boltzmann machines. In: Advances in Neural Information Processing Systems (2012)
14. Feng, F., Wang, X., Li, R.: Cross-modal retrieval with correspondence auto encoder. In: ACM International Conference on Multimedia (2014)
15. Feng, F., Wang, X., Li, R.: Effective multimodal retrieval based on stacked auto encoders. In: International Conference on Very Large Data Bases (2014)
16. Chatfield, K., Simonyan, K., Vedaldi, A., Zisserman, A.: Return of the devil in the details: delving deep into convolutional nets. BMVC (2014)
17. Huiskes, M.J., Lew, M.S.: The MIR flickr retrieval evaluation. In: MM. ACM (2008)
18. Chua, T.-S., Tang, J., Hong, R., Li, H., Luo, Z., Zheng, Y.: NUS-WIDE: a real-world web image database from national university of Singapore. In: CIVR (2009)
19. Lin, Z., Ding, G., Hu, M., Wang, J.: Semantics-preserving hashing for cross-view retrieval. In: CVPR (2015)
20. Wang, D., Gao, X., Wang, X., He, L.: Semantic topic multimodal hashing for cross-media retrieval. In: IJCAI (2015)
21. Zhang, D., Li, W-J.: Large-scale supervised multimodal hashing with semantic correlation maximization. In: AAAI (2014)
22. Ding, G., Guo, Y., Zhou, J.: Collective matrix factorization hashing for multimodal data. In: CVPR (2014)
23. Li, C., Deng, C., Li, N., Liu, W., Gao, X., Tao, D.: Self-supervised adversarial hashing networks for cross-modal retrieval. In: CVPR (2018)

Applying NSGA-II to a Multiple Objective Dial a Ride Problem

Pedro M. M. Guerreiro[1], Pedro J. S. Cardoso[1,2(✉)],
and Hortênsio C. L. Fernandes[3]

[1] ISE, University of the Algarve, Faro, Portugal
pmguerre@ualg.pt
[2] LARSyS, University of the Algarve, Faro, Portugal
pcardoso@ualg.pt
[3] Yellowfish Travel, Lda, Albufeira, Portugal
hortensio@yellowfishtransfers.com

Abstract. In Dial-a-Ride Problem (DARP) customers request from an operator a transportation service from a pick-up to a drop-off place. Depending on the formulation, the problem can address several constraints, being associated with problems such as door-to-door transportation for elderly/disabled people or occasional private drivers. This paper addresses the latter case where a private drivers company transports passengers in a heterogeneous fleet of saloons, estates, people carriers and minibuses. The problem is formulated as a multiple objective DARP which tries to minimize the total distances, the number of empty seats, and the wage differential between the drivers. To solve the problem a Non-dominated Sorting Genetic Algorithm-II is hybridized with a local search. Results for daily scheduling are shown.

Keywords: Dial a Ride ·
Non-dominated Sorting Genetic Algorithm-II (NSGA-II) ·
Multiple objective optimization · Private drivers scheduling

1 Introduction

In recent years there has been a significant increase in the number of companies transporting tourists in the Algarve (Portugal), mainly between Faro's Airport and the various accommodation units in the region. Many companies, including Yellowfish Travel, Ltd. (YFT) provide a private transportation service into the region and surrounding areas (e.g., Lisbon, Alentejo and southern Spain). The company has a heterogeneous fleet of almost 100 vehicles, with capacities ranging from 4 to 8 passengers. The service philosophy of the company is based on a close relationship with each of the customers and business partners in order to meet all the requirements and needs in a personalized way, included in provision of services by private chauffeurs. This type of product distinguishes itself from the shuttle buses services as each request is carried out separated from the others.

© Springer Nature Switzerland AG 2019
J. M. F. Rodrigues et al. (Eds.): ICCS 2019, LNCS 11540, pp. 55–69, 2019.
https://doi.org/10.1007/978-3-030-22750-0_5

Customers are currently ordering services through a web platform, by email, or via telephone, with advance appointments that can go from several days to a few hours before the service. The requests are stored in a database and the daily scheduling of these, by vehicles/drivers, are made at the end of the previous day. In the present, the scheduling is built manually which implies a considerable human effort and a strict dependence on the capacity and availability of one or more operators. In addition, it is a process that has to be carried out by specialized people due to some required knowledge of key variables, such as the location of the accommodation or the time of transport/travel between pick-up and drop-off places, in order to articulate the various services. The set of variables to be taken into account for the distribution of services is, however, much broader, including the vehicles' capacities, vehicles' availabilities (e.g., more generally the availability of drivers and vehicles), traffic restrictions, the pick-up/drop-off time, the number and type of people to be transported, the type and size of the luggage, the number and age of the children, etc.

The aforementioned scheduling consists in the assignment of a set of transport services by drivers/vehicles, which must be fitted in such a way as to avoid large journeys and high waiting times between the end of a service and the beginning of the other, thus optimizing the resources and costs. The usage of "appropriate" vehicles is also an asset, avoiding, for instance, the usage of the vehicles with large capacities to transport a small number of passengers, as smaller vehicles generally provide more comfort and are cheaper to operate. Other factors, not directly linked to transport, are also taken into account, such as the balanced distribution of services by drivers, both for safety reasons (avoiding sharp wear and tear by providing rest to drivers) and for wage equity issues. The two aspects are related since driver earning come from service commissions, which depend on the number of services and drove distances.

Taking into account all of the above, the problem is formulated as a multiple objective DARP which tries to minimize the total distances, the number of empty seats, and the wage differential between the drivers.

DARP [5, 22] consider transportation requests, each associated with an origin and a destination, resulting in paired pick-up and drop-off points, but with the constraint that passengers are transported service by service. DARP combines Scheduling and Vehicle Routing Problems (VRP) [17, 26, 28] as it tries to find the best set of routes that a fleet of vehicles must carry out in order to deliver goods to a certain set of customers from a certain set of depots, using a (possibly) heterogeneous set of vehicles. Variants of the VRP adapt it to fulfill constraints, such as: the Capacitated VRP (CVRP) where capacity constraints are added to the vehicles, the VRP with time windows (VRP-TW) where each delivery must be performed during a certain time interval, the VRP with Pick-Up and Delivering (VRP-PD) where it is admitted that not all services are delivery but also the pick-up of goods is considered, the VRP with Multi-Depot (VRP-MD) in which routes do not start in the same depot, the VRP with Last In First Out (VRP-LIFO) in which the last commodity to be loaded is the first one to be delivered, etc.

Several adaptations and techniques were proposed over the last decades to solve the variants of DARP problem. For example, a branch-and-price-and-cut algorithm for heterogeneous DARP with configurable vehicle capacity is proposed in [23] to solve a route planning problem arising at a senior activity center. In [4] is introduced a mixed-integer programming formulation of the problem and a branch-and-cut algorithm capable of solving small to medium-size instances of the formulated problem. The pickup and delivery problem with time windows (PDPTW) and a DARP along with some families of inequalities used within branch-and-cut algorithms were formulated in [25], being those inequalities tested on several instance sets of the PDPTW and the DARP. At the level of metaheuristics there is also a wide set of solutions that solve several variants of the problem in question, e.g., Genetic Algorithms (GA), Ant Colony Algorithms (ACO), or Tabu Search (TS) [9]. For example, in [29] an ACO is used to minimize the fleet size required to solve a DARP. In [6] is presented a study to develop and test different GAs in the aim of finding an appropriate encoding and configuration, specifically for the DARP problem with time windows. Parallel implementations of the TS are described and compared in [1] when applied to a static DARP. Other algorithmic solutions include the use of hybrid method such as in [15], where it is implemented a two-stage hybrid meta-heuristic method (uses ACO and TS) for vehicle routing problems with constraints of simultaneous pickups and deliveries and time windows (VRP-SPDTW). In [27] is proposed a parallel approach for solving the VRP-SPD. The parallel algorithm is embedded with a multi-start heuristic which consists of a variable neighborhood descent procedure, with a random neighborhood ordering, integrated in an iterated local search framework. The hybrid algorithm for the dynamic dial-a-ride problem in [2] combines an exact constraint programming algorithm and a TS heuristic. In [3] the authors address the problem of lack of transport service in sparsely inhabited areas as a Demand responsive transport problem (DRTP), comparing the Non-dominated Sorting Genetic Algorithm II (NSGA-II), the Strength Pareto Evolutionary Algorithm 2 (SPEA-2), and the Indicator Based Evolutionary Algorithm (IBEA). Furthermore, improvements using an iterative local search (ILS), added in the mutation operator, are used to select the best approach in a solution capable of producing answers in a short period of time. Another application of NSGA-II to a multiple objective variation of the DARP problem considering disruptive scenarios (e.g., accidents with the transporting vehicles, vehicle breakdown and traffic jams) is presented in [14]. NSGA-II is also used to solve another bi-objective formulation of DARP [10] consisting in the determination of routes to be performed by a fleet of vehicles available to serve a set of geographically dispersed customers (corresponding to patients). Please refer to [5, 12] for two surveys on modeling and algorithms for the DARP.

From a modeling point of view, in addition to other constraints, the problem proposed in this paper ends up combining some of the above mentioned variants, since the vehicles have limited capacities (CVRP), involve the collection and delivery of customers (VRP-PD) with a LIFO of size 1 (VRP-LIFO), vehicles do not leave and have a final stop in the same place (VRP-MD), and there are temporary windows for pick-up and drop-off (VRPTW). The problem will be formulated in Sect. 2 as a Multiple Objective DARP (MO-DARP), adapting the work presented in [19]. The difficulty of obtaining efficient methods, capable of giving the optimal or near-optimal scheduling in a timely manner, will be increased due to the dealing with problems with multiple objectives [8,20]. As stated, the objective function to be minimized in all its components, includes (a) the total distance run by the vehicles during a day work, (b) the total number of empty seats during a day work, and (c) the difference between the commissions earned by each driver and a reference value (promoting drivers working and earnings equity).

This paper shows an approach to solve the real problem proposed by the company. Solutions are computed based on a NSGA-II [8] hybridized with a local search (see Sect. 3). Furthermore, the implementation takes into account the restrictions imposed by the company, optimizing not only the operations cost but also the welfare of its workers. The prototype implemented in Python allows to obtain solution for a day's work in bearable times, provided the service request in proper advance. Solutions will be presented showing that distinct concerns from decision maker (relative to the objectives) are allowed.

The paper is structured as follows. The next section formulates the MO-DARP problem in study. Section 3 outlines the NSGA-II algorithm and how it was applied to solve the addressed MO-DARP problem, including the local search operator description and the system's architecture. Results are reported in Sect. 4 and a conclusion and future work is presented in the last section.

2   Problem Formulation

Adapted from [19], the problem is formulated in a directed graph $G = (V, A)$, where V is the set of vertices/nodes, $A \subset V \times V$ is the set of edges/arcs, and an arc (g, h) has associated a distance $\delta_{g,h}$ and a transversing time $t_{g,h}$.

Each request $i \in R = \{1, 2, \ldots, n\}$ is characterized by a pick-up node $p_i^+ \in P^+ = \{1, 2, \ldots, n\}$, a drop-off node $p_i^- \in P^- = \{n+1, \ldots, 2n\}$, a load $q_i = \sum_{l \in J} q_{i,l}$ where J is the type of passengers set (e.g., baby, child, adult) and $q_{i,l}$ is the number of passenger of type l in request i, and either a pick-up time e_i or a drop-off time d_i. The pick-up vs drop-off time specified in the request is related with its type, which is given by the binary parameter T_i ($T_i = 0$ if request i refers to a pick-up and 1 if it is a drop-off).

Moreover, the set of nodes is defined by $V = P^+ \cup P^- \cup D$, where $D = \{2n + 1, \ldots, 2n+m\}$ is the set of depot points. In the proposed model there are as many depots as drivers (a total of m), since it is considered that a vehicle starts and ends at the driver's home, being $d_k^+, d_k^- \in D$ the start and end nodes of vehicle k, respectively. Each vehicle $k \in K$ of capacity C_k is allocated to one driver, information given by the binary parameter $\omega_{j,k}$, where $j \in M = \{1, 2, \ldots, m\}$. Drivers' working time window $[\alpha_j, \beta_j]$ and maximum workload Max_j are also parameters of the model. Two binary variables

$$x_{g,h}^k = \begin{cases} 1 \text{ if arc } (g,h) \text{ is performed by the vehicle } k \in K \\ 0 \text{ otherwise} \end{cases}$$

and

$$y_{i,j}^k = \begin{cases} 1 \text{ if request } i \text{ immediately precedes request } j \text{ in vehicle } k \in K \\ 0 \text{ otherwise} \end{cases},$$

associate the vehicles to going through an arc and the vehicle's services order, respectively.

Regarding scheduling issues, v_g^k indicates what time vehicle k starts serving node $g \in V$. In addiction, the model defines $\Delta_i^k = C_k - q_i$ as the number of unused seats on vehicle k while serving request i (Δ_i^k is equal to 0 when i is not served by k), S_j the cumulative salary of the driver j, and μ_S the drivers average salary. In order to calculate S_j the set of services for the vehicle k is needed, A_k.

The goal of the problem is expressed by a multi-objective function which comprises (a) the minimization of total distance made by the vehicles

$$f_1 \equiv \sum_{k \in K} \sum_{g \in V} \sum_{h \in V} \delta_{g,h} x_{g,h}^k \tag{1}$$

(b) the minimization of the drivers' wages difference

$$f_2 \equiv \sum_{j \in M} (S_j - \mu_S)^2, \tag{2}$$

and (c) the minimization of the total number of empty seats while satisfying all requests

$$f_3 \equiv \sum_{i \in R} \sum_{k \in K} \Delta_i^k. \tag{3}$$

It can be observed that f_1 and f_3 are linear functions, but f_2 is non-linear.

The problem constraints can be defined as follows. The first constrains ensure that requests are satisfied by a single vehicle

$$\sum_{k \in K} x_{p_i^+, p_i^-}^k = 1, i \in R$$

and that the vehicle which picks-up the customer is the same that drops him off

$$x_{p_i^-, p_j^+}^k = y_{i,j}^k, k \in K, i, j \in R.$$

The next equations ensure that a vehicle starts and ends its daily service at its depot

$$\begin{cases} \sum_{h \in P^+} x_{d_k^+, h}^k = 1, k \in K, d_k^+ \in D \\ \sum_{g \in P^-} x_{g, d_k^-}^k = 1, k \in K, d_k^- \in D. \end{cases}$$

The vehicles' capacity restrictions for each service are verified by satisfying

$$x^k_{p^+_i,p^-_i}\left(\sum_{l\in J} q_{i,j} + \Delta^k_i\right) = x^k_{p^+_i,p^-_i} C_k, k \in K, i \in R$$

Time specifications ensure that none of the requests of a vehicle overlap in time

$$y^k_{i,j}(v^k_{p^+_i} + t_{p^+_i,p^-_i} + t_{p^-_i,p^+_j}) \leq v^k_{p^+_j}, k \in K, i,j \in R.$$

The next restrictions address if a customer either defines a pick-up time or drop-off time from a certain place:

$$\begin{cases} v^k_{p^+_i} + t_{p^+_i,p^-_i} \leq d_i T_i, k \in K, i \in R \\ v^k_{p^+_i} \geq e_i(1 - T_i), k \in K, i \in R. \end{cases}$$

The last constraints ensure that a driver only works within a previously defined time window

$$\begin{cases} \omega_{j,k}(v^k_{d^-_k} - v^k_{d^+_k}) \leq Max_j, j \in M, k \in K, d^-_k, d^+_k \in D \\ \omega_{j,k} \min(v^k_g) \geq \alpha_j, k \in K, g \in V, j \in M \\ \omega_{j,k} \max(v^k_g) \leq \beta_j, k \in K, g \in V, j \in M \end{cases}$$

and rev_i the drivers' revenue associated to each request

$$\omega_{j,k} S_j = \sum_{i \in A_k} rev_i, k \in K, j \in M.$$

3   Computational Solution

3.1   General NSGA-II

Section 2 formulated MO-DARP, a nonlinear multi-objective optimization problem, derived from the single customer DARP. Non trivial multi-objective problems are characterized by having a conflicting set of objective functions, which implies that no single solution is capable of optimizing all objectives at once. In this sense, the solution of a MO problem is a set of trade-off solutions, called Pareto set [7,18], being its representation called Pareto Front (PF). In the Pareto (or efficiency) order relation, a solution R is said to dominate solution S, $R \prec S$, when R is not worse than S for all objectives and there is at least one on which it is strictly better, i.e., given an objective function with n objectives, $(f_1, f_2, \ldots, f_n)$, where all objectives are to be minimized, then

$$R \prec S \Leftrightarrow \begin{cases} \forall_{i\in\{1,2,\ldots,n\}} : f_i(R) \leq f_i(S) \\ \exists_{i\in\{1,2,\ldots,n\}} : f_i(R) < f_i(S). \end{cases} \tag{4}$$

Using the dominance definition, the Pareto set is defined as $\mathcal{P} = \{S \in \Omega | \nexists_{R\in\Omega} : R \prec S\}$, where Ω is the problem's admissible solutions set. Without preferential information, all solution in the Pareto set are considered equally good, as they cannot be ordered completely, leaving the decision of which solution to use to a decision maker. Furthermore, as the Pareto set can be a continuous or discrete set (with possibly infinite cardinal), many times the goal is to find a representative set of optimal solutions. This can be even further relaxed as, under complex problems, it might be acceptable a representative set of quasi-optimal solutions. These representative sets are called approximation sets.

To measure the approximation sets quality, several metrics where analyzed in the past [24]. This paper will use the Hypervolume (HV) [30] which uses a reference point to measure the size of the objective space covered by the approximation set. HV considers accuracy, diversity and cardinality, being the only unary metric with this capability.

The method elected to solve the proposed MO-DARP was NSGA-II [8]. As the name suggests, NSGA-II is a multiple objective Genetic Algorithm [13]. GAs evolve a population of candidate solutions throughout a number of generations, until a stopping criteria is reached (e.g, number of generations, number of evaluations of the objective, computational time, etc). In each generation, the candidate solutions are subjected to two genetic operators, namely: crossover and mutation. The crossover operator creates the offspring by combining randomly selected (according to some given crossover probability, p_c) candidate solutions, and the mutation operator applies a mutation to each offspring, again according to a given mutation probability, p_m. While the first operator retains the parents characteristics to the offspring, the second preserve and maintain the diversity of the population, preventing it from becoming trapped in some local minima. After applying the crossover and mutation operators, parents and offspring are joined and every solution is evaluated to determine their respective fitness. That fitness is used to select the best solutions to maintain for the next generation, discarding the others.

A few differences distinguish the NSGA-II procedure from the generic GA, such as the selection of the candidate solutions which are preserved for the next generation. Since in a non-dominated set of solutions of a MO optimization problem there is no "best" solution, but a set of trade-off solutions (approximation set), NSGA-II algorithm ranks each solution according to their respective layer in the approximation set, and uses the crowding distance to maintain the diversity, so that when needed, the solutions discarded are those with lower rank or in a more "crowded" space (refer to [8] for a detailed explanation of the process). Here, given an approximation set $\mathcal{AS}$, the referred solution's layer corresponds to a construction where the first layer is given by

$$\mathcal{L}_0 = \{S \in \mathcal{AS} | \nexists_{R \in \mathcal{AS}} : R \prec S\} \tag{5}$$

and the next layers are recursively given by

$$\mathcal{L}_{i+1} = \{S \in \mathcal{AS}_i | \nexists_{R \in \mathcal{AS}_i} : R \prec S\}, \tag{6}$$

where $\mathcal{AS}_i = \mathcal{AS} - \cup_{k=0}^{i} \mathcal{L}_k$ for $i = 0, 1, \ldots$.

3.2   Applying NSGA-II to MO-DARP

Returning to the MO-DARP, the solutions of the problem can be seen as assignments of drivers/vehicles to services. In this sense, to implement the NSGA-II method, vectors of integers were used to represent solutions, where each column represents a service, and the value is the vehicle assigned to it, as the two parents sketched in Fig. 1 – top-left. For instance, Parent 1 represents a solution where service 1 is assigned to vehicle 3, service 2 is assigned to vehicle 2, etc.

The algorithm can be summarized as follows. (i) The initial population is generated by randomly attributing vehicles to services. (ii) Then the fitness of each individual in the population is computed, using Eqs. (1)–(3) and the number of violated restrictions, implementing a restriction-violation non-dominated sorting. In other words, solution which do not violate any restriction are ranked/sorted as in Eqs. (5) and (6). The ranking process is then repeated with the solutions which violate one restriction, then with two restrictions, etc. This means that the process is allowed to work with solutions which are not valid (e.g., a vehicle has two services which overlap in time). (iii) In the next step, pair of parents are selected proportionally to their rank (called fitness proportionate selection or roulette wheel selection) and the crossover operator is applied with probability, p_c. Figure 1 – top represents a crossover operation where Parent 1 and 2 are combined to generate Offspring 1 and 2. The represented combination is done by randomly selecting three cutting points followed by swap between the parents of the blocks defined by those cutting points. The number of cutting point, κ, is a parameter of the algorithm. (iv) Mutation is then randomly applied to each offspring vehicle/service assignment with probability p_m. In our case, the operator was implemented by randomly changing the vehicle assigned to the service, as the example sketched in Fig. 1 – bottom, where Offspring 1 is mutated by changing the vehicles assigned to services S3 and S7, obtaining Offspring 1*, which replaces Offspring 1. (v) The use of local search operators to improve the offspring is optional, but generally it allows to improve the algorithm performance in terms of solutions fitness for a given time/number of iterations/number of objective function evaluations (see Sect. 3.3 for the considered local search). Finally, (vi) the offspring are evaluated and from the set of parents and offspring, taking into consideration their fitness and the solutions crowding distance, a new population is selected, with the original size, returning to step (iii) if the stopping criteria is not met. When the stopping criteria is met, the feasible non-dominated solutions in the last population are returned as the proposed solution for the MO-DARP. An empty set is returned if no feasible solution was found.

3.3   Local Search

In order to improve overall performance, a local search (LS) operator was implemented and applied to the offspring resulting from the NSGA-II's crossover and mutation operators as follows. Let us consider an offspring, such as the ones presented in Fig. 1. This offspring represents a solution by assigning services to vehicles. For each service, the local search operator starts by checking if there are vehicles satisfying the service's constraints (e.g., number of seats, luggage restrictions, etc.) – to speed up this step, a pre-computation associates admissible vehicles to the services. Secondly, if the set of admissible vehicles (except the assigned one) is not empty, for each of them, it is verified if assigning the service to the vehicles violates any of the time restrictions, namely if overlapping of services occurs, if the vehicles' working hours are satisfied, if it is possible to go from the previous drop-off to the service's pick-up point in time and if after the service is still possible to reach the next pick-up location in time. If the previous conditions are all satisfied then it is computed the variation in the total distance that the change of the service from one vehicle to the other would induce. If the total distance would be improved in any of the admissible changes then the one with the greatest gain is executed.

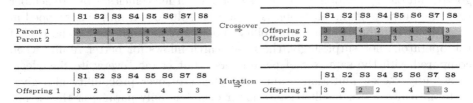

Fig. 1. Solution representation using vectors of integers. Example of the crossover (top) and mutation (bottom) operators.

3.4 System's Architecture

The proposed architecture comprises 4 main components: database, routes server, the MO-DARP optimizer, and visualization. The (relational) database stores the data associated with the problem, such as: services (pick-up/drop-off places, pick-up/drop-off times, etc.), vehicles (availability, number of places, etc.), drivers, etc. To compute the routes between the places, an instance of the Open Source Routing Machine (OSRM) [16] was deployed in a local server. OSRM runs on OpenStreetMap data [21] and, among other things, furnishing an Web application programming interface (HTTP API) which replies to requests for nearest street matches (snaps coordinates to the street network and returns the nearest matches), route computation (finds the fastest route between coordinates), duration or distances tables (computes the duration or distances of the fastest route between all pairs of supplied coordinates), etc. The MO-DARP optimizer was prototyped over the Platypus (Multiobjective Optimization in Python) evolutionary computing framework [11]. Platypus provides optimization algorithms and analysis tools, allow to define constrained and unconstrained problems, and currently supports around ten MO algorithms.

4 Computational Results

This section presents some results of the usage of the proposed algorithms. Two operation days were chosen: Day 1 with 54 services and 22 cars available, and

Day 2 with 197 services and 38 vehicles. In both days it is considered that vehicles can do pick-ups between 00:00 and 24:00. It should be noticed that there might be more than one driver associated to a vehicle, i.e., it is not usual that the same driver does pick-ups early and late in the day – for the chosen days that data was not yet systematized in the database. The algorithm was run 10 times for each set of parameters and the results are summarized in Table 1, which presents for each set of parameters the Hypervolume values of the normalized (to the interval [0, 1]) objective function values and cardinality of the Pareto Front. Regarding the Hypervolume, in the results for Day 2 it can be observed that the most influential parameter is the mutation probability, and the lower its value the better are the results obtained. What this means is that as the algorithm is reaching good solutions, the mutation in reality is "destroying" those good solutions by randomly moving services to other vehicles – but remember that the purpose of the mutation is to prevent solutions from being trapped in some local minima, and its probability should not be null. As for the rest of the operators, not surprisingly, the local search allows for better results when compared with the same set of parameters – but cannot overwrite the effects of a badly chosen mutation – and an higher crossover probability also gives systematically better results than a lower one. The only parameter that is not "coherent" is κ, the number of points for the crossover. If for a lower mutation probability ($p_m = 0.01$), a 4-points crossover always gives better results, when this probability is higher ($p_m = 0.05$), using the classic 2-points crossover yields better results, when compared with using an higher crossover probability, as can be seen in the lower half of the table. The results for Day 1 are a little different, but it can be observed that all the values are very similar, indicating that this problem is "simple" to solve, and the values of the parameters do not influence much the results obtained. Similar results can be observed with respect to the cardinality of the PF, seeable from the fact that the maximum number of solution in the PF (equal to the defined population size) is always reached when $p_m = 0.01$, establishing again mutation as the most influential parameter.

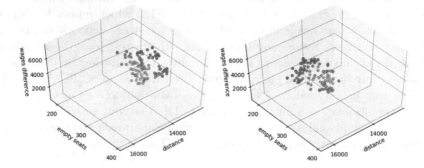

Fig. 2. Pareto fronts for $\kappa = 4, p_c = 0.9$ and $p_m = 0.01$, with (left column) and without (right column) local search operator.

Table 1. Average of Hypervolume indicator and cardinality of the Pareto Front, for 10 executions of each set of parameters, ordered by the values of Day 2's Hypervolume.

κ	p_c	p_m	LS	Hypervolume		# PF	
				Day 1	Day 2	Day 1	Day 2
4	0.9	0.01	✓	0.80	0.81	100	100
2	0.9	0.01	✓	0.80	0.80	100	100
4	0.7	0.01	✓	0.77	0.76	100	100
2	0.7	0.01	✓	0.77	0.73	100	100
4	0.9	0.01	–	0.81	0.66	100	100
2	0.9	0.01	–	0.80	0.65	100	100
4	0.7	0.01	–	0.80	0.65	100	100
2	0.7	0.01	–	0.80	0.64	100	100
2	0.9	0.05	✓	0.80	0.37	99.8	70.5
2	0.7	0.05	✓	0.80	0.36	100	76.9
4	0.9	0.05	✓	0.80	0.36	100	63.5
4	0.7	0.05	✓	0.79	0.35	100	59.5
2	0.9	0.05	–	0.76	0.30	99.9	71.7
2	0.7	0.05	–	0.75	0.29	100	81.2
4	0.9	0.05	–	0.75	0.29	99.6	64.4
4	0.7	0.05	–	0.75	0.27	100	79.2

In Fig. 2 are depicted the Pareto fronts for the best parameters for Day 2 ($\kappa = 4, p_c = 0.9$ and $p_m = 0.01$), with the local search (left) and without the local search (right). It is observable that, without the local search (right), the values for f_1 (distance) are centered around 14000 km, but using the local search (left), all solutions found are well below that value. Regarding f_2 (the number of empty seats) and f_3 (the difference of the commissions) the difference between using or not the local search is negligible. This is not a surprise, as the local search is about finding better solutions considering only the first objective (f_1).

Figure 3 plots two Gantt charts for the schedulings of Day 2, obtained from a single run of the proposed algorithm using as parameter $\kappa = 4$, $p_c = 0.9$, $p_m = 0.01$, and applying the local search. Each line of the plot is associated to a vehicle and the blue bars depicts connection times, i.e., the time to go from the drop-off place of one service to the pick-up place of the next, from the vehicles' starting place to the first pick-up place, or from the last vehicles' drop-off to its ending place. On the other hand, the red bars depict services times, i.e., the time each service takes from the pick-up to the drop-off place. The absence of blue bar between services (red bars) means that the drop-off place of the previous service coincides with the pick-up place of the next one. The scheduling represented in the top of the figure was chosen from the approximation set for being the one which minimizes the number of empty seats: the objective function values are

$f_1 = 13782.37$ (kms), $f_2 = 190$ (empty seats) and $f_3 = 2762, 41$ (wage equity). On the other hand, the figure in the bottom corresponds to scheduling which minimizes the total distance: the objective function values are $f_1 = 12516.58$, $f_2 = 267$, and $f_3 = 2212.65$. Comparing plots it is observable that the second one has a smaller density of blue bars and, although possibly not so discernible, is less "dense". These observation follow the expected behavior since to diminish total distance vehicles should be allowed to have longer waiting times avoiding the movement between drop-off and pick-up points.

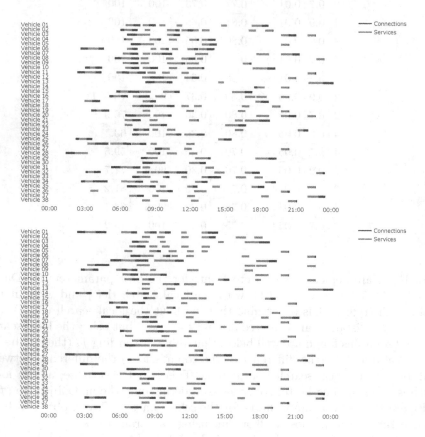

Fig. 3. Two examples of scheduling proposed from a single run of the algorithm using parameter $\kappa = 4$, $p_c = 0.9$, $p_m = 0.01$, and local search. On top the scheduling which minimizes the number of empty seats (objective function values $f_1 = 13782.37$, $f_2 = 190$ and $f_3 = 2762, 41$ and on the bottom the scheduling which minimizes the total distance (objective function values: $f_1 = 12516.58$, $f_2 = 267$, and $f_3 = 2212.65$). (Color figure online)

5   Conclusion and Future Work

The number of companies transporting tourists in the Algarve and in many touristic regions has recently increased dramatically. Between those companies are the ones which provide private transportation services in a heterogeneous fleet of vehicles. These companies have to schedule the daily services taking into consideration constraints (e.g., vehicles and divers availability, vehicles vs. services characteristics, etc.) and objectives. Among these objectives are the operational costs, the appropriate usage of the vehicles fleet, and welfare of the drivers. The problem can be formulated as a multiple objective DARP.

This paper presents the usage of a NSGA-II with a local search to solve the MO-DARP problem originated by those transportations. The proposed system uses data collected by a company (e.g., services, vehicles, drivers, etc.) and a geographical information system (namely, OSRM) to compute the routes between the pick-up and drop-off points. The returned approximation set gives the company a set of scheduling from which a decision maker can choose one, and do the arrangements for the next day.

In the future, the authors intend to improve the algorithm by including other pre and post operators. The comparison with other heuristics is also an objective, looking for opportunities to further hybridize and consequently improve the computational process. Another major asset will be the adaptation of the algorithmic solution to allow a dynamic environment where new services, delays, vehicles/drivers unavailabilities, traffic, etc. are taken into consideration.

Acknowledgments. This work was supported by the Portuguese Foundation for Science and Technology (FCT), project LARSyS (UID/EEA/50009/2013).

References

1. Attanasio, A., Cordeau, J.F., Ghiani, G., Laporte, G.: Parallel tabu search heuristics for the dynamic multi-vehicle dial-a-ride problem. Parallel Comput. **30**(3), 377–387 (2004)
2. Berbeglia, G., Cordeau, J.F., Laporte, G.: A hybrid tabu search and constraint programming algorithm for the dynamic dial-a-ride problem. INFORMS J. Comput. **24**(3), 343–355 (2012). https://doi.org/10.1287/ijoc.1110.0454
3. Chevrier, R., Liefooghe, A., Jourdan, L., Dhaenens, C.: Solving a dial-a-ride problem with a hybrid evolutionary multi-objective approach: Application to demand responsive transport. Appl. Soft Comput. **12**(4), 1247–1258 (2012). https://doi.org/10.1016/j.asoc.2011.12.014
4. Cordeau, J.F.: A branch-and-cut algorithm for the dial-a-ride problem. Oper. Res. **54**(3), 573–586 (2006)
5. Cordeau, J.F., Laporte, G.: The dial-a-ride problem: models and algorithms. Ann. Oper. Res. **153**(1), 29–46 (2007)
6. Cubillos, C., Rodriguez, N., Crawford, B.: A study on genetic algorithms for the DARP problem. In: Mira, J., Álvarez, J.R. (eds.) IWINAC 2007. LNCS, vol. 4527, pp. 498–507. Springer, Heidelberg (2007). https://doi.org/10.1007/978-3-540-73053-8_50

7. Deb, K.: Multi-objective Optimization Using Evolutionary Algorithms. Wiley, Chichester (2001)
8. Deb, K., Pratap, A., Agarwal, S., Meyarivan, T.: A fast and elitist multiobjective genetic algorithm: NSGA-II. IEEE Trans. Evol. Comput. **6**(2), 182–197 (2002)
9. Gendreau, M., Potvin, J.Y.: Handbook of Metaheuristics, vol. 2. Springer, Boston (2010). https://doi.org/10.1007/978-1-4419-1665-5
10. Haddadene, S.A., Labadie, N., Prodhon, C.: NSGA-II enhanced with a local search for the vehicle routing problem with time windows and synchronization constraints. IFAC-PapersOnLine **49**(12), 1198–1203 (2016)
11. Hadka, D.: Platypus - Multiobjective Optimization in Python (2015). https://platypus.readthedocs.io. Accessed 08 Feb 2019
12. Ho, S.C., Szeto, W., Kuo, Y.H., Leung, J.M., Petering, M., Tou, T.W.: A survey of dial-a-ride problems: literature review and recent developments. Trans. Res. Part B Methodol. **111**, 395–421 (2018). https://doi.org/10.1016/j.trb.2018.02.001
13. Holland, J.: Adaptation in Natural and Artificial Systems. University of Michigan Press, Ann Arbor (1975)
14. Issaoui, B., Khelifi, L., Zidi, I., Zidi, K., Ghédira, K.: A contribution to the resolution of stochastic dynamic dial a ride problem with NSGA-II. In: 13th International Conference on Hybrid Intelligent Systems (HIS 2013), pp. 54–59 (2013)
15. Lai, M., Tong, X.: A metaheuristic method for vehicle routing problem based on improved ant colony optimization and tabu search. J. Ind. Manage. Optim. **8**(2), 469–484 (2012). https://doi.org/10.3934/jimo.2012.8.469
16. Luxen, D., Vetter, C.: Real-time routing with openstreetmap data. In: Proceedings of the 19th ACM SIGSPATIAL International Conference on Advances in Geographic Information Systems, GIS 2011, pp. 513–516. ACM, New York (2011). https://doi.org/10.1145/2093973.2094062
17. Männel, D., Bortfeldt, A.: A hybrid algorithm for the vehicle routing problem with pickup and delivery and three-dimensional loading constraints. Eur. J. Oper. Res. **254**(3), 840–858 (2016)
18. Miettinen, K.: Nonlinear Multiobjective Optimization. Kluwer Academic Publishers, Boston (1999)
19. Morais, A.C., Torres, L., Dias, T.G., Cardoso, P.J.S., Fernandes, H.: A combined data mining and tabu search approach for single customer dial-a-ride problem. In: 7th International Conference on Metaheuristics and Nature Inspired Computing, Marrakech, Morocco, pp. 121–123, October 2018
20. Murata, T., Itai, R.: Multi-objective vehicle routing problems using two-fold EMO algorithms to enhance solution similarity on non-dominated solutions. In: Coello Coello, C.A., Hernández Aguirre, A., Zitzler, E. (eds.) EMO 2005. LNCS, vol. 3410, pp. 885–896. Springer, Heidelberg (2005). https://doi.org/10.1007/978-3-540-31880-4_61
21. OpenStreetMap contributors: Planet dump (2017). https://planet.osm.org. https://www.openstreetmap.org
22. Parragh, S.N., Doerner, K.F., Hartl, R.F.: A survey on pickup and delivery problems. Journal für Betriebswirtschaft **58**(1), 21–51 (2008)
23. Qu, Y., Bard, J.F.: A branch-and-price-and-cut algorithm for heterogeneous pickup and delivery problems with configurable vehicle capacity. Transp. Sci. **49**(2), 254–270 (2014)
24. Riquelme, N., Lücken, C.V., Baran, B.: Performance metrics in multi-objective optimization. In: 2015 Latin American Computing Conference (CLEI), pp. 1–11, October 2015. https://doi.org/10.1109/CLEI.2015.7360024

25. Ropke, S., Cordeau, J.F., Laporte, G.: Models and branch-and-cut algorithms for pickup and delivery problems with time windows. Networks Int. J. **49**(4), 258–272 (2007)
26. Salhi, S., Imran, A., Wassan, N.A.: The multi-depot vehicle routing problem with heterogeneous vehicle fleet: formulation and a variable neighborhood search implementation. Comput. Oper. Res. **52**, 315–325 (2014)
27. Subramanian, A., Drummond, L., Bentes, C., Ochi, L., Farias, R.: A parallel heuristic for the vehicle routing problem with simultaneous pickup and delivery. Comput. Oper. Res. **37**(11), 1899–1911 (2010). https://doi.org/10.1016/j.cor.2009.10.011
28. Toth, P., Vigo, D.: Vehicle Routing: Problems, Methods, and Applications. SIAM (2014)
29. Tripathy, T., Nagavarapu, S.C., Azizian, K., Ramasamy Pandi, R., Dauwels, J.: Solving dial-a-ride problems using multiple ant colony system with fleet size minimisation. In: Chao, F., Schockaert, S., Zhang, Q. (eds.) UKCI 2017. AISC, vol. 650, pp. 325–336. Springer, Cham (2018). https://doi.org/10.1007/978-3-319-66939-7_28
30. Zitzler, E., Thiele, L.: Multiobjective evolutionary algorithms: a comparative case study and the strength Pareto approach. IEEE Trans. Evol. Comput. **3**(4), 257–271 (1999)

Smart Campus Parking – Parking Made Easy

Amanda Vieira[1], Iolanda Rosa[1], Ivo Santos[1,2],
Tiago Paulo[1], Nuno Costa[1,2], Marisa Maximiano[1,2],
and Catarina I. Reis[1(✉)]

[1] School of Technology and Management, Polytechnic of Leiria, Leiria, Portugal
{2160832, 2161477, 2160837, 2161352}@my.ipleiria.pt,
{nuno.costa, marisa.maximiano,
catarina.reis}@ipleiria.pt
[2] Computer Science and Communication Research Centre,
Polytechnic Institute of Leiria, Leiria, Portugal

Abstract. The number of users of the parking lots from the campus of the Polytechnic of Leiria, a higher education institution in Portugal, has been increasing each year. It is becoming a major concern to the organization to address the high demand for a free parking spot on campus. In order to ease this problem, this paper proposes the design of a smart parking system that can help users to easily find a free parking spot, using an integrated system that includes sensors and a mobile application.

The system is based on the information about the occupation status of parking lots generated by parking sensors. This information is available in the mobile application that consumes a REST webservice and is presented to end-users, thus contributing to the decrease of time wasted on the quest of finding a free spot. The software architecture consists on a set of decoupled modules that compute and share the information generated by sensors. This architectural approach is noteworthy because it maximizes the system scalability and responsiveness to change. It allows the system to expand with the integration of new applications and perform updates on the existing ones, without an overall impact on the operations of the other system modules.

Keywords: IoT · Sensors · Smart parking · Android · API · REST

1 Introduction

1.1 Context

The development of the economy leads to a continuous growth in the number of available motor vehicles. Nowadays, the demand for a free parking spot versus the number of available spots is a daily concern [1]. Therefore, drivers that need to park their vehicles have the, sometimes troublesome, task to find a free parking slot. Usually drivers waste their time and end up parking in an available space on the streets that they find through a struck of luck or, in a worst case scenario, they fail to find a free parking slot [2]. Parking issues also show a close relation with traffic congestion, accident, and environment pollution [3].

© Springer Nature Switzerland AG 2019
J. M. F. Rodrigues et al. (Eds.): ICCS 2019, LNCS 11540, pp. 70–83, 2019.
https://doi.org/10.1007/978-3-030-22750-0_6

This reality is starting to become a challenge even for higher educational institutions that want to provide the best parking conditions to their users, and the Polytechnic of Leiria is not an exception. The increasing number of enrollments (shown in Table 1) and the high percentage of vehicle ownership, lead to a decrease of supply of free parking spots, which results in triggering blockage of vehicles, traffic congestion, wastage of time and fuel [4] and to high levels of pollution.

Table 1. Evolution of the total number of enrolled students and employees of the Polytechnic of Leiria for the academic years 2015–2018 [5].

Academic year	2015/2016	2016/2017	2017/2018
Total students	10417	10472	11026
Total employees	1128	1162	1245

Therefore, the development of a smart parking system can play an important role to help with this scenario. According to a report, smart parking can result in 220 000 gallons of fuels saving till 2030 and approx. 300 000 gallons of fuels saved by 2050, when implemented successfully (shown in Fig. 1) [6].

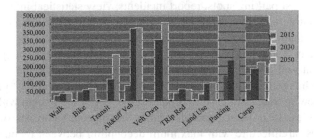

Fig. 1. Assumption of saving fuel consumption.

In addition to the reasons described above, there are other advantages resulting from the usage of smart parking systems that also establish a motivation for this work:

- Efficiently manage the use of parking lots decreasing the traffic flow. Since drivers do not need to circle around to find an available parking spot, this leads to an improvement of the time that a driver takes to find the best spot, while minimizing the occurrence of incidents;
- Improve the environment by reducing pollutants' emission resulting from fuel wasted on the quest of searching for a parking spot;
- Allow users to access real-time information about the parking lots occupancy rate (information also available for any time of the day);
- Allow the integration with other functionalities such as obtaining user's vehicle location when it's parked, see the most popular parking spot, the last reported incidents on a parking spot, among others;

- Conduct various statistical analysis. For instance, explore the periods with most traffic congestion and see if there's a possible relation with the occurrence of incidents.

1.2   Motivation

This paper aims to present the design and development of a smart parking system for the efficient management of parking lots in the Campus 2 of the Polytechnic of Leiria. The main goal is to provide guidance to drivers about the availability of parking spots at the campus, delivering them with up-to-date information obtained from sensors in the parking lots.

The work relies on the integration of several technologies and applications that aim to collect, compute and persist data generated by parking sensors in real-time. End-users can access this data using an Android application that communicates with a REST webservice in order to obtain a response to the user's requests.

The operation of the system, shown in Fig. 2, can be described as a group of sensors detecting the occupancy status of the parking spots and sending this data to a set of integrated applications that unify and persist the information on a centralized database. On the other hand, there is a REST webservice that will expose this data to answer to the received requests from end-user applications. View parking information, use and find the best parking spots, report incidents, view statistical information among other administrative tasks, are some of the features provided by the web service.

One of the main concerns while developing the system was its scalability considering the premise that this platform should easily allow the future integration of new applications that will also receive and share information. Besides, the system must be responsive to change, and thus, it should be easy to expand in order to integrate new parking lots, new services and perform updates in a part of the system without affecting the other modules. In order to meet this goal, the approach consists of an interconnected modules' architecture with maximum decoupling between them.

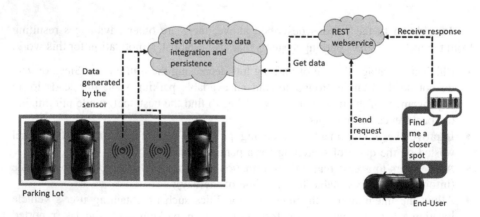

Fig. 2. General outline of system operation.

There are some important features that should be considered when choosing real sensor devices to use in the system:

- Low installation and replacing time to reduce the costs;
- Resistance to atmospheric conditions specially in parking lots located outdoors;
- Compatibility with wireless networks;
- Optimized for low-power consumption;
- Simple for developers to work;
- Allow remote management of several node parameters at the same time.

A good example of a sensor device with all these features is the one of Libelium shown in Fig. 3(a).

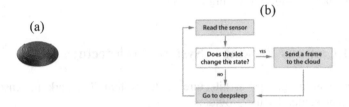

Fig. 3. (a) Libelium smart parking sensor device and (b) the basic working loop diagram [7].

This sensor can be configured remotely using wireless access and supports bidirectional communication. It works in a very efficient way as shown in Fig. 3(b). Reads the sensor data and sends a frame when the parking spot changes its status. Then, it sleeps a desired time before starting the loop again [7].

Currently, and regarding the work here described, the Polytechnic of Leiria does not have sensors on its parking spots due to the initial economic investment that is necessary to deploy such a system. Thus, to mitigate this limitation, two applications that mimic the data generated by real sensors were created (providers).

1.3 Related Work

Related systems already exist. We started by identifying related mobile applications available in the Apple Store [8] and Play Store [9]. Their main functionalities include obtaining the location of an available parking slot in an area, reserve it and pay it via app.

Next, we present some examples of applications with the referred functionalities:

- **JustPark Parking:** Android application that includes around 20 000 parking locations that are available for booking [10].
- **ParkMobile Parking:** Android application that allows the users to have favorite parking zones, update personal details, change payment method, download their invoice, view parking history, etc. [11].

- **ParkNow Parking:** Android application that works on-street and off-street and the payment due is charged monthly. It also allows a user to choose a parking spot based on distance and price [12].
- **PayByPhone:** Smart parking application available on iPhone, Android, Blackberry and Windows. Allows the user to find parking lots, pay the parking fee through the application and even extend the time the user wishes to be parked for [13].
- **BestParking:** iPhone application that allows the user to find an available spot at the best price [14].
- **SmoothParking:** iPhone application that notifies the user when the legal parking time is up. It's available in English and Spanish [15].
- **Parking Hero:** iPhone application that notifies the user about promotions, and shows the locations of special spots like taxis, pickup and delivery points, tourist buses or truck parking on the highway [16].

2 Smart Campus Parking – System Architecture

Figure 4 presents the general architecture of the system. The parking sensors on our system are represented by the providers.

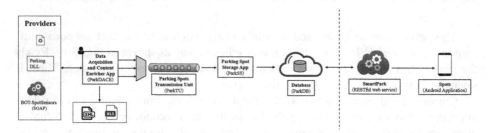

Fig. 4. General service and Spots app communication layer architectural schema.

2.1 Application Programming Interface – Tailored Services

The Spots app (see Sect. 3) will obtain the parking lots information using a REST webservice named SmartPark. The REST architectural style [17] was adopted in the development of the SmartPark API because it is stateless, which is a fundamental aspect for the scalability of the system allowing many clients to connect to the API without degradation of the response time. Thus, the SmartPark role is to provide endpoints and supply data to client applications such as:

- Parking lots general information: name, operating hours, total of spots, etc.;
- Information about occupation status of parking spots and sensor's battery status;
- Instant occupation rate value and perform historical queries about parking occupation at a specific time or during a time period, among others.

This service acquires information from a database where data is placed by the ParkSS application which, on its side, is a part of a system formed by a layer of modules whose purpose is to receive and persist the information generated by the sensors in a unified format.

2.2   Sensors Data Collector and Transformation Modules

As previously mentioned, there are no sensors installed on the parking lots of the Polytechnic of Leiria, so in order to develop the system two providers that mimic the behavior of real sensors were made. The ParkingSensorNodeDLL is a library that generates and pushes data directly to the receiver application in a preset time interval. The BOT-SpotSensors, on other way, is a SOAP webservice [18] that generates data in xml and data contract formats. BOT-SpotSensors follows a "pull" approach meaning that the applications who want to obtain the sensors' data are responsible to request this information to the service. Both providers generate data about occupation status and sensor battery, in a random way. This approach is not the ideal, but it's the simplest and fastest way to implement and, accurately, test the system.

The data generated by these providers needs to be integrated with other information (geolocation and parking lots data) into a unified format and this is done by the ParkDACE application. Therefore, the ParkDACE goal is to receive and collect the data from different sources, validate and integrate it in order to send this information to other application that is responsible for the data persistence.

The sharing of the data is made using a messaging system working with the MQTT [19] connectivity protocol (see Sect. 2.3). Under this protocol, ParkDACE plays the role of a publisher that sends messages with the unified information to the broker. The broker is responsible for distributing the data to all of the applications that subscribed the publisher topics. This system allows multiple subscriptions from various applications to the same topic, without service degradation.

The ParkSS application is a subscriber that will receive the information sent by ParkDACE and will persist the data into a relational database, named ParkDB.

All the information described above can be seen in Fig. 4.

2.3   Broker – Sensors Integration

To make parks and sensors' information available to interested applications it was used a message broker system. We used the broker Eclipse Mosquitto, which is an open source broker, [21] that implements the MQTT protocol, an open protocol that uses the publish-subscribe model, being the application ParkDACE responsible for publishing information about the parking lots and sensors.

The MQTT protocol is lightweight and, ultimately, a good choice for sensors that have limited resources, since it allows an efficient transmission of information.

The ParkTU application plays the essential role of the broker since it is the mediator that controls the connection and message transmission between applications.

Regarding the Quality of Service (QoS) we used a Level 2 – Exactly Once – since it ensures that each message is only received once. It is considered the most reliable level in terms of guarantees for the application because it delivers all the messages. If we have used a Level 0 – At Most Once – the analysis of the state of the parks and sensors, such as future statistical analyzes would not be reliable. Similarly, on Level 1 – At Least Once – the duplication of messages could occur, being then necessary an additional processing by the applications that would receive the information to manipulate the potential duplicated messages. Nevertheless, this level could be considerable advantageous if the performance of the transmission was indeed a crucial factor for the application that receives the data.

In the message broker, two topics were created for the transmission of information:

- Parks – information of parking lots;
- Spots – information about the sensors' status.

When using the message broker with the publish-subscribe model, the application becomes mores scalable. Thus, any application that wants to receive information provided by the ParkDACE can simply subscribe to the desired topics, according to the information that it wants to receive.

3   Spots

The Spots mobile application was developed for the Android operating system and allows users to obtain information about the occupation state of the available parks.

The mobile application consumes data from a Firebase instance and from the RESTful API described in Sect. 2. It also uses the Google API (including Maps), specific georeferentiation libraries from the mobile device (GPS and Gallery) and a specific library to create charts. The architecture can be visualized in Fig. 5 and, additionally, more details can be found on Appendix A.

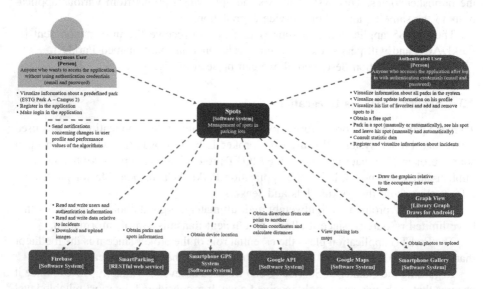

Fig. 5. Spots application architecture.

This application can be used by both authenticated and unauthenticated users. Unauthenticated users can access to the initial screens where they: see information about the occupation state of a specific park, and where they can register or authenticate in the application.

Figure 6(2) presents the parking layout where the green spots correspond to the geolocation of available slots, the total number of free and occupied slots on the park and the last update date for the information presented. It's also possible to unauthenticated users to register or authenticate themselves on the application. To register and authenticate on the Spots app, an email address and password are required.

For authenticated users, the Spots app provides a set of functionalities displayed on an always-present slide-drawer menu. Authenticated users can see the occupation state for all the available parking lots of the system (Fig. 6).

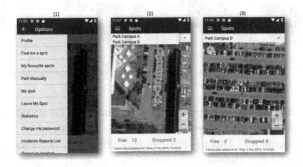

Fig. 6. Initial menu functionalities available for authenticated users (1), initial screen displaying Parking A occupation status on Campus 2 and the selection of menu with all parks available (2), Parking D with all slots occupied (3).

The most relevant features are:

- **Profile:** Users can see the information of their profile and update it. They have the possibility to manage their favorite parking spots' list;
- **My Spot:** When the user is parked the application shows the vehicle location on the parking lot (the spot);
- **Find a me a spot:** The aim of this functionality is to provide the location of an available spot on the user's selected parking according to his preference. There are three search algorithms that can be applied in order to find a free place: select the spot with closest location according to user's GPS location; select the best rated spot from all the spots in a parking lot; and, select the best rated place from the current user's favorites list.

The user can preset a preference for the "Find me a spot" option in the update profile screen and the application will choose the best spot according to user's preference. Otherwise, if the user does not have a preset preference, the application will ask him what option to use. After that, the application will present an available spot; will use the Google Maps service where the geolocation spot is shown, and the user could get the directions to reach that location;

- **Park/Leave My Spot:** The Spots application has an automatic detection mechanism. When there is a change in the occupancy state of a parking, it will be checked if there is any logged user parking or leaving the spot on that sensor location (Fig. 7). However, it also has options in the menu to allow a user to manually specify where he intends to park or when he's leaving a spot. Whenever the user leaves a parking spot, he can rate it and add it to his favorites list if that spot isn't already part of that list.

Fig. 7. Park Manually operation (1) and leave my spot manually (2).

- **Occupation rate/Statistics/Algorithm Performance:** The occupation rate screen shows the evolution of the average occupation rate during a time period in a graphical representation. There is also some additional information regarding users and application usage statistics, as shown in Fig. 8. The algorithm performance screen shows the average computation time in milliseconds that each algorithm takes to give a response to users. All this information is updated in real-time.

Fig. 8. Statistics (1), algorithm performance (2) and occupation rate during time (3) charts/graphical representation.

- **Incidents:** Allows the user to report incidents, see the list of all reported incidents and their details (see Fig. 9).

Fig. 9. Incidents report list (1), incident details (2) and report an incident (3) screens.

The development of the Spots application rests in some general principles of agile methodologies like Scrum [22], XP [23] and Kanban [24]. In this app the features where developed in an incremental and iterative way which allows developers to get constant feedback and improvement of the application.

For automated testing, the Cucumber tool [25] was used and it's based on the XP principle of Test First Programming [23], which leads the team to have a clearly understanding of the requirements before their actual implementation. The acceptance tests were written using the Gherkin language [26] - a "natural" language that contributes to part of the documentation. For the specific implementation of the "steps" we used the Espresso framework [27], the standard for automated testing for Android.

4   Conclusion and Future Work

This paper aims to present the design of a platform for an efficient, cost-effective smart parking system to the Polytechnic of Leiria and thereby enhance the parking quality to the students and employees working on this institution. This system manages data received from sensors with focus on providing real-time information regarding the availability of parking spots to end-users. The main goal is achieving the maximum decoupling between the several modules that compose the system, in order to maximize scalability and minimize the failure of the entire system just because a single module becomes unavailable.

From the analysis of related work, we conclude that a smart parking system has many advantages: decreases traffic congestion; minimizes fuel consumption; reduces the number of incidents on parking lots, and optimizes the time wasted by the drivers to search for a parking spot. Although there are still no sensors implemented on the Polytechnic of Leiria parking lots, in the future, when it becomes a reality, the timely planning of the platform presented on this paper will speed up the process of implementation of the smart park system.

There is future work that could be done in order to improve this system like the upgrade of the authentication mechanism from the client application allowing the users to have other options besides the email to authenticate themselves, like authentication by username or with their personal institutional identification number.

Also, it should be considered the development of an administration platform for the parking lots and other client applications, alike the proposed on this paper, and that work on distinct operating systems. Besides, the option to incorporate activity detection and even environmental factors would also allow us to predict, with some accuracy, the occupancy rate and the other significant statistics of the platform.

In the future, it would also be interesting to allow remote users to previously book an available spot for a short amount of time, before they get there, and park. This mechanism is more difficult to implement as many changes would be required since nowadays the Polytechnic of Leiria offers free parking. The solution would require, at least, an access control mechanism to the parks and the count of available spots excluding the ones that have been booked. In addition to this, it should also be considered the implementation of a security mechanism that ensures the use of the parking lots only by students and employees from the institution.

Acknowledgements. This work is financed by national funds through the FCT - Foundation for Science and Technology, I.P., under project UID/CEC/04524/2016.

Appendix A

See Figs. 10, 11, 12, 13, 14 and 15.

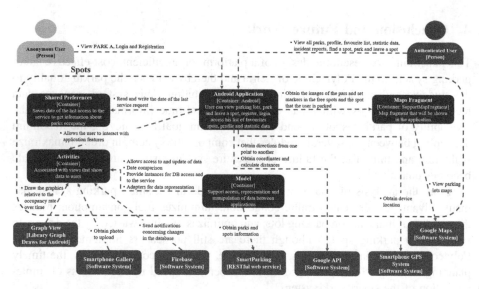

Fig. 10. Spots application containers details architectural scheme.

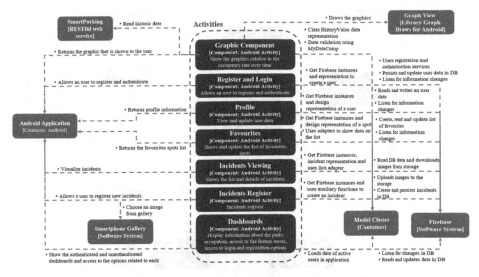

Fig. 11. Components of activities container of Spots application.

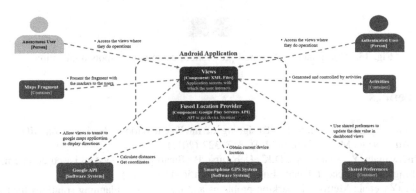

Fig. 12. Components of android application container of Spots application.

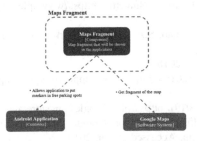

Fig. 13. Components of maps fragment container of Spots application.

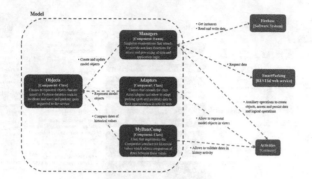

Fig. 14. Components of model container of Spots application.

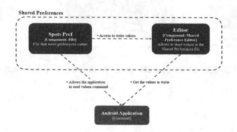

Fig. 15. Components of shared preferences container of spots application.

References

1. Wang, Z., Lian, Z., Han, K.: Designing of intelligent parking lot based on MQTT. Int. J. Adv. Netw. Monit. Control. **2**(4), 317–322 (2017)
2. Vishwanath, Y., Aishwarya, D.K., Debarupa, R.: Smart parking system based on Internet of Things. Int. J. Recent Trends Eng. Res. **2**(3), 156–160 (2016)
3. Liu, Y., et al.: Metropolis parking problems and management planning solutions for traffic operation effectiveness. Math. Probl. Eng. **2012**, 6 (2012)
4. Lingam, C.: University Car Parking Application. Texas (2018)
5. Polytechnic of Leiria: Factos e Números. https://www.ipleiria.pt/ipleiria/ipl-em-numeros/. Accessed 07 Feb 2019
6. Ahad, A., Khan, Z.R., Ahmad, S.A.: Intelligent parking system. World J. Eng. Technol. **4**, 160–167 (2016)
7. Libelium: Smart Parking (2018)
8. Apple Inc.: App Store (2019). https://www.apple.com/pt/ios/app-store/. Accessed 13 Feb 2019
9. Google: Google Play (2019). https://play.google.com/store/apps. Accessed 13 Feb 2019
10. JustPark: JustPark Parking – Apps on Google Play. https://play.google.com/store/apps/details?id=com.justpark.jp. Accessed 29 Jan 2019
11. ParkMobile: Parkmobile Parking – Apps on Google Play. https://play.google.com/store/apps/details?id=com.parkmobile. Accessed 29 Jan 2019
12. ParkNow: ParkNow Parking – Apps on Google Play. https://play.google.com/store/apps/details?id=com.parknow.app. Accessed 29 Jan 2019

13. PayByPhone: Start new parking. PayByPhone. https://m2.paybyphone.com/parking/start/location. Accessed 29 Jan 2019
14. BestParking: BestParking: Find and Book Parking Anywhere. https://www.bestparking.com/. Accessed 29 Jan 2019
15. SmoothParking Inc.: SmoothParking. https://itunes.apple.com/us/app/smoothparking/id615418863?mt=8. Accessed 07 Feb 2019
16. Parking Hero: Smart Parking - Parking Hero. https://itunes.apple.com/us/app/smart-parking-parking-hero/id1249834332?mt=8. Accessed 06 Feb 2019
17. Lange, K.: The Little Book on REST Services. Copenhagen (2016)
18. Tidwell, D., Snell, J., Kulchenko, P.: Programming Web Services with SOAP, 1st edn. O'Reilly, pp. 21–36 (2001)
19. MQTT: MQTT.org (2014). http://mqtt.org. Accessed 13 Feb 2019
20. Fawcett, J., Quin, L.R.E., Ayers, D.: Beginning XML, 5th edn. (2012)
21. Eclipse Foundation: Eclipse Mosquitto. https://mosquitto.org. Accessed 13 Feb 2019
22. Schwaber, K., Sutherland, J.: The Scrum Guide: The Definitive the Rules of the Game. Scrum.org and Scrum Inc., p. 19 (2017)
23. Beck, K., Andres, C.: Extreme Programming Explained - Embrace Change, 2nd edn. John Wait, United States (2004)
24. Anderson, D.J., Carmichael, A.: Essential Kanban Condensed, 1st edn. Lean Kanban University Press, Seattle (2016)
25. Wynne, M., Hellesoy, A., Tooke, S.: The Cucumber Book - Behaviour-Driven Development for Testers and Developers, 2nd edn. Pragmatic Programmers, United States (2017)
26. Gherkin Reference. https://docs.cucumber.io/gherkin/reference/. Accessed 13 Feb 2019
27. Espresso (2019). https://developer.android.com/training/testing/espresso/. Accessed 13 Feb 2019

The Network Topology of Connecting Things: Defence of IoT Graph in the Smart City

Marta Chinnici[1(✉)], Vincenzo Fioriti[2], and Andrea Arbore[3]

[1] ENEA-ICT Division, C.R. Casaccia Via Anguillarese 301,
00123 Rome, Italy
marta.chinnici@enea.it
[2] ENEA-ISER Division, C.R. Casaccia Via Anguillarese 301,
00123 Rome, Italy
vincenzo.fioriti@enea.it
[3] SiliconDev SpA, 00131 Rome, Italy
arborean@hotmail.com

Abstract. The Internet of Things (IoT) is a novel paradigm based on the connectivity among different entities or "things". IoT environment in the form of interconnected smart "things" represents a great potential in terms of effective and efficient solutions related to urban context (e.g., system architecture, design and development, human involvement, data management and applications). On the other hand, with the introduction of the IoT environment, devices and network security have become a fundamental and challenging issue. Moreover, growing number of users connected via IoT system necessitates focusing on the vulnerability of complex networks and defence challenges at the topological level. This paper addresses these challenges from the perspective of graph theory. In this work, the authors introduce a novel AV11 algorithm to identify the most critical and influential IoT nodes in a Social IoT (SIoT) network in a smart city context using ENEA Portici CRESCO infrastructure.

Keywords: IoT · Malware · Complex networks · Graph theory ·
Infrastructure · Smart city · Social Internet of Things (SIoT) network ·
Risks · Big data

1 Introduction

In a smart city context, the introduction of the Internet of Things (IoT) paradigm has become a fundamental concept: devices become pervasive and blend with human beings. It is the state wherein there is no distinguishable difference between the operation of devices surrounding us and our actions. Consequently, devices play an interactive role in the IoT systems and enhance the human experience. There is a seamless integration between us and the things around us. Various devices communicate intelligently with one another with minimal human intervention. Thus, devices of the IoT system are interconnected; they communicate with one another, transfer and retrieve data, intelligently respond to requests and trigger actions [1].

© Springer Nature Switzerland AG 2019
J. M. F. Rodrigues et al. (Eds.): ICCS 2019, LNCS 11540, pp. 84–96, 2019.
https://doi.org/10.1007/978-3-030-22750-0_7

Hence, the introduction of the IoT in smart cities context involves the consideration of a vast number of aspects. These include system architecture, design and development, use of embedded devices, communication technology, human involvement, data management and applications, security and privacy concerns. According to the reports by Cisco [2], Gartner [4] and Ericsson [3], 50 billion of heterogeneous devices will be connected to the Internet by 2020. These devices are getting increasingly smarter, connected to the global networks and among themselves, giving value to the IoT paradigm that is generating an unprecedented volume and variety of data. In [2] and [5], authors have estimated that the data caused by physical objects would reach 507.5 Zettabytes (ZB) per year (42.3 ZB per month) by 2019, which is 269 times more than the amount of data transmitted to Data Centers (DCs) from end-user devices and 49 times higher than total DCs traffic [13]. This brings up crucial discussion regarding all the data generated, stored or transmitted by IoT devices as well as its security and how this relates to the privacy of the users. Every approach of IoT system must have a secure network and provide a necessary level of control and privacy to the users. To accomplish these goals and then the success of the IoT systems' implementation, a baseline to build secure IoT network is one of the central aspects achievable with IoT graph topology.

Motivated by such challenges and inspired by graph theory, we aim to identify the most critical and influential IoT nodes in a Social IoT (SIoT) network [6] to address the way towards building a secure IoT environment for smart city context. This paper discusses security issues at the baseline level of IoT system and proposes topology of a graph, to target security challenges and vulnerabilities of different IoT systems. In detail, in this work the authors utilize a revised version of the AV11 algorithm [15] in a real SIoT network [21] to identify crucial infected nodes in the graph that corresponds to a set of key IoT systems (e.g., smartphone, laptop, computer, tablet, home sensors, etc.). The procedure allows extracting information from the graph topology regarding the best k nodes (namely the "budget") to immunise or remove; thus, the remaining network is more robust to attacks. In IoT networks, the concept of attack handling can be translated to the identification of the IoT nodes that become "infected" in a specific configuration and hence, to defend the graph discovering these infected nodes become a crucial issue. Besides the defence concerns, the procedure also allows controlling virality of the network, identifying the most influential nodes (the influencers).

In summary, this work aims to investigate the network risk security based on the rapid deployment of IoT systems around the digital world. For IoT network, risk assessment is complex due to its vast deployment and diversity in terms of devices. Thus, traditional risk assessment frameworks do not adequately address the risk related to the topology of the network and then the risk assessment of IoT to be completed needs also to include a risk framework at a graph level.

The following objectives will support this aim:

(a) Apply the algorithm AV11 for real IoT network [21] and provide the assessment risk framework;
(b) Analyse topology structure of the network in terms of infective IoT nodes;
(c) Perform data analysis to calculate statistics associated with the infective IoT nodes and hence, the graph topology in terms of devices;

(d) Evaluate stability of the IoT graph with IoT devices configuration;

(e) Assess the risk based on (a–d) in a real IoT network in a smart city context;

(f) *Versus* risk assessment and mitigation of IoT dynamic network and future association of energy consumption of IoT device in the configuration network.

The paper is organized as follows: Sect. 1 – Introduction; Sect. 2 – Background: Related Works and AV11 Algorithm presentation; Sect. 3 – Methodology; Sect. 4 – Assessment of an IoT graph; Sect. 5 – Conclusions and Future Works.

2 Background: Related Work and AV11 Algorithm

In recent years, significant research efforts and technological developments have been devoted to IoT paradigm [8] targeting energy efficiency of IoT systems. Indeed, these "things" enable new computing applications and represent the base of the vision of a global infrastructure composed of complex networked physical objects. According to [7], IoT systems represent the principal source of big data and, consequently, are the drivers of the plethora of applications, e.g. in smart cities [14]. Due to the inherently diverse nature of IoT paradigm, it attracts lots of risks in various forms. Therefore, understanding the processes and mechanisms involved in the evolution of complex networks for the IoT is a significant challenge.

Undeniably, in the IoT paradigm in smart cities context, mathematics plays an essential role in understanding complex networks. To address the networks with mathematical models, one is naturally led to dynamical systems, in which the graph describing the network is also a dynamical variable. The graph's dynamics is coupled with that of other variables not explicitly considered in the model. Analysis of such dynamics requires development of some new tools inspired by the graph theory [16–18]. Historically, the study of networks has been mainly a branch of discrete mathematics known as graph theory [9] that proves useful in understanding complex networks. The network structure is irregular, complex and dynamically evolving in time, with the main focus moving from the analysis of small networks to that of systems with thousands or millions of nodes, and with renewed attention to the properties of networks with dynamical units.

The complexity of the network structure poses significant challenges of capturing risks associated with the topological structure of the graph and risk assessment. In [10–12] a summary of the current literature related to fundament, kernel, methods, environment for IoT and associated risks is provided. Even if many efforts are addressed to IoT paradigm architecture, no investigation into the security aspects associated with the IoT graph in terms of complex network topology is present. Nowadays, IoT is missing security and in particular at the topologic level; for this reason, the security has emerged as a significant challenge for the IoT. Therefore, in this paper, the authors investigate topological and functional structures of an IoT network - where IoT system is represented by the nodes - to analyse the system and the specific infection nodes and malicious propagation attack. This study is based on the application of the algorithm AV11 [15, 22, 23] which intends to immunise or remove chosen nodes and make the rest of the network more robust and resilient to attacks. With this analysis, we are also able to figure out intelligent and robust characters of IoT.

The application of our AV11 algorithm at the topological structure level of the graph is the baseline for the dynamic network analysis. In the paper, the authors also explore those intrinsic structures that are independent of data and methodology. The AV11 algorithm is a topological vulnerability tool, meaning that the vulnerability is taken into consideration according to a particular position of a node in a graph representing a technological network, such as an IoT network. In straightforward cases, such as the star, it is clear that the node in the centre is the most important/critical/influential one, but usually, for more complex topologies, the problem is not that trivial. The spreading of dangerous malware (malicious software) in networks of electronics devices has raised deep grave concern because infections may propagate from the ICT networks to other Critical Infrastructures producing a well-known domino effect. There are two diffusion strategies: targeted intrusion and cooperative search. The first strategy foresees a direct conventional approach to the actual target, while the second one demands a distributed control system, a sophisticated communication scheme and a consensus-like decision-making process. As a side effect of the cooperative search, the malware will spread in the network like a disease (the "epidemic" spreading). Actually, any worm follows the epidemic spreading, but a standard worm will attempt to invade the maximum number of machines as quickly as possible. In contrast, a sophisticated malware adopting a cooperative search strategy or even a simpler network aware approach will infect (relatively) few machines during an extended period. In any case, both seem to propagate following the epidemic spreading model, at least during the initial phase of the attack. Understanding this model may help to counteract the spreading at its very beginning when the cost of defence is more affordable. Researchers are attempting to develop a high-level analysis of malware propagation, discarding software details, to generalise to the maximum extent the defensive strategies. Since the maximum eigenvalue of the adjacency matrix of the network acts as a threshold for the malware's spreading [15], spectral analysis of the graph's adjacency matrix has a relevant role.

In this section, a brief description of the AV11 algorithm development by the authors is presented; the application of the algorithm to a specific IoT network will be shown in the next paragraph.

2.1 Description of AV11 Algorithm

The problem to be faced is: find k best nodes (the "budget") of an IoT network to immunise/remove them with the intention to make the whole network more resilient to malware attacks. Malware is malicious software designed to damage an ICT (Information and Communications Technology) network, often called viruses. Today viruses are net-aware in the sense that they can exploit vulnerabilities of the network, carefully selecting the critical nodes. Something similar could describe the spreading of faults, a well-known domino effect or cascade failure.

A defensive strategy should protect the most critical or influential nodes. Since, unfortunately, available resources are limited, to safeguard at most only k nodes of the network we should select them to maximise the probability to stop or reduce the spread of the malware or the fault in the network. This is what we mean by "find the best k nodes". This task is not trivial since it is not clear what nodes are the most important in a network, because the interdependencies among them are often counter-intuitive. Thus, to identify these k nodes many algorithms have been used in the past years: degree, closeness, betweenness, Estrada indices, most infected, k-core, dynamical importance, and other [15, 19, 22, 23]. Standard topological centrality indicators such as degree, closeness, clustering, etc., are relevant quantities for assessing useful information. These parameters can also be employed to provide the first, non-trivial understanding of the network's dependencies, but are insufficient to unveil more subtle relations among the nodes, and as a matter of fact, spectral methods usually perform better. Even worse is the performance of the "most infected" strategy. According to this procedure, influential nodes that provide significant support to the epidemic spreading are the nodes that get infected more frequently during simulations. Our experiments demonstrate the weak points of this strategy. In the complex environment of the IoT, relying on such an approach could be extremely dangerous.

In this paper, we propose to solve the issue related to finding the best k nodes with our AV11 algorithm, which follows a combinatorial spectral paradigm [19] and has proved to be the most effective [15]. We use standard notation and terminology of graph theory and refer to the *network* as a *graph*, the *node* as a *vertex* and the *link* as an *edge* hereafter. Let G be the graph for it. It is well known that the largest eigenvalue λ_1 of a graph is related to a threshold for the epidemic propagation of a fault or a malware in the network [22]. If the ratio probability of infection, i.e. probability to cure is under this threshold, the spreading does not take place and vice versa [22]. Here, to "cure" means to provide an antivirus or some other kind of protection to the node/device. In practice, AV11 removes a set of k nodes and finds a sub-optimal decrease of the largest eigenvalue of the graph G as indicated in the pseudocode below. A brute-force strategy would be impossible to use even for small graphs, because of the vast number of combinations given that the problem is NP-complete [22]. Nevertheless, our suboptimal algorithm AV11 reduces the algorithmic complexity to $O(k \cdot n^3 \cdot \log n)$. However, it should be noted that even such a complexity is by far too cumbersome for a PC. Therefore, we have used the ENEA' infrastructure (CRESCO) to determine the most critical nodes from the adjacency matrix representing an IoT network [21]. Out of 16216 devices (nodes), 3300 were identified as the most critical ones. These 3300 nodes are those to be immunized somehow, since they guarantee to provide the maximum protection. It has to be considered that immunizing a node involves a non-negligible cost and that the available resources are scarce. In our case-study, the number of 3300 was chosen large enough to test the CRESCO potential, but actually, to determine the number of nodes to be immunized it would be necessary to run some Monte Carlo simulations to know the minimum number of nodes able to stop the spreading of the malware. Therefore 3300 it is crucial to refer only to verify the CRESCO calculation capabilities.

The AV11 pseudocode (see [15] for more details) is:

Input: the adjacency matrix A and an integer $0 < k < n$
Output: a set S with k nodes
Algorithm:
1. Calculate eigenvalues of the adjacency matrix A; let λ_1 be the largest eigenvalue;
2. Initialize: S to empty; $Z = I_n$; $D = \left(1 - \min_{i \in [1,n]} \mathbb{Re}(\lambda_i)\right) I_n$; $node = 0$;
3. **for** $i = 0$ to k **do**
4. $P = (Z * A * Z + D)^p$;
5. $node = index\ of\ \max\left(\mathrm{diag}\,(P)\right)$;
6.   Add $node$ to S;
7.   Set $Z\,[node, node] = 0$;
8. **end for;**
9. **return** S.

Where, I_n is the identity matrix of order n and $0 < p \le n$ is a parameter based on the longest cycle of the graph.

3 Methodology

This work focuses on IoT network graph assessment through a topology structure analysis using the AV11 algorithm [15, 22], particularly, on the evaluation of infected IoT-nodes [21]. To address this challenge, in this paper, we analyse real data of the IoT network consisted of 16216 IoT devices (nodes) provided by the authors in [21]. In detail, real IoT devices are available in the city of Santander and categorized in [6, 21] with respect to typologies and data model for objects introduced in the FIWARE data models [24]. As we have already mentioned, an experimental campaign which consists of the application of the AV11 algorithm to this real network is conducted by the ENEA infrastructure, on the Cluster named CRESCO4 (hosted by ENEA R.C. Portici). The principal goal is to calculate critical nodes of IoT devices and explore their characteristics. In this work, improvements have been made to previous studies conducted on the IoT environment concerning the topology and graph control. In particular, the authors provide an assessment of dynamical properties of the network through spectral eigenvalue analysis and also more in-depth knowledge of the IoT devices. Results are also expressed in terms of statistical data that could be generalized and applied in a real smart city context.

Briefly, the cluster CRESCO4 consists of 38 Supermicro F617R3-FT chassis, each hosting 8 dual CPU nodes. Each CPU, specifically an Intel E5-2670, hosts in its turn 8 cores, for a total number of 4864 cores. These CPUs operate at a clock frequency of 2.6 GHz. Moreover, the system is provided with a RAM memory of 4 GB per core. Computing nodes access a DDN storage system, for a total storage amount of 1 Pbyte. The connection between computing nodes is realized via an Infiniband 4xQDR QLogic/Intel12800-180 switch (432 ports, 40 Gbps).

3.1   IoT Graph in the Smart City

As aforementioned, the authors consider as IoT network the case-study provided by the Santander testbed [20, 21] of real IoT objects.

The Santander testbed considered in this current work as IoT network is composed of several thousand devices that comply with IEEE 802.15.4 standard (10-m communications range with a transfer rate of 250 kbit/s), 200 GPRS modules and 2000 joint RFID, positioned at static locations (e.g., streetlights, bus stops) or mobile location such as on-board of vehicles (e.g., buses, taxis), in order to provide environmental monitoring, outdoor parking area management, mobile environmental monitoring, traffic intensity monitoring, guidance to free parking lots, parks and gardens irrigation, and participatory sensing. The general idea is to develop an architecture to support the smart city concept.

Briefly, we present characteristics of Santander IoT network used in this manuscript to apply the AV11 algorithm. The IoT objects are categorized with typologies and data model for objectives introduced by FIWARE Data Model and comprehend a total of 16216 devices, of which 14600 from private users and 1616 from public services. The following form has been used to describe network objects: id_device, id_user (owner id of device), device_type (category associated with a device in the form of code to differentiate between public and private devices from 1 to 16), device_brand (each device is assigned with a number from 1 to 12 encoding a brand), device model (it is associated with each device, a number from 1 to 24). The device_type code is provided to every object by the global Web Index 2017 [27] that identifies both the status (*static:* home sensors, PC, etc., or *mobile*: smartphone, car, etc.) and the type of device being private or public (Table 1).

Besides, the adjacency matrix is compiled by notions of Social Internet of Things (SIoT): social relationships between the nodes are established by a disjunction (OR) of five elementary relationships [6]. Mobile devices are carried with the users during their movements, while static objects are left in the users' home.

To simulate the user's movements, authors in [6] rely on a mobility model called Small World in Motion (SWIM). In this way, the authors of [6] obtained the estimate of the relationships between the devices on a given day, producing the 16216 × 16216 adjacency matrix of the Santander case study. In Fig. 1 a Gephi [26] visualization of this network is showed. These datasets are freely downloadable from [21]; they are among very few ones available for IoT offering a valuable estimation of real-life phenomena.

Table 1 summarizes a legend of groups of devices: a code number identifies each group, also classified as public or private and static or mobile.

Table 1. The first column is the code number defined by the device type and mobility characteristic provided in the second and third columns respectively.

Device type	IoT-device	Kind
1	Smartphone	Mobile
2	Car	Mobile
3	Tablet	Mobile
4	Smart Fitness	Mobile
5	Smartwatch	Mobile
6	PC	Static
7	Printer	Static
8	Home Sensors	Static
9	Point of Interest (specific point location that a user may find useful or interesting)	Static
10	Environment Weather	Static
11	Transportation (Vehicles, taxis or buses)	Mobile
12	Indicator (Digital signage to display information)	Static
13	Garbage Truck	Mobile
14	Street Light	Static
15	Parking (Location designed for parking)	Static
16	Alarms (Security supervisor or traffic monitoring)	Static

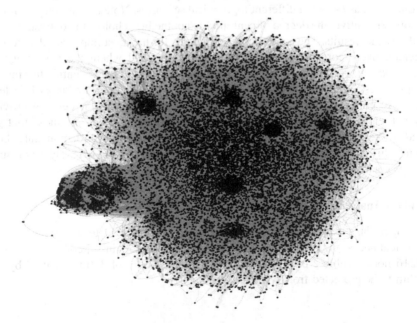

Fig. 1. Gephi visualization of the graph representing the network of the Santander case study with all 16216 nodes.

4 Assessment of an IoT Graph

As we have already mentioned, among topological analytical tools, the spectral description of networks deserves special attention. By its means, it is possible to infer insights on dynamical properties based on "static" parameters. Following Restrepo, Ott and Hunt [19] the authors focus on the largest eigenvalue of the graph adjacency matrix. This parameter (under a set of commonly accepted hypotheses) is closely related to the epidemic spreading of viruses or failures. The dynamical importance of a node (or an edge) [19] is defined as a normalised amount by which the maximum eigenvalue of the graph decreases if the node is removed. Therefore, if we remove a node, we can infer a measure of its "dynamical dependence" for all the other nodes employing the variation of the maximum eigenvector (the eigenvector of the maximum eigenvalue). Thus, if we consider a node (link), its dynamical importance will define its "importance" for the graph. The purpose of a network defence technique is to intervene in certain elements (the "budget") of the network to limit the impact of an attack or the spread of a virus or information.

Our algorithm AV11 selects a subset of k nodes all at once according to spectral combinatorial methods, to immunize or remove them to make the remaining network more robust to attacks. In particular, AV11 classifies all the nodes of a graph according to a spectral parameter that describes dynamical properties of the graph's nodes in terms of the node influence on the network. A selected subset may be optimal or suboptimal, as a result of the brute-force method, and is not unique.

In [15] it is shown that the AV11 algorithm performs better than other selection techniques such as k-core in different topological scenarios. Note that although it might seem counterintuitive, the degree of centrality is not the best choice. In contrast, it often gives the wrong results. The same supposedly holds for the most important degree-like parameters, like k-core presented by the articles [15, 23] where the percentage of infected nodes is presented after the application of techniques to identify the immunizations nodes. Even more critical is the inferior performance of the "most infected" strategy. According to this procedure, influential nodes, i.e. those nodes that provide significant support to the epidemic spreading or the cascade effect, are those that get infected more frequently during simulations. Our experiments demonstrate how deceiving this idea is [22, 23]. In the complex environment of the IoT, relying on such a strategy could be extremely dangerous.

4.1 Data Analysis: Results and Discussion

In this paper, the authors have used the spectral algorithm AV11 to determine the first 3300 critical nodes on the IoT network created by the authors in [6] composed in total by 16216 nodes. Table 2 presents the first "best" ten nodes of 3300 classified by the algorithm to be protected from malware.

Table 2. The first ten nodes/devices of the IoT graphs from the AV11 analysis; they are all type 1 device, i.e. smartphones. The AV11 algorithm provides the value that is used to rank the importance of the node (third column); thus, these nodes are recognized as the most influential and need to be protected in the first place.

Classification	Node	Max_Eingevalue
1	247	364.837169
2	1407	363.841058
3	1521	362.846303
4	593	361.856203
5	728	360.910795
6	274	359.966537
7	240	359.021899
8	3631	358.075309
9	3322	357.135808
10	2314	356.211899

The algorithm goes like this: choosing a set of k nodes (in our case study $k = 3300$) simultaneously and removing them from the graph G, we obtain a new graph, and therefore we can calculate a new λ_1, smaller with respect to the old λ_1. Since the infection threshold depends on the inverse of λ_1, the lower λ_1 the stronger the resilience to the infection spreading. Thus, we are interested in choosing k nodes that will provide the minimum possible λ_1. Moreover, AV11 provides an individual ranking for each of the k selected nodes (Table 2, third column AV11 value).

Then we analyse the major device groups (see Table 2) in order to provide some group statistics (Table 3).

For each type of group device, parameters taken into consideration are the number of devices in the group, the percentage of the total number of devices, mean and variance of the eigenvalues per group, eigenvalue sum per group and the percentage related to the total sum of the 3300 eigenvalues.

Therefore, within the subset of the 3300 nodes, the relevance of each device group concerning the transmission of any interaction to the whole network is represented by the sum per group of the eigenvalues of the group. For example, as shown in Table 3, the sum of the 375 eigenvalues of the type 2 devices accounts for 12.50% of the 3300 eigenvalues sum. Then the type 2 group is placed among the most influential of the network. In other words, since any physical action in the real world somehow finds a counterpart in the graph spectral analysis, the of contribution of each group type to the eigenvalue sum (see Table 3, last column) may be read as the relative relevance in the virtual world of the graph produced by real phenomena in the physical world.

The most relevant effect on the total eigenvalue sum is caused by smartphones, accounting for 27% of the total number of devices and 31.50% of the total eigenvalue sum. The second massive contribution comes from the PCs (25.70%), the third one from cars (12.50%). Smartphones, cars, tablets, printers and PCs represent 86% of the total sum of the eigenvalues, therefore are by far the most critical part of the network. However, it is immediately evident that some devices, namely 6, 7, 12, 13 (PCs, printers, indicators, trucks) even if not the most numerous, have high eigenvalue mean values, meaning that their impact on the network is important, despite the total number of devices.

It is particularly interesting to note the importance of indicators and printers since at present they are the favourite targets of hackers. Moreover, if we associate a cost to a device, it would be possible to produce a cost function describing the trade-off between the risk reduction provided by a security measure applied to a specific group of devices and the economic cost of the measure itself. Not surprisingly in this sense, it may be more affordable to spend more on the security of PCs, printers, indicators and trucks than on smartphones.

Table 3. In the first column codes of devices are indicated, then for each type of device the following statistics are shown: the absolute number of that type of device present in the network, percentage of the device type related to the total number of devices, the average and the variance of the AV11 values per type, and finally the percentage of the eigenvalues of a type_device related to all the AV11 values.

Type_device	Nodes	% Nodes	AV11 value Mean	AV11 value Var	% AV11 Total
1	887	27	114	11399	31.50
6	535	16	155	9432	25.70
2	375	11.30	107	4654	12.50
15	361	10.90	33	208	3.80
7	317	9.60	99	3644	9.80
3	252	7.60	85	2519	6.60
14	141	4.30	54	1003	2.39
11	128	3.90	69	944	2.80
4	127	3.80	52	661	2.00
8	68	2.10	35	168	0.75
9	50	1.50	39	1136	0.61
10	27	0.82	93	9775	0.79
5	15	0.45	21	9	0.10
12	10	0.30	98	1799	0.31
13	7	0.21	138	15	0.30

In essence, the IoT environment will continue to increase in size and complexity. Therefore, a domino effect due to net-aware malware has to be considered probable and not only possible. Inherent vulnerability of IoT devices and structures exacerbate the problem to the point that does not allow an ex-post reaction of the attack, and it is not even viable to trust the enhancement of the IoT resilience. Thus, it is mandatory to rely on a formal ex-ante defence analysis, able to provide cost-effective countermeasures.

5 Conclusion and Future Works

In this paper, we proposed the AV11 algorithm and shown its applicability on a real IoT network example with the goal to provide an assessment framework based on the graph theory for the IoT network in a smart city context. The IoT network has been modelled as an undirected graph, and the nodes/devices have been classified according to a spectral technique able to identify the most influential nodes, i.e. the nodes that can propagate malware through the network. Since it was demonstrated [22, 25] that immunizing a small set of properly selected nodes can prevent or at least reduce the spreading, it is mandatory to identify these nodes in advance. Meanwhile, it is not a viable solution to immunise a large part of an IoT network, because of the vast number of devices. The economic advantages of the proposed algorithm are thus clear. For example, instead of concentrating the defence efforts on smartphones, it could be equally efficient but more economical to protect static assets such as computers, printers, indicators. Further, as part of future work, we aim to approach an IoT network dynamic risk assessment and mitigation and the association of the energy consumption of each device in the configuration network. Since limited research is devoted to topological structures of other complex networks for IoT in comparison with the investigation of IoT software technology and also to the ongoing investigation on the energy efficiency of the IoT devices, the authors aim to address in the future works these results on dynamic IoT network in smart cities context.

Acknowledgments. The authors would like to express their gratitude to the HPC research group at ENEA R.C. Portici to using ENEA Infrastructure for calculations.

References

1. Mendez, D.M., Papapanagiotou, I., Yang, B.: Internet of things: survey on security. Inf. Secur. J. Glob. Persp. **27**(3) (2018). https://doi.org/10.1080/19393555.2018.1458258
2. Dave, E.: The Internet of Things How the Next Evolution of the Internet Is Changing Everything. White Paper, Cisco, April 2011
3. Hans, V.: CEO to Shareholders: 50 Billion Connections 2020. White Paper, Ericsson, April 2010
4. Gartner: Gartner says 6.4 billion connected "things" will be in use in 2016, up 30 percent from 2015. http://www.gartner.com/newsroom/id/3165317. Accessed 06 Dec 2016
5. Forbes: 152,000 smart devices every minute in 2025: IDC outlines the future of smart things. http://www.forbes.com/sites/michaelkanellos/2016/03/03/152000-smart-devices-every-minute-in-2025-idc-outlines-the-future-of-smart-things/#34bf983369a7. Accessed 06 Dec 2016
6. Atzori, L., Iera, A., Morabito, G., Nitti, M.: The Social Internet of Things (SIoT) - when social networks meet the Internet of Things: concept, architecture and network characterization. Comput. Netw. **56**(16), 3594–3608 (2012)
7. Chen, M., Mao, S., Zhang, Y., Leung, Victor C.M.: Big Data: Related Technologies, Challenges and Future Prospects. SCS. Springer, Cham (2014). https://doi.org/10.1007/978-3-319-06245-7

8. Chinnici, M., De Vito, S.: IoT meets opportunities and challenges: edge computing in deep urban environment. In: Dependable IoT for Human and Industry, Chapter 11, pp. 241–272. River Publishers (2018)
9. Kranenburg, R.V., et al.: The Internet of Things. Paper Prepared for the 1st Berlin Symposium on Internet and Society, 25–27 October 2011
10. Yao, B., et al.: Applying graph theory to the internet of things. In: Proceedings of IEEE International Conference on High Performance Computing and Communications & IEEE International Conference on Embedded and Ubiquitous Computing (2013)
11. Shivraj, V.L.: A graph theory based Generic Risk assessment framework for Internet of Things (IoT). In: IEEE International Conference on Advanced Networks and Telecommunications Systems (ANTS) (2017)
12. Mendez, D., et al.: Internet of Things: survey on security and privacy. Inf. Secur. J. Glob. Persp. **27**(3), 162–182 (2018)
13. Chinnici, M., Capozzoli, A., Serale, G.: Measuring energy efficiency in data centers. In: Pervasive Computing Next Generation Platforms for Intelligent Data Collection, Chapter 10, pp. 299–351 (2016)
14. Dey, N., et al.: Internet of Things and Big Data Analytics. Series: Studies in Big Data, vol. 30. Springer (2018). ISSN 2197-6503
15. Arbore, A., Fioriti, V., Chinnici, M.: The topological defense in SIS epidemic models. Chaos, Solitons Fractals **86**, 16–22 (2016). ISSN: 0960-0779
16. Fioriti, V., Chinnici, M., Palomo, J.: Predicting the sources of an outbreak with a spectral technique. Appl. Math. Sci. **8**(135), 6775–6782 (2014). ISSN: 1312885X
17. Fioriti, V., Chinnici, M.: Identifying sparse and dense sub-graphs in large graphs with a fast algorithm. Euro Phys. Lett. **108**(5), 50006 (2014). ISSN: 0295-5075
18. Fioriti, V., Chinnici, M.: Node seniority ranking in networks. Stud. Inf. Control **26**(4), 397–402 (2017). ISSN: 1220-1766
19. Restrepo, J., Ott, E., Hunt, B.: Characterizing the dynamical importance of network nodes and links. Phy. Rev. Lett. **97**, 094102 (2006)
20. http://www.smartsantander.eu/index.php/testbeds/item/132-santander-summary
21. http://www.social-iot.org/index.php?p=downloads
22. Arbore, A., Fioriti, V.: Topological protection from the next generation malware: a survey. IJCIS **9**(1/2), 52–73 (2013)
23. Arbore, A., Fioriti, V.: Sub-optimal topological protection strategy from advanced malware. CRITIS **81–92**, 2011 (2011)
24. https://www.fiware.org/developers/data-models/
25. Chakrabarti, D., Wang, Y., Wang, C., Leskovec, J., Faloutsos, C.: Epidemic thresholds in real networks. ACM Trans. Inf. Syst. Secur. **10**(4), 1–26 (2008)
26. Bastian, M., Heymann, S., Jacomy, M.: Gephi: an open source software for exploring and manipulating networks. In: International AAAI Conference on Weblogs and Social Media (2009)
27. https://cdn2.hubspot.net/hubfs/304927/Downloads/Trends-17.pdf

SILKNOWViz: Spatio-Temporal Data Ontology Viewer

Javier Sevilla$^{(\boxtimes)}$ ⓘ, Cristina Portalés ⓘ, Jesús Gimeno ⓘ,
and Jorge Sebastián ⓘ

Universitat de València, Blasco Ibáñez 13, 46013 València, Spain
{javier.sevilla, cristina.portales, jesus.gimeno,
jorge.sebastian}@uv.es

Abstract. Interactive visualization of spatio-temporal data is a very active area that has experienced remarkable advances in the last decade. This is due to the emergence of fields of research such as big data and advances in hardware that allow better analysis of information. This article describes the methodology followed and the design of an open source tool, which in addition to interactively visualizing spatio-temporal data that are represented in an ontology, allows the definition of what to visualize and how to do it. The tool allows selecting, filtering and visualizing in a graphical way the entities of the ontology with spatiotemporal data, as well as the instances related to them. The graphical elements used to display the information are specified on the same ontology, extending the VISO graphic ontology, used for mapping concepts to graphic objects with RDFS/OWL Visualization Language (RVL). This extension contemplates the data visualization on rich real-time 3D environments, allowing different modes of visualization according to the level of detail of the scene, while also emphasizing the treatment of spatio-temporal data, very often used in cultural heritage models. This visualization tool involves simple visualization scenarios and high interaction environments that allow complex comparative analysis. It combines traditional solutions, like hypercube or time-animations with innovative data selection methods.

Keywords: Ontology · Visualization · Spatio-temporal data

1 Introduction

Data visualization, their analysis and interpretation, are areas of high interest to the scientific community [1, 2]. In our Big Data era, massive amounts of data are processed all the time, some of them leading to scientific discoveries [3]. Therefore, visualization, in order to detect patterns and relationships, has become a research avenue of great importance.

Spatio-temporal data are the main data sets that are analysed; they are related to the evolution of information in two key variables, space and time. By analysing these data, can be known where and when they fluctuate. There are many applications and studies [4, 5] related to the visualization of this type of data, which require special strategies for the graphical representation of information with multiple dimensions.

© Springer Nature Switzerland AG 2019
J. M. F. Rodrigues et al. (Eds.): ICCS 2019, LNCS 11540, pp. 97–109, 2019.
https://doi.org/10.1007/978-3-030-22750-0_8

In the 2014 Ontology Summit [6] it was concluded that the use of ontologies to represent information models and infer knowledge has played a key role in the development of new semantic technologies. However, the collaboration between the Semantic Web and Application of Technologies communities has not been as active as one might expect. In this Summit, the existing problems and means to solve them were defined. Due to the application of these solutions and the need to understand the meaning of the data, in order to process them in areas such as Big Data, the problem is being solved [7, 8] and there is an increasing number of projects in this area that use ontologies [9, 10].

The term ontology is defined for the first time as: "An ontology defines the basic terms and relationships that comprise the vocabulary of a thematic area, as well as the rules for combining terms and relationships to define the extensions of the vocabulary" [11].

Most data visualization software tools are standalone applications, which handle small data sets [12]. In addition, they do not usually use data that are in an ontology, neither to show them nor to know how to do it.

Traditionally, speaking about graphic visualization of ontologies has been synonymous with visualization of graphs, tree maps and similar techniques that have developed a multitude of standalone and online tools. These tools are used to see the ontology and navigate through it to see its instances [13]. Recent projects use ontologies in order to know how to show the content of an ontology [14, 15]. One of the most advanced is the one proposed by the VISO ontology [16]. This ontology, together with the RVL [17], defines how to visualize the information and interact with it in a user-definable graphical environment.

In this paper a wide related work section is depicted, where are described the advances in the research lines involved in the design of SILKNOWViz. Next, in the motivation section, the limitations in these research lines are listed, and ways to alleviate these problems are introduced. SILKNOWViz, a tool developed within the SILKNOW project, is proposed as a way to achieve this objective. In the section on design and methodology, some considerations on the design of the tool are described, as well as the graphic interface and the first developments. Finally, the conclusions of this work are exposed.

2   Related Work

The visualization of information helps the process of obtaining knowledge and a better understanding of the information through the graphical representation of massive data sets [4]. In the case of spatio-temporal models, the represented data fluctuate in space and time. They are data sets that are difficult to visualize, especially the more dimensions there are, in addition to space and time [5]. In order to achieve this objective, the data have been structured in different levels in the last two decades. A series of operations are defined for each level to be performed. These classifications define what can be queried to the system and what can be obtained in each query. These studies have been carried out in conjunction with the evolution of the tools and user interface elements that allow consulting data, as well as viewing them. Next, is depicted an analysis of some of the most widespread classification techniques, as well as the different types of developments.

2.1   Classification of Spatio-Temporal Data

Peuquet talks about the existence of three main parts in spatio-temporal data: where, when and what [18]. These parts describe a location, a time and the data to be represented. On the same type of task-oriented model, more extensions were added; for example, Andrienko [19] classified the tasks to analyze the information consulting by points, regions and trajectories. With the evolution of technology the data has changed, the size of the information is much larger and they have many more attributes to consider. It is important to consider objects and their attributes. Nowadays, with massive amounts of available data, more attributes can be considered, helping with the identification of patterns.

For this reason, new ways of classifying data arise, such as that proposed by Guo [12], which adds an additional part to the three proposed by Peuquet, the "who", which refers to objects, or series of objects. In addition, Guo associates the "what" part with the attributes of these objects. The four parts can be combined around four levels, performing queries with more to fewer variables, so that the fewer variables used, the more information will be shown, reducing the certainty of the analysis. A table with the different tasks for each level can be seen in Fig. 1. These new ways of classifying data generate new results that allow better analysis of the data and discover relationships, or patterns, in an increasingly simple and rapid manner. However, since reality is very complex, new visualization demands appear as the system improves.

Level	Tasks contained in each task level
1	where + when + who → what
	where + when + what → who
	where + who + what → when
	who + what + when → where
2	where + when → what + who
	what + when → who + where
	where + what → who + when
	when + who → what + where
	where + who → when + what
	who + what → where + when
3	where → what + when + who
	what → who + when + where
	when → where + who + what
	who → where + what + when
4	NULL → what + where + when + who

Fig. 1. List with the tasks contained in each task level. Source: Guo [12]

A new requirement appears when objects are both objects and attributes. This happens when an object is related to other objects. In this case, the related objects are new objects and attributes of the source object. An example would be manufacturers and manufactured objects that evolve throughout space and time. It is very interesting to see where these objects have been manufactured, but also where they have been used and what they are related to. If we want to see the evolution of this evolution in space and/or time, the problem is even greater and requires further improvement of techniques.

2.2   Visualization Tools

Since the beginning of this research, many software tools have been developed to visualize information [1, 2, 4]. In the last two decades, due to technological advances, the use of three-dimensional representations with a high level of interactivity has increased.

Some developments show information (what + where + when) over a bi-dimensional or three-dimensional region. These tools use various techniques such as colour maps, colour temperature patterns and clustering techniques [22, 23]. Colour, cluster size, the number of dimensions (2D/3D), elevation, and even the use of additional labels, are the visual elements that are used to represent information in the graph (see Fig. 2).

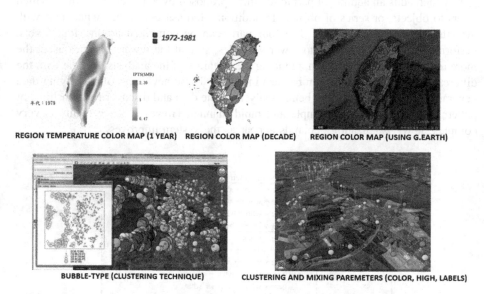

Fig. 2. At the top, screen shots of a web application that uses coloured regions and colour temperature maps, to visualize the incidence of cancer mortality in Taiwan. Source: [20]. Below, screenshots from tools that use clustering techniques to represent the information. Source [21].

In addition to visualizing information and displaying common form controls in the user interface, innovative controls have also been created. These controls are used in order to define and filter the time in which the information to be displayed is represented. One of the most used interactive devices is the timeline [24, 25], which contains information on the different time levels. It starts from a broad level, usually in the top part, to go down interactively, until selecting a concrete period, in the lower zone. It is often used in cultural heritage visualization tools [26]. See Fig. 3 for examples of timelines, specifically, a Time Wheel control [19], which allows the selection of specific months, as well as days and a time interval in a simple way.

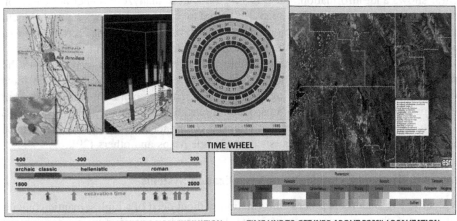

TIME LINE TO GET INFO ABOUT ARCHEOLOGICAL EXCAVATION   TIME LINE TO GET INFO ABOUT FOSSIL LOCALIZATION

Fig. 3. Left and right show screenshots of spatio-temporal data visualization applications in cultural heritage application. Source: [24, 26]. Screenshot of a Time Wheel control in the upper central part. Source: [19]

Analysing how information evolves over time can be the main goal in determinate information systems. In this case, it is important to see the date in different time intervals. To solve this problem, the scientific community has also developed several solutions. Some are based on three-dimensional spaces, like space-time cubes [5], or spirals [27], which displays a 3D spiral on the objet, where each lap represents a period of time.

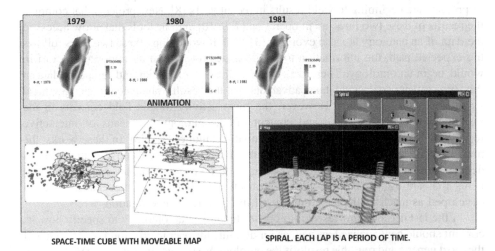

SPACE-TIME CUBE WITH MOVEABLE MAP   SPIRAL. EACH LAP IS A PERIOD OF TIME.

Fig. 4. The upper part of the image shows the different frames, one per year, of the incidence of terminal cancer in Taiwan, resulting in an animation. Source: Adapted from [20]. In the lower left zone a space-time cube is displayed, where a horizontal plane (map) can be moved. Source: [19] To the right, two screenshots of an application that use 3D spirals, where each lap is a period of time and has additional information related with the appropriate year. Source: [27].

Spiral shows a lot information, but it is very difficult to analyze it. Other are based on two-dimensional spaces, like flow maps [28], used to represent trajectories, or movement of objects in time. Finally, animations are also used, where each frame corresponds to a specific period. See some applications that use these techniques in Fig. 4.

The advance in the use of these techniques has been important, especially in the graphic quality and the possibilities of interaction, because many of these ideas have been developed several decades ago. One of the main problems in this area is that developments are often linked to a particular application. For this reason, it is difficult to reuse the tool and frequently the only reused components are the web services used to show maps, or satellite images, such as Google Maps and others [29].

2.3   Application of Ontologies in Visualization Systems

Typically, Big Data tools process information without need of a high semantic content. For instance, there are image-processing tools, where the content of images is analysed using neural networks, looking for patterns in the image, and taking as a reference a large sample of correctly classified images [30]. Probably, if semantic information was available, the process would be faster, because this information could contribute to avoid higher computational cost processes.

Ontologies are used to represent complex models of information and to infer knowledge from them, because a high semantic content can be defined in the representation. In 2014 Ontology Summit [6] it was stated that although ontologies have played a key role in Semantic Web technologies, their application in new lines, such as Big Data, is lower than expected. It seems logical that both lines of research should have had a greater relationship, but this has not happened, in fact. Seemingly, the measures proposed in the 2014 Ontology Summit to increase its application in this type of projects are beginning to yield results in recent years [8]. New projects that combine the results of these two lines are in operation [7, 9, 10] and the tools that allow access to the data of an ontology are also evolving [31, 32]. If work along these two lines follows the expected path, the information to be visualized, or at least an important part of it, would be in an ontology. Therefore, the visualization of spatial and temporal data in this type of projects could take advantage of the results obtained in the graphical visualization of ontologies. However, the main activity of this field consists in the visualization of the structure and content of the ontology by means of interactive representations of graphs, treemaps and similar techniques [13, 33]. In these works, the main objective is usually not the analysis of content in order to discover patterns, or to deduce complex relations. For this reason, these investigations may provide components that are useful in the study of spatio-temporal data, but they have not been developed as much as the techniques used in this line of investigation.

In the last decade, new research proposes the use of an ontology to specify how its content should be visualized, as well as for defining the user interaction [14–16]. One of the most generic and reusable results is the ontology Visualization Ontology (VISO) [16]. This ontology allows the definition of its concepts considering how to visualize them and how to interact with them. This ontology was developed as the basis of the RDFS/OWL Visualization Language (RVL) [17].

The problem with the VISO ontology is that it involves close consideration of visualization aspects in a two-dimensional environment, but hardly delves into 3D visualization components. It would be necessary to create new classes and extend others to consider three-dimensional representation, which is fundamental in many spatio-temporal data visualization techniques.

3 Motivation

After reviewing previous publications, the following issues and conclusions arise regarding the visualization and analysis of spatio-temporal data:

- More work needs to be done on the relationships between different objects, objects that are at the same time objects and attributes.
- It is necessary to create more open and multiplatform tools that can be used to represent and analyse spatio-temporal data. These tools must be capable to be embedded in other projects, especially in web applications.
- The use of ontologies to represent the information to be visualized should be encouraged. In addition to values, data can contain semantic information, allowing new possibilities of interaction and visualization.
- Generic tools have been created in order to define how the information is to be displayed. VISO ontology, one of the most advanced, needs to be extended to contemplate more deeply the needs of graphics in three-dimensional environments.

In order to satisfy these needs, we propose to carry out the following tasks:

- To eliminate limitations in the three-dimensional environment of the VISO ontology and the RVL language, as well as including activities and interactions in order to implement techniques for the representation of spatial and temporal data.
- To define a way to visualize relationships between objects, which allows visualizing the evolution of these relationships in different moments of time, simultaneously.
- If the data are represented by an ontology, use the VISO ontology and the RVL language to define how to visualize the information.
- To develop an open source software tool, in a multi-platform development system that allows to visualize the results of any project, especially on web platforms.

These tasks will be developed within the SILKNOW project, by designing and developing the SILKNOWViz tool.

The SILKNOW project is an interdisciplinary project, funded by the H2020 Programme of the European Union, in order to preserve and promote the heritage of silk textiles. SILKNOW will take digital silk textiles' data from the databases of several institutions, from online catalogues and from APIs (where available). These data will be analysed and processed with advanced text analytics and image-based deep learning techniques in order to homogenise their content, automatically retrieve semantic information, complete poorly tagged data, and translate the text into four languages.

The CIDOC Conceptual Reference Model [34] is used in SILKNOW. The selected implementation is the Erlangen CRM/OWL [35], a RDFS/OWL ontology.

One of the results of SILKNOW is the implementation of a public web site where the data of the project can be searched. In this website, the queries' results can be visualized through a 2D/3D component with an interactive map. This component is the SILKNOWViz tool. Next, we describe how to manage relationships between different, related objects in a spatio-temporal data model. Taking into account these considerations, the design of the tool user interface will be defined, as well as the necessary extensions to represent the required visualization on the VISO ontology.

4   SILKNOWViz Tool

4.1   Object Relationships

In the visualization of spatio-temporal data, the representation of different objects related to each other is one of the most complex activities. It is not possible to take advantage of the work carried out in the visualization of graphs, or trees, at least not in a general way, because in this model, location and time are fixed values, and therefore it is not possible to move them to improve their visualization. When there are many objects, the classic solution consists of developing clusters of objects, visualizing the relationship of an object with a cluster of objects, or relationships between different groups.

Flow maps [28] usually include trajectories. These visualizations systems have worked with this type of techniques in the relations between objects. This approach reduces the simultaneous number of connections and facilitates visualization and comprehension. However, in spatio-temporal data, groupings usually have a spatial base, and despite using these techniques, the number of connections to be visualized usually prevents a correct visualization. A typical solution is interaction: selecting an object, or group of objects, and displaying only the relationships of the selected object. An application that solves this problem by visualizing the objects relations interactively is shown in Fig. 5.

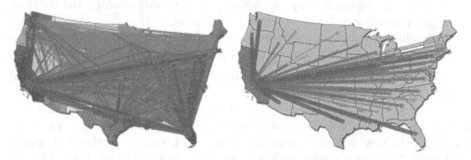

Fig. 5. Visualizing all the relationships generates images that are very difficult to analyze. Visualizing the relationships of a single object is more readable. Source: [28].

Using interaction has an important limitation: previous lack of knowledge whether an object is related or not, nor the number of its relationships. However, investigating those objects with an extreme number of relationships, different from the average, is generally more interesting. This problem becomes more difficult to handle when we want to study several elements simultaneously, as will be detailed in Sect. 2.

Is proposed the use of marks around the object, or clusters of objects, that identify the level of relations with the average. The mark would consist of a ring that would enclose the object, divided into as many sectors as objects it is related to. Longer or shorter lines will be drawn on this sector, depending on the number of relations. On the other hand, the length of the sector is associated with the time interval. Thus, the beginning of the sector will be the beginning of the period, and the end of the sector, the end of the interval. Each sector is a histogram that is enconded into a circle. A diagram showing the ring marker and the final appearance on a map can be seen in Fig. 6. This marker will be implemented in the tool and mapped using VISO.

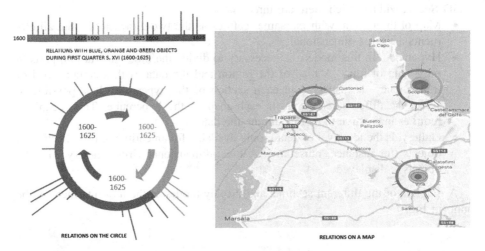

Fig. 6. Ring mark for showing information about the number of relations that each object/cluster has with other objects/clusters during the selected time. To see the relationships you would have to select any of the objects. (Own image)

The tool must present two implementations, because it is designed to be integrated in web pages, like a plugin, but also to run like a standalone application. One of the versions will be limited, displaying only the visualization zone. Another one will include all the features of the tool. The input of the application consists of access to the ontology, by means of a RESTFul API SPARQL, and a query to show the results. The query is optional because the queries may be constructed or extended from the tool itself, if the user requires so. The ontology will need to have the extensions VISO + RVL.

As far as the user interface is concerned, the tool has a display section and a query definition section. The display section displays the query results, allowing the user to change the point of view and select items.

The query definition section has the following elements:

- A three-dimensional graph with the ontology classes that can be displayed, and their relationships.
- A dynamic section with a form, whose data depends on the class selected in the graph, to specify its properties and filter the query.
- A timeline control to set the time interval.
- A window to visualize the selected elements properties.

There are two main scenes:

- 2D Scene: a map of the region, Europe being the largest area, allowing you to move interactively and zoom in to city level. All the elements that appear in the map are two-dimensional, icons, clusters, etc. Levels of detail are available in this visualization. From a certain level, clusters are disaggregated and objects are represented with icons.
- 3D Scene, within which there are three classes:
  - Map of the region, with the same features as the two-dimensional, but all elements are three-dimensional.
  - Hypercube. In this mode it is necessary to divide the time interval into parts (a maximum of four). A map of the region and the data, with a certain level of transparency, is displayed in each section of the hypercube. It is possible to select the different elements of each map of the hypercube. This technique requires interactive tools to examine the data.
  - Gallery. Is the top level of detail, only available for an entity with a collection of objects (for instance, museums). A three-dimensional virtual gallery with the objects is displayed.

A diagram of the different sections and display modes of the tool interface can be found in Fig. 7.

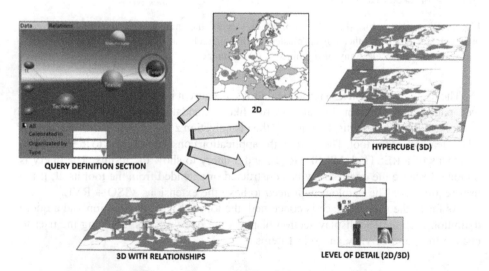

Fig. 7. At the top left appears an outline of the query definition area, the rest are representations of the different display modes. (Own Image)

4.2 VISO Extensions

The VISO ontology and the RVL language form a very powerful set to define how the data of an ontology should be visualized, and to specify the interaction between them. It is a useful tool to define how to represent and interact with the classes and relationships of an ontology. However, it has two limitations for the purposes of our tool. First, since it is very generic, it lacks tools to define complex scenes, shaped by other graphical elements, in a specific order and position. Secondly, it is very oriented to representation in two dimensions, so it is necessary to create extensions for three-dimensional visualizations.

In order to solve these problems, the next extensions are created:

- Creation of the 3D_Object class, which represents a generic three-dimensional object. Creation of the Scene class, which represents a scene, as well as relations with 3D_Object, to define the structure of the scene, following a philosophy similar to the one used by Swing. Is also defined the LOD (Level Of Detail) class, in order to manage the changes in function of the proximity of the viewer.
- Extensions of VISO objects have been created, with 3D_Object as superclass. In this way, existing objects can be represented in three dimensions in a simple way. For example, the 3D_Map class will extend to 3D_Object and the Map class, which indicates that a Map has to be displayed. In this way, the 3D_Map class will indicate that this class has to be represented with a three-dimensional map.
- Classes have been created for the special controls: Ring_Mark, Hypercube and Timeline.

5 Conclusions

In this work, on the one hand, we have reviewed the state of the art in tools for displaying spatio-temporal data. On the other hand, the use of ontologies in the visualization of this type of data and its relationship with Big Data has been analyzed. In addition, ontologies that depict how to visualize and interact with their own data have been reviewed.

The development of an open source tool has also been described, as a way to solve part of the existing problems in these lines of research and their practical application. This tool is designed for the visualization of spatio-temporal data contained in an ontology.

The most important details of this tool, as well as the user interface, have been explained. This open source tool is based on the use of the VISO ontology and the RVL language. This ontology has also been extended, to satisfy the tool requirements.

This tool will be fully developed during 2019 and it will be evaluated by a set of experts in cultural heritage from France, Italy and Spain. It will be available at https://github.com/silknow/silknowviz.

Acknowledgments. The research leading to these results takes place within the project "SILKNOW. Silk heritage in the Knowledge Society: from punched cards to big data, deep learning and visual/tangible simulations", which has received funding from the European Union's Horizon 2020 research and innovation program under grant agreement No. 769504.

References

1. Kehrer, J., Hauser, H.: Visualization and visual analysis of multifaceted scientific data: a survey. IEEE Trans. Visual Comput. Graphics **19**, 495–513 (2013)
2. Liu, S., Cui, W., Wu, Y., Liu, M.: A survey on information visualization: recent advances and challenges. Visual Comput. **30**, 1373–1393 (2014)
3. Lavalle, S., Lesser, E., Shockley, R., Hopkins, M.S., Kruschwitz, N.: Big data, analytics and the path from insights to value. MIT Sloan Manag. Rev. **52**, 21–32 (2011)
4. Zhong, C., Wang, T., Zeng, W., Müller Arisona, S.: Spatiotemporal visualisation: a survey and outlook. In: Arisona, S.M., Aschwanden, G., Halatsch, J., Wonka, P. (eds.) Digital Urban Modeling and Simulation. CCIS, vol. 242, pp. 299–317. Springer, Heidelberg (2012). https://doi.org/10.1007/978-3-642-29758-8_16
5. Bach, B., Dragicevic, P., Archambault, D., Hurter, C., Carpendale, S.: A review of temporal data visualizations based on space-time cube operations. In: Eurographics Conference on Visualization (EuroVis 2014), pp. 23–41 (2014)
6. Gruninger, M., et al.: Ontology Summit 2014 Communique: Semantic Web and Big Data Meet Applied Ontology (2014)
7. Bennett, M., Baclawski, K.: The role of ontologies in linked data, big data and semantic web applications. App. Ontol. **12**, 189–194 (2017)
8. Baclawski, K., et al.: Ontology Summit 2018 Communiqué: Contexts in Context (2018)
9. Konys, A.: Ontology-based approaches to big data analytics. In: Kobayashi, S.-y., Piegat, A., Pejaś, J., El Fray, I., Kacprzyk, J. (eds.) ACS 2016. AISC, vol. 534, pp. 355–365. Springer, Cham (2017). https://doi.org/10.1007/978-3-319-48429-7_32
10. Dou, D., Wang, H., Liu, H.: Semantic data mining: a survey of ontology-based approaches. In: Proceedings of the 2015 IEEE 9th International Conference on Semantic Computing (IEEE ICSC 2015), pp. 244–251 (2015)
11. Neches, R., et al.: Enabling technology for knowledge sharing. AI Mag. **12**, 36–56 (1991)
12. Guo, D., Du, Y.: A visualization platform for spatio-temporal data: a data intensive computation framework. In: 2015 23rd International Conference on Geoinformatics, pp. 1–6 (2015)
13. Dudáš, M., Lohmann, S., Svátek, V., Pavlov, D.: Ontology visualization methods and tools: a survey of the state of the art. Knowl. Eng. Rev. **33**, e10 (2018)
14. Nazemi, K., Burkhardt, D., Ginters, E., Kohlhammer, J.: Semantics visualization – definition, approaches and challenges. Procedia Comput. Sci. **75**, 75–83 (2015)
15. Falconer, S.M., Bull, R.I., Grammel, L., Storey, M.: Creating visualizations through ontology mapping. In: 2009 International Conference on Complex, Intelligent and Software Intensive Systems, pp. 688–693 (2009)
16. Polowinski, J., Voigt, M.: VISO: a shared, formal knowledge base as a foundation for semi-automatic InfoVis systems. In: CHI '13 Extended Abstracts on Human Factors in Computing Systems (CHI EA 2013). ACM (2013). https://doi.org/10.1145/2468356.2468677

17. Polowinski, J.: Towards RVL: A declarative language for visualizing RDFS/OWL data. In: Proceedings of the 3rd International Conference on Web Intelligence, Mining and Semantics, pp. 38:1–38:11. ACM, New York (2013)
18. Peuquet, D.J.: It's about time: a conceptual framework for the representation of temporal dynamics in geographic information systems. Ann. Assoc. Am. Geogr. **84**, 441–461 (1994)
19. Andrienko, N., Andrienko, G., Gatalsky, P.: Exploratory spatio-temporal visualization: an analytical review. J. Visual Lang. Comput. **14**, 503–541 (2003)
20. Ku, W.-Y., et al.: An online atlas for exploring spatio-temporal patterns of cancer mortality (1972–2011) and incidence (1995–2008) in Taiwan. Medicine **95**, e3496–e3496 (2016)
21. Hengl, T., Roudier, P., Beaudette, D., Pebesma, E.: plotKML: scientific visualization of spatio-temporal data. J. Stat. Softw. **63**, 1–25 (2015)
22. Clauset, A., Newman, M.E.J., Moore, C.: Finding community structure in very large networks (2005)
23. Jänicke, S., Heine, C., Scheuermann, G.: GeoTemCo: comparative visualization of geospatial-temporal data with clutter removal based on dynamic delaunay triangulations. In: Csurka, G., Kraus, M., Laramee, Robert S., Richard, P., Braz, J. (eds.) VISIGRAPP 2012. CCIS, vol. 359, pp. 160–175. Springer, Heidelberg (2013). https://doi.org/10.1007/978-3-642-38241-3_11
24. Kraak, M.J.: Timelines, temporal resolution, temporal zoom and time geography (2005)
25. Lee, C., Devillers, R., Hoeber, O.: Navigating spatio-temporal data with temporal zoom and pan in a multi-touch environment. Int. J. Geogr. Inf. Sci. **28**, 1128–1148 (2014)
26. Wang, C., Ma, X., Chen, J.: Ontology-driven data integration and visualization for exploring regional geologic time and paleontological information. Comput. Geosci. **115**, 12–19 (2018)
27. Hewagamage, K., Hirakawa, M., Ichikawa, T.: Interactive Visualization of Spatiotemporal Patterns Using Spirals on a Geographical Map (1999)
28. Guo, D.: Flow mapping and multivariate visualization of large spatial interaction data. IEEE Trans. Visual Comput. Graphics **15**, 1041–1048 (2009)
29. Google Maps (2019). https://cloud.google.com/maps-platform/
30. Zerdoumi, S., et al.: Image pattern recognition in big data: taxonomy and open challenges: survey. Multimedia Tools Appl. **77**, 10091–10121 (2018)
31. Zhao, B., et al.: Ontobee: a linked ontology data server to support ontology term dereferencing, linkage, query and integration. Nucleic Acids Res. **45**, D347–D352 (2016)
32. Verhodubs, O.: Realization of Ontology Web Search Engine (2017)
33. Dudáš, M., Zamazal, O., Svátek, V.: Roadmapping and navigating in the ontology visualization landscape. In: Janowicz, K., Schlobach, S., Lambrix, P., Hyvönen, E. (eds.) EKAW 2014. LNCS (LNAI), vol. 8876, pp. 137–152. Springer, Cham (2014). https://doi.org/10.1007/978-3-319-13704-9_11
34. CIDOC Documentation Standards Group: CIDOC Conceptual Reference Model (CRM) – ISO 21127:2006 (2006)
35. Erlangen CRM/OWL. CIDOC-CRM Implementation (2013). http://erlangen-crm.org/

Ontology-Driven Automation of IoT-Based Human-Machine Interfaces Development

Konstantin Ryabinin[✉] [ID], Svetlana Chuprina [ID], and Konstantin Belousov [ID]

Perm State University, Bukireva Str. 15, 614990 Perm, Russia
kostya.ryabinin@gmail.com, chuprinas@inbox.ru, belousovki@gmail.com

Abstract. The paper is devoted to the development of high-level tools to automate tangible human-machine interfaces creation bringing together IoT technologies and ontology engineering methods. We propose using ontology-driven approach to enable automatic generation of firmware for the devices and middleware for the applications to design from scratch or transform the existing M2M ecosystem with respect to new human needs and, if necessary, to transform M2M systems into human-centric ones. Thanks to our previous research, we developed the firmware and middleware generator on top of SciVi scientific visualization system that was proven to be a handy tool to integrate different data sources, including software solvers and hardware data providers, for monitoring and steering purposes. The high-level graphical user SciVi interface enables to design human-machine communication in terms of data flow and ontological specifications. Thereby the SciVi platform capabilities are sufficient to automatically generate all the necessary components for IoT ecosystem software. We tested our approach tackling the real-world problems of creating hardware device turning human gestures into semantics of spatiotemporal deixis, which relates to the verbal behavior of people having different psychological types. The device firmware generated by means of SciVi tools enables researchers to understand complex matters and helps them analyze the linguistic behavior of users of social networks with different psychological characteristics, and identify patterns inherent in their communication in social networks.

Keywords: Ontology engineering · Internet of Things ·
Human-machine interaction · Tangible interfaces · Firmware ·
Middleware · Scientific visualization · Visual analytics

1 Introduction

Universal computer input devices like keyboard and mouse build the habitual environment for user and normally suit everyday needs in terms of human-machine

The reported study is supported by Ministry of Education and Science of the Russian Federation, State Assignment No. 34.1505.2017/4.6 (Research Project of Perm State University, 2017–2019).

interaction. However specific cases may require specific input hardware and specific software reactions to enable needed user experience. Moreover, real-world experience shows that the today's special cases are the common ones in the future. Nowadays high-level tools adapting the factory-assembled input devices to the specifics of particular task or hardware and software infrastructure are often missing.

When it comes to virtual reality applications, the user's eyes are covered with the head-mounted display, so it becomes very uncomfortable to use keyboard and special hardware control elements (like Oculus Touch) are required. Different vehicle simulators require special hardware control panels including steering wheel and pedals to provide full immersion and enable training function. Some content generation tasks also require special tools, for example, the digital painting requires drawing tablets to increase ergonomics. Special environment conditions may also require space saving, so touchscreens and trackpads are also very popular nowadays. Next, it is often required to bring the benefits of the Internet of Things (IoT) to legacy systems, to combine various old edge devices and sensors with smart ones, or to put the human into the IoT ecosystem. In our opinion, in the near future, putting human centric innovation to drive a composition of IoT systems, in which their constituent systems are individually discovered, becomes the everyday task.

In this context, Internet of Things become a promising branch of computer science and engineering. The technologies and design patterns developed for IoT can be applied to the human-machine interaction bringing it to the new customization level in a unified form. Different sensors provide the user with quite unlimited possibilities to control the computers via so-called tangible interfaces [8], and in the same time provide the automation systems with the same amount of possibilities to monitor the human activity. The problem here is the user's qualification: complex interconnections, sensor-based networks and device making normally require knowledge in both electronics and programming. In contrary, high-level software tools and handy interoperable IoT modules can be easy building blocks ready to be used even by the beginners. The leaders of IoT industry already provide particular solutions for typical use cases like, for example, the organization of Smart Home. But there is still a lack of high-level tools for building arbitrary IoT devices and middleware to turn them into human-machine interface (HMI). Thereby not only the legacy systems can get new HMI by means of IoT, but also the IoT systems themselves can be turned into human-centric ones.

We propose ontology-driven approach to build smart systems, which can automatically generate the firmware for IoT devices and the middleware to connect these devices with third-party software in adaptable way. Thereby this system is a way to automate the creation of software part of IoT-based HMI. Regarding the hardware part we suggest using particular electronic components meeting the price/quality balance. To demonstrate the proposed approach we present the new capabilities of SciVi platform[1] to tackle the problems mentioned above.

---

[1] https://scivi.tools/.

As follows from the description of our system presented in [17], the SciVi platform was initially intended to be the scientific visualization tool adaptable to different types of software solvers and input data formats. Our innovation presented in this paper is using SciVi platform as IoT-based HMI generator without changing the SciVi core. In this case IoT devices play the solvers' role. Moreover, the SciVi architecture allows using its modules separately to solve wide range of tasks, not limited by scientific visualization. For example, if third-party system already has its own visualizer, SciVi can be used as the middleware to provide new HMI capabilities including gesture-based ones.

2   Key Contributions

In this paper we describe the ontology-driven smart system, which enables automation of IoT-based HMI development. The following key results of conducted research are presented.

1. The ontology-driven approach is proposed to automate the firmware and middleware generation for the cases when IoT devices are intended to be used as a part of HMI to the third-party software.
2. The proposed approach is implemented within a smart system SciVi-Middle, which is built upon the adaptive scientific visualization toolset SciVi we developed previously.
3. The recommendations are made for choosing IoT-based HMI hardware with respect to price/quality ratio.
4. The discussed approach is used to tackle real-world problems of social network users behavior analysis.

3   Related Work

The essence of the IoT is the constant monitoring of the environment and allowing humans the remote management in order to ensure the symbiosis of real and digital worlds [7]. Being the basis of ubiquitous computing [7], IoT provides great possibilities to monitor the human activity [21]. This, in turn, opens up the new horizons of human-machine interaction utilizing more specific actions than traditional pushing of buttons. As stated in [7], "in an IoT environment, the human and computer interaction happens through objects containing embedded technology to sense or interact with their internal state and external environment". Human presence (detected by passive infrared sensor), gestures (detected by some kind of motion capture system, like for example [9]), voice and even physical condition (detected by wearable medical sensors) could be turned into the commands for the computing systems whereby multimodal HMI can be provided [24]. Moreover, IoT ensures so-called tangibility of HMI [8], enabling seamless coupling between physical objects and virtual data. As a result, smart systems can be created, which respect human skills and knowledge, as well as the dynamic user behavior [7].

However, achieving the desired user experience within IoT ecosystem is a challenging task. It is necessary to integrate diverse independent components into a one-stop solution, saving functionality and reliability of individual building blocks and ensuring seamless interoperability [10]. To tackle this challenge, both high-level user interface (UI) frameworks and middleware frameworks are demanded.

Middleware is a "glue" to stick together the existing programs and devices that were not originally designed to be connected. Detailed survey of modern IoT middleware made by Farahzadi et al. [6] shows the variety of platforms aimed to ensure integration of different hardware into a single ecosystem. The popular middleware helps to efficiently solve the interoperability issues, however the tasks of building versatile and ergonomic UI for this ecosystem still do not have unified solutions. This is why specific circumstances force developers to create related software from scratch.

The middleware should be very flexible, extensible and adaptable to ensure both interoperability of devices and versatile HMI capabilities [6]. One of the most adequate ways to achieve these features is a model-driven approach, when the system providing middleware functionality is governed by some formal model. In this case, the modification of the model will modify the system's behavior, while the code of system's core remains unchanged. If the used model maintains semantics, the system possesses context-awareness and thereby can be treated as intelligent one.

Well-proven semantic models are provided by ontology engineering. Since 2012 ontologies are widely used in IoT domain to tackle different key issues like heterogeneity of devices, interoperability and scalability problems, monitoring tasks, etc. [2]. The main point to use ontologies is their ability to be close to human conceptual level, while understandable to machines [11]. This is why we decided to use ontology engineering as a core methodology in automating the IoT-based HMI development.

In our previous research we studied the ways to automate monitoring and calibration of IoT devices based on ontology engineering methods and scientific visualization tools [20]. We implemented the corresponding capabilities within scientific visualization system SciVi. The current paper describes the next step of SciVi development generalizing its capabilities from visualization system to HMI platform. It was an intention to build HMI platform upon a visualization system, because visualization is considered as an essential part of human-machine interaction [13].

Research work related to the versatile HMI software for IoT has already justified ontology engineering in this field. As stated by Calderon et al., IoT is still too device-centric, but should be more human-centric, and the semantic technologies are one of the clues to achieve this human-centricity [5]. Ontologies were successfully used by Luz et al. to specify the human-machine computation workflow retaining knowledge and semantics of atomic tasks [11]. The system called Perci developed by Broll et al. relies on application model ontologies to compose an application interface and UI [4]. The research by Steinmetz et al.

is dedicated to using ontologies to enhance IoT middleware and personalize the visualization available for end users [23]. Several projects, for example the one by Noguera-Arnaldos et al. [1] or by Petnik et al. [14], utilize ontologies to provide users with natural language interface to IoT devices.

The distinctive features of our approach are the following:

1. We propose the platform for developing HMI *based on* IoT devices, not *to* the IoT ecosystem. Of course, the methods we suggest can be used to build interactive environments like Smart Home, similar to the ones described in [14] and [13]. However, the main intention behind our research was to enable device makers creating the custom gadgets to control third-party software.
2. We propose using cognitive interactive graphics [12] capabilities provided by scientific visualization and visual analytics tools to support HMI development and functioning.
3. We utilize complex extensible ontology-driven mechanism of data processing called semantic filtering [20] to enable customizable conversion of data representations and formats. The SciVi platform ontology-driven capabilities to adapt to specifics of different application domains and to tackle different problems of scientific visualization in a uniform way are proven in a number of use cases [18]. In this paper we describe the new interdisciplinary use case tackling multimodal content management.

4   Proposed Solution

4.1   Background

While researching the scientific visualization methods and means we found out, that there is a lack of high-level tools providing easy-to-use capabilities to integrate with third-party data sources. This was a motivation to create the adaptive multiplatform scientific visualization SciVi. The detailed description of this system and underlying integration mechanisms can be found in our previous work [19]. SciVi relies on ontology engineering methods, its behavior is completely governed by the set of ontologies describing the supported visualization techniques, data processing algorithms and data sources. In addition, SciVi provides the data flow diagram editor enabling the user to describe the visualization task in terms of data processors (nodes) and data links (arcs). The aforementioned capabilities can be easily extended by enrichment of corresponding ontologies. SciVi was designed to adapt to third-party data sources including software and hardware data generators (so-called solvers). In this regard, we were able to use it as a monitoring tool for lightweight electronic devices within IoT ecosystem [20]. In the present work we introduce SciVi-Middle, the smart system built upon SciVi. SciVi-Middle is dedicated to generalize the integration capabilities of SciVi, to enable it not only integrating with IoT devices, but also interconnecting them with third-party software. The main purpose of SciVi-Middle is to allow inexperienced users to build custom HMI for different systems using IoT technologies.

4.2 SciVi-Middle Functioning Principles

The high-level pipeline of using SciVi with IoT devices as solvers is shown in the Fig. 1. The *Adaptation Stage* catches the data from the IoT device and, if needed, converts them to the form suitable for SciVi internal mechanisms. For example, the conversion may involve parsing the network packages, etc. The *Data Filtering Stage* performs custom data processing that user can tune by data flow diagram editor. Being in fact optional, this stage allows different context-aware data transformation by semantic filters described in the SciVi knowledge base. For example, the filtering may reduce sensor noise of IoT device, etc. The *Rendering Stage* is responsible to synthesize and present the image according to the user's settings and data caught. This stage also provides UI to organize optional feedback for the IoT device. For example, this may be calibration functionality.

The usage of SciVi-Middle is depicted in the Fig. 2. As seen, in this case only the "middle" stages of the system's pipeline are utilized. The internal adaptation mechanisms of SciVi are applied as middleware to ensure needed connection of the hardware device to the third-party software. Also, the new *Control Emission Stage* is introduced that is responsible to steer the desired application and catch its feedback. The architecture of SciVi-Middle is shown in Fig. 3.

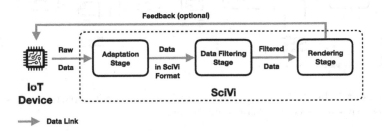

Fig. 1. Functioning pipeline of SciVi with IoT devices as solvers (introduced in the previous work).

The core of SciVi-Middle is its knowledge base that controls the behavior of all the internal modules. It contains 6 ontologies (stored in the standard OWL format) explained below. Let us denote as *Dev* the IoT device we want to use as hardware HMI, and denote as *App* the corresponding third-party application we want to steer by *Dev*.

Programming Languages I/O Structure Ontology (*L*) takes part in automating the process of integration SciVi-Middle with IoT device and third-party application. It is used if the source code of *Dev* firmware or source code of *App* is available to the user. In this case, the *Integration Module* of SciVi-Middle can automatically build the source code parser to extract the input and output

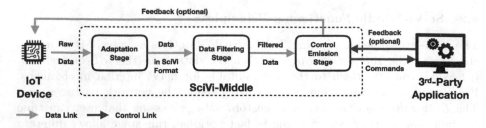

Fig. 2. Functioning pipeline of SciVi-Middle (introduced in the current work).

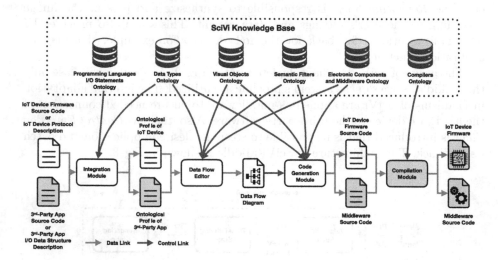

Fig. 3. SciVi-Middle Architecture (SciVi core is white, updated part is yellow, newly introduced parts are green). (Color figure online)

data structures description and thereby compose appropriate descriptions (so-called *Ontological Profiles*) of *Dev* and/or *App*. The corresponding mechanism is described in our previous work [17].

Data Types Ontology (T) describes all the supported data types, for example, numbers (both integer and real), vectors, matrices, quaternions, strings, etc. The description comprises the type name, the values domain, the type cast rules and some service information needed for internal mechanisms (for example, the color of corresponding typed arcs in data flow diagram). Almost every internal SciVi-Middle module requires knowledge stored in this ontology.

Visual Objects Ontology (U) is the most important one for scientific visualization system SciVi, because it describes the available visual objects, graphical scenes and rendering algorithms. But in SciVi-Middle it plays rather accessory role and may be used in some special cases like supplementary monitoring of the *Dev* activity while debugging, or some kind of additional graphical indication for the result IoT-based HMI.

Semantic Filters Ontology (F) provides knowledge about the supported data processing mechanisms. It may be extremely useful, because the raw data from *Dev* may be inappropriate to steer the *App* directly. For example, denoising of sensor data is often required, or, in more specific cases, some complex data transformation is indispensable like fusing the gyroscope and accelerometer measurement results into the representation of orientation. The principles of ontology-driven data filtering are described in details in [19].

Electronic Components and Middleware Ontology (C) is built upon the ontology of electronic components introduced in [20]. It is the most important knowledge base part in the SciVi-Middle. As it is clear from its name, it has two purposes. Firstly, it provides description of different programmable controllers (together with their built-in modules like analog-to-digital converter, etc.), chips, sensors, actuators and so on to give the user semantic overview of the *Dev*. Secondly, it provides description of different programming interfaces (parts of middleware), which may be used to access the *App*. Despite these purposes are quite different, both categories of knowledge are merged into the single ontology, because they require very similar interpretation.

First of all, the ontology C together with U and F governs the *Data Flow Editor*. As mentioned above, this module enables the user to compose visual representation of the task being solved in terms of data flow paradigm. Parsing the ontologies U, F and C the *Data Flow Editor* automatically builds the palette of available data filters, visual objects, graphical scenes, electronic components and middleware parts. User can choose the appropriate items from this palette and add them to the data flow diagram. These items are depicted as nodes having input and output sockets. After that, user can connect sockets building data links and thereby define the concrete processing algorithm. The usage of data flow paradigm to visually program the task solving algorithms within SciVi is described in [19].

Next, the ontologies U, F and C provide necessary knowledge for *Code Generation Module*. This module was first introduced in [20] as the tool generating firmware for IoT devices. In the present work it was enhanced and generalized to enable both firmware and middleware source code generation. This module traverses *Data Flow Diagram* (DFD) received from *Data Flow Editor*. Each node of DFD is linked to the corresponding concept in the ontology U, F or C. This concept is in turn connected to the set of concepts representing corresponding code snippets, function calls, library dependencies, etc. Assembling these building blocks, *Code Generation Module* composes the program source for the firmware and middleware.

It is worth noting, that both *Dev Firmware Source Code* and *Middleware Source Code* are actually optional. In some cases the firmware already exists and SciVi-Middle adapts to it, so no its modification (no generation) is required. The same is applied to the middleware: if the *App* provides necessary interface that can directly connect to the *Dev*, no middleware is created to minimize mediators. One more important point is that both firmware and middleware source codes, if generated, are made "human-friendly": readable coding style is followed. If the user is skilled enough, he/she can make some changes in this code, for example, implement some additional functionality or just inspect the code in learning purposes.

Compilers Ontology (K) is first introduced in the present work and enables to connect compilers directly to SciVi-Middle to build the firmware/middleware from the generated sources. This simplifies the toolchain needed to get the IoT-based HMI creation task solved. The user can obtain not just source code, but the ready-to-run application. The ontology K describes makefile-templates for building the source and, in case of the firmware, deploying it to the device. The make utility is assumed to be used to automate the building process. The makefile is generated by the *Compilation Module* and afterwards make is invoked. As a result, SciVi-Middle provides fully functional toolchain to automate creation of all the corresponding software to organize custom HMI for the *App* based on the *Dev*.

4.3   Key Points of SciVi-Middle Usage

The SciVi-Middle smart system retains the multiplatform portability of SciVi it is based on. It has two versions: desktop application written in C++ using Qt (while some ontology parsing and code generation parts are written in Python) and web application written in Python (server side) and JavaScript (client side). This enables two usage scenarios: standalone application and SaaS (Software as a Service).

Creating the hardware tangible HMI for some application, user should first of all get the device. Our suggestions of hardware components are listed in the below subsection. Here are two variants possible: the device already exists, or user assembles it from scratch. In both cases the firmware for the device either already exists (as a source code or in a binary form), or should be generated. If no generation is required, SciVi-Middle needs ontological profile of the firmware. In case the source code is available, this profile can be created automatically by *Integration Module*. Otherwise, the user should describe the device communication protocol manually using high-level graphical UI provided by *Integration Module*.

The next step is to get the application, which should be controlled via hardware HMI. Again, two variants are possible: the user either has an access to the source code, or not. In case the source code is available, the ontological profile of the application is built automatically. Else, the user should describe input and (optionally) output data structures of the application, or, alternatively, skip this step relying on the built-in middleware components presented in the ontology C. The idea is that the ontology C describes some presets for different use cases, allowing, for example, emulate the mouse movements or certain keyboard buttons. If the required communication is that simple, no special adaptation to the given application is required. So, the ontological profile of the application is optional.

After that, the user should build the data flow diagram describing the actions of IoT-based HMI. For this task high-level visual editor is provided. On this step the user can find out, that there are not enough items in the data flow editor's palette, for example, some needed data filter is missing. In this case SciVi-Middle functionality can be extended by enrichment of the corresponding ontologies without core source code modification. It is worth noting, that the set of available filters, visual objects, electronic components and middleware parts can be personalized: if there are too many of them described in the ontologies U, F and C, some excerpt of these ontologies can be used.

Next step is code generation that takes part automatically according to the data flow diagram composed. On this stage the user can obtain the results and use them for further development, or, alternatively, can proceed to compilation phase. In the latter case, the binaries of the firmware and middleware are created. The user can obtain the ready-to-run middleware (if it was required on previous steps) and deploy the firmware (if necessary) right to the device. The only limitation is, that for now the deploying works for desktop SciVi-Middle only; in case of Web-based version the firmware as well as the middleware can be just downloaded in the binary form.

4.4 Hardware Recommendations

Despite the diversity of microcontrollers, which may be used to assemble IoT devices, now we suggest using ESP8266 with the WiFi module on board. This controller is fairly cheap yet powerful (running at 80 MHz clock frequency and providing 80 Kb RAM) and energy efficient (energy consumption is about 80 mA at 3.3 V). In particular, we propose using WeMos D1 mini – the compact and cheap circuit board containing ESP8266 microcontroller, needed service circuits (like, for example, power and reset circuits) and USB-to-UART interface. This board is very handy in terms of firmware uploading, because it has built-in USB interface. Also, it already contains all needed service circuits to ensure stable work of ESP8266. The ontological description of this microcontroller is already included in the Electronic Components ontology mentioned above.

To simplify the device assembly, we propose using *shields*. Shields are extension circuit boards carrying sensors, displays, plugs for other electrical components (including actuators), etc. Shields require neither soldering nor any special instruments to connect. If they are described in the ontology C, the programming of corresponding firmware become as easy as connecting the components together: the user just adds the related nodes to the data flow diagram and visually builds the data processing chain linking these nodes in an appropriate order.

5 Use Case

The project titled "Socio-Cognitive Modeling of Social Networks Users Verbal and Non-Verbal Behavior Based on Machine Learning and Geoinformation Technologies" aims to identify behavior patterns of Social Network Services (SNS) users through a comprehensive multiparameter analysis of social, behavioral, psychological and linguistic parameters of the individuals. The information from a SNS user profile, such as gender, age, education, sphere of interests, communities, etc. is considered as social parameters. User preferences such as publications marked as favorite, posts, etc. are considered as behavioral. Psychological parameters are revealed as a result of a psychological survey based on BFI (Big Five Inventory) [16] with an adapted Russian-language version of BFI [22]. Linguistic parameters are found out through linguistic analysis of 19161 automatically collected replicas of 340 users who have passed a psychological survey. As a result of linguistic analysis, 163 categories related to deixis, modality, stylistics, graphic means, etc. were singled out (in particular, the list of language genres included 41 nominations). Main goal of the research is to discover the dependencies between language categories and personality types according to BFI methodology. To achieve this, different statistical methods are used. The research is conducted within information system Semograph [3] that provides automation the research work of applied linguists. The large volume of heterogeneous parameters and the revealed numerous dependencies between them pose the problem of scientific data visualization for analyzing the obtained results of a multiparameter SNS users description. These problems are tackled within SciVi.

First, SciVi provides graph-based interactive visualization of the results[2]. Secondly, to simplify semantic filtering SciVi enables special IoT-based HMI. In particular, SNS users' messages with spatial and temporal deictic semantics were considered as language categories: (1) here/there, (2) right/left, (3) up/down, (4) now/then (and also in past/future) and (5) the speed of movement/change (fast/slow). An additional parameter of the analysis was the representation of one or several selected categories in SNS users' texts. The analysis is very complex because there are 144 variants of filtering available (filtering is exposed by hierarchy of 6 scales, each one having 2 or 3 variants). The advanced visualization tools and appropriate filter management interface are demanded. This is why we suggest to use not only traditional slider-based scales to filter the data, but also gesture interface as an alternative. For this purpose, the 6 categories mentioned above are mapped to gestures as described in the Table 1. Combinations of different gestures are possible. They are interpreted as combinations of their semantics. In case of an open palm, speed of gesture is ignored. In case of a feast, speed is mapped to the corresponding speed semantics.

Table 1. Verbal-gestural correspondences (bold font: **gesture**; italic font: *meaning*).

		Vertical move		Horizontal move		Depth move	
		Down	**Up**	**Left**	**Right**	**Back**	**Forth**
Hand back	**Up**	*Down*	*Up*	*Left*	*Right*	*Here*	*There*
	Down	*Past*	*Future*	*Past*	*Future*	*Past*	*Future*

---

[2] https://graph.semograph.org/cgraph/psycho/index.html?lang=en.

To detect gestures, we assembled glove-style wearable device similar to the ones presented in [21] or [9]. The distinctive feature of our approach is to use ontology-based SciVi tools to automate the generation of firmware for the device and middleware to turn this device into HMI. The hand gesture interface enables to apply the filtering with the single gesture instead of searching for the needed options within 6 separated filtering scales. That is why the efficiency of visual analysis is increased: the single gesture takes approximately 1 s, while interacting 6 sliders takes about 11 s.

The Fig. 4 demonstrates two roles of SciVi: scientific visualizer and IoT-based HMI provider. SciViCGraph [20] visualizer is used to establish structures of dominant connections between variables for different states of the graph that represent informants' social parameters, for example, gender. Glove-style IoT device simplifies managing semantic filtering needed to interactive visual analysis of multiparameter SNS users description. That helped the researchers to prove the hypothesis of the impact of gender differences on the behavior of users with the same psychological traits.

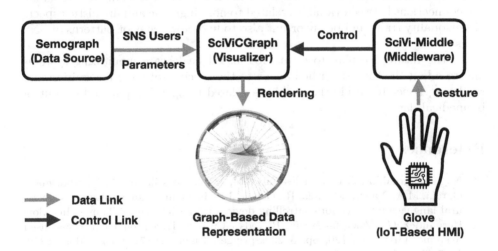

Fig. 4. Roles of SciVi in the research of SNS users' parameters.

6   Conclusion

In this paper, we present the unified high-level solution to build custom IoT-based HMI by means of a self-service smart system that has been developed to automate IoT devices firmware and middleware generation. On the one hand this system allows user to build a new custom device, program it and embed in the existing hardware and software ecosystem. On the other hand our solution is suitable to build the IoT ecosystem from scratch.

Nowadays such kind of solutions may be useful for solving special tasks, which require custom hardware controls to improve ergonomics compared to traditional mouse, keyboard, etc. In the near future, by evolving of virtual and augmented reality tools, the ability to have a hardware and software interface wizard will become more and more demanded. For example, gesture-based interfaces, tangible interfaces, wearable electronics and sensor network based interfaces are already indicated as today's rising trends [15].

The distinctive features of our approach are ontology-driven architecture that enables high-level adaptivity to specifics of different application domains and usage of cognitive graphics (initially based on scientific visualization engine) that enables building complex 2D and 3D monitoring and control graphical interface.

We tested our approach within the real-world interdisciplinary project named "Socio-Cognitive Modeling of Social Networks Users Verbal and Non-Verbal Behavior Based on Machine Learning and Geoinformation Technologies" and successfully improved visual analytics tools to discover correlations between the verbal behavior characteristics and psychological parameters of SNS users. The tools of visual analytics used in the study made it possible not only to streamline the connections between variables related to non-linguistic and linguistic aspects of personality traits and behavior but also to find new behavior patterns of SNS users.

In the future, we plan to create IoT-based healthcare monitoring systems and to adapt the tools described above to the specifics of voice recognition and rendering issues to tackle the problems related to speech input and output of biomedical data.

References

1. Noguera-Arnaldos, J.Á., Paredes-Valverde, M.A., Salas-Zárate, M.P., Rodríguez-García, M.Á., Valencia-García, R., Ochoa, J.L.: im4Things: an ontology-based natural language interface for controlling devices in the Internet of Things. In: Alor-Hernández, G., Valencia-García, R. (eds.) Current Trends on Knowledge-Based Systems. ISRL, vol. 120, pp. 3–22. Springer, Cham (2017). https://doi.org/10.1007/978-3-319-51905-0_1
2. Bajaj, G., Agarwal, R., Singh, P., Georgantas, N., Issarny, V.: A Study of Existing Ontologies in the IoT-Domain. arXiv (2017). https://arxiv.org/abs/1707.00112. Accessed 14 Feb 2019
3. Belousov, K., Erofeeva, E., Leshchenko, Y., Baranov, D.: "Semograph" information system as a framework for network-based science and education. In: Uskov, V.L., Howlett, R.J., Jain, L.C. (eds.) SEEL 2017. SIST, vol. 75, pp. 263–272. Springer, Cham (2018). https://doi.org/10.1007/978-3-319-59451-4_26
4. Broll, G., Rukzio, E., Paolucci, M., Wagner, M., Schmidt, A., Hussmann, H.: Perci: pervasive service interaction with the Internet of Things. IEEE Internet Comput. 13(6), 74–81 (2009). https://doi.org/10.1109/MIC.2009.120
5. Calderon, M., Delgadillo, S., Garcia-Macias, A.: A more human-centric Internet of Things with temporal and spatial context. Procedia Comput. Sci. 83, 553–559 (2016). https://doi.org/10.1016/j.procs.2016.04.263

6. Farahzadi, A., Shams, P., Rezazadeh, J., Farahbakhsh, R.: Middleware technologies for cloud of things: a survey. Digit. Commun. Networks **4**, 176–188 (2018). https://doi.org/10.1016/j.dcan.2017.04.005

7. Ferrari, M.I., Aquino, P.T.: Human interaction and user interface design for IoT environments based on communicability. In: Amaba, B. (ed.) Advances in Human Factors, Software, and Systems Engineering. AISC, vol. 492. Springer, Cham (2016). https://doi.org/10.1007/978-3-319-41935-0_10

8. Ishii, H.: Tangible bits: beyond pixels. In: Proceedings of the 2nd International Conference on Tangible and Embedded Interaction, pp. XV–XXV (2008). https://doi.org/10.1145/1347390.1347392

9. Kumar Mummadi, C., Philips, F., Deep Verma, K., Kasireddy, S., Scholl, P., Kempfle, J., Van Laerhoven, K.: Real-time and embedded detection of hand gestures with an IMU-based glove. Informatics **5**, 28 (2018). https://doi.org/10.3390/informatics5020028

10. Lazarevich, K.: 5 keys to designing great UX for IoT products. IoT For All (2018). https://medium.com/iotforall/5-keys-to-designing-great-ux-for-iot-products-d10bda51842e. Accessed 14 Feb 2019

11. Luz, N., Pereira, C., Silva, N., Novais, P., Teixeira, A., Oliveira e Silva, M.: An ontology for human-machine computation workflow specification. In: Polycarpou, M., de Carvalho, A.C.P.L.F., Pan, J.-S., Woźniak, M., Quintian, H., Corchado, E. (eds.) HAIS 2014. LNCS (LNAI), vol. 8480, pp. 49–60. Springer, Cham (2014). https://doi.org/10.1007/978-3-319-07617-1_5

12. Nechaev, Y.I., Degtyarev, A.B., Boukhanovsky, A.V.: Cognitive computer graphics for information interpretation in real time intelligence systems. In: Sloot, P.M.A., Hoekstra, A.G., Tan, C.J.K., Dongarra, J.J. (eds.) ICCS 2002. LNCS, vol. 2329, pp. 683–692. Springer, Heidelberg (2002). https://doi.org/10.1007/3-540-46043-8_69

13. Nuamah, J., Seong, Y.: Human machine interface in the Internet of Things (IoT). In: 12th System of Systems Engineering Conference, pp. 1–6 (2017). https://doi.org/10.1109/SYSOSE.2017.7994979

14. Petnik, J., Vanus, J.: Design of smart home implementation within IoT with natural language interface. IFAC-PapersOnLine **51**, 174–179 (2018). https://doi.org/10.1016/j.ifacol.2018.07.149

15. Poh, M.: 8 Next-Generation User Interface That Are (Almost) Here. Hongkiat (2018). https://www.hongkiat.com/blog/next-gen-user-interface/. Accessed 14 Feb 2019

16. Goldberg, L.R.: The development of markers for the big five factor structure. Psychol. Assess. **4**, 26–42 (1992). https://doi.org/10.1037/1040-3590.4.1.26

17. Ryabinin, K., Chuprina, S.: Development of ontology-based multiplatform adaptive scientific visualization system. J. Comput. Sci. **10**, 370–381 (2015). https://doi.org/10.1016/j.jocs.2015.03.003

18. Ryabinin, K., Chuprina, S.: Using scientific visualization tools to bridge the talent gap. Procedia Comput. Sci. **51**, 1734–1741 (2015). https://doi.org/10.1016/j.procs.2015.05.376

19. Ryabinin, K., Chuprina, S.: High-Level toolset for comprehensive visual data analysis and model validation. Procedia Comput. Sci. **108**, 2090–2099 (2017). https://doi.org/10.1016/j.procs.2017.05.050

20. Ryabinin, K., Chuprina, S., Kolesnik, M.: Calibration and monitoring of IoT devices by means of embedded scientific visualization tools. In: Shi, Y., et al. (eds.) ICCS 2018. LNCS, vol. 10861, pp. 655–668. Springer, Cham (2018). https://doi.org/10.1007/978-3-319-93701-4_52

21. Sawasaki, N., Ishihara, T., Mouri, M., Murase, Y., Masui, S., Nakamoto, H.: Front-end device technology for human centric IoT. Fujitsu Sci. Tech. J. **52**, 61–67 (2016)
22. Shchebetenko, S.: Reflexive characteristic adaptations explain sex differences in the big five: but not in neuroticism. Personality Individ. Differ. **111**, 153–156 (2017). https://doi.org/10.1016/j.paid.2017.02.013
23. Steinmetz, C., Rettberg, A., Ribeiro, F.G.C., Schroeder, G., Soares, M.S., Pereira, C.E.: Using ontology and standard middleware for integrating IoT based in the industry 4.0. IFAC-PapersOnLine **51**, 169–174 (2018). https://doi.org/10.1016/j.ifacol.2018.06.256
24. Wu, J., Grimsley, R., Jafari, R.: A robust user interface for IoT using context-aware bayesian fusion. In: IEEE 15th International Conference on Wearable and Implantable Body Sensor Networks, pp. 126–131 (2018). https://doi.org/10.1109/BSN.2018.8329675

Towards Parameter-Optimized Vessel Re-identification Based on IORnet

Amir Ghahremani[✉], Yitian Kong, Egor Bondarev, and Peter H. N. de With

Video Coding and Architectures Group (VCA), Eindhoven University of Technology, Eindhoven, The Netherlands
a.ghahremani@tue.nl

Abstract. Reliable vessel re-identification would enable maritime surveillance systems to analyze the behavior of vessels by drawing their accurate trajectories, when they pass along different camera locations. However, challenging outdoor conditions and varying viewpoint appearances combined with the large size of vessels limit conventional methods to obtain robust re-identification performance. This paper employs CNNs to address these challenges. In this paper, we propose an Identity Oriented Re-identification network (IORnet), which improves the triplet method with a new identity-oriented loss function. The resulting method increases the feature vector similarities between vessel samples belonging to the same vessel identity. Our experimental results reveal that the proposed method achieves 81.5% and 91.2% on mAP and Rank1 scores, respectively. Additionally, we report experimental results with data augmentation and hyper-parameters optimization to facilitate reliable ship re-identification. Finally, we provide our real-world vessel re-identification dataset with various annotated multi-class features to public access.

Keywords: Re-identification of vessels · CNNs ·
Maritime surveillance · Vessel re-identification dataset

1 Introduction

Camera-based maritime surveillance systems monitor harbors and waterways to increase the safety and security against unknown pathless watercrafts, prevent out-of-region fishery, manage urban transportation, and control the cargo flow. Recently, vessel-behavior analysis is also an expected function for such systems, since this ability can drastically improve the efficiency of an automated surveillance system. To this end, keeping track of vessels over consecutive camera locations is of a vital importance. This requires a reliable vessel re-identification approach, which aims at the successful detection of the identity of a specific vessel at different camera locations. This concept is visualized in Fig. 1 with image samples captured by different cameras.

The outdoor maritime environment poses considerable challenges to camera-based surveillance systems by precipitation, sunshine reflection, fog, water waves,

© Springer Nature Switzerland AG 2019
J. M. F. Rodrigues et al. (Eds.): ICCS 2019, LNCS 11540, pp. 125–136, 2019.
https://doi.org/10.1007/978-3-030-22750-0_10

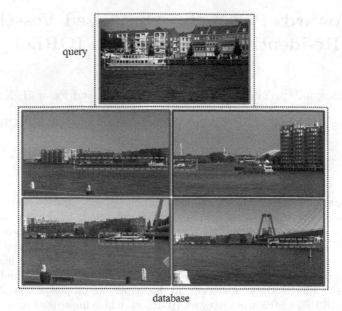

Fig. 1. Vessel re-identification is about finding the query vessel in the existing database images. In this example, the blue box in the top image represents the query image. The red box in the upper-right database image indicates the same vessel re-identified at a clearly different location. (Color figure online)

etc. In addition to these typical problems, a vessel re-identification method has to overcome its task-specific challenges. For instance, surveillance cameras at different locations often capture vessels from varying viewpoints. Since vessels are large objects, their appearances (including color, shape, hull textures, etc.) can be entirely different from alternative viewpoints. Moreover, vessels captured at different camera locations are surrounded by diverse types of backgrounds. Furthermore, illumination changes caused by different weather conditions and daytimes also deteriorate the vessel re-identification performance.

To the best of our knowledge, an in-depth study on the vessel re-identification problem is virtually absent in literature. However, with the emergence of Convolutional Neural Networks (CNNs), the related field of pedestrian re-identification already presents methods with promising performance [1–6]. Since vessel re-identification is conceptually similar to pedestrian re-identification, this paper addresses the vessel re-identification problem by extending a triplet-based pedestrian re-identification approach to the maritime surveillance domain. However, unlike pedestrian re-identification, in a maritime environment vessels of the same model but different identity, may still have extremely similar appearance, which makes the vessel re-identification even more challenging. Figure 2 illustrates this problem by a few example vessel images.

In this paper, first, we attempt to solve vessel re-identification by introducing a new identity-oriented loss function for learning the vessel identity. Second,

Fig. 2. Illustration of four images of similar vessels, which belong to different vessel identities, although they are made by the same vessel manufacturing company and have the same model.

since there is no public dataset available for exploring the vessel re-identification problem, we provide our annotated vessel re-identification dataset, which was captured at various locations in several harbor cities and suburbs in the Netherlands (Amsterdam, Rotterdam), to open public access [7]. Third, this paper investigates the efficiency of high-performing human re-identification techniques for the vessel re-identification problem (e.g. data augmentation, a different number of training iterations, etc.). These experiments lead to a parameter-optimized re-identification of vessels.

The sequel of the paper is organized as follows. Section 2 provides an overview of the related work. Section 3 explains the proposed method. Section 4 presents the experimental results and validation. Section 5 concludes the paper.

2   Related Work

As already mentioned, due to the absence of literature on vessel re-identification, we commence with addressing the widely investigated pedestrian re-identification from several research works.

Pedestrian re-identification approaches attempt to re-identify the same person at different camera locations. These methods typically search for the best match of a query image among previously captured database images. Two common research directions for such methods are (1) to attempt to improve the image discrimination in feature space and/or (2) introduce better distance metrics [8]. Generally, pedestrian re-identification methods are divided into three main categories: (a) verification models [1,2,9,10], (b) identification models [5,6,11–13], and (c) combinational models [8,14,15], which are all briefly discussed below.

Figure 3(a) illustrates a common architecture of verification models. These models re-identify the vessel samples belonging to the same identity by assessing the feature vector similarities between their input images. The work in [1] employs a patch-matching technique, which finds the mid-level feature similarity of pairwise images, to modify a Siamese network. In [9], the method uses matching gates to improve a Siamese network. These gates predict the critical point of a Siamese network in higher layers by inspecting its low-level features.

A common architecture of identification models is presented in Fig. 3(b). These methods investigate a single input image to determine the person's identity. The work in [12] engages handcrafted features in a network for fine-tuning

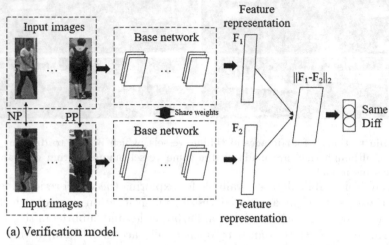

(a) Verification model.

(b) Identification model.

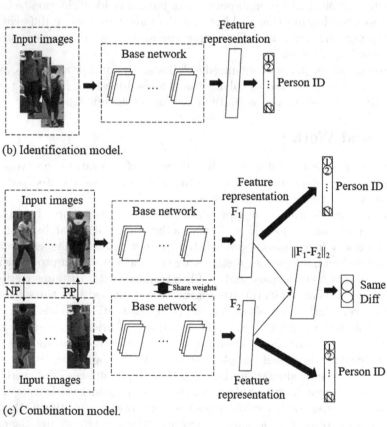

(c) Combination model.

Fig. 3. Architecture of three generic pedestrian re-identification models. The numbers in the circles at the output represent different identities, where N is the total number of person identities in the database. Additionally, NP and PP represent negative and positive pairs, respectively.

a procedure to improve the re-identification performance. The method in [6] achieves better fine-tuning performance by employing a pedestrian-attribute dataset. This work uses the data disparity between practical datasets (which have low quality) and the ImageNet [13] dataset. The work from [11] proposes to use a reliable classification model, which is obtained by combining several pedestrian re-identification datasets for person identity recognition.

Figure 3(c) presents the architecture of a typical combinational model. These architectures incorporate both the identification and verification loss-functions to optimize the performance. In [8], the proposed method improves a Siamese architecture by comparing the feature representations of input images using a square layer. The work in [14] combines two identification subnets and one verification subnet with a Siamese network to provide robust pedestrian re-identification performance.

This paper modifies the pedestrian re-identification concept towards the vessel re-identification problem. Additionally, we introduce a new triplet-based loss function to increase the feature vector similarity between vessels belonging to the same vessel identity. Here, we focus on re-identifying vessels in harbors and waterways. Additionally, we provide an annotated vessel re-identification dataset, which includes 4,616 real-world images. These images were captured with two cameras under different weather, lighting, and timing conditions and at different locations with variable backgrounds. All annotated vessels have been labeled by a unique ID and appear in several images. Moreover, we have also annotated the bounding box, vessel type and vessel orientation of each vessel for potential further experiments.

3   Architecture Pipeline

This section describes the proposed vessel re-identification architecture, which is depicted in Fig. 4. The visualized method (IORnet) includes three modules. The first one is the feature extraction module, which receives a set of three images per vessel identity and transforms them into the feature space. The second module is the triplet subnet. This module calculates the triplet loss of the input, aiming to pull samples closer when they originate from the same vessel, while increasing the distance to different vessels/objects. The third module is the identification subnet, which increases the feature vector similarity between vessels belonging to the same vessel identity. After extracting the feature vectors from the first module, the second and third module operate in parallel to calculate the loss function. Then, the base-network weights are updated according to the calculated loss value. The following subsections discuss the three modules in detail.

3.1   Feature Extraction Module

This module consists of three basic CNNs to transform the input images into feature vectors. Here we employ ResNet50 [16] as the basis architecture. The

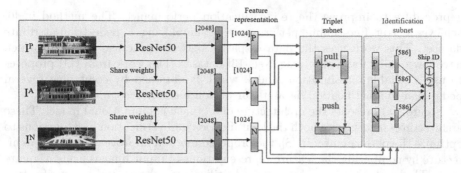

Fig. 4. Identity-Oriented Re-identification network (IORnet). This method receives three input images, which are Anchor image together with its Positive and Negative pairs. Then, the triplet and identification modules calculate the loss function and update the feature extraction CNNs.

extracted feature vectors have a dimension of 2,048 elements. Then, we append a batch normalization layer to speed up the convergence and optimize the deep convolution networks [17]. Finally, this module resizes the feature vectors to 1,024 elements. The three input images submitted to the basis CNNs are denoted by IA, IP, and IN, while IA and IP belong to the same vessel identity (positive pair) and IN represents another vessel identity (negative sample).

3.2 Triplet Module

Here, the conventional triplet model and its limitations are first briefly reviewed. The triplet loss was introduced in [18] to improve face re-identification performance. The objective of the triplet loss is to pull image features belonging to the same class closer to each other, while pushing the features of different image classes away from that cluster. For more clarification, we assume that A, P, and N denote the Anchor, Positive, and Negative image samples, while A and P contain the same object identity and N contains another identity. The triplet loss is then expressed by:

$$D_{AP} - D_{AN} \geq \alpha, \tag{1}$$

where D_{XY} represents the distance between images X and Y in feature space and α is the distance margin. By iteratively optimizing this process over the whole dataset, positive pairs converge into a single cluster, while distantiating that cluster from the negative samples. In this work, we employ the TriNet [4] framework as the triplet subnet. This method uses a variant of the triplet loss to perform end-to-end deep metric learning and achieves reliable results in pedestrian re-identification.

Unfortunately and as a limitation, the triplet architecture only considers two different identities at a time. This can push a negative object sample against its cluster [19], which can lead to having dissimilar feature representations for vessel samples belonging to the same vessel identity. This drawback also increases the convergence time.

3.3 Identification Module

As just discussed, the triplet loss function may generate dissimilar feature vectors for object samples belonging to the same object identity. In order to solve this problem, we propose an Identity Oriented Re-identification network (IORnet). In IORnet, we add an identification subnet to the triplet network, as illustrated in Fig. 4. In this subnet, we consider all the samples belonging to the same identity as a unique label and perform multi-class detection learning. To this end, the feature representations extracted by the basis networks are supplied into a new fully-connected layer. Then, the softmax function is used to normalize these feature vectors. By adding the identification subnet, the final loss function can be formulated now as follows:

$$L = \gamma.L^{\text{triplet}} + (1 - \gamma) \cdot L^{\text{identification}}, \tag{2}$$

where

$$L^{\text{triplet}} = \alpha + D_{AP} - D_{AN},$$

and $L^{\text{identification}}$ represents the softmax loss function. Trade-off parameter γ is defined in the unity interval. With $\gamma = 0$, the final loss becomes the identification loss function, while $\gamma = 1$ changes the equation into the pure triplet loss function. This proposed loss function restricts the whole system to provide more similar feature representations for the image samples belonging to the same vessel identity.

4 Empirical Validation

This section starts with a dataset overview and then discusses the training parameters and analyzes the performance of the proposed method.

4.1 Vessel Re-identification Dataset

In order to train the vessel re-identification model, we have recorded several videos from various locations in the Netherlands. These videos were captured using two cameras during different daytimes. The videos contain a vast variety of different viewpoints on vessels. Additionally, several vessel types with divergent sizes and distances to the camera are found in this dataset. Finally, challenging scenarios including vessel occlusion/truncation are also annotated.

The dataset contains 4,616 images with 733 different vessel identities. Each vessel identity is represented by several images. Additionally, we have labeled each vessel with a bounding box, its vessel type, and vessel orientation (i.e. the approximate positioning angle towards the camera) to facilitate future research. The vessel type range contains 10 classes: sailing vessel, container ship, passenger ship, fishing vessel, tanker, river cargo, small boat, yacht, tug, and taxi vessel. The vessel orientations are described with the following 5 orientation labels: front view, front-side view, side view, back-side view, and back view. Besides this, we have provided a unique ID to each vessel and have cropped each vessel from the whole image by an annotated bounding box. Then the dataset is split into training and test datasets. The training dataset contains 3,651 images with 586 unique vessel identities, while the test dataset includes 965 images with 147 unique vessel identities.

4.2   Training Procedure

In this work, we have employed ResNet50 [16] pre-trained on ImageNet [13] as the basis network for feature extraction. The Adam optimizer [20] is used with default hyper-parameters. We have set the initial learning rate to 0.0003, which is exponentially decayed after 35,000 iterations. The proposed re-identification CNN is trained for 50,000 iterations. We have selected 18 vessel identities and 4 images per identity to form a mini-batch of size 72. Furthermore, we have added a dropout layer [21] to reduce the risk of overfitting. The trade-off parameter γ in Eq. (2) is empirically set to $\gamma = 0.6$.

4.3   Validation Results

This subsection evaluates our Identity Oriented Re-identification network (IORnet). We have trained both the state-of-the-art triplet-based network (TriNet) and the IORnet on the published training dataset. Table 1 compares the performance of these two methods on our test dataset. In this table, the methods are evaluated according to mAP (mean Average Precision), Rank1, Rank5, and Rank10 metrics. According to Table 1, TriNet provides mAP and Rank1 scores of 78.4% and 88.4%, respectively. These values indicate that extending the triplet concept to the vessel re-identification problem by training this network on an annotated vessel re-identification dataset provides robust results. Additionally, IORnet improves the mAP and Rank1 results of TriNet by approximately 3% and 2%, respectively. Evidently, the proposed loss function provides a higher performance due to increasing the similarities between feature vectors belonging to the same identity.

4.4   Discussion on Parameters and Data Augmentation

In order to explain the parameter-optimized vessel re-identification model in depth, this subsection provides additional discussion on the proposed method.

Figure 5 illustrates the mAP scores provided by IORnet for different iterations. It can be observed that the mAP results are not stable within the first 35,000 iterations. This occurs because the learning rate is relatively large and the network skips some local optimizations. After 35,000 iterations (while decreasing the learning rate), the calculated scores tend to become more stable and also slightly improve.

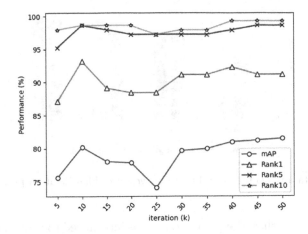

Fig. 5. Impact of training iterations on IORnet performance.

Generally, data augmentation improves the performance of pedestrian re-identification methods. Therefore, we have tested this technique also on the vessel re-identification problem. To this end, our vessel images are augmented with random cropping and horizontal flipping in the training phase, similar to pedestrian re-identification work in [1,17]. More specifically, the image size is first increased by 9/8 with the same aspect ratio. Then, the image is randomly copped to obtain the original size. We have also performed the horizontal image flipping on randomly chosen images. Figure 6 compares the method performance with and without data augmentation. It appears that the data augmentation technique deteriorates the vessel re-identification performance. For instance, the mAP rate is decreased from 81.46% to 76.68%. We conclude that training on random parts of vessels does not improve the re-identification model, since for large objects, random fragments do not provide a reliable statistical base for identity retrieval.

Additionally, during our experiments, we have noticed that many failures in vessel re-identification are the outcome of performing re-identification on vessel samples captured from different orientations (camera viewing angle to the ship). This happens also because vessels are large objects, having very different appearances from varying orientations. To address this problem, the orientation information can be integrated into the re-identification method in our future work.

Table 1. Vessel re-identification performance.

Models	mAP	Rank1	Rank5	Rank10
TriNet	78.4%	88.4%	97.3%	98.6%
IORnet	81.5%	91.2%	98.6%	99.3%

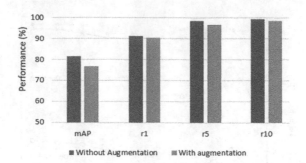

Fig. 6. Data augmentation influence on vessel re-identification performance.

As mentioned earlier, the parameter γ was introduced to control the trade-off between the identification module and the triplet module. Tuning this parameter to the optimal value is of high importance. Here, the re-identification performance is tested with different values of the parameter to discover its influence on the performance. The results are illustrated in Fig. 7. It can be deduced that for $\gamma < 0.36$ and $\gamma > 0.6$ the re-identification performance deteriorates. For this reason, we have chosen $\gamma = 0.6$, since this value achieves the highest mAP and Rank1 scores.

Finally, to pursue real-time vessel re-identification, we have calculated the average time of identity retrieval. The tests were performed on a workstation with E5-1620 CPU, 16 GB of memory and a GTX-1080 GPU. There are 147 vessel identities in our query set and 818 images in the database. The total retrieval time measured for all 147 query images was 558.7 ms. Hence, for a single query image, it takes 3.8 ms to perform the re-identification procedure and return the ranking list, which would satisfy real-time execution.

It is also important to mention that the original TriNet decreases the feature vector size from 2,048 to 128. However, according to our experiments, this small

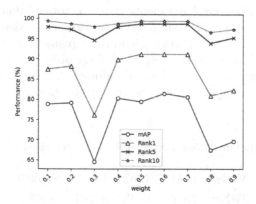

Fig. 7. Influence of γ parameter on IORnet performance.

feature vector size does not allow the identification module to achieve the desired results. Therefore, we have adopted 1,024 as the feature vector size at the output of the feature extraction module. By doing so, the training time increases from 0.3 s/iteration to 0.56.

5 Conclusion

This paper has proposed a robust vessel re-identification method to track the identity of a specific vessel throughout a network with different camera locations. The proposed CNN-based method extends the TriNet loss function with an identification method. The improved architecture, called IORnet, concentrates on enhancing the similarities between vessel images belonging to the same vessel identity in feature space. This approach also leads to a better discrimination from other vessels. Experimental results have shown that our approach achieves 81.5% and 91.2% on mAP and Rank1 scores, respectively. Additionally, experiments were conducted for vessel re-identification using several re-identification techniques with proven value for pedestrian re-identification (like data augmentation, training parameters, etc.). This supplementary inspection has resulted into a parameter-optimized re-identification of vessels. As an important contribution, we have also developed a vessel re-identification dataset, which is annotated with bounding boxes, vessel identities, vessel categories, vessel orientations, and vessel capturing status (whether a vessel is truncated and/or occluded). This dataset includes images with annotated vessels, captured at different locations under varying weather conditions and with variable backgrounds and has become available for public access [7].

Acknowledgement. The authors gratefully acknowledge the project PASSANT, funded by the H2020 Interreg program, for supporting the research work.

References

1. Ahmed, E., Jones, M., Marks, T.K.: An improved deep learning architecture for person re-identification. In: CVPR, pp. 3908–3916 (2015)
2. Ding, S., Lin, L., Wang, G., Chao, H.: Deep feature learning with relative distance comparison for person re-identification. Pattern Recognit. **48**(10), 2993–3003 (2015)
3. Guindel, C., Martin, D., Armingol, J.M.: Joint object detection and viewpoint estimation using CNN features. In: ICVES, pp. 145–150. IEEE (2017)
4. Hermans, A., Beyer, L., Leibe, B.: In defense of the triplet loss for person re-identification, arXiv preprint arXiv:1703.07737 (2017)
5. Lin, Y., Zheng, L., Zheng, Z., Wu, Y., Yang, Y.: Improving person re-identification by attribute and identity learning, arXiv preprint arXiv:1703.07220 (2017)
6. Matsukawa, T., Suzuki, E.: Person re-identification using CNN features learned from combination of attributes. In: ICPR, pp. 2428–2433. IEEE (2016)
7. http://vca.ele.tue.nl/

8. Zheng, Z., Zheng, L., Yang, Y.: A discriminatively learned CNN embedding for person reidentification. TOMM **14**(1), 13 (2017)
9. Varior, R.R., Haloi, M., Wang, G.: Gated siamese convolutional neural network architecture for human re-identification. In: Leibe, B., Matas, J., Sebe, N., Welling, M. (eds.) ECCV 2016. LNCS, vol. 9912, pp. 791–808. Springer, Cham (2016). https://doi.org/10.1007/978-3-319-46484-8_48
10. Cheng, D., Gong, Y., Zhou, S., Wang, J., Zheng, N.: Person re-identification by multi-channel parts-based CNN with improved triplet loss function. In: CVPR, pp. 1335–1344 (2016)
11. Xiao, T., Li, H., Ouyang, W., Wang, X.: Learning deep feature representations with domain guided dropout for person re-identification. In: CVPR, pp. 1249–1258. IEEE (2016)
12. Wu, S., Chen, Y.-C., Li, X., Wu, A.-C., You, J.-J., Zheng, W.-S.: An enhanced deep feature representation for person re-identification. In: WACV, pp. 1–8 (2016)
13. Deng, J., Dong, W., Socher, R., Li, L.-J., Li, K., Li, F.-F.: Imagenet: a large-scale hierarchical image database. In: CVPR, pp. 248–255. IEEE (2009)
14. Geng, M., Wang, Y., Xiang, T., Tian, Y.: Deep transfer learning for person re-identification, arXiv preprint arXiv:1611.05244 (2016)
15. Chen, w., Chen, X., Zhang, J., Huang, K.: A multi-task deep network for person re-identification. In: AAAI, vol. 1, p. 3 (2017)
16. He, K., Zhang, X., Ren, S., Sun, J.: Deep residual learning for image recognition. In: CVPR, pp. 770–778 (2016)
17. Ioffe, S., Szegedy, C.: Batch normalization: Accelerating deep network training by reducing internal covariate shift, arXiv preprint arXiv:1502.03167 (2015)
18. Schroff, F., Kalenichenko, D., Philbin, J.: Facenet: a unified embedding for face recognition and clustering. In: CVPR, pp. 815–823 (2015)
19. Zhang, Y., Liu, D., Zha, Z.-J.: Improving triplet-wise training of convolutional neural network for vehicle re-identification. In: ICME, pp. 1386–1391 (2017)
20. Kingma, D.P., Ba, J.: Adam: A method for stochastic optimization, arXiv preprint arXiv:1412.6980 (2014)
21. Srivastava, N., Hinton, G., Krizhevsky, A., Sutskever, I., Salakhutdinov, R.: Dropout: a simple way to prevent neural networks from overfitting. J. Mach. Learn. Res. **15**(1), 1929–1958 (2014)

Towards Low-Cost Indoor Localisation Using a Multi-camera System

Oualid Araar[1]([⊠]), Saadi Bouhired[2], Sami Moussiou[1], and Ali Laggoune[1]

[1] Ecole Militaire Polytechnique, Bordj El Bahri, Algiers, Algeria
o.araar@emp.mdn.dz
[2] Institut de Recherche DAT, Algiers, Algeria

Abstract. Indoor localisation is a fundamental problem in robotics, which has been the subject of several research works over the last few years. Indeed, while solutions based on fusion of inertial and global navigation satellite system (GNSS) measurements have proved their efficiency in outdoor environments, indoor localisation remains an open research problem. Although commercial motion tracking systems offer very accurate position estimation, their high cost cannot be afforded by all research laboratories. This paper presents an indoor localisation solution based on a multi-camera setup. The proposed system relies on low-cost sensors, which makes it very affordable compared to commercial motion-tracking systems. We show through the experiments conducted that the proposed approach, although being cheap, can provide real-time position measurements with an error of less than 2 cm up to a distance of 2 m.

Keywords: Indoor localisation · Multi-camera system ·
3D ground-truth position · Motion tracking

1 Introduction

A fundamental problem in robotics is how to obtain the position and orientation of the robot with a sufficient accuracy. Localisation solutions can be used for navigation purposes, in order to make a robot follow a desired path or trajectory. They can also be employed for obtaining the ground-truth to evaluate the accuracy of other positioning systems.

While localisation in outdoor environments has been successfully solved since GPS signal was made available to the public, solving the problem for indoor space remains a challenging task. Indeed, techniques based on the fusion of inertial and GNSS measurements have proved their efficiency for both ground and aerial vehicles navigation [1,3,5]. The poor quality or even absence of GNSS signal in indoor environments, however, makes the accuracy of these techniques very limited.

This limitation has motivated the research for other localisation techniques, which do not rely on GNSS signal. Existing solutions can be classified into

© Springer Nature Switzerland AG 2019
J. M. F. Rodrigues et al. (Eds.): ICCS 2019, LNCS 11540, pp. 137–148, 2019.
https://doi.org/10.1007/978-3-030-22750-0_11

two main classes, depending on whether the system is carried by the vehicle or installed externally.

While the first class offers the advantage of making the vehicle independent of any external infrastructure, it generally exhibits a drift problem due to error accumulation. The second class, on the other hand, offers drift-free measurements since no integration process is included in the position estimation. They require making modifications to the environment, which may be inconvenient in some applications.

Among commercial solutions which have been used for the purpose of indoor localisation, motion-tracking systems are the most accurate option. Such systems, which were originally developed for the purpose of tracking the joints of the human body, were quickly adopted for getting the pose of all kind of robots.

Besides their high precision which can achieve sub-millimeter accuracy, motion-tracking systems operates at high sampling rates which can attain several hundreds of FPS. Such sophisticated characteristics, however, are far from what is really needed in most robotic applications. This makes the justification of the high cost of such solutions often questionable.

The purpose of this paper is to develop a localisation approach which presents a better trade-off between performance and cost. The proposed system rely on ordinary video-cameras, which makes it very affordable.

2   Related Work

Among research works which addressed the problem of indoor position estimation using multiple cameras, the one in [11] employed a ceiling-mounted system for localising a ground robot. The overlap between pairs of cameras was exploited to get extrinsic and intrinsic parameters of each camera through stereo-calibration. A chessboard marker with side circular pattern were used for marking the robot. Given the large size and weight of the calibration chessboard, the latter cannot be carried by a robot with a limited payload. A multiple ceiling-mounted camera system was also used in [10] for tracking the position of multiple unmarked ground-robots, using a white floor to facilitate the detection of the vehicles.

Shim and Cho [12] exploited a video-surveillance system composed of multiple cameras for localising a service robot navigating in an indoor environment. In [7], a multi-agent tracking system for marked and unmarked ground-robots was presented. The cameras were mounted vertically which limits its use to only ground robots, moving on a floor perfectly parallel to the cameras plan.

Among research works which have dealt with indoor localisation of aerial vehicles the one in [6] investigated the localisation of a quadrotor UAV. This was achieved by installing a big number of markers in the laboratory and estimating the pose of the drone using an on-board camera. In [8], a multi-camera setup was used for controlling the position of a quadrotor. The vehicle was marked with four balls of different colours. In addition to the important weight of the markers, the proposed approach can only be used for localising a single agent.

Besides its low cost, the solution discussed in this paper presents the following advantages:

- Modularity: Both the hardware and software implementations are modular, which makes our solution independent of the number of sensors.
- Simplicity: The system relies on passive tags which can be easily obtained and printed on ordinary paper.
- Real-time performance: The parallel implementation of the processing algorithms allows the system to operate on full frame-rate (30 FPS).
- Opensourceness: All the third-party software used in this work is open-source, which means anyone can use it free of charge.

In the rest of this paper a detailed description of the localisation system is presented, and learned lessons from each experience are discussed. We start by a description of the different hardware components we used, followed by the optimal placement of the cameras. After that, we discuss the patterns employed for marking the object of interest as well as the algorithms allowing their detection. The estimation of the 3D position of the markers is then presented, where two techniques are implemented and compared. The last part covers the pipeline of the processing algorithms and its real-time implementation.

3   Proposed Solution

3.1   Hardware Architecture

Our localisation system uses four commercial video-cameras, Fig. 1, which provide a resolution of 640 × 480 pixels at a frame rate of 30 FPS. The cameras are interfaced to the processing laptop through a router, where each one is defined by a static IP address. Therefore, from a software point of view, adding a new sensor to the system is a matter of adding its IP address to the list of the already existing ones.

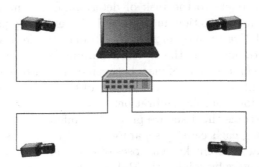

Fig. 1. Hardware components of the localisation solution.

3.2   Camera Placement

Using a multi-camera system permits enhancing the quality of the estimated position on the one hand and increasing the covered volume on the other hand. The free software "IP Video System Design Tool" was used in this work to visualise the volume covered by each sensor. In order to maximise the workspace, the sensors were placed in such a way that only two cameras cover the same space at the time. Figure 2 illustrates this configuration, where the light-green regions are covered by a single camera while at each point of the darker-green regions, the object is seen by two cameras. In this region, the position of the object can be estimated using two approaches, while in the light-green one it can only be obtained by solving the PnP problem as will be discussed in what follows.

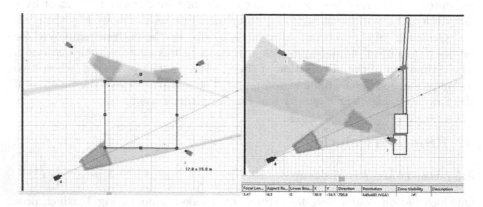

Fig. 2. Optimal placement of the four cameras obtained using the software "IP Video System Design Tool". (Color figure online)

3.3   System Calibration

Camera calibration refers to the task of determining the intrinsic parameters which appears in its projection matrix, and the extrinsic parameters which defines its pose relative to another reference frame. While intrinsic calibration was straightforward, obtaining the extrinsic parameters was a challenging task. The reason was the limited field of view between each pair of cameras, constraining the different configurations one can obtain of the calibration pattern—thing which is necessary for a precise calibration.

In order to overcome this issue we proceeded differently to the classical approach, by calibrating each camera separately, see Fig. 3. This was achieved by placing a pattern at different known poses of each camera, and then obtaining the extrinsic parameters by solving the PnP problem. Figure 4 presents a sample of calibration results using patterns of different sizes and confirms the advantages of using a big pattern.

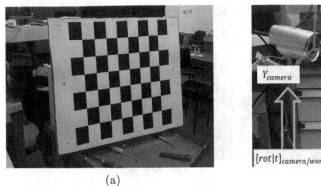 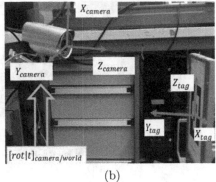

(a)   (b)

Fig. 3. System calibration: (a) stereo-calibration using a chessboard pattern; (b) calibrating each camera separately using tags.

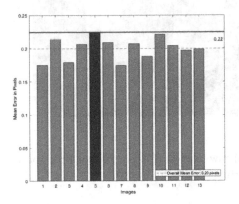 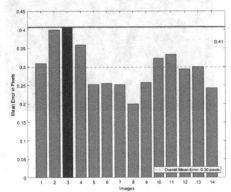

Fig. 4. Calibration results obtained using different pattern sizes. (left) big pattern, (right) smaller one.

3.4 Object Detection and Identification

In order to facilitate the detection of an object of interest and determine its identity in case of multi-object application, two kinds of artificial markers were employed. The first one is a coloured marker while the second is a tag from the AprilTag library. In what follows, we discuss the detection and identification of each of the two markers.

Coloured Marker. For detecting the coloured marker, the images are first transformed from the RGB to the HSV space where the detection thresholds are specified by the user. Pixels within these thresholds represent eventual candidates and are thus retained for the next operation (Fig. 5b). After a dilation/erosion step only regions with the colour of interest are conserved (Fig. 5c). The centroid of the object of interest is calculated based on image-moment, Fig. 5d.

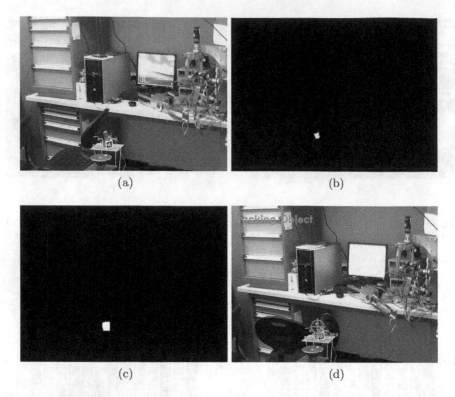

Fig. 5. Coloured marker detection steps: (a) raw images; (b) detection in HSV space; (c) noise suppression after erosion/dilation step; (d) centroid calculation.

Tag Markers. Among existing visual tag systems, one can mention the ARToolkit [4], Studier-stube Tracker [13], ARTag [2], and the more recent April-Tag system [9]. The latter is chosen in this work, as it enables encoding a large number of distinct ids. Furthermore, its encoding scheme enables codewords with a low hamming distance, which results in a lower misidentification rate.

The AprilTag system works in two steps: detection and decoding. The detector begins by clustering pixels based on their gradient magnitude and direction, to detect line segments. Sequences of lines that form a 4-sided shape (quad) are then identified, see Fig. 6.

The decoding stage reads the bits from each tag's payload field, in order to identify it. This stage begins by transforming each bit field into image coordinates using the Homography matrix estimated from the four corners of the detected quad. The resulting pixels are then classified using a spatially varying model of the intensities of the white and black colours.

Fig. 6. Detection steps of the tags from the AprilTag library.

3.5   Position Estimation

The detection and identification steps allow obtaining the 2D coordinates of the object of interest in the image space. In order to get the 3D position of the object relative to the scene, two approaches can be considered. In the first, measurements from at least two cameras are combined in order to get the depth of the object, using what is commonly referred to as triangulation. In the second approach, the position is estimated from each camera separately by solving what is known as the perspective n point (PnP) problem. Both approaches are detailed in what follows.

Triangulation. Unlike RGBD cameras which provides the full 3D coordinates of a point of interest, only the direction of the object can be obtained from a video-camera. If measurements of the same object are available from a second sensor, its 3D position can be obtained from the intersection of the two lines of sight, Fig. 7. This approach is employed for estimating the coloured marker position, since just the 2D coordinates of its centroid are available and thus its 3D position can only be estimated using at least two cameras.

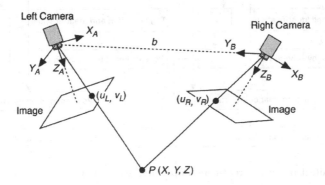

Fig. 7. Object position estimation using Triangulation.

Solving PnP Problem. Unlike the coloured marker, the identified tags provide the 2D coordinates of four points from each tag. Given the prior knowledge about the dimension and shape of the tags, one can estimate the pose of the marked object from a single camera. This problem, which is commonly coined PnP, can be solved if a minimum of four points is available from the object of interest.

3.6   Real-Time Implementation

In order to ensure real-time performance, all the previous steps where implemented in C++, based on the Open Computer Vision (OpenCV) library. A serial implementation of the previously discussed processing algorithms turned-out to be impractical. In fact, even without any processing, i.e only the acquisition and display of the images, an important lag among the four cameras was observed. To overcome this issue, we proceeded to a parallel implementation, using the Microsoft TBB library. An overview of the architecture of this implementation is given in Fig. 8. Using this approach, we were able to run the code in full frame rate on a laptop with an *i*7 2.5 GHz processor and 8 GB of RAM.

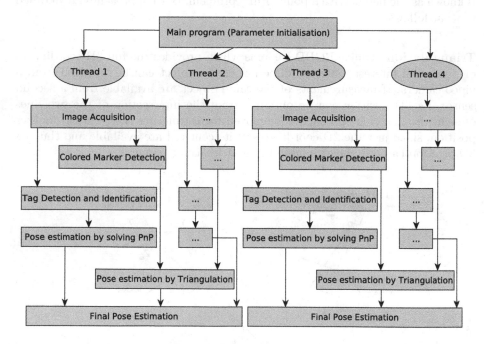

Fig. 8. Architecture of the real-time implementation of the processing algorithms.

4   Experimental Results

In order to validate the system, the experimental set-up of Fig. 9 was realised. The four cameras were installed in the optimal configuration discussed previously.

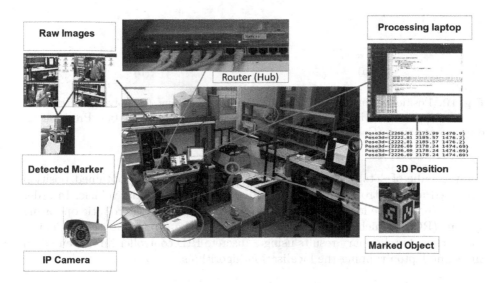

Fig. 9. Experimental set-up used for the localisation of a nano-drone CrazyFly.

Figure 10 presents an evaluation of the estimation error for the position obtained by solving the PnP problem. Figure 10a compares two estimation approaches when multiple tags are detected for the same object of interest. The first solution (green) solves the PnP problem for each tag separately and then calculates the object location as the average of the detected tags positions. In the second approach, all the detected tags are combined in a single vector and the PnP problem is then solved for the combined tags.

Figure 10b depicts the performance of the position estimation by solving the PnP problem using different sizes of the marking tags. As was expected, the precision of the estimation increases with the size of the tag. The detection algorithm was incapable of detecting the small tag, from distances exceeding two meters with an error around 5 cm for such a distance. The biggest tag on the other hand, was detectable even at a distance of more than 6 m, with an average error of less than 1 cm up to a depth of two meters, and less than 9 cm for a distance not exceeding 10 m.

Figure 11a presents a comparison of the precision of the triangulation and PnP methods. As this figure shows, the triangulation algorithm gave better estimation for all the tested positions in the scene.

To test the solution on a real platform the nano-drone CrazyFly was used. The latter, although having a very limited payload of merely 4 g, was able to

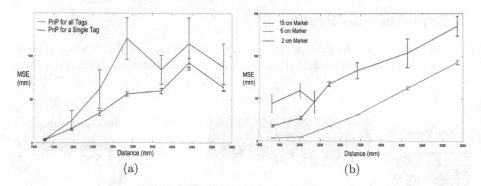

Fig. 10. Position estimation error using the PnP solution: (a) Solving PnP for all detected tags vs. solving PnP for each tag separately; (b) Estimation Precision for different tag sizes. (Color figure online)

carry a cube of 4 tags as well as the coloured marker. The obtained position was used as a control feedback to stabilise the altitude of the vehicle. In order to achieve that, the localisation solution was integrated to the robotic operating system (ROS), which facilitated the interface to the drone. Figure 11b depicts the obtained preliminary results using a discrete PID controller implemented on the same laptop running the localisation algorithms.

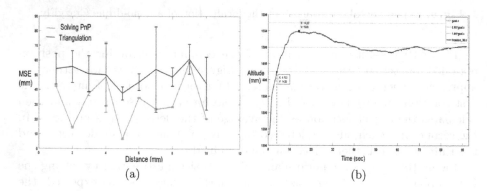

Fig. 11. Experimentation results: (a) Comparison of position estimation precision using triangulation and PnP methods; (b) Results of altitude visual servoing for the nano-drone CrazyFly.

5   Conclusion

Motivated by the actual high cost of motion tracking systems, this work targeted the development of a low-cost localisation solution for indoor environments. The proposed solution was based on a multi-camera system. We showed, through the experiments conducted, that a system composed of four ordinary IP cameras, can provide a precision of a couple of centimetres up to a distance of two meters. Challenges related to the real-time operation of the localisation system were overcome using the adequate implementation. This allowed processing the images of the four cameras in full rate on an ordinary laptop.

Future works target a more thorough evaluation of the positioning errors, in order to achieve an optimal fusion of the measurement obtained from the four cameras. Another aspect, which should enhance the precision and further reduce the processing time, is the introduction of a filtering approach which takes into account the previous position of the object of interest in both the detection and position estimation steps.

References

1. Caron, F., Duflos, E., Pomorski, D., Vanheeghe, P.: GPS/IMU data fusion using multisensor Kalman filtering: introduction of contextual aspects. Inf. Fusion **7**(2), 1–19 (2006). http://www.sciencedirect.com/science/article/pii/S1566253 50400065X
2. Fiala, M.: ARTag, a fiducial marker system using digital techniques. In: IEEE Computer Society Conference on Computer Vision and Pattern Recognition, pp. 590–596 (2005). http://ieeexplore.ieee.org/xpls/abs_all.jsp?arnumber=1467495
3. Groves, P.D.: Principles of GNSS, Inertial, and Multisensor Integrated. Artech House, California (2008)
4. Kato, H., Billinghurst, M.: Marker tracking and HMD calibration for a video-based augmented reality conferencing system. In: Proceedings of the 2nd IEEE and ACM International Workshop on Augmented Reality (IWAR 1999), pp. 85–94 (1999)
5. Kendoul, F.: Survey of advances in guidance, navigation, and control of unmanned rotorcraft systems. J. Field Robot. **29**(2), 315–378 (2012). https://doi.org/10.1002/rob. http://onlinelibrary.wiley.com/doi/10.1002/rob.20414/full
6. Lee, G.H., Achtelik, M., Fraundorfer, F., Pollefeys, M., Siegwart, R.: A benchmarking tool for MAV visual pose estimation. In: 11th International Conference on Control, Automation, Robotics and Vision, ICARCV 2010, vol. 1, pp. 1541–1546 (2010). https://doi.org/10.1109/ICARCV.2010.5707339
7. Lochmatter, T., Roduit, P., Cianci, C., Correll, N., Jacot, J., Martinoli, A.: SwisTrack-a flexible open source tracking software for multi-agent systems. In: IEEE/RSJ International Conference on Intelligent Robots and Systems, IROS 2008, pp. 4004–4010 (2008)
8. Oh, H., Won, D.Y., Huh, S.S., Shim, D.H., Tahk, M.J., Tsourdos, A.: Indoor UAV control using multi-camera visual feedback. J. Intell. Robot. Syst. Theory Appl. **61**(1–4), 57–84 (2011). https://doi.org/10.1007/s10846-010-9506-8
9. Olson, E.: AprilTag: a robust and flexible visual fiducial system. In: IEEE International Conference on Robotics and Automation, pp. 3400–3407. IEEE, May 2011. https://doi.org/10.1109/ICRA.2011.5979561. http://ieeexplore.ieee.org/lpdocs/epic03/wrapper.htm?arnumber=5979561
10. Visvanathan, R., et al.: Mobile robot localization system using multiple ceiling mounted cameras. In: IEEE SENSORS, pp. 1–4 (2015)
11. Schmidt, A., Kasiński, A., Kraft, M., Fularz, M., Domagała, Z.: Calibration of the multi-camera registration system for visual navigation benchmarking. Int. J. Adv. Rob. Syst. **11**(1) (2014). https://doi.org/10.5772/58471

12. Shim, J.H., Cho, Y.I.: A mobile robot localization using external surveillance cameras at indoor. Procedia Comput. Sci. **56**, 502–507 (2015). https://doi.org/10.1016/j.procs.2015.07.242. http://linkinghub.elsevier.com/retrieve/pii/S1877050915017238
13. Wagner, D., Schmalstieg, D.: Making augmented reality practical on mobile phones, Part 2. IEEE Comput. Graphics Appl. **29**, 6–9 (2009). http://ieeexplore.ieee.org/xpls/abs_all.jsp?arnumber=5167481

A New Shape Descriptor and Segmentation Algorithm for Automated Classifying of Multiple-morphological Filamentous Algae

Saowanee Iamsiri[1], Nuttha Sanevas[2], Chakrit Watcharopas[1],
and Pakaket Wattuya[1(✉)]

[1] Department of Computer Science, Kasetsart University, Bangkok, Thailand
{saowanee.ai,chakrit.w,pakaket.w}@ku.th
[2] Department of Botany, Kasetsart University, Bangkok, Thailand
fscintsv@ku.ac.th
http://cs.sci.ku.ac.th/, http://botany.sci.ku.ac.th/

Abstract. In our previous work on automated microalgae classification system we proposed the multi-resolution image segmentation that can handle well with unclear boundary of algae bodies and noisy background, since an image segmentation is the most important preprocessing step in object classification and recognition. The previously proposed approach was able to classify twelve genera of microalgae successfully; however, when we extended it to work with new genera of filamentous algae, new challenging problems were encountered. These difficulties arise due to a variety of the forms of filamentous algae, which complicates both image segmentation and classification processes, resulting in substantial degradation of classification accuracy. Thus, in this work we propose a modified version of our multi-resolution segmentation algorithm by combining them in such a way that the strengths of both algorithms complement each other's weaknesses. We also propose a new skeleton-based shape descriptor to alleviate an ambiguity caused by multiple morphologies of filamentous forms of algae in classification process. Effectiveness of the two proposed approaches are evaluated on five genera of filamentous microalgae. SMO is used as a classifier. Experimental result of 91.30% classification accuracy demonstrates a significant improvement of our proposed approaches.

Keywords: Microalgae image classification · Image segmentation ·
Shape descriptor · Skeleton

1 Introduction

Microalgae are important aquatic life forms and live in most aquatic ecosystems. Because of responding strongly to environmental changes, they have long been

Supported by Kasetsart University Research and Development Institute under Grant No. 146.58.

© Springer Nature Switzerland AG 2019
J. M. F. Rodrigues et al. (Eds.): ICCS 2019, LNCS 11540, pp. 149–163, 2019.
https://doi.org/10.1007/978-3-030-22750-0_12

used as biological indicators for an assessment of water quality [21]. In recent studies of diversity of microalgae [13, 14] the researchers found that some species of microalgae can tolerate to contaminated water resources and can be used to phytoremediate such contaminated sites. Thus, in water resource management, a regular assessment of water conditions allows early warning of deteriorating conditions that can be toxic to both humans and animals. Rapid, accurate identification of microalgae is therefore one of the most important issues in water resource management. While the need of identification of algae increases, only a few biological experts with taxonomy competence are available. Moreover, a task of identifying the species of algae is also time consuming. Thus, an automated algae image classification is essential.

An image segmentation is a process of separating objects of interest from an image background, e.g., detecting algae bodies. It is the most important preprocessing step in an automated image classification task, especially in algae image classification. Since a shape of algae is a key characteristic used for taxonomical identification of its genus, we expect the segmentation algorithm capable of producing segmented results with property that generates boundaries of algae bodies lying as close as possible to the true edges of algae bodies, and preserves important morphological features of algae as many as possible. Completeness of morphological features of the extracted shape will contribute to high quality of shape descriptors, which finally contributes to high classification accuracy. In our previous work on automated microalgae image classification [12] we proposed a segmentation algorithm (i.e. the multi-resolution edge detection) that can efficiently deal with problems of unclear boundary of algae bodies and noisy background. The proposed algorithm worked well with twelve genera (both non-filamentous and filamentous forms) of microalgae. Unfortunately, when applied to new genera of filamentous algae, the algorithm produced unexpected results.

Five genera of microalgae studied in this work are ones of the most commonly found in water resources of Thailand (as shown in Fig. 1). The two genera, i.e. *Anabaena* and *Oscillatoria*, are from the previous work, and the three new genera, i.e. *Spirogyra*, *Spirulina*, and *Anabaenopsis*, are newly collected. All five genera are filamentous algae. In biology, the filamentous forms of these algae are called trichomes. Trichomes can be found in many characteristics. In this work we categorize them into three groups according to their silhouette shape (as shown in Fig. 2) as following:

1. Smoothed boundary: trichomes in this group are composed of cylindrical cells. Thus, they have a smoothed contour along their elongate body. Microalgae in this group are *Oscillatoria* and *Spirogyra* (see Fig. 2b).
2. Crenate boundary: trichomes in this group are composed of spherical cells. So, they have notches at joints of spherical cells along their contours, and hence, the contour appears crenate along their length. Microalgae in this group is *Anabaena* (see Fig. 2a).
3. Spirally coiled boundary: trichomes in this group can be short and regularly spirally coiled like *Anabaenopsis* or irregularly spirally coiled, such as *Spirulina* (see Fig. 2d, e).

(a) *Anabaena* (b) *Oscillatoria* (c) *Spirogyra* (d) *Spirulina* (e) *Anabaenopsis*

Fig. 1. Five genera of filamentous algae in the study.

(a) Crenate   (b) Smoothed   (c) Smoothed   (d) S.Coiled   (e) S.Coiled

Fig. 2. Silhouette of algae in Fig. 1.

The main purpose of inventing the multi-resolution edge detection is that we want to extract algae body from a background, where its boundary lies as close as possible to the true edges of algae, so that the shape features of algae body will be preserved as many as possible. However, when facing with new microalgae of filamentous genera, the detected boundary of algae seems not quite fit enough to efficiently capture small details of boundary in each group. The following examples show the difficulties we faced when the detected boundary of algae lies further away from the true edges of algae: (i) when size of *Anabaena* in an input image is quite small or the trichomes are composed tightly of small spherical cells (see Fig. 3a), notches along its crenate boundary in the segmentation result are shallow. As a result, its boundary appears more like smoothed contour than crenate contour. Consequently, *Anabaena* is more likely to be misclassified to *Spirogyra* or *Oscillatoria*, and (ii) when segmenting a tiny spirally coiled algae (*Spirulina*), curls occurring along a spirally coiled boundary will appear like notches (see Fig. 3b), resulting in ambiguous classification between the two genera (*Spirulina* and *Anabaena*). Finally (iii), the multi-resolution edge detection was not designed to handle trichomes of spirally coiled form that curl in a circle (or a ring shape), and hence segmentation results of algae of this morphology wrongly turn into a coccoid form (see Fig. 3c). Thus, in this work we propose a new modified version of the multi-resolution edge detection so that segmentation results of algae with ring shape remains a ring shape, and notches along crenate boundary and curls of spirally coiled boundary are distinct. The experimental results show that our new segmentation algorithm does benefit a classification process by bringing the differences of the boundaries in the three groups clearer.

In addition, in order to alleviate an ambiguity of classifying ambiguous multiple morphologies of the trichomes forms, we propose a new shape descriptor computed based on a skeleton of a shape. The experimental results show that by using our new shape descriptor, the classifier can efficiently cope with several cases of ambiguity, for example, (i) discriminate shapes with crenate boundary with tiny notches from smoothed boundary shapes, and (ii) discriminate shape with spirally coiled boundary with tiny curls from shapes of crenate boundary.

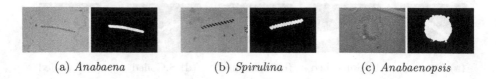

(a) *Anabaena*   (b) *Spirulina*   (c) *Anabaenopsis*

Fig. 3. Incorrect segmentation results of the old approach [12].

2 Related Works

Microalgae segmentation as well as automated algae classification are a very difficult task due to various factors: not only a wide variation of characteristics of algae in nature, but also the poor conditions of microscopic images (e.g. low contrast due to non-uniform illumination conditions, strong background noise arising from lights scattered, presence of various extraneous objects in the background). Therefore, algae image classification and its associated problems become challenging for applications of computer vision [16]. Various image processing and machine learning approaches have been proposed to tackle these problems.

Edge-based segmentation approach is one of the most popular approaches that widely uses in microalgae image segmentation because of its robustness to noise. Li et al. [7] proposed to use Sobel edge detection for semi-automated microorganism segmentation. Since Sobel edge detection often produces discontinuous boundaries, Santhi et al. [16] and Promdaen et al. [12] proposed to use Canny edge detection on a result of Sobel edge detection in order to improve segmentation results. A Canny edge detection was also used by [8,9]. Because most of edge detection methods naturally requires postprocessing for linking discontinuities of edges to form closed contours of objects, deformable model approaches become more attractive since they produce results as a closed contour. Giraldo-Zuluaga et al. [5] proposed to used active contour, while Borges et al. [2] proposed to use level set for segmenting microalgae images. However, edge-based active contours are sensitive to noise and may omit minute features and blurry boundaries during energy minimizing. Therefore, this approach is not suitable for our problem at hand, in which input images may contain both blurry-boundary objects and noisy background. Besides edge-based approach,

Zheng et al. [24] proposed an interesting intensity-based image segmentation approach, namely GSDAM. GSDAM separates an object from a background by considering the connectivity and directionality of lineal structures of intensity information. It can handle well with low contrast as well as noisy background. Unfortunately it is particularly designed for segmenting objects with thin and long shape.

Feature extraction and description are also of important for automated algae classification. Shape descriptors are inevitable primary features for this task, while size and texture depend on the nature of the problem. For example, the works of [1, 7, 9, 16, 22] could use size of objects as one of key features because their input images came from a single source, so size of algae in an image are unvarying. In contrary to our work where input images come from several sources with unknown imaging conditions, the size of algae are varying in a wide range, and thus using size as a key feature merely complicates the problem. Commonly used shape descriptors are such as geometric shape descriptors [7, 9, 16, 22], Fourier descriptors [12], and moment invariant [5, 22]. A major drawback of moment invariants is that higher-order moments are very sensitive to noise. Texture is also one of the most important characteristics used for identifying algae. Most commonly used texture descriptors in this area of problems are Haralick [5, 8] and local binary pattern (LBP) [5, 22]. Because of the way LBP captures texture information (i.e. considering difference of pixel intensity, while neglecting magnitude of intensity), its major drawback is a limitation of capturing the texture information.

For a classification task the most commonly used classifiers are support vector machine [5, 7, 12, 22] and an artificial neural network [5, 8, 9, 16]. For more review about microorganism classification we refer to an interesting work of Li et al. [7]. They provide a comprehensive review of 240 related works on the applications of content-based microscopic image analysis in microorganism classification domains.

3   A New Segmentation Algorithm

In our previous work [12] we proposed two segmentation algorithms, namely the single-resolution edge detection and the multi-resolution edge detection. The single-resolution edge detection was purposely designed for segmenting microalgae of coccoid form, while the multi-resolution edge detection was purposely designed for segmenting microalgae of filamentous and tube-like forms. The rationale behind our idea of treating the two groups of algae differently is that algae of coccoid form require a small degree of image smoothing in an edge detection process in order to preserve their spines or flagella, while filamentous and tube-like algae require high degree of image smoothing so that an edge detector can extract an algae body whose boundary lies as close as possible to the true algae edges. In this work we still adhere to this idea and attempt to extract an algae body whose boundary fits with the true boundary of algae as much as possible.

Since the multi-resolution edge detection already performs well with filamentous algae with smoothed boundary and has strong ability to deal with a noisy background, our plan is to combine a segmentation result of the multi-resolution edge detection with one of a new proposed algorithm. The new segmentation algorithm now combines the previous abilities with one that can handle a wide variety of morphology of filamentous forms. The new algorithm is called a generalized multi-resolution segmentation algorithm, since it can deal with a wide range of filamentous forms. The new algorithm is composed of four main steps. The details of each step are described as follow.

Step 1: Preprocessing

A preprocessing is a process of preparing an input image to have suitable properties for successive processes, e.g. image segmentation and feature extraction. The preprocessing that performs in this work is similar to one described in the previous work [12]. Only a process of converting an RGB input image into a grayscale image is reconsidered. In this work we found that edges of algae in a blue channel of input RGB image appear more distinguishable from a background than those in a grayscale image obtained by a color-to-gray transformation. Thus, in this work we propose to use a blue channel of input image as a grayscale version of input image, instead of a grayscale image obtained by a color-to-gray transformation. And since edges of algae play an important role in a segmentation process, we perform edge enhancement on a grayscale image using the Contrast-Limited Adaptive Histogram Equalization (CLAHE) [11] to improve its acutance.

Step 2: Foreground/Background Segmentation

The objective of this step is to extract algae body from an image background in a way that the contours of algae bodies lie as close as possible to the true edges of algae bodies. To achieve this we hence decide to perform following.

1. We use a thresholding method instead of edge detection. The adaptive thresholding method is applied to a grayscale image, transforming the grayscale image into a binary image, where regions of algae bodies are assigned as a foreground (pixels with value 1), leaving background regions to remain as a background (pixels with value 0). After thresholding, a few number of postprocessing tasks are required to improve a quality of this intermediate segmentation result.
2. Small noisy objects are removed using morphological opening operator (we note here that the result of this stage will be used again in Step 3).
3. We then can see that thresholding method has classified some parts of the algae body as background pixels, causing small holes in foreground regions. Small holes are filled (turning background pixels from 0 to 1) by performing morphological closing operator [18] with a small-size structuring element

(SE). Using a small-size SE in this step is preferable to a large-size SE because we want to prevent unwanted objects from being connected to regions of algae bodies.

4. The multi-resolution edge detection proposed in our previous work plays an important role in this task. We utilize its ability to cope with noisy background. Unwanted objects are removed from the segmentation result by intersecting them with the segmentation result produced by the multi-resolution edge detection. As a result, segmentation results with pleasantly clean background is obtained.

5. Finally, large holes within foreground regions must be filled. Note that these large holes do not commonly occur. Only algae with transparent gaps in their body, e.g. *Oscillatoria* and *Spirogyra*, can produce these holes. In order to fill large holes, the result from multi-resolution edge detection is required. Firstly, we would like to note that the multi-resolution edge detection is an edge-based image segmentation approach, while the new algorithm is a region-based image segmentation approach. Edge-based segmentation segments an image by considering a discontinuity of image pixels to identify edges (or boundary) of objects, while region-based segmentation considers a similarity of image pixels and groups them together into regions (or objects). Hence, the results of these two approaches are complement to each other in a way that one's imperfect result may be completed by a result of another [10]. In our situation an advantage of the new segmentation algorithm is the ability to produce segmentation results in which boundary of algae body lie very closely to the true algae boundary. However, this ability comes with an unavoidable drawback; it often yields an indented region of algae body when segmenting algae with transparent gaps in their body. On the other hand, an advantage of our old segmentation algorithm is that it often yields a full regions of algae body, but boundary of extracted algae body lies a bit farther away from the true algae edges. We thus utilize the advantages of both algorithms to complement their drawbacks by combining the two results together by means of morphological reconstruction, i.e. reconstruction by erosion. Reconstruction by erosion involves two images, i.e. a marker image used as a starting condition of transformation and a mask images used to constrain the transformation. So we use a segmentation result of the old algorithm as a marker image and a result of the new algorithm as a mask image. Then a process of reconstruction by erosion begins by iteratively eroding away the boundaries of foreground regions (algae bodies) in a marker image until they fit a foreground of a mask image or reach a predefined number of iterations. We finally obtain a final segmentation result in which a full region of algae body is obtained and the boundary of algae body lies very closely to the true boundary of algae.

The whole process in this step is shown in Fig. 4.

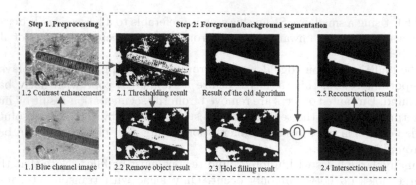

Fig. 4. The whole process of Step 2. Both horizontal and vertical-line SE of size 3 pixels are used in morphological operators.

Step 3: Background Removing (for a Ring-Like Form)

In this final step only images of filamentous algae whose bodies curl in a circle (a ring-liked shape) are involved. In this stage we can see that a background region inside a ring shape (called *area A* for short) is incorrectly assigned as part of algae body. We thus have to remove this area to form a correct ring shape. This process begins with extracting an area A and then removing it from a final segmentation result obtained from the previous step. Area A can be extracted by performing an image subtraction of the final segmentation result and the intermediate result obtained from Step 2. A subtraction result (area A) is shown in Fig. 5. We then enhance its quality by performing a hole filling operation before subtracting it from the final segmentation result, in order to achieve a new correct one (i.e. a ring-shape algae).

Since only images of ring-shape filamentous algae are involved in this step, we have to discriminate algae of ring-liked forms from the others. We simply accomplish this by using a global shape measure called *Eccentricity* (e.g. the ratio of the length of major axis to the length of minor axis of the shape). A value of eccentricity is 0 if a shape is a circle; and the value of 1 indicates a shape of a line segment. Eccentricity is computed on the final segmentation result obtained from Step 2.

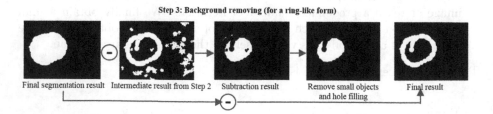

Fig. 5. The whole process of Step 3. 2-pixels disk SE are used.

4   A New Skeleton-Based Shape Descriptor

In general, the accuracy of classification and recognition system depends heavily on the quality of relevant object features (or descriptors) used in a learning process. Conventionally, characteristics of algae, such as size, shape, colors, and texture, are routinely used for taxonomical recognition. However, size and color features do not seem very useful for our problem at hand. In our task microscopic images received from users are produced from several imaging systems with different settings. Consequently, sizes of algae (even in the same genus) are varied in a wide range depending on a magnifying power used in an image acquisition process. Similar situation also happens with the colors of algae in microscopic image. Colors of algae not only typically vary in nature depending on their growing environment, but also vary significantly according to illumination adjustment during imaging process. Hence, in this work shape and texture characteristics are the key features used in a learning process.

Global shape measures, such as geometric features (e.g. compactness, rectangularity, etc.) are a convenient way to describe shapes and are simple to compute [23]. Unfortunately, these features are unable to capture a small difference along a boundary of two different shapes, e.g. shapes with crenate boundary with tiny notches and shapes with wavy boundary with tiny curls. A more sophisticated shape descriptor such as Fourier Shape Descriptors also requires high computational time because a high number of Fourier coefficients must be used, in order to be able to distinguish small differences along the boundary of two different shapes.

In this paper we propose a new shape descriptor that utilizes the weakness of skeleton. The rationale behind our idea is that skeleton is intrinsically sensitive to small changes in a shape's boundary. A small change in the shape's boundary can cause a large change in the skeleton [4,19]. We hence utilize the sensitivity of skeleton to capture small differences along the boundary of two different shapes. A new skeleton-based shape descriptor is computed on a binary image obtained from a segmentation process. The boundary of algae body in a binary image needs to be smoothed (e.g. by morphological opening operator) before we compute a skeleton, in order to avoid excessive amount of unwanted spurs in the resulting skeleton. Unavoidable spurs will be suppressed afterwards by using morphological erosion. Then, three quantities of skeleton-based shape descriptor that measure a wavy degree of the skeleton can be computed as following.

1. *Waviness.* Waviness is a ratio between the length of skeleton and the length of skeleton displacement. A value of waviness is equal to 1 if the skeleton is a straight line, and closer to 0 when the skeleton is highly wavy.
2. *Number of Peaks.* Skeletons of algae with spirally coiled shape will look like sinusoidal wave. Peaks are sinusoidal peaks or location of local maxima of the curve. We count them as peaks if the height of them are higher than a predefined threshold (e.g. 3 pixels).
3. *Average Height of Peaks.* Average height of peaks is a ratio between the total height of all peaks and the number of peaks.

5   Experiments

5.1   Algae Image Dataset

Microscopic algae images of five genera, namely Anabaena, Oscillatoria, Spirogyra, Spirulina, and Anabaenopsis, used in the experiments are collected from various sources. The main sources are the study area of Bung Borapet fresh water source, Nakhon Sawan province [20] and the study area of Khlong Kamphuan watershed, Ranong province [17] in the research projects conducted by Department of Botany, Kasetsart University, Thailand. Other sources include Metropolitan Waterworks Authority and online algae image database from the internet. The data set comprises of 300 images, 60 images per genus.

5.2   Automated Algae Image Classification Framework

An automated algae image classification proposed in the previous work [12] is based on a supervised learning approach that takes an algae image as input and classifies it into one of multiple genera. The classification system composes of three main components, namely image segmentation, feature extraction, and classifier (prediction model) training process. The system works as follows. After separating an algae body from an image background using a segmentation algorithm, multiple features (or descriptors) are computed from the algae body. Multiple features are then formed into a feature vector (one feature vector represents one input image). Finally, a training process takes a set of feature vectors as an input. For supervised learning a training process needs additional input, i.e. a list of labels (names of genus) associated with each input image. A classifier will then be trained based on these sets of inputs. Once the classifier is trained it can be used to predict a genus of an unknown algae image in the future. Details of image features and a classifier used in a learning process will be next described.

5.3   Image Features

As we discussed earlier, only shape and texture features will be used in the learning process. It is interesting to note that both shape and texture features play an important role in this task. Shape features will play a key role when texture features are unavailable, for example, when a size of algae in an input image is very small or when an input image with out-of-focused algae is obtained. On the other hand, texture features will play a key role when shape features are ineffective, for example, when we have to discriminate two algae of the same shape (e.g. *Oscillatoria* and *Spirogyra*). Hence a suggested approach is to combines shape features with texture features to form more powerful discriminating features used for training a prediction model. Shape descriptors used in this work are three geometric shape features plus three quantities of our new skeleton-based shape features. The three geometric shape features [23] are *Solidity, Eccentricity*, and *Convexity* (For more details of these descriptors we refer the reader to the original work). Similar to our previous work [12], texture descriptors used in this work are the thirteen texture descriptors proposed by [6]. The thirteen descriptors are derived from a *gray level co-occurrence matrix* (GLCM). This matrix characterizes texture of an object (or image) by calculating the frequencies of pixel pairs with specific values and in a specified spatial relationship occurred in an image. Based on our preliminary experiments, it suggests that the most effective number of gray-levels used for computing GLCM is 256 and a direction of calculation is set to 0°. The thirteen descriptors include *Angular Second, Contrast, Correlation, Sum of Squares, Inverse Difference Moment, Sum Average, Sum Variance, Sum Entropy, Entropy, Difference Variance, Difference Entropy*, and *Information Measure of Correlation 1, 2*. Hence, the total number of features we used in a training process are 19 features.

5.4   Classifiers

As suggested from our previous work [12], we use the Sequential Minimal Optimization (SMO) algorithm for training a support vector classifier using scaled polynomial kernels in a classification process since it has been successfully applied in many application domains such as medical image analysis, handwritten character recognition, and speech recognition. We set a value of complexity parameter of SMO to 4. Due to a limited data sample, the k-fold cross-validation method is applied in an evaluation process. In all experiments the value of k is set to 4. So, our image dataset is randomly partitioned into 4 subsets of equal size (i.e., 15 images per subset). Of the 4 subsets, a single subset is retained as the test data used for testing the trained classifier, and the remaining 3 subsets are used as training data. The training process repeatedly runs 4 times, each time with different subset of data. The 4 classification results will be averaged to produce a single classification accuracy.

5.5   Classification Results

In this section the three experiments with three different experimental settings are reported. The first experiment is a baseline for evaluating an improvement of the new proposed approaches. The old version of algae image classification [12] is applied with new algae dataset (using the old segmentation algorithm and the old set of shape features), and we let its result be the baseline's performance. The second experiment is conducted by employing the new segmentation algorithm; however, only 16 features are used in a training process. The three features of the new skeleton-based shape descriptor are omitted. Finally, the third experiment is conducted using the new segmentation algorithm and all 19 features (including the three new features). The experimental results of the three experiments are shown in Tables 1, 2, and 3 respectively.

The average classification accuracy of our old approach is 84.00% (Table 1). It can be seen that algae of the first four classes are highly misclassified over each other's classes. The main cause of these mistakes is a consequence of inaccurate image segmentation. When examining the segmentation results of *Anabaena*, we found that crenate boundaries of algae's shape barely appeared. Consequently, shape of *Anabaena* becomes more similar to those of *Oscillatoria* and *Spirogyra*. Similarly, shape of *Spirulina* with tiny curls looks more like shape of *Anabaena* than those of the self-genus. These ambiguities cause a classification task to be more complicated.

Experimental results of the second experiment are shown in Table 2. The average classification accuracy is 86.30%. Even though the average accuracy does not significantly improve from the baseline performance, the ambiguity among the three classes, i.e. *Anabaena*, *Oscillatoria*, and *Spirogyra*, significantly decreases. A group of algae with straight boundary (*Oscillatoria* and *Spirogyra*) is much better discriminated from a group of algae with wavy boundaries (*Anabaena* and *Spirulina*). Only within a group of algae with wavy boundaries (*Anabaena*, *Spirulina*, and *Anabaenopsis*) are confusedly classified to other genera. This classification result explicitly demonstrates the advantage of using efficient segmentation algorithm - unsophisticated case of ambiguity can simply be alleviated.

The third experiment demonstrates that a more sophisticated case of ambiguity (i.e. an ambiguity among a group of shapes with wavy boundaries.) can be alleviated by means of involving a discriminative shape descriptor in the learning process. Experimental results of the third experiment are shown in Table 3. The average classification accuracy is significantly improved to 91.30%. This result strongly demonstrates the effectiveness of the proposed skeleton-based shape descriptor.

5.6 Discussion

In the third experiment, fewer algae in each class are misclassified into incorrect genus. After examining the segmentation results of these samples, we found nothing wrong with the results (example of them are shown in Fig. 6). We thus presume that because a number of images in our dataset is quite limited, hence the training data are not a good representative of algae in a particular genus. On the other hand, texture descriptors, considering as a statistical approach, used in a learning process are probably not sufficient to capture a high variation of texture features of algae in each genus. In addition, shape descriptors may not be effective enough to capture very small detail of object boundary. Finally, it

Table 1. A confusion matrix of the old version of algae image classification [12].

Genus	Anabaena	Oscillatoria	Spirogyra	Spirulina	Anabaenopsis	Accuracy
Anabaena	53	3	4	0	0	88.30%
Oscillatoria	10	42	6	2	0	70.00%
Spirogyra	5	7	48	0	0	80.00%
Spirulina	8	0	0	50	2	83.30%
Anabaenopsis	0	0	0	1	59	98.30%
Average accuracy						84.00%

Table 2. A confusion matrix of the new approach with 16 features (without the new shape descriptor).

Genus	Anabaena	Oscillatoria	Spirogyra	Spirulina	Anabaenopsis	Accuracy
Anabaena	48	4	2	6	0	80.00%
Oscillatoria	5	53	2	0	0	88.30%
Spirogyra	4	4	52	0	0	86.70%
Spirulina	8	0	0	50	2	83.30%
Anabaenopsis	0	1	0	3	56	93.30%
Average accuracy						86.30%

Table 3. A confusion matrix of the new approach with the new shape descriptor.

Genus	Anabaena	Oscillatoria	Spirogyra	Spirulina	Anabaenopsis	Accuracy
Anabaena	52	3	1	4	0	86.70%
Oscillatoria	4	54	2	0	0	90.00%
Spirogyra	3	2	54	1	0	90.00%
Spirulina	4	0	0	54	2	90.00%
Anabaenopsis	0	0	0	0	60	100.00%
Average accuracy						91.30%

is interesting to note that our input images are collected from various sources. Hence size, colors, and illumination of algae in these input images are varied in a wide range. Nevertheless, our new segmentation algorithm is capable of segmenting them successfully in most case, which indicates the effectiveness of the proposed algorithm.

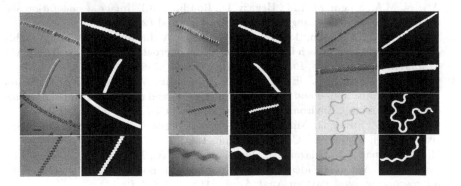

Fig. 6. Example segmentation results of incorrect classification.

6   Conclusion

This paper presented a new image segmentation algorithm called a generalized multi-resolution that provides several advantages. Firstly, it can produce segmentation results in which boundary of algae fits well with the true boundary of algae. Consequently, important morphological features of algae, even in cases of very small algae or very small detail of algae boundary, are preserved. This allows a shape description method to compute high accurate shape descriptors, resulting in high classification accuracy. Secondly, the proposed segmentation algorithm can efficiently handle multiple morphologies of filamentous forms of algae. Lastly, the proposed segmentation algorithm is simple to compute and requires low computational time. So it can be used in a real-time environment effectively. In addition, we also proposed a new skeleton-based shape descriptor that effectively alleviates an ambiguity of multiple morphologies of filamentous forms of algae in a classification process. The proposed shape descriptor is also simple to compute; thus, it can also be applied in a real-time system.

It's also worth noting that biological images are often far more difficult to be processed and recognized than images of daily-life [3]. Image features and object descriptors are of important, especially in the domain where availability of image data is limited. Learning for selecting relevant image features to achieve a good recognition accuracy is still vital [15]. Even a deep learning on limited data still requires a special treatment to improve the accuracy of detection. Image augmentation, however, may help increase the amount of relevant data for deep learning, but in our case augmenting data couldn't be beneficial since algae exhibit diverse morphology.

References

1. Atteya, M.A., Salem, M.A.M., Hegazy, D., Roushdy, M.I.: Image segmentation and particles classification using texture analysis method (2016)
2. Borges, V.R.P., Hamann, B., Silva, T.G., Vieira, A.A.H., Oliveira, M.C.F.: A highly accurate level set approach for segmenting green microalgae images. In: SIBGRAPI Conference on Graphics, Patterns and Images, pp. 87–94 (2015)
3. Coltelli, P., Barsanti, L., Evangelista, V., Frassanito, A.M., Gualtieri, P.: Water monitoring: automated and real time identification and classification of algae using digital microscopy. Environ. Sci. Processes Impacts **16**, 2656–2665 (2014)
4. Cornea, N.D., Silver, D., Min, P.: Curve-skeleton properties, applications, and algorithms. IEEE Trans. Vis. Comput. Graph. **13**(3), 530–548 (2007)
5. Giraldo-Zuluaga, J., Salazar, A., Diez, G., Gomez, A., Martínez, T., Vargas, J.F., Peñuela, M.: Automatic identification of scenedesmus polymorphic microalgae from microscopic images. Pattern Anal. Appl. **21**(2), 601–612 (2018)
6. Haralick, R.M., Shanmugam, K., Dinstein, I.: Textural features for image classification. IEEE Trans. Syst. Man Cybern. **3**(6), 610–621 (1973)
7. Li, C., Shirahama, K., Grzegorzek, M.: Application of content-based image analysis to environmental microorganism classification. Biocybern. Biomed. Eng. **35**(1), 10–21 (2015)

8. Luo, Q., Gao, Y., Luo, J., Chen, C., Liang, J., Yang, C.: Automatic identification of diatoms with circular shape using texture analysis. JSW **6**(3), 428–435 (2011)
9. Mosleh, M.A.A., Manssor, H., Malek, S., Milow, P., Salleh, A.: A preliminary study on automated freshwater algae recognition and classification system. BMC Bioinformatics **13**, S25 (2012)
10. Pavlidis, T., Liow, Y.: Integrating region growing and edge detection. IEEE Trans. Pattern Anal. Mach. Intell. **12**(3), 225–233 (1990)
11. Pizer, S.M., et al.: Adaptive histogram equalization and its variations. Comput. Vis. Graph. Image Process. **39**(3), 355–368 (1987)
12. Promdaen, S., Wattuya, P., Sanevas, N.: Automated microalgae image classification. Procedia Comput. Sci. **29**, 1981–1992 (2014)
13. Renuka, N., Sood, A., Ratha, S.K., Prasanna, R., Ahluwalia, A.S.: Evaluation of microalgal consortia for treatment of primary treated sewage effluent and biomass production. J. Appl. Phycol. **25**(5), 1529–1537 (2013)
14. Renuka, N., Sood, A., Ratha, S.K., Prasanna, R., Ahluwalia, A.S.: Nutrient sequestration, biomass production by microalgae and phytoremediation of sewage water. Int. J. Phytorem. **15**(8), 789–800 (2013)
15. Rezaeilouyeh, H., Mollahosseini, A., Mahoor, M.H.: Microscopic medical image classification framework via deep learning and shearlet transform. J. Med. Imaging **3**(4), 044501 (2016)
16. Santhi, N., Pradeepa, C., Subashini, P., Kalaiselvi, S.: Automatic identification of algal community from microscopic images. Bioinform. Biol. Insights **7**, BBI.S12844 (2013)
17. Sarabol, S., Vajrodaya, S., Ngernsaengsaruay, C., Sanevas, N.: Diversity of algae in Khlong Kamphuan watershed, Kamphuan sub district region, Suk Samran district, Ranong province. Thai J. Bot. **2**(Special Issue), 33–45 (2010)
18. Soille, P.: Morphological Image Analysis: Principles and Applications, 2nd edn. Springer, New York (2003). https://doi.org/10.1007/978-3-662-05088-0
19. Spitzner, M., Gonzalez, R.: Shape peeling for improved image skeleton stability. In: IEEE International Conference on Acoustics, Speech and Signal Processing, pp. 1508–1512 (2015)
20. Sudthang, P.: Diversity of algae and water quality assessment in sediment areas at Bueng Boraphet. Master's thesis, Botany Department, Kasetsart University (2011)
21. Sudthang, P., Vajrodaya, S., Suwanwong, S., Sanevas, N.: Diversity of algae in Bueng Boraphet, Nakhon Sawan province. Thai J. Bot. **2**(Special Issue), 21–31 (2010)
22. Tao, J., Cheng, W., Boliang, W., Jiezhen, X., Nianzhi, J., Tingwei, L.: Real-time red tide algae recognition using SVM and SVDD. In: IEEE International Conference on Intelligent Computing and Intelligent Systems, vol. 1, pp. 602–606 (2010)
23. Yang, M., Kpalma, K., Ronsin, J.: A survey of shape feature extraction techniques. In: Yin, P.Y. (ed.) Pattern Recognition, pp. 43–90. IN-TECH (2008)
24. Zheng, G., Zhao, H., Sun, X., Gao, H., Ji, G.: Automatic setae segmentation from Chaetoceros microscopic images. Microsc. Res. Tech. **77**, 684–690 (2014)

Application of Hierarchical Clustering for Object Tracking with a Dynamic Vision Sensor

Tobias Bolten[1]([✉]), Regina Pohle-Fröhlich[1], and Klaus D. Tönnies[2]

[1] Institute of Pattern Recognition, Hochschule Niederrhein - University of Applied Sciences, Krefeld, Germany
{tobias.bolten,regina.pohle}@hs-niederrhein.de
[2] Department of Simulation and Graphics, University of Magdeburg, Magdeburg, Germany
klaus@isg.cs.uni-magdeburg.de

Abstract. Monitoring public space with imaging sensors to perform an object- or person-tracking is often associated with privacy concerns. We present a Dynamic Vision Sensor (DVS) based approach to achieve this tracking that does not require the creation of conventional grey- or color images. These Dynamic Vision Sensors produce an event-stream of information, which only includes the changes in the scene.

The presented approach for tracking considers the scenario of fixed mounted sensors. The method is based on clustering events and tracing the resulting cluster centers to accomplish the object tracking. We show the usability of this approach with a first proof-of-concept test.

Keywords: Object tracking · Dynamic Vision Sensor · Event clustering

1 Introduction

In the field of computer-vision automated object detection and tracking are challenging topics. Over the past decades, various approaches have been developed. In [1–3] methods under evaluation of the optical flow are considered, whereas in [4–6] variations of the Kalman filter and in [7–9] techniques of deep-learning based approaches are utilized.

However, these approaches are using conventional, frame-based (grey-value) images captured by classical CCD- or CMOS imagers [10]. Depending on the domain of application, this type of recording can quickly lead to problems with the privacy awareness of potential users (especially in in-home environments) or in the case of public places in complex legal issues [11].

This work is part of the project "plsm" which is founded by the European Regional Development Fund under the grant number EFRE-0801082.

J. M. F. Rodrigues et al. (Eds.): ICCS 2019, LNCS 11540, pp. 164–176, 2019.
https://doi.org/10.1007/978-3-030-22750-0_13

The described use case of object tracking in this paper is part of a project whose goal is to improve the planning of public open space by including the specific user behavior in the basic urban design process. For this purpose, it is planned to construct a distributed, sensor-based system in order to automatically derive various parameters of the considered area. In the first step, we focus on the task of object detection and tracking to derive information about the number of users and their movements.

To overcome privacy concerns and restrictions by laws, we suggest the utilization of an alternative image sensor, the so-called Dynamic Vision Sensor (DVS). This type of sensor is biological inspired and works not in a frame-based manner. Instead it transmits the changes within a scene in an asynchronous way when they happen.

The paper is structured as follows: In Sect. 2 the DVS and its functionality are described. A filtering and clustering approach for object tracking based on a DVS is presented in the subsequent section. In Sect. 4 a simple proof-of-concept comparison of the DVS solution to a classical image-processing solution is presented. Section 5 concludes with a short summary.

2   Dynamic Vision Sensor

CCD- or CMOS imagers typically operate at a fixed frame rate and produce a constant data stream independent of changes in the considered scene. This can lead to high redundancies in the individual captured frames. In contrast to this, the pixels of a Dynamic Vision Sensor operate independently and asynchronously, based on relative light intensity changes in the scene. This approach is, as a part of neuromorphic engineering, borrowed from biology. For this reason, DVSs are also called "silicon retinas".

Each pixel of a DVS only transmits an information (called an *event*) when a change in intensity greater than a pre-defined threshold is measured. As a consequence, a static scene generates no sensor output at all.

The output of this sensor is typically encoded as a sparse stream in an Address-Event Representation (AER). Each event in this stream includes [12]:

(x, y)-**Coordinate:**
The pixel coordinate in the sensor array that triggered the event.
Timestamp:
The time of the occurrence of the event. Typically, in a resolution range of milliseconds.
Meta information:
Depending on a specific sensor model, e.g. the polarity of an event (ON: change from dark $\rightarrow$ bright, OFF: change from bright $\rightarrow$ dark) or the brightness value at the moment of the event generation (greyscale value).

Lichtsteiner et al. mentions in [12] that the first sensor of this type was developed in the mid-1990s. An overview of subsequent developments can be found in [13]. In the scope of this work, we used the "CeleX-IV" sensor, which is developed by Hillhouse Technology [14]. This sensor offers a 768×640 pixel array resolution, a high-speed AER output with 200Meps (events per second) and a high dynamic range of ≈ 120 dB.

Figure 1 shows an example scene captured with this sensor. In Fig. 1a the scene is displayed as a greyscale image, whereas Fig. 1b shows the visualization of a 60 ms time window of event data as a binary image. Each pixel, where at least one event occurred in the time window, is set to white. Figure 1c illustrates the spatiotemporal information within the stream of events. Each of the six colors in this figure represents a time window of 60 ms (total 360 ms) of events. The burst of events at the position of the moving human and the tree waving in the wind are clearly visible in these visualizations.

(a) Greyscale reference

(b) Binary event visualization in a 60ms time window

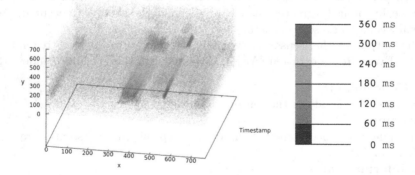

(c) Spatiotemporal visualization of six continuous 60ms time windows (color-coded)

Fig. 1. Example visualizations of AER data captured with a CeleX IV-DVS (Color figure online)

3   Event-Clustering as a Basic Tracking Approach

Based on the inherent properties of the event-based vision sensor, we propose the processing-chain in Fig. 2 to achieve a tracking of moving objects. For this we use a neighborhood-based event filter as a pre-processing step, followed by a hierarchical clustering and a tracing of cluster centroids. These steps are explained in the following sub-sections. Our implementation is based on slicing the continuous event-stream in non-overlapping blocks of a fixed time length (following referred as sliding time window) and the processing of each of these blocks.

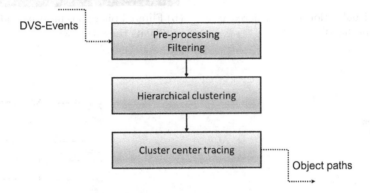

Fig. 2. Suggested processing-chain for object tracking

In addition to the privacy benefits (no grey- or color-value information of the scene is needed) offered by the sparsely populated event stream of a DVS, this approach offers the possibility to achieve a solution with little need of computational and power resources. An important point is, that the static background of the scene does not have to be considered. Especially in the context of sensor networks this can be a great advantage.

3.1   Event-Filtering

Figure 1b and c clearly shows that there is significant sensor noise in the recorded signal, which prevents a sensible use of clustering approaches. Therefore, we suggest a simple filtering step exploiting the spatial and temporal neighborhood for pre-processing. For each event, the number of other events in the von-Neumann neighborhood (4-neighborhood) within the current sliding time window is calculated as

$$f(\text{event}_x, \text{event}_y) = \sum_{t \in \text{time window}} \text{count}(\text{event}_x \pm 1, \text{event}_y \pm 1, t) \qquad (1)$$

Figure 3a clarifies the considered spatio-temporal neighborhood for an event. An event is rejected when $f(\text{event}_x, \text{event}_y) < \text{threshold}$.

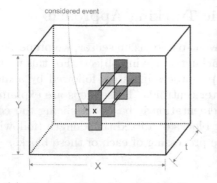

(a) Visualization of considered filter-neighborhood

(b) Filtered binary event visualization in a 60ms time window

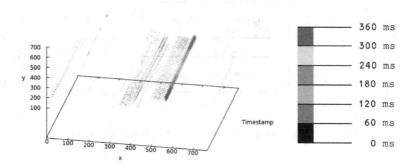

(c) Result of the filter applied to data from Figure 1c

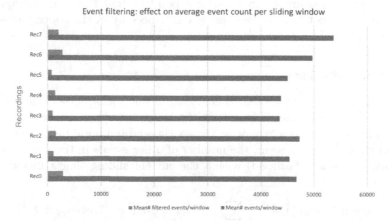

(d) Filter effect on the event count per sliding time window

Fig. 3. Visualization of the event filtering step (Color figure online)

We suggest setting the threshold value depending on the width of the underlying sliding time window. We have chosen the value empirically and set it to 1/8 of the sliding window in ms. The filtering result is shown in Figs. 3b and c (compare with the unfiltered version in Fig. 1).

This filtering drastically reduces the number of events which must be processed in the next step, while preserving most of the events from the desired objects. The effect on the average number of events per sliding window is shown exemplarily in Fig. 3d based on various recordings (compare with Sect. 4). Within this practical example a reduction of about 96% on the average event count per sliding window was achieved.

3.2   Hierarchical Clustering

The next step in the processing chain consists of clustering the pre-filtered events to get semantic related groups of events. As the number of clusters (moving objects) in the scene is not known a priori, a hierarchical clustering approach is used.

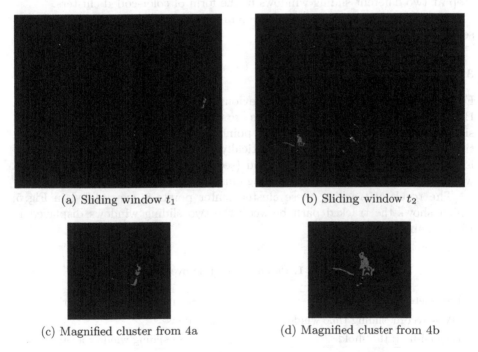

(a) Sliding window t_1 (b) Sliding window t_2

(c) Magnified cluster from 4a   (d) Magnified cluster from 4b

Fig. 4. Color-coded result of clustering at different sliding windows (Color figure online)

We use a hierarchical *single-link* clustering (provided by the "fastcluster"-library [15]) based on the euclidean distance of the (x, y)-coordinates of the pre-filtered events. The clustering break point is controlled by a pre-defined

Fig. 5. Highlighted path of tracked center points between sliding window t_1 and t_2 (compare with Fig. 4) (Color figure online)

cutoff-distance. Only clusters consisting of a minimum number of events are considered for further processing. Figure 4 illustrates the result of the clustering step at two different sliding windows in the form of color-coded clusters.

The Table 1 summarizes all parameters and their selected values of our presented approach.

3.3   Cluster Path Tracking

For each cluster resulting from the previous step, the center point is calculated. Based on this center point, the objects are traced over the time, i.e. over successive sliding time windows. Two center points from clusters in consecutive sliding time windows are considered as semantically linked when their euclidean distance is smaller than a pre-defined threshold (see Table 1). If there is no other point within this distance, the corresponding cluster is interpreted as a new object.

The result of tracking these cluster center points is exemplified in Fig. 5, which shows the tracked path between the two sliding windows displayed in Figs. 4a and b.

Table 1. Parameter setting overview

Parameter	Setting used in our experiments
DVS-Record: sliding time window	60 ms
Event-Filter: threshold	$1/8 \cdot$ sliding window length [ms]
Clustering: cutoff distance	50 px
Clustering: minimal cluster size	100 events
Cluster path tracing: maximum center distance	50 px

Table 2. Scene description within recorded data

Record name	Description of scene	# of sliding time windows
Rec 0	Pedestrian crossing, partially obscured by a tree	310
Rec 1	Pedestrian crossing, partially obscured by a tree	359
Rec 2	Pedestrian crossing, partially obscured by a tree	292
Rec 3	Pedestrian crossing	343
Rec 4	Cyclist crossing, partially obscured by a tree	259
Rec 5	Pedestrian crossing, partially obscured by a tree	303
Rec 6	Cyclist crossing, partially obscured by a tree	225
Rec 7	Riding lawn mower crossing	502

4   Proof of Concept

The presented cluster-based tracking approach on event-based vision information focuses currently on the special use case of a fixed mounted sensor and
moving objects in the monitored scene. Due to the fact that the research area of
event-based computer vision is fairly new, there is a lack of well-known standard
databases covering various use cases and different DVS-sensor resolutions and
characteristics.

Hu et al. [16] are describing an approach to convert "classical" frame-based
vision datasets into synthetic event-based datasets. But the converted databases
are not addressing the described use case of object tracking and the conversion
tries to simulate a DVS sensor with a smaller sensor resolution than the one
used in our practical experiments (see Sect. 2). Hence, creating synthetic converted data for our specific sensor will produce artefacts. Thus, we decided to
use a small, self-recorded database for the first *proof-of-concept* of the proposed
approach. Table 2 briefly summarizes the considered scenes within this dataset.

The following subsections present an alternative tracking approach using
a frame-based imaging technique which is compared with the proposed DVS-
clustering method.

4.1   Comparative Approach: Difference Image

Due to privacy concerns that need to be considered (compare with the project
description in the introduction), it is not possible to use "classical" greyscale or
color images to monitor the desired space. One possible option from the field of
image processing is the approach of using difference images and binarization.

For this purpose, a recording of the background (scene without any objects,
see Fig. 6a) is taken. Each frame of the actual recording (see Fig. 6b) is compared with this background in that the difference between these two images is
calculated. To ensure the privacy concerns this difference image can be binarized
(see Fig. 6c), so that no restoration of color- or greyscale values is possible.

Similar to the described filtering of DVS-events this approach allows also the
use of a filtering as an additional step. In this case, the use of morphological

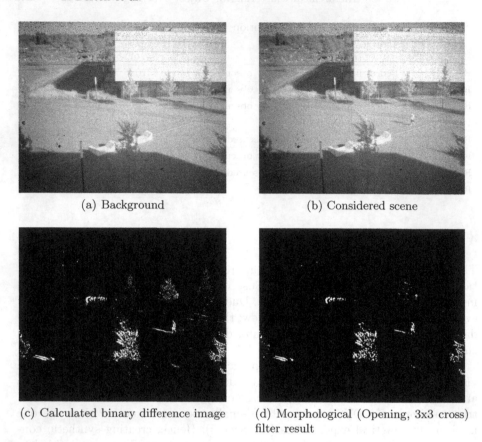

(a) Background

(b) Considered scene

(c) Calculated binary difference image

(d) Morphological (Opening, 3x3 cross) filter result

Fig. 6. Visualization of the described privacy aware 'classical' imaging approach

operations is one possible way. Figure 6d shows the filtered result which arises when using a morphological opening operation with a 3 × 3 cross-structure kernel element.

Based on these images the use of well-known computer vision object tracker is possible. For comparison we used the implementations[1] in the openCV-library [17].

4.2   Comparison: Event-Clustering and OpenCV-Tracker

Compared to the presented clustering procedure on the DVS event data, the implementations of the openCV trackers require a bounding box, which includes the object to be tracked, as input parameter. Due to this fact, we decided to compare the two approaches on the basis of the tracked path of this selected object. This means, that in terms of this proof of concept comparison a *single*

---

[1] See https://docs.opencv.org/3.4.2/d0/d0a/classcv_1_1Tracker.html.

object tracking is performed, although the DVS clustering-based approach could track multiple objects in a scene.

For the two approaches the algorithmically determined object center is compared to a manually defined ground-truth position. In case of the DVS-clustering this object center is the cluster center point and for the openCV-tracker the

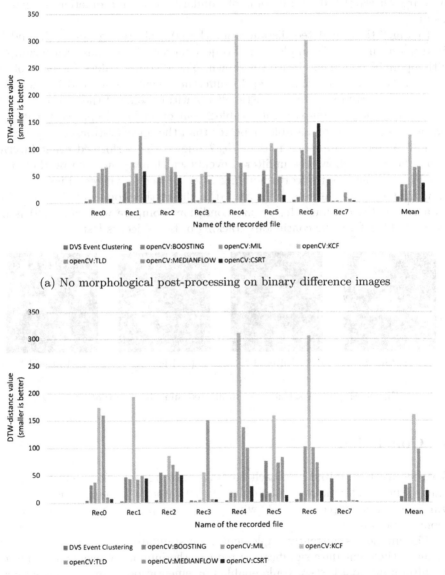

(a) No morphological post-processing on binary difference images

(b) With morphological opening as post-processing on binary difference images

Fig. 7. Calculated path distances for DVS-clustering and six different object trackers implemented by openCV

center point from the points within the returned object bounding box is used. By estimating this center point over continuous sliding time windows (or the generated and filtered binary images), an object path is determined. An example is given by the red line in Fig. 5.

For the quantitative comparison of these paths in comparison with the ground-truth path the dynamic time warp distance is used [18]. This distance measure allows the determination of similarity even for different lengths of results.

In Fig. 7 the calculated distances for the DVS-clustering approach, and for six different in openCV implemented object tackers (for details please compare with openCV documentation) are shown. The distance values for each of the openCV trackers is averaged over 15 different executions with identical initial parameters to compensate stochastic effects within some of the trackers. In the considered footage (compare with Table 2) our event-based approach, with one major exception, is comparable or better than the openCV-trackers.

The presented approach fails in *Rec7* due to the performed event filtering step. The Fig. 8a shows the unfiltered events which are generated by the object and Fig. 8b contains the corresponding filtered events, whereas Figs. 8c and d show that too many events are removed in the further course of the recording. As a result, the minimal cluster size condition (compare with Sect. 3.2) is not reached. Therefore, the continuous tracking of the object is lost.

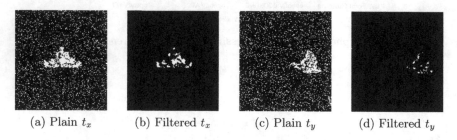

(a) Plain t_x (b) Filtered t_x (c) Plain t_y (d) Filtered t_y

Fig. 8. Magnified selection of unfiltered and filtered events in *Rec7*

5   Conclusion

We presented an initial approach to track moving objects by clustering and tracing their center points based on event data from a Dynamic Vision Sensor. Our method is simple and fast, while respecting possible privacy concerns of depicted persons.

The method is currently implemented on standard PC hardware. Since filtering of the event data significantly reduces the number of events, outsourcing the filter stage to a FPGA would enable implementation on less powerful microprocessors.

Another important aspect for further research is the creation and publication of a larger event-based database with corresponding ground truth annotations to allow a systematic evaluation that goes beyond the presented proof-of-concept test.

Also, the improvement of the presented approach itself is an aspect for further research. The time information for each event (which has a resolution of milliseconds) is mostly unused in the current approach, which represents potential for additional improvement.

References

1. Ranftl, A., Alonso-Fernandez, F., Karlsson, S.: Face tracking using optical flow. In: International Conference of the Biometrics Special Interest Group (BIOSIG), Darmstadt, pp. 1–5 (2015)
2. Liu, Y., Lu, Y., Shi, Q., Ding, J.: Optical flow based urban road vehicle tracking. In: Ninth International Conference on Computational Intelligence and Security, Leshan, pp. 391–395 (2013)
3. Dan, L., Dai-Hong, J., Rong, B., Jin-Ping, S., Wen-Jing, Z., Chao, W.: Moving object tracking method based on improved lucas-kanade sparse optical flow algorithm. In: International Smart Cities Conference (ISC2), Wuxi, pp. 1–5 (2017)
4. Bukey, C.M., Kulkarni, S.V., Chavan, R.A.: Multi-object tracking using Kalman filter and particle filter. In: IEEE International Conference on Power, Control, Signals and Instrumentation Engineering (ICPCSI), Chennai, pp. 1688–1692 (2017)
5. Mu, X., Che, J., Hu, T., Wang, Z.: A video object tracking algorithm combined Kalman filter and adaptive least squares under occlusion. In: 9th International Congress on Image and Signal Processing, BioMedical Engineering and Informatics (CISP-BMEI), Datong, pp. 6–10 (2016)
6. Najafzadeh, N., Fotouhi, M., Kasaei, S.: Object tracking using Kalman filter with adaptive sampled histogram. In: 23rd Iranian Conference on Electrical Engineering, Tehran, Iran, pp. 781–786 (2015)
7. Redmon, J., Farhadi, A.: YOLOv3: an incremental improvement. Technical report arXiv:1804.02767 (2018)
8. Mocanu, B., Tapu, R., Zaharia, T.: Single object tracking using offline trained deep regression networks. In: Seventh International Conference on Image Processing Theory, Tools and Applications (IPTA), Montreal, pp. 1–6 (2017)
9. Behrendt, K., Novak, L., Botros, R.: A deep learning approach to traffic lights: detection, tracking, and classification. In: IEEE International Conference on Robotics and Automation (ICRA), Singapore, pp. 1370–1377 (2017)
10. Mehta, S., Patel, A., Mehta, J.: CCD or CMOS image sensor for photography. In: International Conference on Communications and Signal Processing (ICCSP), Melmaruvathur, pp. 0291–0294 (2015)
11. Mahmood Rajpoot, Q., Jensen, C.: Video surveillance: privacy issues and legal compliance. In: Promoting Social Change and Democracy through Information Technology. IGI Global (2015). ISBN 9781466685024
12. Lichtsteiner, P., Posch, C., Delbruck, T.: A 128 × 128 120 dB 15 µs latency asynchronous temporal contrast vision sensor. IEEE J. Solid-State Circuits $43(2)$, 566–576 (2008)

13. Delbrück, T., Linares-Barranco, B., Culurciello, E., Posch, C.: Activity-driven, event-based vision sensors. In: Proceedings of 2010 IEEE International Symposium on Circuits and Systems, Paris, pp. 2426–2429 (2010)
14. Guo, M., Huang, J., Chen, S.: Live demonstration: a 768×640 pixels 200Meps dynamic vision sensor. In: IEEE International Symposium on Circuits and Systems (ISCAS), Baltimore, p. 1 (2017)
15. Müllner, D.: Fastcluster: fast hierarchical, agglomerative clustering routines for R and Python. J. Stat. Softw. **53**(9), 1–18 (2013)
16. Hu, Y., Liu, H., Pfeiffer, M., Delbruck, T.: DVS benchmark datasets for object tracking, action recognition, and object recognition. J. Front. Neurosci. **10**, 405–410 (2016)
17. Bradski, G.: The OpenCV library. DR DOBBS J. Softw. Tools **25**, 120–125 (2000)
18. Müller, M.: Dynamic time warping. In: Information Retrieval for Music and Motion, pp. 69–84. Springer, Heidelberg (2007). https://doi.org/10.1007/978-3-540-74048-3_4. ISBN 9783540740483

Binarization of Degraded Document Images with Generalized Gaussian Distribution

Robert Krupiński[ID], Piotr Lech[ID], Mateusz Tecław[ID],
and Krzysztof Okarma[(✉)][ID]

Department of Signal Processing and Multimedia Engineering,
Faculty of Electrical Engineering, West Pomeranian University of Technology,
Szczecin, Sikorskiego 37, 70-313 Szczecin, Poland
{rkrupinski,piotr.lech,mateusz.teclaw,okarma}@zut.edu.pl

Abstract. One of the most crucial steps of preprocessing of document images subjected to further text recognition is their binarization, which influences significantly obtained OCR results. Since for degrades images, particularly historical documents, classical global and local thresholding methods may be inappropriate, a challenging task of their binarization is still up-to-date. In the paper a novel approach to the use of Generalized Gaussian Distribution for this purpose is presented. Assuming the presence of distortions, which may be modelled using the Gaussian noise distribution, in historical document images, a significant similarity of their histograms to those obtained for binary images corrupted by Gaussian noise may be observed. Therefore, extracting the parameters of Generalized Gaussian Distribution, distortions may be modelled and removed, enhancing the quality of input data for further thresholding and text recognition. Due to relatively long processing time, its shortening using the Monte Carlo method is proposed as well. The presented algorithm has been verified using well-known DIBCO datasets leading to very promising binarization results.

Keywords: Document images · Image binarization ·
Generalized Gaussian Distribution · Monte Carlo method ·
Thresholding

1 Introduction

Document image binarization is an active area of research in computer vision due to high demands related to the robustness of the thresholding algorithms. As it is one of the most relevant steps in document recognition applications, considering both machine printed and handwritten text documents, many algorithms have been proposed for this purpose. Many of them were presented at Document Image Binarization COmpetitions (DIBCO) held during International Conferences on Document Analysis and Recognition (ICDAR) and H-DIBCO during

© Springer Nature Switzerland AG 2019
J. M. F. Rodrigues et al. (Eds.): ICCS 2019, LNCS 11540, pp. 177–190, 2019.
https://doi.org/10.1007/978-3-030-22750-0_14

International Conferences on Frontiers in Handwriting Recognition (ICFHR). Due to the presence of challenging image distortions, DIBCO datasets [20], used for performance evaluation of the submitted algorithms, became the most popular ones for the verification of newly proposed binarization methods.

The motivation of research related with document image binarization and recognition is not only the possibility of preserving the cultural heritage and discovering some historical facts, e.g. by the recognition of ancient manuscripts, but also potential applications of the developed algorithms in some other areas of industry. Considering the rapid development of Industry 4.0 solutions, similar algorithms may be useful in self-localization and navigation of mobile robots based on machine vision as well as modern autonomous vehicles. Capturing the video data by cameras the presence of some similar distortions may be expected both for natural images and degraded document images. Nevertheless, document image datasets containing also ground truth binary images are still the best tool for verification purposes and therefore the method proposed in the paper is considered using the images from DIBCO datasets.

During the last several years many various approaches to image thresholding have been proposed, outperforming the classical Otsu method [18], including adaptive methods proposed by Niblack [12], Sauvola [22], Feng [3], Wolf [27], or Bradley [1] and their modifications [23], being the most useful for document image binarization purposes. Nonetheless, one of the main issues of such adaptive methods is the necessity of analysis of the neighbourhood of each pixel, increasing the computational effort. Recently, some applications of local features with the use of Gaussian mixtures [11], as well as the use of deep neural networks [24] have been proposed as well. However, to obtain satisfactory results, most of such approaches require multiple processing stages with background removal, median filtering, morphological processing or the time-consuming training process.

Nevertheless, the motivation of the paper is not the direct comparison of the proposed approach with the state-of-the-art methods, especially based on recent advances of deep learning, but the increase of the performance of some known methods due to the application of the proposed approach to image preprocessing.

2 The Basics of the Proposed Approach

2.1 Identification and Definition of the Problem

Handwritten and machine printed documents usually are subject to slow destruction over time influencing their readability. Some characteristic examples of this process are ancient books and old prints, however digital restoration methods allow for reading of even heavily damaged documents. Assuming that the original text was distorted by its summing with a noisy image of a normal distribution, analysing the histograms, it can be noticed that the original information was hidden (blurred) and the histogram of the resulting image is a distorted version of a histogram of a "purely" noisy image.

This similarity is preserved also for the real scanned images of historical documents. Therefore, it can be assumed that potential removing of partial

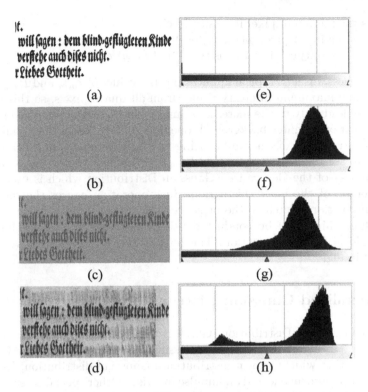

Fig. 1. Illustration of the similarity of histograms for noisy text image and scanned real documents: (a) ground truth binary image, (b) noisy image, (c) text image combined with Gaussian noise, (d) real scanned document images, and their histograms (e)–(h), respectively.

information related with noise should improve the quality of the image being the input for further processing. An illustration of this phenomenon is shown in Fig. 1.

2.2   The Basic Idea of Text Image Reconstruction

Assuming that the real text image is approximately the combination of ground truth (GT) binary image with Gaussian noise, being the most widely present one in nature, the readability of text may be initially improved by normalization of pixel intensity levels according to classical formula

$$q(x, y) = \left| \frac{(p(x, y) - I_{min}) \cdot 255}{I_{max} - I_{min}} \right|, \tag{1}$$

where:

- $0 \leq I_{min} < I_{max} \leq 255$,
- I_{max} is the maximum pixel intensity level,

- I_{min} is the minimum pixel intensity level,
- $p(x, y)$ is the input pixel intensity level at (x, y) coordinates,
- $q(x, y)$ is the output pixel intensity level at (x, y) coordinates.

Nevertheless, in most typical applications the values of I_{min} and I_{max} are the minimum and maximum intensity values from all image pixels, so this normalization may lead only to the increase of image contrast. Assuming the presence of a dark text on brighter background, one may remove safely the detailed data related to brighter pixels not influencing the text information. Therefore, we proposed to set the $I_{max} = \mu_{GGD}$ and $I_{min} = 0$ where the μ_{GGD} is the location parameter of the Generalized Gaussian Distribution which is used for the approximation of the image histogram. Such operation causes the removal of partial information related to the presence of distortions and is followed by the thresholding which may be conducted using one of the typical methods, e.g. classical global Otsu binarization. The illustration of the consecutive steps is presented in Fig. 2.

3   Generalized Gaussian Distribution

Generalized Gaussian Distribution (GGD) is very popular tool in many research areas related to signal and image processing. Its popularity comes from the coverage of other widely known distributions: Gaussian distribution, Laplacian distribution, a uniform one and an impulse function. Other special cases were also considered in literature [5, 6]. Many different methods were designed to estimate the parameters of this distribution [28].

This distribution was also extended to cover the complex variable [13] and multidimensional [19]. GGD was used to design many different models, for instance, to model the tangential wavelet coefficients for compressing three-dimensional triangular mesh data [8], the image segmentation algorithm [25], to generate an augmented quaternion random variable with GGD [7], the natural scene statistics (NSS) model to describe certain regular statistical properties of natural images [29], to approximate an atmosphere point spread function (APSF) kernel [26].

The probability density function of GGD is defined by the equation [2]

$$f(x) = \frac{\lambda \cdot p}{2 \cdot \Gamma\left(\frac{1}{p}\right)} e^{-[\lambda \cdot |x|]^p}, \tag{2}$$

where p is the shape parameter, $\Gamma(z) = \int_0^\infty t^{z-1} e^{-t} dt, z > 0$ [17] and λ is connected to the standard deviation σ of the distribution by the equation $\lambda(p, \sigma) = \frac{1}{\sigma} \left[\frac{\Gamma(\frac{3}{p})}{\Gamma(\frac{1}{p})} \right]^{\frac{1}{2}}$. The parameter $p = 1$ corresponds to Laplacian distribution and $p = 2$ corresponds to Gaussian distribution. When $p \to \infty$, the GGD density function becomes a uniform distribution and when $p \to 0$, $f(x)$ approaches an impulse function. Some examples are shown in Fig. 3.

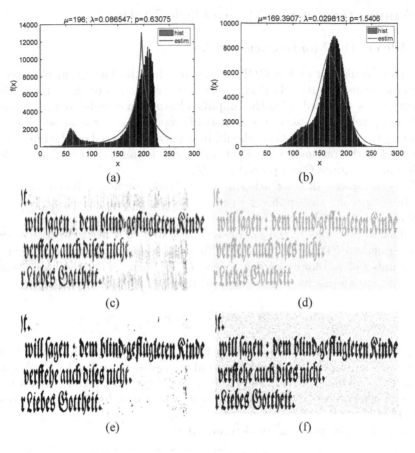

(a)   (b)

(c)   (d)

(e)   (f)

Fig. 2. Illustration of the consecutive steps of the document image processing and obtained GGD distributions - from top: histogram and its GGD approximation for the real document image (a) and GT combined with Gaussian noise (b), results of proposed normalization with the use of μ_{GGD} (c) and (d), respectively, and the results of further Otsu thresholding (e) and (f), respectively.

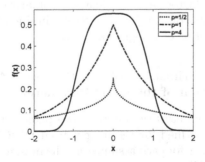

Fig. 3. Density function of GGD with $\lambda = 1$ for three selected exponents $p = 1/2$, $p = 1$ and $p = 4$.

4   Application of the Monte Carlo Method

4.1   Idea of the Monte Carlo Method

Since the calculation of the GGD parameters for the histogram obtained for the whole image is relatively slow, a significant reduction of the computational burden may be achieved using the simplified histogram calculated for the limited number of pixels. To preserve the statistical properties of the analysed image the randomly chosen pixel locations should be evenly distributed on the image plane and therefore the random number generator with uniform distribution should be applied in the Monte Carlo procedure [15].

The general idea of the statistical Monte Carlo method is based on the random drawing procedure applied for the reshaped one-dimensional vector consisting of all $M \times N$ pixels from the analysed image. Then, n independent numbers, equivalent to positions in the vector, are generated by a pseudo-random generator of uniform distribution with possibly good statistical properties. Next, the total number of randomly chosen pixels (k) for each luminance level is determined used as an estimate of the simplified histogram, according to:

$$\hat{L}_{MC} = \frac{k}{n} \cdot M \cdot N, \tag{3}$$

where k is the number of drawn pixels for the specified luminance level in randomly chosen samples, n denotes the total number of draws and $M \times N$ stands for the total number of samples in the entire image. In general, the estimator $\hat{L}_{MC}$ may refer to any defined image feature which may be described by binary values 0 and 1.

The estimation error can be determined as:

$$\varepsilon_\alpha = \frac{u_\alpha}{\sqrt{n}} \cdot \sqrt{\frac{K}{M \cdot N} \cdot \left(1 - \frac{K}{M \cdot N}\right)}, \tag{4}$$

assuming that K represents the total number of samples with specified luminance level and u_α denotes the two-sided critical range.

For such estimated histogram some classical binarization methods may be applied leading to results comparable with those obtained for the analysis of full images [9,16], also in terms of recognition accuracy.

4.2   Experimental Verification of the Proposed Approach for the Estimation of the GGD Parameters

The influence of the number of randomly drawn pixels on the obtained parameters of the GGD was verified for the images from DIBCO datasets with each drawing repeated 30 times for each assumed n. The minimum, average and maximum values of the four GGD parameters: shape parameter p, location parameter μ, variance of the distribution λ, and standard deviation σ were then determined, according to the method described in the paper [4], without the necessity of

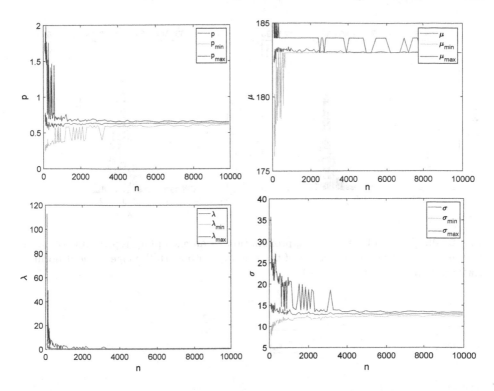

Fig. 4. Illustration of convergence of the Monte Carlo method used for estimation of the GGD parameters p, μ, λ and σ using n randomly chosen samples for an exemplary representative image.

using of more sophisticated estimators based on maximum likelihood, moments, entropy matching or global convergence [21]. The illustration of convergence of the parameters for an exemplary representative image from DIBCO datasets using different numbers of drawn samples (n) is shown in Fig. 4.

Nonetheless, it should be noted that for each independent run of the Monte Carlo method the values of the estimated parameters may differ, especially assuming a low number of randomly chosen samples (n). One of the possible solutions of this issue is the use of the predefined numbers obtained from the pseudorandom number generator with a uniform distribution. Therefore an appropriate choice of n is necessary to obtain stable results. Some local histogram peaks may be related with the presence of some larger smears of constant brightness on the image plane (considered as background information). Since the histogram of a natural image should be in fact approximated by a multi-Gaussian model, a limitation of the analysed range of brightness should be made to obtain a better fitting of the GGD model.

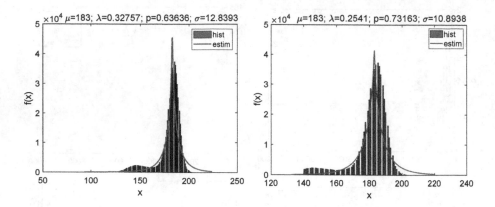

Fig. 5. Illustration of histogram approximation for an exemplary representative image before (a) and after (b) limitation of the brightness range (full image without normalization, $x_{min} = 139$, $x_{max} = 224$).

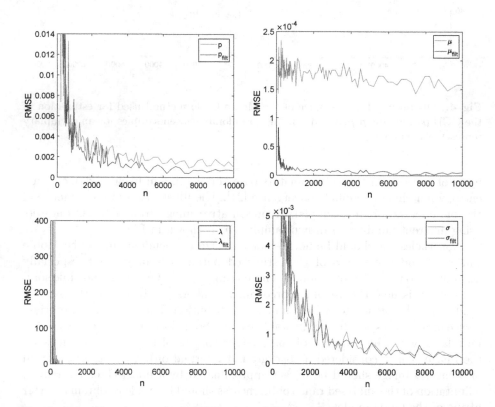

Fig. 6. Approximation errors (RMSE) of the GGD parameters p, μ, λ and σ for various numbers of samples (n) used in the Monte Carlo method for an exemplary representative image.

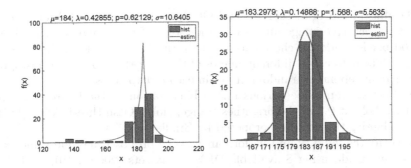

Fig. 7. Illustration of histograms obtained in two steps of the proposed method for an exemplary representative image using $n = 100$ randomly drawn samples.

Determination of the limited brightness range is conducted as follows:

- determination of the simplified histogram using the Monte Carlo method for n samples (e.g. $n = 100$),
- estimation of the GGD using the simplified histogram,
- setting the lower boundary as x_{min}, such that $P(x = x_{min}) = 1/n$,
- setting the upper boundary as x_{max}, such that $P(x = x_{max}) = 1 - 1/n$.

Therefore, the brightness values with probabilities lower than the probability of an occurrence of a single pixel $P(x) = 1/n$ are removed on both sides of the distribution. An example based on the histogram determined for the full image ($M \cdot N$ samples used instead of n) is shown in Fig. 5.

Additionally, the calculation of the Root Mean Squared Error (RMSE), was made to verify the influence of the number of samples n on the approximation error. Since the histograms of natural images are usually "rough", the additional median filtering of histograms with 5-elements mask was examined. Nevertheless, the obtained results were not always satisfactory, as shown in Fig. 6 and therefore this filtration was not used to prevent additional increase of computation time.

5   Proposed Two-Step Algorithm and Its Experimental Verification

On the basis of the above considerations, the following procedure is proposed:

- determination of lower boundary x_{min} and upper boundary x_{max} with the use of the Monte Carlo histogram estimation and GGD approximation,
- limiting the brightness range to $\langle x_{min}; x_{max} \rangle$,
- restarted GGD approximation of the histogram for the limited range with the use of Monte Carlo method,
- estimation of the location parameter μ_{GGD} for the histogram with the limited range,
- limiting the brightness range to $\langle 0; \mu_{GGD} \rangle$ and normalization,
- binarization using one of the classical thresholding methods.

The illustration of histograms and GGD parameters obtained for an exemplary representative image after two major steps of the proposed algorithm for $n = 100$ pixels randomly chosen according to the Monte Carlo method is shown in Fig. 7. The noticeably different shapes of the left (a) and right (b) histograms result from independent random draws in each of two steps.

In the last step three various image binarization methods are considered: fixed threshold at 0.5 of the brightness range, global Otsu thresholding [18] and locally adaptive thresholding proposed by Bradley [1].

To verify the validity and performance of the proposed method some experiments were made using 8 available DIBCO datasets (2009, 2010, 2011, 2012, 2013, 2014, 2016 and 2017). For all of these databases some typical metrics used for the evaluation of binarization algorithms [14] were calculated for five different values of samples (n) used in the Monte Carlo method. The executions of the Monte Carlo method were repeated 30 times and the obtained results were compared with the application of three classical thresholding methods mentioned above without the proposed image preprocessing based on the GGD histogram approximation. Detailed results obtained for the fixed threshold (0.5), Otsu and Bradley thresholding are presented in Tables 1, 2 and 3, respectively. Better results are indicated by higher accuracy, F-Measure, specificity and PSNR values, whereas lower Distance-Reciprocal Distortion (DRD) values denotes better quality [10]. All the metrics marked with (GGD) were calculated for the proposed GGD based approach with the Monte Carlo method setting the n value at 5% of the total number of pixels (about 1000–5000 depending on image resolution).

As can be observed analysing the results presented in Tables 1, 2 and 3, the proposed approach, utilising the GGD histogram approximation with the use of Monte Carlo method for image preprocessing, leads to the enhancement of binarization results for Otsu and Bradley thresholding methods, whereas its application for the binarization with a fixed threshold is inappropriate. Particularly significant improvements can be observed for DIBCO2012 dataset with the use of Otsu binarization, however the advantages of the proposed approach can also be observed for the aggregated results for all datasets (weighted by the number of images they contain). A visual illustration of the obtained improvement is shown in Fig. 8 for an exemplary H10 image for DIBCO2012 dataset, especially well visible for Otsu thresholding.

It is worth noting that the results shown in Fig. 8 were obtained using the proposed method applied with random drawing of only $n = 120$ samples using the Monte Carlo method. The obtained improvement of the accuracy value due to the proposed preprocessing is from 0.7765 to 0.9748 for Otsu method and from 0.9847 to 0.9851 for Bradley thresholding. The respective F-Measure values increased from 0.4618 to 0.8608 for Otsu and from 0.9220 to 0.9222 for Bradley method. Nevertheless, depending on the number of randomly drawn pixels the values achieved for the proposed method may slightly differ.

Proposed application of the GGD based preprocessing combined with the Monte Carlo method leads to the improvement of the binarization results which are comparable with the application of adaptive thresholding or better for some

Table 1. Results of binarization metrics obtained for DIBCO datasets using the classical binarization with fixed threshold (0.5) without and with proposed GGD preprocessing.

DIBCO dataset	2009	2010	2011	2012	2014	2016	2017	All
Accuracy	0.9621	0.9574	0.8870	0.9728	0.9531	0.9638	0.9482	0.9491
Accuracy (GGD)	0.9534	0.9485	0.9532	0.9561	0.9330	0.9570	0.9414	0.9481
F-Measure	0.7999	0.5461	0.6988	0.7799	0.7123	0.7614	0.6698	0.7052
F-Measure (GGD)	0.6888	0.4538	0.6228	0.5612	0.5305	0.6818	0.4647	0.5597
Specificity	0.9704	0.9988	0.8938	0.9957	0.9985	0.9914	0.9719	0.9741
Specificity (GGD)	0.9966	0.9997	0.9919	0.9987	0.9995	0.9987	0.9985	0.9978
PSNR	15.5993	14.4964	12.9377	16.1540	14.8660	16.8802	13.7190	14.8101
PSNR (GGD)	14.2591	13.5879	13.8179	14.7964	13.1314	16.0966	13.0014	13.9737
DRD	11.4208	9.5613	38.6952	7.1093	8.2956	7.2295	11.9349	13.2896
DRD (GGD)	9.2247	12.3666	9.5546	12.1856	12.0275	8.4014	12.2352	11.0136

Table 2. Results of binarization metrics obtained for DIBCO datasets using Otsu binarization without and with proposed GGD preprocessing.

DIBCO dataset	2009	2010	2011	2012	2014	2016	2017	All
Accuracy	0.9426	0.9816	0.9491	0.9132	0.9844	0.9729	0.9427	0.9538
Accuracy (GGD)	0.9493	0.9819	0.9588	0.9797	0.9831	0.9787	0.9532	0.9674
F-Measure	0.7860	0.8537	0.7894	0.7417	0.9162	0.8659	0.7723	0.8127
F-Measure (GGD)	0.7931	0.8530	0.8021	0.8571	0.9101	0.8745	0.7924	0.8349
Specificity	0.9447	0.9930	0.9543	0.9149	0.9953	0.9788	0.9448	0.9590
Specificity (GGD)	0.9566	0.9947	0.9693	0.9904	0.9960	0.9884	0.9600	0.9771
PSNR	15.3070	17.4839	15.0603	14.4945	18.7161	17.7851	13.9273	15.8619
PSNR (GGD)	15.3389	17.5737	15.3613	17.1587	18.3658	18.1447	14.5766	16.4100
DRD	22.5705	4.0850	11.2789	30.4799	2.6494	5.5839	15.9928	13.5486
DRD (GGD)	19.7113	3.9453	8.8428	5.9430	2.8434	4.5330	12.6115	8.8330

Table 3. Results of binarization metrics obtained for DIBCO datasets using Bradley binarization without and with proposed GGD preprocessing.

DIBCO dataset	2009	2010	2011	2012	2014	2016	2017	All
Accuracy	0.9540	0.9730	0.9483	0.9660	0.9778	0.9557	0.9456	0.9584
Accuracy (GGD)	0.9627	0.9753	0.9577	0.9718	0.9771	0.9620	0.9531	0.9642
F-Measure	0.7700	0.8278	0.7702	0.8124	0.8885	0.7604	0.7590	0.7938
F-Measure (GGD)	0.7945	0.8350	0.7940	0.8326	0.8852	0.7815	0.7800	0.8107
Specificity	0.9543	0.9806	0.9498	0.9682	0.9880	0.9570	0.9455	0.9613
Specificity (GGD)	0.9670	0.9843	0.9630	0.9756	0.9898	0.9664	0.9565	0.9701
PSNR	14.0103	16.5663	14.0493	15.8350	17.4899	13.8843	13.1672	14.7915
PSNR (GGD)	14.9064	16.8107	14.8448	16.2801	17.5676	14.5243	13.8088	15.3381
DRD	18.0002	7.6067	13.5784	11.4206	4.3950	16.7249	14.8879	12.6598
DRD (GGD)	15.1632	6.6653	10.8255	9.2529	4.3959	14.5060	12.6725	10.7451

images. In most cases its application for adaptive thresholding allows for further slight increase of binarization accuracy.

Original image

Ground truth binary image

Result of global Otsu binarization

Result of adaptive Otsu binarization with GGD

Result of adaptive Bradley binarization

Result of adaptive Bradley binarization with GGD

Fig. 8. Illustrations of the obtained improvement of binarization results for an exemplary image from DIBCO2012 dataset.

6   Summary and Future Work

Although the obtained results may be outperformed by some more complex state-of-the-art methods, especially based on deep CNNs [24], they can be considered as promising and confirm the usefulness of the GGD histogram approximation with the use of the Monte Carlo method for preprocessing of degraded document images before binarization and further analysis. Since in the proposed approach, only one of the GGD parameters (location parameter μ) is used, a natural direction of our future research is the utilisation of the other parameters for the removal of additional information related to contaminations.

Our future research will concentrate on further improvement of binarization accuracy, although an important limitation might be the computational burden. However, due to an efficient use of the Monte Carlo method, the overall processing time may be shortened and therefore our proposed approach may be further combined with some other binarization algorithms proposed by various researchers.

References

1. Bradley, D., Roth, G.: Adaptive thresholding using the integral image. J. Graph. Tools **12**(2), 13–21 (2007). https://doi.org/10.1080/2151237X.2007.10129236
2. Clarke, R.J.: Transform Coding of Images. Academic Press, New York (1985)
3. Feng, M.L., Tan, Y.P.: Adaptive binarization method for document image analysis. In: Proceedings of the 2004 IEEE International Conference on Multimedia and Expo (ICME), vol. 1, pp. 339–342 (2004). https://doi.org/10.1109/ICME.2004. 1394198
4. Krupiński, R.: Approximated fast estimator for the shape parameter of generalized Gaussian distribution for a small sample size. Bull. Polish Acad. Sci. Tech. Sci. **63**(2), 405–411 (2015). https://doi.org/10.1515/bpasts-2015-0046
5. Krupiński, R.: Reconstructed quantized coefficients modeled with generalized Gaussian distribution with exponent 1/3. Image Process. Commun. **21**(4), 5–12 (2016)
6. Krupiński, R.: Modeling quantized coefficients with generalized Gaussian distribution with exponent $1/m$, $m = 2, 3, \ldots$. In: Gruca, A., Czachórski, T., Harezlak, K., Kozielski, S., Piotrowska, A. (eds.) ICMMI 2017. AISC, vol. 659, pp. 228–237. Springer, Cham (2018). https://doi.org/10.1007/978-3-319-67792-7_23
7. Krupiński, R.: Generating augmented quaternion random variable with Generalized Gaussian Distribution. IEEE Access **6**, 34608–34615 (2018). https://doi.org/ 10.1109/ACCESS.2018.2848202
8. Lavu, S., Choi, H., Baraniuk, R.: Estimation-quantization geometry coding using normal meshes. In: Proceedings of the Data Compression Conference (DCC 2003), p. 362, March 2003. https://doi.org/10.1109/DCC.2003.1194027
9. Lech, P., Okarma, K.: Optimization of the fast image binarization method based on the Monte Carlo approach. Elektronika Ir Elektrotechnika **20**(4), 63–66 (2014). https://doi.org/10.5755/j01.eee.20.4.6887
10. Lu, H., Kot, A.C., Shi, Y.Q.: Distance-reciprocal distortion measure for binary document images. IEEE Signal Process. Lett. **11**(2), 228–231 (2004). https://doi. org/10.1109/LSP.2003.821748
11. Mitianoudis, N., Papamarkos, N.: Document image binarization using local features and Gaussian mixture modeling. Image Vis. Comput. **38**, 33–51 (2015). https:// doi.org/10.1016/j.imavis.2015.04.003
12. Niblack, W.: An Introduction to Digital Image Processing. Prentice Hall, Englewood Cliffs (1986)
13. Novey, M., Adali, T., Roy, A.: A complex Generalized Gaussian Distribution - characterization, generation, and estimation. IEEE Trans. Signal Process. **58**(3), 1427–1433 (2010). https://doi.org/10.1109/TSP.2009.2036049
14. Ntirogiannis, K., Gatos, B., Pratikakis, I.: Performance evaluation methodology for historical document image binarization. IEEE Trans. Image Process. **22**(2), 595–609 (2013). https://doi.org/10.1109/TIP.2012.2219550
15. Okarma, K., Lech, P.: Monte Carlo based algorithm for fast preliminary video analysis. In: Bubak, M., van Albada, G.D., Dongarra, J., Sloot, P.M.A. (eds.) ICCS 2008. LNCS, vol. 5101, pp. 790–799. Springer, Heidelberg (2008). https:// doi.org/10.1007/978-3-540-69384-0_84

16. Okarma, K., Lech, P.: Fast statistical image binarization of colour images for the recognition of the QR codes. Elektronika Ir Elektrotechnika **21**(3), 58–61 (2015). https://doi.org/10.5755/j01.eee.21.3.10397
17. Olver, F.W.J.: Asymptotics and Special Functions. Academic Press, New York (1974)
18. Otsu, N.: A threshold selection method from gray-level histograms. IEEE Trans. Syst. Man Cybern. **9**(1), 62–66 (1979). https://doi.org/10.1109/TSMC.1979.4310076
19. Pascal, F., Bombrun, L., Tourneret, J.Y., Berthoumieu, Y.: Parameter estimation for multivariate Generalized Gaussian Distributions. IEEE Trans. Signal Process. **61**(23), 5960–5971 (2013). https://doi.org/10.1109/TSP.2013.2282909
20. Pratikakis, I., Zagori, K., Kaddas, P., Gatos, B.: ICFHR 2018 competition on handwritten document image binarization (H-DIBCO 2018). In: 16th International Conference on Frontiers in Handwriting Recognition (ICFHR), pp. 489–493, August 2018. https://doi.org/10.1109/ICFHR-2018.2018.00091
21. Roenko, A.A., Lukin, V.V., Djurović, I., Simeunović, M.: Estimation of parameters for generalized Gaussian distribution. In: 6th International Symposium on Communications, Control and Signal Processing (ISCCSP), pp. 376–379, May 2014. https://doi.org/10.1109/ISCCSP.2014.6877892
22. Sauvola, J., Pietikäinen, M.: Adaptive document image binarization. Pattern Recogn. **33**(2), 225–236 (2000). https://doi.org/10.1016/S0031-3203(99)00055-2
23. Saxena, L.P.: Niblack's binarization method and its modifications to real-time applications: a review. Artif. Intell. Rev. 1–33 (2017). https://doi.org/10.1007/s10462-017-9574-2
24. Tensmeyer, C., Martinez, T.: Document image binarization with fully convolutional neural networks. In: 14th IAPR International Conference on Document Analysis and Recognition, ICDAR 2017, Kyoto, Japan, 9–15 November 2017, pp. 99–104. IEEE (2017). https://doi.org/10.1109/ICDAR.2017.25
25. Wang, C.: Research of image segmentation algorithm based on wavelet transform. In: IEEE International Conference on Computer and Communications (ICCC), pp. 156–160, October 2015. https://doi.org/10.1109/CompComm.2015.7387559
26. Wang, R., Li, R., Sun, H.: Haze removal based on multiple scattering model with superpixel algorithm. Signal Process. **127**, 24–36 (2016). https://doi.org/10.1016/j.sigpro.2016.02.003
27. Wolf, C., Jolion, J.M.: Extraction and recognition of artificial text in multimedia documents. Formal Pattern Anal. Appl. **6**(4), 309–326 (2004). https://doi.org/10.1007/s10044-003-0197-7
28. Yu, S., Zhang, A., Li, H.: A review of estimating the shape parameter of generalized Gaussian distribution. J. Comput. Inf. Syst. **21**(8), 9055–9064 (2012)
29. Zhang, Y., Wu, J., Xie, X., Li, L., Shi, G.: Blind image quality assessment with improved natural scene statistics model. Digit. Signal Process. **57**, 56–65 (2016). https://doi.org/10.1016/j.dsp.2016.05.012

Nonlinear Dimensionality Reduction in Texture Classification: Is Manifold Learning Better Than PCA?

Cédrick Bamba Nsimba[✉] and Alexandre L. M. Levada

Department of Computer Science,
Federal University of São Carlos, São Carlos, SP, Brazil
cedrick.bamba@ifsp.edu.br, alexandre@dc.ufscar.br

Abstract. This paper presents a comparative analysis of algorithms belonging to manifold learning and linear dimensionality reduction. Firstly, classical texture image descriptors, namely Gray-Level Co-occurrence Matrix features, Haralick features, Histogram of Oriented Gradients features and Local Binary Patterns are combined to characterize and discriminate textures. For patches extracted from several texture images, a concatenation of the image descriptors is performed. Using four algorithms to wit Principal Component Analysis (PCA), Locally Linear Embedding (LLE), Isometric Feature Mapping (ISOMAP) and Laplacian Eigenmaps (Lap. Eig.), dimensionality reduction is achieved. The resulting learned features are then used to train four different classifiers: k-nearest neighbors, naive Bayes, decision tree and multilayer perceptron. Finally, the non-parametric statistical hypothesis test, Wilcoxon signed-rank test, is used to figure out whether or not manifold learning algorithms perform better than PCA. Computational experiments were conducted using the Outex and Salzburg datasets and the obtained results show that among twelve comparisons that were carried out, PCA presented better results than ISOMAP, LLE and Lap. Eig. in three comparisons. The remainder nine comparisons did not presented significant differences, indicating that in the presence of huge collections of texture images (bigger databases) the combination of image feature descriptors or patches extracted directly from raw image data and manifold learning techniques is potentially able to improve texture classification.

1 Introduction

Texture analysis plays an important role in the computer vision area. It is responsible for extracting meaningful information from texture images. Texture is considered as an essential attribute, among all characteristics present in an image, which can be used as a rich source of information in many application areas, such as object recognition, remote sensing, content-based image retrieval and so on.

This study was financed in part by the Coordenação de Aperfeiçoamento de Pessoal de Nível Superior - Brasil (CAPES) - Finance Code 001.

© Springer Nature Switzerland AG 2019
J. M. F. Rodrigues et al. (Eds.): ICCS 2019, LNCS 11540, pp. 191–206, 2019.
https://doi.org/10.1007/978-3-030-22750-0_15

In the computational context, texture analysis can be defined as a set of techniques capable of processing texture information in every perspective or in certain regions of interest (ROIs) of an image. According to Materka, Strzelecki et al. [12], this set is basically composed of four major categories, namely: feature extraction, texture classification, texture segmentation and shape reconstruction from texture. Feature extraction is the first stage of texture analysis. It is responsible for translating the correlation between visual patterns contained in an image into quantitative values. The methods responsible of performing this function are known as texture descriptors. The purpose of texture classification is to associate textured samples with two or more classes according to a given criteria of similarity [8,12,21]. Texture segmentation, unlike classification, focuses on the delimitation of regions based on texture information. In this case there is no need for prior knowledge of the characteristics of each surface meaning that unsupervised grouping techniques can be used [12,21,22]. Shape reconstruction through textures is directly linked to the recomposition of three-dimensional surfaces by means of texture information [12,13,21,22]. This work is confined mainly to texture classification which is composed of two main processes, regarding feature extraction and classification of patterns.

Many methods of texture analysis have been developed over the years, each one exploring a novel approach to extract the image's texture information. For instance, we have classical methods based on second-order statistics (such as Co-occurrence Matrices (GLCM)), Haralick, Histogram of Oriented Gradients (HoG) and Local Binary Patterns (LBP). Over many years, these four descriptors have gained a large amount of interest in many computer vision researching groups. The features extracted using the aforementioned descriptors have proven to be discriminative in classifying texture patterns.

By using those four descriptors to choose a subset of features that really represent a texture image separately, some texture information can be lost which may result in decreased recognition performance. According to this paper, in order to compensate for lost texture information and to fill the gaps left by each descriptor separately, the features extracted by those four descriptors are combined. Feature vector representing the patches are then taken through dimensionality reduction methods which give an improvement in recognition performance. On the other hand, we also propose to consider as feature vector a concatenation of 32×32 patches extracted directly from an image raw data.

Finally, we use this effective combination of the features of those four descriptors and also those ones extracted directly from an image raw data (after concatenating them and applying dimensionality reduction algorithms) with four classifiers (KNN [10], Naive Bayes [23], Decision Trees [14] and Multilayer Perceptron [5]) on Outex [19] and Salzburg [11] datasets. Finally, a comparative analysis is achieved using the Wilcoxon test [24] to figure out if nonlinear dimensionality reduction methods are better than the linear ones using the two proposed approaches.

The rest of this paper is organized as follows. We give a brief overview of PCA and manifold learning methods in Sects. 2 and 3 respectively. Section 4 is dedicated to present our proposal. Computational experiments and results for this work are presented in Sects. 5 and 6 respectively. Finally, we conclude the paper's work in Sect. 7.

2   Principal Component Analysis

High-dimensional datasets (i.e datasets with many features) may present difficulty in visualization process, may require high computational performance or may contain noise. To overcome these issues, there exist methods related with linear dimensionality reduction with the purpose of shrinking datasets by transformation and/or selection of features, while minimizing information loss. In this section, we will focus particularly on PCA [9].

It is common that some datasets follow certain distributions that are majorly embedded in a few orthogonal components, where a component is the result of a linear combination of the original features. PCA is a statistical technique that aims at transforming a n-dimensional dataset X to a m-dimensional dataset Y, where, obviously, $m \leq n$. Moreover, the dimensions of Y are captured in the direction in which the variance of samples in X is maximum and should necessarily be orthogonal components, commonly known as principal components [9]. In Fig. 1, the orange and purple arrows are the principal components of the synthetic dataset K, with 1000 samples and 2 features (2 principal components). In addition, the Algorithm 1 summarizes in a few simple steps the idea behind PCA method.

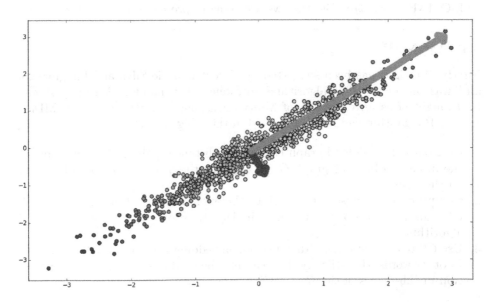

Fig. 1. The principal components of dataset K (orange and purple arrows). (Color figure online)

Data: a dataset D with n samples and f features and m as being the number of dimensions desired for the reduced dataset [17,15].

Result: the matrix consisting of the principal components

1. Find $X = HD$, where H is the centering matrix defined as $H = I_n - \frac{1}{n}11^T$, where I_n and 1 are, respectively, the identity matrix of order n and the column vector of 1's.
2. Calculate the covariance matrix $\sum_X$.
3. Use singular value decomposition to find the eigenvalues $\sum = \{\sigma_i\}$ and eigenvector $V = \{v_i\}$ of $\sum_X$.
4. Sort the eigenvalues by their absolute value in descending order and select the first m ones and their respective eigenvectors.

Algorithm 1. Simple idea behind PCA Algorithm

3   Manifold Learning

Manifold learning is a popular approach to nonlinear dimensionality reduction. Algorithms for this task are based on the idea that the dimensionality of many datasets is only artificially high; though each data point consists of perhaps thousands of features, it may be described as a function of only a few underlying parameters. That is, the data points are actually samples from a low-dimensional manifold that is embedded in a high-dimensional space. Manifold learning algorithms attempt to uncover these parameters in order to find a low-dimensional representation of the data. Manifold learning algorithms include methods such as ISOMAP, LLE, Lap. Eig. that we are going to present briefly in this section.

3.1   ISOMAP

Firstly, ISOMAP has been suggested by Tenenbaum, de Silva and Langfor [20] and was one of the first algorithms introduced for manifold learning. It can be thought of as an extension of Multidimensional Scaling (or simply MDS). ISOMAP algorithm can be summarized in three big steps:

1. Construct the weighted graph G from the distances pairwise for all points in the input and find the graph G' by applying the nearest-neighbor algorithm on the graph G.
2. Compute the shortest path graph G'' between all pairs of nodes from graph G'. This might be done by the all-pairs Dijkstra's [6] or by the Floyd Warshall algorithm [1].
3. Use G'' to construct the p-dimensional embedding using the MDS algorithm. In other words, the MDS method can now be used to construct a representation in sub-spaces of the R^n.

3.1.1   Multidimensional Scaling (MDS)

Isomap algorithm finds points whose Euclidean distances equal the geodesic distances that were calculated in steps 1 and 2. In fact, such points exist and are unique, since the manifold is isometrically embedded. Multidimensional Scaling [7] is a classical technique that may be used to find such points. The main goal of the MDS algorithm can be formalized as follows: Let $\{x_i, y_i\}$ be the input dataset for $i = 1, 2, ..., n$ where x_i denotes the feature vector that represents the ith sample and y_i denotes the class or category to which the vector belongs, being generally an integer greater than zero. Given a Euclidean pairwise distances matrix, retrieve which are the coordinates of $x_r \in R^k$, $r = 1, 2, ..., n$, where k is defined by the user (plan, space 3D, etc.). A distance matrix is given by $D = \{d_{rs}^2\}$, $r, s = 1, 2, ..., n$ where d_{rs}^2 is a distance between x_r and x_s vectors. Its expression is given by

$$d_{rs}^2 = \|x_r - x_s\|^2.$$

Now, let B be the matrix of inner products defined as $B = \{b_{rs}\}$, where $b_{rs} = x_r^T x_s$. Based on these configurations, the MDS method seeks to address two main problems:

(i) From the distance matrix D find a matrix B.
(ii) From B retrieve the coordinates $x_r \in R^p$ where p is the rank of B.

To address the first problem, MDS method assume that points are mean-centered—i.e. $\sum_{r=1}^{n} x_r = 0$. After some algebraic operations on D and using the initial form of matrix B, the final expression of matrix B is as $B = HAH$ such that $A = -\frac{1}{2}D$ and $H = I - \frac{1}{n}11^T$.

Finally, for retrieving the coordinates of the points, some properties of matrix B can be used. In fact B has three important properties, namely B is symmetric, the rank of B is p (this corresponds to the maximal number of linearly independent rows/columns of B that generates a base in R^p) and B is positive semidefinite. This implies that the matrix B has p non-negative eigenvalues and $n-p$ eigenvalues are null. Thus, by the spectral decomposition of B one can write $B = V' \wedge' V'^T$ where $\wedge' = diag(\lambda_1, \lambda_2, ..., \lambda_n)$ is a diagonal matrix of eigenvalues of B and V' with $n \times p$ dimension is defined as

$$\begin{bmatrix} | & | & \cdots & | \\ | & | & \cdots & | \\ v_1 & v_2 & \cdots & v_p \\ | & | & \cdots & | \\ | & | & \cdots & | \end{bmatrix}$$

However, as B can be written in these following format $B = X_{n \times p} X_{p \times n}^T$, thus, the observed $X_{n \times p}$ can be expressed as $X_{n \times p} = V'_{n \times p} \wedge'^{\frac{1}{2}}_{p \times p}$ where $\wedge'^{\frac{1}{2}} = diag(\sqrt{\lambda_1}, \sqrt{\lambda_2}, ..., \sqrt{\lambda_p})$.

Each row of $X_{n \times p}$ will have the coordinate of a vector $x_i \in R^p$, where p is a parameter that controls the number of dimensions of the output space: if we want a $2D$ plot, $p = 2$, in case of a $3D$ plot, $p = 3$. The Algorithm 2 summarizes the process of MDS.

Data: $D = \{d_{rs}^2\}$
Result: points in low-dimensional (k) Euclidean space whose interpoint distances match the geodesic distances found in step 1 of ISOMAP
1. Do $A = -\frac{1}{2}D$
2. Do $H = I - \frac{1}{n}11^T$
3. Compute $B = HAH$
4. Find the eigenvalues and the eigenvectors of B
5. Take the k eigenvectors associated with the largest k eigenvalues of B and set V'_{nxk} and $\wedge'^{\frac{1}{2}} = diag(\sqrt{\lambda_1}, \sqrt{\lambda_2}, ..., \sqrt{\lambda_k})$
6. Compute $X_{nxk} = V'_{nxk}\wedge'^{\frac{1}{2}}_{kxk}$

Algorithm 2. MDS Algorithm

It should be noted that the ISOMAP algorithm is totally unsupervised, in the sense that it does not use any information about class distribution. It attempts to find a more compact representation of the original data only by preserving the distances.

3.2   LLE

LLE has been suggested by Saul and Roweis [16] and was introduced at about the same time as ISOMAP, but it is based on different idea than that of ISOMAP. The idea comes from visualizing a manifold as a collection of overlapping coordinate patches. If the neighborhood sizes are small and the manifold is sufficiently smooth, then these patches will be approximately linear. Moreover, the chart from the manifold to R^d (d is the dimensionality of the manifold that the input is assumed to lie on and, accordingly, the dimensionality of the output i.e. $y_i \in R^d$) will be roughly linear on these small patches. The idea is to identify these linear patches, characterize the geometry of them, and find a mapping to R^d that preserves this local geometry and is nearly linear. It is assumed that these local patches will overlap with one another so that the local reconstructions will combine into a global one [3]. LLE algorithm can be summarized in two big steps:

1. Model the manifold as a collection of linear patches and attempts to characterize the geometry of these linear patches.
2. Find a configuration in d-dimensions (the dimensionality of the parameter space) whose local geometry is characterized well by W (W_i is a characterization of the local geometry around x_i of the manifold). Unlike ISOMAP, d must be known a priori or estimated.

3.3   Laplacian Eigenmaps

The idea behind the Lap. Eig. [2] is to use spectral graph theory in order to capture local information about the manifold. Given a graph and the corresponding matrix of edge weights, W, the Laplacian graph is defined as $L = D - W$ in which D represents the diagonal matrix with elements $D_{ii} = \sum_j W_{ij}$.

Here, a matrix W is used as a local similarity in order to measure the degree to which points are near to one another. Thus $W_{ij} = e^{-\|x_i - x_j\|^2 / 2\sigma^2}$ if x_j is one of the k-nearest neighbors of x_i and equals 0 otherwise. The main goal is to use W in order to find output points $y_1, ..., y_n \in R^d$ that are the low-dimensional analogs of input points $x_1, ..., x_n \in R^D$ where $d < D$. Lap. Eig. algorithm can be summarized in two big steps:

1. Given $x_1, ..., x_n \in R^D$, d, k, σ as input, define a local similarity matrix W_{ij}.
2. Let U be defined as the matrix whose columns are the eigenvectors of $Ly = \lambda Dy$ associated to the non-zero eigenvalues. Here L represents Lagrange multipliers. Y is then given by the $n \times d$ matrix whose columns are these d eigenvectors and whose rows are the embedded points $(y_1, ..., y_n \in R^d)$.

4   Proposed Method

This section presents a dimensionality reduction based approach for texture classification. Our method is firstly inspired by the observation of how dimensionality reduction techniques can effectively learn relevant information from the raw data, and secondly, by a wide variety of characteristics that present the feature vectors generated by each one of the classical descriptors adopted in this work. Based on this, we propose a straightforward solution which concatenates all these feature vectors (referred as first approach) and also those feature vectors constructed from patches extracted directly from the raw data (referred as second approach) and makes a feature selection through PCA and manifold learning techniques to choose the most effective features for class separability.

The idea consists in creating an approach for texture classification strongly based on the application of dimensionality reduction techniques. According to the block diagram of the proposed method (refer to Fig. 2), the two stages in Fig. 2(a) are part of the first approach and consist in extracting HoG, GLCM, Haralick and LBP features followed by their concatenation. We concatenated these four feature vectors aiming at constructing one big feature vector to serve as an input for dimensionality reduction using PCA and manifold learning algorithms. The stage in Fig. 2(b) is part of the second approach and consists in extracting 32 × 32 patches from an image which will then be concatenated to form a feature vector of the underlying image. In the next stage of the diagram flow, feature selection is achieved through PCA, ISOMAP, LLE and Lap. Eig. algorithms during which thresholds are used to maintain only the most effective features among all belonging to the concatenated feature vector. From this process, we have as an output a discriminative and reduced feature vector that contains only the best elements (the most relevant informations) of the input descriptors. Then, the selected feature vector is taken through the classification process using KNN, Naive Bayes, Decision Trees and ML Perceptron. Finally, a comparative analysis is achieved using the Wilcoxon test to figure out if nonlinear dimensionality reduction methods are better than the linear ones using the two proposed approaches.

5   Experimental Setup

5.1   Image Datasets

Two texture image datasets were used in the experiments:

- **Salzburg** [11] contains a collection of 476 color texture images that have been captured around Salzburg (Austria) and, herein, depicted in Fig. 4. For each texture class (from total of 10 classes used in this paper), there were 128×128 source images, of which 80% was used for training the classifier, while the other 20% was used for testing.
- **Outex** [19] has a total of 960 ($24 \times 20 \times 2$) image instances of illuminant "inca". The training set consists of 480 (24×20) images and the other 480 (24×20) images are part of the test set. The test suite for Outex used in this work is Outex_TC_00011-r. Some of the texture instances for each class are depicted in Fig. 5.

5.2   Description of Experiments

In Fig. 2 we present the flow diagram of the proposed method, including details of the pipeline, and the methods used in the experiments. Note that our contribution is to perform a statistical comparative analysis to show which approach is better than other between PCA and manifold techniques and how these approaches can be used to reduce the dimensionality of the feature space in texture classification task.

In this paper, we perform the following two types of experiments:

- **Experiments using the full feature vector created by concatenating all descriptors,** followed by classification with feature selection;
- **Experiments using the feature vector created by extracting 32 × 32 patches,** followed by classification with feature selection.

The main topic under investigation is the effect of different methods in reducing dimensionality for concatenation of descriptors and 32×32 image patches-based feature vector in texture classification. In our experiments we analyze if PCA reduction produces better results than manifold algorithms for both feature extraction approaches adopted in this paper. It is for this reason that we used the Wilcoxon test to find which method has significant difference. A significance level of $\alpha = 0.05$ was used in all tests. In this case the comparison is pair-wise. In addition, all experiments were performed with a stratified $k - fold$ cross validation setting. The accuracy proved to be a suitable measure to evaluate the classification performance, since all datasets are balanced and also the sampling for the cross validation is stratified [18].

6   Results and Discussion

6.1   Full Feature Space Experiments

The accuracies for the first set of experiments are shown in Table 1 and Table 3. They present a general perspective of the results. For each dataset and dimensionality reduction method among PCA, ISOMAP, LLE and Lap. Eig., we displayed ten accuracies, corresponding to the dimensionality reduction using 1, 2, 3, 4, 5, 6, 7, 8, 9 and 10 features (attributes).

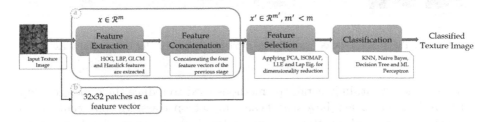

Fig. 2. Block diagram of the proposed method.

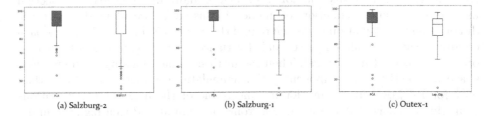

Fig. 3. Accuracies of PCA, ISOMAP, LLE and Lap. Eig. for the Salzburg dataset using the 32×32 image patch-based feature vector (a), Salzburg dataset using the concatenated feature vector (b) and Outex dataset using the concatenated feature vector (c). Boxplots in gray correspond to significances when compared to the accuracy of other method with p-value < 0.05. Here, the name Salzburg-2 (a) refers to the second approach applied to Salzburg dataset. Analogously, the names Salzburg-1 (b) and Outex-1 (c) refer to the first approach applied to Salzburg and Outex datasets, respectively.

Fig. 4. A sample from Salzburg's texture database.

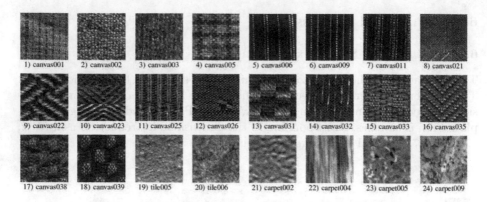

1) canvas001 2) canvas002 3) canvas003 4) canvas005 5) canvas006 6) canvas009 7) canvas011 8) canvas021

9) canvas022 10) canvas023 11) canvas025 12) canvas026 13) canvas031 14) canvas032 15) canvas033 16) canvas035

17) canvas038 18) canvas039 19) tile005 20) tile006 21) carpet002 22) carpet004 23) carpet005 24) carpet009

Fig. 5. 128 × 128 blocks from the 24 texture classes of the Outex database.

It is evident that, in general, the methods used to reduce the dimensionality of the full feature space have significant impact on classification accuracy. In this way, the choice of d (target feature vector size) can improve or hamper the discriminative capacity of texture descriptors. While the brute force search for the best value of d seems a major drawback of many works in feature selection using dimensionality reduction techniques, in this paper, we adopted an intrinsic dimensionality estimation approach to find the intrinsic value of d. This approach computes the accumulated percentual of variance of the ith component from a list of n components (dimensions). Then the intrinsic dimensionality of the reduced space will be the first component with accumulated variance ratio greater than 95% for ISOMAP algorithm, for instance. One of the important observations from this result is that the reduction from the computed d to a higher number of dimensions often maintains the accuracies and sometimes can even slightly increase them. Figure 6 shows the observed value of d, varying from 0 to 100, based on the percentage of variance retained in the attributes. It also shows that the intrinsic dimensionalities of Salzburg and Outex datasets for ISOMAP algorithm are 10 and 24, respectively.

Since the results showed in Tables 1 and 3 are only an overall view, we performed a deeper analysis using the following pairs of methods: PCA with ISOMAP; PCA with LLE and PCA with Lap. Eig. Also, because of the comparison must be carried out in pairs and under no assumption about the distribution of the data, the Wilcoxon's test was performed to compare the accuracies obtained by the reduction methods showed in Tables 1 and 3. A significance level of $\alpha = 0.05$ was used in all tests. To observe the results in more detail, boxplots for Salzburg and Outex datasets using PCA with LLE and PCA with Lap. Eig are shown in Figs. 3 (b) and (c), respectively. The boxplots shaded in gray correspond to data with significant difference when compared to the accuracy of the other method, obtaining p-value < 0.05.

We obtained significantly better results for PCA when compared with the LLE method for Salzburg (see Fig. 3 (b)) and with the Lap. Eig method for

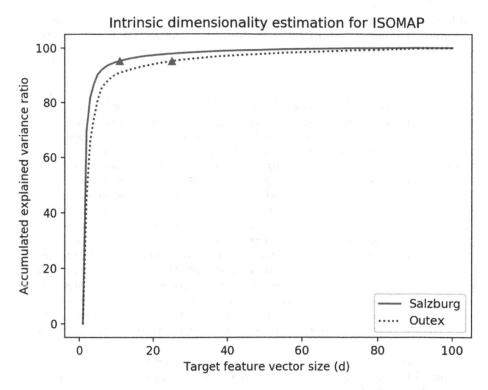

Fig. 6. Intrinsic dimensionality estimation for ISOMAP for Salzburg and Outex datasets using the full feature space.

Outex (see Fig. 3 (c)), according to the Wilcoxon's statistical test. On a more general note, it can be observed that the results were not significantly worse, except for the value of d equals to 1 for all the methods in the two datasets.

6.2   32 × 32 Patch-Based Feature Space Experiments

An overview of the results for the second set of experiments is shown in Tables 2 and 4. For each data set and dimensionality reduction method, we showed eleven accuracies, corresponding to the dimensionality reduction with the value of d equals to 5, 10, 15, 20, 25, 30, 35, 40, 45, 50 and 100. Here, the idea is to test if the extraction of patches directly from a texture image raw data can improve or maintains the results, and also to compare the use of PCA and manifold learning methods in this way.

The only result that significantly presented difference among the pair-wise comparisons was between PCA and ISOMAP for Salzburg data set as shown in Fig. 3 (a). PCA significantly presented better accuracies than ISOMAP for salzburg data set, according to the Wilcoxon's statistical test for p-value < 0.05. More importantly, on a more general perspective (i.e., for all accuracies), the results were not significantly worse.

Table 1. Quantitative results when comparing PCA method with four techniques of manifold learning in Salzburg dataset. Here we considered the full feature vector created by concatenating all descriptors.

		Number of features									
		1	2	3	4	5	6	7	8	9	10
PCA	KNN	58.33	88.89	88.89	100.00	100.00	100.00	100.00	100.00	100.00	100.00
	N.B.	80.56	86.11	97.22	100.00	97.22	100.00	97.22	100.00	100.00	100.00
	MLP	38.89	77.78	80.56	94.44	97.22	91.67	97.22	100.00	100.00	100.00
	DT	52.78	86.11	88.89	94.44	94.44	88.89	94.44	94.44	97.22	88.89
ISOMAP	KNN	86.11	100.00	100.00	100.00	100.00	100.00	100.00	100.00	97.22	100.00
	N.B.	75.00	100.00	100.00	97.22	100.00	97.22	100.00	94.44	88.89	100.00
	MLP	44.44	77.78	91.67	91.67	94.44	94.44	100.00	97.22	97.22	100.00
	DT	75.00	97.22	88.89	94.44	97.22	94.44	83.33	83.33	91.67	83.33
LLE	KNN	66.67	69.44	94.44	100.00	97.22	97.22	100.00	100.00	100.00	97.22
	N.B.	41.67	68.75	78.13	83.33	82.29	82.29	94.79	92.71	89.58	90.63
	MLP	30.56	16.67	55.56	55.56	47.22	50.00	69.44	47.22	94.44	83.33
	DT	63.89	75.00	91.67	94.44	91.67	97.22	97.22	88.89	94.44	94.44
Lap. Eig.	KNN	72.22	83.33	100.00	100.00	97.22	100.00	100.00	94.44	100.00	100.00
	N.B.	72.22	83.33	100.00	97.22	100.00	97.22	97.22	94.44	100.00	97.22
	MLP	11.11	55.56	91.67	100.00	91.67	94.44	97.22	91.67	100.00	97.22
	DT	63.89	80.56	100.00	100.00	94.44	94.44	94.44	94.44	91.67	91.67

Table 2. Quantitative results when comparing PCA method with four techniques of manifold learning in Salzburg dataset. Here a 32×32 patch of a texture image is considered as the feature vector.

		Number of features									
		5	10	15	20	25	30	35	40	45	50	100
PCA	KNN	100.00	100.00	100.00	100.00	100.00	100.00	100.00	100.00	100.00	100.00	100.00
	N.B.	53.82	68.14	70.33	72.68	71.99	75.23	78.95	79.15	79.15	80.78	83.24
	MLP	91.11	100.00	100.00	100.00	100.00	100.00	100.00	100.00	100.00	100.00	100.00
	DT	100.00	100.00	100.00	100.00	100.00	100.00	100.00	100.00	100.00	100.00	100.00
ISOMAP	KNN	100.00	100.00	100.00	100.00	100.00	100.00	100.00	100.00	100.00	100.00	100.00
	N.B.	44.08	45.85	52.97	51.37	54.35	56.01	54.48	56.24	58.69	60.03	68.89
	MLP	88.46	98.76	100.00	100.00	100.00	100.00	100.00	100.00	100.00	100.00	100.00
	DT	100.00	100.00	100.00	100.00	100.00	100.00	100.00	100.00	100.00	100.00	100.00
LLE	KNN	100.00	100.00	100.00	100.00	100.00	100.00	100.00	100.00	100.00	100.00	100.00
	N.B.	22.91	36.01	41.08	43.50	48.04	49.51	49.58	51.80	52.75	53.89	63.33
	MLP	43.46	52.55	54.80	57.78	63.89	68.82	69.12	71.86	72.42	77.25	90.20
	DT	100.00	100.00	100.00	100.00	100.00	100.00	100.00	100.00	100.00	100.00	100.00
Lap. Eig.	KNN	99.97	99.97	100.00	100.00	100.00	99.97	100.00	100.00	100.00	100.00	100.00
	N.B.	47.78	46.54	54.64	52.97	55.33	56.34	55.39	56.67	54.41	56.70	61.01
	MLP	66.86	76.18	80.36	86.44	91.57	92.88	95.26	98.56	98.14	99.08	99.74
	DT	99.90	99.93	99.93	99.93	99.87	99.84	99.97	99.97	99.84	99.87	99.80

Table 3. Quantitative results when comparing PCA method with four techniques of manifold learning in Outex dataset. Here we considered the full feature vector created by concatenating all descriptors.

		Number of features									
		1	2	3	4	5	6	7	8	9	10
PCA	KNN	20.83	77.08	94.79	90.63	95.83	95.83	98.96	97.92	96.88	97.92
	N.B.	25.00	73.96	91.67	87.50	91.67	92.71	91.67	95.83	96.88	98.96
	MLP	13.54	59.38	81.25	84.38	88.54	88.54	94.79	95.83	96.88	98.96
	DT	20.83	68.75	88.54	83.33	84.38	85.42	86.46	88.54	92.71	92.71
ISOMAP	KNN	31.25	83.33	89.58	92.71	92.71	96.88	97.92	97.92	98.96	95.83
	N.B.	31.25	70.83	81.25	88.54	84.38	94.79	92.71	96.88	94.79	94.79
	MLP	17.71	60.42	82.29	87.50	88.54	91.67	92.71	92.71	94.79	95.83
	DT	22.92	69.79	85.42	83.33	78.13	90.63	88.54	84.38	87.50	86.46
LLE	KNN	40.63	67.71	79.17	88.54	89.58	92.71	98.96	97.92	100.00	96.88
	N.B.	41.67	68.75	78.13	83.33	82.29	82.29	94.79	92.71	89.58	90.63
	MLP	30.56	16.67	55.56	55.56	47.22	50.00	69.44	47.22	94.44	83.33
	DT	63.89	75.00	91.67	94.44	91.67	97.22	97.22	88.89	94.44	94.44
Lap. Eig.	KNN	46.88	84.38	88.54	91.67	88.54	95.83	92.71	88.54	88.54	91.67
	N.B.	50.00	73.96	75.00	78.13	77.08	82.29	82.29	82.29	83.33	86.46
	MLP	10.42	42.71	58.33	58.33	66.67	67.71	73.96	69.79	77.08	86.46
	DT	43.75	69.79	77.08	83.33	83.33	85.42	88.54	88.54	88.54	88.54

Table 4. Quantitative results when comparing PCA method with four techniques of manifold learning in Outex dataset. Here a 32×32 patch of a texture image is considered as the feature vector.

		Number of features									
		5	10	15	20	25	30	35	40	45	50	100
PCA	KNN	56.51	62.24	61.98	60.42	58.33	54.69	53.65	52.60	48.96	45.83	37.50
	N.B.	61.98	70.05	73.95	76.30	79.43	78.91	79.69	80.47	79.95	80.47	80.47
	MLP	63.80	73.70	77.08	75.26	76.82	71.88	69.79	67.97	63.80	63.02	49.23
	DT	53.39	56.25	61.20	58.33	57.81	58.85	58.33	55.73	57.81	57.29	53.13
ISOMAP	KNN	70.31	76.82	76.56	76.30	75.78	76.56	76.56	76.56	75.78	75.50	72.14
	N.B.	55.99	68.23	72.66	73.18	73.18	73.70	73.96	72.92	73.96	74.48	73.70
	MLP	51.82	60.42	65.63	63.28	66.41	62.76	60.94	57.81	63.02	56.25	54.95
	DT	58.07	65.36	66.93	66.41	64.58	62.50	63.02	61.20	61.46	63.02	64.32
LLE	KNN	59.11	68.23	66.15	70.57	72.14	70.57	71.88	71.88	73.44	72.40	65.10
	N.B.	58.59	64.32	68.23	71.88	75.00	75.52	77.08	79.17	78.13	79.95	82.55
	MLP	37.24	37.50	42.19	42.71	44.53	45.83	48.18	46.88	46.09	47.66	53.91
	DT	53.13	53.65	50.26	58.33	59.11	55.73	59.11	56.77	53.65	51.82	52.34
Lap. Eig.	KNN	61.46	69.27	69.27	71.61	73.70	73.96	76.56	75.78	74.48	74.48	67.45
	N.B.	47.66	58.59	63.02	61.72	64.58	66.93	67.71	68.23	68.49	68.49	67.45
	MLP	59.38	67.71	74.22	71.09	71.88	75.26	74.22	74.22	74.22	73.96	67.97
	DT	58.85	63.28	62.76	62.50	61.98	65.10	66.41	64.32	63.80	62.50	60.42

7   Conclusions

This research presented an important contribution regarding the use of dimensionality reduction techniques for texture classification task. We stated that an original feature vector of D dimensions could be reduced to only d features (where $d < D$) that can better represent the data, achieving similar or higher accuracies. To produce an higher feature vector size to be used as input of the dimensionality reduction methods, we proposed two approaches: one using the full feature vector created by concatenating HoG, LBP, GLCM and Haralick descriptors; and other using the feature vector created by extracting 32×32 patches of each texture image. In this paper we especially carried out investigations through Wilcoxon's statistical test with regard to figure out whether or not manifold learning algorithms perform better than PCA. In this case for each data set, we performed six pair-wise comparisons, namely PCA with ISOMAP; PCA with LLE and PCA with Lap. Eig such as each comparison was done for both approaches. Thus, we had twelve pair-wise comparisons in total for all experiments.

In the first place, it is important to note that the use of the two approaches (full feature vector and 32×32 patches-based feature vector) followed by feature selection using both PCA and manifold learning produced, on average, superior results than the individuals descriptors for texture classification task. In the second one, among all the pair-wise comparisons involved in the experiments, only three results (from a set of twelve ones) were significantly differents, according to the boxplots in Fig. 3. These results clearly indicated that PCA overperformed the ISOMAP in Salzburg using the second approach, also overperformed the LLE in Salzburg using the first approach, and finally, overperformed the Lap. Eig in Outex using the first approach.

Since the results for the rest of nine comparisons did not significantly show the difference between the accuracies of PCA and each one of the manifold learning method, we believe that increasing the dimension of the feature vectors of both approaches by respectively adding more descriptors and by extracting 8×8 patches from texture image, may promote better performance in favour of manifold learning methods. The reason for this is that manifold learning methods are more efficient in dimensionality reduction than PCA in the presence of huge collections of texture images (bigger databases) and especially for texture classification task [4, 25].

Future studies can take into account new approaches for estimating the intrinsic dimensionality value of d for each manifold learning method, can also explore different manifold learning methods and variations of the supervised classifier's algorithms in order to produce better accuracies, as well as investigate the subspaces generated by such methods in order to understand better their discriminative power.

Acknowledgment. This study was financed in part by the Coordenação de Aperfeiçoamento de Pessoal de Nível Superior - Brasil (CAPES) - Finance Code 001.

References

1. Chumbley, J.Y.A., Moore, K., et al.: Floyd-warshall algorithm. https://brilliant. org/wiki/floyd-warshall-algorithm/. Accessed Nov 2018
2. Belkin, M., Niyogi, P.: Laplacian eigenmaps and spectral techniques for embedding and clustering. In: Dietterich, T.G., Becker, S., Ghahramani, Z. (eds.) Advances in Neural Information Processing Systems, vol. 14, pp. 585–591. MIT Press (2002)
3. Cayton, L.: Algorithms for manifold learning (2005)
4. Chen, J., et al.: Updating initial labels from spectral graph by manifold regularization for saliency detection. Neurocomputing **266**, 79–90 (2017)
5. Cireşan, D.C., Meier, U., Gambardella, L.M., Schmidhuber, J.: Deep big multilayer perceptrons for digit recognition. In: Montavon, G., Orr, G.B., Müller, K.-R. (eds.) Neural Networks: Tricks of the Trade. LNCS, vol. 7700, pp. 581–598. Springer, Heidelberg (2012). https://doi.org/10.1007/978-3-642-35289-8_31
6. Rivest, R.L., Stein, C., Cormen, T.H., Leiserson, C.E.: Introduction to Algorithms. The MIT Press, Cambridge (2009)
7. Cox, T., Cox, M.: Multidimensional scaling. In: Chen, C.-H., Härdle, W.K., Unwin, A. (eds.) Handbook of Data Visualization. Springer Handbooks of Computational Statistics. Springer, Heidelberg (2008). https://doi.org/10.1007/978-3-540-33037-0_14
8. Dash, S., Chiranjeevi, K., Jena, U.R., Trinadh, A.: Comparative study of image texture classification techniques, pp. 1–6, January 2015
9. Dunteman, G.H.: Principal Components Analysis. Sage Publications, Newbury Park (1989)
10. Guo, G., Wang, H., Bell, D., Bi, Y., Greer, K.: KNN model-based approach in classification. In: Meersman, R., Tari, Z., Schmidt, D.C. (eds.) OTM 2003. LNCS, vol. 2888, pp. 986–996. Springer, Heidelberg (2003). https://doi.org/10.1007/978-3-540-39964-3_62
11. Kwitt, R., Meerwald, P.: Salzburg texture image database. http://www.wavelab. at/sources/STex/. Accessed Feb 2018
12. Materka, A., Strzelecki, M.: Texture analysis methods - a review. Technical report, Institute of Electronics, Technical University of Lodz (1998)
13. Mirmehdi, M., Xie, X., Suri, J.: Handbook of Texture Analysis. Imperial College Press, London (2009)
14. Quinlan, J.R.: Induction of decision trees. Mach. Learn. **1**(1), 81–106 (1986)
15. Raschka, S.: Principal Component Analysis (PCA) Step by Step, April 2014
16. Saul, L.K., Roweis, S.T.: Think globally, fit locally: unsupervised learning of low dimensional manifolds. J. Mach. Learn. Res. **4**, 119–155 (2003)
17. Smith, L.I.: A tutorial on principal components analysis. Technical report, Cornell University, USA, 26 February 2002
18. Sun, Y.: Cost-sensitive boosting for classification of imbalanced data. Pattern Recogn. **40**, 3358–3378 (2007)
19. Pietikainen, M., Viertola, J., Kyllonen, J., Ojala, T., Maenpaa, T., Huovinen, S.: Outex - new framework for empirical evaluation of texture analysis algorithms, vol. ICPR/1, pp. 701–706 (2002)
20. Tenenbaum, J.B., de Silva, V., Langford, J.C.: A global geometric framework for nonlinear dimensionality reduction. Science **290**(5500), 2319 (2000)
21. Tuceryan, M., Jain, A.K.: Texture Analysis, pp. 207–248 (1998)
22. Wankhade, P.D.: A review on aspects of texture analysis of images. Int. J. Appl. Innov. Eng. Manag. (IJAIEM) **3**, 229–232 (2014)

23. Webb, G.I.: Naïve Bayes, pp. 713–714. Springer, Boston (2010). https://doi.org/10.1007/978-0-387-30164-8
24. Wilcoxon, F.: Individual comparisons by ranking methods. In: Kotz, S., Johnson, N.L. (eds.) Breakthroughs in Statistics. Springer Series in Statistics (Perspectives in Statistics), pp. 196–202. Springer, New York (1992). https://doi.org/10.1007/978-1-4612-4380-9_16
25. Yang, C., Zhang, L., Lu, H., Ruan, X., Yang, M.-H.: Saliency detection via graph-based manifold ranking. In: Proceedings of the 2013 IEEE Conference on Computer Vision and Pattern Recognition, CVPR 2013, pp. 3166–3173. IEEE Computer Society, Washington, DC (2013)

Event-Oriented Keyphrase Extraction Based on Bi-clustering Model

Lin Zhao[1,2], Liangjun Zang[1(✉)], Longtao Huang[1], Jizhong Han[1,2],
and Songlin Hu[1,2]

[1] Institute of Information Engineering, Chinese Academy of Sciences,
Beijing, China
{zhaolin,zangliangjun,huanglongtao,hanjizhong,husonglin}@iie.ac.cn
[2] School of Cyber Security, University of Chinese Academy of Sciences,
Beijing, China

Abstract. Keyphrase extraction, as a basis for many natural language processing and information retrieval tasks, can help people efficiently discover their interested information from vast streams of online documents. Previous methods are mostly proposed in general purpose, where keyphrases that represent the main topics are extracted. However, such keyphrases can hardly distinguish events from massive streams of long text documents that share similar topics and contain highly redundant information. In this paper, we address the task of keyphrase extraction for event-oriented retrieval. We propose a novel bi-clustering model for clustering the documents and keyphrases simultaneously. The model consequently makes the extracted keyphrases more specific and related to the event. We conduct a series of experiments on a real-world dataset. The experimental results demonstrate the better performance of our approach than other unsupervised approaches.

Keywords: Event-oriented · Keyphrase extraction ·
Simultaneous learning · Bi-clustering model · Information retrieval

1 Introduction

With tremendous amounts of documents on trending and breaking news generated by various Internet media providers, it becomes increasingly difficult for people to digest such a great many streaming news information. Search engines retrieve documents from large corpora based on users' input queries that specify their interested events. However, the quality of the retrieval results depends on people's capability of refining proper keyphrases about the events. Keyphrase extraction, a task of automatically extracting descriptive phrases or concepts that represent the main topics of a document, can help people achieve more

Supported by the National Key Research and Development Program of China (No. 2017YFB1010000) and the National Natural Science Foundation of China (No. 61702500).

representative phrases to issue their queries. However, most existing researches on keyphrase extraction did not take the retrieval performance into consideration. This causes the extracted keyphrases fail to retrieve events from massive streams of long text documents that share similar topics and contain highly redundant information. To this end, this paper addresses the task of event-oriented keyphrase extraction. The goal is to automatically extract keyphrases that represent the specific events and distinguish with others.

Accurately identifying event-oriented keyphrases from documents will benefit many downstream applications such as event-oriented information retrieval, event monitoring, event tracking, news recommendation, text summarization. For example, censors look for solutions to monitor online current affairs hotspots from blogs, forums and news site for possible criminal activities and event-oriented public opinion analysis. The key to extract event-oriented keyphrases is to find more documents about the specific events and decrease the possibility of finding documents describing other events.

Generally, previous studies about automatic keyphrase extraction fall into two categories: the supervised keyphrase extraction and the unsupervised keyphrase extraction. The supervised keyphrase extraction task is usually treated as a binary classification problem [8,14]. In this approach, keyphrases and non-keyphrases are labeled by human judges in training documents and a classifier is trained by using training documents annotated with keyphrases, then the classifier is applied to determine whether a candidate phrase is a keyphrase in test documents. On the contrary, the unsupervised methods do not require labeled training data. Instead, some external statistic information are explored to identify the keyphrases [18,34]. In this paper, we mainly focus on extracting keyphrases with an unsupervised method. However, the existing unsupervised keyphrase extraction methods suffer from two drawbacks:

- **Ignore users' specific needs.** The document collection may contain several events (or several aspects of an event). For different applications, the users may be interested in different events or aspects, thus they need different keyphrases to reflect their needs. Taking the theme of earthquake from Sina news as an example (see Sect. 4.1), there are some earthquake events, e.g. "四川康定发生6.3级地震" (Sichuan earthquake), "尼泊尔发生8.1级强震" (Nepal earthquake)", "earthquake" is a keyphrase for the entire theme while it should not be the keyphrase when we only consider the rescue aspect in Nepal earthquake.
- **Fail to extract event-oriented keyphrases.** Considering different events with simi-lar topics are likely to share keyphrases, which makes existing methods fail to distinguish a specific event from others with similar topics. For example, topic models (e.g. Latent Dirichlet Allocation) can simultaneously cluster documents and generate representative words for each topic, but topic-based methods are likely to put the documents about "Sichuan earthquake" and those about "Nepal earthquake" into the same cluster because both events share many topical keywords like "坍塌" (collapse), "救援" (rescue) and "重建" (rebuild).

Existing work rarely focuses on automatically extracting keyphrases for a particular event. In this paper, we address the task of event keyphrase extraction, which is a specific task to obtain keyphrases for users about their interested events. Actually, the task of event keyphrase extraction comprises two sub-tasks, i.e. event identification and keyphrase extraction. Event identification aims to cluster all documents into different groups, each of which corresponds to one specific event. Keyphrase extraction aims to extract keyphrases from the documents for each event. A simple and straightforward strategy is to combine them in a pipelined way, but it suffers the problem that the errors raised by event identification will lead to poor performance of keyphrase extraction. Another more reasonable strategy is to simultaneously address the two subtasks. It provides the opportunity of mutual boosting, i.e., the two sub-tasks can potentially benefit from each other, therefore we adopt the way of simultaneous learning. For evaluating our approach, we implemented a retrieval system to evaluate the effectiveness of event-oriented keyphrases in the application of information retrieval. The major contributions of this paper are summarized as follows:

(1) We address the task of event-oriented keyphrase extraction, which can benefit many downstream applications.
(2) We take users' specific needs and event-oriented characteristics into consideration and propose the novel method based on the bi-clustering model.
(3) Extensive experiments over real data show that our method achieves better performance compared with the other unsupervised methods.

The rest of the paper is organized as follows. In Sect. 2, we give some background information and related works. Section 3 presents the methodology, including problem formulation, keyphrase extraction, and document retrieval. In Sect. 4, we describe the experiment and analysis, including experiment setting and results analysis. Section 5 presents conclusions and a discussion of further research.

2   Related Work

Automatic keyphrase extraction, event detection, and document retrieval are three lines of studies that are related to our work. Automatic keyphrase extraction selects the important and topical keywords from documents automatically [32] and its goal is to extract phrases that represent or are related to the topics discussed in the given documents [5,10,19,30]. Event detection is a basic problem of the information extraction, and it is also a sub-field in information retrieval, which aims to detect real events from multiple document streams [7,31,37,39]. Document retrieval is a process of matching the user's request against a collection of documents [21]. In this paper, we apply document retrieval to evaluate the performance of event-oriented keyphrases.

2.1 Keyphrase Extraction

Existing methods for keyphrase extraction can be divided into supervised and unsupervised methods [22]. Most supervised methods formalize keyphrase extraction as a binary classification problem. However, supervised methods require training data with labeled keyphrases, which is extremely expensive and time-consuming for do-main-oriented application scenarios and hard to adapt to other domains.

Existing unsupervised methods for keyphrase extraction can be categorized into several groups. First, rank-based methods build a word co-occurrence graph from the input document and then ranking all nodes on the graph [12,18,33]. Such methods often suffer the problem of information loss. For example, if two words never occur together within a predefined window size, there will be no edge to connect them in the co-occurrence graph. Second, topic-based methods aim to group the candidate keyphrases in a document into topics, such that each topic is composed of all and only those candidate keyphrases that are related to that topic [15,19]. Third, external resource based methods explore external text corpus to enhance the performance of keyphrase extraction [28,29]. However, such methods often bring about the information overload problem, e.g. the real meanings of words in the document may be overwhelmed by a large amount of introduced external corpus. Finally, knowledge-based methods combine semantic similarity clustering with knowledge graphs to help discover hidden semantic relations in documents [6]. Such a method does not consider the relevance of news articles belong to the same event, which demonstrates efficiency only in single-document keyphrase extraction task.

2.2 Event Detection

Many tasks are put forward to develop and evaluate technologies for event detection, e.g. topic detection and tracking (TDT), automatic content extraction (ACE), and text analysis conference knowledge base population (TAC KBP). In addition, there has been a lot of research work in event detection. Neural network models have been the most successful methods for event detection. For instance, Chung et al. [1] introduce a DAG-GUR architecture that captures the syntactic and context information through a bi-directional reading of the text with dependency parse relationships. Feng et al. [7] combined a convolutional neural network (CNN) with a bi-directional long short-term memory (Bi-LSTM) to create a hybrid network. The hybrid network was fed to a linear model for capturing sequence and chunk information from specific contexts.

2.3 Document Retrieval

The main concern of document retrieval is to retrieve documents from a corpus based on the specific user query. Different probabilistic model and language model have been proposed over the past decade for document retrieval, such as BM25 [26], probabilistic retrieval model for semi-structured data [17], mixture of language models [24], and machine learning-based methods [23,38]. However, early methods mentioned above are difficult to meet the needs of users. They assume bag-of-words document representation and match phrases directly in queries and documents, which suffers from the issue of lexical gap, when similar concepts are expressed using different words in queries and relevant documents. The success of deep learning has revitalized research on text matching recently. Several neural architectures have been proposed for document retrieval [2,3,25,35,36], which can be categorized into two classes according to their model architecture. One is representation-based model, such as DSSM [13], CDSSM [27], ARC-I [11] and DCNN [16], which builds a good representation for a text with deep neural network, and then conducts matching between text representations. The other is the interaction-based model, such as DeepMatch [20], ARC-II [11], DRMM [9] and K-NRM [35], which builds local interactions between two texts, and then learns hierarchical interaction patterns for matching with deep neural networks.

3   Methodology

The framework of our event-oriented keyphrase extraction is shown in Fig. 1. It consists of the following three steps: (1) problem formulation, (2) keyphrase extraction, and (3) document retrieval. We collect a lot of news documents, each document corresponds to the specific event and each event includes many documents. Then we utilize the bi-clustering model to extract keyphrases from news documents. Finally, we use the keyphrases as queries to retrieve news documents, in order to evaluate the ability of keyphrases to represent events.

3.1   Problem Formulation

Event-Oriented Keyphrases. We define event-oriented keyphrase as a continuous sequence of keywords, which is highly important and relevant to the event. For instance, "地震" (earthquake), "康定县" (Kangding County), "四川省" (Sichuan Province), and "倒塌" (collapse) are the keyphrases of the example document in Fig. 2(a). "防御中心" (Prevention Center), "监测" (monitoring), "调查" (investigation), and "损失" (damage) are the keyphrases of the example document in Fig. 2(b). Note that "地震" (earthquake) and "倒塌" (collapse) are not the keyphrases in the event of Nepal earthquake, because (1) the event of Nepal earthquake expresses the rescue phase rather than report the earthquake in this example, and (2) they are shared in both the "Sichuan earthquake" and "Nepal earthquake", which cannot distinguish different events with similar topics.

Keyphrases Extraction. Given the event set E and news set N includes D news subset, each news subset d_j corresponds to an event e_i. Our goal is to extract event-oriented keyphrases of each event from the corresponding collection D_i, where

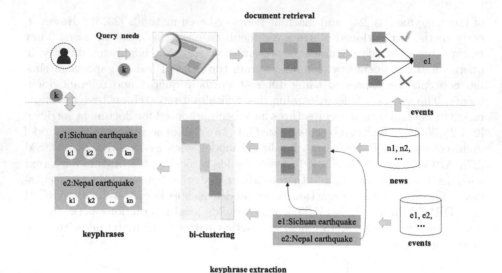

Fig. 1. The process of event-oriented keyphrase extraction

- $E = \{e_i | 1 \leq i \leq |E|\}$ denotes the collection of specific events.
- $D = \{d_j | 1 \leq j \leq |D|\}$ denotes the news subsets.
- $N = \sum\limits_{i=1}^{|E|} D_i$ denotes the news data set grouped by events.

Keyphrase Evaluation. We evaluate event-oriented keyphrases through event-oriented retrieval tasks. Given the event-oriented keyphrases K by keyphrase extraction mentioned above. We issues $K \in e_i$ as queries to retrieve news in search engine (e.g. Google Search). The search engine returns a collection of news set R. The goal is to evaluate the effectiveness of K by judging whether set R belongs to e_i.

- $K = \{k_1, k_2, ..., k_{|K|}\}$ denotes the collection of event-oriented keyphrases are extracted from D_i.
- $R = \{r_1, r_2, ..., r_{|R|}\}$ denotes the retrieved news collection related to the event e_i.

3.2   Keyphrases Extraction

Feature Selection. Different from the keyphrase extraction in document level, we consider the correlation of keyphrases as event features in specific events and propose the following hypothesis:

Assume that an event is represented by a set of key elements. Event features are used to enrich these key elements. There are the following relationships:

22日16时55分在**四川**省甘孜藏族自治州康定县发生6.3级**地震**。住在**康定县**城的一位居民告诉记者，当地房屋摇晃厉害，震感很强烈，但他目击范围之内没有看到房屋**倒塌**。(An **earthquake** of magnitude 6.3 occurred in **Kangding County**, Ganzi Tibetan Autonomous Prefecture, **Sichuan Province** at 16:55 on the 22nd. A resident living in Kangding County told reporters that the local houses shook badly and felt a strong shock, but he did not see the **collapse** of the houses within the scope of his witness.)

(a) Sichuan earthquake

地震发生1小时，房屋倒塌十分严重，西藏自治区地震局派出了由西藏地震灾害**防御中心**主任XX带队的第一支现场工作队，赶赴灾区开展震情**监测**、灾情**调查**和震害**损失**评估等现场工作。(One hour after the earthquake, houses collapsed seriously. The Seismological Bureau of Tibet Autonomous Region dispatched the first on-site team led by XX, director of the Tibet Earthquake Disaster **Prevention Center**, to carry out on-site work such as earthquake **monitoring**, disaster **investigation** and **damage** assessment.)

(b) Nepal earthquake

Fig. 2. Example of event-oriented keyphrases from "Sichuan earthquake" (a) and "Nepal earthquake" (b) Shared keyphrases are red font and underlined in "Nepal earthquake". (Color figure online)

(1) Key elements extracted from similar topics are similar. (2) Event features are to enrich a group of key elements. For instance, the three key elements of the event of earthquake = "四川" (Sichuan province in China), "地震" (earthquake), "记者" (journalist). The features are "康定县" (Kangding country), "应急" (emergency), "刘忠俊" (Zhongjun Liu), which are to describe "四川" (Sichuan province in China), "地震" (earthquake), "记者" (journalist) respectively.

Based on the above hypothesis, the method of extracting features include two steps: (1) Find the set of key elements. First, we select candidate elements with named entity recognition from each news. Then we calculate the word frequency of each candidate element in the entire news dataset. Finally, the target key elements with maximum word frequency are selected. (2) Find corresponding features. We obtain candidate features using a stop word list to remove stop words and filtering words with certain part-of-speech tags (e.g., nouns, verbs). Then we exploit the co-occurrence relation between features and key elements. The feature is selected if this feature and an element appear simultaneously in a sentence.

Bi-clustering Modeling. Bi-clustering consists of simultaneous partitioning of the set of samples and the set of their attributes (features) into subsets (classes). Samples and features classified together are supposed to have high relevance to each other [6]. Motivated by this work, we employ the bi-clustering process to analyze the relationship between news documents and event features. A basic premise behind our method is a duality of features and document clustering. That is, feature clustering induces document clustering while document clustering induces feature clustering. The goal is to find the best bi-clusters with correlation higher than those in the corresponding other rows and columns.

Generally, given news and features to bi-clustering, the process can be summarized as follows:

(1) Each of the news documents n to be clustered is represented by a p-dimensional feature vector. The entire documents are represented by a matrix of shape $n \times p$. We calculate TF-IDF values of each of the features, and a word frequency matrix (denoted as $A_{n \times p}$) is obtained by values vectorizing. We apply log normalization to normalize the matrix $A_{n \times p}$. The log of the data matrix is computed with $L = log A_{n \times p}$. The final matrix is computed according to the formula:

$$K_{ij} = L_{ij} - \overline{L_{i\cdot}} - \overline{L_{\cdot j}} + \overline{L_{\cdot\cdot}} \tag{1}$$

where $\overline{L_{i\cdot}}$ denotes the column mean, $\overline{L_{\cdot j}}$ denotes the row mean and $\overline{L_{\cdot\cdot}}$ denotes the overall mean of L.

(2) We take a matrix K as input to bi-clustering and get the bi-cluster partitions with Dhillon's Spectral Co-Clustering algorithm [4]. Ideally, the rearranging the matrix reveals the bi-clusters on the diagonal because each row and each column belongs to exactly on bi-cluster. But there are some noise data in actual news, so the diagonal structure is not perfect.

(3) We filter out the noise data from sub-matrix in the diagonal as the best features bi-clusters. Each of the bi-clusters corresponds to an event. The best bi-clusters indicate subsets features used more often in those subsets news. For instance, the bi-clustering results of raw data and filtering data are shown in Fig. 3(a) and (b), respectively, where the raw data come from our experimental dataset. First, we select 100 news documents and corresponding 300 features, Fig. 3(a) presents the bi-clustering result. There are five bi-cluster partitions in Fig. 3(a), where the first and the fourth bi-clusters contain noise data. Then we filter out the noise data from the two bi-clusters and run the bi-clustering process again, the result is shown in Fig. 3(b).

3.3 Document Retrieval

Our task of event-oriented keyphrases extraction is to retrieve events from massive streams of long text documents. Hence, we can issue event-oriented keyphrases as queries to the document retrieval system, and evaluate the quality of event-oriented keyphrases by returning the results of the retrieved document (see Fig. 4).

It is ambiguous for using raw keyphrase to search event directly in a corpus containing lots of similar topic collections because it may appear in multiple events. For instance, we use "地震" (earthquake) as the query to search in a search engine, the returned retrieval lists will be much news about the different event. Hence, preparing for information retrieval tasks, we need to generate

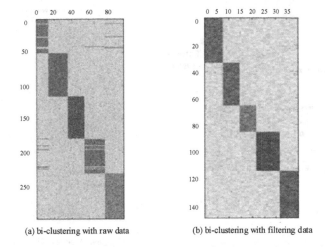

(a) bi-clustering with raw data   (b) bi-clustering with filtering data

Fig. 3. An example of bi-clustering before and after filtering the raw data

query suggestions based on the extracted keyphrases. The method of generating suggestion is the combination of keyphrases. Keyphrases combination can narrow down the scope of the search and help users to find more relevant news. The detailed processes are as follows:

(1) Query suggestion generation. Each query corresponds to a user's need for event queries. Query suggestions are generated based on different events. We apply a query parser to map the relationship between the query and the event. The mapping method is to calculate the news numbers of different events and select the event with maximum news numbers as the target event.
(2) We issue different queries (e.g. event title, query suggestion) into a document retrieval system based on BM25 in order to compare the quality of different queries.
(3) We calculate the accuracy, recall, and F1-measure based on the retrieved results returned by different queries. The higher the three metrics, the better the quality of the keywords.

4   Experiment and Analysis

In this section, we describe the process of experiment and evaluation result of testing our bi-clustering model on the dataset. We first describe the dataset collected by crawling from Sina news, then we introduce the evaluation tasks and metrics for the upcoming event-oriented retrieval. Finally, we compare and analyze the results from the bi-clustering model and baselines in evaluation tasks.

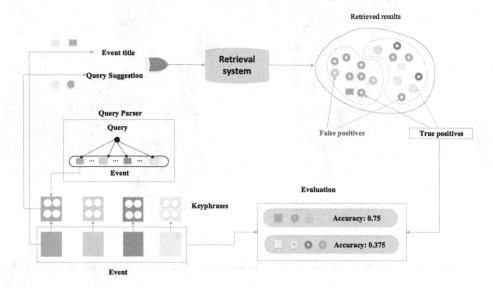

Fig. 4. The process of document retrieval based on event-oriented keyphrases

4.1 Experiment Setting

Data Description. To our knowledge, there are no datasets available to satisfy our task currently. We picked 52 events ranged from 2014 to 2016 from Sina News, including 5932 news. The criterion we used for selecting events is that they should contain multiple categories and have similar topics in different categories. These events are involved in the categories of disaster, accident, conference, current affairs, and military. Table 1 shows some examples of the selected events and corresponding category. We hired several annotators to create truth data. Annotators were asked to (1) crawl news collection from Sina News, (2) classify news documents into events corresponding to the above-mentioned categories. Finally, we extract event-oriented keyphrases from 80% news of each category and the remaining 20% news are used for event-oriented retrieval tasks.

Evaluation Metrics. For evaluating the quality of event-oriented keyphrases, we consider the following queries generated by different methods for performance comparisons: (i) event titles as queries directly, (ii) query suggestions based on the topic model as queries, and (iii) query suggestions based on the bi-clustering model. We collect the top-K news documents ($K = 10$) of the query results. The metrics for comparison are precision, recall, and F1-measure. For the returning results of each event, these metrics were defined as follows:

$$p_e = \frac{\sum_{q \in Q} p_q}{|Q|} \tag{2}$$

$$r_e = \frac{\sum_{q \in Q} r_q}{|Q|} \tag{3}$$

Table 1. The distribution of selected events and category.

Category	Number	Event sample
Disaster	10	四川省康定县6.3级地震 (A 6.3 magnitude earthquake in Kangding, Sichuan)
Accident	10	俄罗斯客机在埃及坠毁 (The Russian airliner crashed in Egypt)
Conference	5	全国人民代表大会常务委员会第十一次会议 (The eleventh meeting of the Standing Committee of the 12 National People's Congress)
Current affairs	19	习近平访问瑞士 (Xi Jinping's visit to Switzerland)
Military	8	中国人民解放军军改 (Military reform of the Chinese Liberation Army)

$$f_{measure-e} = \frac{2 \times p_e \times r_e}{p_e + r_e} \tag{4}$$

where Q is the number of query suggestions in each event, p_q is the precision of each query suggestion and r_q is the recall of each query suggestion.

4.2 Results Analysis

We present the results of precision, recall, and F1-measure based on queries generated by different methods in Table 2.

We have the following observations from Table 2: (1) From the results on all categories, bi-clustering model outperforms the other two both on precision and recall. The topic model performs better than the event titles on precision and recall. In our consideration, event titles as queries can find key elements and corresponding features in order to match target news accurately, while other methods are missing important event features. For instance, there is returned seldom results using the title "2016各地两会" (The National People's Congress and Chinese People's Political Consultative Conference around the nation in 2016). The reason is that "各地" (around the nation) does not appear in news, it was expressed using the name of different cities, such as "四川" (Sichuan province in China), "成都" (Chengdu city in Sichuan province). (2) From the results on each category, we can find the disaster category performs the best and conference category performs the worse than other categories. This is because the news described different events is a higher similarity in the conference, so most of the key elements are the same. For instance, "全国人民代表大会常务委员会第十一次会议" (the eleventh meeting of the Standing Committee of the 12 National People's Congress) and "全国人民代表大会常务委员会第十一次会议" (the tenth meeting of the Standing Committee of the 12 National People's Congress) are shared the same keyphrases, so it is difficult to distinguish such events.

Table 2. Performance comparisons of different methods, the highest values are in bold.

Category	Metric	Event titles	Topic model	Bi-clustering model
Overall	P	0.665	0.809	**0.854**
	R	0.443	0.538	**0.568**
	F	0.531	0.647	**0.683**
Disaster	P	0.633	0.953	**0.980**
	R	0.422	0.635	**0.652**
	F	0.507	0.762	**0.784**
Accident	P	0.825	0.890	**0.940**
	R	0.549	0.592	**0.626**
	F	0.659	0.712	**0.752**
Conference	P	0.400	0.553	**0.647**
	R	0.264	0.368	**0.430**
	F	0.318	0.442	**0.517**
Current affairs	P	0.700	0.813	**0.833**
	R	0.466	0.542	**0.555**
	F	0.559	0.651	**0.667**
Military	P	0.766	0.840	**0.873**
	R	0.516	0.556	**0.581**
	F	0.613	0.672	**0.698**

5   Conclusion

In this paper, we have implemented the task of event-oriented keyphrases and proposed a novel method for event-oriented keyphrase extraction, which uses the bi-clustering model to cluster news documents and keyphrases simultaneously. Also, we implement a simple document retrieval system based on BM25 to evaluate the quality of the extracted keyphrases. The experimental results showed that the event-oriented keyphrases greatly improve the performance of event retrieval.

Our future work may include the following directions. First, we only evaluated the performance of the proposed model by conducting experiments on a single Chinese news dataset for engineering needs in this paper. Therefore, in the future, we will try to conduct experiments on English dataset (e.g. crawl events from Wikipedia). Then, we can extract event-oriented keyphrases by deep learning model. Finally, we plan to utilize external knowledge to compensate for event information. For example, integrate knowledge graph into the task of event-oriented keyphrases extraction to improve the performance of event retrieval.

References

1. Chung, J., Gülçehre, Ç., Cho, K., Bengio, Y.: Empirical evaluation of gated recurrent neural networks on sequence modeling. CoRR abs/1412.3555 (2014)
2. Dai, Z., Xiong, C., Callan, J.P., Liu, Z.: Convolutional neural networks for soft-matching n-grams in ad-hoc search. In: WSDM (2018)
3. Dehghani, M., Zamani, H., Severyn, A., Kamps, J., Croft, W.B.: Neural ranking models with weak supervision. In: SIGIR (2017)
4. Dhillon, I.S.: Co-clustering documents and words using bipartite spectral graph partitioning. In: KDD (2001)
5. Ding, Z., Zhang, Q., Huang, X.: Keyphrase extraction from online news using binary integer programming. In: IJCNLP (2011)
6. Farzindar, A., Khreich, W.: A survey of techniques for event detection in twitter. Comput. Intell. **31**, 132–164 (2015)
7. Feng, X., Huang, L., Tang, D., Ji, H., Qin, B., Liu, T.: A language-independent neural network for event detection. Sci. China Inf. Sci. **61**, 1–12 (2016)
8. Frank, E., Paynter, G.W., Witten, I.H., Gutwin, C., et al.: Domain-specific keyphrase extraction. In: Proceedings of the Sixteenth International Joint Conference on Artificial Intelligence, pp. 668–673. Morgan Kaufmann Publishers (1999)
9. Guo, J., Fan, Y., Ai, Q., Croft, W.B.: A deep relevance matching model for ad-hoc retrieval. In: CIKM (2016)
10. Hasan, K.S., Ng, V.: Automatic keyphrase extraction: a survey of the state of the art. In: Proceedings of the 52nd Annual Meeting of the Association for Computational Linguistics (ACL 2014) (2014)
11. Hu, B., Lu, Z., Li, H., Chen, Q.: Convolutional neural network architectures for matching natural language sentences. In: NIPS (2014)
12. Huang, C., Tian, Y., Zhou, Z., Ling, C.X., Huang, T.: Keyphrase extraction using semantic networks structure analysis. In: Sixth International Conference on Data Mining (ICDM 2006), pp. 275–284 (2006)
13. Huang, P.S., He, X., Gao, J., Deng, L., Acero, A., Heck, L.P.: Learning deep structured semantic models for web search using click through data. In: CIKM (2013)
14. Hulth, A.: Improved automatic keyword extraction given more linguistic knowledge. In: Proceedings of the 2003 Conference on Empirical Methods in NLP, pp. 216–223 (2003)
15. Joorabchi, A., Mahdi, A.E.: Automatic keyphrase annotation of scientific documents using wikipedia and genetic algorithms. J. Inf. Sci. **39**, 410–426 (2013)
16. Kalchbrenner, N., Grefenstette, E., Blunsom, P.: A convolutional neural network for modelling sentences. In: ACL (2014)
17. Kim, J., Xue, X., Croft, W.B.: A probabilistic retrieval model for semistructured data. In: ECIR, pp. 228–239 (2009)
18. Liu, Z., Huang, W., Zheng, Y., Sun, M.: Automatic keyphrase extraction via topic decomposition. In: EMNLP (2010)
19. Liu, Z., Li, P., Zheng, Y., Sun, M.: Clustering to find exemplar terms for keyphrase extraction. In: EMNLP, pp. 257–266 (2009)
20. Lu, Z., Li, H.: A deep architecture for matching short texts. In: Advances in Neural Information Processing Systems, pp. 1367–1375 (2013)
21. Manning, C.D., Raghavan, P., Schütze, H.: Introduction to information retrieval (2008)
22. Mihalcea, R., Tarau, P.: TextRank: bringing order into text. In: EMNLP (2004)

23. Nikolaev, F., Kotov, A., Zhiltsov, N.: Parameterized fielded term dependence models for ad-hoc entity retrieval from knowledge graph. In: SIGIR (2016)
24. Ogilvie, P., Callan, J.P.: Combining document representations for known-item search. In: SIGIR (2003)
25. Onal, K.D., Altingövde, I.S., Senkul, P., de Rijke, M.: Getting started with neural models for semantic matching in web search. CoRR abs/1611.03305 (2016)
26. Robertson, S.E., Zaragoza, H., Taylor, M.J.: Simple BM25 extension to multiple weighted fields. In: CIKM (2004)
27. Shen, Y., He, X., Gao, J., Deng, L., Mesnil, G.: A latent semantic model with convolutional-pooling structure for information retrieval. In: CIKM (2014)
28. Shi, T., Jiao, S., Hou, J., Li, M.: Improving keyphrase extraction using wikipedia semantics. In: 2008 Second International Symposium on Intelligent Information Technology Application, vol. 2, pp. 42–46 (2008)
29. Shi, W., Zheng, W., Yu, J.X., Cheng, H., Zou, L.: Keyphrase extraction using knowledge graphs. In: Chen, L., Jensen, C.S., Shahabi, C., Yang, X., Lian, X. (eds.) APWeb-WAIM 2017, Part I. LNCS, vol. 10366, pp. 132–148. Springer, Cham (2017). https://doi.org/10.1007/978-3-319-63579-8_11
30. Tomokiyo, T., Hurst, M.: A language model approach to keyphrase extraction. In: Proceedings of ACL Workshop on Multiword Expressions, pp. 33–40 (2003)
31. Tu, W., Cheung, D.W.L., Mamoulis, N., Yang, M., Lu, Z.: Real-time detection and sorting of news on microblogging platforms. In: PACLIC (2015)
32. Turney, P.D.: Learning algorithms for keyphrase extraction. Inf. Retr. **2**, 303–336 (2000)
33. Wan, X., Xiao, J.: Exploiting neighborhood knowledge for single document summarization and keyphrase extraction. ACM Trans. Inf. Syst. **28**, 8:1–8:34 (2010)
34. Wan, X., Yang, J., Xiao, J.: Towards an iterative reinforcement approach for simultaneous document summarization and keyword extraction. In: ACL (2007)
35. Xiong, C., Dai, Z., Callan, J.P., Liu, Z., Power, R.: End-to-end neural ad-hoc ranking with kernel pooling. In: SIGIR (2017)
36. Yang, L., Ai, Q., Guo, J., Croft, W.B.: aNMM: ranking short answer texts with attention-based neural matching model. In: CIKM (2016)
37. Yang, M., Cui, T., Tu, W.: Ordering-sensitive and semantic-aware topic modeling. In: AAAI (2015)
38. Zhiltsov, N., Kotov, A., Nikolaev, F.: Fielded sequential dependence model for ad-hoc entity retrieval in the web of data. In: SIGIR (2015)
39. Zhu, J., Xu, C., Li, Z., Fung, G.P.C., Lin, X., Huang, J., Huang, C.: An examination of on-line machine learning approaches for pseudo-random generated data. Cluster Comput. **19**, 1309–1321 (2016)

Track of Solving Problems with Uncertainties

Path-Finding with a Full-Vectorized GPU Implementation of Evolutionary Algorithms in an Online Crowd Model Simulation Framework

Anton Aguilar-Rivera[✉]

Barcelona Supercomputing Center, Barcelona, Spain
anton.aguilar@bsc.es

Abstract. This article introduces a path-finding method based on evolutionary algorithms and a fully vectorized GPU implementation of it. The algorithm runs on real-time and it can handle dynamic obstacles in maps of arbitrary size. The experiments show the proposed approach outperforms other traditional path-finding algorithms (e.g. A*). The conclusions present further improvement possibilities to the proposed approach like the application of multi-objective algorithms to represent full crowd models.

1 Introduction

Crowd behavior has been widely studied in the literature [6], being crowd models especially important to the digital entertainment industry, where animation of large groups of characters are desired. Moreover, crowd models are also important to architectonic design and emergency planning. For example, the number and position of facility exits should be carefully selected to minimize evacuation time in the case of an emergency [8].

Crowd simulation is a complex problem. Both map size and number of agents are concerns when devising scalable simulators. Besides, a natural behavior of agents is usually desirable, specially for visualization use. Also, researchers working with crowd simulations may be interested in the over-all effect of subtle changes in the model, specially in social sciences studies [13].

In their survey, Ijaz, Sohail, and Hashish [6] made a classification of the different types of crowd models. They can be classified by resolution in the following manner: Macroscopic, mesoscopic, and microscopic. Macroscopic models describe the general motion of the crowd only. Mesoscopic models are based on cellular automaton and provide better resolution; movement rules are applied to the grid instead to agents. On the other hand, in microscopic models the crowd is composed by individual agents who make their own decisions. As expected, computational cost increases with an increase of resolution.

Microscopic models should provide information about the agent's position, speed, acceleration, intended stops, and others. Algfoor, Sunar, and Kolivand [1]

© Springer Nature Switzerland AG 2019
J. M. F. Rodrigues et al. (Eds.): ICCS 2019, LNCS 11540, pp. 223–236, 2019.
https://doi.org/10.1007/978-3-030-22750-0_17

remark the importance of path-finding for crowd simulation problems. In a microscopic setting, a path from each agent's current position to their goals should be defined. The problem can include dynamic obstacles as well. Besides, agents are usually programmed to avoid collisions with each other. Realism level and terrain texture are also possible restrictions to the path-finding problem.

This article is concerned with path-finding in the context of crowd simulation. This means the proposed approach should be able to provide paths to a large number of agents and be fast enough to be used for real-time visualization. A zone-based hybrid approach is proposed to allow studying of psychological factors and other subtle elements of the model. Also, this work intends exploring an evolutionary computation approach to path-finding problems. Further arguments in favor of the proposed approach are explained below.

1.1   Background

A general classification of solution approaches will be explained first to later discuss the proposed method under the light of the current state-of-the-art. Ijaz et al. [1] classify the crowd simulation approaches in the following categories: Zone-based models, layer-based models, and sequential models. Each of these categories may use a combination of methods.

Zone-based models divide the map and apply different resolution levels to each part of it. This approach allows handling large maps efficiently. Although, space restriction also restrict the possible solutions to the problem. Also, this means the optimal path could change if the zones are defined in a different manner.

Layer-based models apply both macro models and micro models simultaneously but they are applied in different layers. This allows to separate global planning from local navigation. Techniques like cellular automaton are used to model the global level, using simple rules to guide agents, while the refined movement is computed in other layers. Their problem is psychological factors are not included in the global level and the approach is still dependent of crowd density because its application of microscopical models to individual agents.

Finally, sequential models apply both macro models and micro models, one after another. The macroscopic model is applied until more refined movement is needed. Then, the system switches to a microscopic model. Synchronization is important while applying this approach. These models are better suited to cases where crowd movement is stable.

In regards of path-finding techniques, Algfoor et al. [1] classify techniques in two categories: Terrain-based methods and hierarchical techniques. The former is further divided into regular or irregular grids. Regular grids differ in their geometric shape (square, triangular, etc). They mention visibility graphs, mesh navigation, and waypoints to be techniques to create irregular grids. Some examples of hierarchical techniques are probabilistic road maps, quadtrees, and rapidly explored random trees.

The use of genetic algorithms (GA) for crowd simulation and path-finding problems is mentioned in the literature. Most of the works using GAs have limited their use to secondary parts of their respective solution approaches. For example, Johansson and Helbing [8] have used GAs to define the number of position of facility exits to minimize evacuation time. Vigueras, Lozano, Orduna, and Grimaldo [18] presented a zone-based crowd simulation model where GAs are used to determine map partitions. Zones were defined using Convex-hulls. Junior, Musse, and Jung [9] proposed using neural networks to estimate crowd density in subway stations. Optimization of the method was performed using GAs, among other techniques. Zhong et al. [20] used GAs to calibrate their crowd simulation model. Bera and Manocha [2] also used GAs, this time to allow a model to learn crowd movement patterns from data.

Direct application of GAs to solve path-finding problems seems to be limited in the literature, with less publications than the approach explained above. Naderan-Tahan and Manzuri-Shalmani [11] presented a GA specifically designed to solve path-finding problems. In their approach, each gene represented a point in the map space, and assumed that points were joined by linear segments to form the path. The chromosomes could have variable length. The first and the last genes of the chromosome were always the initial point and the goal. Some of this approach's drawbacks are the initial population should be obtained using other methods besides random initialization, which could bias the result towards premature convergence. Besides, the reported times are too high to be applicable to online visualization, although, their method was intended for robot navigation instead.

Song, Wang, and Sheng [16] proposed improvements. They used Bézier curves to define the paths instead of linear segments. The former approach is better suited to provide smoother paths than the latter. Besides, the paths are fully-derivable. In their approach, the map is divided by a fixed grid, where tile centers become potential control points to Bézier curves (i.e. paths). The GA searches for clear paths using these control points only. This feature simplifies the search, but it limits path resolution; there could be cases where the available control points are not enough to find a clear path for some complicated parts of the map.

Their method solves the global path-finding problem (i.e. the full map) using a one-level microscopic approach. Therefore, a global optimization process should be applied for each agent. This will have an impact on performance in large maps. On the other hand, their experiments show grids of 16×16 only and execution times were not reported, therefore, the performance of the approach is unknown. The reported setup suggests their approach was not intended for online simulations. Although, the main focus of that article was the use of GAs to path-finding problems instead of high performance.

This analysis seems indicate a novel implementation approach is necessary to make GAs useful for online path-finding. The references where GAs are explicitly used for path-finding are few because their execution times. Nevertheless the use of hybrid computation technologies (e.g. GPUs) open new possibilities to the application of GAs.

The combination of GPUs with machine learning techniques is a popular trend. Neural networks and deep learning are some examples [4]. Although, the evolutionary computation community has made efforts to accelerate their methods with GPUs and other distributed technologies [5]. The problem of evolutionary algorithms is they are inherently sequential, being necessary the population of the last generation to compute the next one. Although, state-of-the-art implementation have been reported in the literature.

For example, Pospichal, Jaros, and Schwarz [15] presented an implementation for NVIDIA GPUs using CUDA. Nowotniak and Kucharski [12] reported a GPU-based GA inspired on quantum systems. Wang and Sheng [19] introduced a GPU GA for task planning. Jaros [7] reported a multi-GPU island-based genetic algorithm for solving the knapsack problem.

Therefore, the approach proposed in this work is an extension of the current effort to use GAs for path-finding problems [11,16]. This approach introduces an GPU-based GA implementation for path-finding-problems. The implementation was designed with performance in mind, and the details of how this goal is reached are explained in sections below. It is a dynamic zone-based model-free method, where macroscopic and microscopic models are managed in layers. The implementation is able to handle maps of any size and generate paths at frame-rate time even when dynamic obstacles are present in the implementation. The implementation is tested against standard path-finding implementations and suitable experimental results are presented to show the performance improvement of the proposed approach.

The rest of the article is organized in the following sections: Sect. 2 introduces the approach, covering both the path-finding algorithm and the GPU, full-vectorized, GA implementation. Section 3.1 explains the experiments. Section 3.2 presents the results. Section 4 is the discussion, and the conclusions appear on Sect. 5.

2 Proposed Approach

This article follows the trend of the references mentioned above [11,16]. In a similar manner, Bézier curves are used to describe paths because they can be handled by GAs easily and because of their mathematical properties. Bézier curves are defined in the following manner:

$$\mathbf{B}(t) = \sum_{i=1}^{r} \binom{n}{i}(1-t)^{r-i}t^i\mathbf{P}_i. \tag{1}$$

Where t is a parametric variable in the range $[0, 1]$ and $\mathbf{P}_i$ are the control points. An explicit version of Eq. 1 is

$$\mathbf{B}(t) = (1-t)^r\mathbf{P}_0 + \binom{r}{1}(1-t)^{r-1}t\mathbf{P}_1 + \cdots + \binom{r}{r-1}(1-t)t^{r-1}\mathbf{P}_{r-1} + t^r\mathbf{P}_r. \tag{2}$$

GAs encode these control points, allowing them handling complete curves using a few parameters only. Smooth transition between path segments can be achieved when the derivatives of the curve are included in the optimization process. Further detail can be found in the references [16].

2.1   GA-GPU Implementation

The implementation both considers the encoding of Bézier curves and a GA implementation on GPU. This approach makes use of the Julia language [3]. The implementation also uses ArrayFire [10], a parallel computing library that interfaces with either CUDA or OpenCL. The library frees the user from the usual burdens of managing GPUs dedicated hardware. Although, high performance with ArrayFire can only be achieved by using full-vectorized code. Therefore, vectorization is a priority of the present approach.

The implementation proposes a traditional GA where selection, crossover, mutation, and evaluation of individuals will be performed for a fixed number of generations. The sequential nature of these operation avoids a generation-wise, parallel implementation. Parallelization is performed at population level, processing a large number of individuals simultaneously.

Coding. Individuals are encoded using virtual genes [17]. Differing from the references, This implementation uses 2 values to represent the control points, one for the x coordinate and other for the y coordinate. The values of the coordinates are random numbers in the range $[0, 2^b - 1]$, where b is the number of bits. In our case $b = \log_2(N)$, where N is the size of the map. The implementation is intended to work with $N \times N$ map tiles. The values of the current position and goal position should be passed to the GPU and appended to the chromosomes to perform further operations. We will call this variable g_d from now on. Finally, we assume the population has n individuals and m genes. Therefore, the population is of size $n \times m$.

Selection. The vectorized implementation is better expressed in equations. This operator is a variation of tournament selection. Let us define f_s to be the vector of fitness values of the population and f_{sp} the vector of fitness values of the shuffled population. Shuffling is performed using the Julia rand() command to create a AFArray with values from 1 to N. Also, we have i_x, which contains the original indices of the individuals, and i_{xs} are the shuffled indices. Assuming minimization, we define the following variables:

$$\Delta_1 = \text{sgn}\left(\text{sgn}(f_{sp} - f_s) + 1\right), \tag{3}$$

$$\Delta_2 = \text{sgn}\left(\text{sgn}(f_s - f_{sp}) + 1\right). \tag{4}$$

Where sgn is the sign operator. The indices of the selected individuals are

$$i_{xn} = i_x .* \Delta_1 + i_{xs} .* \Delta_2. \tag{5}$$

Where the .* operator denotes element-wise multiplication. Δ_1 is always 0 when $f_s > f_{sp}$ and the contrary is true for Δ_2. In this way, it is easy to discriminate the tournament winner. Individuals with index i_{xn} will reach the next generation and will be subject to further operations.

Crossover. Crossover operation involves 2 tasks. One is swapping the chromosome around the crossover gene, the other is applying inter-bit crossover to it. The first part requires splitting the population into left, right, and center parts. Let us assume c_r, is a matrix of random values in the range $[1, m]$ (Julia uses 1-based indexing) and c_c is a matrix with the cumulative sum of c_r, column-wise. Then, the linearized crossover indices should be

$$x_{\text{vec}} = v_m + n. * x_{r_{0,m-1}} \tag{6}$$

Where v_m is a vector with consecutive values from 1 to m, and $r_{0,m-1}$ is a vector of random values in the range $[0, m-1]$. Therefore, the c_c value at the crossover point should be $x_{\text{ix}} = c_c [x_{\text{vec}}]$. Let us now assume x_{mix} is a matrix composed by m copies of x_{ix}. The discrimination values m_d of the left, right, and center part of the chromosome are

$$s_{\text{gs}} = \text{sgn} (c_c - x_{\text{mix}}) . \tag{7}$$

s_{gs} will be 0 at the crossover point, the left part will be -1 and the right part will be 1. We can create a mask for each part with the following equations:

$$m_X = 1 - |s_{\text{gs}}| , \tag{8}$$

$$m_L = \|0.5(-s_{\text{gs}} + 1)\| - m_X \tag{9}$$

$$m_R = \|0.5(s_{\text{gs}} + 1)\| - m_X \tag{10}$$

Where m_L, m_R, and m_X are the masks for the left, right, and center parts, respectively. $\| * \|$ denotes the round operation. g_d can be multiplied by any of these masks to extract the corresponding part of the chromosome for all the individuals in the population simultaneously.

The second task is performing inter-bits crossover at the selected gene. Valenzuela-Rendón [17] explains the needed integer value to extract the low parts from the binary string representing the crossover gene from parents p_1 and p_2 with virtual genes is

$$\mathcal{X}_m (p_1, p_2) = p_1 \text{mod} 2^{m_c} - p_2 \text{mod} 2^{m_c} . \tag{11}$$

Where m_c is a random number between $[1, m]$ (assuming 1-based indexing). The post-crossover population is finally computed in the following manner:

$$g_d = gd_1. * m_L + g_{d_2}. * m_R + (g_{d_1} - \mathcal{X}_m). * m_X \tag{12}$$

Where g_{d_1} and g_{d_2} denote the original population and a scrambled copy of it.

Mutation. Mutation follows the same approach used on crossover. This time, 2 masks are created to separate the mutation gene from the rest of the chromosome. In this case c_m is a matrix of size $n \times m$ of random numbers in the range $[0, 1]$. Then the mask would be

$$m_m = \text{sgn} (\text{sgn}(c_m - p_m) + 1) \tag{13}$$

Where p_m is mutation probability. To obtain the mutated gene, we use the following mask:

$$m_{m_1} = 1 - m_m \tag{14}$$

Finally, the post-mutation population is

$$g_d = g_{d_1}.\ast m_m + g_{d_r}.\ast m_{m_1} \tag{15}$$

Where g_{d_1} is the original population, and g_{d_r} is a matrix of random genes.

Evaluation. The particular implementation is focused on finding clear paths, although, other optimality criteria could be used as well. Evaluation is composed by 2 tasks: Computation of Bézier curves and computation of the fitness function.

We need g_d and the values of t to compute the curves. Equation 2 is used for an efficient computation on the device (i.e. GPU). Let us assume t_d is a matrix of size $(r+1) \times (1/\Delta)$, where Δ is the resolution of t. t_d holds the terms of Eq. 2 for each value of vector t. t is defined from 0 to 1 with a step size Δ. Therefore $g_d \times t_d$ will compute the points of the curve.

In the actual implementation, the columns of t_d are duplicated to manage separately the x coordinates from the y coordinates. This means t_d has the form

$$t_d = \begin{bmatrix} (1 - t_{0x})^r & (1 - t_{0y})^r & \dots & t_{0x}^r & t_{0y}^r \\ (1 - t_{1x})^r & (1 - t_{1y})^r & \dots & t_{1x}^r & t_{1y}^r \\ \vdots & \vdots & \dots & \vdots & \vdots \\ (1 - t_{\Delta x})^r & (1 - t_{\Delta y})^r & \dots & t_{\Delta x}^r & t_{\Delta y}^r \end{bmatrix} \tag{16}$$

To allow an effective multiplication a pair of masks are defined: st_x and st_y. Where

$$\mathrm{st}_x = \begin{bmatrix} 1 & 0 & \dots & 1 & 0 \\ \vdots & \vdots & & \vdots & \vdots \\ 1 & 0 & \dots & 1 & 0 \end{bmatrix}, \tag{17}$$

while st_y would be

$$\mathrm{st}_y = |\mathrm{st}_x - 1|. \tag{18}$$

These matrices have successive columns of ones and zeros to avoid multiplication of the wrong values. Therefore, the values of the path will be

$$\mathbf{B}_x(t) = (g_d.\ast \mathrm{st}_x)\, t_d, \tag{19}$$

$$\mathbf{B}_y(t) = (g_d.\ast \mathrm{st}_y)\, t_d. \tag{20}$$

The fitness function to evaluate individuals is based on clearance. To compute the metric we need the Euclidean distance between the origin and any other point in the path. Let us call this variable d_{ab}. Also, we need to compute collisions along the path. Let us assume M_d is the map loaded on the device. We can use linearized indices from $\mathbf{B}_x(t)$ and $\mathbf{B}_y(t)$ to handle it

$$\mathbf{B}_{xy}(t) = \mathbf{B}_x(t) + N\left(\mathbf{B}_y(t) - 1\right). \tag{21}$$

Collisions will be simple the indexing

$$c_d = M_d\left[\mathbf{B}_{xy}(t)\right] \tag{22}$$

Using these variables, we can compute the fitness value in the following manner:

$$f_d = (c_d. * d_{ab})\, V_1. \tag{23}$$

Where V_1 is a vector of ones used to perform the final multiplication step.

Initialization. Data transfer between the host and the device should be used carefully because it has a direct impact on performance. Relevant variables should be created on GPU memory to avoid unnecessary overload. These variables are initialized before the run to save computation time. g_d, f_s, t_d, st_x, st_y, V_1 are all suitable candidates to initialization. Also, the algorithm returns the best-so-far individual instead of the final population to avoid unnecessary overhead time.

2.2   Path-Finding Algorithm

Path-finding is divided in a static stage and a dynamic stage. Figure 1 shows a flowchart with the path-finding algorithm.

Static_Field() refers to the static path-finding stage that is computed off-line to save time. Breadth-first search is used to create a gradient field on the map. This approach was preferred because it is able to provide paths to any number of agents. Other methods could have been used as well.

Initialization() refers to the setup mentioned in Sect. 2.1. The necessary variables are loaded on the device beforehand to save time. Besides, these variables will be used constantly along the run, therefore, overhead is reduced drastically by applying this step.

Static_Path() deals with providing a static path to a particular agent. Given the field, this step is straightforward. The static path is used to guide the agent in a global scale. We say it is a macroscopic path in the sense it provides the general path movement of the agent, which will be subject to corrections when dynamic obstacles appear in the way.

The actual path-finding algorithm starts with the Move() function. This function moves the agent subject to different conditions, depending on the type of simulation desired. While moving, the agent is constantly applying Is_Goal() and Is_Obstacle() functions. Their names are self descriptive, but we can say Is_Obstacle() search for obstacles s_g steps ahead in the path. The agent moves freely as long as the path remains clear, until the goal is reached.

The path-finding algorithm is called when an obstacle is on sight on the path. Tiling() is a function that takes a piece of the map of size $2s_g \times 2s_g$. The current position of the agent becomes the local starting point and the closest path point to the tile edges will become the local goal. Tile size should be large enough to allow the necessary freedom to the agent to find an alternative path. Also, dynamic tiling is necessary because we cannot guarantee the tile will contain a clear path to the local goal because the static field was created without obstacles information. Dynamic tiling overcomes this occurrence. Also, tiling centers the obstacle to the tile to avoid obstructions with the edges. Dynamic tiling size could be used to guarantee a path will be found. In the experiments, it was found dynamic tiling was enough to prevent blockage.

Path-Finding() refers to the application of GAs to find clear paths at frame-rate times for the given tile. Tiling is necessary to guarantee GPU's memory is not overwhelmed by large maps. The device capabilities allows to run large populations (in the order of thousands) to speed-up the search. According with GAs theory, larger populations contain a much larger schemata contents and they have more probability to contain the optimal ones. Given speed is our main concern, we need to work with the largest population possible. Also, the number of generations is limited, because the sequential execution has direct impact on performance. Also, the number of control points should be the lowest possible to keep performance up and avoid the path to be unnecessarily complicated. Although, if the GAs cannot find a clear path with the current number of control points, this one is increased and the GA is run again. This will happen until a clear path is found.

Finally, once the alternative path is found, the global path is updated with the new segment. The agent will follow the new path until it finds a new obstacle in sight or the goal is reached.

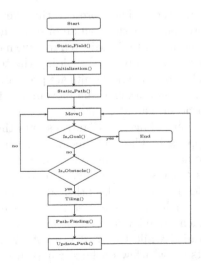

Fig. 1. Path-finding algorithm

3   Experiments

Experiments were conducted to test the performance of the proposed approach. In this case, we used Voronoi maps [14], which were algorithmically created for the experiments. Maps have a size of 2048 × 2048, while tile size is $2s_g = 32$. Initial and goal points were randomly selected, but they were chosen in a way the goal is reachable from the initial point. Also, we assume obstacles are previously inflated, therefore, there is no potential collision risk involved if paths stick to them.

The approach assumes the necessary time for agents to move along the path is much longer than the time needed to generate alternative paths, therefore, we can assume the algorithm works with snapshots of the world. The path-finding algorithm is activated when Is_Obstacle() function returns true.

3.1   Experiments Description

The experiments tested 100 different maps where the system should provide paths to agents. Dynamic obstacles appear randomly in the path, and the agent system should provide an alternative path for each obstacle. The approach is compared against traditional path-finding methods: Breadth-first search, Dijkstra algorithm, and A*. All the algorithms were run against the same path-finding problems, and the same obstacles. The average running time per obstacle is computed from each method. The experiments were run on a computer equipped with an Intel® Core™ i7-7700HQ CPU @ 2.80 GHz @ 8 processor, 7,7 GiB of RAM, and NVIDIA GeForce GTX 1050/PCIe/SSE2 GPU.

3.2   Experiments Results

The results are shown in Fig. 3. They present the average time it takes the algorithms to find an alternative path when an obstacle has been detected. The number of obstacles is random, but it is adjusted to have more or less one obstacle per tile size. The average time is obtained dividing the total computation time by the number of obstacles solved. The results are presented on frames/obstacle, where a frame is 1/60 s. We can assume the movement of agents and obstacles is much more slower to be measured on frames, but a faster algorithm will be able to serve a larger number of agents. Figure 3 shows the comparison between A* and the proposed approach. The experiments using breadth-first search and Dijkstra's algorithm are not shown because they are order of magnitude slower than A*. This happens because the number of revised nodes by A* is much more less than the nodes revised by the other algorithms (Fig. 2). This result was expected. On the other hand, we can see the average of the distribution of GA's running time is much more lower than the average of A*. Fig. The median of GA is less than 1 frame, while the median of A* is around 80 frames per obstacle. Both box plots show a few outliers, where we can assume particular conditions caused a delay for the algorithms.

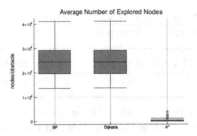

Fig. 2. Average number of nodes explored by search algorithms.

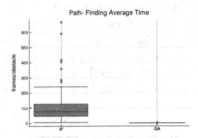

Fig. 3. Average execution time to solve path-finding with dynamic obstacles. Results shown in frames/obstacle, (one frame $= 1/60\,\text{s}$)

4   Discussion of Results

Figure 4 shows examples of paths found by different algorithms. We can see breadth-first tends to find paths that are relatively far away from the obstacle, this occurs due the particular manner this algorithm expands the frontier. On the other hand, A* has a tendency to stick to obstacles to minimize the deviation form the optimal path. In the case of GAs, paths are conditioned to the particular tile used by the algorithm. This happens because both the path and the control points must be located inside the tile, causing the rest of the map to be off-limits to the algorithm. There is a possibility a better path could be found if we change the tile size or position. Therefore, an optimal global path could be rather different from the paths computed using a global approach. Nevertheless, optimizing the whole map could be impossible because of hardware limitations (e.g. GPU memory).

We can observe a slightly erratic movement on the paths generated using Dijkstra's algorithm. We have to say this algorithm is designed to find the less cost path in weighted graphs, which is not the case of this problem. Therefore, cost does not provide guidance to the path-finding process. Finally, we observe the GA has a tendency to generate smoother curves than other algorithms because of the nature of Bézier curves. By the moment, clearance (i.e Eq. 23) is the only fitness metric considered. Of course, we can include minimum distance, smoothness of curves, deviations from a constant speed or acceleration, etc. The integral inclusion of all these metrics could probably be achieved with multi-objective evolutionary algorithms.

In regards to outliers, it was found there were cases where the GA ran out of time and it could not find a clear path. This happened when the particular tile was specially complicated. The initial number of control points is 1 and it was increased when the GA failed to find a clear route. It was found sometimes even 5 control points were necessary to find a clear path. We wish to keep the number of control points the minimum possible because of two main reasons: One is to minimize execution time. Also, a Bézier curve has a tendency to become unnecessarily complicated with a large number of control points. Having a way to estimate the necessary control points beforehand will have a positive impact on performance.

In relation to the classification explained in Sect. 1.1, the proposed approach is inherently zone-based because it uses tiling to solve the path-finding problem with dynamic obstacles. Also we can say the method is layered when we consider the global path-finding to be on the macroscopic level and obstacles avoidance belong to the microscopic level. Also it is a terrain-based technique where octagonal nodes are used to allow both straight and diagonal motion.

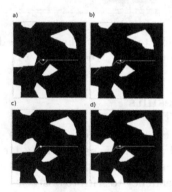

Fig. 4. Example of application of path-finding algorithms: (a) Breadth-first search, (b) Dijkstra's algorithm, (c) A*, (d) GAs. The small white square is a dynamic obstacle.

5   Conclusions

This article presented a novel approach to path-finding problems using evolutionary algorithms. The approach is intended to be applied to crowd simulation applications, where online path-finding is desired. The main problem for the application of evolutionary algorithms was the required speed to serve a large number of agents online. The article presented an evolutionary algorithm based on Bézier curves to handle the path-finding problem, and a vectorized GPU implementation that allows managing large populations at frame-rate times. The experiments showed the proposed approach is faster than traditional path-finding methods, being A* the most remarkable one.

There are many possibilities of future work for the proposed approach. For example, a multi-objective version of the algorithm can be devised to simultaneously include minimum distance, clearance, and social forces. The advantage of this approach is it will generate a Pareto front of optimal solutions, providing with variety to the simulations. This will happen because the agent will have the possibility of choosing one of the solutions based on its own preference, allowing the inclusion of "personality" to them.

The current method is subject to further improvement. It was explained above how execution of GA was affected by the fact the number of control points should be gradually incremented until the algorithm can generate flexible enough paths to traverse the tile. A method could be devised to estimate this value beforehand, and a good estimation will have a positive impact on performance. The integration of this improvement with a multi-objective version of the algorithm will be able to online full crowd modeling. Also, the algorithm can be compared against other custom path-finding methods and be subject of further improvement.

The current implementation was made using the ArrayFire library. Probably a lower level implementation could be desirable (e.g. using CUDA). This vectorized approach has the advantage porting should be straightforward, because most of the operations used are matrix operations: A task GPUs are specially useful. Also, other approaches to path-finding could be combined with evolutionary algorithms to find novel approaches. This implies the exploration of sequential approaches, hierarchical techniques, and others.

Acknowledgments. The author thanks Consejo Nacional para la Ciencia y Tecnología (CONACyT) for the postdoctoral fellowship at Barcelona Supercomputing Center.

References

1. Algfoor, Z.A., Sunar, M.S., Kolivand, H.: A comprehensive study on pathfinding techniques for robotics and video games. Int. J. Comput. Games Technol. **2015**, 7 (2015)

2. Bera, A., Manocha, D.: REACH - realtime crowd tracking using a hybrid motion model. In: 2015 IEEE International Conference on Robotics and Automation (ICRA), pp. 740–747. IEEE (2015)

3. Bezanson, J., Edelman, A., Karpinski, S., Shah, V.B.: Julia: a fresh approach to numerical computing. SIAM Rev. **59**(1), 65–98 (2017)

4. Cui, H., Zhang, H., Ganger, G.R., Gibbons, P.B., Xing, E.P.: GeePS: scalable deep learning on distributed GPUs with a GPU-specialized parameter server. In: Proceedings of the Eleventh European Conference on Computer Systems, p. 4. ACM (2016)

5. Gong, Y.-J., et al.: Distributed evolutionary algorithms and their models: a survey of the state-of-the-art. Appl. Soft Comput. **34**, 286–300 (2015)

6. Ijaz, K., Sohail, S., Hashish, S.: A survey of latest approaches for crowd simulation and modeling using hybrid techniques. In: 17th UKSIMAMSS International Conference on Modelling and Simulation, pp. 111–116 (2015)

7. Jaros, J.: Multi-GPU island-based genetic algorithm for solving the knapsack problem. In: 2012 IEEE Congress on Evolutionary Computation (CEC), pp. 1–8. IEEE (2012)
8. Johansson, A., Helbing, D.: Pedestrian flow optimization with a genetic algorithm based on Boolean grids. In: Waldau, N., Gattermann, P., Knoflacher, H., Schreckenberg, M. (eds.) Pedestrian and Evacuation Dynamics 2005, pp. 267–272. Springer, Heidelberg (2007). https://doi.org/10.1007/978-3-540-47064-9_23
9. Jacques Jr., J.C.S., Musse, S.R., Jung, C.R.: Crowd analysis using computer vision techniques. IEEE Signal Process. Mag. 27(5), 66–77 (2010)
10. Malcolm, J., Yalamanchili, P., McClanahan, C., Venugopalakrishnan, V., Patel, K., Melonakos, J.: ArrayFire: a GPU acceleration platform. In: Modeling and Simulation for Defense Systems and Applications VII, vol. 8403, p. 84030A. International Society for Optics and Photonics (2012)
11. Naderan-Tahan, M., Manzuri-Shalmani, M.T.: Efficient and safe path planning for a mobile robot using genetic algorithm. In: IEEE Congress on Evolutionary Computation, CEC 2009, pp. 2091–2097. IEEE (2009)
12. Nowotniak, R., Kucharski, J.: GPU-based tuning of quantum-inspired genetic algorithm for a combinatorial optimization problem. Bull. Pol. Acad. Sci. Tech. Sci. 60(2), 323–330 (2012)
13. Pan, X., Han, C.S., Dauber, K., Law, K.H.: A multi-agent based framework for the simulation of human and social behaviors during emergency evacuations. Ai Soc. 22(2), 113–132 (2007)
14. Papadopoulou, E., Zavershynskyi, M.: The higher-order Voronoi diagram of line segments. Algorithmica 74(1), 415–439 (2016)
15. Pospichal, P., Jaros, J., Schwarz, J.: Parallel genetic algorithm on the CUDA architecture. In: Di Chio, C., et al. (eds.) EvoApplications 2010, Part I. LNCS, vol. 6024, pp. 442–451. Springer, Heidelberg (2010). https://doi.org/10.1007/978-3-642-12239-2_46
16. Song, B., Wang, Z., Sheng, L.: A new genetic algorithm approach to smooth path planning for mobile robots. Assem. Autom. 36(2), 138–145 (2016)
17. Valenzuela-Rendón, M.: The virtual gene genetic algorithm. In: Cantú-Paz, E., et al. (eds.) GECCO 2003, Part II. LNCS, vol. 2724, pp. 1457–1468. Springer, Heidelberg (2003). https://doi.org/10.1007/3-540-45110-2_18
18. Vigueras, G., Lozano, M., Orduna, J.M., Grimaldo, F.: A comparative study of partitioning methods for crowd simulations. Appl. Soft Comput. 10(1), 225–235 (2010)
19. Wang, K., Shen, Z., et al.: A GPU-based parallel genetic algorithm for generating daily activity plans. IEEE Trans. Intell. Transp. Syst. 13(3), 1474–1480 (2012)
20. Zhong, J., Hu, N., Cai, W., Lees, M., Luo, L.: Density-based evolutionary framework for crowd model calibration. J. Comput. Sci. 6, 11–22 (2015)

Analysing the Trade-Off
Between Computational Performance
and Representation Richness
in Ontology-Based Systems

Salvatore F. Pileggi[1(✉)], Fabian C. Peña[2], Maria Del Pilar Villamil[2],
and Ghassan Beydoun[1]

[1] School of Information, Systems and Modelling,
University of Technology Sydney, Ultimo, Australia
{SalvatoreFlavio.Pileggi,Ghassan.Beydoun}@uts.edu.au
[2] Systems and Computing Engineering Department, School of Engineering,
Universidad de los Andes, Bogota, Colombia
{fc.pena,mavillam}@uniandes.edu.co

Abstract. As the result of the intense research activity of the past decade, Semantic Web technology has achieved a notable popularity and maturity. This technology is leading the evolution of the Web via interoperability by providing structured metadata. Because of the adoption of rich data models on a large scale to support the representation of complex relationships among concepts and automatic reasoning, the computational performance of ontology-based systems can significantly vary. In the evaluation of such a performance, a number of critical factors should be considered. Within this paper, we provide an empirical framework that yields an extensive analysis of the computational performance of ontology-based systems. The analysis can be seen as a decision tool in managing the constraints of representational requirements versus reasoning performance. Our approach adopts synthetic ontologies characterised by an increasing level of complexity up to OWL 2 DL. The benefits and the limitations of this approach are discussed in the paper.

Keywords: Semantic web · Semantic technology · Ontology ·
Computational performance

1   Introduction

The Semantic Web [3] has achieved a notable popularity as a mature technological environment. In big part, this is due the intense research activity of the last 15 years, and the efforts of W3C[1] to promote a standardisation process for the different languages and their underlining models. Semantic technologies are leading the evolution of the Web via interoperability by providing structured

---

[1] World Wide Web Consortium (W3C) - https://www.w3.org.

© Springer Nature Switzerland AG 2019
J. M. F. Rodrigues et al. (Eds.): ICCS 2019, LNCS 11540, pp. 237–250, 2019.
https://doi.org/10.1007/978-3-030-22750-0_18

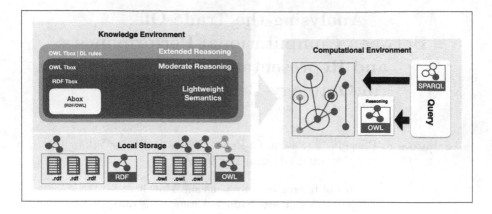

Fig. 1. Computational model.

metadata. Because of the adoption of rich data models on a large scale to support the representation of complex relationships among and standard reasoning, the computational performance of ontology-based systems may be hard to evaluate, as it may change significantly depending on the application context. A number of critical and key factors should be considered. Firstly, the Semantic Web technology provides a technological ecosystem composed of several languages. These languages are characterised by an increasing complexity to support different data modelling spaces. Secondly, applications may propose very different behaviours and may consequently adopt the technology in different ways.

This paper provides a performance evaluation framework for ontology-based systems supported by empirical measurements. The proposed framework takes into account the perennially conspicuous trade-off between computational performance and representational capabilities. Our analysis is limited to decidable technology, including *lightweight semantics* based on RDF [5] reasoning, *moderate reasoning* equivalent to OWL-Lite [2] reasoning and *extended reasoning* corresponding to OWL-DL [2]. OWL ontologies are implemented in OWL 2.

2   Related Work

The analysis of the trade-off between computational performance and representation richness is a classic topic extensively reported in literature. A number of OWL benchmarks are compared in [16], where also the specification of a set of requirements for an ideal OWL benchmark is provided.

Similar approaches are followed also to compare different reasoners in other contributions (e.g. [1, 4, 6, 7]). An interesting comparison between two of the most relevant computation techniques (tableau and hyper-tableau calculus) is proposed in [13]. One of the most popular OWL benchmarks is LUBM [10], which provides advanced analysis features to evaluate systems characterized by different reasoning capabilities and storage mechanisms. It addresses generic OWL data-spaces and approaches performance analysis by providing global metrics suitable to compare different systems. More recently, a competition based on a testing framework agreed within the community has been arranged [14].

Our work differs from those mentioned as we provide an environment suitable to multi-dimensional analysis in which the performance of a given system may be evaluated as the function of the ontology complexity and its scale (population). These two dimensions are addressed by generating synthetic ontologies (Sect. 3.4) which enable fine-grained analysis. Our approach assures a generic and an application-independent performance analysis that relies on the specification of complexity ranges (Sect. 3) and on computational experimentation. Furthermore, we aims at providing domain and architecture agnostic results by introducing a number of simplifications (see Sect. 3.1). For instance, we don't take into account architectural (e.g. storage system) and network factors, as well as we adopt a query-independent approach. Those simplifications allow a more focused, direct and understandable analysis framework. Last but not the least, our framework is extensible, meaning that further dimensions of analysis may easily be addressed.

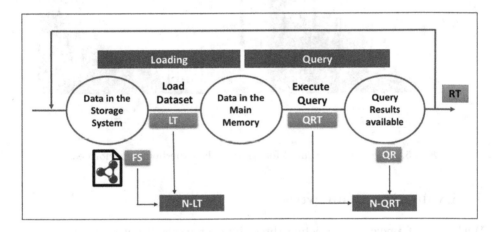

Fig. 2. Experiment phases and metrics associated.

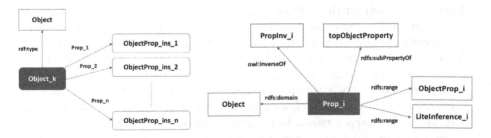

Fig. 3. Synthetic object.   **Fig. 4.** Synthetic property.

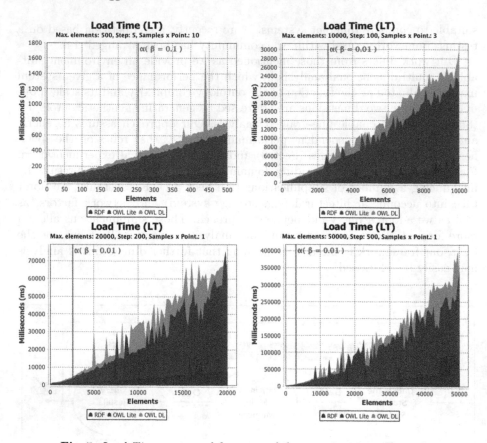

Fig. 5. *Load Time* measured for some of the experiments performed.

3   Evaluation Framework

Within our framework, we define three different levels of ontology complexity (Fig. 1) as follows:

- **Lightweight semantics.** We associate lightweight semantics with a minimal set of knowledge representation requirements and, therefore, with the best computational performance. We assume RDF [5] structures and reasoning.
- **Moderate reasoning.** The most immediate extension for lightweight semantics as previously defined is to provide more extended reasoning capabilities, to uncover non-explicit relations among basic concepts. We associate this level of complexity with OWL and, more concretely, with OWL-Lite [2] complexity. Such a step forward introduces additional constructs and abstractions (e.g. data and object property), structural relations (e.g. class/sub-class and property/sub-property), constraints (e.g. class disjointedness, functional property), relations among properties (e.g. inverse properties) and basic inference on properties (domain and range).

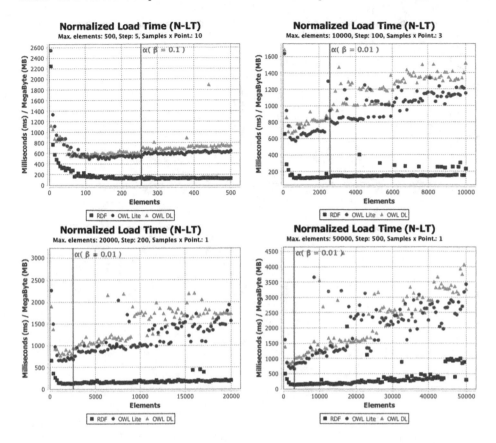

Fig. 6. *Normalized Load Time* measured for some of the experiments performed.

– **Extended reasoning capabilities.** The highest level of complexity that we consider within this work corresponds to OWL-DL [2], which assures the maximum expressiveness maintaining computational completeness and decidability. This level of complexity extends the previous one by providing the capability to define inference rules according to a Description Logic. This extension results in more advanced reasoning capabilities.

3.1   Assumptions and Simplifications

A comprehensive study on the computational performance of ontology-based systems should consider several factors. For simplicity sake, we consider a simplified, still in our opinion exhaustive, environment, adopting the following assumptions:

– **Local storage (file system).** The Semantic Web is a distributed environment by definition. In a Web context, data can be potentially retrieved from multiple, eventually remote, sources. Moreover, target data could be stored in files, normally accessible by URLs, as well as in common databases or even

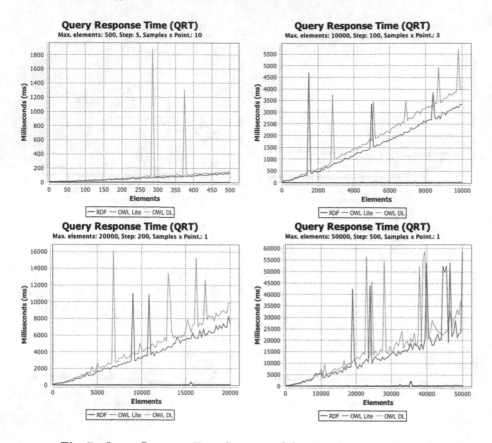

Fig. 7. *Query Response Time* for some of the experiments performed.

specialised data-stores (triple-stores). In this work, we consider uniquely local storage in the file system. This allows an analysis independent from the performance of the storage system.

– **Query-independent evaluation.** A study that takes into account the complexity of the query would be very interesting. However, it would add a significant complexity. To assure a query-independent evaluation, we consider the generic SPARQL query below:

```
PREFIX onto: ourPrefix
SELECT ?x ?y
WHERE {    ?x a ?y .
      FILTER regex( str( ?x ), ourPrefix ) .
      FILTER regex( str( ?y ), ourPrefix )
}
```

This query can be applied to both RDF and OWL environments and returns all the elements that are member of some class. As will be later explained, this allows to clearly identify the contribution of inference to the query outcome.

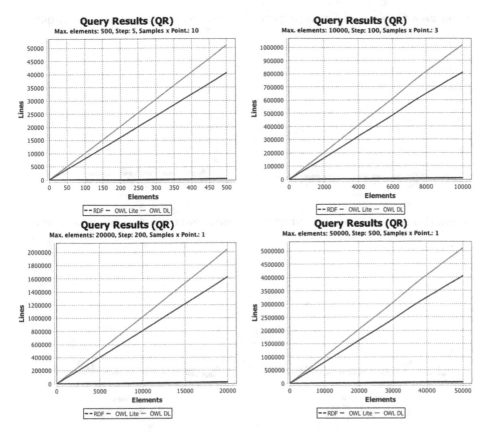

Fig. 8. *Query Response* for some of the experiments performed.

- **Ontology size increases by increasing its population only (Abox).**
  We consider two main dimension of analysis: the ontology complexity and
  the ontology size. The former, associated with the Tbox including inference
  rules, is defined by a number of templates as a kind of static configuration.
  The latter is associated with the Abox and is dynamically addressed.
- **Agnostic approach to software components.** We consider the semantic
  engine (reasoner) as a black-box. That is, we have designed our framework
  on the basis of macro-operations common to all common APIs in semantic
  technology. For our experiments, without loss of generality, we only use HermiT [15]. Naturally, we can perform the same experiments using any other
  semantic engines supporting RDF and OWL2 DL.
- **Synthetic Ontology.** In order to provide a fine-grained analysis, we opt for
  an environment which produces synthetic ontologies according to common
  approaches [12]. We believe the experimentation on real ontologies is not a
  relevant factor within the scope of this work. However, as briefly discussed
  later on, it may introduce some uncertainty in the analysis.

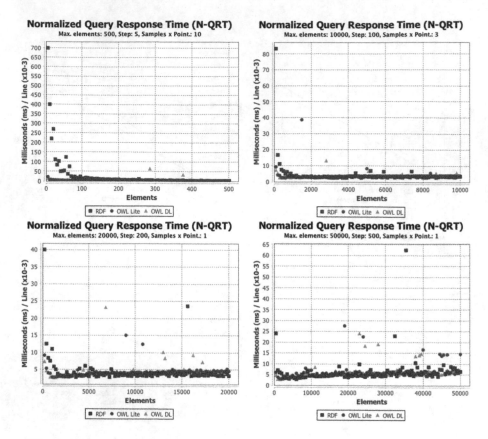

Fig. 9. *Normalized Query Response Time* for some of the experiments performed.

3.2   From Lightweight Semantics to Complex Reasoning

We approach the computational performance evaluation of ontology-based systems according to a classic perspective, which takes into account two major macro-operations: *loading the information* into the semantic engine and *executing a query* on the information available (Fig. 2).

Control Mechanism for the Input Dataset. In order to control eventual gaps between logical and physical representations (e.g. RDF and OWL implementations), we introduce a simple control mechanism for input datasets. The *convergence point* (α) is defined as the function of a parameter, the *convergence threshold* (β); for instance, $\alpha_{(\beta=1\%)} = 100000$ means that, for a scale higher than 100000 atomic elements, the difference in size for the considered representations is within the 1% of the smallest size. β is normally expressed as a percentage of the shorter representation size and is normally supposed to be a small value. Such a metric may be very relevant for experiments involving small datasets. Indeed, it expresses the end of the transitory and the beginning of the stationary condition for a given experiment: while those experiments at a lower scale than α (transitory) are affected by the difference in size of the considered representations, the experiments at an higher scale (stationary condition) are assumed not affected by such overheads.

3.3   Metrics

The experiment performed are modelled as an iterative process (Fig. 2). After each iteration, the size of the input dataset is increased. Each iteration is composed of two different phases: in the first, a file of size FS is loaded from the storage systems into the main memory (loading phase); in the second, a query on the available dataset is executed (query phase).

To assess the computational performance in each phase, we consider the three following metrics:

- **Load Time (LT)** is the time needed to load the dataset from the storage systems into the main memory.
- **Query Response Time (QRT)** is the execution time for a query. QRT assumes the target dataset already loaded in the main memory.
- **Query Results (QR)** is the number of rows of the result set returned by a given query.

In order to have a concise assessment of the performance, we define two further normalized metrics based on the three defined above:

- **Normalized Load Time (N-LT)** is defined as LT/FS. Within our evaluation framework, $N - LT$ concisely expresses loading performance because the load time is considered as the function of the dataset size.
- **Normalized Query Response Time (N-QRT)** is defined as QRT/QR. $N - QRT$ reflects the query performance: it provides an understanding of the query response time as the function of the query result set size.

For completeness, we also define a global metric, *Response Time (RT)*, which is the sum of the Load Time and of the Query Response Time ($LT + QRT$). However, In order to assure a consistent analysis, RT should be considered both with and in the context of the normalized metrics previously discussed.

3.4   Synthetic Patterns

As earlier mentioned, the scale of the ontologies adopted in the experiments is enlarged by increasing its population (Abox). The synthetic object adopted to populate the Abox is represented in Fig. 3: it is declared to be an instance of the class *Object* and it is related to n static objects through an equivalent number of properties.

We provide moderate reasoning by defining a number of Object Properties according to the model depicted in Fig. 4. Each object property is a sub-property

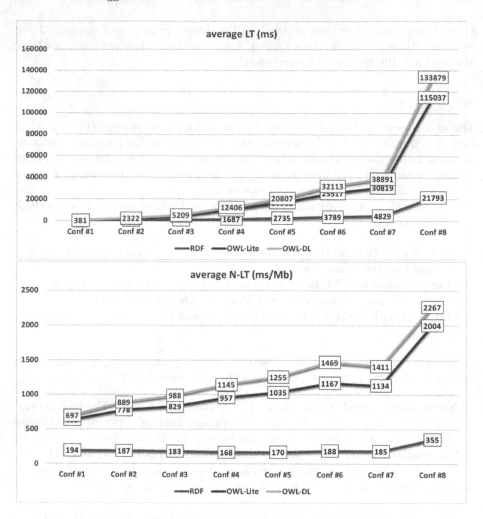

Fig. 10. Loading performance (average values).

of *topObjectProperty*. Its domain and range are defined according to RDF-S specifications. For each object property, an additional property is defined and declared as an inverse of the considered property, according to OWL specifications.

Finally, extended reasoning capabilities are provided by a set of DL statements that define equivalent classes according to the pattern reported in Eq. 1. The DL statement is composed of two different sub-statements adopting the operator *some*; those two sub-statements are related by the operator OR or AND.

$$(Prop_i \text{ some } ObjectProp_i) \quad AND|OR \quad (Prop_{i+1} \text{ some } ObjectProp_{i+1}) \quad (1)$$

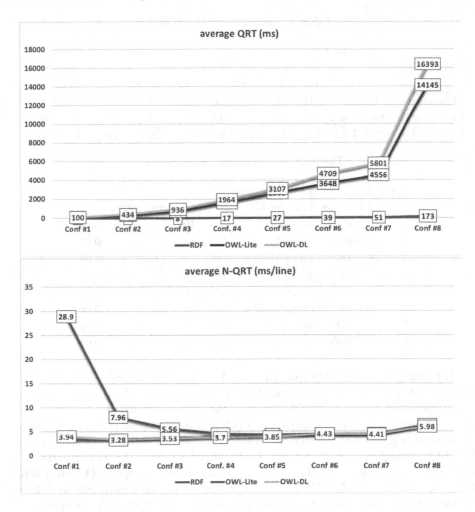

Fig. 11. Query performance (average values).

4   Experimental Performance Evaluation

We adopt commonly accepted metrics from Protege [8] to measure the complexity of the ontologies considered. The different configurations corresponding to non-populated ontologies (Tbox) are reported in Table 1. We select three different perspectives by considering *axioms*, *classes* and *object properties*. We distinguish between logical and declaration axioms. We also report the number of sub-classes and equivalent classes defined by DL statements. For properties, we consider the number of inverse properties and the statements associated with the object property domain and range.

Our experimentation consists of a number of empirical measurements for the different ontologies by increasing the Abox size as in the configurations reported in Table 2.

Table 1. Ontology configuration.

Ontologies configuration (Tbox)									
Ont. complexity	Axioms			Classes			Object properties		
	Tot	Log.	Decl.	Tot	SubC	Equiv.	Inv.	Dom.	Ran.
Lightweight (≈ RDF)	21	1	0	**1**	0	**0**	0	0	0
Moderate (≈ OWL-Lite)	301	196	105	**43**	40	**0**	20	20	40
Extended (≈ OWL-DL)	362	236	126	**64**	60	**20**	20	20	40

Table 2. Experiments configuration.

Conf.	Experiment configuration				
	Elements	File size (MB)	Step	Samples x point	β
1	0-500	up to 1	5	10	0.1
2	0-2500	up to 5	25	5	0.05
3	0-5000	up to 10	50	5	0.05
4	0-10000	up to 20	100	3	0.01
5	0-15000	up to 30	150	2	0.01
6	0-20000	up to 40	200	1	0.01
7	0-25000	up to 50	250	1	0.01
8	0-50000	up to 100	500	1	0.01

For example, the experiment #1 considers files with logical elements in the range 0–500, namely files of a size up to 1 MB; the experiment starts with a file of the minimum size (5 elements in this case); the file is increased of a step of 5; each measure reported is the average over 10 independent samples; β is 0.1. We adopt a software engine based on Hermit [9,15] as a reasoner and OWL-BGP[2] [11] as SPARQL wrapper. The framework is developed in Java. All the experiments reported in the paper have been executed on a common laptop (1.8 GHz Intel Core i5, 8 GB 1600 MHz DDR3, macOS Sierra). For each experiment as defined in Table 2, the measurements are executed sequentially without re-booting.

The *Load Time (LT)* measured for some of the experiments performed is reported in Fig. 5. Likewise, the *Normalized Load Time (N-LT)* is shown in Fig. 6. Similarly we report the metrics for query performance evaluation: Figs. 7, 8 and 9 show respectively the *Query Response Time (QRT)*, the *Query Response (QR)* and the *Normalized Query Response Time (N-QRT)*.

As previously explained, normalised metrics take into account both the size of the dataset imported and the size of the result set. Therefore, the metrics provide a concise measure of loading and query performance, allowing comparison among the different languages. We design our experiments to minimize the impact of size variations for the input datasets (Sect. 3.2). Indeed, *LT* and *N-LT* present a similar pattern. Average values for all experiments are reported in Fig. 10. Looking exclusively at loading performance, RDF clearly outperforms OWL. That is because the reasoner adopted implements hyper-tableau calculus. A minor difference between OWL-Lite and OWL-DL is also detected.

---

[2] OWL-BGP - https://github.com/iliannakollia/owl-bgp.

Looking at query performance, average values for QRT and N-QRT are reported in Fig. 11. QRT highlights variations in performance across the different models considered throughout the range of experiments. The pattern detected for N-QRT is quite interesting as it clearly shows the computation performance of hyper-tableau calculus in normalized conditions. Indeed, according to this normalized metric that takes into account the contribution of inference in terms of query output, at a significant scale there is no difference of performance among the three levels of complexity.

5 Main Limitations and Uncertainty

The simplifications introduced in the framework (Sect. 3.1) have allowed a systematic, fine-grained and relatively simple analysis in stable and normalized conditions. On the other hand, such an approach may introduce a number of possible uncertainty factors to the key question on performance evaluation of real ontology-based systems.

The very first factor of uncertainty is the use of synthetic ontologies. They are designed around a number of typical design patterns. Real ontologies may include those patterns or a part of them, as well as they may propose completely different ones. Additionally, the distributed approach to modern systems may introduce key trade-offs beyond network factors, such as between cloud and edge computing along a wide range of hybrid solutions (e.g. fog computing). Similar considerations affect key architectural components, such as the storage system. Indeed, the design of the whole architecture needs to be considered depending on the data consistency model (e.g. weak and strong consistency).

6 Conclusions and Future Work

In this paper, we define a performance evaluation framework for ontology-based systems in which ontology complexity and ontology size are considered as the main dimensions for performance analysis. In essence, our framework can be used to ensure the right mix of responses to the constraints imposed by the trade-off between reasoning computation and knowledge expressiveness requirements (ranging from lightweight semantics, moderate reasoning and extended reasoning capabilities respectively).

We introduced a number of simplifying assumptions (discussed in Sect. 3.1) to enable a relatively simple environment for the analysis of computational performance. We believe however that our metrics reflect relatively realistic working conditions. They combine simplicity and coverage leading to a direct and viable analysis. We have performed a number of experiments at a relatively low scale, involving files up to 100 MB. Without loss of generality, we fixed the semantic engine across all experiments and opted for uniform use of a single reasoned. Comparing various reasoners has been undertaken elsewhere. We thus focused on the definition of a framework which allows generic and effective performance analysis by considering increasing capabilities in terms of representation and reasoning. With the increasing prominence of IoT and AI based applications, the trade off between complexity in representation and performance is a more pressing concern for many innovations. This is particularly true for distributed settings. Hence, in future work, we will consider distributed environments, namely data sets imported by different remote sites and performance analysis as the function of the ontology Tbox.

References

1. Abburu, S.: A survey on ontology reasoners and comparison. Int. J. Comput. Appl. **57**(17), 33–39 (2012)
2. Bechhofer, S.: OWL: web ontology language. In: Liu, L., Özsu, M.T. (eds.) Encyclopedia of Database Systems, pp. 2008–2009. Springer, Boston (2009). https://doi.org/10.1007/978-0-387-39940-9
3. Berners-Lee, T., Hendler, J., Lassila, O., et al.: The semantic web. Sci. Am. **284**(5), 28–37 (2001)
4. Bock, J., Haase, P., Ji, Q., Volz, R.: Benchmarking OWL reasoners. In: ARea2008-Workshop on Advancing Reasoning on the Web: Scalability and Commonsense. Tenerife (2008)
5. Decker, S., et al.: The semantic web: the roles of XML and RDF. IEEE Internet Comput. **4**(5), 63–73 (2000)
6. Dentler, K., Cornet, R., Ten Teije, A., De Keizer, N.: Comparison of reasoners for large ontologies in the OWL 2 EL profile. Semant. Web **2**(2), 71–87 (2011)
7. Gardiner, T., Tsarkov, D., Horrocks, I.: Framework for an automated comparison of description logic reasoners. In: Cruz, I., et al. (eds.) ISWC 2006. LNCS, vol. 4273, pp. 654–667. Springer, Heidelberg (2006). https://doi.org/10.1007/11926078_47
8. Gennari, J.H., et al.: The evolution of Protégé: an environment for knowledge-based systems development. Int. J. Hum. Comput. Stud. **58**(1), 89–123 (2003)
9. Glimm, B., Horrocks, I., Motik, B., Stoilos, G., Wang, Z.: HermiT: an OWL 2 reasoner. J. Autom. Reason. **53**(3), 245–269 (2014)
10. Guo, Y., Pan, Z., Heflin, J.: LUBM: a benchmark for OWL knowledge base systems. In: Web Semantics: Science, Services and Agents on the World Wide Web, vol. 3(2), pp. 158–182 (2005)
11. Kollia, I., Glimm, B.: Optimizing SPARQL query answering over OWL ontologies. J. Artif. Intell. Res. **48**, 253–303 (2013)
12. Link, V., Lohmann, S., Haag, F.: OntoBench: generating custom OWL 2 benchmark ontologies. In: Groth, P., et al. (eds.) ISWC 2016, Part II. LNCS, vol. 9982, pp. 122–130. Springer, Cham (2016). https://doi.org/10.1007/978-3-319-46547-0_13
13. Motik, B., Shearer, R., Horrocks, I.: Hypertableau reasoning for description logics. J. Artif. Intell. Res. **36**(1), 165–228 (2009)
14. Parsia, B., Matentzoglu, N., Gonçalves, R.S., Glimm, B., Steigmiller, A.: The OWL reasoner evaluation (ORE) 2015 competition report. J. Autom. Reason. **59**(4), 455–482 (2017)
15. Shearer, R., Motik, B., Horrocks, I.: HermiT: a highly-efficient OWL reasoner. In: OWLED, vol. 432, p. 91 (2008)
16. Weithöner, T., Liebig, T., Luther, M., Böhm, S.: What's wrong with OWL benchmarks. In: Proceedings of the Second International Workshop on Scalable Semantic Web Knowledge Base Systems (SSWS 2006), pp. 101–114. Citeseer (2006)

A Framework for Distributed Approximation of Moments with Higher-Order Derivatives Through Automatic Differentiation

Michel Schanen(✉), Daniel Adrian Maldonado, and Mihai Anitescu

Mathematics and Computer Science Division,
Argonne National Laboratory, Lemont, IL, USA
{mschanen,maldonadod,anitescu}@anl.gov

Abstract. We present a framework for the distributed approximation of moments, enabling the evaluation of the uncertainty in a dynamical system. The first and second moment, mean, and variance are computed with up to third-order Taylor series expansion. The required derivatives for the expansion are generated automatically by automatic differentiation and propagated through an implicit time stepper. The computational kernels are the accumulation of the derivatives (Jacobian, Hessian, tensor) and the covariance matrix. We apply distributed parallelism to the Hessian or third-order tensor, and the user merely has to provide a function for the differential equation, thus achieving similar ease of use as Monte Carlo-based methods. We demonstrate our approach using with benchmarks on Theta, a KNL-based system at the Argonne Leadership Computing Facility.

Keywords: Uncertainty · Method of moments ·
Automatic differentiation

1 Introduction

Mathematical models are an approximation of real life systems and their validity resides in how well the outputs of the model agree with measured data. Often, the input or parameters of the model are uncertain because data is unavailable or inaccurate; for these cases, one typically performs an uncertainty quantification (UQ) analysis to determine how much the outputs vary with the input parameters of the model. The uncertainty in the outputs can be quantified as a range of values, but also as a probability distribution function (pdf). Several methods, for example Monte Carlo computation and polynomial chaos, try to solve the problem of computing the probability distribution of the output given parameters defined as pdf's.

A field that has experienced renewed interest in these techniques is energy systems engineering. The electrical power grid with the adoption of renewable

J. M. F. Rodrigues et al. (Eds.): ICCS 2019, LNCS 11540, pp. 251–260, 2019.
https://doi.org/10.1007/978-3-030-22750-0_19

energy has tied its behavior to stochastic weather fluctuations requiring the use of UQ techniques to predict its performance. However, the scale of these problems is such that conventional methods are not satisfactory from a computational point of view. Monte Carlo methods can be thought of as the first-line tools for UQ; with sufficient sampling they are able to quantify the uncertainty regardless of the input distribution or the nonlinearities of the system. However, Monte Carlo methods suffer from slow convergence, which has led to the search for alternative approximations [8]. Recently, the method of moments sparked new interest as one alternative [5].

The method of moments is an approximating technique that works with the moments of probability distributions instead of their density functions. The main idea is to use a Taylor expansion of the function and write the moments of the output distribution as a polynomial function of the moments of the input distribution. Depending on the characteristics of the function, only a few terms of the Taylor expansion might be enough to achieve enough precision. As noted in [9], one of the main issues with the method of moments is that although its accuracy increases with the degree of the Taylor polynomial, computing higher-order derivatives poses serious technical challenges, leading to mostly linearization techniques for acquiring sensitivities [3].

In this paper we present ADUPROP[1], a framework developed at Argonne National Laboratory that combines the automatic differentiation (Sect. 2), method of moments (Sect. 3), uncertainty quantification, and distributed parallelism (Sect. 4) into an easy to use tool that is able to quantify uncertainty of dynamical systems using the method of moments at an unprecedented scale. We use automatic differentiation (AD) by overloading through a C++ template library. This flexible technique allows a straightforward augmentation of C++ codes for computing higher-order derivatives. By exploiting the structure of this approach, we implement a scheme that parallelizes both the accumulation of the derivative information and the computation of the covariance based on the derivative values.

2   Algorithmic Differentiation

Automatic differentiation [2] allows one to differentiate computer programs by applying differential calculus at a program's statement level. It uses compilers or language-based approaches to transform an implementation of a multivariate vector function $y = g(x), \mathbb{R}^n \mapsto \mathbb{R}^m$ into Jacobian vector products $y^{(1)} = J(x) \cdot x^{(1)}$ (tangent-linear model) or transposed Jacobian vector products $x_{(1)} = J^T \cdot y_{(1)}$ (adjoint model), where $x^{(1)}$, $y^{(1)}$ denotes the tangents and $x_{(1)}$, $y_{(1)}$ denotes the adjoints. The tangent-linear mode is equivalent to the finite difference method with the additional advantage of providing derivative information up to machine precision with no truncation or cancellation errors.

In this paper we solely rely on the tangent-linear or forward mode where the transformed code computes the product of the Jacobian J at point x times a

---

[1] https://gitlab.com/aduprop/aduprop.

directional derivative $x^{(1)}$, yielding the output tangent. The directional derivatives, denoted with a superscript order of differentiation, are defined as a partial derivative of y and x with respect to an auxiliary variable s. For readability we use Spivak's notation for derivatives $y^{(1)} = \frac{\partial y}{\partial s} = \frac{\partial y}{\partial x} \cdot \frac{\partial x}{\partial s} = Dg(x) \cdot x^{(1)} \in \mathbb{R}^m$. Letting $x^{(1)}$ go over the Cartesian basis vectors of the implementation $J \cdot x^{(1)}$ yields, column by column, the entire Jacobian $J = Dg(x) \in \mathbb{R}^{m \times n}$. Thus, for the *accumulation* of the full Jacobian we need to rerun the tangent-linear code n (number of columns) times. For higher-order derivative models we use the inner product $<>$ notation introduced in [6] where the tangent-linear model is written as a projection of the Jacobian onto the tangent:

$$y = g(x), \, y^{(1)} = < Dg(x), x^{(1)} > = Dg(x) \cdot x^{(1)} \, . \tag{1}$$

Note that in general an implementation transformed by an automatic differentiation (AD) tool computes both $g(x)$ and the Jacobian vector product. Reapplying an AD tool to an already first-order differentiated code yields a second-order forward over forward (FoF) code computing (2):

$$
\begin{aligned}
y \quad &= g(x), & y^{(2)} \quad &= < Dg(x), x^{(2)} > \\
y^{(1)} &= < Dg(x), x^{(1)} >, & y^{(1,2)} &= < D^2 g(x), x^{(1)}, x^{(2)} > + < Dg(x), x^{(1,2)} > \, .
\end{aligned}
\tag{2}
$$

The superscript $^{(2)}$ denotes the second order of differentiation. Rerunning this FoF model and letting $x^{(1)}$ and $x^{(2)}$ each go over the Cartesian basis vectors, we obtain all the entries of the Hessian $D^2 g \in \mathbb{R}^{m \times n \times n}$ evaluated at x. Here $< D^2 g(x), x^{(1)}, x^{(2)} >$ is the projection of $x^{(1)}$ onto the Hessian followed by the projection of $x^{(2)}$; $x^{(1,2)}$ must be set to zero. (For a detailed definition of Jacobian, Hessian and tensor projections, please refer to [6]). Following this logic, we reapply the tangent-linear model to acquire third order derivatives using the forward over forward over forward model (FoFoF):

$$
\begin{aligned}
y \quad &= g(x), & y^{(3)} \quad &= < Dg(x), x^{(3)} > \\
y^{(2)} &= < Dg(x), x^{(2)} >, & y^{(2,3)} &= < D^2 g(x), x^{(2)}, x^{(3)} > + < Dg(x), x^{(2,3)} > \\
y^{(1)} &= < Dg(x), x^{(1)} >, & y^{(1,3)} &= < D^2 g(x), x^{(1)}, x^{(3)} > + < Dg(x), x^{(1,3)} > ,
\end{aligned}
$$

and the last term capturing the third-order tensor D^3:

$$
\begin{aligned}
y^{(1,2)} \quad &= < D^2 g(x), x^{(1)}, x^{(2)} > + < Dg(x), x^{(1,2)} >, \\
y^{(1,2,3)} &= < D^3 g(x), x^{(1)}, x^{(2)}, x^{(3)} > + < D^2 g(x), x^{(1,3)}, x^{(2)} > \\
&\quad + < D^2 g(x), x^{(1)}, x^{(2,3)} > + < D^2 g(x), x^{(1,2)}, x^{(3)} > + < Dg(x), x^{(1,2,3)} > \, .
\end{aligned}
\tag{3}
$$

The original code using one variable x went up to two for the tangent-linear model, four for the FoF model and eight for the FoFoF model. To accumulate the third-order tensor $D^3 g(x) \in \mathbb{R}^{m \times n \times n \times n}$ in $y^{(1,2,3)}$ we have to let $x^{(1)}$, $x^{(2)}$, and $x^{(3)}$ go over the Cartesian basis vectors, thus requiring n^3 reruns of the model. The remaining tangents of x must be set to zero. These properties will translate directly to the implementation described in Sect. 5.

3   Method of Moments

What is the distribution of y if we let $y = g(x)$ be a function of a random variable x with known properties? Computing this analytically is often difficult; and, in particular, obtaining the pdf of y is not possible in general. An alternative approach is to consider the moments of the distributions. Depending on the shape of the pdf, the first few moments of the pdf can be sufficient to capture relevant behavior. More concretely, consider $g(\boldsymbol{x})$ where $\boldsymbol{x}$ is a random variable with density $f(x)$. If $g(x)$ is sufficiently smooth, given the mean value theorem, we can write [7]

$$\mathbb{E}\left[g(\boldsymbol{x})\right] = \int_\infty^\infty g(x)f(x)dx \approx g(\mu) \int_\infty^\infty f(x)dx = g(\mu)\,, \tag{4}$$

where $\mu = \mathbb{E}\left[\boldsymbol{x}\right]$. Using a third order Taylor expansion for a function $g(x)$ around $\mu = \mathbb{E}\left[\boldsymbol{x}\right]$ and considering the expressions for the mean $\mu^g = \mathbb{E}\left[g(x)\right]$ and covariance $c_{pq}^g = \mathbb{E}\left[(g_p(\boldsymbol{x}) - \mu_p)(g_q(\boldsymbol{x}) - \mu_q)\right] = \mathbb{E}\left[g_p(\boldsymbol{x})g_q(\boldsymbol{x})\right] - \mathbb{E}\left[g_p(\boldsymbol{x})\right]\mathbb{E}\left[g_q(\boldsymbol{x})\right]$, we obtain

$$\mu_p^g = g(\mu) + \frac{1}{2}\sum_{i,j=1}^n \mathrm{D}_{ij}^2 g_p \cdot c_{ij} \tag{5}$$

and

$$
\begin{aligned}
c_{pq}^g = \tfrac{1}{2!}\sum_{i,j=1}^n & \left(\mathrm{D}_j g_p \cdot \mathrm{D}_i g_q + \mathrm{D}_j g_q \cdot \mathrm{D}_i g_p\right) c_{ij} \\
+ \tfrac{1}{4!}\sum_{i,j,k,l=1}^n & \left(\mathrm{D}_i g_p \cdot \mathrm{D}_{jkl}^3 g_q + \mathrm{D}_j g_p \cdot \mathrm{D}_{ikl}^3 g_q + \mathrm{D}_k g_p \cdot \mathrm{D}_{ijl}^3 g_q \right.\\
& + \mathrm{D}_l g_p \cdot \mathrm{D}_{ijk}^3 g_q + \mathrm{D}_{ij}^2 g_p \cdot \mathrm{D}_{kl}^2 g_q + \mathrm{D}_{ik}^2 g_p \cdot \mathrm{D}_{jl}^2 g_q \\
& + \mathrm{D}_{il}^2 g_p \cdot \mathrm{D}_{jk}^2 g_q + \mathrm{D}_{jk}^2 g_p \cdot \mathrm{D}_{il}^2 g_q + \mathrm{D}_{jl}^2 g_p \cdot \mathrm{D}_{ik}^2 g_q \\
& + \mathrm{D}_{kl}^2 g_p \cdot \mathrm{D}_{ij}^2 g_q + \mathrm{D}_{jkl}^3 g_p \cdot \mathrm{D}_i g_q + \mathrm{D}_{ikl}^3 g_p \cdot \mathrm{D}_j g_q \\
& \left. + \mathrm{D}_{ijl}^3 g_p \cdot \mathrm{D}_k g_q + \mathrm{D}_{ijk}^3 g_p \cdot \mathrm{D}_l g_q\right) c_{ijkl} \\
- \tfrac{1}{2!}\sum_{i,j,k,l=1}^n & \left(\mathrm{D}_{ij}^2 g_p \cdot \mathrm{D}_{kl}^2 g_q\right) c_{ij} c_{kl}\,.
\end{aligned}
\tag{6}
$$

In the derivation of these formulas we assume a Gaussian distribution, which results in the cancellation of the odd terms of the expansion.

4   Tensor Decomposition

In addition to combining the aforementioned methods and tools in a novel way to compute the moments, our main contribution is the distributed parallelization described in this section. The FoF model always computes one projection $y^{(1,2)}$ of $D^2 g \in \mathbb{R}^{m \times n \times n}$ onto $x^{(1)}, x^{(2)} \in \mathbb{R}^n$ (see Fig. 1). Hence we cannot decompose the Hessian along the entries of $y^{(1,2)}$. By parallelizing over the entries of $x^{(1)}$ or $x^{(2)}$, we can restrict the Cartesian basis vectors (see Sect. 2) to the local indices and thus distribute the Hessian accumulation over all processes. The same holds true for the computation of the four-dimensional tensor $D^3 g$ (3) where we can parallelize over the entries of $x^{(1)}$, $x^{(2)}$, or $x^{(3)}$ and thus distribute the tensor over all processes.

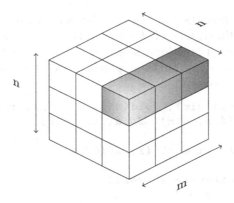

Fig. 1. Hessian D^2g

The computation of the mean in (5) is then parallelized in a straightforward way over either the index i or j. Each process thus has a local copy of μ that needs to be *allreduced* at the end of the summation. We parallelize the computation of C over p or q and perform *allgather* to share C among all processes. Solving the linear system for the time stepper has a runtime complexity of at most $\mathcal{O}\left(n^3\right)$, or lower in case of sparsity in the inner Jacobian. Accumulating the tensor D^3g has a runtime complexity of $\mathcal{O}\left(n^2\right)\cdot cost(g)$. Thus, the total run-time complexity for accumulating the tensor is $\mathcal{O}\left(n^6/p\right)$, where p is the number of processes. The covariance computation in (6) yields the same complexity of $\mathcal{O}\left(n^6/p\right)$, which is our global runtime complexity.

5 Implementation

ADUPROP (AD for Uncertainty Propagation) is a C++ implementation of the propagation of moments. In particular, it implements the concepts of Sect. 3 in the context of differential equations. ADUPROP is a template-based code that allows easy computation of higher-order derivatives. The library provides vector, matrix, and tensor data structures using a template type T instead of double. For our implementation of ADUPROP we chose the AD tool CoDiPack, which is based on operator overloading. To create an $n + 1$ derivative type t3s we recursively apply differentiation on an n order type.

```
1 typedef RealForwardGen <RealForwardGen <RealForwardGen <double
        > > > t3s;
```

This maps exactly to the notation in Sect. 2. Accessing, for example, x.gradient ().gradient().value() of the third-order type t3s of a variable x is equivalent to accessing $x^{(2,3)}$. The variable type T in the implementation of f allows us to instantiate the function using the types double, t1s, t2s, and t3s.

As an example, we implement UQ of a differential equation system with ADUPROP. When the system is discretized ($x_k = \phi(x_{k-1})$), the procedure is identical to the one described in Sect. 3. The default integration scheme that we use is backward Euler. One of the main advantages of using AD is that we can differentiate through functions, loops, or other complex functions in which obtaining explicit derivatives might be practically challenging and tedious. The integration loop is written as follows

```
1   pVector<T> xold(dim), yold(dim), y(dim), res(dim);
2   pMatrix<T> J(dim, dim), Jold(dim, dim);
3   xold = x;
4   sys->residual_beuler<T>(x, xold, y);
5   do {
6       sys->jac_beuler<T>(x, xold, J);
7       yold = y; Jold = J;
8       adlinsolve<T>(J, y);
9       res = Jold * y - yold;
10      x = x - y;
11      sys->residual_beuler<T>(x, xold, y);
12  } while (y.norm() > eps);
```

The object sys, defined by the user, has to contain the residual function residual_beuler and the Jacobian jac_beuler. The function adlinsolve provides an interface to linear solvers. Currently we support BLAS and Eigen for dense and sparse linear systems, respectively. For an order of differentiation k we differentiate $Ax = b$. Let $A(s) \in \mathbb{R}^{n \times n}$, $b(s) \in \mathbb{R}^n$, and $x(s) \in \mathbb{R}^n$ with s being some input dependency. We define $\frac{\partial^k A}{\partial s^k} = A_k$, $\frac{\partial^k b}{\partial s^k} = b_k$, and $\frac{\partial^k x}{\partial s^k} = x_k$. With $Ax = b$, we have

$$c_k \cdot A \cdot x^{(k)} = b^{(k)} - c_0 \cdot A^{(k)} \cdot x - c_1 \cdot A^{(k-1)} x^{(1)} - \ldots - c_{k-1} \cdot A^{(1)} \cdot x^{(k-1)}. \quad (7)$$

In summary, we have to solve 2, 4, and 8 linear systems for first-, second-, and third-order derivatives, respectively. With these three basic blocks in place, a time stepper, residual function, Jacobian, and linear system, we can compute the Jacobian, Hessian, and third order derivative tensor using the logic described in Sect. 2.

6   Scalability

A prototype sequential implementation of this method in Julia was tested on power system dynamics in [5]. To assess the scaling and computational capabilities of ADUPROP, we resort to a well-known dampened nonlinear dynamical system used in the weather simulation community and elsewhere [4,10,11]:

$$\dot{x}_i = x_{i-1}(x_{i+1} - x_{i-2}) - x_i + F, \; i = 1, \ldots, n > 3. \quad (8)$$

This system is known to show chaotic behavior when the *forcing term* $F \geq 8$, having an equilibrium $(F, \ldots, F)$ that becomes unstable for all $n \geq 4$. The system transitions from a damped, constant-valued system to a traveling wave with a

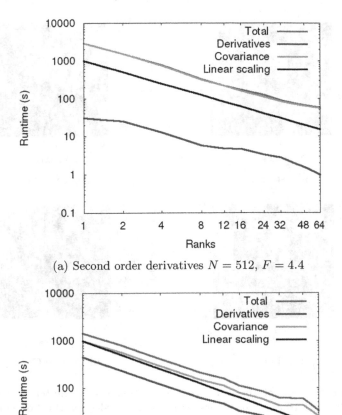

(a) Second order derivatives $N = 512$, $F = 4.4$

(b) Third order derivatives $N = 64$, $F = 4.4$

Fig. 2. Strong scaling

periodic attractor, and eventually to chaotic behavior, all adjustable through the selection of F.

The scalability study was done on Theta at the Argonne Leadership Computing Facility. Theta is composed of 1.3 GHz Intel Xeon Phi 7230 SKU nodes with 64 cores each. Our goal was to achieve strong scaling on a single node with up to 64 MPI processes. Our focus is on the strong scaling of the covariance computation using Hessians and third-order derivative tensors. We use the same $F = 4.4$ forcing but increase the dimension to $N = 64$ for third-order derivatives and $N = 512$ for second-order derivatives. The time horizon is irrelevant

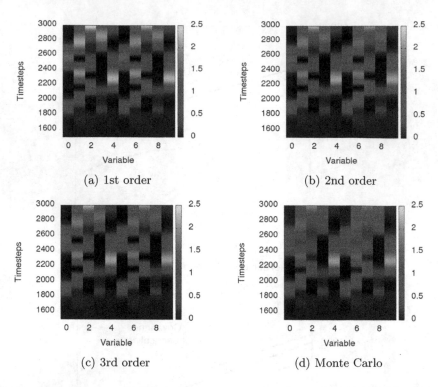

Fig. 3. Approximation of the variance propagation from timestep 1,500 to 3,000 (h = 0.001) for $N = 10$ $F = 5$ using first-order, second-order and third-order derivatives in addition to fully converged Monte Carlo at 1,000 samples.

to the scaling, since there is no parallelization in time. With the timestep set to 1 we achieved the strong-scaling results in Fig. 2. The black line serves as a reference point for linear scaling. We show that our implementation scales nearly linearly with up to 64 cores, with both second- and third-order derivatives. As anticipated by our complexity analysis in Sect. 4, the covariance computation dominates the runtime with second-order derivatives, whereas with third-order derivatives both the tensor accumulation and covariance computation are much closer.

Each KNL node has 64 cores, limiting the strong scaling to one row or projection of the derivative tensor. Our code is able to run beyond a single node, but the runtime cost of computing the derivative tensor becomes too high. In the future we will investigate low-rank approximations to compress the derivative tensor and decrease its computational cost [1].

We validate the approximation of the variance propagation in Fig. 3 with dimension $N = 10$ and in a nonlinear regime with $F = 5$. The higher-order derivatives allow us to better capture the effects of nonlinearity in power systems [5]. The numerical aspects of this research will be subject of future research.

7   Conclusion

This paper describes a distributed parallel framework for using the method of moments backed with AD. The extension to third-order derivatives and the parallelization over the derivative and covariance accumulation is unprecedented at this scale and speed. The distribution of the Hessian and tensor is chosen such that AD and the covariance computation benefit from the parallelism. Nonetheless, the computational cost grows at a factor of $\mathcal{O}\left(N^6\right)$, N being the dimension, times the original simulation evaluation using third-order derivatives. This factor is reduced to $\mathcal{O}\left(N^4\right)$ while using only the Hessian. In both cases we have shown the scalability on the current Intel KNL architecture. As opposed to Monte Carlo, this algorithm provides an analytical propagation workflow that showed promising results in [5]. To scale beyond a single KNL node, future research will focus on exploiting Hessian, tensor and covariance matrix structure in order to apply sampling methods. This approach has shown promising results in machine learning [12] and we plan to integrate this approach in our software. While being highly problem dependent, it has the potential to significantly reduce the complexity of the covariance computation. In particular, at higher dimensions, this method of propagating uncertainties may become a valuable alternative to the Monte Carlo-based sampling methods.

Acknowledgment. This material was based upon work supported by the U.S. Department of Energy, Office of Science, under Contract DE-AC02-06CH11347.

Government License (will be removed at publication): The submitted manuscript has been created by UChicago Argonne, LLC, Operator of Argonne National Laboratory ("Argonne"). Argonne, a U.S. Department of Energy Office of Science laboratory, is operated under Contract No. DE-AC02-06CH11357. The U.S. Government retains for itself, and others acting on its behalf, a paid-up nonexclusive, irrevocable worldwide license in said article to reproduce, prepare derivative works, distribute copies to the public, and perform publicly and display publicly, by or on behalf of the Government. The Department of Energy will provide public access to these results of federally sponsored research in accordance with the DOE Public Access Plan. http://energy.gov/downloads/doe-public-access-plan.

References

1. Abdel-Khalik, H.S., Hovland, P.D., Lyons, A., Stover, T.E., Utke, J.: A low rank approach to automatic differentiation. In: Bischof, C.H., Bücker, H.M., Hovland, P., Naumann, U., Utke, J. (eds.) Advances in Automatic Differentiation. LNCSE, vol. 64, pp. 55–65. Springer, Heidelberg (2008). https://doi.org/10.1007/978-3-540-68942-3_6

2. Griewank, A., Walther, A.: Evaluating Derivatives: Principles and Techniques of Algorithmic Differentiation, vol. 105, 2nd edn. SIAM, Philadelphia (2008)

3. Hiskens, I.A., Alseddiqui, J.: Sensitivity, approximation, and uncertainty in power system dynamic simulation. IEEE Trans. Power Syst. **21**(4), 1808–1820 (2006). https://doi.org/10.1109/TPWRS.2006.882460

4. Lorenz, E.: Predictability: a problem partly solved. In: Seminar on Predictability, 4–8 September 1995, vol. 1, pp. 1–18. ECMWF, Shinfield Park, Reading (1995)

5. Maldonado, D.A., Schanen, M., Anitescu, M.: Uncertainty propagation in power system dynamics with the method of moments. In: 2018 IEEE Power Energy Society General Meeting (PESGM), pp. 1–5, August 2018. https://doi.org/10.1109/PESGM.2018.8586023
6. Naumann, U.: The Art of Differentiating Computer Programs: An Introduction to Algorithmic Differentiation, vol. 24. SIAM, Philadelphia (2012)
7. Papoulis, A.: Probability, Random Variables and Stochastic Processes. McGraw-Hill, New York (1965)
8. Preece, R., Milanović, J.V.: Efficient estimation of the probability of small-disturbance instability of large uncertain power systems. IEEE Trans. Power Syst. **31**(2), 1063–1072 (2016). https://doi.org/10.1109/TPWRS.2015.2417204
9. Smith, R.C.: Uncertainty Quantification: Theory, Implementation, and Applications, vol. 12. SIAM, Philadelphia (2013)
10. Sterk, A.E., van Kekem, D.L.: Predictability of extreme waves in the Lorenz-96 model near intermittency and quasi-periodicity. Complexity (9419024) (2017). https://doi.org/https://doi.org/10.1155/2017/9419024
11. Wilks, D.S.: Effects of stochastic parametrizations in the Lorenz '96 system. Q. J. R. Meteorol. Soc. **131**(606), 389–407. https://doi.org/10.1256/qj.04.03, https://rmets.onlinelibrary.wiley.com/doi/abs/10.1256/qj.04.03
12. Xu, P., Roosta-Khorasan, F., Mahoney, M.W.: Second-order optimization for non-convex machine learning: An empirical study. (2017) arXiv:1708.07827

IPIES for Uncertainly Defined Shape of Boundary, Boundary Conditions and Other Parameters in Elasticity Problems

Marta Kapturczak$^{(\boxtimes)}$ (iD) and Eugeniusz Zieniuk (iD)

University of Bialystok, Białystok, Poland
{mkapturczak,ezieniuk}@ii.uwb.edu.pl

Abstract. The main purpose of this paper is modelling and solving boundary value problems simultaneously considering uncertainty of all of input data such as: shape of boundary, boundary conditions and other parameters. The strategy is presented on the basis of problems described by Navier-Lamé equations. Therefore, the uncertainty of parameters here, means the uncertainty of the Poisson's ratio and Young's modulus. For solving uncertainly defined problems we use implementation of interval parametric integral equations system method (IPIES). In this method we propose modification of directed interval arithmetic for modeling and solving uncertainly defined problems. We consider an examples of uncertainly defined, 2D elasticity problems. We present boundary value problems with linear as well as curvelinear (modelled using NURBS curves) shape of boundary. We verify obtained interval solutions by comparison with precisely defined (without uncertainty) analytical solutions. Additionally, to obtain errors of such solutions, we decided to use the total differential method. We also analyze influence of input data uncertainty on interval solutions.

Keywords: Boundary problems · Uncertainty · Interval arithmetic · Parametric integral equations system

1 Introduction

Modeling of uncertainty is a very important problem and it generates considerable interest among researchers. However direct application of existing mathematical apparatuses is often useless in practice. In this paper we present interval parametric integral equations system (IPIES) [8,10] for solving uncertainly defined boundary value problems on examples of 2D elasticity problems. The parametric integral equations system method (PIES) was previously developed and widely tested for precisely (exactly) defined problems [7,11]. Many studies

The scientific work is founded by resources of Ministry of Science and Higher Education for research, granted to the Faculty of Mathematics and Informatics, University of Bialystok.

J. M. F. Rodrigues et al. (Eds.): ICCS 2019, LNCS 11540, pp. 261–268, 2019.
https://doi.org/10.1007/978-3-030-22750-0_20

have confirmed PIES advantages over other methods, such as well known FEM and BEM methods.

Mentioned methods have also corresponding interval methods, such as IFEM [1] and IBEM [4]. However, in these methods accuracy of solutions depends on (finite or boundary) elements number. So, discretization increases the number of interval data and results in solutions overestimation. Therefore, the IFEM and IBEM researchers have mainly focused on modeling uncertainty of boundary conditions and other parameters only.

The article presents the impact of all uncertainly defined input data (necessary to define the problem) on IPIES solutions, obtained based on implemented program of the method. We consider 2D elasticity problems modeled by Navier-Lamé equations and we define uncertainty of the shape of boundary, boundary conditions and other parameters (Poisson's ratio and Young's modulus). We model uncertainty using modified directed interval arithmetic (applied in IPIES). To verify obtained interval solutions we use analytical solutions with errors obtained using total differential method [3]. Additionally we test an impact of change in data uncertainty on interval solutions.

2   Mathematical Foundations of IPIES

Till now, PIES was applied for precisely defined boundary problems [7,11]. It is an analytical modification of boundary integral equations. Now, to include uncertainly defined input data, we can present PIES (on example of Navier-Lamé equation) using intervals:

$$
0.5\boldsymbol{u}_l(s_1) = \sum_{j=1}^{n} \int_{\widehat{s}_{j-1}}^{\widehat{s}_j} \left\{ \boldsymbol{U}_{lj}^*(s_1,s)\boldsymbol{p}_j(s) - \boldsymbol{P}_{lj}^*(s_1,s)\boldsymbol{u}_j(s) \right\} J_j(s)ds, \qquad (1)
$$

where $l = 1, 2, ..., n$, $\widehat{s}_{l-1} \le s_1 \le \widehat{s}_l$, $\widehat{s}_{j-1} \le s \le \widehat{s}_j$, where $\widehat{s}_{l-1}, \widehat{s}_{j-1}$ are the beginnings and $\widehat{s}_l, \widehat{s}_j$ are the ends of segments with index l or j. Function $J_j(s) = [\underline{J}_j(s), \overline{J}_j(s)]$ is the Jacobian for interval curve segment $\boldsymbol{S}_j = [\boldsymbol{S}_j^{(1)}(s), \boldsymbol{S}_j^{(2)}(s)]^T$ with index j, where $\boldsymbol{S}_j^{(1)}(s) = [\underline{S}_j^{(1)}(s), \overline{S}_j^{(1)}(s)], \boldsymbol{S}_j^{(2)}(s) = [\underline{S}_j^{(2)}(s), \overline{S}_j^{(2)}(s)]$. Integral functions $\boldsymbol{p}_j(s) = \left\{ [\underline{p}_j^{(1)}(s), \overline{p}_j^{(1)}(s)], [\underline{p}_j^{(2)}(s), \overline{p}_j^{(2)}(s)] \right\}, \boldsymbol{u}_j(s) = \left\{ [\underline{u}_j^{(1)}(s), \overline{u}_j^{(1)}(s)], [\underline{u}_j^{(2)}(s), \overline{u}_j^{(2)}(s)] \right\}$ are the interval parametric boundary functions on corresponding boundary segments $\boldsymbol{S}_j$ (on which the boundary was theoretically divided). One of these functions will always be defined as uncertain (interval) boundary conditions on segment $\boldsymbol{S}_j$, then the other will be obtained as a result of numerical solution of IPIES (1).

Including uncertainty in the first integral $\boldsymbol{U}_{lj}^*(s_1,s)$ for plane state of strain we obtained following interval matrix:

$$
\boldsymbol{U}_{lj}^*(s_1,s) = -\frac{1}{8\pi(1-\boldsymbol{\nu})\boldsymbol{\mu}} \begin{bmatrix} (3-4\boldsymbol{\nu})\ln(\eta) - \frac{\eta_1^2}{\eta^2} & -\frac{\eta_1\eta_2}{\eta^2} \\ -\frac{\eta_1\eta_2}{\eta^2} & (3-4\boldsymbol{\nu})\ln(\eta) - \frac{\eta_2^2}{\eta^2} \end{bmatrix}, \qquad (2)
$$

where $l, j = 1, 2, ..., n$, $\mu = 0.5 \cdot E/(1 + \nu)$ is an interval Lamé parameter, $\nu = [\underline{\nu}, \overline{\nu}]$ is an interval Poisson's ratio, $E = [\underline{E}, \overline{E}]$ is an interval Young's modulus and the formulas to obtain $\eta = [\underline{\eta}, \overline{\eta}]$, $\eta_1 = [\underline{\eta}_1, \overline{\eta}_1]$ and $\eta_2 = [\underline{\eta}_2, \overline{\eta}_2]$ are:

$$\eta = [\eta_1^2 + \eta_2^2]^{0.5}, \qquad \eta_1 = S_l^{(1)}(s_1) - S_j^{(1)}(s), \qquad \eta_2 = S_l^{(2)}(s_1) - S_j^{(2)}(s). \quad (3)$$

The second integral $P_{lj}^*(s_1, s)$ we also defined using intervals:

$$P_{lj}^*(s_1, s) = -\frac{1}{4\pi(1 - \nu)\eta} \begin{bmatrix} P_{11} & P_{12} \\ P_{21} & P_{22} \end{bmatrix}, \quad (4)$$

where $l, j = 1, 2, ..., n$ and $P_{ik} = [\underline{P}_{ik}, \overline{P}_{ik}]$ (where i, k = 1, 2) are defined as:

$$P_{ii} = \left\{ (1 - 2\nu) + 2\frac{\eta_i^2}{\eta^2} \right\} \frac{\partial\eta}{\partial n}, \quad (5)$$

$$P_{ik} = \left\{ 2\frac{\eta_i \eta_k}{\eta^2} \frac{\partial\eta}{\partial n} - (1 - 2\nu) \left[\frac{\eta_i}{\eta} n_k(s) + \frac{\eta_k}{\eta} n_i(s) \right] \right\}, \quad (6)$$

$$\frac{\partial\eta}{\partial n} = \frac{\eta_1}{\eta} n_1(s) + \frac{\eta_2}{\eta} n_2(s). \quad (7)$$

We use directed intervals for modeling uncertainty and modified directed interval arithmetic for calculations:

$$x \cdot y = \begin{cases} x_s \cdot y_s - x_s \cdot y_m - x_m \cdot y_s + x_m \cdot y_m & \text{for } x \leq 0, y \leq 0 \\ x_s \cdot y - x_m \cdot y & \text{for } x > 0, y \leq 0 \\ x \cdot y_s - x \cdot y_m & \text{for } x \leq 0, y > 0 \\ x \cdot y & \text{for } x > 0, y > 0 \end{cases}, \quad (8)$$

where for any interval number $a = [\underline{a}, \overline{a}]$ we define $a_s = a + a_m$ and $a_m = \begin{cases} |\overline{a}| & \text{for } \overline{a} > \underline{a} \\ |\underline{a}| & \text{for } \overline{a} < \underline{a} \end{cases}$, where $a > 0$ means $\underline{a} > 0$ and $\overline{a} > 0$ i $a \leq 0$ means $\underline{a} < 0$ or $\overline{a} < 0$ then multiplication ($\cdot$) is an interval multiplication. Research on such modification was widely discussed in [8].

3   Verification of IPIES on Examples

To obtain interval solutions, presented mathematical apparatus was implemented as computer program of IPIES method. We decided to verify obtained interval solutions using analytical solutions with errors obtained by total differential method (used directly to define errors in arithmetic operations). This method allows us to obtain error of the function, when the errors of all function arguments are known. If the function $u = f(x_1, x_2, ..., x_n)$ is differentiable and we define $|\Delta x_i|(i = 1, 2, ..., n)$ as absolute errors of function arguments, then the general formula to obtain an absolute error of the function is [3]:

$$\Delta u = \sum_{i=0}^{n} \left| \frac{\partial f}{\partial x_i} \right| |\Delta x_i|, \tag{9}$$

where in boundary value problem the function f correspond to analytical solution.

Example 1. The first of considered problems is known as the Lamé problem [6]. We define simultaneously uncertainty of the boundary shape and boundary conditions (Fig. 1). We analyze thick-walled pipe, which length of the internal radius ($a = 10$ cm) and external radius ($b = 25$ cm) is defined with the width of uncertainty band (the width of interval) $\varepsilon_a = \varepsilon_b = 1$. So defined pipe is subjected to a uniform internal pressure p_a. Therefore, we define the uncertainty of the boundary condition by interval value $p_a = [99, 101]$. The problem is defined in plane state of strain and the material parameters are defined as degenerate intervals with values: $E = 2 \cdot 10^5$ MPa i $\nu = 0.25$. We use NURBS curves [5,9] of second degree (defined using interval points) to model uncertainty of the shape of boundary.

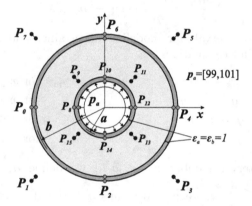

Fig. 1. Modelling uncertainty of the shape of boundary and boundary condition.

As already mentioned, we decided to compare obtained interval solutions with analytical ones [6]:

$$\sigma_x = \frac{p_a a^2 - p_b b^2}{b^2 - a^2} - \frac{(p_a - p_b)a^2 b^2}{r^2(b^2 - a^2)}, \qquad \sigma_y = \frac{p_a a^2 - p_b b^2}{b^2 - a^2} + \frac{(p_a - p_b)a^2 b^2}{r^2(b^2 - a^2)}, \tag{10}$$

where r is the one of polar coordinates $r^2 = x^2 + y^2$, $a < r < b$. We obtain solution error using total differential method, where as the error of function arguments ($|\Delta x_i|$ from the (9)) we assume the half of the interval width ($\Delta a = |\overline{a} - \underline{a}|/2$) for uncertainly defined shape of boundary $a = [\underline{a}, \overline{a}]$, $b = [\underline{b}, \overline{b}]$ and boundary condition $p_a = [\underline{p_a}, \overline{p_a}]$, $p_b = [\underline{p_b}, \overline{p_b}]$. Additionally we assume $p_b = [0,0]$ and omit the uncertainty of material constant in this example, because the Young's modulus and Poisson's ratio are not defined in analytical solution (10).

We present interval solutions and analytical solutions with errors obtained using total differential method in Table 1 in cross section: $x = 12, 14, ..., 24$ and $y = 0$. To allow direct comparison, we present interval solutions using the middle of interval $(\text{mid}(a) = (\overline{a} + \underline{a})/2)$ and the half of interval width $(\Delta a = |\overline{a} - \underline{a}|/2)$. Both obtained middle values of interval solutions and analytical solutions, as well as obtained halves of the width of interval solutions and analytical solutions errors are almost equal. Average relative error is ca one percent, so the example confirms correctness of proposed strategy.

Table 1. Interval solutions in domain with compare to analytical ones.

x	Analytical solution				Interval solution (IPIES)			
	σ_x	σ_y	$\Delta\sigma_x$	$\Delta\sigma_y$	$mid(\sigma_x)$	$mid(\sigma_y)$	$\Delta\sigma_x$	$\Delta\sigma_y$
12	-63.624	101.720	8.488	14.664	-64.074	102.686	8.535	14.714
14	-41.691	79.786	5.824	11.666	-41.980	80.573	5.833	11.708
16	-27.455	65.551	4.096	9.721	-27.670	66.223	4.101	9.758
18	-17.695	55.791	2.911	8.387	-17.861	56.385	2.914	8.421
20	-10.714	48.810	2.063	7.433	-10.846	49.347	2.066	7.464
22	-5.549	43.644	1.436	6.727	-5.649	44.135	1.441	6.754
24	-1.620	39.716	0.959	6.190	-1.629	40.180	1.003	6.230
Average relative error [%]					0.95	1.06	0.87	0.43

Example 2. In the next example we decided to examine the influence of input data uncertainty on interval solutions of IPIES method. We consider 2×2 square plate presented in [2] and uncertainly defined in Fig. 2. The shape of boundary is defined with width of uncertainty band $\varepsilon = 0.1$. Material constants are defined as follow: Young's modulus $\boldsymbol{E} = [0.9, 1.1]$ and Poisson's ratio $\boldsymbol{\nu} = [0.29, 0.31]$. In Fig. 2, we also present interval force $\boldsymbol{p}$ acting to one side of the plate. We obtain solutions of the problem in plane state of strain.

Analytical solutions, of above mentioned example, are defined exactly (without uncertainty) as follow [2]:

$$u_x = -0.195x^2 - 0.455(y-1)^2, \qquad u_y = 0.91x(y+1), \qquad (11)$$

and corresponding stress [2]:

$$\sigma_x = 0, \qquad \sigma_y = x. \qquad (12)$$

We obtain interval solutions in cross-section presented on Fig. 2, where $y = 0$ and x changing from -1 to 1. Results are presented in Fig. 3. We denote the width

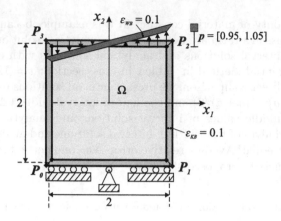

Fig. 2. Uncertainly defined boundary value problem.

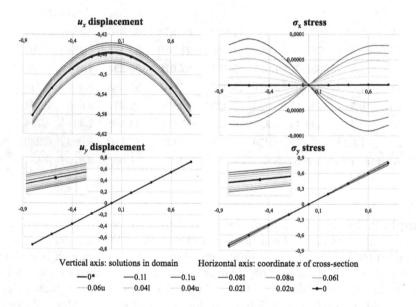

Fig. 3. Interval solutions in domain.

of uncertainty band as values 0.1 – 0. For example, value 0.08 means the width of uncertainty band equal to $\varepsilon_{KB} = 0.08$ for the shape of boundary, $\varepsilon_{WB} = 0.08$ for boundary conditions, $\varepsilon_E = 0.16$ for Young's modulus and $\varepsilon_\nu = 0.016$ for Poisson's ratio. Interval solutions are presented as lower (l) and upper (u) bound.

So, we can note, that the strategy works well for problems where all of input data are defined uncertainly. As we expect, by increasing the width of interval input data, the width of interval solutions is also increased. Additionally presented IPIES solutions with the width of uncertainty band $\varepsilon = 0$ and analytical ones 0* (for precisely defined problem) are almost equal and are located inside of each interval solution.

4 Conclusions

We presented mathematical foundation of IPIES with uncertainly defined all of input data simultaneously. Based on such mathematical model the program of IPIES method has been implemented. We test the program on examples of the problems described by Navier-Lamé equations. Therefore we modeled uncertainly defined shape of boundary, boundary conditions and parameters: Young's modulus and Poisson's ratio. To verify interval solutions we decided to use analytical solutions and its errors obtained by total differential method. Obtained solutions are almost equal. We also tested an impact of the width of interval input data on width of interval solutions. We noted, that the width of interval solution, as expected, increases with the width of interval input data. Additionally exact solutions of PIES (almost equal to the analytical ones) are located inside all of considered interval solutions. Therefore, in conclusion, it is very difficult and time-consuming to define exactly what kind of problems can be solved by IPIES, but we present a high potential of the method, in solving problems with all of input data defined uncertainly, for further investigations.

References

1. Muhanna, R.L., Mullen, R.L., Rao, M.V.R.: Nonlinear finite element analysis of frames under interval material and load uncertainty. ASCE-ASME J. Risk Uncertain. Eng. Syst. Part B Mech. Eng. **1**(4), 041003 (2015)
2. Panzeca, T., Fujita Yashima, H., Salerno, M.: Direct stiffness matrices of BEs in the Galerkin BEM formulation. Eur. J. Mech. Solids **20**, 277–298 (2001)
3. Peng, F.Y., Ma, J.Y., Wang, W., Duan, X.Y., Sun, P.P., Yan, R.: Int. J. Mach. Tools Manuf. **70**, 53–62 (2013)
4. Piasecka-Belkhayat, A.: Interval boundary element method for 2D transient diffusion problem using the directed interval arithmetic. Eng. Anal. Bound. Elem. **35**(3), 259–263 (2011)
5. Piegl, L., Tiller, W.: The NURBS Book, 2nd edn. Springer, Heidelberg (1997). https://doi.org/10.1007/978-3-642-59223-2
6. Timoshenko, S.P., Goodier, J.N.: Theory of Elasticity. McGraw-Hill, Tokyo (1970)
7. Zieniuk, E.: A new integral identity for potential polygonal domain problems described by parametric linear functions. Eng. Anal. Bound. Elem. **26**(10), 897–904 (2002)
8. Zieniuk, E., Kapturczak, M., Kużelewski, A.: Concept of modeling uncertainly defined shape of the boundary in two-dimensional boundary value problems and verification of its reliability. Appl. Math. Model. **40**(23-24), 10274–10285 (2016)
9. Zieniuk, E., Kapturczak, M.: Modeling the shape of boundary using NURBS curves directly in modified boundary integral equations for Laplace's equation. Comput. Appl. Math. **37**(4), 4835–4855 (2018)

10. Zieniuk, E., Kapturczak, M., Kużelewski, A.: Modification of interval arithmetic for modelling and solving uncertainly defined problems by interval parametric integral equations system. In: Shi, Y., et al. (eds.) ICCS 2018, Part III. LNCS, vol. 10862, pp. 231–240. Springer, Cham (2018). https://doi.org/10.1007/978-3-319-93713-7_19

11. Zieniuk, E., Kużelewski, A.: GPU-based acceleration of computations in elasticity problems solving by parametric integral equations system. Adv. Eng. Softw. **79**, 27–35 (2015)

Enabling UQ for Complex Modelling Workflows

Małgorzata J. Zimoń[1(✉)], Samuel Antão[1(✉)], Robert Sawko[1], Alex Skillen[2], and Vadim Elisseev[1]

[1] IBM Research, Daresbury, UK
{malgorzata.zimon,rsawko}@uk.ibm.com,
{samuel.antao,vadim.v.elisseev}@ibm.com
[2] Scientific Computing Department, STFC, Daresbury, UK
alex.skillen@stfc.ac.uk
http://research.ibm.com/labs/uk
http://www.scd.stfc.ac.uk

Abstract. The increase of computing capabilities promises to address many scientific and engineering problems by enabling simulations to reach new levels of accuracy and scale. The field of uncertainty quantification (UQ) has recently been receiving an increasing amount of attention as it enables reliability study of modelled systems. However, performance of UQ analysis for high-fidelity simulations remains challenging due to exceedingly high complexity of computational workflows. In this paper, we present a UQ study on a complex workflow targeting a thermally stratified flow. We discuss different models that can be used to enable it. We then propose an abstraction at the level of the workflow specification that enables the modeller to quickly switch between UQ models and manage underlying compute infrastructure in a completely transparent way. We show that we can keep the workflow description almost unchanged while benefitting of all the insight the UQ study provides.

Keywords: Uncertainty quantification · Workflows · Modelling · High-performance computing

1 Introduction

With the increase of computational power, scientific and engineering simulations are tackling increasingly higher complex systems. Modelling is being used as a strategic tool to provide key insights for predictions and decision making. Therefore, it is important to address the presence of uncertainty in the model data, such as inexact knowledge of initial conditions, and incorporate probabilistic behaviours into analysis. Uncertainty quantification (UQ) methods can be used in model certification, parameter estimation, and inverse problems [10].

This work was supported by the STFC Hartree Centre's Innovation Return on Research programme, funded by the Department for Business, Energy & Industrial Strategy.

J. M. F. Rodrigues et al. (Eds.): ICCS 2019, LNCS 11540, pp. 269–281, 2019.
https://doi.org/10.1007/978-3-030-22750-0_21

Although the need for confidence intervals in modelling predictions is apparent, the inclusion of UQ into complex engineering studies is non-trivial. Not only can the model have a high degree of sophistication but also the underlying high-performance computing (HPC) resources can be challenging to manage. HPC is usually associated with UQ as an enabler, in the sense that the modelling applications typically require the memory, parallelisation ability, and compute power only found in top-tier HPC resources. Also, the need to replay simulations for multiple sets of inputs increases the amount of computation needed. Other property of UQ that stresses the HPC infrastructure is how the runtime requirements can change between runs, particularly with multi-resolution methods [17].

As scientific simulations and modelling become more complex, they tend to be organised in workflows as opposed to single monolithic applications. During the process of designing an orchestrated set of applications, the requirements and the way data is used may change. Additionally, the number of applications and their dependences can vary. Therefore, the end-user would appreciate a flexible and interactive environment. The modeller should be able to focus exclusively on improving the computational experiment and avoid having to dive into challenging infrastructure details. In addition, the assimilation of the UQ study within the workflow in a integrated way would bring an increased benefit. This can be used to drive the workflow design itself and enable defining trade-offs at an early stage, e.g. the accuracy of the UQ model versus the computational cost. Workflows also add an auditability dimension to the modelling work which is relevant for many organisations. A platform that delivers on these requirements would democratise the use of UQ (and HPC for that matter) across less experienced users who might avoid performing stochastic modelling due to a steep learning curve.

In this work, we present an example of a complex UQ analysis needed to estimate the influence of parametric variability in the transient simulation of conjugate heat transfer. We demonstrate the use of a non-intrusive, adaptive method by assessing the propagation of thermal shock within a u-shaped bend. This case is relevant to nuclear applications, where hot-cold cycles can lead to thermal fatigue and material failure. The influence of shock-magnitude on wall temperatures is assessed.

Supported on this analysis, we then discuss an abstraction in which the end-user can express UQ studies cleanly as part of its own simulation/modelling workflow and with them have a common infrastructure that handles these two components jointly as they were a single one. Furthermore, we explore possibilities to enhance reusability of workflow components by multiple end-users.

This paper expands on [17]. Therein, an architecture to easily deploy UQ studies as workflows is proposed. That solution uses an as-a-service approach, integrating a mix of cloud and traditional HPC technology as well as the components to manage a dynamic and scalable infrastructure. That architecture not only provides the requirements of UQ studies but also reduces the load on the users, as it is able to *smart* schedule jobs and predetermine important runtime parameters (domain partitions, compute nodes, threading) with an artificial intelligence engine, providing the baseline infrastructure for the abstraction proposed herein.

The paper is organised as follows. Section 2 describes the general polynomial chaos method and its non-intrusive, multi-element variant; Sect. 3 discusses current infrastructures to enable UQ, while Sect. 4 describes our proposed abstraction; Sect. 5 demonstrates an application of the abstraction to perform a complex UQ study for heat transfer flow modelling; finally, Sect. 6 draws some conclusions.

2 Non-intrusive Methods for Uncertainty Quantification

In this work, the general polynomial chaos (gPC) approach has been applied to quantify uncertainty in the simulation of thermally stratified flow. The following section describes a variant of generalised polynomial chaos, non-intrusive spectral projection (NISP). We discuss an extension of the method which utilises discretisation of the stochastic space for low-regularity problems and long-time integration. This modification of gPC was first proposed by Wan and Karniadakis [12] as multi-element generalised polynomial chaos (ME-gPC), which yields approximate local statistics for uniformly distributed variables. The formulation was then extended to deal with stochastic inputs with arbitrary probability measures [13]. Non-intrusive multi-element surrogate modelling was later described by Foo et al. [5]. This approach performs the probabilistic stochastic collocation in a decomposed random space. In this article, we apply the same logic as in the multi-element probabilistic collocation method (ME-PCM), but instead of using a Lagrangian interpolant to obtain the surrogate, we perform a pseudo-spectral (discrete) projection [15].

Let $(\Omega, \mathcal{F}, \mu)$ be a probability space, where Ω is the sample space, $\mathcal{F}$ is the σ-algebra of subsets of Ω, and μ is a probability measure. Given an uncertain $\mathbb{R}^d$-valued input, $\boldsymbol{\xi} = (\xi_1, \ldots, \xi_d)$, gPC seeks a representation of a quantity of interest (QoI), $Y(\omega) \in L_2(\Omega, \mathcal{F}, \mu)$, as a sum of weighted polynomials:

$$Y(\omega) = \sum_{k=0}^{\infty} y_k \Psi_k(\boldsymbol{\xi}(\omega)), \tag{1}$$

where ω is a random event, the basis functions $\Psi_k(\boldsymbol{\xi}(\omega))$ are orthogonal with respect to μ and $\{y_k\}_{k=0}^{\infty}$ is a set of expansion coefficients. The list of polynomials whose weights are associated with a particular random variable can be found in [16]. The representation in Eq. 1, including the series truncated at $N_p + 1$ elements, enables the calculation of the stochastic moments, i.e. expected value, $\mathbb{E}$, and variance, $\mathbb{V}$, in terms of polynomial coefficients:

$$\mathbb{E}[Y]\langle \Psi_0 Y \rangle = y_0, \quad \mathbb{V}[Y] = \mathbb{E}\left[|Y - \mathbb{E}[Y]|^2\right] = \sum_{k=0}^{N_p} y_k^2 \langle \Psi_k^2 \rangle, \tag{2}$$

where $\langle . \rangle$ denotes the inner product; number of terms $N_p + 1 = \frac{(p+d)!}{p!d!}$ depends on p and d representing polynomial order and stochastic dimension, respectively.

In the intrusive gPC approach, the polynomial expansion of QoI is introduced in the solver, resulting in a coupled system of equations which is often non-trivial to derive. On the other hand, stochastic collocation methodologies treat the simulator as a *black box* and only use the simulation outputs to estimate the polynomial coefficients. Therefore, the sampling methods are attractive if the UQ study has to be performed with large, complex codes and workflows which are not amenable to changes.

Non-intrusive algorithms comprise the following steps: (1) choose a set of inputs, (2) run the deterministic code for each value, and (3) using the set of solutions construct the expansion of summed polynomials, either through interpolation or discrete projection. The approach is independent of the solver, and executions of the simulator can be made concurrently. In the NISP approach, a surrogate model $\tilde{Y}$ is constructed using the realisations of Y:

$$\tilde{Y} = \sum_{k=0}^{N_p} \tilde{y}_k \Psi_k \approx Y, \tag{3}$$

where $\tilde{y}_k = \frac{1}{\langle \Psi_k^2 \rangle} \sum_{q=1}^{Q} y\left(\xi^{(q)}\right) \Psi_k\left(\xi^{(q)}\right) w^{(q)}$ and $\left(\xi^{(q)}, w^{(q)}\right)$ are the prescribed quadrature nodes and their corresponding weights.

In the multi-element variant of stochastic collocation, the parametric space is discretised into N_e non-overlapping subsets $\left\{B^j\right\}_{j=1}^{N_e}$. For each element, the set of collocation points (quadrature nodes) is prescribed and local surrogates are estimated. Over each element B^j, a change of variable is applied in order to have a mapping on $[-1, 1]$. Construction of orthogonal polynomials for arbitrary PDFs can be performed with a three-term recurrence relation. After the global approximant is reconstructed from multiple models, the approximate global stochastic moments can be assembled from the local statistics with the Bayes' formula [5]. In [5,13], the level of stochastic discretisation is decided adaptively by assessing the contribution of the highest polynomial terms to total variance.

3   Existing Infrastructure for Workflow-Based UQ Studies

UQ models are typically designed disconnected from the management of the compute infrastructure (execution engine). On the UQ model side, DAKOTA is arguably one of the most popular frameworks, it comprises a vast range of UQ models that it exposes through textual specification with its own syntax or through a C++ application programming interface (API). OpenTURNS is another popular framework that exposes many UQ models through a Python API. None of these frameworks offer a way to easily specify infrastructure-related requirements unless the user explicitly integrates them in his/her workflow specification. There are some other projects, e.g. Sandia Analysis Workbench (SAW) [6] for DAKOTA and Salome [4] for OpenTURNS, that provide wrappers that enable some of this integration.

On the compute infrastructure side, there are many execution engines that support the execution of workflows, some of them comprise the architecture proposed in [17]. These execution engines can be more cloud-oriented (e.g. ICP [9]), others more traditional batch-job oriented (e.g. CWLEXEC [3] with IBM Spectrum LSF) [8], and others somewhere in between (e.g. Flux [1]).

Approaches to represent workflows based on programatic languages have also been proposed in [14] (XML formats) and [2] (Python). Both focus on decorating the data types so that the execution engine can optimise data management.

A characteristic of these works is that they all consider that the logic to perform UQ (or other kind of parametric studies) will have to be appended to the baseline workflow, making it more complex.

4 Abstraction for UQ Studies Specification

In this section, we propose an abstraction to allow users to easily specify a workflow and identify where quantities of interest for a UQ study are produced and execute all the required logic on a distributed HPC system seamlessly, with minimum user intervention. The scientific/modelling workflow is the central element, the UQ study is built/derived from it.

4.1 A UQ Case Study

We motivate the discussion of the proposed abstraction with a UQ case study around the thermal properties of a flow in a u-shaped bend. We aim to estimate the temperature distribution in a pipe as a function of the Froude number to gain better understanding of stratified flow development. We measure the wall temperature at rings shown in Fig. 1a.

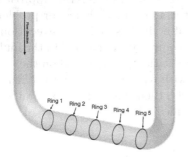

(a) Sketch of ring locations. Rings are located at $x/D \approx 1.5, 3, 4.5, 6$ and 7.5 for rings 1 to 5 respectively, and have radius $0.525D$ (i.e. in the middle of the wall).

(b) Temperature distribution in a u-shaped bend at non-dimensionalised time $\tilde{t} = 40$ for $Fr \approx 0.621$.

Fig. 1. U-shaped bend geometry and simulation result sample.

The Reynolds-averaged Navier Stokes (RANS) in our target simulation is based upon that of Viollet [11]. The thermal and material properties of the system are determined from the following dimensionless groups:

- Reynolds number: $Re = 10000$, based on pipe diameter, D.
- Reduced Froude number: $Fr \equiv \dfrac{U}{\sqrt{g \frac{\delta \rho}{\rho} D}} = 0.67 \pm 40\%$.
- Peclet number: $Pe \equiv \frac{UD}{\alpha} = 6 \times 10^4$.

where U, g, ρ and α are the bulk velocity, gravitational acceleration, fluid density, and thermal diffusivity, respectively.

The flow is fully developed at the inlet, and is allowed to reach a (statistically) steady isothermal state before commencing the thermal transient. A hot shock is introduced at the inlet at time t_0, and increases linearly until time t_1 where it remains at the maximum temperature. The duration of the ramp is $t_1 = t_0 + 7.5U/D$. The ratio of thermal diffusivities is based on water flowing within a steel pipe, i.e. $\alpha_{solid}/\alpha_{fluid} = 144.8$.

The surrogates are built at non-dimensionalised $\tilde{t} \equiv Ut/D = 40$; at this time a stable stratification occurs as shown in Fig. 1b. We aim to show the benefit of using a multi-element variant of NISP to build an accurate input-output representation relationship. At first, we approach the problem with a standard NISP method. We discuss the strengths of the method and identify its limitations which can be addressed with local polynomial bases. We estimate the stochastic response using global orthogonal bases. As the function of the simulator is unknown, the convergence is analysed based on results from consecutive polynomial resolutions. The approximation error of expected value E obtained with p-th polynomial order is defined here as $|E_p - E_{p-1}|/|E_{p-1}|$, where $|.|$ denotes L_2 norm.

In this study, we treat the Froude number $Fr(\xi)$ to be a random variable, where ξ has a uniform distribution $Fr \sim U(0.402, 0.938)$. Both NISP approaches use Gaussian quadrature approximation as it is an optimal choice for problems with a small stochastic dimension. However, if a single simulation execution is expensive, nested quadratures may reduce the computational effort. In the multi-element variant of NISP, we perform the discretisation purely based on the value of the last term of the polynomial expansion with respect to the first coefficient. If the ratio is larger than a prescribed threshold, then the term has a significant influence and the random space should be divided.

4.2   Workflow Abstraction for UQ

Figure 2a shows how a workflow underlying the UQ study specified in Sect. 4.1 would look like. In the following, we propose a set of directions to abstract the way users can specify a UQ study based on a given workflow.

The Data. Data can have a myriad of different formats depending on the tool/domain. If data needs to be filtered the user needs to specify an adaptor,

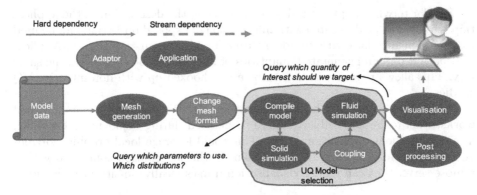

(a) Workflow specification. Hard and stream connector denote dependencies between applications and data.

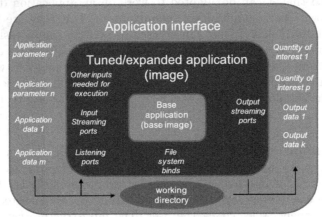

(b) Application specification.

Fig. 2. Abstraction used to define the workflow.

see below. Connection to external data repositories can also be specified, either by valid URLs that can be used directly by applications, or by adaptors that access that data and mirror to a file system. Data persistence/replication can be controlled and minimised by using hashing techniques similar to what exists already today in modern data repositories, e.g. Git or DockerHub.

The Application. At the heart of every workflow is the application. Like the data, applications can be very diverse and require a specific environment to execute correctly. This environment can also depend on other applications. This kind of hierarchic dependence of applications is addressed by several cloud technologies through containerisation with Docker [7] being arguably the most popular. Therefore, here we assume that the application is defined through means of an image whose instances can be deployed as a container.

Similar to what happens today to containers, the description of the application will have to be decorated with information of exposed ports that can be used to stream in/out data without going through a file system - see Fig. 2b. Application parameters and generated quantities of interest can be exposed in the same way. This allows parametrisation to be easily hooked up with non-intrusive UQ studies.

Applications can also be visualisation tools. We consider the user owned workstation to be a possible resource to be used during deployment for which a matching client application exists. This could leverage local graphics for rendering a stream of data or deploy a client in the workstation connected with a remote server that can render visualisation frames locally (in-situ visualisation).

Adaptors. Adaptors are just like applications, except that they are meant to change data so that it can be adapted to work on an application downstream in the workflow. The assumption on the adaptors is that they are lightweight and can be replayed safely. They can be used to perform simple tasks like serialisation/deserialisation of data or to generate plots. Adaptors are designed to address a problem that limits the adoption of workflow design tools, which is the tweaking of an application for a slightly different format of data, or set of inputs. They make the components of the workflows more modular and therefore more reusable.

UQ Specification. UQ can be seen as a more sophisticated form of parameterisation study. As discussed above, applications and adaptors are configured to expose parts of the internal environment as parametrised entries. For a given workflow, one can potentially perform a UQ study on a connected subset of workflow components selected by the user. The infrastructure can detect the inputs and quantities of interests from the applications/adaptors in that subset. For that, it only has to evaluate the connections to other components that are not part of that subset (UQ model selection area in Fig. 2a). This can be done automatically based on the application/adaptor specification. The user only has to select the type of UQ model. Once the set of components and UQ model is known, the user is notified of which parameters and quantities were identified. Then, he/she can select which should be targeted, and which distributions should be used. This makes it almost trivial to incorporate a UQ study as part of the design of the workflow.

This approach has the property of being suitable for nesting, i.e. one can make UQ studies within UQ studies, by selecting subsets. The infrastructure would repeat the process above as needed.

5   Results

We executed the workflows based on the abstraction proposed in the Sect. 4 using the standard NISP approach. Initially the UQ study is performed on the temperature distribution in ring 2 as shown in Fig. 1a.

As shown in Fig. 3a and b, for given threshold values, we reach statistical convergence above $p = 10$. For the sake of the study, we validate the surrogate with $p = 13$ (which was obtained with $p + 1 = 14$ runs) against a large number of simulation results. Figure 4 shows that the polynomial expansion reproduces well the mean temperature in the second ring for varying Froude numbers.

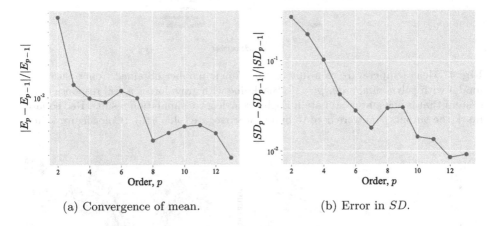

(a) Convergence of mean.   (b) Error in SD.

Fig. 3. Error in expected value and standard deviation of temperature. Statistics were obtained from surrogate models built with increasing polynomial order.

The resulting mean temperature distribution and standard deviation are plotted in Fig. 5a. Such representation of results suggests that each value of temperature within the uncertainty range is equally probable to occur. However, the PDFs of the simulation outputs in Fig. 5b give an interesting insight – the distribution of QoIs is bimodal in nature, i.e. extreme values of temperature are more likely to occur. This observation highlights misinterpretation of the system response in finite order statistics and stresses the need for performing the UQ analysis. If classical Monte Carlo approach was used, at least an order of magnitude more calculations would be required to obtain the same result as with polynomial expansion.

For the approximation, a fairly large polynomial order is used. Surrogates built with smaller number of quadrature nodes allow to accurately describe stochastic moments. However, if the approximant is to be treated as a replacement of high-fidelity simulation, then better representations might be needed. More polynomial terms are retained due to the complexity of the functions that describe input-output relationship. This is particularly visible when analysing the temperature sampled at the first ring. Although the convergence of the first moments is reached, even for $p = 15$ the model does not return all of the simulation points (see Fig. 6a). Using polynomials of higher degree could improve the result but at significant computational cost.

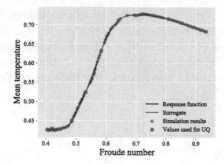

Fig. 4. Mean temperature as a function of Froude number obtained from a surrogate model with polynomial order $p = 13$. Black line with gray dots is a real response of the system that is being approximated. Each circle denotes simulation results. Red squares mark the outputs that were used to build the surrogate (blue line). (Color figure online)

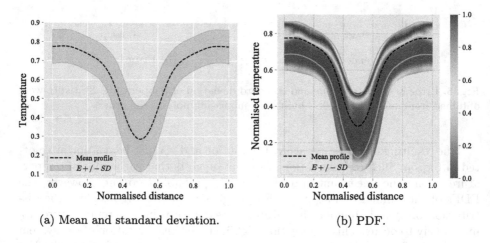

(a) Mean and standard deviation.   (b) PDF.

Fig. 5. Statistics derived from the PC-based surrogate with $p = 13$.

To circumvent this limitation, we define an analysis which uses piecewise low-degree polynomial basis [5,12]. The random space refinement (increasing number of elements N_e) is performed adaptively based on the behaviour of the last polynomial coefficient. If its value with respect to the zeroth polynomial term is larger than a prescribed threshold, then the random space is halved.

We should stress here that changing between methods does not require modifications to the defined workflow. Instead, we only adjust the selection of the UQ model to use. This highlights the benefits of using an abstraction for the parametric study and improved modelling.

Figures 6b and c showcase the improvement in surrogate modelling when using piece-wise smooth polynomials. By adaptively dividing the random space into sections and constructing partial model response functions, we can obtain a better approximation as with global basis. The individual polynomial order of each surrogate was kept low, $p = 3$. Therefore, a better model approximation can be achieved with discretisation ($N_e > 1$) at comparable computational cost of a global surrogate. The mean of relative errors, $\delta = \frac{1}{N} \sum_{n=1}^{N} |Y(Fr_n) - \tilde{Y}(Fr_n)|/|Y(Fr_n)|$ for the partial surrogates was $\delta = 0.0018$ and $\delta = 0.0011$ for $N_e = 4$ and $N_e = 5$, respectively, while the error of the global approximant with 16 simulation runs was $\delta = 0.0053$.

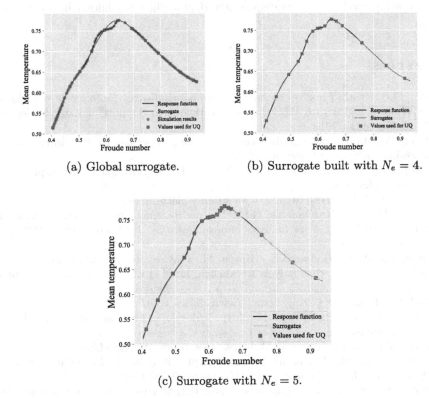

(a) Global surrogate.   (b) Surrogate built with $N_e = 4$.

(c) Surrogate with $N_e = 5$.

Fig. 6. Comparison between a surrogate for the 1st ring's temperature constructed with a global basis and a set of locally supported polynomials. In (b) the same number of points was used as for the solution shown in (a). The global surrogate with the same number of points as $N_e = 5$ would have 3× higher error.

6   Conclusions

We have shown that the knowledge extracted from the modelled system can be maximised by using surrogate-based UQ analysis, allowing the prediction of the influence of varying conditions on QoI. This can be done without changing existing simulation codes, and at much smaller cost than using a brute-force approach.

We have applied a non-intrusive polynomial-based approach to model parametric uncertainty associated with conjugate heat transfer modelling. The study of response functions shows that in this case piece-wise polynomials are better suited for constructing accurate surrogates.

The analysis is facilitated by the introduction of an abstraction that comprises a way to specify the scientific application along with high-throughput workflow. The solution derives the necessary details for the UQ study from the existing application specification. The paper has highlighted the need for flexibility allowing modellers to transparently deploy their workflows in a dynamic fashion.

References

1. Ahn, D.H., et al.: Flux: overcoming scheduling challenges for exascale workflows. In: WORKS 2018: 13th Workshop on Workflows in Support of Large-Scale, pp. 10–19, November 2018
2. Babuji, Y., et al.: Parsl: scalable parallel scripting in python. In: 10th International Workshop on Science Gateways (2018)
3. Computing, I.S.: CWLEXEC (2018). https://github.com/IBMSpectrum-Computing/cwlexec
4. EDF: SALOME (2018). https://www.salome-platform.org/
5. Foo, J., Wan, X., Karniadakis, G.E.: The multi-element probabilistic collocation method (ME-PCM): error analysis and applications. J. Comput. Phys. **227**(22), 9572–9595 (2008)
6. Friedman-Hill, E., Hoffman, E., Gibson, M., Clay, R.: Incorporating workflow for V&V/UQ in the sandia analysis workbench. In: NAFEMS World Congress, June 2015
7. GmbH, D.I.: Docker (2018). https://www.docker.com/
8. IBM: IBM Spectrum LSF (2017). https://developer.ibm.com/storage/products/ibm-spectrum-lsf
9. IBM: IBM Cloud Private (2018). https://www.ibm.com/cloud/private
10. Sullivan, T.J.: Introduction to Uncertainty Quantification. TAM, vol. 63. Springer, Cham (2015). https://doi.org/10.1007/978-3-319-23395-6
11. Viollet, P.: Observation and numerical modelling of density currents resulting from thermal transients in a non rectilinear pipe. J. Hydraul. Res. **25**(2), 235–261 (1987)
12. Wan, X., Karniadakis, G.E.: An adaptive multi-element generalized polynomial chaos method for stochastic differential equations. J. Comput. Phys. **209**(2), 617–642 (2005)
13. Wan, X., Karniadakis, G.E.: Multi-element generalized polynomial chaos for arbitrary probability measures. SIAM J. Sci. Comput. **28**(3), 901–928 (2006)
14. Wu, S., Mortveit, H.S.: A general framework for experimental design, uncertainty quantification and sensitivity analysis of computer simulation models. In: 2015 Winter Simulation Conference (WSC), pp. 1139–1150, December 2015
15. Xiu, D.: Numerical Methods for Stochastic Computations: A Spectral Method Approach. Princeton University Press, Princeton (2010)

16. Xiu, D., Karniadakis, G.E.: The Wiener-Askey polynomial chaos for stochastic differential equations. SIAM J. Sci. Comput. **24**(2), 619–644 (2002)
17. Zimoń, M., Elisseev, V., Sawko, R., Antão, S., Jordan, K.: Uncertainty quantification-as-a-service. In: Proceedings of the 28th Annual International Conference on Computer Science and Software Engineering, CASCON 2018, pp. 331–337. IBM Corp., Riverton (2018)

Ternary-Decimal Exclusion Algorithm for Multiattribute Utility Functions

Yerkin G. Abdildin[✉] [iD]

Nazarbayev University, 53 Kabanbay Batyr Ave.,
Nur-Sultan, 010000, Kazakhstan
yerkin.abdildin@nu.edu.kz

Abstract. We propose methods to eliminate redundant utility assessments in decision analysis applications. We abstract a set of utility assessments such that the set is represented as a matrix of ternary numbers. To achieve efficiency, the matrix is converted to a decimal vector for further processing. The resulting approach demonstrates excellent performance on random sets of utility assessments. The method eliminates the redundant questions for the decision maker and can serve for consistency check.

Keywords: Uncertainty · Decision analysis · Decision maker · Multiattribute utility problem · Redundant utility assessments

1 Introduction

In multi-objective decisions under uncertainty, the decision maker's trade-offs over the different attributes (e.g. increase output, but decrease expenses [1]) can be expressed through a multiattribute utility function (MUF). In complex decision problems, e.g. power plant siting [2–4], the construction of a MUF [5] becomes difficult with the number of attributes. For example, Keeney and Raiffa [6] discuss the siting of nuclear power facilities utilizing ten attributes while Keeney and Sicherman [7] model alternatives for generating electricity using 14 attributes. To simplify the assessment of a MUF from the decision maker [8], it can be decomposed into lower-order terms [9]. The lower-order terms of MUF may contain redundancies. The elimination of possible redundancy in utility assessments saves time significantly for the decision maker.

Although in different context, the task to eliminate redundant expressions in a program may arise in compiler optimization. The example of a redundant expression here could be a computation that is performed twice. Different algorithms have been discussed to address this problem [10–12]. Redundancy is also met in biomedical science. In particular, if some biological functions (or terms) in the gene ontology (GO) hierarchy do not provide additional information, then they are redundant and must be excluded from the GO lists [13].

In this paper, we consider the problem of determining the minimal set of utility assessments (lower-order terms) needed to construct a MUF from a given set of assessments. We introduce a novel and efficient method for eliminating duplicate and redundant terms in a MUF. The efficiency of the method is achieved by transferring the

© Springer Nature Switzerland AG 2019
J. M. F. Rodrigues et al. (Eds.): ICCS 2019, LNCS 11540, pp. 282–292, 2019.
https://doi.org/10.1007/978-3-030-22750-0_22

computational burden from a two-dimensional input (ternary matrix) to a decimal vector. The algorithm sorts the vector, eliminates possible duplicates and redundancies, and returns the unique set of utility assessments. The method can help to exclude redundant questions for the decision maker. It can also be used to check the consistency of the utility assessments, which are required for a construction of multiattribute utility functions.

2 Preliminaries

Consider a decision problem with a set of n attributes $X = \{X_1, X_2, \ldots, X_n\}$, where the domain of the attribute X_i is represented by instantiations: x_i^0, x_i^*, x_i, such that the former two are the subsets of the latter and distinct. Namely, x_i^0 and x_i^* are the *least* and *most* preferred values of attribute X_i, and x_i denotes *all* values. Let $U = \{U_1, U_2, \ldots, U_m\}$ be a given set of utility assessments, where U_j is a function of X. The goal is to return U without redundant assessments.

For example, consider the five utility assessments presented in Fig. 1(a). The first two terms, $U(x_1, x_2, x_3^0, x_4^0)$ and $U(x_1^*, x_2, x_3, x_4^0)$, are *redundant* due to the terms $U(x_1, x_2, x_3^0, x_4)$ and $U(x_1^*, x_2, x_3, x_4)$, respectively. The fifth term is a *duplicate* of $U(x_1^*, x_2, x_3, x_4)$. Therefore; a unique set of utility assessments will contain only two terms.

To simplify the encoding and decoding of utility assessments, Abdildin and Abbas [14] introduced the ternary matrix of utility assessments, so that the given set can be encoded as a 5×4 ternary matrix M as shown in Fig. 1(b).

Definition 1 [14]. *The ternary matrix of utility assessments, M, is an $m \times n$ matrix storing integers 0, 1, and 2, representing respectively the least preferable value, the most preferable value, and all values of an attribute, where m is the number of utility assessments, and n is the number of attributes of the decision-making problem.*

(a)
$$U(x_1, x_2, x_3^0, x_4^0),$$
$$U(x_1^*, x_2, x_3, x_4^0),$$
$$U(x_1^*, x_2, x_3, x_4),$$
$$U(x_1, x_2, x_3^0, x_4),$$
$$U(x_1^*, x_2, x_3, x_4).$$

(b)

2	2	0	0
1	2	2	0
1	2	2	2
2	2	0	2
1	2	2	2

Fig. 1. (a) Utility assessments; (b) Ternary matrix M.

To eliminate duplicate and redundant terms from the ternary matrix M in a brute-force approach, every row of M has to be compared with all other rows elementwise. The brute-force approach can be implemented in different ways, the pseudo-code of the one utilized in this paper is given in the appendix. Another approach [14] is to compare only parts of the rows called twos-complement. The twos-complement part represents the set of columns in a row containing ones and zeros, i.e. complements of twos. We now propose a novel and more efficient approach for eliminating redundant utility assessments.

3 Proposed Approach

3.1 Main Idea

The main idea of the *ternary-decimal exclusion algorithm* [15] can be described by small example when $n = 2$. In this case, the nine values of the ternary number, T, can vary from [0 0] to [2 2] (see Table 1).

We can observe that:

- If ternary number [2 2] exists in the input matrix (or $max = 3^n - 1 = 8$ in decimal), then all other numbers must be excluded from the ternary matrix.
- If ternary number [2 1] (or $max - 1 = 7$ in decimal) is present, then decimal numbers 1 and 4 can be eliminated (note, a modulo 3 division (%) returns the remainder 1). Similarly, 6 covers 0 and 3, because, for example, 6 % 3 = 0.
- The decimal 2 covers all numbers below it and 5 covers 3 and 4.
- In addition, 2 and 5 divide the decimal numbers into three sub-intervals of equal length.

Table 1. Coverage table for $n = 2$.

Ternary number T for $n = 2$	Decimal number D corresponding to T	Coverage: What can D eliminate?								
2 2	8	0	1	2	3	4	5	6	7	8
2 1	7		1			4			7	
2 0	6	0			3			6		
1 2	5				3	4	5			
1 1	4					4				
1 0	3				3					
0 2	2	0	1	2						
0 1	1		1							
0 0	0	0								

Therefore, we will convert the input ternary matrix $M(m, n)$ into a decimal vector $V(m)$, sort the vector, eliminate duplicate numbers, and then (recursively) eliminate redundancies using the observations discussed above. The recursion utilized in the proposed approach improves the performance of the algorithm. It recursively divides the sorted vector into three sub-vectors using the right bounds: $lb = \lfloor min + (max - min)/3 \rfloor$ and $mb = \lfloor min + 2(max - min)/3 \rfloor$, for the left and mid sub-vectors, respectively, where initially $min = 0$ and $max = 3^n - 1$.

Note that the ternary numbers in Table 1 are unique and their number grows fast with n.

Definition 2. *For a decision problem with n attributes, the cardinality of the ternary matrix of utility assessments is the total number of possible unique ternary numbers that it can contain and is equal to 3^n.*

So, the amount of unique ternary numbers (i.e. utility assessments) that can be formed from the rows of matrix $M(m, n)$ is limited by its cardinality. The cardinality of the ternary matrix of utility assessments will be used later in our simulations.

3.2 Pseudocodes

The algorithm consists of the following steps. First, we convert the ternary matrix $M(m, n)$ to a decimal vector $V(m)$. Recall, a d-digit ternary number, T, can be converted to a decimal as follows: $D = t_d R^d + t_{d-1} R^{d-1} + \ldots + t_1 R^1 + t_0 R^0$, where R denotes radix (base 3), and t_d denotes the digits of T. As an example, $T = 102$ in decimal is $D = 1 * 3^2 + 0 * 3^1 + 2 * 3^0 = 11$.

Next, we sort the decimal vector $V(m)$ and eliminate duplicates. One can use any efficient algorithm (see comparisons in [16]) to sort the decimal vector and then eliminate the duplicates in linear time.

Then, we recursively exclude redundancies from the vector. The pseudocode of the **exclude_redundant** subroutine is presented in Table 2. A modulo division is denoted by %, an empty vector by [], and elements of vector V from 1 to idx by V(1:idx). The **exclude_redundant** procedure uses the **search_bound** subroutine (Table 3).

Now, we convert the decimal vector to the ternary matrix. Recall, a decimal number D can be converted to a ternary T by continually dividing D by three to have a quotient and a remainder until the former is equal to zero. Then the remainder is read in a reverse order.

Complex interdependence conditions in a decision problem may lead to specific redundancies in utility assessments. The last step of the proposed algorithm, **exclude_specific** subroutine (Table 4), is needed to eliminate such redundancies.

Table 2. Pseudocode of the **exclude_redundant** subroutine.

Procedure exclude_redundant(V, min, max):
Input: A vector of m sorted unique nonnegative decimal integers, V[m], and the minimal and maximal possible values of m in V[m].
Output: A new vector of sorted unique positive decimal integers, V.

```
m ← length(V)
if m < 2
   return V    // base condition
if (V(m) = max)
   return V ← max
k ← 1
if (V(m) = max - 1) AND (V(m – 1) = max - 2)
   while k < m - 1
      if (V(k) % 3) = 1 OR (V(k) % 3) = 0
         delete V(k)
elseif (V(m) = max - 1)
   while k < m
      if (V(k) % 3) = 1
         delete V(k)
elseif (V(m) = max - 2)
   while k < m
      if (V(k) % 3) = 0
         delete V(k)
m ← length(V)
lb ← floor(min + (max-min)/3)       // left bound
mb ← floor(min + 2*(max-min)/3)    // mid bound
idxL ← search_bound(V, 1, m, lb)   // find elm ≤ lb to create Left sub-vector
if idxL = 0
   L ← []
elseif V(idxL) = lb
   L ← lb
else
   L ← V(1 : idxL)
idxM ← search_bound(V, idxL + 1, m, mb) // find elm ≤ mb to create Mid sub-vector
if idxM = idxL
   M ← []
elseif V(idxM) = mb
   M ← mb
else
   M ← V(idxL + 1 : idxM)
if m = idxM     // create Right sub-vector
   R ← []
else
   R ← V(idxM + 1 : m)
[L] ← exclude_redundant(L, min, lb)
[M] ← exclude_redundant(M, lb, mb)
[R] ← exclude_redundant(R, mb, max)
[V] ← [L M R]    // merge L M R
return V
```

Table 3. Pseudocode of the **search_bound** subroutine.

Procedure search_bound(V, left, right, x):
Input: A sorted vector of unique nonnegative decimal integers, V, its first and end indices, and number x to search.
Output: idx, an index of x if it exists in V, index of element lower than x, otherwise.
if left > right
idx ← right
return idx
else
mid ← floor(left + (right - left)/2)
if x < V(mid)
idx ← search_bound(V, left, mid - 1, x)
elseif x > V(mid)
idx ← search_bound(V, mid + 1, right, x)
else
idx ← mid
return idx

Table 4. Pseudocode of the **exclude_specific** subroutine.

Procedure exclude_specific(M):
Input: Ternary matrix M.
Output: Matrix M without redundant rows.
r ← 1
while r < size(M, 1) + 1
tc ← find(M(r, :) ≠ 2) // indices of 0s and 1s in row r
i ← r + 1
while i < size(M,1) + 1
if isequal(M(r, tc), M(i, tc)) // compare elms of r and i by indices in tc
M(i, :) ← [] // delete row i
i ← i - 1
i ← i + 1
r ← r + 1
return M

3.3   An Illustrative Example

Let us consider a three-attribute decision problem in which attributes X_1 and X_2 are utility independent [6] from attribute X_3, and attribute X_3 is utility independent from attributes X_1 and X_2. These partial utility independence conditions imply that:

- $U(x_1|x_2,x_3) = U(x_1|x_2,x_3^0)$
- $U(x_2|x_1,x_3) = U(x_2|x_1,x_3^0)$
- $U(x_3|x_1,x_2) = U(x_3|x_1^0,x_2^0)$.

The multiattribute utility function (MUF) can be decomposed into lower-order terms by expanding it through the attributes, which present some utility independence. For example, the expansion of a MUF [9, 14] through the first attribute leads to the following functional form: $U(x_1, x_2, x_3) = U(x_1^*, x_2, x_3)U(x_1|x_2, x_3) + U(x_1^0, x_2, x_3) \bar{U}(x_1|x_2, x_3)$, where $\bar{U}(x_1|x_2, x_3) = 1 - U(x_1|x_2, x_3)$. A further expansion of the MUF through the remaining two attributes and incorporation of the partial utility independence conditions (see details in [9, 14]) will result in the functional form with multiple lower-order terms some of which are duplicate or redundant.

The utility terms arising from the decomposition of the MUF for this decision problem can be encoded as an input ternary matrix for our algorithm (see Fig. 2). Here, the input matrix is converted to a decimal vector (#1), the vector is sorted, and duplicates are eliminated (#2), redundancies are excluded (#3), and the vector is converted to a ternary matrix, which is then returned as an output (#4-5-Output). The output represents the five terms, which are required for construction of the multiattribute utility function, namely, $U(x_1, x_2, x_3^0)$, $U(x_1^*, x_2^*, x_3^*)$, $U(x_1^*, x_2^0, x_3^*)$, $U(x_1^0, x_2^*, x_3^*)$, $U(x_1^0, x_2^0, x_3)$. These utility terms should be assessed [8] from the decision maker. Thus, the proposed algorithm eliminates redundant questions for the assessor.

Input			#1	#2	#3	#4-5-Output		
1	1	1	13	24	24	2	2	0
2	2	0	24	15	13	1	1	1
1	2	0	15	13	10	1	0	1
0	0	2	2	12	4	0	1	1
1	1	0	12	10	2	0	0	2
0	0	2	2	9				
1	0	1	10	6				
1	0	0	9	4				
0	1	1	4	3				
2	2	0	24	2				
0	2	0	6	1				
0	1	0	3	0				
0	2	0	6					
0	0	1	1					
0	0	0	0					

Fig. 2. Illustration of the proposed algorithm on a three-attribute decision problem.

4   Simulation Results

The simulation results demonstrate very good performance of the proposed algorithm compared to the brute-force approach. The simulations were done in MATLAB R2017b on a machine with 8 GB of RAM and Intel(R) Core(TM) i5-6400 CPU @2.70 GHz 2.71 GHz under Windows 10 Enterprise. Figure 3 illustrates performance of the proposed algorithm relative to the brute-force approach for different numbers of functions (utility terms) and fixed number of attributes for 10,000 runs of randomly generated input ternary matrices. The average running times of the brute-force approach were set to unit. Comparing to the brute-force approach, the ternary-decimal exclusion algorithm demonstrates excellent performance for various values of m for a fixed value of n.

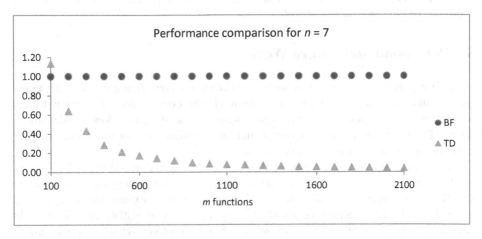

Fig. 3. Performance of the ternary-decimal exclusion algorithm (TD) relative to the brute-force approach (BF) for different numbers of functions.

The proposed algorithm also shows very good performance for different values of n (see Fig. 4). Note that the value of m is proportional to n in this test and is equal to half of the cardinality of the ternary matrix. One limitation of the proposed approach is the case when m is disproportionately very low for a given n. For example, if $m < floor(0.05(3^n))$, then the efficiency of the proposed approach decreases. This gives a motivation for future work.

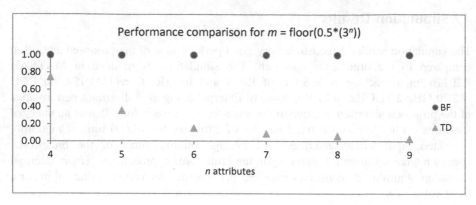

Fig. 4. Performance of the ternary-decimal exclusion algorithm (TD) relative to the brute-force approach (BF) for different numbers of attributes.

5   Conclusion and Future Work

For solving problems under uncertainty, the decision maker often must consider many factors and use multiple criteria (i.e. attributes). The construction of a multiattribute utility function, in this case, may require assessments of various lower-order utility terms. The exclusion of utility terms, which are duplicate or redundant, can be quite challenging in complex decision problems.

This paper introduced a novel method for excluding redundant terms in a multi-attribute utility function. The ternary-decimal exclusion algorithm demonstrated excellent performance on random sets. Future work may extend the analysis and possibly find better recurrence relations. The proposed algorithm can be used in decision support systems and can serve as a consistency check for the utility assessments.

Appendix

See Table 5.

Table 5. Pseudocode of the brute-force approach.

Algorithm brute-force(M):
Input: Ternary matrix M.
Output: Matrix M without redundant rows.

```
r ← 1
while r < size(M, 1) + 1
  i ← 1
  while i < size(M, 1) + 1
    if i ≠ r
      if M(r, :) ≥ M(i, :)  // if all elms of row r ≥ the corresponding elms of row i
        redundant ← true  // then row i can be redundant due to row r
        for j ← 1:size(M, 2)
          if M(r, j) > M(i, j) AND M(r, j) = 1
            redundant ← false
            break
        if redundant = true
          M(i, :) ← []  // delete row i
          if i < r
            r ← r - 1
          i ← i - 1
    i ← i + 1
  r ← r + 1
return M
```

References

1. Abdildin, Y.G., Abbas, A.E.: Canonical multiattribute utility functions: enumeration, verification, and application. Procedia Comput. Sci. **18**, 2288–2297 (2013). https://doi.org/10.1016/j.procs.2013.05.400

2. Gros, J.: Power plant siting: a paretian environmental approach. Nucl. Eng. Des. **34**, 281–292 (1975). https://doi.org/10.1016/0029-5493(75)90125-9

3. Gros, J., Avenhaus, R., Linnerooth, J., Pahner, P.D., Otway, H.J.: A systems analysis approach to nuclear facility siting. Behav. Sci. **21**(2), 116–127 (1976). https://doi.org/10.1002/bs.3830210206

4. Solomon, B.D., Haynes, K.E.: A survey and critique of multiobjective power plant siting decision rules. Socio Econ. Plann. Sci. **18**(2), 71–79 (1984). https://EconPapers.repec.org/RePEc:eee:soceps:v:18:y:1984:i:2:p:71-79

5. Abbas A.E.: Constructing multiattribute utility functions for decision analysis. In: Hasenbein, J. (ed.) INFORMS Tutorials in Operations Research, pp. 62–98. Hanover, Maryland (2010). https://doi.org/10.1287/educ.1100.0070

6. Keeney, R.L., Raiffa, H.: Decisions with Multiple Objectives: Preferences and Value Tradeoffs. Wiley, New York (1976)
7. Keeney, R.L., Sicherman, A.: Illustrative comparison of one utility's coal and nuclear choices. Oper. Res. **31**(1), 50–83 (1983). https://doi.org/10.1287/opre.31.1.50
8. Abdildin, Y.G., Abbas, A.E.: Analysis of decision alternatives of the deep borehole filter restoration problem. Energy **114**, 1306–1321 (2016). https://doi.org/10.1016/j.energy.2016.08.034
9. Abbas, A.E.: General decompositions of multiattribute utility functions with partial utility independence. J. Multicriteria Decis. Anal. **17**, 37–59 (2010). https://doi.org/10.1002/mcda.452
10. Knoop, J., Ruthing, O., Steffen, B.: Lazy code motion. ACM SIGPLAN Not. **27**(7), 224–234 (1992). https://doi.org/10.1145/143095.143136
11. Morel, E., Renvoise, C.: Global optimization by suppression of partial redundancies. Commun. ACM **22**(2), 96–103 (1979). https://doi.org/10.1145/359060.359069
12. Paleri, V.K., Srikant, Y.N., Shankar, P.: Partial redundancy elimination: a simple, pragmatic, and provably correct algorithm. Sci. Comput. Program. **48**, 1–20 (2003). https://doi.org/10.1016/S0167-6423(02)00083-7
13. Jantzen, S.G., Sutherland, B.J.G., Minkley, D.R., Koop, B.F.: GO trimming: systematically reducing redundancy in large gene ontology datasets. BMC Res. Notes **4**(267), 1–9 (2011). https://doi.org/10.1186/1756-0500-4-267
14. Abdildin, Y.G., Abbas, A.E.: An algorithm for excluding redundant assessments in a multiattribute utility problem. Procedia Comput. Sci. **9**, 802–811 (2012). https://doi.org/10.1016/j.procs.2012.04.086
15. Abdildin, Y.G.: Multiattribute utility functions for the deep borehole filter restoration problem. Ph.D. dissertation, University of Illinois at Urbana-Champaign (2014). http://hdl.handle.net/2142/50440
16. Biggar, P., Nash, N., Williams, K., Gregg, D.: An experimental study of sorting and branch prediction. ACM J. Exp. Algorithmics **12**, 1–39 (2008). https://doi.org/10.1145/1227161.1370599. Article 1.8

Sums of Key Functions Generating Cryptosystems

Nataliya Kalashnykova[1] ⓘ, Viktor V. Avramenko[4] ⓘ,
and Viacheslav Kalashnikov[2,3,4(✉)] ⓘ

[1] Universidad Autónoma de Nuevo León (UANL), av. Universidad S/N,
66455 San Nicolás de los Garza, NL, Mexico
[2] Tecnológico de Monterrey (ITESM), av. Eugenio Garza Sada 2501 Sur,
64849 Monterrey, NL, Mexico
kalash@tec.mx
[3] Central Economics and Mathematics Institute (CEMI), Nakhimovsky pr. 47,
Moscow 117418, Russia
[4] Sumy State University (SumDU), Rimsky-Korsakov str. 2,
Sumy 40007, Ukraine

Abstract. The paper develops an algorithm based on derivative disproportion functions (DDF) for modeling a cryptosystem for transmitting and receiving devices. The transmitted symbols are encoded with the aid of sums of at least two of those functions weighted with random coefficients. Some important properties of the derivative disproportion functions are also discussed. Numerical experiments demonstrate that the algorithm is quite reliable and robust.

Keywords: Cryptosystems · Derivative disproportion functions (DDF) · Decoding algorithms

1 Introduction

In the modern competitive world, the significance and price of information are steadily growing up. Hence, the information is often encrypted when transmitted by various ways from a transmitter to a receiver in order to prevent it from unauthorized access. The latter necessity strongly ushers one to the use of cryptographic techniques within information systems, the most popular of which are the Data Encryption Standard (DES) [1], Advanced Encryption Standard (AES) [2], and the Rivest-Shamir-Adleman (RSA) cryptosystem [3]. However, the new super-computers and the technologies of network and neural computing that have appeared in the 2000th, conduct to the reevaluation of the previous cryptographic systems that had been thought to be highly reliable. Because of that, the development of new principles for the generation of new cryptosystems is very reasonable.

This research was financially supported by the SEP-CONACYT Grants CB-2013-01-221676 and FC-2016-01-1938.

© Springer Nature Switzerland AG 2019
J. M. F. Rodrigues et al. (Eds.): ICCS 2019, LNCS 11540, pp. 293–302, 2019.
https://doi.org/10.1007/978-3-030-22750-0_23

At present, the majority of cryptosystems exploit integer numbers for their keys. Moreover, the longer the key, the more hacking-proof is the cryptosystem because it becomes more difficult to fit the key by solving an appropriate factorization problem. The shift from integers to reals, or even more, to real functions makes the task of code breaking (hacking) much more complicated, which promises to enhance the cryptosystem's reliability (resistance property).

In this paper, we study such a new tool for classifying and declassifying both an analog signal and a signal in the form of a sequence of symbols from the specified alphabet [5]. Such a cryptosystem is built with the use of derivative disproportion functions (DDF) [5, 6]. The input symbols are encrypted by the sum of real functions (keys) weighted with randomly selected coefficients. Owing to the derivative disproportion functions, one has a possibility of recognizing which functions had been involved to encrypt the signal. The latter permits the receiver to decode the encrypted symbols *even though* the randomly selected weights of the key functions are *unknown*. The latter fact allows the considered topic to fit well into the area of solving problems under *uncertainty*.

Moreover, the derivative disproportion functions (DDF) are accepted in a steadily growing list of areas of applications under uncertainty, such as the identification of quasi-stationary dynamic objects [7, 8], and pattern recognition [9].

The rest of the paper is organized as follows. Sections 2 and 3 introduce the derivative disproportion functions (DDF) thus specifying the problem. The decoding algorithm is presented in Sect. 4, while Sect. 5 considers numerical examples and the results of numerical experiments. Sections 6 and 7 argue the proposed cryptosystem's robustness and the requirements to the disproportion key functions, respectively. The paper is finished with the conclusions (Sect. 8) and the list of references.

2 Derivative Disproportion Functions (DDF)

The brand-new ways of classifying information can be generated with the use of derivative disproportion functions (DDF). Disproportion functions related to the derivatives and to the values were developed and examined by the authors in the previous publications [5, 6].

The derivative disproportion functions (DDF) are exercised in order to identify (label, tag) relevant real functions. The DDFs permit to quantitatively estimate the degree of a deviation of a numerical function from the specified functions (like, e.g., power function $y = k \cdot x^n$) for any fixed value of the argument, regardless of the associated parameters (like, for example, multiplier k for the power function). Here, $n \geq 1$ is an integer.

Definition 1. The *derivative disproportion function (DDF) of order n* of the function $y = y(x)$ with respect to x $(x \neq 0)$ is defined as follows:

$$@d_x^{(n)}y = \frac{y}{x^n} - \frac{1}{n!} \cdot \frac{d^n y}{dx^n}. \tag{1}$$

In the particular case of $n = 1$ (order 1), Eq. (1) of the derivative disproportion is easily reduced to:

$$@d_x^{(1)}y = \frac{y}{x} - \frac{dy}{dx}. \tag{2}$$

As one could expect, for the linear function $y = kx$, its DDF of order 1 is *zero* for any value of the coefficient k. The symbol @ is chosen to designate the operation of determination of disproportion. The symbol "d" is selected to refer to the function's derivative as the main object of disproportion calculated. Finally, the left-hand side of Eq. (2) reads "*at d one y with respect to x*".

If a function is reported in a parametric form, the n-th order derivative disproportion function (DDF) defined by Eq. (1) is determined by applying the rules of calculation of $d^n y/dx^n$ under the parametric dependence of y on x. In particular, the first-order derivative disproportion of the function defined parametrically as $y = \psi(t)$ and $x = \varphi(t)$ (where t is the parameter and $\varphi(t) \neq 0$, $\varphi'(t) \neq 0$ for all t) has the form

$$@d_x^{(1)}y = @d_{\varphi(t)}^{(1)}\psi(t) = \frac{y}{x} - \frac{y'_t}{x'_t} = \frac{\psi(t)}{\varphi(t)} - \frac{\psi'(t)}{\varphi'(t)}. \tag{3}$$

It is clear that if $\psi(t) = k\varphi(t)$ for some constant k, its derivative disproportion defined by Eq. (3) equals *zero* on the (shared) domain of the functions $y = \psi(t)$ and $x = \varphi(t)$.

Lemma 1. Each derivative disproportion function (DDF) of order n has the following properties:

1. Multiplying the function y by any scalar m results in multiplying its DDF by the same scalar.
2. The order n derivative disproportion function (DDF) of a sum (difference) of functions equals the sum (difference) of their DDFs.
3. For the linear function $y = kx$, its derivative disproportion of order 1 is zero for any value of the coefficient k.

Proof. It is readily verified by simple algebraic manipulations with the use of Definition 1.

Remark 1. In other words, the operator $@d_x^{(n)}$ defined on the space $C^n(\Omega)$ of n times continuously differentiable real functions is *linear* on this space.

3 The Problem's Statement

Examine a communication system (channel) transmitting symbols (signals) encoded with a cryptosystem $\mathcal{K}$ based on the key functions $f_i = f_i(t)$, each defined on a (time) interval $t \in [0, T_i]$, $T_i > 0$, $i = 1, \ldots, m$. The functions are assumed to be smooth and n times (continuously) differentiable. A symbol transmitted at the time moment t is encrypted with the sum of (at least two) key functions with possible time delays (shifts) $\tau_i \in [0, T_i], i = 1, \ldots, m$.

For example, if the transmitted symbol is encrypted by the (weighted) sum of two key functions f_p and f_q, $1 \leq p, q \leq m$, the signal transmitted through the communication channel has been encoded as

$$y(t) = k_p f_p (t + \tau_p) + k_q f_q (t + \tau_q), k_p > 0, k_q > 0. \tag{4}$$

We assume that an invader (intruder, hacker, etc.) who may have found unauthorized access to the channel is **not aware** of either the key functions f_i or their time delays (shifts) τ_i, or the coefficients k_i, $i = p, q$.

On the receiver's side of the communication system (channel), the complete list of key functions and their delays is known, but which of them (and with what weights) are involved in the received signal coded as in Eq. (4) is to be yet detected. The identification of these functions and their weights in Eq. (4) allows one to work out the received symbol $y(t)$.

The problem of identifying both the key functions and their weights in Eq. (4) is solved by the algorithm presented in the next section.

4   The Algorithm's Description

The problem in question is by no means easy to solve because the key functions and their weights can be detected only approximately (uncertainty environment). The received message $y(t)$ is unfolded in time, so the exact or approximate derivatives of this function are necessary. When one works with discrete data, e.g., $y(t) = (y(t_0), y(t_1), \ldots, y(t_{N-1}))^T$, then the desired approximate "derivative" of the (discrete) function $y(t)$ is estimated by a special approximation method, similar to that by Gregory-Newton (*cf.*, [4]).

The initial first version of our algorithm is technically quite burdensome, and due to the space restriction, we present here its description only for $m = 3$ (the complete version can be found in [6] and other publications of the authors).

The main idea of the general algorithm is as follows: if the key functions' delays (shifts) $\tau_i, i = 1 \ldots, m$, are known, we may represent the received message $y(t)$ as the sum of *all* key functions with (yet unknown) weights k_i:

$$y(t) = \sum_{i=1}^{m} k_i f_i(t + \tau_i). \tag{5}$$

Next, we have to identify these weights at the present moment t. The coefficients will be equal to zero for those functions that are **not really** involved in the encoded signal Eq. (5).

As we mentioned above, the algorithm will be described only for the case $m = 3$. In general, the algorithm consists of m steps (that is, 3 in our case).

Step 1. Select randomly one of the key functions, for instance, the first one $f_1 = f_1(t + \tau_1)$. By making use of Eq. (3), estimate the derivative disproportion function (DDF) value for the signal $y(t)$ and denote it as $F_{01}(t) := @d_{f_1}^{(1)}y(t)$. Besides, the DDF values $F_{21}(t)$ and $F_{31}(t)$ are calculated for the key functions $f_2(t + \tau_2)$ and $f_3(t + \tau_3)$ **with respect to** $f_1(t + \tau_1)$. Owing to the linearity of operator $@$ on the space $C^n(\Omega)$ (*see*, Remark 1), Eq. (5) yields (for $m = 3$):

$$
\begin{aligned}
F_{01}(t) &\equiv @d_{f_1}^{(1)}y(t) = \frac{y(t)}{f_1(t + \tau_1)} - \frac{y'(t)}{f_1'(t + \tau_1)} = k_1 \cdot 0 + k_2 \left[\frac{f_2(t + \tau_2)}{f_1(t + \tau_1)} - \frac{f_2'(t + \tau_2)}{f_1'(t + \tau_1)} \right] \\
&+ k_3 \left[\frac{f_3(t + \tau_2)}{f_1(t + \tau_1)} - \frac{f_3'(t + \tau_2)}{f_1'(t + \tau_1)} \right] = k_2 @d_{f_1}^{(1)}f_2(t + \tau_2) + k_3 @d_{f_1}^{(1)}f_3(t + \tau_2) \quad (6) \\
&\equiv k_2 F_{21}(t) + k_3 F_{31}(t).
\end{aligned}
$$

Here, the first term on the right-hand side of the upper line of Eq. (6) is zero due to Assertion 3 of Lemma 1.

Step 2. Again, pick up randomly one of the remaining DDFs $F_{21}(t)$ and $F_{31}(t)$; let it be, for instance, $F_{21}(t)$. Now, we calculate the derivative disproportions of the functions $F_{01}(t)$ and $F_{31}(t)$ with respect to $F_{21}(t)$; denote them as $F_{0121}(t)$ and $F_{3121}(t)$, respectively.

Applying the operator of the derivative disproportion of order 1 to both sides of Eq. (6), then making use of its linearity and Assertion 3 of Lemma 1, one easily comes to the equalities

$$
F_{0121}(t) \equiv \frac{F_{01}(t)}{F_{21}(t)} - \frac{F_{01}'(t)}{F_{21}'(t)} = k_2 \cdot 0 + k_3 \left[\frac{F_{31}(t)}{F_{21}(t)} - \frac{F_{31}'(t)}{F_{21}'(t)} \right] \equiv k_3 F_{3121}(t). \quad (7)
$$

Step 3. The relationship given by Eq. (7) shows the linear dependence of the function F_{0121} on the function F_{3121}. Again, on the ground of Assertion 3 of Lemma 1, we conclude that the DDF $F_{01213121}(t)$ of the function F_{0121} with respect to F_{3121} is *zero* for all feasible t:

$$
F_{01213121}(t) \equiv @d_{F_{3121}}^{(1)}F_{0121}(t) = \frac{F_{0121}(t)}{F_{3121}(t)} - \frac{F_{0121}'(t)}{F_{3121}'(t)} = k_3 - k_3 = 0.
$$

Now, one can use relationships of Eqs. (6) and (7) in the converse order and compute the desired values of the weights k_i. Indeed, first from Eq. (7), one readily obtains

$$
k_3 = \frac{F_{0121}}{F_{3121}}; \quad (8)
$$

the latter, in its turn, combined with Eq. (6) implies:

$$
k_2 = \frac{F_{01} - k_3 F_{31}}{F_{21}}. \quad (9)
$$

Finally, by substituting the just found weights k_2 and k_3 in Eq. (5), one deduces the value of k_1:

$$k_1 = \frac{y(t) - k_2 f_2(t + \tau_2) - k_3 f_3(t + \tau_3)}{f_1(t + \tau_1)}. \tag{10}$$

The algorithm stops after having decoded the received message $y(t)$ by having identified the (unknown) weights associated with the participating key functions. All the weights related to the idle (non-participating) key functions are *zero*.

Remark 2. As one can smoothly infer, the explicit list of basic key functions and their possible delay (shift) values τ_i is **indispensable** for the implementation of this simplified version of the decoding algorithm. The more sophisticated procedures that may be needed to decipher the received message in the lack of such important information, that is, under more profound uncertainty, are described in [6].

5 Numerical Examples and Experiments

In order to illustrate the cryptosystem's operation, let us consider the binary coding in the form of an arbitrary sequence of symbols "0", "1", space "_", and a transition to the new line (paragraph return) "\". For this model, only three real key functions are employed. The symbols being transmitted are encoded by the (weighted) sum of at least two of these functions multiplied by random factors (weights). The time delays (shifts) of the standard functions with respect to the current time t are assumed to be *zero*. The communication system (TV or radio channel) can transmit only binary code symbols. Therefore, if there appears any other symbol apart from those listed above, it is perceived as a paragraph return.

To develop the numerical methods calculating the approximate derivatives, it is necessary to control the signal $y(t)$ within the interval containing at least 10 (discrete) points of the time variable t. In fact, the number of points in this interval may vary (the greater this number of points, the higher the cryptosystem's stability (resistance)), but in our case, it is selected constant and equals 75 (*cf.*, again, [4]).

In order to simulate the operations of the cryptosystem, the following three functions are employed as key functions:

$$f_1(t) = 100 \sin((\alpha_1 - \beta_1)t) \cos(15\beta_1 t);$$
$$f_2(t) = 100 \exp(-0.1\alpha_2 t) \sin(10\beta_2 t) \cos((\alpha_2 + \beta_2)t);$$
$$f_3(t) = 100 \exp(-0.1\alpha_3 t) \sin(400\beta_3 t),$$

where $\alpha_1 = 1$; $\alpha_2 = 0.12$; $\alpha_3 = 0.5$; $\beta_1 = 0.1$; $\beta_2 = 1.2$; $\beta_3 = 0.7$.

The weights k_1, k_2, and k_3, with which the key functions encode the signal $y(t)$ by Eq. (5) before its transmission, have been selected randomly by making use of a pseudo-random generator with the uniform distribution from *zero* to 10 (for each symbol). However, only when encoding a symbol '1', $y(t)$ includes the entire (weighted) sum of all three key functions and therefore, their coefficients k_1, k_2, and k_3

are not equal to zero. When encrypting '0', we put $k_1 = 0$, and while encoding a space, we set $k_3 = 0$. Finally, if another symbol or the paragraph return is encoded, then $k_2 = 0$.

At any given time moment, the receiver tries to identify the involved key functions and calculate the unknown weights (coefficients) k_i, $i = 1, 2, 3$, by making use of the formulas from Eqs. (8)–(10). Thereafter, the received message is decoded.

When an arbitrary text is encrypted with the application of derivative disproportion functions, it is always recommended to introduce at least two random letters before the transmission of the binary code.

Figures 1, 2, and 3 show the diagrams of the signal $y(t)$ transmitted via the communication channel. Various examples of the cryptosystem operation when the same symbols are transmitted, as well as when the binary symbols are alternated, were treated.

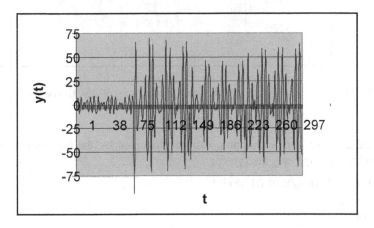

Fig. 1. The signal corresponding to the serial transmission of four symbols '0'.

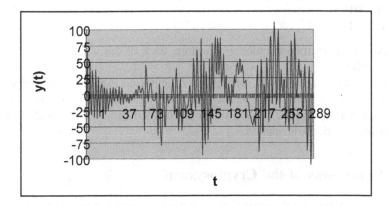

Fig. 2. The signal corresponding to the serial transmission of four symbols '1'.

Besides, the case when ASCII- codes of symbols A, B, C, D, O are transmitted, was tested. The corresponding codes were as follows:

01000001 01000010 01000011
01000100 01001111.

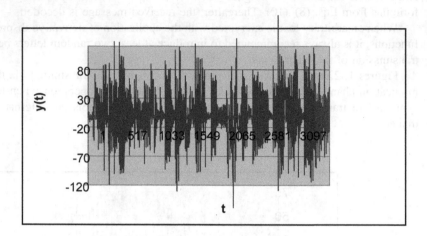

Fig. 3. The signal corresponding to the serial transmission of ASCII- codes of symbols A, B, C, D, O.

The cryptosystem's operation can be illustrated by the last (third) example. The following message was encrypted:

01000001 01000010 01000011
01000100
01001111

The following message was obtained after decoding:

01000001 01000010 01000011
01000100
01001111

According to the operational algorithm, each of the letters led to the appropriate paragraph return.

In all cases, the received message was deciphered exactly as what was transmitted. At the same time, as it can be seen from the figures, it is quite difficult to reveal a message born by the transmitted signal through the communication channel unless the decoding algorithm is applied.

6   Robustness of the Cryptosystem

The cryptosystem's robustness (stability) depends on the choice of the key functions as well as on their total number. The more components are involved in the signal, the more difficult becomes the task of deciphering (in case it's been intercepted as a result of a hacker attack). Obviously, it is necessary *not only* to identify the type and the number of key functions *but also* to fit the weights involved.

How difficult it is to fit their values can be judged from the fact that in the given example, it suffices to apply $sin\,(9.9999\,\beta_2 t)$ instead of the present $sin\,(10\,\beta_2 t)$ in $f_2(t)$, or to select 400.0001 instead of the current 400 in $f_3(t)$, so that the code word consisting of four consecutive 0's is "decoded" as four 1's. This simple example confirms that any attempts on part of a hacker to "guess" the coded word by an exhaustive search for the coefficients (weight) even after having detected the key functions used, is almost always doomed to fail.

Another instance: the replacement of $\alpha_1 = 1$ with $\alpha_1 = 0.99$ in $f_1(t)$ has resulted in the distorted reception of the sole line 11000000011010000001101000000000000100000 000000 without breaks in contrast to the three-line original message boasting with spaces as well.

The cited examples show that it is quite difficult just to fit the weights by a simple guess, to say nothing about the necessity to determine the number of functions and to fit their types.

It should be also noted that the same symbol is encoded differently depending on its position (location). Besides, one should pay attention to the fact that in this case, the frequency analysis cannot be applied for unauthorized access and decoding.

All the above-mentioned facts show that the cryptosystems based on the (weighted) sum of real key functions are sufficiently resistant to hacking (cryptographically secure).

7 Requirements for the Key Functions

1. Each (one real variable) key function has to be real-valued and sufficiently smooth (n times continuously differentiable).
2. The key function and its derivatives up to order n must *not* be constant.
3. The key functions should *not* asymptotically approach a constant value within its domain (e.g., like the function $x^{-\alpha}$, $\alpha > 0$, for the large values of x).

4. The collection of key functions must be selected so that to exclude the possibility that the value of one function at some point be negligible (too small by its absolute values) as compared to the values of other functions at the same point; that is, every function has to make a quite significant "contribution" to the (weighted) sum of all key functions.

5. The key functions cannot be identical.

8 Concluding Remarks

We develop a cryptosystem where (one-variable) real functions are used as the keys. An example is presented to illustrate the operation of such a system where symbols are encrypted by the (weighted) sum of the key functions with random coefficients. The decoding is conducted with the aid of the first order derivative disproportion functions (DDF) calculated for the received signal and the key functions.

For a practical application of such cryptosystems, one should bear in mind that in the process of computing the weights during deciphering, a division by small numbers, or a ratio of two numbers both close to zero may happen. This can lead to certain information distortion. Therefore, the encrypted message must be decoded before it is transmitted via a communication channel. If necessary, the message should be encrypted once again with other coefficients (weights) generated randomly for every key function.

Acknowledgments. The authors would like to express their profound gratitude to the two anonymous referees whose valuable critical and constructive comments and proposals have helped a lot in improving the paper.

This work was financially supported by the SEP-CONACYT grants CB-2013-01-221676 and FC-2016-01-1938, as well as by Tecnológico de Monterrey (ITESM), Campus Monterrey, Strategic Group for Optimization and Data Science.

References

1. U.S. Department of Commerce/National Institute of Standards and Technology Data Encryption Standard (DES) Federal Information, Processing Standards Publication, 46-3, 25 October (1999). http://csrc.nist.gov/publications/fips/fips46-3/fips46-3.pdf
2. Federal Information Processing Standards Publication 197, 26 November 2001, Specification for the ADVANCED ENCRYPTION STANDARD (AES) (2001). http://crsc.nist.gov/publications/fips/fips197/fips197.pdf
3. Rivest, R., Shamir, A., Adleman, L.: A method for obtaining digital signatures and public–key cryptosystems. Commun. ACM **21**(2), 120–126 (1978). https://doi.org/10.1145/359340.359342
4. Khan, I.R., Ohba, R., Hozumi, N.: Mathematical proof of closed-form expressions for finite difference approximations based on Taylor series. J. Comput. Appl. Math. **150**(3), 303–309 (2003)
5. Avramenko, V.V., Zabolotny, M.I.: A Way of Data Coding, Patent UA H04L 9/00 №42957, Ukraine (2009)
6. Avramenko, V.V., Karpenko, A.P.: Recognition of fragments of given standards in an analyzed signal with the aid of disproportionality functions. Trans. Sumy State Univ. (SumDU) **34**(1), 96–101 (2002)
7. Kalashnikov, V.V., Avramenko, V.V., Kalashnykova, N.I., Kalashnikov Jr., V.V.: A cryptosystem based upon sums of key functions. Int. J. Comb. Optim. Prob. Inf. **8**(1), 31–38 (2017)
8. Kalashnikov, V.V., Avramenko, V.V., Slipushko, NYu., Kalashnykova, N.I., Konoplyanchenko, A.E.: Identification of quasi-stationary dynamic objects with the use of derivative disproportion functions. Procedia Comput. Sci. **108**(C), 2100–2109 (2017)
9. Kalashnikov, V.V., Avramenko, V.V., Kalashnykova, N.I.: Derivative disproportion functions for pattern recognition. In: Watada, J., Tan, S.C., Vasant, P., Padmanabhan, E., Jain, L.C. (eds.) Unconventional Modelling, Simulation, and Optimization of Geoscience and Petroleum Engineering, pp. 95–104. Springer, Heidelberg (2018). Chapter 7

Consistent Conjectures
in Globalization Problems

Nataliya Kalashnykova[1] ⓘ, Mariel A. Leal-Coronado[2] ⓘ,
Arturo García-Martínez[2] ⓘ, and Viacheslav Kalashnikov[2,3,4(✉)] ⓘ

[1] Universidad Autónoma de Nuevo León (UANL), av. Universidad S/N,
66455 San Nicolás de los Garza, NL, Mexico
[2] Tecnológico de Monterrey (ITESM), av. Eugenio Garza Sada 2501 Sur,
64849 Monterrey, NL, Mexico
kalash@tec.mx
[3] Central Economics and Mathematics Institute (CEMI), Nakhimovsky pr. 47,
Moscow 117418, Russia
[4] Sumy State University (SumDU), Rimsky-Korsakov str. 2,
Sumy 40007, Ukraine

Abstract. We study the effects of merging two separate markets each originally
monopolized by a producer into a globalized duopoly market. We consider a
linear inverse demand with cap price and quadratic cost functions. After globalization, we find the consistent conjectural variations equilibrium (CCVE) of
the duopoly game. Unlike in the Cournot equilibrium, a complete symmetry
(identical cost functions parameters of both firms) does not imply the strongest
coincident profit degradation. For the situation where both agents are low-
marginal cost firms, we find that the company with a technical advantage over
her rival has a better ratio of the current and previous profits. Moreover, as the
rival becomes ever weaker, that is, as the slope of the rival's marginal cost
function increases, the profit ratio improves.

Keywords: Duopoly game · Conjectural variations equilibrium · Cap price ·
Globalization

1 Introduction

The purpose of the present paper is to investigate a market with two competing producers of an identical commodity. We consider two stages: before globalization
(separate markets) and after globalization (united market). Before globalization, each
producer satisfies the separate demand of the market that it monopolizes. After globalization, both firms compete in a united market. This model is often said to have the
structure of a pure (classic) duopoly where both companies satisfy the complete market
demand.

This research was financially supported by the SEP-CONACYT Grants CB-2013-01-221676 and
FC-2016-01-1938.

© Springer Nature Switzerland AG 2019
J. M. F. Rodrigues et al. (Eds.): ICCS 2019, LNCS 11540, pp. 303–314, 2019.
https://doi.org/10.1007/978-3-030-22750-0_24

One can find numerous studies on the effects of combining two or more markets in the literature. According to [1], there are two types of global markets: (a) the free trade market which allows the existence of n different markets with a separate supplier; and (b) a single integrated market in which all producers compete.

Since the 1980s, there has been a lot of research on the role of imperfect competition. This was pointed out in [2], which deals with global markets of type (a). In fact, there are several works which models correspond to these type of markets. Some examples are found in [3–5], too. On the other hand, [1] analyzes a globalized market of type (b), through a Nash-Cournot equilibrium model, whereas in [1], the authors examine cases where all producers' profits are degraded in the same manner. For each producer, they use the ratio of the profit obtained after globalization to the profit before globalization to represent the degree of the profit degradation, and the largest of the ratios among the producers is a measure of coincident degradation. They found that under a complete symmetry, i.e. when the values of parameters of cost and demand functions are equal, all producers have profit degradation coincidently. For the model they use which boasts linear demand functions for the separated markets and the globalized market, as well as linear cost functions, under Nash-Cournot conjectures, the value of the measure of coincident degradation is the lowest (the worst) when the firms are identical.

The present paper also discusses the situation of type (b). As in [1], we use the ratio of the profit obtained after globalization to the profit before globalization to represent the degree of profit degradation or improvement. However, our purpose is to analyze the effects of globalization considering the diverse values that can take the parameters of the cost functions of the companies, which in our case are quadratic. We reveal that is possible that one producer loses while the other one gains; or both lose.

Next, at the stage of globalization, when competition takes place, we raise a kind of equilibrium with consistent conjectural variations (CCVE). Conjectural Variations Equilibria (CVE) were introduced by Bowley in 1924 [6] and Frisch in 1933 [7] as another possible solution concept in static games. According to this concept, agents behave as follows as was stated in [8]: each agent chooses her most favorable action taking into account that every rival's strategy is a conjectured function of her own strategy. In [8, 9], we studied mixed oligopoly models with consistent conjectural variations (CCV), which correspond to the market price variations due to the change in the output level of a producer. Concepts such as exterior and interior equilibrium were introduced, and proofs of existence and uniqueness of equilibrium were presented in the above-mentioned papers. We apply these concepts in our present paper, too.

Since any conjectures inevitable bear doubts about whether they will be followed by all the players at all, this topic has a direct link to the area of solving problems with uncertainties. The conjectures accepted by each agent can be considered as attempts to evaluate the robustness of the model's solution subject to the players' preferences and market powers.

The paper is organized by follows. In Sect. 2, we describe the mathematical model and specify the assumptions to accept for each stage. This section also shows the optimal output levels produced by each firm before globalization. The consistent conjectural variations equilibrium price and production volumes are justified in this section as well. In Sect. 3, we define two types of agents: *low-marginal* and *high-marginal cost* firms (abbreviated as LMCF and HMCF, respectively). As we study a market with 2 agents, we have four feasible situations corresponding to the possible combination of types of firms. We define the profit ratio and compute it for each situation in terms of the parameters in order to analyze the effect of the cost parameters on this ratio. To do so we use the concept of technical advantage introduced by [10]. In that section, we also display an example showing that, unlike the Nash-Cournot case [1], a complete symmetry does not necessarily render the worst-case ratio under consistent conjectures. Finally, in Sect. 4, we present our conclusions and outline our future work. The list of references and acknowledgments finish the paper.

2 The Model Specification

We assume that before globalization, there exist two monopolistic markets. Each monopoly faces an active demand D_i, $i = 0, 1$, which does not depend on market price, and current demand $G_i = G_i(p_i)$, $i = 0, 1$, whose argument p_i is the market clearing price. We will also assume that in every market, the price value $p_i = \bar{P}$ is the *cap price*. This means that the demand functions have a discontinuity point (a break) and for prices higher than $\bar{P}$ the demand is zero. Therefore, the company i output volume, $q_i \geq 0$, will satisfy the following inequality if the market is "balanced":

$$g_i(p_i) + D_i \leq q_i \leq G_i(p_i) + D_i, \; i = 0, 1. \tag{1}$$

Here, $g_i(p_i)$ is the right limit of the function G_i, $i = 0, 1$, and it may happen that $g_i(p_i) < G_i(p_i)$ for some price p_i, whereas the left limit of the current demand function at each point is assumed to coincide with its proper value.

After globalization, both firms compete in the common market. The consumers' (current) demand is described by a demand function $G = G(p_w)$, whose argument $p_w \geq 0$ is the (common) market clearing price. An active demand value D is nonnegative and does not depend on the market price. Here we take for granted that after globalization, the cap price will be the same as before globalization. Since the demand function has a point of discontinuity (a break at the cap price $\bar{P}$), the balance between the demand and supply for a given price $p_w \geq 0$ is described by the following ("balance") inequality:

$$g(p_w) + D \leq Q \leq G(p_w) + D. \tag{2}$$

Here again, $g = g(p_w)$ is the right limit of the function $G = G(p_w)$ at any point $p_w \geq 0$ while $Q = q_0 + q_1$.

2.1 The Model's Assumptions

Accept the following assumptions about the demand and cost functions in order to study the effects of globalization.

2.1.1 Before Globalization

A1.1. The inverse demand function for each firm $i, i \in \{0, 1\}$, is defined as follows:

$$p_i(\theta_i) = \begin{cases} \bar{P}, & \text{if } 0 \leq \theta_i \leq \bar{Q}/2; \\ c - d \cdot \theta_i, & \text{if } \bar{Q}/2 < \theta_i \leq c/d. \end{cases} \tag{3}$$

Here c and d are positive values, and $\bar{P} = c - d\bar{Q}/2$. The total quantity demanded in the market i at the price p_i is θ_i, which includes the passive and the active quantities demanded.

A1.2. For each $i \in \{0, 1\}$, the cost function $f_i(q_i)$ is quadratic, i.e., $f_i(q_i) = 1/2 a_i q_i^2 + b_i q_i$, where $a_i > 0$ and $0 \leq b_i \leq c$, $i = 0, 1$.

A1.3. Also, assume that $\bar{P} > \max_{i=0,1}\{a_i\bar{Q}/2 + b_i\}$.

2.1.2 After Globalization

A2.1. The market inverse demand function is defined as follows:

$$p_w(\theta) = \begin{cases} \bar{P}, & \text{if } 0 \leq \theta \leq \bar{Q}; \\ c - d \cdot \theta/2, & \text{if } \bar{Q} < \theta \leq 2c/d. \end{cases} \tag{4}$$

Here c, d, and $\bar{P}$ are defined as in **A1.1**. The variable θ is the total quantity demanded (including both the passive and active demands).

Assumption **A1.2** about the cost function is also made; the cost structure won't change after globalization. As a consequence of **A1.3**, if $q_0 + q_1 < \bar{Q}$ then $\bar{P} > a_i q_i + b_i$ for at least one $i, i \in \{0, 1\}$.

2.2 Objective Functions of the Companies

2.2.1 Before Globalization

Recall that before globalization, there exists a single company in each market commercializing the commodity. Firm $i, (i \in \{0, 1\})$ chooses its output volume so as to maximize its net profit function: $\pi_i(q_i) = p_i(q_i)q_i - f_i(q_i)$.

Note that assumption **A1.3** implies that the output value that maximizes the benefits cannot be lower than $\bar{Q}/2$. Because of that, we can rewrite the maximization problem of any firm with sub-index i as follows:

$$\max_{q_i \geq \bar{Q}/2} \pi_i(q_i) \equiv p_i(q_i) - f_i(q_i), \tag{5}$$

which can be easily replaced with Karush-Kuhn-Tucker (KKT) equations (*cf.*, [13, p. 26]). The optimal output value $\bar{q}_i$ for private firm $i, i \in \{0, 1\}$ is found as:

$$\bar{q}_i = \begin{cases} \frac{c - b_i}{2d + a_i}, & \text{if } \frac{c - b_i}{2d + a_i} > \frac{\bar{Q}}{2}; \\ \frac{\bar{Q}}{2}, & \text{otherwise.} \end{cases} \tag{6}$$

2.2.2 After Globalization

After globalization, there is an integrated market where both companies compete in a classic duopoly. The price at this stage is determined in the global market, so it obeys the inverse demand function Eq. (4) cited in assumption **A2.1**.

The problem of each private company i is to maximize its net profit

$$\xi_i(q_i) \equiv p_w(Q)q_i - f_i(q_i), \ i = 0, 1. \tag{7}$$

The output level by each company under the assumptions made is found using the theory from [11]. As in [11], we also claim that the output volume chosen by a producer influences the market price. This can be described by a conjectured function of the variations of the price upon variations of the production volume. Then, the first order maximum condition to define the equilibrium would have the form for each $i, i \in \{0, 1\}$:

$$\frac{\partial \xi_i}{\partial q_i} \equiv p_w(Q) + \frac{\partial p_w(Q)}{\partial q_i} \cdot q_i - a_i q_i - b_i \begin{cases} = 0, & \text{if } q_i > 0; \\ \leq 0, & \text{if } q_i = 0. \end{cases} \tag{8}$$

As in [11], the (negative of the) rate of the price function p_w variation implied by a possible variation of output conjectured by agent $i(i = 0, 1)$ is denoted as $v_i = -\partial p_w(Q)/\partial q_i$. In order to describe each agent's behavior, we need to estimate v_i. The conjectured dependence of p_w on q_i must account for the (local) concavity of the i-th agent's objective function; otherwise, one cannot guarantee that the output volumes found via the *first order optimality conditions* Eq. (8) *maximize* (but not minimize) the profit functions. For instance, it suffices to assume that the coefficient v_i (from now on referred to as the i-th agent's *influence coefficient*) is nonnegative and constant, for $i, i \in \{0, 1\}$.

In [11, 12], we defined the concept of *exterior equilibrium*, i.e., *conjectural variations equilibrium* (CVE) with the influence coefficients fixed in an exogenous model. As the competition after globalization has been described with the model presented in [11], the equilibrium would be found exactly as before. Theorem 1 in [11] establishes the existence and uniqueness of the exterior equilibrium $(p_w; \tilde{q}_0, \tilde{q}_1)$ under assumptions **A1.2** and **A2.1**, and also provides the left and right derivatives of the equilibrium price $p_w = p_w(D, v_0, v_1)$ with respect to D. This theorem serves as a base for the concept of *interior equilibrium*, which was defined in [11] as the exterior equilibrium with *consistent* conjectures (influence coefficients). Under the above assumptions, according to Theorem 2 in [11], there exists *interior equilibrium* after globalization. Namely, we define the following concept.

2.2.3 Consistent Conjectural Variations Equilibrium (CCVE)

Let us define the following auxiliary parameter

$$\tau = \begin{cases} -\infty, & \text{if } p_w = \bar{P}; \\ -2/d, & \text{if } p_w < \bar{P}. \end{cases}$$

Given the previous results obtained in [11], the *consistent* (justified) influence coefficient of agent $i, i \in \{0, 1\}$, after globalization is found by solving the following (nonlinear) equation system:

$$v_i = \frac{1}{\frac{1}{v_{-i} + a_{-i}} - \tau}, \; i = 0, 1, \tag{9}$$

where the symbol $(-i)$ represents the competitor's sub-index.

The CVE with the consistent conjectures (9) is called *interior* equilibrium. In Eq. (9), $\tau \in [-\infty, 0)$. When $\tau = -\infty$, system (9) has the unique solution $v_i = 0, i \in \{0, 1\}$. The latter result corresponds to the perfect competition equilibrium (*cf.*, [11]).

The following result was already derived and published as Theorem 3 in [11] and Theorem 4.3 in [12], including for the case of a mixed oligopoly (competition among a public firm and several private companies).

Theorem 2.1 ([11, 12]). *Under assumptions A1.2 and A2.1, for any $\tau \geq 0$, Eq. (9) has a unique solution $v_i = v_i(\tau)$, $i = 0, 1$, continuously depending upon τ. Furthermore, $v_i(\tau) \to 0$ when $\tau \to -\infty$, and strictly increases and tends up to $v_i(0) > 0$ as $\tau \to 0, i = 0, 1$.*

In our case, the solution of the system formed by equations Eq. (9) for the firm i's influence coefficient is:

$$v_i = \begin{cases} -\frac{a_i}{2} + \sqrt{\frac{a_i^2}{4} + \frac{\Gamma}{K_{-i}}}, & \text{if } \tau = -\frac{2}{d}; \; i = 0, 1, \\ 0, & \text{if } \tau = -\infty, \end{cases} \tag{10}$$

where $\Gamma = a_i + a_{-i} + 2a_i a_{-i}/d$ and $K_{-i} = 2(2 + \frac{2}{d} a_{-i})/d$.

For the interior equilibrium price $p_w \geq \bar{b} = \max\{b_1, b_2\}$, Theorems 1 and 2 from [11] imply that relationship Eq. (8) defines uniquely the equilibrium production volumes $\tilde{q}_i$, $i = 0, 1$ (taking into account that $p_w = \bar{P}$ implies $v_i = 0$):

$$\tilde{q}_i = \begin{cases} \frac{p_w - b_i}{v_i + a_i}, & \text{if } p_w < \bar{P}; \\ \frac{p_w - b_i}{a_i}, & \text{if } p_w = \bar{P}, \end{cases}, i = 0, 1. \tag{11}$$

In the particular case when $b_0 = b_1$ assumption **A1.3** entails that the total output level given by Eq. (11) at $p_w = \bar{P}$, is greater than $\bar{Q}$. However, at this price the quantity demanded is at most $\bar{Q}$, which means that the market is *not* balanced. Hence, in this particular case, the equilibrium can be reached only when $p_w < \bar{P}$. From now onward, for simplicity, we restrict ourselves to this case with $b_0 = b_1 = b$.

In the equilibrium when $p_w < \bar{P}$, the total supply output equals the demand in the market. Then, from **A2.1**, $p_w = c - dQ/2$, where $Q = \tilde{q}_0 + \tilde{q}_1$. Plugging in this in the equilibrium outputs Eq. (11), it is easy to obtain both the total output and the equilibrium price p_w.

3 The Effect of Globalization for the Profits

To find the effects of globalization on profits we look for the ratio of benefits. We determine conditions involving the parameters under which these ratios are greater or smaller than 1. If the profit ratio is greater than 1 for company $i, i \in \{0, 1\}$, we say that globalization is *beneficial* for this firm, and it is *not* otherwise, that is, if the profit ratio is less than 1. In order to do that, we first introduce the properties of companies being *low-marginal*, or vice versa, *high-marginal* cost firms.

Definition 3.1. We say that agent i is a *low-marginal cost firm (LMCF)* if the marginal cost $f_i'(q_i)$ evaluated at $\bar{Q}/2$ is less than the cap price minus a proportion d of the quantity $\bar{Q}/2$, that is, $f_i'(\bar{Q}/2) < \bar{P} - d\bar{Q}/2$. Conversely, agent i is a *high-marginal cost firm (HMCF)* if $f_i'(\bar{Q}/2) \geq \bar{P} - d\bar{Q}/2$.

Before globalization, the output level produced by firm i to supply to a separate market depends on the value of the corresponding parameters. On the one hand, if firm i is an ***LMCF***, it produces $\bar{q}_i = \frac{c-b}{2d+a_i}$. Finally, if it is an ***HMCF*** it supplies $\bar{q}_i = \frac{\bar{Q}}{2}$. Because of that, before globalization, four situations in total are feasible depending on the characteristics of the firms of both markets. These situations are described in Table 1, which shows the optimal outputs and the profits for both firms.

Table 1. Possible situations before globalization.

	Before globalization		
Situation	Types of firms	Outputs: $\bar{q}_i, i = 0, 1$	Profits: $\pi_i(\bar{q}_i)$, $i = 0, 1$
1	Both are LMCFs	$\bar{q}_0 = \frac{c-b}{2d+a_0}; \bar{q}_1 = \frac{c-b}{2d+a_1}$	$\pi_0(\bar{q}_0) = \frac{(c-b)^2}{2(2d+a_0)}; \pi_1(\bar{q}_1) = \frac{(c-b)^2}{2(2d+a_1)}$
2	Agent 0 is an LMCF and agent 1 is a HMCF	$\bar{q}_0 = \frac{c-b}{2d+a_0}; \bar{q}_1 = \frac{\bar{Q}}{2}$	$\pi_0(\bar{q}_0) = \frac{(c-b)^2}{2(2d+a_0)} \pi_1(\bar{q}_1) = \frac{1}{8}\bar{Q}[4(c-b) - \bar{Q}(2d+a_1)]$
3	Agent 0 is a HMCF and agent 1 is an LMCF	$\bar{q}_0 = \frac{\bar{Q}}{2}; \bar{q}_1 = \frac{c-b}{2d+a_1}$	$\pi_0(\bar{q}_0) = \frac{1}{8}\bar{Q}[4(c-b) - \bar{Q}(2d+a_0)] \pi_1(\bar{q}_1) = \frac{(c-b)^2}{2(2d+a_1)}$
4	Both are HMCFs	$\bar{q}_0 = \bar{q}_1 = \frac{\bar{Q}}{2}$	$\pi_0(\bar{q}_0) = \frac{1}{8}\bar{Q}[4(c-b) - \bar{Q}(2d+a_0)] \pi_1(\bar{q}_1)$ $= \frac{1}{8}\bar{Q}[4(c-b) - \bar{Q}(2d+a_1)]$

Let R_i denote the profit ratio of company $i, i \in \{0, 1\}$ and be given by:

$$R_i \equiv \frac{\xi_i(\bar{q}_i)}{\pi_i(\bar{q}_i)}, \ i = 0, 1. \tag{12}$$

Formula (12) would take different values according to the situation encountered.

3.1 Measure of Coincident Profit Degradation

Globalization may improve or degrade the profits of the companies. However, [1] study the cases when coincident profit degradation occurs, that is, both firms have smaller profits after globalization than before. In the above-mentioned work, the profit ratio of a producer after globalization to that before globalization is proposed as the degree of profit degradation for the producer due to globalization. They utilize the largest of the ratios of profit degradation among producers as a measure of *coincident degradation*.

According to [1], the reason is: a smaller value of the measure is supposed to indicate stronger coincident degradation. The situation where only one of the producers suffers profit degradation cannot be considered as a coincident producer profit degradation, as far as the other producer enjoys profit improvement. The measure of coincident profit degradation used in [1] for a duopoly is defined in the following terms:

$$k_R = \max\{R_0, R_1\}. \tag{13}$$

The main result obtained in [1] is that the worst-case ratio of coincident profit degradation for all producers due to globalization is reached by a market system if, and only if the system is in a complete *symmetry*. In the next subsection, we show how this result is not necessarily true in the case of the equilibrium with consistent conjectural variations (CCVE), at least for a system of two firms with quadratic cost functions. We also describe the effect of the cost parameters values, a_i and a_{-i} on the profit ratios. In order to do so, we use the concept of *technical advantage*. According to the definition of technical advantage introduced in [10], a firm has a technical advantage over its rival if it can produce the same output that its rival produces at lower marginal and total costs than its rival. Therefore, we say that firm i has the technical advantage over the other firm $(-i)$ if $a_i < a_{-i}$. The proofs of the propositions are too long and will be published elsewhere.

3.2   Situation 1

Situation 1 stated in Table 1 refers to the case when both agents are low-marginal cost firms (LMCF). Substitute the profits at the equilibrium and the optimal profits (Table 1) into formula (12), and after some algebraic manipulations obtain:

$$R_i = \frac{2d + a_i}{2v_i + a_i} \cdot \left(\frac{2v_i}{d}\right)^2, i = 0, 1, \tag{14}$$

where $v_i = -\frac{a_i}{2} + \sqrt{\frac{a_i^2}{4} + \frac{\Gamma}{K_{-i}}}$, $i = 0, 1$, according to Eq. (10).

Proposition 3.1. *There is a degradation of the profits of private firm i, $i = 0, 1$, if and only if*

$$\lambda_1(a_i, a_{-i}, d) + \lambda_2(a_i, a_{-i}, d) > 1, \tag{15}$$

where $\lambda_1(a_i, a_{-i}, d), \lambda_2(a_i, a_{-i}, d) \in (0, 1)$ are defined as follows:

$$\lambda_1(a_i, a_{-i}, d) = \left[2\sqrt{a_i^2/4 + \Gamma/K_{-i}} \Big/ (2d + a_i)\right]^{1/2},$$

$$\lambda_2(a_i, a_{-i}, d) = 1 - 2\left(-a_i/2 + \sqrt{a_i^2/4 + \Gamma/K_{-i}}\right) \Big/ d, i = 0, 1. \tag{16}$$

The simultaneous degradation of the benefits occurs when inequality (15) is valid for both $i = 0, 1$. The degradation or increase of company i's profit depends not only on the cost parameters of the same company but also on the cost parameters of the other agent.

Proposition 3.2. *The profit ratio of competitor i increases if a_{-i} grows.*

Proposition 3.2 states that the larger the coefficient of the quadratic term of the rival's cost function, the lower the profit degradation for producer i (or, its profits may even increase). The proof of Proposition 3.2 shows that the increase of the parameter a_{-i} of the cost function of the rival affects positively the profit ratio of player i, as expected.

For the current situation, the profits of the weaker firm $(-i)$ are degraded after globalization. Another important fact is that if a firm has the technical advantage over the other, the degradation of its own profit due to globalization (if the latter happens at all) is lower than the profit degradation of the other firm. Even more, the profits of firm i can increase. These results are summarized in Proposition 3.3.

Proposition 3.3. *If $a_i < a_{-i}$, i.e., competitor i has technical advantage over $(-i)$, then*

(a) $R_{-i} < 1$;
(b) $R_{-i} < R_i$.

Note that if we consider the case where there is coincident degradation of the profits, the measure of the latter in this case would equal $k^R = R_i$, i.e., the profit degradation of the firm with the technical advantage.

Under the complete symmetry, both producers suffer coincident profit degradation. This result is the same as in [2] and is stated in Proposition 3.4:

Proposition 3.4. *If the firms are identical $(a_i = a_{-i} = a)$ the ratio of both firms is given by*

$$R = \frac{2d + a}{2\sqrt{\frac{a^2}{4} + \frac{ad}{2}}} \cdot \left(\frac{a}{\frac{a}{2} + \sqrt{\frac{a^2}{4} + \frac{ad}{2}}} \right)^2. \tag{17}$$

This resulting value is less than 1 for any positive values of a and d, which means that globalization degrades profits for each company when both firms face the same costs.

In contrast to [1], the latter is *not* necessarily the worst case under consistent conjectures. We introduce a numerical example to show it. In the following examples, we compute, together with the consistent conjectural variations equilibrium (CCVE), the equilibrium under Nash-Cournot conjectures considering the quadratic cost functions. In [1], the cost function is linear. Nevertheless, our examples with quadratic costs show that a complete symmetry implies the worst-case ratio under Nash-Cournot conjectures, too.

Table 2. Example 1: $c = 50, d = 10, \bar{P} = 30, \bar{Q} = 4, b = 1$ and $a_0 = 0.1$

a_1	v_0	v_1	R_0^{CV}	R_1^{CV}	k_R^{CV}	R_0^{Nash}	R_1^{Nash}	k_R^{Nash}
0.06	0.585	0.602	0.217	0.230	**0.230**	0.886	0.894	0.894
0.08	0.623	0.632	0.232	0.239	0.239	0.888	0.892	0.892
0.10	0.659	0.659	0.246	0.246	0.246	0.890	0.890	**0.890**
0.12	0.693	0.684	0.260	0.253	0.260	0.893	0.888	0.892
0.14	0.725	0.708	0.273	0.260	0.273	0.895	0.886	0.894

Example 1. Consider a duopoly with $c = 50, d = 10, \bar{P} = 30, \bar{Q} = 4, b = 1$, and $a_0 = 0.1$. Table 2 simulates Situation 1 for the above-given values of the parameters and different values of the parameter a_1, starting with $a_1 = 0.06$ and increasing with a mesh of 0.02. The above-mentioned table shows the influence coefficients in the case of CCVE, while the values of the influence coefficients at the Nash-Cournot equilibrium are always $v_0 = v_1 = -d/2$. For the consistent CVE, the minimal value of the degradation measure $k_R^{CV} = 0.230$ (among the values presented in Table 2) is achieved when $a_1 = 0.06$, which means that the worst case is *not* the one where the firms are identical, unlike the Nash-Cournot case in which the worst case ratio is obtained when $a_1 = a_0 = 0.10$. That is, a complete symmetry does not necessarily entail the worst case ratio in for consistent CVEs.

3.3   Situations 2 and 3

In Situation 2 from Table 1, there is one *low-marginal cost* firm and the rival is a *high-marginal cost* producer. Plug in the equilibrium and optimal values into formula (12), and simple algebraic manipulations yield (in Situation 3, the results are similar):

$$R_0 = \frac{2d + a_0}{2v_0 + a_0} \cdot \left(\frac{2v_0}{d}\right)^2; R_1 = \frac{(c - b)^2 (2v_1/d)^2}{\bar{Q}(2v_1 + a_1)[c - b - \bar{Q}(2d + a_1)/4]}. \tag{18}$$

Here, Propositions 3.1 and 3.2 are still valid for firm 0 because the formulas of the profit ratio for firm 0 from Eq. (18) are identical to the formulas (14). Firm 0 would face degradations of her profits if, and only if (15) holds. The profit ratio of firm 0 increases with respect to a_1. The larger the value of a_1 the higher is the profit ratio for firm 0. The latter means that if globalization damages the profits of firm 0, the degradation would not be too strong as it would be with a smaller value of a_1. For firm 1, the higher values of the slope of the rival's marginal cost would result in a better profit ratio as stated in Proposition 3.5.

Proposition 3.5. *Profit ratio of competitor 1 increases together with a_0.*

3.4   Situation 4

In Situation 4 from Table 1, both producers are high-marginal cost firms. Substituting the equilibrium and optimal values in formula (12), after simple algebraic manipulations one obtains for $i = 0, 1$:

$$R_i = \frac{(c-b)^2(2v_i/d)^2}{\bar{Q}(2v_i + a_i)[c - b - \bar{Q}(2d + a_i)/4]}. \tag{19}$$

Proposition 3.6. *The profit of competitor i increases together with* a_{-i}.

Therefore, the effect of the increase of the quadratic cost coefficient a_{-i} on the rival's profit (player i) is positive.

4   Conclusions and Future Works

In this paper, we examine consistent conjectural variations equilibrium (CCVE) for a duopoly in a market of a homogeneous product. We study the effects of uniting two separate markets each monopolized by a producer: after globalization, both firms compete in one common market. Our model assumes an inverse demand function with a cap price and quadratic cost functions of both agents. Similar to previous studies, we investigate if the companies lose or gain due to globalization by evaluating their profit ratios, i.e., the ratios of their net profits after and before entering the common market. For the situations where both agents are low-marginal cost firms, we find that the company with a technical advantage over its rival has a better profit ratio. In addition, as the rival becomes weaker, this is, as the slope of the rival's marginal cost function increases, the first agent's profit ratio enhances, too. Moreover, when both agents are low-marginal cost firms, at least the weaker company suffers the degradation of its profits due to the globalization.

Unlike the previous study [1], which considers Nash-Cournot equilibrium, we show with an example that the complete symmetry of the agents does *not* always provide the worst case (the lowest profit ratio) in the case of consistent CVE. As a consequence, we demonstrate that under consistent conjectures it is important to analyze not only the case where firms are identical (although this leads one to deal with more complicated or even intractable problems).

In our forthcoming papers, we are going to analyze a system with a public firm whose maximized objective is distinct from its net profit.

Acknowledgments. The authors would like to express their profound gratitude to the three anonymous referees whose valuable critical and constructive comments and proposals have helped a lot in improving the paper.

This work was financially supported by the SEP-CONACYT grants CB-2013-01-221676 and FC-2016-01-1938, as well as by Tecnológico de Monterrey (ITESM), Campus Monterrey, Strategic Group for Optimization and Data Science.

References

1. Kameda, H., Ui, T.: Effects of symmetry on globalizing separated monopolies to a Nash-Cournot oligopoly. Int. Game Theory Rev. **14**(2), 1–15 (2012)
2. Brander, J.A., Spencer, B.J.: Intra-industry trade with Bertrand and Cournot oligopoly: the role of endogenous horizontal product differentiation. Res. Econ. **69**(2), 157–165 (2015)
3. Brander, J.A.: Intra-industry trade in identical commodities. J. Int. Econ. **11**(1), 1–14 (1981)
4. Brander, J.A., Krugman, P.: A "reciprocal dumping" model of international trade. J. Int. Econ. **15**(3), 313–321 (1983)
5. Yilmazkuday, D., Yilmazkuday, H.: Bilateral versus multilateral free trade agreements: a welfare analysis. Rev. Int. Econ. **22**(3), 513–535 (2014)
6. Bowley, A.L.: The Mathematical Groundwork of Economics. Clarendon Press, Oxford (1924)
7. Frisch, R.: Monopoly - polypoly – the concept of force in the economy. Natl. Int. Econ. Pap. **1**, 23–36 (1951). (Translation by W. Beckerman)
8. Kalashnikov, V.V., Bulavsky, V.A., Kalashnykova, N.I., Castillo, F.J.: Mixed oligopoly with consistent conjectures. Eur. J. Oper. Res. **210**(3), 729–735 (2011)
9. Kalashnykova, N.I., Kalashnikov, V.V., Montantes, M.A.O.: Consistent Conjectures in Mixed Oligopoly with Discontinuous Demand Function. In: Watada, J., Watanabe, T., Phillips-Wren, G., Howlett, R., Jain, L. (eds.) Intelligent Decision Technologies. Smart Innovation, Systems and Technologies, vol. 15, pp. 427–436. Springer, Heidelberg (2012). https://doi.org/10.1007/978-3-642-29977-3_43
10. Flores, D., García, A.: On the output and welfare effects of a non-profit firm in a mixed duopoly: a generalization. Econ. Syst. **40**(4), 631–637 (2016)
11. Kalashnikov, V.V., Bulavsky, V.A., Kalashnikov Jr., V.V., Kalashnykova, N.I.: Structure of demand and consistent conjectural variations equilibrium (CCVE) in a mixed oligopoly model. Ann. Oper. Res. **217**(1), 281–297 (2014)
12. Kalashnykova, N.I., Bulavsky, V.A., Kalashnikov, V.V., Castillo-Pérez, F.J.: Consistent conjectural variations equilibrium in a mixed duopoly. J. Adv. Comput. Intell. Intell. Inform. **15**(2), 425–432 (2011)
13. Isac, G., Bulavsky, V.A., Kalashnikov, V.V.: Complementarity, Equilibrium, Efficiency, and Economics. Kluwer Academic Publishers, Dordrecht/Boston/London (2002)

Verification on the Ensemble of Independent Numerical Solutions

A. K. Alekseev[1,2] , A. E. Bondarev[1] ,
and A. E. Kuvshinnikov[1(✉)]

[1] Keldysh Institute of Applied Mathematics RAS, Moscow, Russia
aleksey.k.alekseev@gmail.com, bond@keldysh.ru,
kuvsh90@yandex.ru
[2] Moscow Institute of Physics and Technology, Dolgoprudny, Russia

Abstract. The element of the epistemic uncertainty quantification concerning the estimation of the approximation error is analyzed from the viewpoint of the ensemble of numerical solutions obtained via independent numerical algorithms. The analysis is based on the geometry considerations: the triangle inequality and measure concentration in spaces of great dimension. In result, the feasibility for nonintrusive postprocessing appears that provides the approximation error estimation on the ensemble of the solutions. The ensemble of numerical results obtained by five OpenFOAM solvers is analyzed. The numerical tests were made for the inviscid compressible flow around a cone at zero angle of attack and demonstrated the successful estimation of the approximation error.

Keywords: Approximation error · Ensemble of numerical solutions · Distances between solutions · Triangle inequality · Measure concentration · Euler equations · Flow around a cone · OpenFOAM

1 Introduction

The approximation (discretization) error $\Delta u^{(k)} = u^{(k)} - \tilde{u}$ is usually considered as a subject of the epistemic uncertainty quantification. The reasons for this statement are the dependence of the approximation error on the truncation error $\Delta u^{(k)} = (u^{(k)} - \tilde{u}) = A_h^{-1} \delta u^{(k)}$ and the fact, that the Lagrange form of the truncation error $\delta u^{(k)}(\alpha_n) = Ch^m \partial^{m+1} u(x_n + \alpha_n)/\partial x^{m+1}$ (n is the grid point number) contains unknown deterministic parameter $\alpha_n \in [0,1]$.

Herein, the CFD system is formally denoted as $A\tilde{u} = f$, the numerical solution (obtained by k-th algorithm) is governed by the discrete operator $A_h u^{(k)} = f_h$. Truncation error $\delta u^{(k)}$ is obtained from the Taylor series of the numerical solution (grid function $u^{(k)}$ inserted to the main system of PDE.

However, there exist works (for example, [1]), which more or less successfully consider the approximation error to be the normally distributed random value. On the other hand, the numerical tests [2] demonstrate the universal (although, non-Gaussian) probability density distribution $P(\alpha_n)$ (for the Lagrange parameter α_n) with the mean value about 1/3 for the heat conduction problem. So, some random features of the

© Springer Nature Switzerland AG 2019
J. M. F. Rodrigues et al. (Eds.): ICCS 2019, LNCS 11540, pp. 315–324, 2019.
https://doi.org/10.1007/978-3-030-22750-0_25

approximation and truncation errors are observed in numerical tests. Thus, there exists a formal contradiction between random (observed) and deterministic (theoretically based) approaches to the approximation error estimation. This paradox may be resolved via the measure concentration effect [3], relating the probability and the geometry of the high dimensional spaces, however, it is far above the present paper scope.

From the stochastic standpoint, the ensemble-based approach may provide a natural alternative in the absence of information on the truncation error probability density distribution. Herein, the element of the epistemic uncertainty quantification concerning the estimation of the discretization error is analyzed from the viewpoint of the ensemble of numerical solutions obtained via independent algorithms governed by a vector valued parameter $\delta u^{(k)}$ (truncation error).

The ensemble of solutions is related to the flow around elongated bodies of rotation (EBR). In the late 80's–early 90's the computational technology allowed to make routine computing for a flow around EBR with a high degree of reliability. For example, the deviation between the numerical and experimental results for aerodynamic drag coefficients did not exceed 2–3%. The essence of this technology was that the aerodynamic drag coefficient Cx was considered as a sum of three components: Cp – coefficient for inviscid flow, Cf – coefficient for viscous friction and Cd – coefficient for near wake pressure. The present work is a part of the general project to create a similar technology [4–6]. The OpenFOAM software package (Open Source Field Operation and Manipulation CFD Toolbox) [7] is used to calculate the aerodynamic characteristics for inviscid flow around the elongated bodies of rotation. OpenFOAM contains a number of solvers [8–12] based on independent numerical algorithms.

This paper presents an analysis of the ensemble of solutions, obtained using five OpenFOAM solvers, for the discretization error estimation. Etalon solutions [13], which have a high accuracy and are used for verification of numerical methods for many years, are employed for the true error estimation. The analysis is performed for the inviscid flow around a cone.

2   The Test Problem

The statement of the CFD problem is presented in accordance with the work [13], where the results of the inviscid flow around cones with different angles at various Mach numbers are addressed. We consider a cone in the uniform supersonic flow of ideal gas at zero angle of attack $\alpha = 0°$ with a Mach number 2. The body under consideration is a cone with the half angle $\beta = 20°$ (Fig. 1). The inflow conditions are denoted by the index "∞", and the outflow ones by the index ξ, since the solution is self-similar and depends on the dimensionless variable. The Euler equations system is used for the flow field calculation. The system is supplemented by the equation of state of the ideal gas.

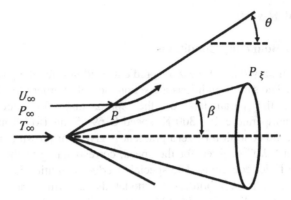

Fig. 1. Flow scheme

3   OpenFOAM Solvers

The following solvers from the OpenFOAM software package were used:

- *rhoCentralFoam* (rCF), based on a central-upwind scheme, which is a combination of central-difference and upwind schemes [8, 9]. The essence of the central-upwind schemes consists in a special choice of a control volume containing two types of domains: around the boundary points, the first type; around the center point, the second type. The boundaries of the control volumes of the first type are determined by means of local propagation velocities. The advantage of these schemes is the possibility to achieve a good resolution for discontinuous solutions in gas dynamics, using the appropriate technique for the numerical viscosity reducing.
- *sonicFoam* (sF), based on the PISO algorithm (Pressure Implicit with Splitting of Operator) [10]. The basic idea of the method is the application of two difference equations to calculate the pressure for the correction of the pressure field obtained from discrete analogs of the equations of moments and continuity. This approach is takes account of the fact that the velocities changed by the first correction may not satisfy the continuity equation, therefore, a second corrector is introduced, which enables us to calculate the velocities and pressures satisfying the linearized equations of momentum and continuity.
- *rhoPimpleFoam* (rPF), based on the PIMPLE algorithm, which is a combination of the PISO and SIMPLE (Semi-Implicit Method for Pressure-Linked Equations) algorithms. In this method, an external loop is added to the PISO algorithm, due to which the method becomes iterative one and allows to count with the Courant number greater than 1.
- *pisoCentralFoam* (pCF), which is a combination of a Kurganov-Tadmor scheme [8] with the PISO algorithm [11].
- *QGDFoam* (QGDF), which is based on the implementation of quasi-gas dynamic equations [12].

4   Numerical Results

4.1   Initial and Boundary Conditions

On the upper boundary ("*top*") the zero gradient condition for the gas dynamic functions is specified. The same conditions are set on the right border ("*outlet*"). On the left border ("*inlet*"), the parameters of the approaching flow are set: pressure P = 101325 Pa, temperature T = 300 K, speed U 694.5 m/s (Mach number = 2). On the cone surface, the condition of zero gradient is given for pressure and temperature, and the condition "*slip*" is given for the speed, corresponding to the non-penetration condition for the Euler equations. The special "*wedge*" condition is used for the front ("*front*") and back ("*back*") borders to model the axisymmetric geometry in the OpenFOAM package. The OpenFOAM package also employs a special "*empty*" boundary condition. This condition is specified in cases when calculations in a given direction are not carried out. In our case, this condition is used for the "*axis*" border.

The number of grid cells is 13200. The initial conditions correspond to the boundary conditions on the inlet edge, that is, the initial conditions are used for the parameters of the inflow stream. The molar mass $M = 28.96$ kg/mol and the specific heat at constant pressure Cp = 1004 were also set.

4.2   Parameters of Solvers

In the OpenFOAM package, there are two options for approximating differential operators: directly in the solver's code or using the fvSchemes and fvSolution configuration files. In order the comparison to be correct, we used the same parameters, where possible. In the fvSchemes file: ddtSchemes – *Euler*, gradSchemes – *Gauss linear*, divSchemes – *Gauss linear*, laplacianSchemes – *Gauss linear corrected*, interpolationSchemes – *vanLeer*. In the fvSolution file: solver – *smoothSolver*, smoother *symGaussSeidel*, tolerance – 1e–09, nCorrectors – 2, nNonOrthogonalCorrectors – 1.

5   Analysis of True Errors on the Ensemble

Tables 1 and 2 show the results of calculations of the L_2 and L_1 norm of "true" error (differences between tested and exact (etalon) solutions):

$$\sqrt{\sum_{m} \left| y_m - y_m^{exact} \right|^2 V_m} \tag{1}$$

$$\sum_{m} \left| y_m - y_m^{exact} \right| V_m \tag{2}$$

Herein, the norm is not considered as the vector norm, rather it should be treated as the grid function norm (discrete analogue of the function norm).

In Tables 1 and 2 y_m is the velocity components U_x and U_y, pressure p and density ρ in the cell. V_m is the cell volume for the cone angle $\beta = 20°$ and Much number $M = 2$. Figures 2 and 3 shows errors from Tables 1 and 2 in the form of spider diagram, where the tabular data are normalized to the maximum values of the corresponding gas-dynamic functions.

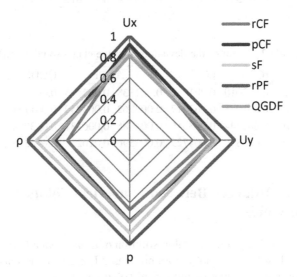

Fig. 2. Norm of the deviation from the etalon solution (L_2)

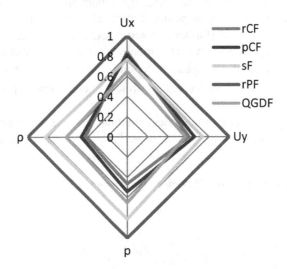

Fig. 3. Norm of the deviation from the etalon solution (L_1)

Table 1. Norm of the deviation from the etalon solution (L$_2$)

	rCF	pCF	sF	rPF	QGDF
U_x	0.007258	0.007367	0.006663	0.008037	0.006500
U_y	0.007895	0.009107	0.008890	0.010691	0.008532
p	0.011566	0.013465	0.015504	0.017360	0.013721
ρ	0.015545	0.018697	0.023374	0.025612	0.019400

Table 2. Norm of the deviation from the etalon solution (L$_1$)

	rCF	pCF	sF	rPF	QGDF
U_x	0.002010	0.001952	0.001838	0.002405	0.001548
U_y	0.001848	0.002017	0.002262	0.003059	0.001741
p	0.002844	0.003316	0.004985	0.006131	0.003984
ρ	0.004322	0.004729	0.008396	0.010201	0.006245

6   Analysis of Distances Between Numerical Solutions on the Ensemble

The above analysis is based on the etalon solution of *a priori* small error. However, the availability of such solution is not a common case. There is some feasibility for error analysis based on the ensemble of numerical solutions [14–16].

Let the relation of the approximation error of these schemes be known *a priori*. We denote the numerical solution as the vector (grid function) $u^{(k)} \in R^N$ (k is the scheme number, N is the number of grid points respectively), the values of unknown exact solution at nodes of this grid (further denoted as exact solution) as $\tilde{u} \in R^N$ and use a discrete (L$_2$ or L$_1$) equivalent norm. The unknown deviation of exact solution values at grid points $\tilde{u} \in R^N$ from the computed solution is assuming the form $\left\| u^{(i)} - \tilde{u} \right\| = r_i$. The numerical solutions $u^{(i)}$ are located at surfaces of concentric hyperspheres with the center at $\tilde{u}$ and unknown radii r_k.

In the simplest event of two numerical solutions $u^{(1)}$ and $u^{(2)}$ with *a priori* known errors relation $r_1 \geq 2r_2$ ($r_1 = \left\| \tilde{u} - u^{(1)} \right\|$, $r_2 = \left\| \tilde{u} - u^{(2)} \right\|$) the following theorem [14] may be stated.

Theorem 1. *Let the norm of difference of two numerical solutions $u^{(1)} \in R^{(N)}$ and $u^{(2)} \in R^{(N)}$*

$$\left\| du_{1,2} \right\| = \left\| u^{(1)} - u^{(2)} \right\| \tag{3}$$

be known from computations and there is a priori information

$$\left\|\tilde{u} - u^{(1)}\right\| \geq 2\left\|\tilde{u} - u^{(2)}\right\| \tag{4}$$

then the norm of approximate solution $u^{(2)}$ error is less than the norm of difference of solutions:

$$\left\|\tilde{u} - u^{(2)}\right\| \leq \left\|du_{1,2}\right\| \tag{5}$$

Unfortunately, *a priori* information (Eq. 4) usually is not available. However, the separation of distances between solutions into clusters of the small error and the large error (i.e., the availability of highly precise solutions concentrated in a vicinity of the exact one) may be considered as evidence of the existence of solutions with significantly different error norms [14–16].

The quantitative criterion for applicability of Theorem 1, based on dimension of first cluster and the distance between clusters, may be stated as the following semi-heuristic criterion:

Criterion. *If the distance between clusters is greater than the size of the cluster of accurate solutions $\delta_2 - \delta_1 > \delta_1$ then $\left\|\tilde{u} - u^{(i)}\right\| \leq \left\|du_{1,i}\right\|$, where $u^{(i)}$ belongs to the cluster of more accurate solutions and $u^{(1)}$ is the maximally inaccurate solution.*

From this standpoint the collection of distances between solutions may contain the important information. The matrix of the distances engendered by L_2 or L_1 norm is provided in the Tables 3 and 4 for U_x. The analysis for other components is omitted for brevity due to their similar behavior.

In the Tables 3 and 4, one can see no visible separation of distances by clusters, which may satisfy the above criterion. So, the above mentioned criterion is not applicable.

Table 3. Norm of the deviation from the etalon solution (L_2)

	rCF	pCF	sF	rPF	QGDF
rCF	0.0	0.003522	0.002741	0.003444	0.003580
pCF		0.0	0.003067	0.003666	0.003511
sF			0.0	0.002529	0.002852
rPF				0.0	0.004785
QGDF					0.0

Table 4. Norm of the deviation from the etalon solution (L_1)

	rCF	pCF	sF	rPF	QGDF
rCF	0.0	0.001336	0.001085	0.001294	0.001411
pCF		0.0	0.001266	0.001462	0.001399
sF			0.0	0.000911	0.001256
rPF				0.0	0.001873
QGDF					0.0

In this situation we consider the maximum value of distance between solutions over all data of Tables 3 and 4 as the upper bound of the error.

Corresponding values $\|du_{i,j}\|_{L_2}$ and $\|du_{i,j}\|_{L_1}$ are provided in Tables 5 and 6 for U_x component as the maximum estimation of error norms. The relations of error norms to the maximum estimate $\|\Delta u\|_{L_2}/\|du_{max}\|_{L_2}$ and $\|\Delta u\|_{L_1}/\|du_{max}\|_{L_1}$ are also presented in the Tables.

Table 5. Norm of the deviation from the etalon solution (L_2)

	rCF	pCF	sF	rPF	QGDF
$\|\Delta u\|_{L_2}$	0.07258	0.007367	0.006663	0.008037	0.006500
$\|du_{max}\|_{L_2}$	0.004785				
$\|\Delta u\|_{L_2}/\|du_{max}\|_{L_2}$	1.517	1.54	1.392	1.68	1.358

Table 6. Norm of the deviation from the etalon solution (L_1)

	rCF	pCF	sF	rPF	QGDF
$\|\Delta u\|_{L_1}$	0.002010	0.001952	0.001838	0.002405	0.001548
$\|du_{max}\|_{L_1}$	0.001873				
$\|\Delta u\|_{L_1}/\|du_{max}\|_{L_1}$	1.073	1.042	0.981	1.284	0.827

The analysis of data provided in Tables 5 and 6 demonstrates the maximum distance between solutions to be a reasonable estimate for the deviation of considered numerical solution from the etalon one. The maximum relation of the true error to estimate of the error $\|\Delta u\|_{L_2}/\|du_{max}\|_{L_2}$ over all solutions is about 1.7 that seems to be quite acceptable value. As mentioned in [16], the L_1 norm is more suitable for the error estimation. Herein, the results of Table 6 also confirm the higher quality of the error estimation via the maximum distance between solutions calculated in the L_1 norm.

If one consider $\|du_{max}\|$ as the error estimator and the value $\|\Delta u\|/\|du_{max}\|$ as the effectivity index of this estimator, its magnitude in the L_1 norm is about unit, that is close to "ideal" estimator [17]. The error estimator $2\|du_{max}\|_{L_2}$ is acceptable for the L_2 norm.

7 Discussion

The above results demonstrate the potential of the ensemble of numerical solutions, obtained by independent algorithms, for a posteriori error estimation and verification. In papers [14–16] the independence of numerical methods was obtained by using schemes of the formally different order of approximation. Herein, the independence is ensured by differences of the numerical algorithms' structure that significantly expands the application domain.

The results demonstrate also that the location of exact solution determined by the distances between solutions along with [14–16] is found to be close to the etalon one. It confirms the high quality of the etalon solution [13].

The evolution of the notion "solution" for the CFD equations ("strong", "weak", "measure-valued" [18]) is far from the final. At present, the ensemble-based option for solutions (statistic [1, 18], measure-valued [18]) seems to be of most current interest. These ensemble-based solutions may provide both the reasonable solution notion for the shocked and turbulent flows and some natural way for the Uncertainty Quantification.

The statistic solution is used in [18] for approximation of the measure-valued solution moments. The key element of this approach is the scalar stochastic parameter directly inserted in the boundary condition. In contrast, in present paper, the stochastic parameter is related with the source term of the differential approximation (truncation error) implicitly. The stochastic properties of the truncation error are caused by the high dimensionality of the problem (about 10^4 nodes) and the independence of the truncation error (source term) for the solvers based on different algorithms. The cumulative action of these features may cause stochastic properties due to the measure concentration effect [3].

8 Conclusion

In above presented numerical tests five OpenFOAM solvers were compared with the etalon solution [13] and with each other in the metrics engendered by L_1 and L_2 norm. All solutions are found to be close with each other and with the etalon one. The discretization error is estimated as the maximum distance between solutions in the ensemble.

The estimated discretization error is close to the "true" errors (between numerical and etalon solutions). Thus, the tests demonstrate that the ensemble of the numerical solutions obtained by different solvers, based on independent algorithms, provides the feasibility for verification of any considered solver with the same quality as the etalon solution.

The above error norm estimation is obtained without conditions listed in [14–16] (a priori information on error norm relation or availability of small error clusters) that enhances the applicability domain for the ensemble based discretization error estimation. Both the code and solution verification may be performed via an ensemble of numerical solutions obtained by independent algorithms.

Acknowledgments. This work was supported by grant of RSF № 18-11-00215.

References

1. Rauser, F., Marotzke, J., Korn, P.: Ensemble-type numerical uncertainty quantification from single model integrations. J. Comput. Phys. **292**, 30–42 (2015). https://doi.org/10.1016/j.jcp.2015.02.043
2. Alekseev, A.K., Makhnev, I.N.: On using the Lagrange coefficients for a posteriori error estimation. Numer. Anal. Appl. 2(4), 302–313 (2009). https://doi.org/10.1134/S1995423909040028

3. Zorich, V.A.: Multidimensional geometry, functions of very many variables, and probability. Theory Probab. Appl. **59**(3), 481–493 (2015). https://doi.org/10.1137/S0040585X97 T987181

4. Bondarev, A.E., Kuvshinnikov, A.E.: Comparative study of the accuracy for OpenFOAM solvers. In: Proceedings of Ivannikov ISPRAS Open Conference (ISPRAS), pp. 132–136. IEEE, Moscow (2017). https://doi.org/10.1109/ISPRAS.2017.00028

5. Bondarev, A.E., Nesterenko, E.A.: Approximate method for estimation of friction forces for axisymmetric bodies in viscous flows. Mathematica Montisnigri **31**, 54–63 (2014)

6. Bondarev, A.E., Kuvshinnikov, A.E.: Analysis of the accuracy of OpenFOAM solvers for the problem of supersonic flow around a cone. In: Shi, Y., et al. (eds.) ICCS 2018. LNCS, vol. 10862, pp. 221–230. Springer, Cham (2018). https://doi.org/10.1007/978-3-319-93713-7_18

7. OpenFOAM. http://www.openfoam.org. Accessed 30 Jan 2019

8. Kurganov, A., Tadmor, E.: New high-resolution central schemes for nonlinear conservation laws and convection-diffusion equations. J. Comput. Phys. **160**(1), 241–282 (2000). https://doi.org/10.1006/jcph.2000.6459

9. Greenshields, C., Wellerr, H., Gasparini, L., Reese, J.: Implementation of semi-discrete, non-staggered central schemes in a colocated, polyhedral, finite volume framework, for high-speed viscous flows. Int. J. Numer. Meth. Fluids **63**(1), 1–21 (2010). https://doi.org/10.1002/fld.2069

10. Issa, R.: Solution of the implicit discretized fluid flow equations by operator splitting. J. Comput. Phys. **62**(1), 40–65 (1986). https://doi.org/10.1016/0021-9991(86)90099-9

11. Kraposhin, M., Bovtrikova, A., Strijhak, S.: Adaptation of Kurganov-Tadmor numerical scheme for applying in combination with the PISO method in numerical simulation of flows in a wide range of Mach numbers. Procedia Comput. Sci. **66**, 43–52 (2015). https://doi.org/10.1016/j.procs.2015.11.007

12. Kraposhin, M.V., Smirnova, E.V., Elizarova, T.G., Istomina, M.A.: Development of a new OpenFOAM solver using regularized gas dynamic equations. Comput. Fluids **166**, 163–175 (2018). https://doi.org/10.1016/j.compfluid.2018.02.010

13. Babenko, K.I., Voskresenskii, G.P., Lyubimov, A.N., Rusanov, V.V.: Three-Dimensional Ideal Gas Flow Past Smooth Bodies. Nauka, Moscow (1964). (In Russian)

14. Alekseev, A.K., Bondarev, A.E., Navon, I.M.: On Triangle Inequality Based Approximation Error Estimation. arXiv:1708.04604 [physics.comp-ph], 16 August 2017

15. Alekseev, A.K., Bondarev, A.E., Navon, I.M.: On Estimation of Discretization Error Norm via Ensemble of Approximate Solutions. arXiv:1704.04994 [physics.comp-ph] 18 April 2017

16. Alexeev, A.K., Bondarev, A.E.: On Exact Solution Enclosure on Ensemble of Numerical Simulations. Mathematica Montisnigri XXXVIII, 63–77 (2017)

17. Repin, S.I.: A Posteriori Estimates for Partial Differential Equations, vol. 4. Walter de Gruyter (2008). https://doi.org/10.1515/9783110203042

18. Fjordholm, U.S., Mishra, S., Tadmor, E.: On the computation of measure-valued solutions. Acta Numerica **25**, 567–679 (2016). https://doi.org/10.1017/S0962492916000088

On the Estimation of the Accuracy
of Numerical Solutions in CFD Problems

A. E. Bondarev$^{(\boxtimes)}$ ⓘ

Keldysh Institute of Applied Mathematics RAS, Moscow, Russia
bond@keldysh.ru

Abstract. The task of assessing accuracy in mathematical modeling of gas-dynamic processes is of utmost importance and relevance. Modern software packages include a large number of models, numerical methods and algorithms that allow solving most of the current CFD problems. However, the issue of obtaining a reliable solution in the absence of experimental data or any reference solution remains relevant. The paper provides a brief overview of some useful approaches to solving the problem, including such approaches as a multi-model approach, the study of an ensemble of solutions, the construction of a generalized numerical experiment.

Keywords: Mathematical modeling · Computational fluid dynamics ·
Accuracy estimation · Multi-model approach · Ensemble of solutions ·
Generalized numerical experiment

1 Introduction

The task of assessing accuracy in mathematical modeling of gas-dynamic processes is of utmost importance and relevance. A huge number of works devoted to this topic, for example [1]. Precisely, accuracy estimation played a key role in the entire history of the development of numerical methods in CFD. Throughout the history of CFD, the main criterion for accuracy and reliability has been a comparison with a physical experiment [2, 3]. The development of numerical methods followed the path of complicating the mathematical models under consideration. At the first stage, the Euler equations were used to model the inviscid flow. To calculate the friction coefficient on the body, the boundary layer equations were used, where the results of calculations of inviscid flow were used as boundary conditions at the upper boundary of the layer. In order to simulate viscous effects (vortices, separation zones), it was already necessary to consider the complete system of Navier-Stokes equations. To simulate turbulent flows, it was necessary to add turbulence models. The history of the development of numerical methods is presented in detail in [4].

At each stage of this development for the construction of numerical methods and algorithms for their implementation, the main criterion for accuracy and reliability was a comparison with a physical experiment. Having a numerical method with the necessary approximation, stability and convergence, it was possible to compare the numerical solution of the simulated problem with the experiment and verify the reliability of the method. However, each task is characterized by a whole set of defining

© Springer Nature Switzerland AG 2019
J. M. F. Rodrigues et al. (Eds.): ICCS 2019, LNCS 11540, pp. 325–333, 2019.
https://doi.org/10.1007/978-3-030-22750-0_26

parameters, such as the Mach number, the Reynolds number, the geometric parameters of the problem, and so on. Having achieved a satisfactory agreement with the experimental data for a specific set of determining parameters, it was assumed by default that with some reasonable variation of them, the solution is obtained quite accurately.

Modern software systems for solving CFD problems, both open and commercial, have now been greatly developed. Such complexes include a large number of numerical methods, turbulence models, methods for parallelizing algorithms. It would seem that now the problem of accuracy and reliability is solved. However, in practice there is a certain kind of paradox. With all the wealth of opportunities provided by modern computing software packages, in these packages there are a large number of tuning parameters. These parameters may vary in certain ranges. On the one hand, this is very good, since it gives the opportunity to customize the algorithms to match the experimental result. But in the absence of experimental data or any reference solution, there are serious problems in evaluation the accuracy of the solution obtained.

In this case, we are dealing with a complex type of uncertainty, where the total error consists of such components as model selection error, numerical method error, error of the algorithm's numerical implementation method, computational grid construction error, and finally inaccuracies associated with setting numerous parameters characterizing the selected turbulence model. Analyzing and evaluating the accuracy for each of these components separately is quite difficult and ultimately inefficient. It is much more expedient to develop integrated approaches for obtaining a reference solution and an assessment of accuracy.

It should also be noted that the question of the method and standard of evaluation also plays a big role. The gas-dynamic fields obtained in the calculation can be compared with the experiment and the reference solution. For example, quite often numerical solutions are compared by the presence of oscillations in a shock wave and the degree of its smearing. The solutions obtained by monotonous schemes look best from this point of view. However, oscillating solutions may converge in the norms of L_1, L_2 better than monotone ones. Another way is to compare commonly used valuable functionals in practice, such as the drag coefficient of an object placed in a stream. Thus, the following question remains relevant: how can one obtain a relatively reliable solution in the absence of data from a physical experiment or a reference solution? The question is of special importance if we are not talking about a single calculation, but about the formulation of mass industrial calculations. Below some possible approaches to solving this problem are considered. The article presents three approaches to solving listed above problems that are developed simultaneously by our team at Keldysh Institute of Applied Mathematics (KIAM RAS). All the results presented in this article were developed at KIAM RAS as part of the development of these three approaches.

2 Approaches to Obtaining a Reference Solution

2.1 Multi-model Approach

This approach is in a sense historical. It was widely distributed at the end of the twentieth century. The approach allowed to carry out with sufficiently high accuracy mass industrial and scientific calculations in a wide class of problems of modeling flows around objects. This computing technology was complex and combined several mathematical models. Each of the models used the results obtained using a different model as boundary conditions. It is here that the body drag coefficient in the flow was used as a valuable functional. This technology was especially effective for elongated bodies of rotation. Here aerodynamic drag coefficient was computed as a sum of three components: coefficient for inviscid flow, coefficient for viscous friction and coefficient for near wake. The results for inviscid flow were used as boundary conditions for computing of viscous friction coefficient. Then the results for viscous friction were used as boundary conditions for computing of near wake problem. This approach was widely used for analyzing the aerodynamic properties of different bodies with high efficiency.

Currently, there are successful attempts to implement this approach at the modern level using parallel computing in the form of a computational pipeline, where data is automatically transferred from the model to the model [5, 6]. To simulate a non-viscous flow around an open software package OpenFOAM (Open Source Field Operation And Manipulation CFD Toolbox) [7] is used. This package has a large number of solvers, both standard and developed by various teams. For a comparative assessment of the accuracy of these solvers, a series of calculations were carried out on the test problem of a flow around a cone at an angle of attack. During the test calculations, the Mach number, the angle of the cone, and the angle of attack were varied. The results allowed to make conclusions about the most appropriate solvers in terms of accuracy [6].

To determine the friction coefficient on the body placed in a flow, a computational technique [5] is implemented, based on an approximate semi-empirical model combining the results of experimental studies and the well-known effective length method. This technique uses the results of calculations of non-viscous flow as input data and allows one to obtain a drag friction resistance coefficient and characteristic boundary layer thicknesses in a wide range of Mach and Reynolds numbers both for the laminar and turbulent regime. To determine the coefficient for near wake pressure, the Navier-Stokes equations are used, where the results obtained in the previous stages are used as boundary conditions.

This approach is not universal. It works well for classes of problems where friction coefficient and coefficient for near wake pressure are small compared with coefficient for inviscid flow. Nevertheless, for many classes of problems, this technology allows obtaining results that can be used as a reference solution in the absence of experimental data.

2.2   Use of Ensemble Solutions

If there is a set of numerical solutions (for example, obtained using various finite-difference schemes) and *a priori* information about the error ranking of these solutions, then we can estimate the neighborhood of the approximate solution containing the exact solution *(exact solution enclosure)*. If an ensemble of numerical solutions can be divided into clusters of "accurate" and "inaccurate" solutions, then the error ranking of values can be performed using an a posteriori analysis of the distances between the numerical solutions. This can serve as a computational proof of the existence of an exact solution in the case of nonlinear problems. This approach is described in detail in [8], where the results of tests for supersonic flows within the framework of the Euler model are presented. A set of solvers with different approximation orders was used. The comparison considered a set of finite-difference schemes with accuracy order from the first up to forth. The results of comparison demonstrated the exact solution enclosure.

This approach can be considered as perspective. Nevertheless, it has evident drawback. For using of this approach one should have a set of solvers with different accuracy order.

2.3   Construction of Generalized Numerical Experiment

This approach is the most interesting from the point of view of the author. The modern development of high-performance computing clusters and the wide distribution of parallel computing technologies open up a number of new opportunities for solving problems of mathematical modeling in computational gas dynamics. These new features include high-grade parametric research and solving optimization analysis problems. Parametric studies suggest multiple solutions to the direct problem of mathematical modeling with variations in the defining parameters of the problem. The defining parameters of the problem include characteristic numbers, such as the Mach number, Reynolds number, Strouhal number, etc., and the geometric parameters of the problem. Each of the defining parameters varies in a certain range of variation with a certain partitioning step. The tasks of optimization analysis are more complex from a computational point of view. At each split point of the space of defining parameters, such problems assume the solution of the inverse problem, which aims to find the extremum of one or another valuable functional (optimal form, minimal drag coefficient, etc.). Parametric studies and optimization analysis tasks are the basis of a generalized computational experiment. A generalized computational experiment allows one to obtain in discrete form a solution not only for one single task, but for a whole class of problems. Here the class of problems is determined by the ranges of change of defining parameters.

The main advantage of a generalized computational experiment is that it allows one to obtain a solution not for one specific problem, but for a class of problems. However, the discrete solution itself cannot provide an understanding of the results obtained. It requires a wide and creative use of the tools of scientific visualization and visual analytics. When visualizing the results of a generalized computational experiment, it is necessary to combine the use of classical methods of visualization and animation of three-dimensional scalar and vector fields with visual analytics tools designed for analyzing multidimensional data. Various aspects of the construction of a generalized computational experiment and its formal description are described in detail in [9].

Let's consider some examples of this approach applied to some practical problems. It is applied in some variations due to different aims for each class of problems [6, 9–11].

The first example is presented in [9]. We consider a nozzle in supersonic viscous flow. Underexpanded supersonic jet exhausts from the nozzle. Jet propagation creates an obstacle in the main flow. For standard case we have flow structure presented in

Fig. 1(a). If we increase velocity of pressure ratio growth in the jet, then we obtain a new flow structure presented in Fig. 1(b). This crucial value of the velocity is used as control parameter. We consider four characteristic numbers (Mach number, Reynolds number, Prandtl number, Strouhal number) as coordinates in the space of defining parameters. Each of these parameters within the range is divided with some specific step. So we have a set of points in a four-dimensional space created by four defining parameters. For each point of this four-dimensional space, we find the value of the crucial velocity at which the flow structure changes. Then one can construct the space of three first principal components for computed data and make visual presentation for crucial velocity in a new system of coordinates (Fig. 1(c)). The form of dependence in question allows to approximate the dependence by plane. So the construction of generalized numerical experiment allowed to obtain desired dependence in an analytical form [9].

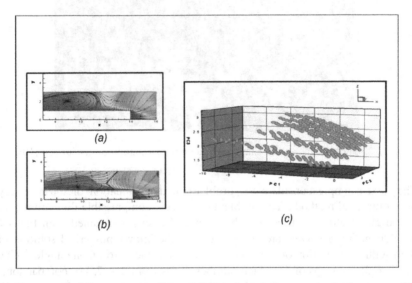

Fig. 1. Two types of flow structure ((a) and (b)) and control parameter in the space of 3 first principal components (c).

The following example of constructing a generalized computational experiment is presented in [10], where the problem of finding the optimal shape for a three-dimensional blade assembly is considered. The blade assembly has a rather complicated configuration and is located in the stream. Note that this blade assembly belongs to the power plant. The target functionals here were chosen the value of the total aerodynamic force acting on the blade, and the amount of torque. As the defining parameters were set two angles, which set the slope of the blade and the transverse width of the blade. These three parameters varied in certain ranges. For the numerical implementation of this generalized computational experiment, a computational technology was constructed that simulates the load on the blade assembly, placed in the air flow at various flow rates. The task was complicated by the fact that the modeling of the flow around and the variation of the geometric parameters were carried out taking into account quantitative restrictions on the moment of inertia and the mass of the blade assembly. These characteristics should not exceed certain boundary values when changing geometric parameters. This problem was solved using parallel computing. Figure 2 shows the resulting shape of the blade assembly. The figure also shows the pressure distribution over the surface of the blade assembly.

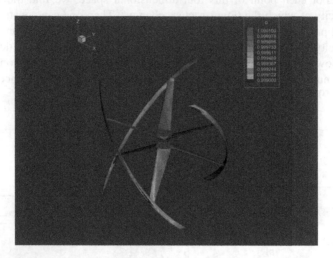

Fig. 2. 3D blade assembly shape and pressure distribution on its surface [10].

The next example considers the problem of the evaluation of the accuracy for different numerical methods. The problem of inviscid compressible flow around a cone at zero angle of attack is used as a base one. The results obtained with the help of various OpenFOAM solvers are compared with the known numerical solution of the problem with the variation of cone angle and flow velocity [6]. Cone angle β changes from 10° to 35° in steps of 5°. Mach number varies from 2 to 7. For comparison, four solvers were selected from the OpenFOAM software package: *RhoCentralFoam, SonicFoam, RhoPimpleFoam, RhoPimpleFoam.* The results of such kind of numerical experiment were presented as errors in the form of an analog of the L2 norm for all solvers. Figure 3 illustrates the results in a form of a change in deviation from the exact solution for pressure depending on the cone angle and the velocity for the solver *rhoCentralFoam.* Such changes were obtained for all solvers.

Figure 3 shows a multidimensional dataset for pressure obtained as a result of parametric calculations in the space of the first three principal components. Yellow shows the results for *rhoCentralFoam* solver, red for *pisoCentralFoam*, green for *sonicFoam* and blue for *rhoPimpleFoam*. Figure 3 shows that the errors for *rhoCentralFoam* and for *pisoCentralFoam* can be roughly approximated by a plane reflecting the dependence of the error on the Mach number and cone angle. The results for *sonicFoam* and especially for *rhoPimpleFoam* are significantly separated from the results for the first two solvers due to their particular numerical characteristics.

This methodical research can serve as a basis for selecting the OpenFoam solver for calculating the inviscid supersonic flow around the elongated bodies of rotation. The results of solvers comparison can also be useful for developers of OpenFoam software content.

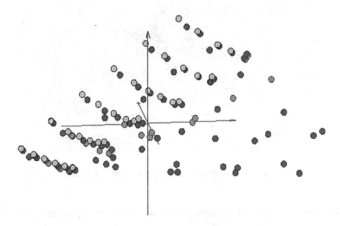

Fig. 3. Errors for different OpenFOAM solvers in the space of principal components (Color figure online)

The following example is devoted to improving the computational properties of finite-difference schemes. The problem of mathematical modelling of the flow in the far wake behind the body is solved. In the general case, in a rectangular computational domain, a viscous compressible heat-conducting gas flow is considered, described by a complete system of time-dependent Navier-Stokes equations. At the input boundary, the distributions of gas-dynamic parameters are given, obtained from calculations of the flow around an axisymmetric body and a portion of the track behind it. The main goal of the generalized computational method was to thoroughly study the properties of artificial viscosity incorporated in the hybrid difference scheme. For this purpose, we studied the properties of the weight coefficients of the hybrid scheme on the example of the problem of flow in the far wake and determined the limitations for the weight coefficients. In this task, the following defining parameters were varied, such as the steps of the grid decomposition in the x and y directions, the weighting coefficients of the difference scheme, the Reynolds number of the problem. As a result of the generalized computational experiment, a limiting surface was constructed for the dependence of the weight coefficient on the other determining parameters of the problem. An example of the limiting surface is presented in Fig. 4. When choosing the value of the weighting factor below the surface, in the numerical solution, non-physical oscillations arise, which can lead to the collapse of the solution. Such surfaces are constructed for non-viscous and viscous flow. In the case of viscous flow, laminar and turbulent regimes are considered.

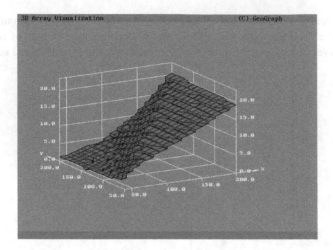

Fig. 4. Limiting surface for inviscid case.

From the point of view of accuracy assessment, the application of this approach allows us to deal with valuable objective functionals constructed for a multidimensional data volume, rather than with gas-dynamic fields. The construction of such a functional as a function of several variables (defining the parameters of the problem) in some cases gives the possibility of representing the function in an analytical form. This will be the solution that interests us for a class of problems, a certain set of external parameters and ranges of changes of these parameters. If inside the multidimensional volume of defining parameters there are some points for which there are experimental data, then we can calculate the average deviation for the resulting objective functional and consider from this point of view the suitability of the constructed functional for practical use.

It should be noted that you need to stay within the framework of the model used. As a rule, there are no clear boundaries of applicable models, there are transition zones. The amendment to these zones requires additional uncertainty in terms of the applicability of the selected model.

3 Conclusion

Evaluation of the accuracy of mathematical modeling problems solutions in CFD is important and relevant. This importance especially increases when it is necessary to carry out calculations in the absence of experimental data for comparison or a reference solution. Three approaches that may be useful in this situation are presented - the multi-model approach, the use of an ensemble of solutions and the construction of a generalized computational experiment. All these approaches are not universal and have their drawbacks, but for different classes of tasks they can be used to complement each other.

Acknowledgments. This work was supported by grant of RSF № 18-11-00215.

References

1. Fjordholm, U.S., Mishra, S., Tadmor, E.: On the computation of measure-valued solutions. Acta Numerica **25**, 567–679 (2016). https://doi.org/10.1017/S0962492916000088
2. Van Dyke, M.D.: Album of Fluid Motion (Parabolic, California, 1982; Mir, Moscow, 1986)
3. Mueller, T.J.: Flow visualization by direct injection. Fluid Mech. Meas. (Hemisphere, 1983) **2**, 367–450 (1983)
4. Bondarev, E.N., Dubasov, V.T., Ryzhov, Y.A., et al.: Aerigidromeckanika. M.: Mashinostroenie (1993)
5. Bondarev, A.E., Nesterenko, E.A.: Approximate method for estimation of friction forces for axisymmetric bodies in viscous flows. Mathematica Montisnigri **31**, 54–63 (2014)
6. Bondarev, A.E., Kuvshinnikov, A.E.: Analysis of the accuracy of OpenFOAM solvers for the problem of supersonic flow around a cone. In: Shi, Y., et al. (eds.) ICCS 2018. LNCS, vol. 10862, pp. 221–230. Springer, Cham (2018). https://doi.org/10.1007/978-3-319-93713-7_18
7. OpenFOAM. http://www.openfoam.org. Accessed 30 Jan 2019
8. Alexeev, A.K., Bondarev, A.E.: On exact solution enclosure on ensemble of numerical simulations. Mathematica Montisnigri **38**, 63–77 (2017)
9. Bondarev, A.E.: On the construction of the generalized numerical experiment in fluid dynamics. Mathematica Montisnigri **42**, 52–64 (2018)
10. Andreev, S.V., et al.: A computational technology for constructing the optimal shape of a power plant blade assembly taking into account structural constraints. Program. Comput. Softw. **43**(6), 345–352 (2017). https://doi.org/10.1134/S0361768817060020
11. Bondarev, A.E., et al.: Design of program tool BURGERS2 for hybrid finite-difference schemes optimization and visualization. Sci. Vis. **5**(1), 26–37 (2013)

"Why Did You Do That?"
Explaining Black Box Models with Inductive Synthesis

Görkem Paçacı[✉], David Johnson, Steve McKeever, and Andreas Hamfelt

Department of Informatics and Media, Uppsala University, Box 513,
75120 Uppsala, Sweden
{gorkem.pacaci,david.johnson,steve.mckeever,andreas.hamfelt}@im.uu.se

Abstract. By their nature, the composition of black box models is opaque. This makes the ability to generate explanations for the response to stimuli challenging. The importance of explaining black box models has become increasingly important given the prevalence of AI and ML systems and the need to build legal and regulatory frameworks around them. Such explanations can also increase trust in these uncertain systems. In our paper we present RICE, a method for generating explanations of the behaviour of black box models by (1) probing a model to extract model output examples using sensitivity analysis; (2) applying *CNPInduce*, a method for inductive logic program synthesis, to generate logic programs based on critical input-output pairs; and (3) interpreting the target program as a human-readable explanation. We demonstrate the application of our method by generating explanations of an artificial neural network trained to follow simple traffic rules in a hypothetical self-driving car simulation. We conclude with a discussion on the scalability and usability of our approach and its potential applications to explanation-critical scenarios.

Keywords: Artificial intelligence · Machine learning ·
Black box models · Explanation · Inductive logic · Program synthesis

1 Introduction

Much of the software industry is characterised by a clear separation of concerns, encouraging users not to be attentive to how a system functions. However with safety critical applications or those that must adhere to tight legal requirements, further assurances are needed. Rigorous software engineering, through formal development or comprehensive testing, has been very successful at supporting complex system development in many exacting fields. However we are now moving into an era where software is generated from examples.

The concept of the *black box* describes a system that can only be viewed with reference to its observed inputs and resulting outputs. Black box models can be validated using well-established software testing methods based on such observations [2]. This validates the intended behaviour of a black box model by providing

© Springer Nature Switzerland AG 2019
J. M. F. Rodrigues et al. (Eds.): ICCS 2019, LNCS 11540, pp. 334–345, 2019.
https://doi.org/10.1007/978-3-030-22750-0_27

us an *interpretation* of the mechanics of the model. Why the observed behaviour occurs is opaque and any *explanation* for its behaviour remains uncertain.

AI and Machine Learning (ML) models are often black boxes. While there are some ML models that exhibit simple relationships that can be straightforwardly explained (e.g. logistic regression) or that inherently display logical structure (e.g. decision trees), complex models that consist of artificial neural networks (ANNs), deep neural networks (DNNs) and random forests are non-intuitive and are not structured with human-readable logic. This makes us unable to posit the question to a model, *"Why did you do that?"*, to seek explanations of the decisions that ANN-based systems make [11].

To provide explanations of black box models is an intersection of technical and human challenges:

- How can we generate an explainable model of a black box?
- How can we ensure these explainable models are human-readable?
- How do we use these explanations of black boxes?

Contributions: In this paper we present a novel method, Rule Induction of CNP Explanations (RICE), that combines sensitivity analysis and logic program synthesis for generating explanation models of black boxes. The explanation models are not intended to be executable, but rather to be human-readable in their explanation of the executable model. Previous work on explaining predictions of ML classifiers, such as the LIME algorithm [21], which uses a local linear approximation of a model's behaviour to highlight what a model has picked out in the input data to make its classification. Such work focuses on explaining classifiers, while our method focuses on explaining the rules and logic learned in a black box.

The paper is structured as follows. In Sect. 2 we position our work within the field of black box interpretability. In Sect. 3 we provide an overview of the RICE methodology. In Sect. 4 we go through a simple traffic light use case, showing how the explainable model is produced using RICE. In Sect. 5 we reflect on our methodology and in Sect. 6 we summarise and list some areas of future work.

2 Related Work

Black box models, and in particular ML models, now exist in many critical domains such as medicine, finance, and even in military applications. There is a clear problem faced by the industrialization of using such models, where the human understanding of how they behave and work is difficult. Model *interpretability* sets out to produce metrics about models, for example, scoring the performance of predictions against ground truth, and to build trust in models by providing examples of when they perform correctly or otherwise [15]. When applied to black box models, for example in supervised learning models, such measures are focused on validating the learning of associations between factors. This of course does not guarantee that the model has learned causal relationships. Associations and causality describe the behaviour of models, however, they do not reveal an *explanation* for why an output was generated nor an explanation for how the black box works.

Interpretable models already exist. Of note, there are a number of classification models that are readily interpretable, for example decision trees, rule-based systems, and nearest-neighbour methods, each with varying levels of usability [7]. Decision trees generate a tree-like structure representing a series of tests on different features in a training dataset where leaf nodes represent various labeled classifications. Rules-based systems explicitly map an input to an action through some explicitly defined series of logical assertions (*if-then* rules). Nearest-neighbour algorithms qualify a classification based on the values of attributes in the immediate neighbourhood of some input.

In [10] the authors produced an "Open the black box taxonomy" to better understand different approaches to explaining black box models. At the top level there are two options: To formulate a black box explanation, and design a transparent box. The latter has been discussed in some depth in [22] where its author argues that explanations are often unreliable and misleading, and therefore black box models should not be used for critical systems. She cites several examples where lack of transparency and accountability of predictive models have had severe consequences. Counter-arguments to interpretable ML models include black boxes holding significant value and intellectual property for the model authors, and that interpretable models can be more costly to construct. They further define the *black box explanation problem* that requires an explanation model that is "globally interpretable"; that is to be able to mimic the behaviour of the black box *and* should be understandable by a human.

Approaches to generating human-readable explanation models include the Automatic Statistician, a system that discovers plausible models from data and automatically presents its findings as figures and as natural language [9,16], LIME (Local Interpretable Model-agnostic Explanations) [21], a method for isolating parts of a given input that contribute most to a classification, and QII (Quantitative Input Influence) [6], a technique for calculating influences of individual inputs or groups of inputs, each provides explanations as summaries of local causal phenomena. A recent survey lists various methods in explainable AI, and notes that due to rise in autonomous systems and complex models, there is even more need for interpretable models [4].

3   Explanations Through Inductive Synthesis

With RICE, we propose a method that combines sensitivity analysis and inductive logic programming. In contrast to LIME and QII, our approach seeks to generate globally interpretable model explanations for rule-based black box by synthesizing human-readable logic programs.

We start with a ML artefact after the learning step is accomplished. Sensitivity analysis consists of a series of methods developed to inspect the input/output

Algorithm 1. Rule Induced CNP Explanation (RICE)

1: **procedure** RICE(training data)
2:     $model \leftarrow$ train(training data)
3:     $observables \leftarrow probe$(model)
4:     $explanation \leftarrow CNPInduce$(observables)

---

relation of a function-like structure and is used in relation to ANNs to identify which input variables are relevant to the model, and which are not. This in turn is used to optimize the learning by reducing the unused dimensions in the training data [24]. We use monothetic analysis, that is, systematically trying different values for one of the inputs while keeping the values for other inputs constant [23]. This is undertaken for each possible assignment of inputs to the model, effectively calculating every partial differential of the model's function. The data points where the partial differential m is greater than an arbitrary e are noted as *critical examples*. Our method relies on the assumption that these extracted critical examples are a good estimate of the model's function. We call this stage of the method *probing*.

For synthesizing logic programs we rely on earlier work on inductive program synthesis, CNPInduce (Induction of CNP) [17,18]. The approach is a meta-interpretative form of inductive logic programming, where the CNPInduce synthesizer is written in Prolog. A meta-interpreter is a higher-level program that executes a program in a language and produces its input/output relation. In RICE, the meta-interpreter of a specially designed target language (CNP) is reversed, and this reversed meta-interpreter is executed with some known examples from its input/output relation thus producing all programs that would produce an input/output relation containing those examples [12,13]. There are other rule-extraction methods that are applicable but they are mostly domain-specific, and require a user strategy to investigate the model. The strength of CNPInduce is its domain-agnostic technique and its ability to synthesize recursive programs [17,18]. The ability to synthesize recursive programs makes it a good candidate to tackle models that deal with vector and matrix data such as audio and video signals. Even though inductive synthesis is a form of ML itself, it cannot perform well with noisy, high volume data; but it can produce a human-readable output. When coupled with a technique such as ANNs, which can deal with noisy input data, the combination gives a novel technique that has the efficiency of ANNs and the human-readable output of program synthesis.

To combine the two techniques, we take the critical examples extracted from the black box model and export them in a format that can be input to CNPInduce, the synthesizer to generate CNP programs. By using critical examples the synthesizer generates a program in the CNP language which satisfies these observables, which constitutes our explanation model. We call this stage of the method simply *synthesis*. Algorithm 1 illustrates this sequence of stages to generate explanations using RICE. Since the synthesized program and the ML artefact are semantically correlated through the critical examples, one can interpret the synthesized program instead of the ML artefact. This allows the inspection and validation of the ML model through the synthesized program. In RICE, this program is expressed in a language specifically developed for this purpose, called Combilog with Named Projection or CNP [17]. Its human-readability is improved compared to other forms of variable-free relational programming languages. Moreover, since it is a pure relational language the programs in this language can be automatically translated to more familiar languages, such as first-order logic, definite clauses [14], or even structured English [8]. We call this stage of the method *validation*. In the next section we demonstrate the RICE method through an example.

4   A Demonstration of Model Validation

To demonstrate our method we devised an experiment based on a simplification of a self-driving vehicle's decision system. We assume a case where through its sensors the vehicle's systems have identified the status of a traffic light (red, amber, green) and the distance to it. The vehicle needs to decide to accelerate or to brake continuously. Existing studies using ANNs on traffic lights focus on detection of the state of the traffic light under complex circumstances, and recent work demonstrate this is achievable in real-time [1]. An action stage which would naturally follow detection is usually left out of the model. For demonstration purposes of RICE we design a model that mimics the action stage. This allows us to focus on the core of the method instead of issues regarding optimization and scale. Let us assume this decision is left to an ANN, the model is trained with data from actual driving sessions, and it seems to function normally. In order to quantify the decision system's reliability, one may want to know precisely how it reacts to specific conditions. In the case of a trained model this is near-impossible since the model consists of binary data. Let us take it from here and show how our method can help us approach validation of this model. The implementation can be accessed in our GitHub repository[1]. We present this demonstration in multiple stages:

1. Training of the model
2. Extract examples from the model (probing)
3. Synthesize a human-readable program from examples (synthesis)
4. Inspect the synthesized program as an indirect representation of source artefact (validation).

4.1   Training of the Model

In our feature vector we encode each light as being on or off with 0 or 1, where position 0 corresponds to the red light, position 1 to the amber light and position 2 to the green light. Our final feature representing the distance of the car from the traffic lights is encoded as a floating point decimal between 0 and 1 representing

---

[1] https://github.com/UppsalaIM/rice/releases/tag/iccs19.

the range of distances from 0 m to 100 m, and is placed at position 3 of the feature vector. Our sample states are therefore:

```
[1, 0, 0, 0.25]  # red and the car is 25m from the lights
[0, 1, 0, 1.0]   # amber and the car is 100m from the lights
[0, 0, 1, 0.0]   # green and the car is 0m from the lights
```

A classification model is implemented and trained using TensorFlow[2]. The network has four layers, the input layer with 4 nodes, two hidden layers with 11 nodes each, and an output layer with 1 node. In the output layer, the value of the single output node represents the action; 1 for *go*, and 0 for *stop*.

Training data was procedurally generated in order to be able to embed algorithmic rules in the training data. While calculating the training labels a 2% noise was introduced. This is important to show that the method works with noisy data, as inductive synthesis methods by themselves would not be able to work with noisy examples. Training data is generated with random values for the state of the traffic light and the distance, within the valid space of input vectors. The labels are calculated according to the following scheme:

- If the light is **red**, and the distance to it is less than 60 m, then stop; otherwise go. For example, if input is $[1, 0, 0, 0.5]$, output is 0; and if the input is $[1, 0, 0, 0.9]$, output is 1.
- If the light is **amber**, and the distance to it is between 10 m and 80m, then stop; otherwise go. The minimum distance of 10 m is for avoiding stopping when it is too close. For example, if input is $[0, 1, 0, 0.2]$, output is 0.
- If the light is **green**, go. For example, if input is $[0, 0, 1, 0.1]$, output is 1.

In the training set there were 45,000 samples, and in the testing set there were 5,000. The data was generated as 50,000 samples and then split into training versus test data. During 10 training runs with 10 epochs each, accuracy up to 99.7% was measured, and it was consistently above 96%.

A model with 99% accuracy was saved to a HDF5 file which contains the model structure and the weights. In the following stage *probing*, we load this model from a file in order to show that the training data is separate from the next stage.

4.2 Probing: Extracting Examples from a Trained Model

The trained model from the previous stage is loaded from a file using TensorFlow with the probing stage implemented in Python. It is evaluated for each combination of the light state, for each possible distance (with a 1/100 granularity). The `distance_sweep` function below displays how the input *sweep* is generated, and the following two lines display how the model is evaluated for the given inputs. Here only the sweep for the red light state is displayed, where the outputs from the model are stored in the variable `red_o`.

---

[2] https://tensorflow.org.

```
def distance_sweep(lights_state):
    return np.array(np.array([lights_state + [x] for
                              x in np.linspace(0, 1, STEPS+1)]))

red_sweep = distance_sweep([1, 0, 0])
red_o = [np.rint(x[0]) for x in model.predict(red_sweep)]
```

Once these input/output sweeps are generated for every possible light state, the ones where the partial differential m is equal to 1 (modulo rounding). This gives all the critical points in the sweep where the output changes from 0 to 1, or 1 to 0. These are the points where the change in input results a dramatic change in the output. These input/output pairs are printed out by the probing stage in the input format of CNPInduce. Along with the input/output pairs, a distinct set of all the constants involved in the input/output pairs are extracted as well, and a synthesis job file is generated, as exemplified below:

```
jobValence([rd:in, am:in, gr:in, dist:in, go:out]).
%% Constants
jobConstant(0.0).
jobConstant(0.08).
jobConstant(0.09).
jobConstant(0.59).
jobConstant(0.6).
jobConstant(0.78).
jobConstant(0.79).
jobConstant(1.0).
%% Observables
jobObservable([rd:1.00, am:0.00, gr:0.00, dist:0.00, go:0.00], true).
jobObservable([rd:1.00, am:0.00, gr:0.00, dist:0.59, go:0.00], true).
jobObservable([rd:1.00, am:0.00, gr:0.00, dist:0.60, go:1.00], true).
jobObservable([rd:1.00, am:0.00, gr:0.00, dist:1.00, go:1.00], true).
jobObservable([rd:0.00, am:1.00, gr:0.00, dist:0.00, go:1.00], true).
jobObservable([rd:0.00, am:1.00, gr:0.00, dist:0.08, go:1.00], true).
jobObservable([rd:0.00, am:1.00, gr:0.00, dist:0.09, go:0.00], true).
jobObservable([rd:0.00, am:1.00, gr:0.00, dist:0.78, go:0.00], true).
jobObservable([rd:0.00, am:1.00, gr:0.00, dist:0.79, go:1.00], true).
jobObservable([rd:0.00, am:1.00, gr:0.00, dist:1.00, go:1.00], true).
jobObservable([rd:0.00, am:0.00, gr:1.00, dist:0.00, go:1.00], true).
jobObservable([rd:0.00, am:0.00, gr:1.00, dist:1.00, go:1.00], true).
```

Each jobObservable line in this synthesis file gives an input/output pair along with a Boolean flag true that indicates if the model *does* produce this output. Examples where the model *does not* produce an output can alternatively be included for the synthesis to be able to eliminate some of the possible programs.

4.3   Synthesis: Synthesizing a Human-Readable Program

In the synthesis stage, the synthesis job file prepared by the previous stage is run through CNPInduce by loading it through the `jobFromLocalFile` command. This command loads the constants and examples (observables) from the file, and initiates a synthesis job with these. In this case, the number of domains in the model goes beyond the synthesizer's current efficiency limits. Therefore we manipulate the synthesis job file to only include the observables where the state of the light is red, which reduces the number of domains to 3, as shown:

```
jobObservable([rd:1.00, dist:0.00, go:0.00], true).
jobObservable([rd:1.00, dist:0.59, go:0.00], true).
jobObservable([rd:1.00, dist:0.60, go:1.00], true).
jobObservable([rd:1.00, dist:1.00, go:1.00], true).
```

When the synthesizer is initiated with these observables, the first program it suggests is the following CNP program:

```
ande(const(rd,1.0),proj(iif(ltValue(a,0.6),0.0,1.0),[a->dist,o->go])))
```

This reveals that the synthesizer found a program that involves all of the given arguments (rd, dist, go). CNPInduce also guarantees that the programs it produces are terminating programs. This means that as long as the inputs are constant values the programs generated will terminate and assign an output value. When a program is suggested, the synthesizer gives the option to stop or to look for other programs. When instructed to look for other programs, the synthesizer continues to find others that are correct, but longer and more complex. These induced logic programs suggested by CNPInduce form our explanation models.

In the next section, let us discuss how CNP programs can be interpreted as explanation models to validate the black box model being studied.

4.4   Validation: Interpreting the Program

CNP is a pure relational language and therefore it may not be straightforward to those who are not familiar to such languages. In definite clause form it can be translated to:

$$\text{model}(\text{Rd}, \text{Dist}, Go) \leftarrow \text{Rd} = 1.0 \wedge \text{Rd} < 0.6 \wedge \text{Go} = 1.0$$
$$\text{model}(\text{Rd}, \text{Dist}, Go) \leftarrow \text{Rd} = 1.0 \wedge \neg(\text{Rd} < 0.6) \wedge \text{Go} = 0.0$$

Or in English it can be written as:

If the red light is 1.0, when the Dist is less than 0.6 assign Go to 1.0, otherwise assign Go to 0.0.

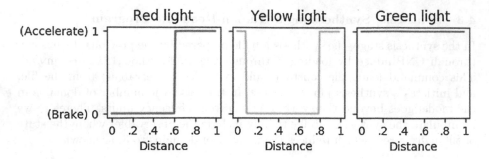

Fig. 1. Sensitivity analysis on the trained model, visualized. (Color figure online)

A visualization of the input/output states of the model can be seen in Fig. 1. The red line shows a shift at $Distance = 0.6$, which is in line with the CNP program. Visualizations like these are useful but not always applicable due to higher dimensions or discrete data. In these cases a program is much more helpful to make sense of.

Once an explanation model expressed in CNP is available for a given black box model, it can be automatically translated to logical rules or to natural language such as structured English. Once in this form, further automated assertions can be made on the explanation model's CNP program. For example, we can use automated testing and validation on the CNP explanations to reason about regulatory and legal compliance, where law may be represented logical axioms [20].

5   Discussion and Future Work

While we have demonstrated that RICE can be used to generate explanation models in this paper, we have identified a number of challenges that need to be addressed to improve its applicability, usability and performance.

A weakness of our method lies in indirection. The probing stage extracts examples and these examples are used for program synthesis, which introduces a level of indirection between the synthesized program (the explanation model) and the actual model (the black box being studied). One way we suggest approaching this is to exploit the fact that both models can be evaluated for any input. By generating randomized samples for inputs and comparing the difference of the explanation model's output to those of the black box model, a difference value can be calculated that would converge to 0 if they are semantically identical.

An opportunity that was discovered during the development of probing phase was that it was just as easy to identify erroneous behaviour in irregular system states (such as red light and green light being on at the same time), and specifically how the errors occur. As the probing stage indifferently extracts all critical data points, it identifies irregular states as well, which are reflected in the synthesized program. This is an opportunity that would be useful for identifying incorrect behaviour trained into ML models that might only use positive examples. Another point that needs improvement is the granularity of probing. In the demonstration presented in this paper, 1/100 granularity was used. Ideally the probing can be much more efficient using techniques such as logarithmic search and dynamic granularity. Using heuristics to make this stage more efficient is an acceptable approach since the final program can always be confirmed with randomized input/output pairs against the actual model.

An advantage of our generic probing stage is that the only manual requirement it needs is the valid value ranges of input/output variables, and a name for each so the synthesized program can refer to these names. The probing stage can be written in a completely generic fashion and since the synthesis stage is also domain-agnostic, the whole method can be considered generic.

We found that the success of synthesis is highly sensitive to the accuracy of the model itself. In the example, the model was measured to have over 99% accuracy. With models under 90% accuracy we did not find the synthesis successful in identifying a program.

Scaling of the RICE method is a significant issue. Currently, the CNPInduce program synthesizer is implemented as a Prolog program, which cannot be parallelized trivially. But the simplicity of the search algorithm behind CNPInduce [17] may allow for a MapReduce implementation that is more amenable to parallelization [5].

Finally, while the usability of CNP has been studied [17], its use as a target language for explanation models needs to be tested. Studies have shown that model interpretability (transparent models vs. explained black box models) and the styles of algorithmic explanations have varying degrees to which they build trust in ML models and there is currently no 'best' approach emerging [3, 19].

6   Conclusions

In this paper we have demonstrated a novel method of assigning meaning to opaque software artifacts whose specification is unknown. Our methodology called RICE, does not try to interfere, steer or model the black box directly. Instead we have adopted a three stage post-processing approach in which we use the artefact as an opaque prototype on which to extract meaning. Our first stage, called *probing*, inspects the generated artefact using sensitivity analysis to create a set of valid input/output pairs. From these pairs we have shown how to generate a logic-based explanation model, using *synthesis* as a second stage, that subsequently allows the artefact to be interpreted. This is achieved through a final stage, called *validation*, that produces a readable version.

This paper presents a proof of concept and much work remains to successfully scale the technique up to larger and more complex examples. Our interests lie not in the popular areas of machine learning, such as image recognition and consumer marketing, but in applications that require legal interpretation and certification.

Finally while the interoperability problem is usually presented as a concern for systems based on machine learning, the scope is far greater. Legacy systems for which there is no longer a readable source code, or agent based systems that have been finely calibrated are two further examples that would benefit from the ability to reverse-engineer an explainable interpretation that our methodology permits.

References

1. Behrendt, K., Novak, L.: A deep learning approach to traffic lights: detection, tracking, and classification. In: 2017 IEEE International Conference on Robotics and Automation (ICRA). IEEE (2017)
2. Beizer, B.: Black-Box Testing: Techniques for Functional Testing of Software and Systems. Wiley, New York (1995)
3. Binns, R., Van Kleek, M., Veale, M., Lyngs, U., Zhao, J., Shadbolt, N.: 'It's reducing a human being to a percentage': perceptions of justice in algorithmic decisions. In: Proceedings of the 2018 CHI Conference on Human Factors in Computing Systems (New York, NY, USA, 2018), CHI 2018, pp. 377:1–377:14. ACM (2018)
4. Biran, O., Cotton, C.: Explanation and justification in machine learning: a survey. In: IJCAI 2017 Workshop on Explainable AI (XAI), vol. 8 (2017)
5. Chu, C.-T., et al.: Map-reduce for machine learning on multicore. In: Advances in Neural Information Processing Systems, pp. 281–288 (2007)
6. Datta, A., Sen, S., Zick, Y.: Algorithmic transparency via quantitative input influence: theory and experiments with learning systems. In: 2016 IEEE Symposium on Security and Privacy (SP), pp. 598–617, May 2016
7. Freitas, A.A.: Comprehensible classification models - a position paper. ACM SIGKDD Explor. **15**(1), 1–10 (2013)
8. Fuchs, N.E., Schwitter, R.: Attempto controlled English (ACE). arXiv preprint cmp-lg/9603003 (1996)
9. Ghahramani, Z.: Probabilistic machine learning and artificial intelligence. Nature **521**(7553), 452–459 (2015)
10. Guidotti, R., Monreale, A., Ruggieri, S., Turini, F., Giannotti, F., Pedreschi, D.: A survey of methods for explaining black box models. ACM Comput. Surv. **51**(5), 93:1–93:42 (2018)
11. Gunning, D.: Explainable artificial intelligence (XAI), Program update November 2017 (2017)
12. Hamfelt, A., Nilsson, J.F.: Inductive metalogic programming. In: Proceedings of the Fourth International Workshop on Inductive Logic programming. Bad Honnef/Bonn GMD-Studien Nr. 237, pp. 85–96 (1994)
13. Hamfelt, A., Nilsson, J.F.: Inductive synthesis of logic programs by composition of combinatory program schemes. In: Flener, P. (ed.) LOPSTR 1998. LNCS, vol. 1559, pp. 143–158. Springer, Heidelberg (1999). https://doi.org/10.1007/3-540-48958-4_8
14. Hamfelt, A., Nilsson, J.F., Vitoria, A.: A combinatory form of pure logic programs and its compositional semantics (1998, Unpublished draft)
15. Lipton, Z.C.: The mythos of model interpretability. Commun. ACM **61**(10), 36–43 (2018)
16. Lloyd, J.R., Duvenaud, D., Grosse, R., Tenenbaum, J.B., Ghahramani, Z.: Automatic construction and natural-language description of nonparametric regression models. In: Proceedings of the Twenty-Eighth AAAI Conference on Artificial Intelligence, AAAI 2014, pp. 1242–1250. AAAI Press (2014)

17. Paçaci, G.: Representation of Compositional Relational Programs. Ph.D. thesis, Uppsala University, Information Systems (2017)
18. Paçaci, G., McKeever, S., Hamfelt, A.: Compositional relational programming with nominal projection and compositional synthesis. In: Proceedings of the PSI 2017: 11th Ershov Informatics Conference (2017)
19. Poursabzi-Sangdeh, F., Goldstein, D.G., Hofman, J.M., Vaughan, J.W., Wallach, H.M.: Manipulating and measuring model interpretability. CoRR abs/1802.07810 (2018)
20. Prakken, H., Sartor, G.: Law and logic: a review from an argumentation perspective. Artif. Intell. **227**, 214–245 (2015)
21. Ribeiro, M.T., Singh, S., Guestrin, C.: "Why should i trust you?": explaining the predictions of any classifier. In: Proceedings of the 22nd ACM SIGKDD International Conference on Knowledge Discovery and Data Mining (New York, NY, USA, 2016), KDD 2016, pp. 1135–1144. ACM (2016)
22. Rudin, C.: Please stop explaining black box models for high stakes decisions. CoRR abs/1811.10154 (2018)
23. ten Broeke, G., van Voorn, G., Ligtenberg, A.: Which sensitivity analysis method should i use for my agent-based model? J. Artif. Soc. Soc. Simul. **19**(1), 5 (2016)
24. Zurada, J.M., Malinowski, A., Cloete, I.: Sensitivity analysis for minimization of input data dimension for feedforward neural network. In: 1994 IEEE International Symposium on Circuits and Systems, ISCAS 1994, vol. 6, pp. 447–450. IEEE (1994)

Predictive Analytics with Factor Variance Association

Raul Ramirez-Velarde, Laura Hervert-Escobar[(✉)],
and Neil Hernandez-Gress

Tecnológico de Monterrey, Monterrey, NL, Mexico
`laura.hervert@tec.com`

Abstract. Organizations are turning to predictive analytics to help solve difficult problems and uncover new opportunities. Nowadays, the processes are saturated in data, which must be used properly to generate the necessary key information in the decision making process. Although there are several useful techniques to process and analyze data, the main value starts with the treatment of key factors. In this way, a Predictive Factor Variance Association (PFVA) is proposed to solve a multi-class classification problem. The methodology combines well-known machine learning techniques along with linear algebra and statistical models to provide the probability that a particular sample belongs to a class or not. It can also give predictions based on regression for quantitative dependent variables and carry-out clustering of samples. The main contribution of this research is its robustness to execute different processes simultaneously without fail as well as the accuracy of the results.

Keywords: PCA · Singular value decomposition · Machine learning

1 Introduction

Machine learning has recently risen as one the most groundbreaking technologies of our times. Many companies are using machine learning and other data science techniques to improve processes and resource allocation improving significantly business value. While many Machine Learning algorithms have been around for a long time, the ability to automatically apply complex mathematical calculations to big data over and over, faster and faster is a recent development. Some of the challenges faced by companies is the high-dimensional data. Dealing with many variables can help in certain situations but also may divert attention from what really matters. Therefore, it is important to check whether the dimensionality can be reduced while preserving the essential properties of the full data matrix. Principal Component Analysis (PCA) is a most widely used tool in exploratory data analysis and in machine learning for predictive models. Moreover, PCA is an unsupervised statistical technique used to examine the interrelations among a set of variables. It is also known as a general factor analysis where regression determines a line of best fit. The main idea of this procedure is to reduce dimensionality of a dataset while preserving as much 'variability' (statistical information) as possible. Preserving variability may sometimes implies the finding of new variables that are linear functions of those in the original dataset, that successively maximize variance and that are uncorrelated

J. M. F. Rodrigues et al. (Eds.): ICCS 2019, LNCS 11540, pp. 346–359, 2019.
https://doi.org/10.1007/978-3-030-22750-0_28

with each other. Finding such new variables, the principal components (PCs), reduces to solving an eigenvalue/eigenvector problem. Mathematical basis of the methodology were presented by Pearson [1] and Hotelling [2]. The advances in technology for data processing have generated a broad study of PCA method. Literature is vast, some of the most substancial books for understanding PCA are [3, 4], and for more specialized applications are [5, 6]. Also, the use of machine learning techniques for prediction models are widely study in literature, Hervert-Escobar [15] present a PCA method for reduction dimensionality combined with a multiple regression analysis to generate econometrics models. Then, such models are optimized to obtain the optimal price of a set of products. The model was tested in a case-study showing favorable results in profits for the pilot stores. Additionally, the literature provide articles that compile the advances and uses of prediction using machine learning prediction techniques. Sharma [7] present a survey of well-known efficient regression approach to predict the stock market price from stock market data based. Buskirk [16] provide a review of some commonly used concepts and terms associated with machine learning modeling and evaluation. The introduction also provides a description of the data set that was used as the common application example different machine learning methods. In this research we present a machine learning technique that given a matrix of explanatory variables value samples can produce predictions for quantitative dependent variables. It can also label the samples allocating one or several classes. This technique was tested to determine prices of articles for sale in convenience stores, to predict pollution contingencies, to determine leisure activities for tourists, to establish the probability of metastasis in cancer patients or the malignity of tumors in breast cancer patients, and other applications.

The procedure starts with a principal component analysis (PCA). This derives in several linear combinations of the original explanatory variables called principal components projections. These linear combinations are used to carry out a least-squares curve fitting to give predictions about variables of interest. Some applications implies a classification of the samples, mainly when the variables of interest have a definition of success or failure (within a thresh-old value). In such cases, point probabilities are computed for every value in the principal component projections. Then, the curve-fitting model will throw approximate probabilities for the value of a given combination of independent input variables.

The rest of the manuscript is organized as follows. The proposed methodology is presented in Sect. 2. The testing and analysis of the procedure are presented in Sect. 3. Finally conclusions are given in Sect. 4.

2   General Procedure

The general outline of the technique follows. We are given a matrix X with data in which rows are samples (m samples) with numerical values and columns are variables (n variables). The explanatory variables are separated into matrix E (o variables) and the dependent variables (p variables), or variables of interest, are separated into matrix D. Furthermore $n = o + p$. Also, from PCA [8] samples scores and variables loadings, dimension reduction and clustering can also be carried out. Figure 1 shows the steps of the procedure.

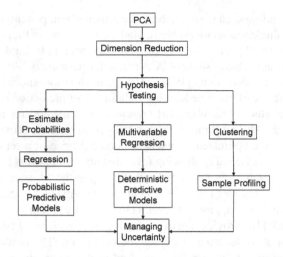

Fig. 1. Predictive factor variance association steps.

2.1   Dimension Reduction

The procedure starts by performing a PCA

1. Carry out PCA for matrix X and find matrices $X = PDQ^t$.
2. Let $F = XQ$ be the principal component scores.
3. Use square cosines and a clustering algorithm over the first two columns of F (First two components) to determine:
   a. Collinearity and dimension reduction. Explanatory variables that are grouped together are collinear and thus eliminate those columns from matrix X. Repeat PCA, go to 1.
   b. Causality relationships. If explanatory variables are grouped with variables of interest then that is evidence of a causality relationship.
   c. The square cosine is the square of the cosine of the angle between the vectors of variable loadings. If variables are close, the cosine will tend to one. If variables are separated, the cosine will approach zero. If $cosine^2(x) > 0.5$ then $cosine > \mp 0.7071$, thus associate variables with squared cosine greater than 0.5.
4. Eliminate variables of interest leaving only matrix E. Carry out PCA over E.
5. Select according to the following cases:
   a. **For predictive analytics** carryout curve fitting. Carry our curve fitting between explanatory variables and variables of interest clustered together in step 3.b. Discard any $R^2 < 0.5$. This indicates which explanatory variables influence the most the variable of interest.
   b. **For multi-label problem** estimate probabilities. Match F_i row by row with variable of interest j (assumed to be categorical with 0 or 1 as value), where i = 1 ... o and j = 1...p. Use window procedure to estimate probabilities.

6. Select according to the following cases:
    a. **For predictive analytics** match F_i row by row with variable of interest j, where $i = 1 \ldots o$ and $j = 1 \ldots p$. Carry out curve fitting. Discard all $R^2 < 0.5$. Each F_i is a linear combination of explanatory variables that potentially has enough information to give good predictions for variables of interest. Thus $\hat{D}_i = f_{ij}(F_j) = f_{ij}(q_1 X_1, q_2 X_2, \ldots, q_o X_o)$ where the coefficients q_i are determined by matrix Q.
    b. **For multi-label problem** carry out curve fitting. Discard all $R^2 < 0.5$. Each F_i is a linear combination of explanatory variables that potentially has enough information to give good predictions for variables of interest. Thus $\hat{D}_i = f_{ij}(F_j) = f_{ij}(q_1 X_1, q_2 X_2, \ldots, q_o X_o)$ where the coefficients q_i are determined by matrix Q. If $\hat{D}_i > 0.5$ then assume a result of 1, and 0 otherwise.
7. Select according to the following cases:
    a. **For predictive analytics** normalized new samples called X' can give predictions on variables, for $F' = X'Q$ and $\hat{D}'_i = f_{ij}\left(F'_j\right)$.
    b. **For multi-label problem** normalized new samples called X' can give predictions on variables, for $F' = X'Q$ and $\hat{D}'_i = f_{ij}\left(F'_j\right)$. Again, if $\hat{D}_i > 0.5$ then assume a result of 1, and 0 otherwise.

2.2 Sorting and Estimating Probabilities

When a multi-label classification solution is required, class probabilities per sample must be calculated [9]. We introduce a new measure called the success probability which is usually interpreted as $P[x_i] > l$, that is the probability that the variable will reach a value above a threshold (although there are many other types of categories). The horizontal axis for this estimate can be any of the explanatory variables and the vertical axis any of the variables of interest which would in this case be categorical with a value of 1 for success or yes, and 0 for failure or no. But the most effective models are usually derived from using an entire factor, that is the linear combinations of explanatory variables obtained as columns of matrix F to predict variables in matrix D.

To estimate this probability, for each factor and each variable of interest, an m-row vector of two entries is created. One entry is vx, or the explanatory linear combination from F (or just single variable) and the other entry is vy, the categorical variable used to estimate probabilities. This vector is sorted by vx. A sample of h items of is taken above and below a given value of vx is taken creating a sliding window of samples. In general, h data points from the plots are lost, $h/2$ points at the beginning of the plot an $h/2$ at the end. We use $h = 21$ and $h = 41$. The samples are taken and then the number of successes are counted. So vy_i is a measure which indicates that horizontal point vx_i is a success $(vy_i = 1)$ or not $(vy_i = 0)$. The probability of success $P_s(i)$ at horizontal point i is estimated by (Eq. 1):

$$P_s(i) = \frac{1}{h} \sum_{j=i-\frac{(h-1)}{2}}^{j=i+\frac{(h-1)}{2}} vy_j \tag{1}$$

In essence, we consider each success result in the sample as a Bernoulli experiment and use the sample mean of a collection of results of the variable of interest as an estimator for the probability, which is a well-known maximum likelihood estimator [10]. This is similar to Krichevsky-Trofimov [12] estimator, which is used as a conditional estimator of the outcome of the next Bernoulli experiment [11]. However, since our dataset is not a time series and our objective is to create a global model for the success probability as a function of some variable, we can use the sample mean of the current windows as seen in [12, 13].

The sample size used to estimate the success probability is small, so tests were made with different values of h such as 11, 41 and even 101 when possible, and it was found that the behaviour of the probability estimate was about the same, showing stability in the estimate. Naturally, the bigger the sample the more precise the estimate will be but the more variability will be lost to averaging.

3 Testing

In this section we present different data sets were the proposed methodology is tested.

3.1 Metastasis on Breast Cancer Patients

This data set consists of 2920 reference values for different genes of 78 breast cancer patients. These patients already have breast cancer tumors. We wish to determine the likelihood of tumors metastasizing. This is a binary classification problem. This data file was obtained from the UCI Machine Learning Repository (Dua and Karra [14]).

The probability estimate procedure with a windows of only 11 samples found a cubic equation with $R^2 = 0.831$. The fitted model with F17 is able to correctly determine if the sample metastasized or not on 75.64% of the samples. See Fig. 2 for a plot of probabilistic model.

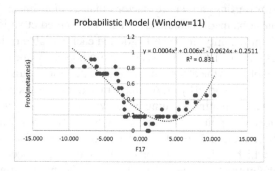

Fig. 2. Probabilistic model determining the probability of metastasis in breast cancer patients.

3.2 Malignity of Tumors in Breast Cancer Patients

This data set consists of 570 samples of information about cancer patients. The variables are information about the tumor such as radius, perimeter, area, texture, softness, concave points, concavity, symmetry and fractal dimension. The objective is to find if the tumor is malignant or benign. This is a binary classification problem. This data file was obtained from the UCI Machine Learning Repository (Dua and Karra [14]).

The fitted probabilistic model using F1 shows a coefficient determination of $R^2 = 0.9434$ and is able to determine correctly if the sample corresponds to a malignant tumor in 91.38% of the samples. See Fig. 3 for a plot of the probabilistic model.

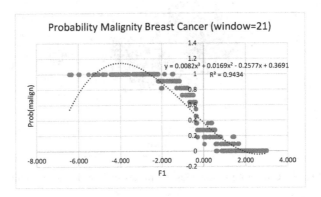

Fig. 3. Probabilistic model determining if a tumor is malignant in breast cancer patients.

Dimension Reduction. PCA on the data set also shows that some of the variables can be eliminated achieving dimension reduction. This is shown in Fig. 4, where some variables can be eliminated. This is very important. If a medical device is being designed to help people heal or prevent a health problem, the less variables that are necessary to control the cheaper the device will be and the more people it will help.

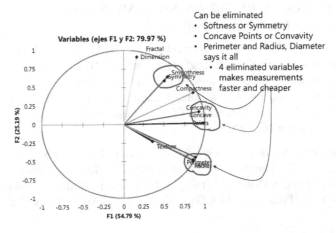

Fig. 4. Dimension reduction to determine of a tumor is malignant

Figure 5 shows how the clustering in the F1 vs F2 biplot of samples agrees with using F1 to separate malignant to benign tumors. As can be seen, there is a clear separation between samples marked as M or B along the F1 axis.

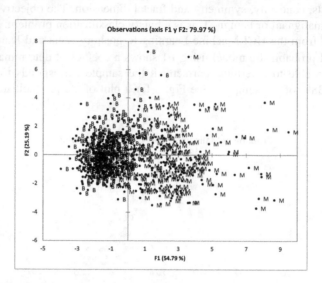

Fig. 5. Clustering of samples. M means malign, B means benign.

3.3   Travel and Activities

This data set corresponds to 87 samples containing answers to a survey. The provided information is the individual psych-socio-economic profile. This data is useful for traveling websites in order to determine what kind of activities a user would prefer. The information contains several features for the socio-economic profile, and for psychological profile.

The socio-economic features are gender, age, education, marital, employment, WEMWBS, PANAS: PA, PANAS: NA, SWLS, SWLS: group.

The features for psychological profile are:

- Personality traits: Extroversion, Agreeableness, Consciousness, Imagination, Neuroticism shown as the columns: 'Big5: Extraversion', 'Big5: Agreeableness', 'Big5: Conscientiousness', 'Big5: Neuroticism', 'Big5: Imagination'
- 3 orientations to happiness: Pleasure, Meaning, and Engagement. The columns: 'OTH: Pleasure', 'OTH: Meaning', 'OTH: Engagement'
- Fear of missing out (FoMO).

The types of activities are (Would the person like to do…?): 'Outdoors-n-Adventures', 'Tech', 'Family', 'Health-n-Wellness', 'Sports-n-Fitness', 'Learning', 'Photography', 'Food-n-Drink', 'Writing', 'Language-n- Culture', 'Music', 'Movements', 'LGBTQ', 'Film', 'Sci-Fi-n-Games', 'Beliefs', 'Arts', 'Book Clubs', 'Dance',

'Hobbies-n-Crafts', 'Fashion-n-Beauty', 'Social', 'Career-n-Business', 'Gardening-n-Outdoor housework', 'Cooking', 'Theatre, Show, Performance, Concerts', 'Drinking alcohol, Partying', 'Sex and Making Love'.

The data file was obtained from online surveys.

This is a multi-label problem. In Table 1 we show which activities are predicted by which factors with and the model R^2.

Table 1. Types of activities by component best able to predict it and model R^2.

Variable	F	R2	Variable	F	R2	Variable	F	R2
1	16	0.748	11	7	0.725	21	3	0.680
2	19	0.609	12	1	0.774	22	2	0.710
3	9	0.674	13	10	0.651	23	11	0.704
4	14	0.769	14	13	0.660	24	15	0.676
5	15	0.670	15	17	0.728	25	19	0.591
6	8	0.711	16	4	0.812	26	13	0.647
7	7	0.727	17	16	0.729	27	19	0.676
8	16	0.715	18	7	0.766	28	13	0.562
9	20	0.690	19	15	0.642			
10	4	0.945	20	7	0.693			

For example Activity 'Language-n-Culture' can be predicted by the linear combination of explanatory variables formed by factor 4, with and $R^2 = 0.945$. The model is able to determine accurately that a person will like this type of activity in 67.81% of the samples. Figure 6 shows the probabilistic model.

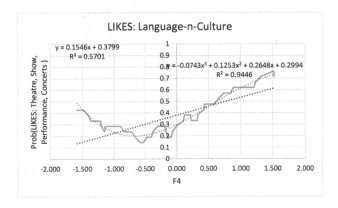

Fig. 6. Probabilistic model for language and culture related activities preferences.

3.4   Convenience Store Pricing Strategy

This is a data set of 6362 samples of sales tickets on different stores of several convenience stores. The data consists of type of merchandise, merchandize identification number, year and year-week, article cost, taxes, units bought, purchase amount, units with discount, amount of discount, articles prize, total profit margin and % of profit margin (from cost), minimum, medium and maximum weather temperatures, and amount of rain.

Principal component analysis determined that profit margin and % of profit margin as well as units sold is dependent only in the units sold with discount and the total amount of discount. See Fig. 7.

For the convenience store, one important goal is to sale articles with profit margin of at least 40%. PFVA determined that F10 of a PCA that excludes variables "profit margin" and "% of profit margin" gave the best prediction about attaining this goal with an $R^2 = 0.753$. The regression model is able to determine if the sales ticket will achieve the goal in 68.86% of the samples.

An important observation is that profit margin and the probability of achieving more than 40% of profit margin are opposite. This is because margin increases with the number of articles sold at discount but the probability of achieving more than 40% profit margin diminishes, as discount articles have lower price at same cost than regular priced articles. That can be seen in Figs. 8a and b for profit margin and Prob(%-margin > 40%) versus discount units sold (usdes) and Figs. 9a and b for profit margin and Prob(%margin > 40%) versus discount total amount (montodes), both for article identifies as sku = 455.

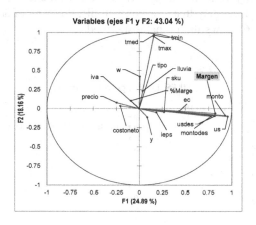

Fig. 7. Principal component analysis of convenience store sales

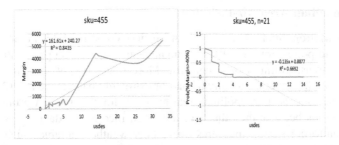

Fig. 8. a. Profit margin versus discount units sold for article sku = 455. b. Prob(%profit margin > 40%) versus discount units sold for article sku = 455

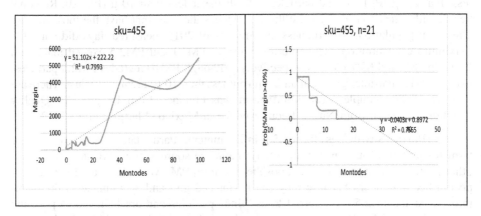

Fig. 9. a. Profit margin versus total discount amount for article sku = 455. b. Prob(%profit margin > 40%) versus total discount amount for article sku = 455

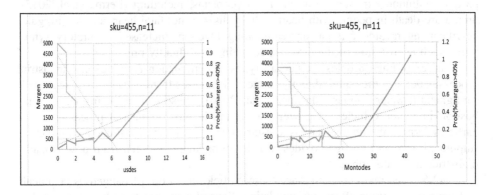

Fig. 10. a. Balance between profit margin versus total margin. Allow at most 5 units sold at discount per sale. b. Balance between profit margin versus total margin. Allow at most 16 currency units discount per sale

3.5 Metropolitan Area Air-Pollution

Monterrey's Metropolitan Area (MMA) in Mexico, consists of 15 municipalities with an approximate population of 4,406,054 inhabitants. As it so happens with other large metropolitan areas, pollution is a concern. More than 60% of days in a year have pollution levels that label air quality as bad or extremely bad. There are 12 monitoring stations throughout the metropolitan area, although for this paper we only used data from five stations because the other had too much incomplete data.

Each monitoring station measures every hour weather variables such as pressure (PRS), temperature (TOUT), relative humidity (HR), solar radiation (SR), rainfall (Rain), wind speed (WSR) and direction (WDV); and pollutants such as carbon oxide (COx), nitrogen oxide (NOx), sulfur oxide (SOx), ozone (O3), particles with diameter less than 2.5 μ (PM2.5) and particles with diameter less than 10 μ (PM10). Regional health authorities consider the last three pollutants the ones that have the most adverse effects on population. Nevertheless, in only 3% of 2015 measured days did daily O3 maximum concentration exceed norms, whereas PM10 and PM2.5 exceed maximum limits in 58% and 63% of days respectively. Although PM10 includes PM2.5 particles, usually PM10 particles greater than 2.5 μ (but less than 10 μ) are mainly dust. There are in total 14,374 samples. The information was obtained from Integral Air Quality Monitoring system from Nuevo Leon province in Mexico (Martinez et al. 2012).

PM2.5 (Particles with Less than 2.5 μ in Diameter). Particulate matter, or PM, is the term for particles found in the air. Many manmade sources emit PM directly or emit other pollutants that react in the atmosphere to form PM (Martinez et al. 2012). PM10 pose a health concern because they can be inhaled into and accumulate in the respiratory system. PM2.5 are referred to as "fine" particles and are believed to pose the greatest health risks. Because of their small size fine particles can lodge in the respiratory system, and may even reach the bloodstream. Exposure to such particles can affect respiratory and cardiovascular systems. Numerous scientific studies have linked particle pollution exposure to a variety of problems, including (Lerma et al. 2013): premature death in people with heart or lung disease, nonfatal heart attacks, irregular heartbeat, aggravated asthma, decreased lung function, increased respiratory symptoms, such as irritation of the airways, coughing or difficulty breathing.

Sources of fine particles include all types of combustion activities and industrial processes but also they are indirectly formed when gases from burning fuels react with sunlight and water vapour (Martinez et al. 2016). These can result from fuel combustion in motor vehicles, at power plants, and in other industrial processes. Most particles form in the atmosphere as a result of complex reactions of chemicals such as sulphur dioxide and nitrogen oxides, which are pollutants emitted from power plants, industries and automobiles.

PM2.5 is the main concern since these particles form out of dangerous chemicals that have very serious effects on population. Several measures have been proposed to reduce PM2.5 pollution such as restricting vehicle transit by license plate number (that is, some vehicles would not be allow to circulate some days of the week) and vehicle verification. As we can see in Figs. 11a and b, although ozone can be directly related to vehicle traffic (as represented by time of day) PM2.5 cannot (Carrera et al. 2015). Therefore, *none of those vehicle related preventive measures would work*.

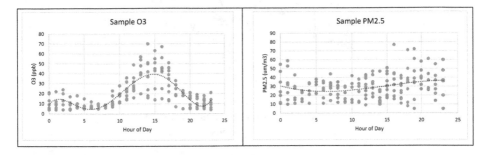

Fig. 11. a. Changes in O3 air concentration according to time of day. O3 is clearly related to daily traffic as represented by time of day. b. Changes in PM2.5 air concentration according to time of day. PM2.5 is clearly NOT related to daily traffic as represented by time of day

Also, weather affects pollution. Our aim is to, given a morning weather forecast, to determine if there will be a pollution contingency, that is, pollution levels higher than allowed by laws and regulations. This would allow regional governments to implement emergency measures to protect the population.

PCA shows (See Fig. 12) that O3 is mainly related to temperature, solar radiation and relative humidity, as it's well known that O3 levels are often high when the day is hot, sunny and dry. Nevertheless O3 levels can also be high at low temperatures.

For O3 we are able to create a deterministic model as shown in Fig. 13. F1 can predict the value of O3 with $R^2 = 0.7801$.

Figure 12 also shows that PM2.5 and PM10 are very weakly related to weather, as the biplot lines are orthogonal from any weather variable. Nevertheless, since we are mainly worried about the daily maximum levels, a new file was created with 365 samples containing daily maximum levels of PM2.5 and maximum, minimum and average daily readings for all weather variables. We found F10 from this new analysis to be best predictor of the class PM2.5 > 40.5. It has $R^2 = 0.9663$ and it's able to correctly predict bad air quality because of high PM2.5 levels in 65.7% of the samples, giving evidence that even though weather does influence pollution, for air suspended particulate neither weather nor traffic are the main determining factors. Since the number of samples was large, we tried windows from 11 samples to 101 samples, finding that the windows from 11 to 51 would have the same percentage of correct predictions, with that percentage dropping slightly at window size 101.

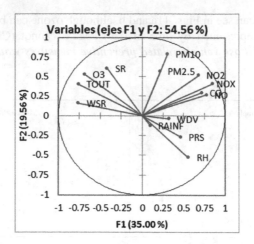

Fig. 12. PCA indicating that O3 is related to weather, specifically temperature, humidity and solar radiation. There seem to be no clear relationship between weather and particles.

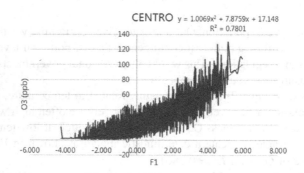

Fig. 13. Predictive model for O3 based on F1.

4 Conclusions

PFVA is a new technique that uses well-known concepts of matrix algebra and probability that is able to solve several problems of data analytics. It provides orthogonal linear combinations of the explanatory variables that can be used to predict the value of a variable of interest given a collection of values for explanatory variables; determine the classes to which a sample belong; and to classify samples into groups with common characteristics.

The main differentiator of this technique is the use of sample window to compute class probabilities that is has proven to be, even though the window can be really small, as seen in the examples presented, to be robust and accurate.

References

1. Pearson, K.: On lines and planes of closest fit to systems of points in space. Phil. Mag. **2**, 559–572 (1901). https://doi.org/10.1080/14786440109462720
2. Hotelling, H.: Analysis of a complex of statistical variables into principal components. J. Educ. Psychol. **24**, 417–441, 498–520 (1993) https://doi.org/10.1037/h0071325
3. Jackson, J.E.: A User's Guide to Principal Components. Wiley, New York (1991)
4. Jolliffe, I.T.: Principal Component Analysis, 2nd edn. Springer, New York (2002). https://doi.org/10.1007/b98835
5. Diamantaras, K.I., Kung, S.Y.: Principal Component Neural Networks: Theory and Applications. Wiley, New York (1996)
6. Flury, B.: Common Principal Components and Related Models. Wiley, New York (1988)
7. Sharma, A., Bhuriya, D., Singh, U.: Survey of stock market prediction using machine learning approach. In: 2017 International Conference of Electronics, Communication and Aerospace Technology (ICECA), Coimbatore, pp. 506–509 (2017)
8. Jolliffe, I.T.: Principal component analysis and factor analysis. In: Principal Component Analysis. Springer Series in Statistics. Springer, New York (2002). https://doi.org/10.1007/0-387-22440-8_7
9. Golub, G.H., Van Loan, C.F.: Matrix Computations, vol. 3. JHU Press, Baltimore (2012)
10. Wilks, S.S.: Mathematical Statistics. Wiley, New York (1962). ISBN 978-0471946502
11. Belyaev, E., Gilmutdinov, M., Turlikov, A.: Binary arithmetic coding system with adaptive probability estimation by "virtual sliding window." In: 2006 IEEE International Symposium on Consumer Electronics, pp. 1–5 (2006). https://doi.org/10.1109/ISCE.2006.1689517
12. Krichevsky, R., Trofimov, V.: The performance of universal encoding. IEEE Trans. Inf. Theory **27**(2), 199–207 (1981). https://doi.org/10.1109/tit.1981.1056331
13. Leighton, F., Rivest, R.: Estimating a probability using finite memory. IEEE Trans. Inf. Theory **32**(6), 733–742 (1986). https://doi.org/10.1109/tit.1986.1057250
14. Dua, D., Karra Taniskidou, E.: UCI Machine Learning Repository. University of California, School of Information and Computer Science, Irvine (2017). http://archive.ics.uci.edu/ml
15. Hervert-Escobar, L., Esquivel-Flores, O.A., Ramirez-Velarde, R.V.: Optimal pricing model based on reduction dimension: a case of study for convenience stores. Procedia Comput. Sci. **108**, 2079 (2017)
16. Buskirk, T.D., Kirchner, A., Eck, A., Signorino, C.S.: An introduction to machine learning methods for survey researchers. Surv. Pract. **11**, 1–10 (2018). https://doi.org/10.29115/SP-2018-0004

Track of Teaching Computational Science

Redesigning Interactive Educational Modules for Combinatorial Scientific Computing

M. Ali Rostami[1] and H. Martin Bücker[1,2](✉)

[1] Institute for Computer Science, Friedrich Schiller University Jena,
07737 Jena, Germany
martin.buecker@uni-jena.de
[2] Michael Stifel Center Jena for Data-driven and Simulation Science,
07737 Jena, Germany

Abstract. Combinatorial scientific computing refers to the field of using combinatorial algorithms to solve problems in computational science and data science. Teaching even elementary topics from this area is difficult because it involves bridging the gap between scientific computing and graph theory. Furthermore, it is often necessary to understand not only the methodologies from mathematics and computer science, but also from different scientific domains from which the underlying problems arise. To enrich the learning process in combinatorial scientific computing, we designed and implemented a set of interactive educational modules called EXPLAIN. The central idea behind EXPLAIN is its focus on describing the equivalence of a problem in terms of scientific computing and graph theory. That is, in EXPLAIN, the scientific computing problem and its graph theoretical representation are treated as two sides of the same coin. The process of solving a problem is interactively explored by visualizing transformations on an object from scientific computing, simultaneously, with the corresponding transformations on a suitably defined graph. We describe the redesign of the EXPLAIN software with an emphasis on integrating a domain-specific scripting language and a hierarchical visualization for recursively defined problems.

Keywords: Combinatorial scientific computing · Sparse matrices · Graph models · EXPLAIN · Recursive nested dissection

1 Teaching Combinatorial Scientific Computing

The term *Combinatorial Scientific Computing* (CSC) refers to an emerging interdisciplinary scientific discipline that is concerned with the development and use of combinatorial techniques for the solution of problems arising from computational and data science. A typical approach in CSC consists of first modeling a problem from scientific computing using graph theory, then solving the combinatorial problem, and finally analyzing the results. If the analysis shows that

© Springer Nature Switzerland AG 2019
J. M. F. Rodrigues et al. (Eds.): ICCS 2019, LNCS 11540, pp. 363–373, 2019.
https://doi.org/10.1007/978-3-030-22750-0_29

the results differ substantially from reality, this approach is refined in an cyclic fashion, starting again from the modeling phase. The reader is referred to [13] and [16] for more details on CSC. Compared to the more traditional disciplines of scientific computing and graph theory, the CSC discipline is young. Therefore, there is currently no textbook that supports students attending a course in CSC. Teaching in this area is also complicated by the fact that the fields of scientific computing and graph theory are typically not meaningfully interwoven through any curricular activities. Moreover, the problems addressed in computational and data science frequently have their roots in various different scientific disciplines requiring the students a comparatively high degree of skills in an interdisciplinary way of thinking.

To enrich, promote, and support teaching and learning effectiveness in the CSC domain, we designed, implemented, and used in the classroom a software called EXPLoring Algorithms INteractively (EXPLAIN). This web-based software is a collection of interactive educational modules for different problems arising from scientific computing that have strong connections to graph theory. More precisely, the current version of EXPLAIN includes the following modules: fill-in of Cholesky factorization [14], different variants of sparse matrix compression [4,5], parallel sparse matrix-vector multiplication [17], and one-way dissection ordering [11]. The primary goal of EXPLAIN is to illustrate the intimate connection between scientific computing and graph theory. The way to accomplish this aim is based on displaying a problem instance in terms of two equivalent representations simultaneously. A representation using an object from scientific computing is displayed immediately adjacent to an equivalent representation using a graph model. EXPLAIN allows students to explore the solution process of a CSC problem interactively. This process corresponds to a sequence of transformations applied to both these representations at the same time.

Although there are a lot of software tools for teaching graph-theoretical topics and graph algorithms, there is currently no other software than EXPLAIN with a focus on the CSC domain. However, we shortly mention some work that is indirectly related to that area. The software packages Gato/CATBox [19] and CABRI-Graph [6] have a focus on animation of graph algorithms. The growing field of data science and network science is a rich source of tools for visualizing, exploring, and analyzing graphs [1,2]. However, none of these graph visualization tools addresses aspects of scientific computing. On the other hand, software tools that teach scientific computing do not involve any aspects from graph theory; see for instance the interactive Java applets and the NCM software associated with the textbooks [12] and [15], respectively.

The novel contribution of the present article is the redesign of EXPLAIN in an attempt to address the issue of increasing the student engagement. More precisely, in the previous version of EXPLAIN, the algorithm, i.e., the transformation that is applied in each step of the solution process, was prescribed by EXPLAIN. The redesigned version now introduces a domain-specific scripting language enabling the student to develop a new algorithm or modify an existing algorithm. To illustrate the redesign we consider the extension of the module addressing the problem of finding a one-way dissection ordering [11].

The outline of this article is as follows. After sketching the well-known concept of a nested dissection ordering in Sect. 2, we start the presentation of the redesign in Sect. 3. Here, we introduce a new simple scripting language used to interactively develop and modify an algorithm within EXPLAIN. As a side effect, this language also allows to add the new feature of algorithm animation. In Sect. 4, we extend our previous work on the one-way dissection ordering [11] to a recursive approach. Finally, in Sect. 5, we describe the extension to larger problem sizes and summarize our findings in Sect. 6.

2   Nested Dissection Ordering

Runtime and storage requirements of direct methods for the solution of systems of linear equations with large sparse coefficient matrices are heavily influenced by the ordering of rows and columns [9]. An ordering with favorable characteristics is a dissection ordering. Let A be a sparse symmetric positive definite coefficient matrix and let P denote a permutation matrix, then finding a dissection ordering is given as the following problem:

Problem 1 (One-Way Dissection Ordering). Given a sparse symmetric positive definite matrix A, find a symmetric permutation

$$P^T AP = \begin{bmatrix} A_1 & 0 & B_1^T \\ 0 & A_2 & B_2^T \\ B_1 & B_2 & C \end{bmatrix} \tag{1}$$

of A such that the size of the last main diagonal (quadratic) block, C, is minimized while the sizes of the two remaining diagonal (quadratic) blocks, A_1 and A_2, are balanced.

The connection of this scientific computing problem to an equivalent graph problem is illustrated by an interactive educational module in EXPLAIN [11]. Let V be a set of vertices and E be a set of edges and let $G = (V, E)$ denote an undirected graph that is associated with the matrix A in the following sense. The adjacency matrix of G has the same nonzero pattern as A. The graph problem that is equivalent to Problem 1 is then as follows:

Problem 2 (Small Vertex Separator). Given the graph G associated with a sparse matrix A, find a disjoint decomposition of the vertices $V = V_1 \cup V_2 \cup S$ with a vertex separator S such that the size of the vertex separator, $|S|$, is minimized while the sizes of the two remaining components, $|V_1|$ and $|V_2|$, are balanced.

In EXPLAIN, a student can determine a vertex separator S by interactively selecting a sequence of vertices from G. Whenever a vertex is selected the layout of the graph and the layout of the matrix are changed simultaneously as follows: (i) The edges incident to that vertex are eliminated from the graph. (ii) The row and column corresponding to that vertex are moved to the last position in the matrix. (iii) The color of that vertex and of its corresponding row and column are changed to the color denoting membership to S.

In an EXPLAIN module, the process of interactively solving a problem instance, such as determining a one-way dissection ordering, is referred to as a *round*. Thus, in each round of this module, the determined ordering produces three sets of vertices, S, V_1 and V_2, that correspond to the matrix blocks C, A_1 and A_2, respectively. The size of these entities characterize the quality of the ordering. To encourage improved participation, EXPLAIN integrates elements of gamification by allowing the student to opt for a sequence of rounds, attempting to improve the quality of an ordering. The history of these rounds is displayed in a *score diagram*. Two excerpts of EXPLAIN's visualization of a problem instance are depicted in Fig. 1. This figure shows the situation when round 4 is completed. The set V_1 and its corresponding block A_1 are printed in blue, while V_2 and A_2 are drawn in red. The color orange is used for the vertex separator S that corresponds to the block C. The score diagram shows that in the first three rounds, the balancing was not achieved. It also indicates that the size of the vertex separator is similar in all rounds.

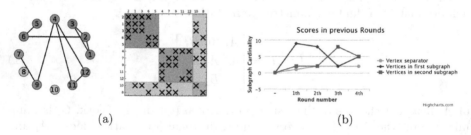

(a)   (b)

Fig. 1. (a) Graph and matrix view. (b) Score diagram. Both figures taken from [11]. (Color figure online)

It is straightforward to apply one-way dissection recursively as follows:

Problem 3 (Nested Dissection Ordering). Given a sparse symmetric positive definite matrix A, find a symmetric permutation $P^T A P$ of A by applying a one-way dissection ordering recursively to the blocks A_1 and A_2 of the matrix (1) such that

- the sum of the sizes of the last diagonal blocks that occur at each of L levels of the recursion is minimized
- while the sizes of all remaining diagonal blocks that occur at these levels of the recursion are balanced.

The corresponding graph problem is then formulated as follows:

Problem 4. Given the graph G associated with a sparse matrix A, find a disjoint decomposition of the vertices V by finding a small vertex separator recursively for the components V_1 and V_2 of the graph such that

- the sum of the sizes of the vertex separators that occur at each of the L levels is minimized
- while the sizes of all the remaining components that occur at these levels of the recursion are balanced.

The module described in [11] illustrates one-way dissection, i.e., the connection between Problems 1 and 2. However, it is not capable of illustrating nested dissection, i.e., the connection between Problems 3 and 4.

3   Redesigning EXPLAIN

When using the interactive educational modules in classroom students asked how, given a certain CSC problem, they can modify an algorithm that is implemented in the module by their own ideas. The previous implementation of EXPLAIN was not designed to allow for interactive modifications of the algorithm. However, students are better integrated into the learning process if they are themselves more profoundly involved in the algorithm design. In particular, a true motivation for a student is to browse and edit the actual algorithm in a pseudo-code form. That is, on the one hand, we do want to engage the students in implementing an algorithm. On the other hand, we do not want to involve students in the details of any programming language which would prevent them from concentrating on the specific CSC topic proposed by the corresponding module. Our new design therefore consists of three elements: a new scripting language, an interactive online editor, and an algorithm animation.

The new scripting language is designed to be user-friendly and is based on the existing JavaScript code in the previous version of EXPLAIN. In an abstract [18], we briefly discuss a preliminary version of this scripting language and give more detail here. In this scripting language, any algorithm is a single (finite) loop over a set of steps. This assumption corresponds to EXPLAIN's concept of a round which solves a CSC problem by clicking on the vertices in the graph view step by step. Further assumptions are as follows: (i) Program variables are divided into three classes: global predefined, global user-defined, and local user-defined variables. Examples of global predefined variables are those variables that store information of the current matrix and graph. (ii) The algorithm consists of a set of simple mathematical operations on a graph and a matrix as well as some predefined visualization functionalities. Table 1 shows a subset of these operations. (iii) The vertices are unique integer values starting from 0. (iv) Arrays are similar to JavaScript arrays.

For this language, we provide an online editor that consists of two sections, denoted by "Globals" and "Code" in Fig. 2. These sections are used to define global variables and implement a step of the algorithm, respectively. Given a global array

Table 1. Subset of operations supported by the new scripting language.

Function interface	Operation
colorVertices(vs, cs)	Color vertices (vs) with colors (cs)
getColors(vs)	Extract colors of vertices (vs)
colorColumn(col, c)	Color column (col) with color (c)
colorRow(row, c)	Color row (row) with color (c)
neighbors(v)	Extract neighbors of vertex (v)
updateMatrix(ord)	Update matrix with ordering (ord)
drawSubMat(c0, c1, r0, r1, c)	Color submatrix (r0:r1, c0:c1) with color (c)

ordering, the algorithm specified in the Code section loops over the values of this array. The array ordering can be either selected from a predefined list of orderings or defined interactively in the Globals section by the student or the teacher.

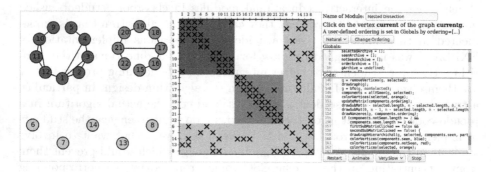

Fig. 2. The online editor is visualized next to the graph and matrix view.

Finally, EXPLAIN supports an animation of the algorithm. That is, the steps of the algorithm are executed one after another, while the current active statement is highlighted and the matrix as well as the graph view change accordingly. For instance, if the active statement carries out a coloring, the colors of the corresponding objects in the matrix and graph view are updated. The student can choose the speed of the algorithm animation. It is also possible to stop this animation at any time or to restart the animation.

To illustrate the use of the scripting language, consider Fig. 3. Here, a part of the nested dissection algorithm is implemented. More precisely, it shows a single step of the algorithm acting on a graph g. From the set of all selected vertices selected, all incident edges are removed and the updated graph is drawn in the graph view. A breadth-first search of the graph g, implemented in the function bfs, followed by the function allVSeen returns the components of g. Then, the matrix view is updated using the function drawSubMat and the nonzeros are drawn in the matrix view. If the determined components

```
var n = numOfVertices(g);
g = removeEdges(g, selected);
drawGraph(g);
g = bfs(g, nonSelected);
components = allVSeen(g, selected);
colorVertices(selected, orange);
drawSubMat(n - selected.length, n - selected.length, 0, n - 1, orange);
drawSubMat(0, n - 1, n - selected.length, n - selected.length, orange);
updateMatrix(components.order);
drawNonzeros(g, components.order);
var seen = components.seen.length;
var nseen = components.notSeen.length;
if(seen < 2 && nseen < 2) return;
drawGraphHierarchichal(g, selected, components.seen, components.notSeen);
colorVertices(components.seen, blue);
colorVertices(components.notSeen, red);
colorVertices(selected, orange);
drawSubMat(0, seen - 1, 0, seen - 1, blue);
drawSubMat(seen, seen + nseen - 1, seen, seen + nseen - 1, red);
```

Fig. 3. Excerpt of the algorithm implementing a step of the nested dissection ordering.

are large enough the algorithm switches to a hierarchical view by the function `drawGraphHierarchichal`. Finally, the components are colored in the graph view and the corresponding matrix blocks are also colored with the same color.

4   Hierarchical Layout of Graph and Matrix View

The JavaScript library D3 (Data-Driven Documents) [3] constitutes not only an important building block for the design of the scripting language introduced in the preceding section, but is also useful to implement visualizations of hierarchical structures. Recall from Sect. 2 that the previous implementation of EXPLAIN is capable of illustrating the connection between the graph and matrix representations of a one-way dissection ordering, cf. Fig. 1, but fails to do so for a nested dissection ordering. To illustrate the recursive application of the dissection idea, the size of the graph/matrix needs to be somewhat larger.

As an example, consider the graph and the matrix depicted in Fig. 4(a). After interactively choosing the vertices 6, 7, 8 and 14 as the vertex separator S_1, the student will get from EXPLAIN a visualization of a one-way dissection depicted in Fig. 4(b). Here, the rows and columns that correspond to the separator S_1 are numbered last. This one-way dissection produces not only S_1 but also the components V_1 and V_2 shown in blue and red, respectively. However, in contrast to the previous implementation, the student can now interactively choose another two vertex separators S_2 and S_3 associated with V_1 and V_2. In Fig. 5(a), the graph and matrix view are given for the separators $S_1 = \{6, 7, 8, 14\}$, $S_2 = \{1, 13\}$ and $S_3 = \{17, 18\}$. In the graph view, these separators are drawn at the bottom of the graph. The separator S_1 from the first level of recursion is visually divided from the separators S_2 and S_3 of the second recursion level by a dashed line. The remaining four components of the graph are shown in circular layout using different colors at the top of this graph view.

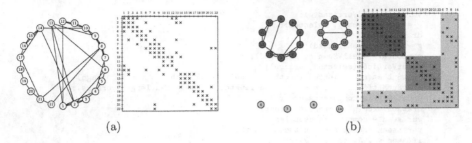

Fig. 4. (a) Initial graph and matrix view. (b) View after $L = 1$ level of recursion. (Color figure online)

The corresponding score diagram is shown in Fig. 5(b). Here, the size of the vertex separator as well as the sizes of the four remaining components are visualized. The separator size is the sum of the sizes of the vertex separators of all recursion levels. The figure illustrates these quantities for a sequence of four rounds. The line chart shows the history of the size of the vertex separator. For each round, a four-bar chart grouped together presents the four sizes of the remaining components. The colors used in the graph view correspond to the matrix view. The goal is to minimize the size of the vertex separator while balancing the sizes of the remaining components. The hierarchical structure of a nested dissection ordering is also visible in the matrix view.

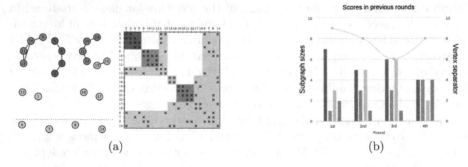

Fig. 5. (a) Graph and matrix view after $L = 2$ levels of recursion. (b) Score diagram with four remaining components, i.e., $L = 2$ recursion levels.

5   Larger Problem Instances

While the basic idea of a specific CSC problem in EXPLAIN is best taught by starting with a small problem instance, the learning process typically also benefits from scaling the problem to larger problem sizes. Increasing the problem size in D3 is straightforward for the matrix view, though some modifications are in order, including hiding the majority of row/column indices on the top and on the left of the matrix as well as replacing the cross symbol denoting a nonzero entry by a small point. However, it is necessary to withdraw from D3 in the graph view. Since visualization in D3 is based on the Scalable Vector Graphics (SVG) technology, the performance is insufficient for displaying larger graphs. The redesign therefore switches to Cytoscape.js [10] which is based on the more efficient canvas element in HTML5 if the problem size is above some threshold. The graph is visualized with the compound layout based on [8].

In Fig. 6(b) the layout of the 65×65 matrix *GD96_c* with 250 nonzeros taken from the SuiteSparse Matrix Collection [7] is shown. As you can see, although the graph is relatively big, the student can find a vertex separator in this layout easily. In Fig. 6(b), the result after selecting a vertex separator is given.

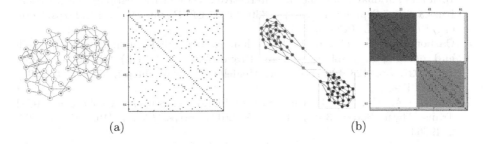

(a)   (b)

Fig. 6. (a) Initial view of *GD96_c*. (b) View after $L = 1$ level of recursion.

6   Concluding Remarks

When teaching elements of combinatorial scientific computing one of the fundamental concepts is the intimate connection between scientific computing and combinatorial analysis. The EXPLAIN software is a web-based set of interactive educational modules designed to illustrate the underlying correspondences between an object from scientific computing and a suitable graph model. Its goal is to enrich, improve, and accelerate the learning process by displaying these two entities next to each other. When an interactive transformation or an iterative algorithm is applied to a problem instance, the resulting modifications on these two entities are dynamically visualized. The redesign of EXPLAIN includes the integration of a simple scripting language enabling the student to solve a problem from combinatorial scientific computing by developing an algorithm using an online editor.

References

1. Auber, D., et al.: TULIP 5. In: Alhajj, R., Rokne, J. (eds.) Encyclopedia of Social Network Analysis and Mining, pp. 1–28. Springer, New York (2017). https://doi.org/10.1007/978-1-4614-7163-9_315-1

2. Bastian, M., Heymann, S., Jacomy, M.: Gephi: an open source software for exploring and manipulating networks. In: Third International AAAI Conference on Weblogs and Social Media, pp. 361–362 (2009)
3. Bostock, M., Ogievetsky, V., Heer, J.: D^3: data-driven documents. IEEE Trans. Vis. Comput. Graph. **17**(12), 2301–2309 (2011). https://doi.org/10.1109/TVCG.2011.185
4. Bücker, H.M., Rostami, M.A.: Interactively exploring the connection between bidirectional compression and star bicoloring. In: Koziel, S., et al. (eds.) International Conference on Computational Science, ICCS 2015, Reykjavík, Iceland, 1–3 June 2015, vol. 51, pp. 1917–1926. Elsevier (2015). https://doi.org/10.1016/j.procs.2015.05.456
5. Bücker, H.M., Rostami, M.A., Lülfesmann, M.: An interactive educational module illustrating sparse matrix compression via graph coloring. In: Proceedings of the 16th International Conference on Interactive Collaborative Learning (ICL), Kazan, Russia, 25–27 September 2013, pp. 330–335. IEEE, Piscataway (2013). https://doi.org/10.1109/ICL.2013.6644591
6. Carbonneaux, Y., Laborde, J.-M., Madani, R.M.: CABRI-graph: a tool for research and teaching in graph theory. In: Brandenburg, F.J. (ed.) GD 1995. LNCS, vol. 1027, pp. 123–126. Springer, Heidelberg (1996). https://doi.org/10.1007/BFb0021796
7. Davis, T.A., Hu, Y.: The University of Florida sparse matrix collection. ACM Trans. Math. Softw. **38**(1), 1:1–1:25 (2011). https://doi.org/10.1145/2049662.2049663
8. Dogrusoz, U., Giral, E., Cetintas, A., Civril, A., Demir, E.: A layout algorithm for undirected compound graphs. Inf. Sci. **179**(7), 980–994 (2009). https://doi.org/10.1016/j.ins.2008.11.017
9. Duff, I., Erisman, A., Reid, J.: Direct Methods for Sparse Matrices. Numerical Mathematics and Scientific Computation, 2nd edn. Oxford University Press, Oxford (2017)
10. Franz, M., Lopes, C.T., Huck, G., Dong, Y., Sumer, O., Bader, G.D.: Cytoscape.js: a graph theory library for visualisation and analysis. Bioinformatics **32**(2), 309–311 (2016). https://doi.org/10.1093/bioinformatics/btv557
11. Bücker, H.M., Rostami, M.A.: Interactively exploring the connection between nested dissection orderings for parallel Cholesky factorization and vertex separators. In: IEEE 28th International Parallel and Distributed Processing Symposium, IPDPS 2014 Workshops, Phoenix, Arizona, USA, 19–23 May 2014, pp. 1122–1129. IEEE Computer Society, Los Alamitos (2014). https://doi.org/10.1109/IPDPSW.2014.125
12. Heath, M.T.: Scientific Computing: An Introductory Survey. Classics in Applied Mathematics, 2nd edn., vol. 80. SIAM (2018)
13. Hendrickson, B., Pothen, A.: Combinatorial scientific computing: the enabling power of discrete algorithms in computational science. In: Daydé, M., Palma, J.M.L.M., Coutinho, Á.L.G.A., Pacitti, E., Lopes, J.C. (eds.) VECPAR 2006. LNCS, vol. 4395, pp. 260–280. Springer, Heidelberg (2007). https://doi.org/10.1007/978-3-540-71351-7_21
14. Lülfesmann, M., Leßenich, S.R., Bücker, H.M.: Interactively exploring elimination orderings in symbolic sparse Cholesky factorization. In: International Conference on Computational Science, ICCS 2010, vol. 1, no. 1, pp. 867–874. Elsevier (2010). https://doi.org/10.1016/j.procs.2010.04.095
15. Moler, C.B.: Numerical Computing with MATLAB. SIAM, Philadelphia (2004)

16. Naumann, U., Schenk, O. (eds.): Combinatorial Scientific Computing. Computational Science Series. Chapman and Hall/CRC, Boca Raton (2012)
17. Rostami, M.A., Bücker, H.M.: An educational module illustrating how sparse matrix-vector multiplication on parallel processors connects to graph partitioning. In: Hunold, S., et al. (eds.) Euro-Par 2015. LNCS, vol. 9523, pp. 135–146. Springer, Cham (2015). https://doi.org/10.1007/978-3-319-27308-2_12
18. Rostami, M.A., Bücker, H.M.: An online scripting language for teaching combinatorial scientific computing. In: Amme, W., Heinze, T.S. (eds.) Programmiersprachen und Grundlagen der Programmierung, 19. Kolloquium, KPS 2017, Weimar, 25–27 September 2015, Tagungsband, pp. 83–85. Jenaer Schriften zur Mathematik und Informatik Math/Inf/02/2017, Friedrich-Schiller-Universität Jena, Jena (2017). Extended abstract. http://www.kps2017.uni-jena.de
19. Schliep, A., Hochstättler, W.: Developing Gato and CATBox with Python: teaching graph algorithms through visualization and experimentation. In: Borwein, J., Morales, M.H., Rodrigues, J.F., Polthier, K. (eds.) Multimedia Tools for Communicating Mathematics, pp. 291–310. Springer, Heidelberg (2002). https://doi.org/10.1007/978-3-642-56240-2_18

A Learner-Centered Approach
to Teaching Computational Modeling,
Data Analysis, and Programming

Devin Silvia[1]([✉]), Brian O'Shea[1], and Brian Danielak[2]

[1] Department of Computational Mathematics, Science, and Engineering,
Michigan State University, East Lansing, MI 48824, USA
{dsilvia,oshea}@msu.edu
[2] Fullstack Academy, New York, NY 10004, USA

Abstract. One of the core missions of Michigan State University's new Department of Computational Mathematics, Science, and Engineering is to provide education in computational modeling and data science to MSU's undergraduate and graduate students. In this paper, we describe our creation of CMSE 201, "Introduction to Computational Modeling and Data Analysis," which is intended to be a standalone course teaching students core concepts in data analysis, data visualization, and computational modeling. More broadly, we discuss the education-research-based rationale behind the "flipped classroom" instructional model that we have chosen to use in CMSE 201, which has also informed the design of other courses taught in the department. We also explain the course's design principles and implementation.

Keywords: Computational science education · Data analysis · Modeling

1 Introduction

In the last few decades, the use of data analysis and computational modeling has become critical to success in a wide variety of careers and the need for these skills is growing. As an example, McKinsey & Company recently released a report arguing that the need for data scientists – that is, workers who can successfully analyze, model, and interpret data, and use it to inform critical business decisions – is going to grow rapidly [8]. A wide range of industries now use computational modeling and data analytics to inform many aspects of their day-to-day practices, encompassing a range of activities that include product design, optimizing manufacturing processes, hiring decisions, and choosing advertising targets. The trend is clear, as are the implications for new employees – being fluent in the tools used to work with data and computational models, as well as being intelligent consumers of the products of these tools, will be a critical and high-demand skillset. Unfortunately, many institutions are relatively slow to provide students the type of computational training that addresses these needs, as indicated by the lack of programs aimed specifically at these areas. And, while not necessarily representative

© Springer Nature Switzerland AG 2019
J. M. F. Rodrigues et al. (Eds.): ICCS 2019, LNCS 11540, pp. 374–388, 2019.
https://doi.org/10.1007/978-3-030-22750-0_30

of all Computer Science departments, the courses offered by the Computer Science department at Michigan State University (MSU) are often heavily oversubscribed and typically do not encompass the range of data analysis and modeling skills that employers desire.

In 2015, MSU formed the Department of Computational Mathematics, Science, and Engineering (CMSE) to address the growing prevalence of computational and data science in academia and industry. The two core missions of the department are to perform cutting-edge, computationally-focused research and to provide a wide range of educational opportunities in computational and data science to MSU's undergraduate and graduate student populations. One particularly important population identified to be under-served in this regard by both the faculty and university administration is undergraduates in the natural and social sciences, as the primary avenues for computational training were computer science courses that do not address the needs outlined above.

In this paper, we describe the creation of CMSE 201, "Introduction to Computational Modeling and Data Analysis." This was the first course created by CMSE and is intended to provide a wide range of undergraduate students with the analytical and technical skills necessary to effectively work with data and computational models. In Sect. 2 we describe the design and implementation of this course. In Sect. 3, we summarize and discuss the limitations of this work as well as future plans, including challenges and opportunities.

2   Course Design and Implementation

2.1   Design Process

The broad goals of "Introduction to Computational Modeling and Data Analysis" (hereafter referred to as 'ICM') come from the process of creating the Department of CMSE. Prior to the MSU faculty endorsing the proposal to create a new department, a series of informal discussions were held with the deans, department chairs, and interested faculty in most of the colleges on campus. As a part of these meetings, faculty were asked "what sort of computational skills would you like your undergraduate and graduate students to have?" Disciplinary jargon aside, the answers were remarkably uniform: faculty would like students in their upper-division classes and in their research groups to be able to take a dataset and manipulate, analyze, and visualize it to extract usable information from it. Similarly, they would like students to be able to create simple models that capture the salient features of a system, and both quantitatively and qualitatively compare those models with the data that they have analyzed. The broad consensus among faculty was that there was not an existing undergraduate course at MSU that met most or all of these goals. A similar set of discussions took place with representatives from companies in a wide range of industries at several conferences (these companies were primarily in the manufacturing and high-tech sectors, and had a significant presence in the Midwest). These discussions had similar outcomes, though a much greater emphasis was put on the ability to *communicate* the process and outcomes of data analysis and modeling to co-workers and supervisors (which is in line with more systematic surveys of the skills desired by employers) [12,13]. The skills identified in all of these discussions are the core of the high-level course learning objectives described below.

An additional goal of the design process was to create a course that exemplified the best practices as set forth in the undergraduate research literature. Our ability to do this was facilitated by several important factors. First, there are a large number of MSU faculty members who participate in disciplinary-based education research and/or successfully implement its outcomes in their classrooms, and thus there is strong support from both fellow faculty and university administration for the creation of such courses. Second, MSU has invested heavily in the creation of "REAL" (Rooms for Engaged and Active Learning) classrooms[1], which are specifically designed to facilitate student and faculty engagement and provide opportunities for innovative learning experiences. Finally, a key factor is that the course was created from a "blank slate," and was not a required course in any major or minor at the time of its creation[2]. This *ab initio* property of the course design meant that there were few institutional preconceptions about what should be taught or how it should be done.

Informed by the goals described above, we spent the Fall 2015 semester designing a new course and creating course content. Using the principles of *backwards course design* [18], we created both a high-level set of course learning goals (Sect. 2.2) and a much more finely-grained set of intended learning and content objectives (Sect. 2.4). We solicited feedback on these goals and objectives from many faculty and students, and the goals were iteratively tuned and improved. After solidifying our curriculum, we decided that we would evaluate student success based on their ability to work with data and implement and modify models. We decided on an assessment structure combining both formative assessment [2] (pre-class assignments and in-class group problem-solving activities) and summative assessment (homework assignments, exams, and semester projects).

2.2   Learning Goals

The over-arching purpose of this ICM course is to help students develop practical technical skills and ways of thinking that allow them to effectively construct, manipulate, visualize, and interpret datasets and computational models. These skills are not only applicable to the physical, life, and social sciences, but are rapidly becoming a necessity in these fields. In addition to developing technical skills and new thought processes, the course strives to promote an understanding of the social and scientific relevance of modern data analysis and computational modeling, which can be critical in recruiting and retaining students from underrepresented groups in STEM [11].

---

[1] https://tech.msu.edu/teaching/real/; see also Sect. 2.6.
[2] Although it is now a requirement in three degree programs and one minor, and a selective or elective course in several other degrees and minors.

As described in Sect. 2.1, this course targets students in the natural and social sciences rather than computer science students. As such, the computer programming skills that are needed within the course are motivated entirely by the goals and applications mentioned previously. Students certainly learn many of the same computer science topics (e.g., variables and arrays, loops, functions, etc.), but they do so in service of building and running simulations and analyzing data. Although this approach ensures that the students learn the skills necessary to ask meaningful questions of and extract answers from computational models and data, it places unavoidable limitations on the amount of computer science topics that can be covered in a single semester. More complex topics, like recursion and object-oriented programming, are left to a subsequent course for those students that are motivated to dive deeper. Finally, to reach a broad cross section of the student population, the only prerequisite for the course is one semester of calculus, and the course content touches on a wide variety of disciplines. We expect that our students have **no prior programming experience**.

The broad goals of the course are as follows. By the end of the semester, we intend that students who have successfully completed this class will be able to:

1. Gain insight into physical, biological, and social system through the use of computational algorithms and tools.
2. Write programs to solve common problems in a variety of disciplines.
3. Identify salient features of a system that can be codified into a model.
4. Manipulate, analyze, and visualize datasets and use to evaluate models.
5. Understand basic numerical methods and use them to solve problems.
6. Synthesize results from a scientific computing problem and present it both verbally and in writing.

2.3 Theoretical Underpinnings and Pedagogical Motivations

After developing the learning objectives for this course, the next steps were to determine the overall course structure, decide on the types of evidence that we would use to identify that learning was taking place, and sketch out specific assessments and assignments that would be used. While there are many theoretical frameworks that could be used to inform our design decisions, we decided to use the principles articulated in the book "How Learning Works: Seven Research-Based Principles for Smart Teaching" [1]. Our motivation for this choice is that the principles articulated in *How Learning Works* are grounded in a theoretical understanding of how people learn that is based on empirical evidence, and that these principles can be practically implemented in courses intended for large numbers of students and facilitated by instructors coming from variety of backgrounds and skill levels with only a modest amount of training. In designing this course, we have engaged most heavily with the following principles:

To develop mastery, students must acquire component skills, practice integrating them, and know when to apply what they have learned. In a given assignment, students are typically engaging with only a small number of new concepts in programming, modeling, or data analysis. The arc of assignments focusing on a given topic (from pre-class, to in-class, to homework) are scaffolded to first give students practice with each new concept individually, and then integrate them with prior skills. For example, students learn to use iteration by first learning to write for loops in the pre-class assignment that perform simple tasks such as incrementing a variable. Then, in class they use the same conceptual structure to do numerical integration using the Trapezoidal Rule and, later, to evolve through time the position and velocity of a projectile acting under the influence of gravity and air resistance. As part of homework assignments and later pre-class assignments, loops are built upon as part of more complex modeling and data analysis tasks, often in scenarios where students have to make critical decisions about their usage.

How students organize knowledge influences how they learn and apply what they know. As described in the example above, new concepts are introduced in a scaffolded manner that assists students in making connections between new pieces of knowledge, and in productively integrating that new knowledge with existing ideas. The ultimate goal here is to help students develop a strong conceptual understanding of computational modeling and data analysis, as well as the tools and techniques for doing so, that they can apply appropriately in new situations. As an example of this scaffolding, students learned to work with variables prior to learning the concept of loops and their implementation, and in following activities learn Boolean logic (which is combined with simple loops to create more sophisticated control flow). After this, functions are introduced, and students integrate functions into loops and Boolean logic as they create increasingly complex models and data analysis workflows.

Goal-directed practice coupled with targeted feedback enhances the quality of students' learning. Each assignment addresses one or more course learning or content objectives. In-class assignments are typically pursued by students in small groups of 3–4 students, with instructional staff facilitating a handful of groups. Group members are encouraged to ask each other questions and the instructional staff regularly check in with all students to ensure that they are on the right track and to help guide the students in the right direction when necessary. Homework assignments and the semester project are accompanied by written feedback, with careful attention paid to struggling students.

Students' motivation determines, directs, and sustains what they do to learn. Students enter into our course with a variety of motivations, and are majoring or interested in a variety of topics. To that end, the topics of assignments are chosen to apply to a wide range of applications and subject areas, and to connect these areas together by the use of common numerical algorithms. Furthermore, we choose applications that are motivated by authentic scientific questions (e.g., disease spread) or social issues (e.g., racial segregation), and when possible, both (e.g., climate change).

To become self-directed learners, students must learn to monitor and adjust their approaches to learning. To this end, we structure assignments that are designed to encourage students to go beyond thinking about the task that they are pursuing, but to think about *what they are doing, why they are doing it*, and *if they are doing it correctly*. We do this by asking students to justify the choices they have made and to explain how what they are doing fits within the context of the course.

Taken together, these principles motivate the course structure (as described in Sect. 2.4 through Sect. 2.7) and the course assignments.

2.4   Content Objectives

Working from our learning objectives and theoretical framework, we then chose the content of the course and its structure to achieve these objectives and to facilitate students learning the broad array of skills described in Sect. 2.1. The course progression is shown in Table 1, which includes a day-by-day breakdown of course activities. Motivated by our learning objectives, we describe a given day's activity in terms of the modeling and/or data analysis concept(s) that are developed during that class period, the context within which that is being taught (i.e., application area), and the new programming practice(s) that are required in order to be able to implement the modeling or data analysis concept.

The temporal progression shown in Table 1 demonstrates how students are gradually introduced to new skills and concepts. We assume that students come into the class with little or no programming experience, and thus early class sessions introduce fundamental programming concepts such as variables, loops, Boolean logic, and functions, and software-related tasks such as creating code flowcharts and writing pseudo-code to represent programs. These programming concepts are introduced using a modeling or data analysis concept that motivates the need for them – for example, in Day 2 of class students learn to use variables in programs to solve order-of-magnitude estimation problems. Motivated by the principles described in Sect. 2.3, we gradually introduce new skills and concepts in a way that facilitates their integration with previous content, and emphasize in class the ways that the new material builds on what has been previously learned. Furthermore, students are given clearly-defined problems in class, which they pursue in small groups with targeted instructor feedback. The modeling and data analysis/visualization concepts are taught within a variety of different contexts or application areas to both show the breadth of potential application areas and to engage students' personal interests.

2.5   Selecting and Training Instructional Staff

The instructional model used to support the learning and content objectives in this course is relatively resource-intensive and requires instructional staff to be carefully chosen and trained. We use a combination of PhD-level instructors, graduate teaching assistants, and undergraduate learning assistants. PhD-level instructors are assigned to the course based on interest and availability, and all faculty in the Department of CMSE are expected to rotate through the course.

Table 1. Course Content and Progression. An overview of the most recent iteration of the ICM course. The first column highlights the main modeling or data analysis concept explored in the pre-class/in-class activities. The second column provides the context in which the concept(s) was explored. The third column indicates the programming practices that were most important for the successful completion of the material.

Day	Modeling/Data analysis concept	Context/Application	Programming practices
1	*Students are provided a course overview*		
2	Order of magnitude estimation	Varied (e.g. population density)	Variable definition and manipulation, simple math
3	Mathematical representations of physical systems	Kinematics, projectile motion	Defining lists, writing loops, variable manipulation
4	Evaluating the state of physical systems	Kinematics, projectile motion	Writing loops, using boolean logic, using if/else statements, writing and using functions
5	Computing costs and optimizing solutions	Designing a ride share service	Writing and using functions, combining functions
6	Visualizing models	Projectile motion and population growth	Using Python modules (e.g. `math` and `Matplotlib`)
7	Loading and visualizing data	Water levels of the Great Lakes	Using the `NumPy` module, loading/reading csv files, saving plots
8	Numerical integration (finite different method)	Kinematics of extreme sports	Defining initial conditions, updating variables in loops, storing data in lists
9	Numerical integration (using pre-existing solvers)	Orbital Mechanics	Defining functions, using the `odeint` function from the `SciPy` module
10	Compartmental models	Epidemiology, rumors, population dynamics	No programming this day (instead: flow-charting and pseudo-coding)
11	Compartmental models	Epidemiology, rumors, population dynamics	Using `odeint`, modifying function parameters and initial conditions
12	Agent-based models	Forest fires and tipping points	Creating, navigating, and manipulating arrays, visualizing 2D arrays
13	Agent-based models	Schelling's segregation model	Creating, navigating, and manipulating arrays, visualizing 1D arrays
14	Data inspection, manipulation, and visualization	Flint water crisis data	Interacting with data using `Pandas`, applying boolean masks, data slicing and visualization
15	Data cleaning, transformation, and interpretation	Flint water crisis data	Manipulating data, computing statistical quantities, using `Pandas`
16	Finding trends (linear regression)	Political approval ratings	Using `NumPy`'s polyfit and poly1d and `SciPy`'s curve_fit to fit data, visualizing model fits
17	Finding trends (auto-regression)	Weather forecasting	Using the lag_plot function from `Pandas`, using fitting functions, visualizing data
18	*Midterm exam took place on this day*		
19	Stochastic models using random numbers	Brownian motion/random walks	Using random number generators, implementing if/elif/else statements
20	Monte Carlo methods	The traveling salesperson problem	Using random number generators, visualizing data

continued

Table 1. *continued*

Day	Modeling/Data analysis concept	Context/Application	Programming practices
21	Markov chain Monte Carlo parameter estimation	Fitting noisy data	Using random number generators, evaluating conditionals, visualizing data
22	*Class time devoted to semester project work*		
23	Manipulating and analyzing varied datasets	Climate Change	Loading data files, manipulating variables, visualizing data
24	Comparing models to data	Climate Change	Writing loops, storing data, loading data files
25	Building compartmental models, coupling models, and comparing to data	Climate Change	Manipulating and updating variables, visualizing data, over-plotting models and data

Graduate teaching assistants (TAs) are chosen based on their expertise, availability, and interest in teaching with this instructional model. Undergraduate learning assistants (LAs) are students who took the course in a previous semester, received a high course grade, and have an interest in teaching.

Each week, the instructional staff meets to debrief their teaching experience, prepare for the coming week of material and, when time allows, discuss educational pedagogy. In the week preceding the start of the semester, all TAs and LAs are required to attend a full-day training workshop that explains the structure of the course, the motivations for the course design, common issues that may arise in class, and methods for effectively managing student-instructor interactions.

As the course has evolved, progressively more emphasis has been placed on ensuring that *all* members of the instructional staff receive sufficient training, including PhD-level instructors. As course enrollment continues to grow and our demand for undergraduates LAs increases, we plan to develop a more formalized TA and LA training program aimed at providing additional professional development and requiring that the students enroll in a formal course to explore educational pedagogies more deeply and present on their teaching experiences.

2.6   Course Technology and Logistics

In service of reaching our learning goals and implementing the described pedagogical techniques, we leverage a variety of digital and physical technologies. There are also certain course logistics necessary to achieve the quality standards that CMSE strives to uphold in its courses. Coordinating all of these aspects of the course are nontrivial and we attempt to outline the key details here.

Students in this course use the Python programming language[3] due to its ubiquity, overall ease of use as an introductory programming language, and the wide range of available and relevant libraries and supplemental software packages. Specifically, we have students install the Anaconda[4] Python distribution

---

[3] https://www.python.org.

[4] https://www.anaconda.com.

on their own laptop computers. Within this distribution, the main piece of software that we use in the ICM course are Jupyter Notebooks. In fact, the Jupyter Notebooks are a critical component of the course as they act as our student code development environment. They allow raw code, the results of its execution, and explanatory narrative to coexist in a single interactive document. The notebooks can also render plots and print data query results inline, making them extremely powerful for doing data analysis.

Jupyter Notebooks are also extremely powerful from a curriculum authoring and pedagogical perspective. We design our instructional materials with embedded YouTube videos of our mini-lectures and links to helpful resources all within a single document. That document then serves as the template that students edit and then submit for credit. This single-document structure means there is less pressure on students to context-switch between lecture notes, video demonstrations, and code playgrounds to try things out. Since notebooks support `<iframe>` cells, we can also embed a Google Form survey to solicit student feedback. Furthermore, Jupyter Notebooks are used both in academia and in industry, so students are working with the same professional-grade tools as experts.

Although students are expected to bring their own laptops to class, this expectation cannot always be met. We have two solutions for this: First, a cart containing laptops with the necessary software is brought to class every day for student use in class. Second, we maintain a JupyterHub[5] instance that allows students to remotely log on and use Jupyter notebooks from any device with a modern web browser. Between these two solutions, virtually any computing issue can be resolved with minimal impact on student and instructor class time.

Beyond the software that facilities the learning environment we strive for in the ICM course, the physical classroom space is crucial for motivating collaborative learning. To encourage productive collaboration, we offer as many sections of the course as possible in MSU's REAL classrooms, mentioned in Sect. 2.1 and shown in Fig. 1. These classrooms are constructed to discourage traditional lectures as students sit at tables in groups of 4–6 such that they are face-to-face with each other, rather than facing the instructor. These tables are equipped power outlets, audio/video cables, and hi-definition televisions, so that students can work on their laptops and share their screens to cooperatively code their problem solutions. Furthermore, the classrooms are set up such that should the instructor wish to share the work or progress of a particular group or individual, they can broadcast that screen to the entire room. This affords the instructor the opportunity to turn a localized learning moment into a classroom-wide discussion. The REAL classrooms also offer plentiful whiteboard space for student teams to write out their algorithm ideas, model components, and data analysis goals, an important part of the in-class activities outlined in Sect. 2.7.

Although the layout and technological enhancement of the REAL classrooms help to promote a collaborative learning environment, we note that teaching this course in classrooms like the REAL classrooms at MSU is not a requirement for course success. As long as the classroom desks are mobile, they can be

---

[5] https://jupyterhub.readthedocs.io/en/stable/.

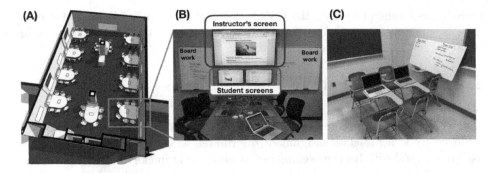

Fig. 1. (A) A rendering of one of MSU's Rooms for Engaged and Active Learning (REAL classroom). The room is designed to foster group work and minimize lecturing. (B) One of the REAL classroom student stations for group work. Students have space for their laptop, power access, and the ability to connect to one of the small screens at the end of the table. The instructor can display content on the large television screen above the tables to be viewed by all students. Students use the whiteboards for tasks like pseudo-coding, flowcharting, brainstorming. (C) One possible alternative setup for classrooms that do not offer the same design advantage of the REAL classroom. This setup achieves much of the same goals for encouraging student interaction and is sufficient for a successful offering of the ICM course.

rearranged into groups and the room can be supplemented with portable white-boards. Although slightly less ideal, this allows for a setup like that shown in Fig. 1c, which still encourages peer-to-peer interaction during class.

During class students work in small groups for discussion, brainstorming, pseudo-coding, and other related activities, but occasionally fragment into pairs when writing code. Group members are expected to be mutually supportive, and are responsible for their own learning and each others' learning. They do so by sharing their ideas and reasoning, asking each other questions, and answering questions posted to them. A positive group environment is created in two ways. First, groups are deliberately constructed – instructors assign students to groups in such a way that groups have similar overall capabilities on average, that group members have a wide variety of majors, and that students from under-represented groups are not overwhelmed by those from the majority group. Groups are re-formed a handful of times per semester, with the instructional staff taking student personalities and abilities into account. Second, students are given clear instructions regarding the goals of the group and receive periodic feedback on how they and their group is performing.

Finally, the size of the course sections and structure of the instructional staff to date as follows. For the semesters discussed in this paper, course sections have included anywhere from 24 to 60 students. We strive to maintain a student-instructor ratio of no greater than 16:1, as going to larger ratios puts unrealistic expectations on our ability to provide sufficient facilitation of student learning [5,10]. For our current average section size of 36 students, we staff each section with one PhD-level instructor (typically a CMSE faculty member), one graduate teaching assistant (TA) from the CMSE doctoral program, and one undergraduate learning assistant (LA). We maintain our desired student-instructor ratio in larger sections by adding additional LAs.

2.7   A Typical Week of Class

A typical week in ICM has the following structure. Individual sections of the course meet for 110 minutes twice a week. Prior to a given class, students are required to do a pre-class assignment distributed as a Jupyter Notebook. This assignment is short, due the evening before class, and typically includes narrative text and/or one or more short videos that introduce them to that day's new topics. After the students watch the videos, they immediately use what they have learned in a series of short programming problems and/or free response questions. Finally, they fill out an embedded Google Form survey that asks them to list any questions that they have about the material – as well as broader questions/issues they may have regarding the class – and then turn in the notebook via the course management system. The content of the Google Form is immediately available to the instructional staff in an easy-to-digest format which allows them to see what students are struggling with or have questions about. These pre-class assignments act as a mechanism for formative assessment and they are assigned grades based on completion rather than on correctness.

Class begins with a classroom-wide discussion of the pre-class assignment. This discussion may include a brief presentation by the instructor or a walkthrough of particular pieces of the assignment. This affords the instructor the opportunity to address any questions or confusion that students may have indicated in the pre-class assignment, which helps put each student on equal footing prior to starting the day's activity. After this discussion, students download the in-class assignment from the course management system (another Jupyter notebook) and begin working on it in small groups. The assignments involve student discussion, calculations, programming, data analysis, and comparing of models to data, all of which is facilitated by the instructional staff. There are usually one or two "check points" during each class that are intended to get students to think critically about the outputs of their data analysis and/or modeling, and often result in whole-class discussions. These assignments also prompt the students to use the available whiteboards to write "pseudo-code," flow-chart their algorithms, or outline their data analysis methods. At the end of each class, students fill out a Google Form survey to give feedback and ask clarifying questions, and then upload a copy of their Jupyter notebook to the course management system. Grading is based on completion and good-faith effort rather than on correctness.

2.8   Summative Assessment

In addition to the various forms of formative assessments we described previously, we also build multiple mechanisms for summative assignment into the course. These take on the form of homework assignments, exams, and semester projects.

After students have explored a topic in class, they receive a homework assignment that builds on that topic, but typically focusing on a different application area. Students are encouraged to discuss the homework with each other, but are required to do the work themselves and turn in the homework individually. This homework is a summative evaluation graded on correctness. In addition to testing students' ability to perform data analysis and implement models, some homework assignments also have an interpretive or descriptive component. These questions provide us with opportunities to explore student understanding around topics that might be overly challenging to code.

Beyond the homework assignments, we have explored a variety of exam formats over the evolution of the ICM course. The most effective format involves giving students Jupyter notebooks that are similar in format to their normal in-class activities with a handful of questions aimed at probing key topics from the course. The questions asked might require: finding a solution to a set of equations to model a time-dependent physical system; loading, manipulating, and visualizing data; using random numbers to carry about numerical integration; or explaining what an agent-based model is, how it can be used to model a real system, and how one might setup the model using code. The exams reach a similar level of complexity as the tasks students normally complete in class, but for these exams they have to work individually. Students are allowed to use the internet to look up documentation for the code they are using or troubleshoot bugs they may run into. Students report that the exams are challenging, but fair and reasonably well-aligned with the course content and design. Averaged across sections, ~80% of students achieve an exam grade of 70% or higher.

The final piece of summative assessment that we use in the ICM course are the semester projects. These projects provide the students with the opportunity to pursue a topic that they are personally interested in and showcase the skills and knowledge they have acquired over the course of the semester.

3   Summary and Discussion

This paper presents the design process, structure, content progression, and assessments mechanisms from the first six semesters of teaching and refining our ICM course. In the four years since the development of this course began in earnest, the Department of CMSE has grown substantially, as has student and faculty interest in the course. This presents both challenges and opportunities.

The primary challenge that we have encountered is rapidly growing demand for this course. Academic advisors and faculty are strongly recommending the course to students from a variety of majors, and enrollment has increased every semester, limited only by course capacity. Since the first offering, course sections have grown in size, with the most recent semester having a maximum section size of 72 students. These large sections are staffed by a faculty instructor, a graduate teaching assistant, and two undergraduate learning assistants. A similarly-structured introductory MSU physics course [9] suggests that is possible to scale up to on the order of 100 students per section without sacrificing the student-centered nature of the course and its associated learning gains, something we will likely have to pursue.

In addition to growing demand from students, undergraduate minors and degree programs are now incorporating the ICM course into their requirements as a prelude to adding computational modeling and data analysis throughout their curriculum. Beyond the capacity increases required to support these programs, it will be critical to reassess the ICM learning goals and detailed curriculum to ensure that it continues to meet the Department's goals, as well as communicate this information to programs requiring the course.

The rapid growth in demand requires the development of a training curriculum for all instructional staff that focuses on methods for facilitating student-centered learning and the development of a more formalized undergraduate learning assistant program [15]. Undergraduate learning assistants are often comparably effective to teaching assistants [4] and may even help to institutionalize course reform and teaching practices [6]. For faculty-level instructors, training in research-based pedagogy is critical as most CMSE faculty are young, with little to no teaching experience and their experience has primarily been in traditional, lecture-style courses. Training faculty in the research-based teaching methods employed in this course is also crucial to its sustainability and continuity.

The success of ICM has led to inter-college discussions about the creation of other courses with complementary goals. For example, faculty in MSU's College of Communication Arts and Sciences are interested in developing a course focused more on data analysis/data science, the emerging field of "data journalism" [7], and the role that reliance on Big Data and algorithms can play in policy and broadly in society (e.g., [14]). The student population for this type of course would be quite different - journalism, marketing, humanities, and social science majors rather than STEM majors - which implies different course prerequisites and different learning goals.

The creation of this course (and its second-semester counterpart) creates a range of research opportunities. Critically, while a set of measurable concepts in programming has been defined in the computer science education research community (see, e.g., [17]), a similar inventory of concepts and a related assessment tool does not exist with regards to computational modeling and data analysis. Such a tool is critical for evaluating individual student assignments and in measuring the impact of curricular changes. Furthermore, the effect of the combination of computer science and modeling/analysis tasks on student learning has received almost no attention, outside of the physics education research community [3,16]. Finally, there is a rich vein of possible work relating to student affect and student identity. Most students who enter the ICM course have already declared their major, and are unlikely to self-identify as computational or data scientists. Exploring the impact of the ICM course on their self-identity and feelings about their major, particularly as they progress through their degree, may prove to be important for student recruitment and retention.

Acknowledgements. The authors thank Nathan Brugnone, Danny Caballero, Andrew Christlieb, Dirk Colbry, Sarah Gady, Nat Hawkins, Morten Hjorth-Jensen, and Luke Stanek for useful discussion and constructive criticism on drafts of this manuscript. We further thank all instructors of CMSE 201 for their enthusiastic participation and thoughtful feedback during the course creation process. We thank the MSU CMSE Department, the Office of the Vice President for Research and Graduate Studies, the College of Natural Science, the MSU Connected Mathematics Endowment (as administered by the MSU Program in Mathematics Education), and the Howard Hughes Medical Institute for their generous support.

References

1. Ambrose, S.A., Bridges, M.W., DiPietro, M., Lovett, M.C., Norman, M.K.: How Learning Works: Seven Research-Based Principles for Smart Teaching. Wiley, Hoboken (2010)
2. Black, P., Wiliam, D.: Inside the black box: raising standards through classroom assessment. Phi Delta Kappan **80**(2), 139–144 (1998)
3. Caballero, M.D., Hjorth-Jensen, M.: Integrating a computational perspective in physics courses. ArXiv e-prints, February 2018
4. Chapin, H.C., Wiggins, B.L., Martin-Morris, L.E.: Undergraduate science learners show comparable outcomes whether taught by undergraduate or graduate teaching assistants. J. Coll. Sci. Teach. **44**(2), 90–99 (2014)
5. Cummings, K., Marx, J., Thornton, R., Kuhl, D.: Evaluating innovation in studio physics. Am. J. Phys. **67**(S1), S38–S44 (1999)
6. Goertzen, R.M., Brewe, E., Kramer, L.H., Wells, L., Jones, D.: Moving toward change: institutionalizing reform through implementation of the learning assistant model and open source tutorials. Phys. Rev. Spec. Top. Phys. Educ. Res. **7**(2), 020105 (2011)
7. Gray, J., Chambers, L., Bounegru, L.: The Data Journalism Handbook: How Journalists Can Use Data to Improve the News. O'Reilly Media Inc., Sebastopol (2012)
8. Henke, N., et al.: The age of analytics: competing in a data-driven world. McKinsey & Company (2016). http://www.mckinsey.com/business-functions/mckinsey-analytics/our-insights/the-age-of-analytics-competing-in-a-data-driven-world
9. Irving, P.W., Obsniuk, M.J., Caballero, M.D.: P3: a practice focused learning environment. Eur. J. Phys. **38**(5), 055701 (2017)
10. Kohl, P.B., Vincent Kuo, H.: Chronicling a successful secondary implementation of studio physics. Am. J. Phys. **80**(9), 832–839 (2012)
11. National Academy of Sciences, Committee on Trends and Opportunities in Federal Earth Science Education and Workforce Development, Board on Earth Sciences and Resources, Division on Earth and Life Studies: Preparing the Next Generation of Earth Scientists: An Examination of Federal Education and Training Programs. National Academies Press. https://www.nap.edu/catalog/18369/preparing-the-next-generation-of-earth-scientists-an-examination-of
12. National Association of Colleges and Employers: Job Outlook 2016: The Attributes Employers Want to See on New College Graduates' Resumes. http://www.naceweb.org/career-development/trends-and-predictions/job-outlook-2016-attributes-employers-want-to-see-on-new-college-graduates-resumes/
13. National Association of Colleges and Employers: Job Outlook 2017. http://www.naceweb.org/store/2017/job-outlook-2017/

14. O'Neil, C.: Weapons of Math Destruction: How Big Data Increases Inequality and Threatens Democracy. Broadway Books (2016)
15. Otero, V., Pollock, S., Finkelstein, N.: A physics department's role in preparing physics teachers: the Colorado learning assistant model. Am. J. Phys. **78**(11), 1218–1224 (2010)
16. Petter Sand, O., Odden, T.O.B., Lindstrøm, C., Caballero, M.D.: How computation can facilitate sensemaking about physics: a case study. ArXiv e-prints, July 2018
17. Qian, Y., Lehman, J.: Students' misconceptions and other difficulties in introductory programming: a literature review. ACM Trans. Comput. Educ. **18**(1), 1:1–1:24 (2017)
18. Wiggins, G.P., McTighe, J.: Understanding by Design. Association for Supervision and Curriculum Development, expanded 2nd edn. https://lccn.loc.gov/2004021131

Enabling Interdisciplinary Instruction in Computer Science and Humanities

An Innovative Teaching and Learning Model Customized for Small Liberal Arts Colleges

William B. Crum Jr.(iD), Aaron Angello(iD), Xinlian Liu(✉)(iD),
and Corey Campion(iD)

Hood College, Frederick, MD 21701, USA
{crum,angello,liu,campion}@hood.edu

Abstract. Infiltration of data-driven computational methods of humanities research has generated mutual interests between the two communities of computer science and humanities. Larger institutions have adopted drastic structural reforms to meet the challenges to bridge the two fields. Successful examples include the integrated major programs launched at Stanford University and the collaborative workshop at Carnegie Mellon University. These types of exploratory experiments require (1) intensive resources as well as (2) strong support of faculty and administration. At a small college, both can be luxuries. We present an innovative model to carry out effective synchronized courses of computational humanities and digital humanities that pulls together efforts between two small programs and needs little additional support. This paper reviews the proposal, design, and delivery of a pair of interdisciplinary graduate courses in the small college setting. We discuss the details of our implementation and provided our observations and recommendations.

Keywords: Interdisciplinary · Synchronized courses ·
Computational digital humanities ·
Interdisciplinary project-based learning · Teaching pedagogy

1 Humanities in the Age of Big Data

Over the past few years, drastically increased data-collecting capability and massively parallel data processing capability have enabled a swift shift in the way research projects are conducted in humanities, arts, and social sciences fields. The emergence of large data collections, e.g., digitized documents including books, government records, images, audio, and video, websites, social media, health, business, and communication records, sensor data, etc., changes the way scholars generate and present meaningful research in the 21st century [7]. Due to the nature of these large corpora, traditional means of analysis and presentation are no longer sufficient. This research requires research teams with expertise in

© Springer Nature Switzerland AG 2019
J. M. F. Rodrigues et al. (Eds.): ICCS 2019, LNCS 11540, pp. 389–400, 2019.
https://doi.org/10.1007/978-3-030-22750-0_31

multiple fields. The trend has been recognized by academia, as higher education institutions are taking steps to offer students interdisciplinary programs which integrate computational and humanities curricula.

Despite the success of such programs in leading institutions such as The *CS+X* Joint majors at the Stanford University and a *Conjoint Model* piloted at Villanova University [21], it remains a challenge in smaller colleges to offer similar programs because of barriers in institutional support, faculty buy-in, and curricula design [22].

Among the challenges confronting liberal arts education today is a fundamental disconnect between the curricula that many institutions offer and the training that many students need. In the early twenty-first century, most colleges and universities still adhere to the model of disciplinary-specialization that developed in the nineteenth century when the pressures of industrialization and globalization led to the expansion of higher education and a need to justify the growing number of tenure lines within an institution. Though an important asset for students and faculty alike, discipline-specific models of higher education struggle to prepare students for the kinds of interdisciplinary collaborations now expected by many employers. Aware of the need for change, many institutions have tried to translate the widespread rhetoric about interdisciplinarity into new programs and curricula that better serve today's students.

With an eye toward providing students in the Master of Arts in Humanities and Master of Science in Computer Science programs the opportunity to practice the interdisciplinary collaboration, the program directors and co-instructors undertook to create an exciting new learning opportunity within the unique budgetary and enrollment environment found at the hundreds of smaller institutions across the United States. From the identification of student learning outcomes and the development of an interdisciplinary syllabus to the logistics of sharing a course across programs and the experiences of both instructors and students, the paper considers both the benefits and challenges of such collaboration and offers recommendations for those looking to undertake similar projects at their own institutions.

2   Design and Architecture of the Courses

2.1   Customization Based on Local Resources

The project began when the director of the M.S. in Computer Science program approached the director of the M.A. in Humanities program about the possibility of offering humanities courses that would enhance the communication skills of computer scientists and help them develop the empathy required to understand the human communities their computer programs serve. Coincidentally, the M.A. director had been looking for ways to help humanities students expand their research and share their ideas with a wider audience through the use of technology. The results of this exchange were the institution's first-ever paired-courses in digital humanities. Unlike traditional dual-listed courses, the paired courses allowed students to explore digital humanities within their own disciplines before coming together to collaborate on shared projects. Students from both fields learned to balance pragmatism and ambition through collaboration and communication. By combining enhanced training in the students' fields with the opportunity for interdisciplinary collaboration, the courses produced an experience that more closely approximated the working environments of today.

To mitigate the barriers to implementing a joint program at a small college, we came to the realization that our approach should:

1. take little additional resources
2. require few curricular changes to existing programs
3. provide maximum support to the instructors.

To gain institutional support, we used recent research findings [12, 15, 20] and media coverage [3, 16] to reason with the administrators that Computer Science could be the *enabler* of liberal arts education, which is the core of our College's education missions.

Typically, computer science students are given assignments related to the specified course in isolation. Even when teamwork is assigned, the teams are made up of computer science students. Neither the project goals nor the team makeup allows students to gain experience in interdisciplinary research or collaboration beyond the field of computer science. The same is often true for graduate students in the humanities, few of whom enjoy opportunities to collaborate with and benefit from the unique skill sets of their colleagues in the natural and social sciences.

As we went about planning the two courses, we confronted sets of problems unique to each group of students. Very simply, most of the students who were a part of the Masters in Humanities program had few technical skills, while the students who were a part of the Masters in Computer Science lacked an understanding of the kinds of research undertaken in humanities fields. These concerns had to be addressed separately, but we also had to keep in mind that one of our primary goals was to bring these two classes together.

The students in the Masters of Humanities program had virtually no previous experience with digital humanities (DH). For the most part, these students are high school teachers or adults who are lifelong learners, interested in expanding their knowledge of topics in the humanities in general. They are not specialists, and have not spent time in research universities with digital humanities programs. There was a general sense of curiosity among the students, a curiosity that presumably inspired them to sign up for the class, but they demonstrated a distinct lack of understanding of the history of DH, of the types of projects that are or ought to be considered DH, or even of what DH, broadly speaking, is.

Furthermore, writing code was a terrifying prospect to most of these students. A few had some web development experience, but none were comfortable even with HTML or CSS. We had to assure the students that practicing DH does not necessarily mean one has to write code (although there are certainly those who have argued otherwise [14]), that our very intention in setting the classes up the way we had was to create an environment in which computer programmers and humanists could collaborate. Different people have different skill sets, and digital humanities are, in no small part, about collaboration. The students would not need to learn to write code in this class, but they would need to have some understanding of how the code works so that they could communicate with the computer programmers.

2.2 Student Learning Outcomes

Upon completion of this course pair, students were able to:

- Demonstrate critical technical skills necessary to conduct interdisciplinary research in computational social sciences and digital humanities.
- Demonstrate the skills necessary to work in interdisciplinary teams.
- Demonstrate effective programming skills necessary to develop natural language processing applications in the humanities (for students attending the Computational Humanities class).
- Speak intelligently regarding topics in computational humanities (for students attending the Computational Humanities class).
- Recognize limitation of technologies and technical methods (for students attending the Digital Humanities class).

2.3 Course Planning

We began the course by trying to define the term "digital humanities." As anyone who works in the field knows, this is a daunting prospect. Even among the field's preeminent practitioners, there is much debate about what this term means. The students read a number of "defining DH" articles, including Matthew Kirschenbaum's "What is the Digital Humanities and What's It Doing in English Departments?" [8] and Patrik Svensson's "The Landscape of Digital Humanities," [17] both articles have practically become canonical in Introduction to DH classes. We also found that Svensson's "Beyond the Big Tent" [18] helped the students engage in the "what counts as DH" debate. For our purposes, when it comes to the "who's in and who's out" debate [14], we found it useful to try to embrace the broadest and most inclusive definition of DH. We tried to adopt Svensson's recommendation that we think of DH not as a "big tent," a field in itself that defines itself by being exclusionary, but rather "as a meeting place, innovation hub, and trading zone," a site that highlights a "commitment to interdisciplinary work and deep collaboration" [18]. Given the students' backgrounds in English, history, and art, and given their trepidation with regard to digital research methods in the humanities, we found that thinking about DH in this way allowed for more fruitful engagement in the class.

Next, we provided a bit of historical contextualization for the field. Students read accounts of early examples of the use of computing in the humanities such as Father Roberto Busa's twenty-five-year project creating a lemmatized concordance of all eleven million words in the works of St. Thomas Aquinas' writings, a project which began in 1949 [1]. The students went on to look at the work of

Vannevar Bush's "As We May Think" [2] and Theodor Nelson's "A File Structure for the Complex, the Changing, and the Indeterminate" [6] so that they could understand the origins of linked file structures which are so ubiquitous today.

The bulk of the class, however, was focused upon making. After analyzing a number of existing projects, the students were able to get hands-on experience in the areas of text analysis (e.g. Voyant Tools), creating digital collections (Omeka), encoding metadata (XML and TEI), network visualization (Gephi and Palladio), and geospatial humanities (CartoDB). The idea was to give the students an overview of some common, powerful tools used in a number of successful research projects that are also relatively unintimidating to the DH novice. Once they had an idea of what was possible, they began to conceive projects related to their particular areas of interest, and they began to work on those projects.

In Computational Humanities (CH), the students were taught natural language processing (NLP) in Python using the Natural Language Toolkit (NLTK) [9]. The only prerequisite was a programming class in Java or Python. Only one student had taken a course in Python. The students in CH were traditional age international students largely from one country. These students had no background in natural language processing or in digital humanities.

This project-based course involved three strands: the Python language, NLTK functions, and natural language processing methods. During the early stage, the instruction interlaced these strands with emphasis on Python language and basic NLTK functions. Initially, instruction in natural language processing focuses on basic concepts and vocabulary. Once students gained a firm footing in the Python language, the students' attention turned primarily to the use of NLTK to investigate various aspects of natural language processing.

Where these classes differed from most other introductory courses, however, was in their emphasis on collaboration across disciplines. These courses ran concurrently. Early in the semester, and again in the middle of the semester, the professors from each class met with the students from the other class. The humanities professor prepared lectures and discussions aimed at helping the computer science students better grasp some of the conceptual and methodological approaches to research in the humanities and its importance. The computer science professor introduced the humanities students to the basics of programming and natural language processing using Python and the Natural Language Toolkit (NLTK). Many of the computer science students had spent their entire academic careers as far from English, art, history, and philosophy as they could, opting to surround themselves with other programmers and "math-brained" people. Similarly, the humanities students, in many cases, winced at the prospect of learning to code. They had simply never been exposed to it.

Both classes did find value in their exposure to these other disciplines, but the real value of running these classes together, and the explicitly stated goal of organizing it this way, was to provide an opportunity for the students to learn to communicate to each other across disciplines, to learn to speak the language of their counterparts.

3   Implementation

Two graduate courses Computational Humanities (Computer Science Department) and Introduction to Digital Humanities (Humanities Department) were run in parallel.

Each course was a normal one semester, 15-week course meeting once a week. Originally, we planned to run the courses on the same day and time. However, due to scheduling issues, the classes ran on the same day but during overlapping time periods. The overlap was during the last half of the Digital Humanities course and the first half of the Computational Humanities course.

During the first three weeks, each course met independently. Students in the Digital Humanities course read articles about the history of humanities computing and about ongoing discussions and debates in the field. They discussed foundational texts, including Vannevar Bush's "As We May Think" [2], Theodor Nelson's "A File Structure for the Complex, the Changing, and the Indeterminate," [6], and the work of Father Roberto Busa [1]. Additionally, they read and discussed a number of prominent articles that address the question of how we are to define the digital humanities [8,14,18]. During this initial three weeks, the students were able to begin situating themselves within the complex field of digital humanities and to start grasping a fundamental understanding of the kinds of work that go in the field. Computational Humanities students began to learn the fundamentals of Python programming and the use of NLTK functions to perform basic natural language processing activities. They developed programs that performed basic activities, such as accessing text corpora, text from the Internet, and other lexical resources, extracting content from web pages, processing raw text, searching text, counting vocabulary, using regular expressions to detect word patterns, creating frequency distributions, and plotting and tabulating the results of frequency distributions. By the end of the first three weeks, students developed skills to write programs in Python and to use NLTK functions. They gained a basic understanding of the vocabulary and concepts of natural language processing.

On the fourth week, the classes met together for part of the session. During this time, the students got to know each other and discuss some preliminary project ideas. Following the joint meeting, the Computational Humanities students met with the humanities professor to begin understanding the myriad ways humanists are using digital tools to expand research opportunities and to begin understanding the value of employing computational methods in humanities research. The Introduction to Digital Humanities students met with the computer science professor to receive some basic instruction in writing Python programs in natural language processing.

During the next three weeks, each course met independently. The Digital Humanities course began to narrow its focus, looking at specific areas of focus, methods, and tools. They began looking at the markup, metadata, and the value of TEI (Text Encoding Initiative). They also read Franco Moretti's *Maps, Graphs, and Trees* [11] and began exploring the potentials of data visualization, both theoretically [4,10,13] and practically, by creating their own small-scale projects using Gephi and Voyant tools. With experience in writing programs that perform basic natural language processing activities, the Computational Humanities students were ready for more advanced activities. Students wrote programs using part-of-speech taggers to process sentences into lists of (word, part-of-speech) pairs and using tagged corpora and Python dictionaries to analyze text. Students wrote programs to evaluate the performance of automatic taggers. Through this work, students gained an appreciation for the processes and problems associated with categorizing and classifying text.

On the eighth week, the classes met together for part of the session. During this time, the students formed research teams and began work on the projects. Following the joint meeting, the Computational Humanities met with the humanities professor to receive additional instruction in digital humanities, this time exploring a number of existing DH projects, while the Introduction to Digital Humanities met with the computer science professor to receive additional instruction in writing Python programs in natural language processing.

During the next four weeks, each course met independently and the interdisciplinary teams formed during week eight worked on the final project outside of class. The Digital Humanities students examined geo-spatial humanities projects and read articles dealing with potentially new ways of presenting scholarly research [5], explored issues related to digital pedagogy, and discussed issues associated with racial and gender identities in the field. In the past, Computational Humanities students examined text at the word level. Now, the students' focus changed to extracting meaningful word groups, such as noun phrases from the text. They wrote programs that analyzed sentence structure and built and parsed simple grammars.

During the last three weeks, the classes met jointly to work on team projects and give presentations of the projects.

4   Outcome and Discussions

To assess the effectiveness of our approach, we assessed students' readiness in terms of grasping of required skill sets to conduct interdisciplinary projects in computational digital humanities. We then estimated the metrics again towards the end of the class. We provided these figures along with the average grades of the joint project and students' semester grade in Table 1.

The semester grade includes non-interdisciplinary projects given throughout the semester as well as the interdisciplinary joint project. Joint project grades are team-based. The instructors met to jointly evaluate each team project. The joint project evaluation criteria include:

1. quality as a digital humanities project
2. the extent was it interdisciplinary
3. quality of the team work
4. progress towards completion as a proof of concept
5. quality of the grant proposal
6. class presentation

Table 1. Student's readiness on skill sets required to succeed in an interdisciplinary project in computational humanities before and after the joint class, and the average grades they received in the joint project as well as the average semester grade.

		Grasp of required skill sets		Average grades received	
		Pre-class	Post-class	Joint project	Semester grade
CS students	NLP skills	20%	80%	96%	91%
	HUM skills	10%	60%		
HUM students	NLP skills	15%	75%	96%	95%
	HUM skills	50%	85%		

4.1 Observations

This was a first step in the development of an interdisciplinary program involving the Computer Science and Humanities departments. On the whole, it was a good first step.

Due to limited time, it was understood that students would be unable to deliver a production-level project. Therefore, students were to make sufficient progress to provide proof of concept and write a grant proposal.

The projects created by the teams were generally satisfactory. There were some issues regarding team interactions. However, they were manageable. Some projects tended to lean more towards the digital humanities side, while others tended to lean more towards the natural language processing side.

One project that seemed to really get the right mix was a team comprised of an art teacher and a computer science student. The team wanted to create an interactive web-based art history timeline. To start, they wanted to transfer the art teacher's 250 item art history PowerPoint presentation to a web page. The art teacher's approach was to copy and paste materials from the PowerPoint presentation. However, the computer science student saw that this process could be automated via a natural language processing program.

He developed programming tools to extract information from the PowerPoint presentation and insert the information into an Excel spreadsheet. The goal was to create a database to use in the construction of a web page.

In developing the program, they discovered the difficulty of writing a program dealing with the complexities for natural languages. For example, the program needed to extract date information. Date information can take many formats, e.g., numerical (100 BCE) or verbal (sixth century), which makes it more difficult to extract. They overcame many obstacles. However, due to limited time, they were unable to solve all the problems by the end of the semester.

4.2 Issues and Recommendations

Scheduling. While the class schedules overlapped, they did not meet at the same time. This meant that during weeks four and eight the time available for both the join meeting component and cross-training was limited to about two-thirds of the normal class time. In reality, both groups needed more time for cross training. The Computation Humanities students only had time to get a basic understanding of the vocabulary and concepts of digital humanities. Likewise, the Digital Humanities students only got a chance to see the natural language processing programs. They did do a little "follow the leader" programming with the professor.

Recommendation: Future course offerings of an interdisciplinary nature should meet at the same days and times.

Cross Training. The scheduling issue limited cross training. Cross training provides a framework for team members to understand their colleagues in other disciplines. In this setting, the Computational Humanities students needed a basic understanding of digital humanities and the types of research it involves. Likewise, the Introduction to Digital Humanities students needed a basic understanding of writing natural language processing programs. With crossing training, team members can more effectively communicate and develop a better solution.

Recommendation: Cross training session should increase both in length and number. Concurrent meeting times should help.

Team Meetings. Many students in Introduction to Digital Humanities are adult learners with full-time jobs. This makes team meetings outside of class difficult.

During the last three weeks of the courses, the Computational Humanities students made an effort to meet together during the Introduction to Digital Humanities' scheduled time. However, some were not able to make the beginning of the class.

Recommendation: Concurrent meeting times should help provide more opportunities for team meetings including additional out-of-class meeting time before and after regular class meetings.

Student Mindsets. Just because students know a class will have an interdisciplinary project does not mean they will want to work in groups. Students sign up for courses for all kinds of reasons.

Students felt most comfortable working with colleagues within their own discipline. Some students did not want to work in teams at all. Therefore, it was necessary to require interdisciplinary teams and to provide guidance to those reluctant students to work with students of a different discipline.

Many of the digital humanities students, i.e., teachers, were experienced in organizing and presenting information and in leadership. In some cases, they did not have the patience or the inclination to delegate tasks to "junior" members of the teams. Likewise, "junior" members did not always demand a greater role in the project. Therefore, it was necessary to provide guidance for students dealing with team dynamics.

Recommendation: Concurrent meeting times should help provide more opportunities for students to get to know their counterparts in the other class. The opportunity for great human connection should help change negative mindsets.

Projects. The final project was the only interdisciplinary project.

Recommendation: Ideally, the development of project ideas and formation of teams should occur earlier in the semester. This allows more time to complete the project and develop the skills necessary to work in interdisciplinary teams.

The development of one or two mini-interdisciplinary projects early in the course could be beneficial to develop skills necessary to work in interdisciplinary teams. This could be professor directed, i.e., the professors choose project topics, scope, and teams.

5   Conclusion

By combining enhanced training in the students' fields of study with the opportunity for interdisciplinary collaboration, the courses produced an experience that more closely approximated the working environments in which many are now employed. The pilot was well accepted by our students from both Computer Science and Humanities sides. We also learned valuable lessons that will help us to improve this model as an effective teaching strategy.

Acknowledgement. We thank the Hood College Graduate School and our corresponding departments for their support. We are indebted to our students who embraced this pilot project and worked with us along the way when we rethought, re-planned, and refined the project.

This project is in part supported by a grant from the National Endowment for the Humanities. The computing resources are made available in part by Extreme Science and Engineering Discovery Environment (XSEDE) [19], which is supported by National Science Foundation grant number ACI-1548562 under Educational Allocation ASC140025.

References

1. Busa, R.: The annals of humanities computing: the index thomisticus. Comput. Hum. **14**(2), 83–90 (1980)
2. Bush, V.: As we may think. Atl. Mon. **176**(1), 101–108 (1945)
3. Davis, J., Albrecht, J., Alvarado, C., Chen, T.Y., Lee, S.: Teaching and research in computer science at liberal arts colleges: myths and reality. Technical report. http://www.forbes.com/2006/07/28/leadership-careers-jobs-cx_tvr_0728admired.html

4. Drucker, J.: Humanities approaches to graphical display. DHQ: Digit. Hum. Q. **5**(1), 1–21 (2011)
5. Fitzpatrick, K.: Planned Obsolescence: Publishing, Technology, and the Future of the Academy. New York University Press, New York (2011)
6. Nelson, T.H.: Complex information processing: a file structure for the complex, the changing and the indeterminate, January 1965. https://doi.org/10.1145/800197. 806036
7. Khan, N., et al.: Big data: survey, technologies, opportunities, and challenges. Sci. World J. **2014** (2014). https://doi.org/10.1155/2014/712826, http:// www.ncbi.nlm.nih.gov/pubmed/25136682, http://www.pubmedcentral.nih.gov/ articlerender.fcgi?artid=pmc4127205
8. Kirschenbaum, M.G.: What is digital humanities and what's it doing in English departments? ADE Bull. **150**, 55–61 (2010)
9. Loper, E., Bird, S.: NLTK: the natural language toolkit. In: Proceedings of the ACL 2002 Workshop on Effective Tools and Methodologies for Teaching Natural Language Processing and Computational Linguistics, ETMTNLP 2002, vol. 1, pp. 63–70. Association for Computational Linguistics, Stroudsburg, PA, USA (2002). https://doi.org/10.3115/1118108.1118117
10. Manovich, L.: What is visualization. PAJ: J. Initiat. Digit. Hum. Media Cult. **2**(1) (2010). https://journals.tdl.org/paj/index.php/paj/article/view/19
11. Moretti, F., Piazza, A.: Graphs, Maps, Trees: Abstract Models for a Literary History. Verso (2005)
12. Pulimood, S.M., Pearson, K., Bates, D.C.: A study on the impact of multidisciplinary collaboration on computational thinking. In: Proceedings of the 47th ACM Technical Symposium on Computing Science Education, SIGCSE 2016, pp. 30–35. ACM, New York (2016). https://doi.org/10.1145/2839509.2844636
13. Ramsay, S.: In praise of pattern. Faculty Publications, Department of English, p. 57 (2005). http://digitalcommons.unl.edu/englishfacpubs/57/. Accessed 15 Feb 2019
14. Ramsay, S.: Who's in and who's out. In: Melissa Terras, J.N., Vanhoutte, E. (eds.) Defining Digital Humanities: A Reader, pp. 239–241. Taylor and Francis Group, Burlington (2013)
15. Stozhko, N., Bortnik, B., Mironova, L., Tchernysheva, A., Podshivalova, E.: Interdisciplinary project-based learning: technology for improving student cognition. Res. Learn. Technol. (2015). https://doi.org/10.3402/rlt.v23.27577
16. Straumsheim, C.: Computer Science as Liberal Arts 'Enabler', February 2016. https://www.insidehighered.com/news/2016/02/23/liberal-arts-colleges-explore-interdisciplinary-pathways-computer-science
17. Svensson, P.: The landscape of digital humanities. Digit. Hum. **4**(1) (2010)
18. Svensson, P.: Beyond the big tent. In: Gold, M.K., Klein, L.F. (eds.) Debates in the Digital Humanities. University of Minnesota Press, Minneapolis (2012)
19. Towns, J., et al.: XSEDE: accelerating scientific discovery. Comput. Sci. Eng. **16**(5), 62–74 (2014). https://doi.org/10.1109/MCSE.2014.80
20. Walker, H.M., Kelemen, C.: Computer science and the liberal arts: a philosophical examination. Trans. Comput. Educ. **10**(1), 2:1–2:10 (2010). https://doi.org/10. 1145/1731041.1731043
21. Way, T., Whidden, S.: A parallel, conjoined approach to interdisciplinary computer science education. In: Proceedings of the 2016 ACM Conference on Innovation and Technology in Computer Science Education, ITiCSE 2016, pp. 363–363. ACM, New York (2016). https://doi.org/10.1145/2899415.2925486

22. Zhang, C.: Interdisciplinary teaching and research: challenges and solutions. In: Proceedings of the 2017 7th International Conference on Education, Management, Computer and Society (EMCS 2017), pp. 160–163. Atlantis Press, Paris, France, March 2017. https://doi.org/10.2991/emcs-17.2017.31, http://www.atlantis-press.com/php/paper-details.php?id=25876027

A Project-Based Course on Software Development for (Engineering) Research

Kyle E. Niemeyer(✉) (iD)

Oregon State University, Corvallis, OR 97331, USA
kyle.niemeyer@oregonstate.edu

Abstract. This paper describes the motivation and design of a 10-week graduate course that teaches practices for developing research software; although offered by an engineering program, the content applies broadly to any field of scientific research where software may be developed. Topics taught in the course include local and remote version control, licensing and copyright, structuring Python modules, testing and test coverage, continuous integration, packaging and distribution, open science, software citation, and reproducibility basics, among others. Lectures are supplemented by in-class activities and discussions, and all course material is shared openly via GitHub. Coursework is heavily based on a single, term-long project where students individually develop a software package targeted at their own research topic; all contributions must be submitted as pull requests and reviewed/merged by other students. The course was initially offered in Spring 2018 with 17 students enrolled, and will be taught again in Spring 2019.

Keywords: Research software · Teaching software development ·
Software best practices

1 Motivation

Nearly all research relies on software—even experimental—but researchers typically do not receive training in best practices during graduate school in the same way as they do for experimental methods. In fact, in two recent surveys the vast majority of academics confirmed that they use software and that their research would be impractical without it: 90%/70% of UK academics surveyed in 2014 [12], and 95%/63% of US postdoctoral researchers surveyed in 2017 [19]. Computational science in particular depends on software and following good, evidence-based practices when working with software and data.

For example, in the Mechanical Engineering graduate curriculum at Oregon State University, the thermal-fluid sciences option (where I teach) requires a course on experimental measurement techniques, but no analogous course on proper techniques on software development or computational science. (We do require a course on numerical methods that focuses on solving differential equations, but this does not extend to software development.) Instead, our program— as in most similar programs around the world—assumes that such practices

© Springer Nature Switzerland AG 2019
J. M. F. Rodrigues et al. (Eds.): ICCS 2019, LNCS 11540, pp. 401–407, 2019.
https://doi.org/10.1007/978-3-030-22750-0_32

are trivial compared with the physical phenomena or mathematical methods and/or can be self-taught. However, in the same way that appropriate measurement techniques and statistical analysis of data are necessary for experimental (and computational) research, good practices ensure reliability and correctness of research results obtained from software developed for both computational (i.e., modeling-based) or experimental (i.e., analysis of results) research.

As the research community has recognized the importance of software and data skills, in recent years Software Carpentry [27] workshops have become a recognized avenue for graduate students and postdocs (and the occasional faculty member) to learn necessary skills for working with Python, the command line, and version-control systems. While these are essential skills for research, researchers who go further to develop software require additional training. This article describes a course aimed at filling this gap by teaching graduate-student researchers practical software development skills, and also exposing them to topics related to open science and reproducibility.

2   Course Design

The course heavily relies and builds on the *Effective Computation in Physics* textbook by Scopatz and Huff [23] (Chaps. 10–22), as well as recommendations by Wilson et al. [28] and Jiménez et al. [13]. All materials for the course are openly available online via GitHub and shared under a Creative Commons Attribution license; the online course syllabus provides links to each lecture [20]. The course combines lectures, hosted on GitHub and presented using `reveal.js` [8], with in-class activities and discussions, as summarized in Table 1. Out-of-class work, described in Table 2, centers around a software development project discussed in Sect. 2.2.

2.1   Course Description and Learning Objectives

The listed course description is

> This course will advance students' understanding of topics related to computational science and engineering, and advance their skills in applying techniques to solve research problems using high-level programming languages. The course will build on existing abilities in computer programming to cover topics related to computational modeling and scientific software development. Students will gain experience in applying available packages and libraries, as well as developing software to solve problems related to their own research interests. Students will also gain experience in working collaboratively and openly on scientific computing projects.

Table 1. Course schedule over ten weeks, with in-class activities.

Topic	In-class activity
Getting started, and version control	Configure Git
Remote version control, licensing, and copyright	Create, clone, and fork repos
Structuring modules, and testing	Create basic structure of module
Test coverage, continuous integration, documentation	Configure Travis CI
Introduction to Julia (guest lecture)	
Introduction to parallel programming	
Classes and objects (in Python)	
Packaging and distributing your software	Create PyPI, Anaconda packages
Optimizing numerical code in Python	
Working with files, command-line inputs in Python	
Open science, software citation, reproducibility	Connect GitHub and Zenodo
Posters, presentations, and technical writing	
Project presentations	

Table 2. Assignments, with week assigned.

Assignment	Week
Join Gitter chat room and create GitHub profile	1
Project proposal	1
Choose open-source license	3
Create tests and submit as PR for review	3
Finish configuring Travis CI	4
Write comments and docstrings	4
Complete PyPI and/or Anaconda packages	6
Write report and make presentation	7

By the end of the course, students should be able to

1. use high-level programming language to analyze and/or solve practical research problems;
2. apply principles of modern computational science and engineering, reproducibility, and open science to their research;
3. evaluate, visualize, write about, and publish computational research results; and

4. develop and share an open-source research software package that solves a problem in their research area.

These are the formal student learning objectives for the course.

2.2   Project

In lieu of standalone homework assignments, all assigned work contributes to a term-long project where students develop a new software package targeted at their own research area. The project initiates with a proposal that students submit via pull request to an open repository on GitHub, and which the instructor merges upon approval (following any changes requested). Then, students create a repository in the course organization for their software package, and fork this to their own accounts. After this, students submit all project contributions as pull requests to the upstream repository. Partners review these and either approve or request changes; only after the code-review partner approves the contribution can the project's owner merge the pull request.

2.3   Methods of Instruction

The course is delivered using a combination of lectures, discussion, and in-class interactive work. Lectures mostly exist as `reveal.js` [9] presentations, which are shared openly on a public-facing syllabus website [20]. Lectures also use practical demonstrations of Python code, shown using either IPython [21] or via Jupyter Notebooks [15]. Nearly all lectures also ask students to follow along on their own computer, either executing example code or advancing their project packages.

3   Results from First Offering

I offered the first iteration of this course in the Spring 2018 term, with the title "Software Development for Engineering Research;" while the course content is not limited to engineering research, I offered the course out of the Mechanical Engineering program with a targeted audience of graduate students in the College of Engineering. 17 students enrolled in the course, with all but one being graduate students; roughly half were in the second year or later of their graduate programs. Approximately 40% of the students came from mechanical engineering (including thermal-fluid sciences and design engineering), 35% were from nuclear engineering, and the remaining came from robotics and chemical engineering. Three quarters of the students had already taken a course on Python programming for engineering applications, while the others had some self-taught Python programming skills. Half expressed comfort working with the Unix command line, and the other half said they had used it but were not as comfortable with command-line operations. None admitted to being command-line ninjas, and none were completely unfamiliar.

The first offering of this 10-week course on software development for engineering research completed successfully in June 2018, with all 17 students releasing the first version of their software developed during the course. At least four of the software packages have been developed further after the conclusion of the course, and at least one package is being prepared for submission to the *Journal of Open Source Software* (JOSS) [24].

The projects covered a wide variety of topics, with functions including simulation, experimental data analysis, and automation: designing detonation tubes [6], using machine learning to extract features from nuclear physics simulations [10], interfacing with an 8-channel digital pulse processor board [16], simulating and analyzing the combustion engine of a Global Formula Racing formula SAE vehicle [14], optimizing and analyzing wind-farm layouts [17], analyzing spin stabilization of solid rocket motors [18], a nodal quasi-diffusion solver for nuclear fission [22], agent-learning for autonomous path finding [25], generating input for a Monte-Carlo radiation transport code [26], calculating solar-energy terms based on location [11], analyzing radioxenon spectra [7], calculating deep-learning layers for multi-agent reinforcement learners [5], analyzing solvent extraction kinetics [4], simulating rapid compression machine experiments [3], calibrating blackbody infrared cameras [2], and simulating transient heat transfer in a microchannel with passive temperature dependent flow control [1].

Although the sample size is small, students rated the course well in their end-of-term evaluations: they rated the course as a whole 5.3/5.5 (mean/median) out of 6.0 and the instructor contribution 5.5/5.8 (mean/median) out of 6.0. Multiple comments discussed the course favorably, and that it should be taken by all students doing research involving software/programming. Suggestions included clarifying expectations for students and proving more feedback; also, one student felt the course was too advanced for their experience level.

4   Conclusions

This article describes a 10-week course given in Spring 2018 teaching skills for developing research software; all lesson and assignment content is available openly [20]. This course will be offered again in Spring term 2019 (10 weeks, April–June). Planned changes include incorporating more in-class activities in more of the topics, and adding new topics such as peer code review and high-performance computing.

In addition, I am developing alternate versions of the course aimed at different lengths of time, such as an afternoon tutorial session or day-long workshop. These lessons and modules will be shared openly for the community to use, adapt, and extend. Furthermore, while the course at Oregon State University is currently offered out of the Mechanical Engineering program, it may fit better offered as an Engineering course or more broadly in the Graduate Education program.

Acknowledgements. This research was supported by the Better Scientific Software Fellowship, part of the Exascale Computing Project (17-SC-20-SC), a collaborative effort of the U.S. Department of Energy Office of Science and the National Nuclear Security Administration.

References

1. Armatis, P.D.: `PAFloCS` v0.1.0. Zenodo (2018). https://doi.org/10.5281/zenodo. 1291277
2. Bean, D.: `IRCal` v0.1.0. Zenodo (2018). https://doi.org/10.5281/zenodo.1291190
3. Behnoudfar, D.: `SimRCM` v0.1.0. Zenodo (2018). https://doi.org/10.5281/zenodo. 1291302
4. Bettinardi, D.J.: `SepKinetics` pre-alpha v0.110. Zenodo (2018). https://doi.org/ 10.5281/zenodo.1291202
5. Brian, M.: `deep_learning_layer_calculator` v1.0. Zenodo (2018). https://doi. org/10.5281/zenodo.1291273
6. Carter, M.: `BeaverDet` v0.1.0. Zenodo (2018). https://doi.org/10.5281/zenodo. 1288009
7. Czyz, S.A.: `radioxenon_ml` v0.5.0. Zenodo (2018). https://doi.org/10.5281/zenodo. 1291208
8. El Hattab, H.: `reveal.js` 3.7.0, December 2018. https://github.com/hakimel/ reveal.js
9. El Hattab, H.: `reveal.js` (2019). https://github.com/hakimel/reveal.js
10. Grechanuk, P.A.: `ML_FeatureAnalysis` v0.0.1. Zenodo (2018). https://doi.org/10. 5281/zenodo.1286598
11. Guymer, N.: `SolarCalc` v0.1.0. Zenodo (2018). https://doi.org/10.5281/zenodo. 1290030
12. Hettrick, S., et al.: UK research software survey 2014 (dataset). University of Edinburgh on behalf of Software Sustainability Institute (2015). https://doi.org/10. 7488/ds/253
13. Jiménez, R.C., et al.: Four simple recommendations to encourage best practices in research software. F1000Research **6**, 876 (2017). https://doi.org/10.12688/ f1000research.11407.1
14. Kittelman, J.S.: `PyPow` v0.0.1. Zenodo (2018). https://doi.org/10.5281/zenodo. 1287325
15. Kluyver, T., et al.: Jupyter notebooks - a publishing format for reproducible computational workflows. In: Loizides, F., Schmidt, B. (eds.) Positioning and Power in Academic Publishing: Players, Agents and Agendas, pp. 87–90. IOS Press (2016). https://doi.org/10.3233/978-1-61499-649-1-87
16. Mannino, M.: `OkPy_Rad`. GitHub (2018). https://github.com/SoftwareDevEng Research/OKPY_Rad
17. Miller, A.: `BigFan` v0.1.2. Zenodo (2018)
18. Morse, M.D.: `SRMspinanalysis` v0.1.0. Zenodo (2018). https://doi.org/10.5281/ zenodo.1288020
19. Nangia, U., Katz, D.S.: Track 1 paper: surveying the U.S. National Postdoctoral Association regarding software use and training in research. In: Workshop on Sustainable Software for Science: Practice and Experiences (WSSSPE 5.1) (2017). https://doi.org/10.6084/m9.figshare.5328442.v3
20. Niemeyer, K.E.: Syllabus for Software Development for Engineering Research course (2018). https://softwaredevengresearch.github.io/syllabus/
21. Pérez, F., Granger, B.E.: IPython: a system for interactive scientific computing. Comput. Sci. Eng. **9**(3), 21–29 (2007). https://doi.org/10.1109/MCSE.2007.53. https://ipython.org
22. Reynolds, A.J.: `NoQuD`. Zenodo (2018). https://doi.org/10.5281/zenodo.1290779

23. Scopatz, A., Huff, K.D.: Effective Computation in Physics: Field Guide to Research with Python. O'Reilly Media, Inc. (2015). http://physics.codes
24. Smith, A.M., et al.: Journal of Open Source Software (JOSS): design and first-year review. PeerJ Comput. Sci. **4**, e147 (2018). https://doi.org/10.7717/peerj-cs.147
25. Sripada, V.: agent_learning. Zenodo (2018). https://doi.org/10.5281/zenodo.1291474
26. Stewart, R.: FRIDGe v0.1.0-alpha. Zenodo (2018). https://doi.org/10.5281/zenodo.1288681
27. Wilson, G.: Software carpentry: lessons learned. F1000Research **3**, 62 (2016). https://doi.org/10.12688/f1000research.3-62.v2
28. Wilson, G., Bryan, J., Cranston, K., Kitzes, J., Nederbragt, L., Teal, T.K.: Good enough practices in scientific computing. PLOS Comput. Biol. **13**(6), e1005510 (2017). https://doi.org/10.1371/journal.pcbi.1005510

Programming Paradigms
for Computational Science: Three
Fundamental Models

Miguel-Angel Sicilia$^{(\boxtimes)}$, Elena García-Barriocanal, Salvador Sánchez-Alonso,
and Marçal Mora-Cantallops

Computer Science Department, University of Alcalá,
Polytechnic Building. Ctra. Barcelona km. 33.6,
28871 Alcalá de Henares, Madrid, Spain
{msicilia,elena.garciab,salvador.sanchez,marcal.mora}@uah.es

Abstract. The widespread of data science programming languages and
libraries have raised new interest in teaching computational science coding in ways that leverage the capabilities of both single-computer and
cluster-based computation infrastructures. Some of the programming
patterns and idioms are converging, yet there are specialized uses and
cases that require learners to switch from one to another. In this paper,
we report on the experience in action research with more than ten
cohorts of mixed background students in postgraduate level data science
classes. We first discuss the key mental models found to be essential
to understanding solution design, and then review the three fundamental paradigms that students must face when coding data manipulation
and their interrelation. Finally, we discuss some insights on additional
elements found important in understanding the specificities of current
practice in data analysis tasks.

Keywords: Computational science · Education · Data science ·
Programming · Mental models

1 Introduction

Computational science requires programming skills that are to a large extent
determined by the languages, libraries, frameworks and applications used for the
various domains of scientific inquiry. These are not only related to the creation
of models themselves but to the broader scope of data manipulation, which
have been captured in the past as part of knowledge discovery and data mining
process frameworks [13]. The success of the concept of a *data scientist* [5,9]
as a professional role that deals with data-intensive problems and has a broad
and hybrid skill set has to some extent predated the idea of a computational
scientist. Arguably, data scientists that are in the domain of some particular
scientific discipline are computational scientists. This idea is reflected in the
EDISON framework for data science education [16], that includes a competence

© Springer Nature Switzerland AG 2019
J. M. F. Rodrigues et al. (Eds.): ICCS 2019, LNCS 11540, pp. 408–420, 2019.
https://doi.org/10.1007/978-3-030-22750-0_33

area that incorporates scientific methods closer to experimental work as it is done in the sciences than to applied data analytics in a business context.

In any case, teaching programming for computational science or data science is a different endeavor than teaching general-purpose programming as it is practiced in undergraduate degrees in the domain of computing. Typically, data science study programs are offered to audiences with a diverse profile [6], and it is expected in them that students with a background different from computing are able to effectively develop software that makes use of existing scientific libraries and computing infrastructures, without them becoming software engineers. Typical languages taught in data science include R, Python (using SciPy libraries) and Julia, and less frequently nowadays, Octave, Matlab, SAS or others. Some of them as Python are actually general-purpose languages that were not originally devised with data analysis as a goal. However, data wrangling and libraries consuming data (as machine learning libraries) in them tend to follow some particular idioms or paradigms that need not be the same as those used in the language. An example is the difference of using NumPy arrays [21] and Pandas dataframes [17] in Python, which is widely different to using regular Python lists, both in logical and in internal representation aspects. There are elements of efficiency and memory handling, typing and even style that make the experience of data analysis in Python very different from programming Python for other purposes, e.g. to develop a Web site.

Here we describe insights from the experience in teaching scientific programming to non-computer science graduate students. Concretely, it reports on the reflections of the experience in ten cohorts of students enrolled in programs related to data science (namely: Data Science, Business Intelligence and Business Analytics and Big Data) that were taught programming for data science in Python (and secondarily classic statistical analysis programming with R) at the University of Alcalá in Madrid in the last five years. The results come from the observation of common pitfalls and difficulties found when approaching the grading of assignments, and from subsequent interviews focused on particular problems in understanding. Incremental inquiry in cycles was done applying action research principles, similar to those that have been applied to teaching programming elsewhere [15]. We have come up with a number of mental models and programming styles or paradigms that require separate attention and are different in relevant aspects. Further, we report on a number of additional elements that were found important in an adequate understanding of the specifics of programming in data science.

The rest of this paper is structured as follows. Section 2 describes the assumptions, background and overall setting of the educational programs that have served as the basis for the discussion. Then, in Sect. 3 we deal with the key mental models identified that students have to develop in order to efficiently code in data science settings. Section 4 describes the three fundamental programming paradigms or models identified that require significant cognitive effort when moving from one to the other. In Sect. 5, we discuss additional elements that have been found as important in providing students the adequate context to understand the tasks at hand. Finally, conclusions and outlook are provided in Sect. 6.

2   Background and Assumptions

We assume here that teaching data science can be considered as a broader concept than that of teaching computational science, as the latter focuses on models for the diverse scientific disciplines, while the former is more inclusive of computational models not necessarily related to the enterprise of science. Actually, there are some experiences reported on explicitly combining both concepts, e.g. Giabbanelli et al. [11] report on a course combining computational models with data science concepts. In other cases, both concepts are considered as overlapping when discussing educational aspects, e.g. in [8]. It should also be noted that data tasks as the application of machine learning are a current active area in online question-answering sites [1] reflecting the importance of the topic.

The typical path for teaching data science starts with exposing the students to introductory lessons on programming, typically using a high level language as Python (as in [6] for example). This entails the usual sequence of introducing first variables, conditional and iterative control structures, then some fundamental data structures: lists and dictionaries. However, data science environments adhere to a paradigm sometimes called *array-oriented programming* in which operations on scalars apply transparently to vectors, matrices, and higher-dimensional arrays. This results in a degree of conciseness, as operations abstract out the number of dimensions of the data, and for most common operations, it is rare the need to use control structures. Instead, vectorized operations are applied on the data, and there are facilities to select and transform data with operators. Examples are boolean indexing in arrays or clauses that specialize in selecting subsets of data and generalize the notion of slicing lists. In consequence, the teaching of basic Python used in our approach does not emphasize algorithm design, but the use of a core of data structures and the application of functions. Particularly, object oriented design (encapsulation, inheritance, etc.) is not included in the teaching, and the understanding of object orientation is limited to the syntax for sending messages to objects (i.e. invoking methods), which is natural in languages as Python.

We assume here that the teaching is directed to a non-computer science student, so that the emphasis is on using libraries and not on developing new algorithms, parallel versions of them or optimized versions of some complex user interfaces, which would require other kind of skills. Also, we assume that database or data store access will be made transparent using some form of SQL-like languages. It should be noted that these assumptions are rather strong and somewhat controversial. For example, in the case of tasks as data acquisition from APIs (application programmer interfaces as those exposed in RESTful interfaces), data wrangling would require some extra abilities in some cases. In any case, in this paper we focus on the core of the daily activities of a data scientist that involve array and dataframe manipulation and application of library functions, even though there is currently a degree of ambiguity on data science team roles and the skills required for different profiles [18].

3   Key Mental Models

Mental models are concise abstract representations that can help shape behaviour and set an approach to solving problems. It has been recognized that mental models are required by novices to learn programming [7]. The assumption of mental models is that we construct mental "small-scale models" of reality, which can be used to anticipate events, to reason, and to underlie explanation. These models are simplified and incomplete, but are helpful in our construction of knowledge on a topic.

Most students attending data science classes already have a mental model consisting of "table like structures" similar to those used in spreadsheets. Many of them had some knowledge of the relational model and the SQL language so that the ideas of tables is natural. However, there are other mental models that have been found to be essential in our experience and that require dedicated attention in educational programs, which are discussed in what follows.

3.1   Array Programming: Vectorization and Broadcasting

Array programming revolves around the idea of having multidimensional arrays, and operations are applied over those irrespective of their size or number of dimensions. The key concept is that of *vectorized* operations. Library functions for example in SciPy are already vectorized, so that they can be applied to arrays of any number of dimensions. Also, it is possible to automatically vectorize functions. The following example is from SciPy, but similar alternatives are available in other languages as R.

```
def _transform(x,y):
    if (x>0.5*y and y<0.3):
        return (sin(x-y))
    elif (x<0.5*y):
        return 0
    elif (x>0.2*y):
        return (2*sin(x+2*y))
    else:
        return (sin(y+x))
transform = np.vectorize(_transform, otypes=[np.float])
transform(a, b) // a, b can be of any number of dimensions
```

It should be noted that such vectorization does in general not provide any performance advantage over using for loop iteration. However, vectorized operations in libraries are typically optimized explicitly or implemented in lower level languages as C or Fortran, so that they should always be preferred.

Broadcasting is an effect related to applying operations on arrays of different sizes. Under some constraints, one of the arrays is broadcast, meaning that it is repeated so that the sizes are compatible. This occurs trivially in expressions as a * 5 where a is an array, and the scalar is broadcast to the same size of the array.

The understanding of vectorization and broadcasting is critical to all operations, as even the selection of elements is based on vectorized logical or relational operators. This sets repetitive control structures (looping) as an exception that is largely not needed for manipulating arrays and matrices (even though it is still required for other tasks, as processing data from APIs or Web pages, but that can be considered as different tasks).

The mental model is then that of functions that can be used with arrays of any dimensions replacing looping, including implicit extension on dimensions. This requires a change in the mental model of students in the case they have experience in non-vectorized programming and tend to recur to loops as a default strategy.

3.2   Memory Hierarchy, parallelism and Task Models

The second important mental model requires basic knowledge of computer architecture, in an abstract way. The fundamental model is that of a hierarchy of memory, typically considering as levels: (a) internal registers in the processor, (b) system RAM and (c) external storage as in disks. This nowadays gets an additional level in that of (d) distributed storage, e.g. in a cluster of computers. The latter is dealt here considering that the programming constructs are transparent to the actual parallel computing infrastructure, instead of dealing with concrete models as in the case of MPI reported elsewhere [10].

The levels can be clearly aligned by students with different volumes of data. Typically, system RAM goes up to the size of several GB, while in disks there is an scaling to near some TB, and then clusters can scale up. Years ago, the complexity of moving from (b) to (c) or (d) required specific programming, but this is becoming largely transparent in some cases thanks to so-called "Big Data" technologies as Apache Spark and others.

The second, related level is that of parallelism, that is tied to the different levels. Processor-level parallelism involves in-core acceleration, and it is specially relevant to matrix computation, since it is at the core of GPU acceleration. Other forms of parallelism involve in-memory (thread or processes) and then distributed computing in a cluster. Programming models as those of Dask[1] or Apache Spark[2] have made the last two levels transparent to a large extent, and in-core parallelism is also independent of the code, as it will be described later.

The transition to (d) can be grasped by some basic understanding of the idea of *task graphs* which are directed acyclic graphs of partial computation. These, as the one in the example in Fig. 1 generated with Dask[3] for the fitting

---

[1] https://dask.org/.

[2] https://spark.apache.org/.

[3] https://jcrist.github.io/dask-sklearn-part-2.html.

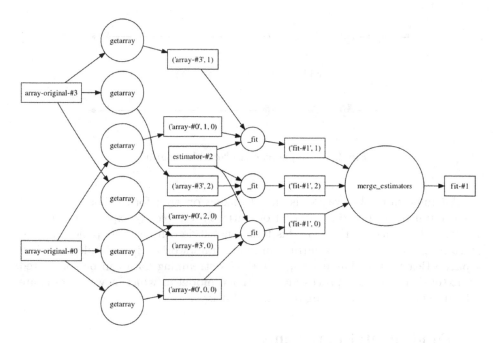

Fig. 1. Examples dask task graph.

of an estimator, are the basis for distributing computation across processes or nodes. Here the mental models needs not be detailed (and as such students are just exposed to simple examples as distributed computations of averages), as they concern internal details of the breakdown of computation. However, a basic understanding of the idea that vectorized computations of arbitrary complexity can be decomposed for distribution in different processing units in that way is required to avoid naive unscientific beliefs about the underlying mechanisms.

3.3 Streams of Data

Dealing with streams of data requires a change in the way of thinking, as data becomes a continuous flow and the programs wait and react to the arrival of data. This is opposite "data applied to functions" to the idea of "functions applied to data" and thus requires separate treatment, typically introduced afterwards. While streaming can be understood as processing in batches of windows [4], the mental model has to focus on the asynchronous and infinite nature of data. That is properly reflected on the use of the dataflow paradigm, and depictions as "marble diagrams" as commonly used in discussions of reactive programming. Figure 2 shows a depiction of an event stream and the temporal transformation effect of a filter function[4].

---

[4] http://reactivex.io/documentation/operators/filter.html.

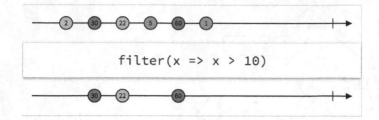

Fig. 2. Example reactive "marble diagram".

An important observation is that thinking on transformations on streams fits the functional paradigm in that operations are typically expressed as purely functional primitives. Further, this functional style can also be applied to non-streaming data thanks to current unifying frameworks [3] as implemented in Apache Beam[5]. For that reason, a model for streaming data can be considered an extension of a data processing functional model in which new concerns are added, e.g. that of windowing or temporal delays.

4   Fundamental Paradigms

A programming paradigm is "an approach to programming a computer based on a mathematical theory or a coherent set of principles" [20]. Each paradigm supports a set of concepts that makes it the best for a certain kind of problem or task. The consideration of the mental models described before is actually orthogonal to the three key programming paradigms we have found as important and are discussed in what follows. This is because all of them abstract out memory hierarchy, are able to somewhat perform automatic parallelism across different levels of that hierarchy, and in some cases allow for dealing with both streaming and non-streaming data in a declarative style. Further, all of them make extensive use of vectorization as a basic primitive. Then, the difference among them is the key building block used in the code, that is discussed in what follows.

4.1   Lower Level Abstractions: Functional Array/bag Processing

The first concept is that of working with homogeneous arrays of data, with one or several dimensions. These can be thought of as arrays or bags (understood as generic sequences of elements), and data science languages, due to vectorization and broadcasting, can be taught in a functional style. In numpy, for example, *ufuncs* have a functional common interface.

```
x = np.arange(1, 6)
np.add.reduce(x)
```

---

[5] https://beam.apache.org/.

The problem with a full functional paradigm is that pure functions are by definition stateless, and efficiency matters require exposing students to in-place modification semantics. An expression such as `a * = 2` corresponds to an in-place operation, where all values of the array are modified. By contrast, `a = a * 2` means that a new array containing the values of `a * 2` is created. The same occurs in `Series` in which the parameter `inplace` is for example used to change the memory management. Also, slices (which can be seen as a filtering operation) are by default views and not copies, again exposing memory management to programming constructs.

A more complete functional paradigm at this level is provided in `bag` libraries of Dask, a multiset library that is able to operate with data out of core and in clusters. The following is an example processing Json data. It is equivalent to Apache Spark RDD interface[6].

```
result = (b.filter(lambda record: record['age'] > 30)
          .map(lambda record: record['occupation'])
          .frequencies(sort=True)
          .topk(10, key=1))
```

As it can be seen, the functional paradigm is a viable alternative for most of array computations in data science frameworks. This is a key practice as leads to more concise code and fits better with other paradigms as dataflow that will be discussed later. The main problem is that current main languages, Python and R still require the student to be aware of buffering, in-memory operations and views, since the languages are imperative and not functional, and writing efficient code requires knowledge of those details.

4.2 Dataframe-Centric Manipulation

The second model introduces the mental model of a *dataframe*. This is essentially a collection of one-dimensional series that share a common index (which is represented typically at the "rows"), and are themselves indexed by name (represented in the "columns"). That is a pervasive notion of data collections, since for example, machine learning libraries assume a tabular representation of this form, with rows as instances and columns as variables or features. However, that notion of dataframe is actually used for holding different forms of data, and the schema that is implicitly required by many machine learning libraries is corresponding with the notion of "tidy data" [23].

Tidy data can be assimilated to normalized relational databases and it is a concept that helps in differentiating the typical structure of a machine learning problem from other types of data encoded in dataframes. As such, the teaching of the dataframe paradigm could be split in several "forms", that of tidy data and others that are untidy and thus may require transformation.

The dataframe paradigm for tidy data then reduces to bag processing when taking a simple column, or to something close to the notion of relational table

---

[6] https://spark.apache.org/docs/latest/rdd-programming-guide.html.

data otherwise. Actually, for students knowing SQL it proves useful to introduce dataframe manipulation using an analogy to SQL first. For example, in pandas: (a) boolean indexing can be assimilated to WHERE clauses, merge() for joins, and groupby() to the corresponding GROUP BY SQL clause. Then, multi-indexes, pivot tables and other advance or more flexible behaviour can be introduced as additional features to what is available in SQL.

A dataframe-centric approach emphasizes thinking on state (collections of tables) instead of on functions, even though the manipulations can be done applying vectorized functions. A typical interaction in an interactive environment as a Jupyter Notebook is typically a sequence of operations that modify, create views or provide data to libraries taking fragments of an in-memory dataframe. Memory hierarchy and parallelism are becoming largely transparent, e.g. with the abstraction of a Dask dataframe that provides an interface that is analogous to that of a regular pandas dataframe, as shown in the example in which compute() triggers the start of the computation of the expression t that is represented as a task graph.

```
import dask.dataframe as dd
df = dd.read_csv('files/2018-*.*.csv', parse_dates=['timestamp'])
t = df.groupby(df.timestamp.dt.hour).value.mean()
t.compute()
```

However, the technicalities and details of those are still surfacing the programming tasks. An example is the need to specify *chunking* parameters, that determine how the underlying task graph divides the data for parallel processing. Another example is the use of different schedulers, i.e. execution frameworks, as can be appreciated in Dask's documentation, e.g.: "The multiprocessing scheduler is an excellent choice when workflows are relatively linear, and so does not involve significant inter-task data transfer as well as when inputs and outputs are both small, like filenames and counts.". This still requires a mental model of the memory hierarchy and some operating system concepts.

4.3   Pipelines of Data Processing

The third model has the notion of a pipeline as the central processing element, and data becomes the input-output of a computation expressed as a directed graph of stateless operations that perform their task as data becomes available. These *dataflow* languages are inherently parallel and can work well in large, decentralized systems.

This paradigm requires a change in thinking from a dataframe-centric approach, since programs are activated by incoming data and do not follow the idea of a typical von-Neumann architecture. Reactive programming is considered a kind of dataflow that emphasizes the fact that the computation is continuous and triggered by external events. The following is an example word count for a stream of data using Apache Beam Python SDK.

```
counts = (lines
| 'split' >> (beam.ParDo(WordExtractingDoFn()))
.with_output_types(unicode))
| 'pair_with_one' >> beam.Map(lambda x: (x, 1))
| beam.WindowInto(window.FixedWindows(15, 0))
| 'group' >> beam.GroupByKey()
| 'count' >> beam.Map(count_ones))
```

As it can be seen, the different stages are composed with pipes, and follow well-known functional primitives. The new element that needs to be added is that of windowing, which introduces concerns for data flowing with some delay. This model can be used at the levels of array/bag processing as in the previous example, but could also be mixed with dataframe style manipulation as in the following Apache Spark example in Scala.

```
people.filter("age > 30")
.join(department, people("deptId") === department("id"))
.groupBy(department("name"), "gender")
.agg(avg(people("salary")), max(people("age")))
```

In both cases, the notion of pipeline is sufficient to introduce streaming elements into the previous paradigms seamlessly. These constructs are sufficient to convey the mental model of a stream of data that can be bursty and is virtually infinite, i.e. data that triggers computation.

5 Other Relevant Aspects

Here we describe additional contextual or related elements that are critical to provide an understanding of some decisions students face when selecting execution frameworks or libraries in particular situations.

5.1 Optimization

Another key decision is that of understanding how immutable structures affect optimization, and the implications of the functional paradigm in so. The two key ideas that require specific instruction are that of static typing and optimization. The idea that statically typed languages as C/C++ provide more information to code optimizers is fundamental and difficult to grasp by students with no CS background. However, it is required to understand why for example in Python, the Cython super-language is used, or even direct call to C or Fortran libraries. This is also key to understand approaches to provide hints for optimization. An interesting example is that of the library numba that translates Python functions to optimized machine code at runtime using the industry-standard LLVM compiler library. A typical decoration of a function with numba is as follows.

```
@vectorize(['float32(float32, float32, float32)',
'float64(float64, float64, float64)'], target='cuda')
def cu_discriminant(a, b, c):
    return math.sqrt(b ** 2 - 4 * a * c)
```

This provides a way of understanding the process of optimization that is related to computer architecture, and links with the idea of parallelism at the microprocessor level.

5.2   Understanding Costs

A typical problem in deciding the computational framework is that of the cost of maintenance and use of the infrastructure. While using a single-computer configuration, the costs are considered transparent, but the decision to switch to a cluster requires an understanding of the total cost of ownership (TCO), which is a complex topic and would require careful consideration of diverse factors [22]. The pricing of cloud computing or *on-premises* equivalents have become more complicated due to the widespread adoption of specialized hardware for particular computing problems, as in the case of training deep learning models [19].

All these hardware and infrastructure needs impact the decision on using some libraries or others, and require that data science students to have some basic knowledge of computer and network architecture. There are reports of using such kind of infrastructures [12] but not about specific instructional designs. At least a basic understanding of complexity notation and "big-O" classes of complexity is required so that students can reason about decisions on choosing one or other algorithm or approach.

5.3   Sequencing and Context

Sequencing is fundamental to instructional design, and it can be argued that the three models may be introduced using different paths. We have experimented an "array *then* dataframe *then* pipeline" approach as a layered approach that presents constructs as compositions of the previously presented ones. A possible alternative route is that of using a streaming-first approach, as it can be argued that data-at-rest is a particular case of data-at-motion [4].

A related aspect is that of providing students with the adequate context. This depends on the discipline or degree, as it should be different if the students are in a bioinfirmatics degree than in a business analytics program. In our case, the context falls in the latter, and required adapting cases and examples so that the training is not disconnect from the rest of the program, as typically real-world or realistic analytic cases come after the training focused on programming, data acquisition, cleaning and preparation. Our experience for our case shows that an approach to profit-driven analytics (see for example a case of churn in [14]) is key from the beginning. A case in which for example, a classifier needs to be selected by considering the cost of the errors provides the context to link business or domain context to technical work. This can also be discussed by situating the training in the framework of a process model as CRISP-DM.

6   Conclusions and Outlook

Teaching programming for computational science using modern data science frameworks requires a careful consideration on mental models, programming paradigms and other aspects that affect the task of coding for data wrangling or analytics.

We have discussed the findings on teaching computational science programming to ten cohorts of students of data science-related programs, coming from diverse, non-computer science backgrounds. We followed an action research framework, that included new elements and variations in different subsequent cohorts and evaluated the effect of these new elements through examining outcomes of assignments and interviewing students. The mental models discussed come from direct experience in facing student difficulties when understanding code and when translating the idea of code running on their laptop to similar code running as a sequence of parallel tasks, possibly in a distributed setting on a cluster.

We argue that there are three key fundamental programming models that should be dealt with separately for a complete understanding and effectiveness in leveraging current data analytics technology and libraries. We have discussed their main implications and relations, and the approach to introduce them as layers that add complexity. Essentially, array/bag models can be taught as purely functional, before introducing the dataframe centric approach and then a pipeline model that fits the additional constructs required to deal with streaming data. It is possible to abstract out the primitives from the concrete language, library or execution framework, which allows for a significant degree of transfer of learning, that is made evident when students move, for example, from a local Python environment to a distributed setting using Apache Spark.

The results discussed as subject to inherent limitations including type of program (oriented to business analytics) and the characteristics of the participants. However, we believe the discussion can be used as a basis for contrasting other experiences and delineating research directions. Future work should deal with specific understanding problems and the transition and effectiveness of the mental models suggested here. Further, controlled experiments should delve into the details of particular programming constructs, idioms or other forms of expression.

References

1. Ahmad, A., Feng, C., Ge, S., Yousif, A.: A survey on mining stack overflow: question and answering (Q&A) community. Data Technol. Appl. **52**(2), 190–247 (2018)
2. Ambler, A.L., Burnett, M.M., Zimmerman, B.A.: Operational versus definitional: a perspective on programming paradigms. Computer **25**(9), 28–43 (1992)
3. Akidau, T., Bradshaw, R., Chambers, C., et al.: The dataflow model: a practical approach to balancing correctness, latency, and cost in massive-scale, unbounded, out-of-order data processing. Proc. VLDB Endow. **8**(12), 1792–1803 (2015)
4. Akidau, T., Chernyak, S., Lax, R.: Streaming Systems: The What, Where, When, and How of Large-Scale Data Processing. O'Reilly Media Inc., Newton (2018)

5. Baškarada, S., Koronios, A.: Unicorn data scientist: the rarest of breeds. Program, **51**(1), 65-74 (2017)
6. Brunner, R.J., Kim, E.J.: Teaching data science. Procedia Comput. Sci. **80**, 1947–1956 (2016)
7. Cañas, J.J., Bajo, M.T., Gonzalvo, P.: Mental models and computer programming. Int. J. Hum. Comput. Stud. **40**(5), 795–811 (1994)
8. Chuprina, S., Alexandrov, V., Alexandrov, N.: Using ontology engineering methods to improve computer science and data science skills. Procedia Comput. Sci. **80**, 1780–1790 (2016)
9. Davenport, T.H., Patil, D.J.: Data scientist. Harv. Bus. Rev. **90**(5), 70–76 (2012)
10. Eijkhout, V.: Teaching MPI from mental models. In: Proceedings of the Workshop on Education for High Performance Computing, pp. 14–18. IEEE Press (2016)
11. Giabbanelli, P.J., Mago, V.K.: Teaching computational modeling in the data science era. Procedia Comput. Sci. **80**, 1968–1977 (2016)
12. Ivica, C., Riley, J.T., Shubert, C.: StarHPC–teaching parallel programming within elastic compute cloud. In: Proceedings of the 31st International Conference on Information Technology Interfaces, pp. 353–356. IEEE (2009)
13. Kurgan, L.A., Musilek, P.: A survey of Knowledge Discovery and Data Mining process models. Knowl. Eng. Rev. **21**(1), 1–24 (2006)
14. Maldonado, S., Flores, Á., Verbraken, T., Baesens, B., Weber, R.: Profit-based feature selection using support vector machines-general framework and an application for customer retention. Appl. Soft Comput. **35**, 740–748 (2015)
15. Malik, S.I.: Improvements in introductory programming course: action research insights and outcomes. Syst. Pract. Action Res. **31**, 1–20 (2018)
16. Manieri, A., Brewer, S., et al.: Data Science Professional uncovered: how the EDISON Project will contribute to a widely accepted profile for Data Scientists. In: IEEE 7th International Conference on Cloud Computing Technology and Science (CloudCom), pp. 588–593. IEEE (2015)
17. McKinney, W.: Data structures for statistical computing in Python. In: Proceedings of the 9th Python in Science Conference, vol. 445, pp. 51–56 (2010)
18. Saltz, J.S., Grady, N.W.: The ambiguity of data science team roles and the need for a data science workforce framework. In: 2017 IEEE International Conference on Big Data, pp. 2355–2361. IEEE (2017)
19. Sze, V., Chen, Y.H., Emer, J., Suleiman, A. Zhang, Z.: Hardware for machine learning: challenges and opportunities. In: 2017 IEEE Custom Integrated Circuits Conference (CICC), pp. 1–8. IEEE (2017)
20. Van Roy, P.: Programming paradigms for dummies: what every programmer should know. New Comput. Parad. Comput. Music **104**, 616–621 (2009)
21. Van Der Walt, S., Colbert, S.C., Varoquaux, G.: The NumPy array: a structure for efficient numerical computation. Comput. Sci. Eng. **13**(2), 22 (2011)
22. Walterbusch, M., Martens, B., Teuteberg, F.: Evaluating cloud computing services from a total cost of ownership perspective. Manag. Res. Rev. **36**(6), 613–638 (2013)
23. Wickham, H.: Tidy data. J. Stat. Softw. **59**(10), 1–23 (2014)

Numerical Analysis Project in ODEs for Undergraduate Students

Sigurdur Hafstein$^{(\boxtimes)}$ (iD)

Science Institute, University of Iceland, Dunhagi 3, 107 Reykjavík, Iceland
shafstein@hi.is

Abstract. Designing good projects involving programming in numerical analysis for large groups of students with different backgrounds is a challenging task. The assignment has to be manageable for the average student, but to additionally inspire the better students it is preferable that it has some depth and leads to them to think about the subject. We describe a project that was assigned to the students of an introductory Numerical Analysis course at the University of Iceland. The assignment is to numerically compute the length of solution trajectories of a system of ordinary differential equations with a stable equilibrium point. While not difficult to do, the results are somewhat surprising and got the better students to get interested in what was happening. We describe the project, its solution using Matlab, and the underlying mathematics in some detail. Further, we discuss the pedagogical aspects of the project and the results in terms of its success and shortcomings.

Keywords: Scientific computing project ·
Ordinary differential equations · Numerical integration ·
Lyapunov functions

1 Background

The project we describe in this paper was assigned to the undergraduate students of Numerical Analysis, a 6 ECTS unit course at the University of Iceland with approximately 150 students. The responsibility for this course is within the Faculty of Physical Sciences in the School of Engineering and Natural Sciences and it is mandatory for all BSc. students of Mechanical-, Industrial-, Chemical-, and Civil and Environmental Engineering as well as for all students of Physics, Engineering Physics, Mathematics, Applied Mathematics, and Mathematics and Mathematical Education. It is an elective course for students of Geophysics, Computer Science, Software Engineering, Electrical and Computer Engineering, and Chemistry.

As can be seen from this long lists the preparation and interests of the enrolled students vary considerably. Usually, the students enroll in the course in the fourth semester of six in total to complete a bachelors degree. Prerequisites for Numerical Analysis are one course in Computer Science (programming in Matlab

© Springer Nature Switzerland AG 2019
J. M. F. Rodrigues et al. (Eds.): ICCS 2019, LNCS 11540, pp. 421–434, 2019.
https://doi.org/10.1007/978-3-030-22750-0_34

or Python), one course in Linear Algebra, and two courses (three recommended) in Calculus. All of the students are thus familiar with applying linear algebra and calculus, but the mathematics students have also studied the theoretical aspects of these disciplines in some detail in the framework of metric spaces, ring theory, etc. Students of Mathematics and Applied Mathematics are required to enroll simultaneously in the course Theoretical Numerical Analysis (2 ECTS units), where rigid mathematical proofs of most of the material covered in the Numerical Analysis course are studied.

The **Course Description** is:

> Fundamental concepts on approximation and error estimates. Solutions of systems of linear and non-linear equations. PLU decomposition. Interpolating polynomials, spline interpolation and regression. Numerical differentiation and integration. Extrapolation. Numerical solutions of initial value problems of systems of ordinary differential equations. Multistep methods. Numerical solutions to boundary value problems for ordinary differential equations.

and the **Learning Outcomes** are:

> Knowledge and understanding: To complete this course the student should be able to
> 1. define, explain and give examples of the main concepts of the course, such as error, matrix factorization, interpolating polynomial, and finite differences,
> 2. state and explain the main results of the course, for example by stating Newton's Method and the Secant Method and estimate the errors, state algorithms to solve boundary value problems using finite differences and verify the degree of the approximation.
>
> Skills: To complete this course the student should be able to
> 1. formulate a simple mathematical problem as a numerical problem, implement it on a computer, and compute an approximate solution,
> 2. estimate the error of numerical solutions,
> 3. use computer software, such as the Anaconda Python platform or Matlab, for programming, computing, and performing numerical experiments,
> 4. validate the results of numerical computations,
> 5. use the concepts and the results of the course to develop and advance algorithms for simple problems the student has not seen before.

In the course two larger group projects count for 30% of the final grade. The assignment we describe here was the third and last part of the second project. The other two parts were to

1. implement adaptive integration using the trapezoidal- and the Simpson's rule and test it for some integrals and
2. implement the shooting method and use it to solve a few boundary value problems.

Adaptive integration and the shooting method are discussed in sufficient detail in the lectures and in the textbook used in the class [20] to make them rather easy to do.

2   The Project and Its Solution

In the project the time-reversed van der Pol oscillator

$$\mathbf{x}' = \mathbf{f}(\mathbf{x}), \quad \text{where} \quad \mathbf{f}(x, y) = \begin{pmatrix} -y \\ -4(1 - x^2)y + x \end{pmatrix} \tag{1}$$

is considered. It has an exponentially stable equilibrium at the origin and an unstable periodic orbit around it; see below for more details. The project has three objectives, which we describe below together with its solution using Matlab.

2.1   Objective I

The first objective of the assignment was to analyze the system (1) by drawing solution trajectories, both forward and backwards in time. This is easily achieved by first defining in f.m

```
1  function y=f(t,x)
2      mu=4.0;
3      y(1)=-x(2);
4      y(2)=-mu*(1-x(1)^2)*x(2)+x(1);
5      y=y';
6  end
```

and then typing in the command window, here using the initial value $(2, 0)^T$ for $\mathbf{x}' = \mathbf{f}(\mathbf{x})$ and $(1, 1)^T$ for $\mathbf{x}' = -\mathbf{f}(\mathbf{x})$ and in both cases integrating over the time-interval $[0, 20]$.

```
1  >> [t,w]=ode45(@f,[0,20],[2,0]);
2  >> plot(w(:,1),w(:,2))
3  >> [t,w]=ode45(@(t,x) -f(t,x),[0,20],[1,1]);
4  >> plot(w(:,1),w(:,2))
5  >> xlabel('X');ylabel('Y');
```

The plots produced are drawn in Fig. 1. A few comments are in order. Usually, one defines the function f as a function of both time t and space x, even though the system is autonomous, i.e. f does not depend explicitly on time. We follow this tradition here, for otherwise we could not use the Matlab solver ode45 directly. The system $\mathbf{x}' = -\mathbf{f}(\mathbf{x})$ is a time-reversion of the time-reversed van der Pol oscillator from (1), i.e. it is the van der Pol oscillator. It is well known that the van der Pol oscillator has a stable periodic orbit and it can be clearly seen in Fig. 1 (right). Since the periodic orbit is stable all "normal" initial values, except

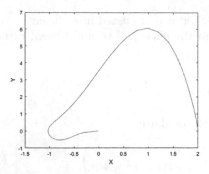 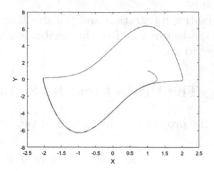

Fig. 1. Left: A solution trajectory of system $\mathbf{x}' = \mathbf{f}(\mathbf{x})$ with $\mathbf{f}$ from (1), starting at $(2,0)^T$ and integrated numerically over the time-interval $[0,20]$ using Matlab's ode45. Right: A solution trajectory of system $\mathbf{x}' = -\mathbf{f}(\mathbf{x})$ with $\mathbf{f}$ from (1), starting at $(1,1)^T$ and integrated numerically over the time-interval $[0,20]$ using Matlab's ode45.

the unstable equilibrium point $(0,0)^T$, will converge to the orbit. The system $\mathbf{x}' = \mathbf{f}(\mathbf{x})$ is the time-reversed van der Pol oscillator and the stable periodic orbit of the van der Pol oscillator becomes an unstable orbit and the unstable equilibrium point at the origin becomes stable. Thus, any initial value inside of the periodic orbit will converge to the equilibrium and outside of the orbit will diverge. Note that the Matlab solver ode45 (and others) reports problems, i.e. "Unable to meet integration tolerances without reducing the step size below the smallest value allowed" in the integration for various initial values and time-intervals. This can be overcome by using a solver with fixed time-steps, to be delivered in the next objective.

2.2 Objective II

The second objective of the project was to program the standard Runge-Kutta method of order 4, commonly abbreviated RK4. In more detail, an implementation for the function

```
1   function [t,w]=RK4(f,xi,a,b,n)
```

should be given, where f is the function $\mathbf{f}$ in $\mathbf{x}' = \mathbf{f}(t, \mathbf{x})$, xi is the initial value at time a, and [a,b] is the time-interval over which the solution is computed, and n is the number of time-steps used. The output [t,w] should be the transposes of the outputs of the Matlab solver ode45 (column vector notation). This is an easy task whose solution is given in the Appendix in Program I. Using RK4 to plot solution trajectories as in Objective I now becomes, using 500 time-steps:

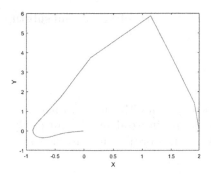 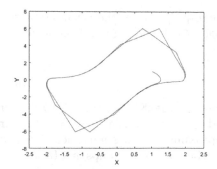

Fig. 2. Left: A solution trajectory of system $\mathbf{x}' = \mathbf{f}(\mathbf{x})$ with $\mathbf{f}$ from (1), starting at $(2,0)^T$ and integrated numerically over the time-interval $[0,20]$ using RK4 and 100 time-steps (step-size $20/100 = 0.2$). Right: A solution trajectory of system $\mathbf{x}' = -\mathbf{f}(\mathbf{x})$ with $\mathbf{f}$ from (1), starting at $(1,1)^T$ and integrated numerically over the time-interval $[0,20]$ again using RK4 and 100 time-steps (step-size $20/100 = 0.2$).

```
1 >> [t,w]=RK4(@f,[2,0]',0,20,500);
2 >> plot(w(1,:),w(2,:))
3 >> [t,w]=RK4(@(t,x) -f(t,x),[1,1],0,20,500);
4 >> plot(w(1,:),w(2,:))
5 >> xlabel('X');ylabel('Y');
```

This delivers trajectories comparable to the ones in Fig. 1. Note, that much fewer time-steps, e.g. 100, results in much less accurate results, see Fig. 2. We now move to the main objective of the problem.

2.3   Objective III

On a uniform 101×101 grid on $[-3,3] \times [-8,8]$, compute the length of the solution trajectories to (1) integrated over a time-interval of length 4 and using RK4 with 100 time-steps. Note that 100 time-steps for a time-interval of length 4 corresponds to the 500 time-steps for a time-interval of length 20 used above, because both have step-size 0.04. The solution to system (1), starting at $\boldsymbol{\xi} \in \mathbb{R}^2$ at time zero, is a function

$$t \mapsto \phi(t,\boldsymbol{\xi}) \text{ fulfilling } \phi(0,\boldsymbol{\xi}) = \boldsymbol{\xi} \text{ and } \phi'(t,\boldsymbol{\xi}) = \frac{d}{dt}\phi(t,\boldsymbol{\xi}) = \mathbf{f}(\phi(t,\boldsymbol{\xi})) \quad (2)$$

for all t in the definition domain of $\phi(\cdot,\boldsymbol{\xi})$ (dependant of $\boldsymbol{\xi}$ either $\mathbb{R}$ or the maximum domain before ϕ becomes infinite). The length of the trajectory $t \mapsto \phi(t,\boldsymbol{\xi})$ on the time-interval $[0,T]$ is well known to be defined as

$$V(\boldsymbol{\xi}) := \int_0^T \|\phi'(\tau,\boldsymbol{\xi})\| d\tau,$$

where $\| \cdot \|$ denotes the Euclidian norm on $\mathbb{R}^2$. Because of (2) we can substitute $\mathbf{f}(\phi(\tau, \boldsymbol{\xi}))$ for $\phi'(\tau, \boldsymbol{\xi})$ and the formula for $V(\boldsymbol{\xi})$ becomes

$$V(\boldsymbol{\xi}) := \int_0^T \|\mathbf{f}(\phi(\tau, \boldsymbol{\xi}))\| d\tau. \tag{3}$$

Since $[\texttt{t,w}]=\texttt{RK4(@f,xi,0,T,n)}$ delivers $\phi(t_i, \boldsymbol{\xi})$ in its ith column, where $\texttt{xi} = \boldsymbol{\xi}$ and $t_i = (i-1)\texttt{T/n}$ for $i = 1, 2, \ldots, \texttt{n + 1}$, the integral in (3) can easily be approximated using numerical integration. In the project it was suggested to use the composite Simpson's Rule

$$\int_0^T g(\tau) d\tau \approx \frac{h}{3} \left(g(t_1) + g(t_{n+1}) + 4 \sum_{i=1}^{n/2} g(t_{2i}) + 2 \sum_{i=1}^{n/2-1} g(t_{2i+1}) \right),$$

where $h = T/n$ is the length of the time-steps. Note that for this formula we need that n is an even number. Since many of the trajectories will be very long, indeed so long that the numerical solver will fail to deliver a numerical value, it is useful to substitute e.g. $V(\boldsymbol{\xi}) \leftarrow \min\{4T, V(\boldsymbol{\xi})\}$ (Matlab's $\texttt{min}$ interprets $\texttt{NaN}$ as larger than any number). The implementation is now simple and is given in the Appendix in Program II. Typing

```
1  >> TraLengths(@f,-3,3,-8,8,4,100)
```

in the command window, where $\texttt{TraLengths}$ is the program from the Appendix, $\texttt{f}$ is the function from $\texttt{f.m}$, $[-3, 3] \times [-8, 8]$ is the the area in the plane where a function is computed and plotted, $\texttt{T = 4}$ is the interval of integration in (3), and $\texttt{n=100}$ is the number of time-steps used in the numerical integration, now delivers after a short time (12.5 s on a computer with an i7-7700K CPU) Fig. 3.

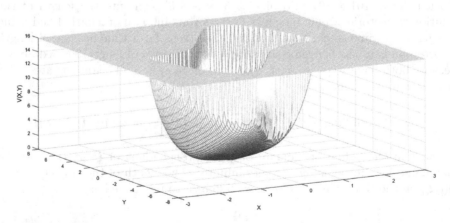

Fig. 3. The length of the trajectories of system (1) plotted as a function of starting position (x, y).

After having produced this figure the less interested students have finished the project. The more interested ones, however, now might start to ask themselves and the instructor why this function looks like it does? Indeed, many of them did ask the instructor.

3   Study of the Results

Let us discuss the results from Objective III and the function V computed in more detail.

Because the origin is a stable equilibrium, solutions $t \mapsto \phi(t, \boldsymbol{\xi})$ slow down close to it. Indeed, they become so slow that the limit $\phi(t, \boldsymbol{\xi}) \to \mathbf{0}$ when $t \to \infty$ is never reached. The reason for this is that since $\mathbf{f}$ is continuous and $\mathbf{f}(\mathbf{0}) = \mathbf{0}$, $\mathbf{f}$ and then also $\mathbf{x}'$ are small close to the origin. Therefore trajectories that start close $\boldsymbol{\xi} \approx \mathbf{0}$ to the origin are short and $V(\boldsymbol{\xi}) \approx 0$.

The function V is a so-called Lyapunov function for the system. This means that it has a local minimum at the stable equilibrium at the origin and that it is decreasing along all solution trajectories in a neighbourhood of the equilibrium. The local minimum is intuitively clear and the decrease can be seen from the following calculations:

$$\frac{d}{dt} V(\phi(t, \boldsymbol{\xi}))\bigg|_{t=0} = \lim_{h \to 0} \frac{V(\phi(h, \boldsymbol{\xi})) - V(\boldsymbol{\xi})}{h} \tag{4}$$

$$= \lim_{h \to 0} \frac{1}{h} \left[\int_0^T \|\phi(\tau, \phi(h, \boldsymbol{\xi}))\| d\tau - V(\boldsymbol{\xi}) \right]$$

and because solution trajectories are unique, i.e. $\phi(\tau, \phi(h, \boldsymbol{\xi})) = \phi(\tau + h, \boldsymbol{\xi})$, we get

$$\int_0^T \|\phi(\tau, \phi(h, \boldsymbol{\xi}))\| d\tau = \int_0^T \|\phi(\tau + h, \boldsymbol{\xi})\| d\tau$$

$$= \int_h^{T+h} \|\phi(s, \boldsymbol{\xi})\| ds$$

$$= \int_0^T \|\phi(s, \boldsymbol{\xi})\| ds + \int_T^{T+h} \|\phi(s, \boldsymbol{\xi})\| ds - \int_0^h \|\phi(s, \boldsymbol{\xi})\| ds$$

$$= V(\boldsymbol{\xi}) + \int_T^{T+h} \|\phi(s, \boldsymbol{\xi})\| ds - \int_0^h \|\phi(s, \boldsymbol{\xi})\| ds.$$

Thus, by (4) and the Fundamental Theorem of Calculus,

$$\frac{d}{dt} V(\phi(t, \boldsymbol{\xi}))\bigg|_{t=0} = \lim_{h \to 0} \frac{V(\phi(h, \boldsymbol{\xi})) - V(\boldsymbol{\xi})}{h}$$

$$= \lim_{h \to 0} \frac{1}{h} \left[\int_T^{T+h} \|\phi(s, \boldsymbol{\xi})\| ds - \int_0^h \|\phi(s, \boldsymbol{\xi})\| ds \right]$$

$$= \|\phi(T, \boldsymbol{\xi})\| - \|\phi(0, \boldsymbol{\xi})\|$$

$$= \|\phi(T, \boldsymbol{\xi})\| - \|\boldsymbol{\xi}\|.$$

Since $\lim_{t\to\infty} \|\phi(t, \boldsymbol{\xi})\| = 0$ for all $\boldsymbol{\xi}$ in a neighbourhood of the origin, we have for all such $\boldsymbol{\xi}$ that $\|\phi(T, \boldsymbol{\xi})\| < \|\boldsymbol{\xi}\|$ for sufficiently large $T > 0$ and $\boldsymbol{\xi} \neq 0$. Thus

$$V'(\boldsymbol{\xi}) := \frac{d}{dt} V(\phi(t, \boldsymbol{\xi})) \bigg|_{t=0} < 0$$

and V is decreasing along solution trajectories in a neighbourhood of the stable equilibrium at the origin, i.e. the mapping $t \mapsto V(\phi(t, \boldsymbol{\xi}))$ is decreasing.

The stability theory of Lyapunov and Lyapunov functions are covered in most textbooks on dynamical systems and/or control theory, e.g. [16,19,21], and the interested student can be pointed to them for additional information. For an attractor, like our stable equilibrium at the origin for the time-reversed van der Pol system, the sublevel sets of a Lyapunov function serve as "traps". Once inside the component of a sublevel set, that includes the origin and does not extend to the boundary of the domain of the Lyapunov function, the solution cannot escape. This comes because the solution is decreasing along solution trajectories and therefore cannot climb over the edge/boundary of the sublevel set.

The theory of complete Lyapunov functions even tells us that every system given by an ODE possesses a *complete Lyapunov function* that goes a long way in characterizing the qualitative behaviour of the system. Indeed, this holds true for very general dynamical systems. A complete Lyapunov function is a scalar-values function from the whole state-space that is non-increasing along along all solution trajectories and strictly decreasing where possible. Note that, e.g. for a periodic orbit, it cannot be strictly decreasing. In general it is strictly decreasing along all solution trajectories on the part of the state-space where the flow is gradient-like and constant on every transitive component of the *chain-recurrent set*. The theory of complete Lyapunov functions was developed by Auslander, Conley, and Hurley [1,6,13–15], see also [18].

Computing Lyapunov functions by using results from converse theorems in dynamical systems, i.e. theorems guarantying the existence of certain kinds of Lyapunov functions for systems with particular stability properties, using integrals or sums over solutions trajectories has been studied in numerous publications [2–5,7–12,17]. A central issue is the verification of the properties of a Lyapunov function for the function computed. One way to do this is to interpolate the values computed over the simplices of a triangulation. Essentially, one demands $\nabla V(\mathbf{x}) \cdot \mathbf{f}(\mathbf{x}) + E_{\mathbf{x}} \|\nabla V(\mathbf{x})\|_1 < 0$ at all vertices of the triangulation, where the error $E_{\mathbf{x}} \geq 0$ assures that not only

$$\frac{d}{dt} V(\phi(t, \mathbf{x})) \bigg|_{t=0} = \nabla V(\mathbf{x}) \cdot \mathbf{f}(\mathbf{x}) < 0$$

at the vertices, but for any $\mathbf{x}$ in the domain of the triangulation. Here $\|\mathbf{x}\|_1 = |x_1| + |x_2|$ and the function V is defined on the simplices by using convex interpolation of the values at the vertices over the whole simplex. A detailed discussion of this method is beyond the scope of this paper, but the interested reader can have a look at [11,12].

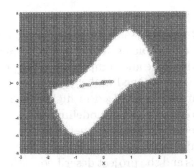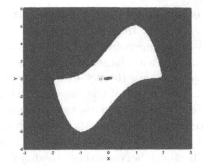

Fig. 4. Verification of the negativity of the orbital derivative of the function V computed in Objective III (left). Areas where it is not negative are plotted in red. Not only is it not negative outside of the periodic orbit, but also in a strip within it. By increasing the spatial resolution the area where the orbital derivative fails to be negative inside of the orbit becomes smaller (right).

By adding to the code for the function `TraLengths` as described in Program III in the Appendix, a verification of the interpolation of the values as described in [11,12] is carried out. If one want to implement this for a different two-dimensional system than (1), only the function `f` in `f.m` and the function `B` in Program III have to be modified accordingly. The modification of `f` is obvious because it is the right-hand side of (1), and the return value for `B(r,s,xm,ym)` should be an upper bound, preferably close, on

$$\max_{\substack{i=1,2 \\ |x_1|\le \text{xm}, |x_2|\le \text{ym}}} \left| \frac{\partial^2 f_i}{\partial x_r \partial x_s}(x_1, x_2) \right|.$$

The results of this verification for our system (1) with the same parameters as in Objective III is plotted in Fig. 4 (left) and with a higher spatial resolution, i.e. `xres=301; yres=301;`, on the right.   By using larger values for the parameters `xres,yres,T,n`, the area within the periodic orbit where the orbital derivative fails to be negative is effectively reduced to an arbitrary small neighbourhood of the equilibrium at the origin, cf. [12, Figs. 1,2]. Using a compiled programming language like C++ also makes these computations very fast.

4   Conclusion

We presented a new project for undergraduate students in Numerical Analysis. It is not difficult to solve the problem and thus the average student should be able to do so without too much difficulties. The solution, i.e. the function plotted, is somewhat surprising and intended to awaken the interest of the better students. The solution to the project is given and the results and the underlying theory are discussed in some detail. Matlab code is given for all objectives of the project together with code for a more sophisticated verification of the results. Further, numerous references to the underlying theory are given.

Let us compare the presented project, in terms of pedagogical value, to the larger projects from the textbook [20] of the course. The textbook contains, apart from numerous Exercises and Computer Problems, eleven so-called Reality Checks, which are larger project of comparable scale and complexity to the project presented. These Reality-Checks are very well designed and include topics such as the kinematics of the Steward Platform frequently used in flight-simulators, positioning using GPS, motion control in computer-aided modeling, and a simple audio codec. These Reality Checks have been used as assignments for the larger projects in the course discussed in the paper with a great success. The difference, however, between those Reality Checks and the project described, is that the Reality Checks inspire the better students to become interested in the technology behind the project, e.g. flight simulators, GPS, audio codecs, but not in mathematics. The teacher of the course has had lively discussion with students about these technological topics during and after the Reality Checks projects, but rarely about mathematics. A possible drawback of the proposed project is that the theory behind the project has little direct connection to technology. While a project on, say the Stewart Platform, might invoke the interest of an engineering student in numerical root-finding of nonlinear equations because she/he finds the steering of a flight simulator fascinating, she/he will just routinely solve the project proposed in this paper without gaining any interest in numerical analysis at all.

The advantage of the proposed project is that it inspires some of the mathematics students to get interested in the theory of dynamical systems, a topic that they had only known indirectly from studying differential equations. Further, as Lyapunov functions are a mathematical extension of the concept of dissipative energy from physics, some of the physics students were particularly interested as well. This allowed to the teacher to discuss attractors, repellers, chaos, and complete Lyapunov functions with interested students.

Appendix

Below is an implementation in Matlab to a solution of the project.
Program I:

```
1  function [t,w]=RK4(f,xi,a,b,n)
2      h=(b-a)/n;
3      t=a:h:b;
4      w=zeros(length(xi),n+1);
5      w(:,1)=xi;
6      for i=1:n
7          wi=w(:,i);
8          s1=f(t(i),wi);
9          s2=f(t(i)+h/2,wi+h/2*s1);
10         s3=f(t(i)+h/2,wi+h/2*s2);
11         s4=f(t(i)+h,wi+h*s3);
12         w(:,i+1)=wi+h/6*(s1+2*s2+2*s3+s4);
13     end
14 end
```

Program II:

```
1  function TraLengths(f,ax,bx,ay,by,T,n)
2    xres=101; yres=101;
3    V=zeros(xres,yres);
4    vx=zeros(n+1,1);
5    x=linspace(ax,bx,xres);
6    y=linspace(ay,by,yres);
7    for ix=1:xres
8      for jy=1:yres
9        xi=[x(ix),y(jy)]';
10       [t,w]=RK4(f,xi,0,T,n);
11       for ivx=1:n+1
12         vx(ivx)=norm(f(t,w(:,ivx)));
13       end
14       V(ix,jy)=T/(3*n)*(vx(1)+vx(n+1)+4*sum(vx(2:2:n))...
15                 +2*sum(vx(3:2:n-1)));
16      end
17    end
18    V=min(V,4*T*ones(size(V)));
19    mesh(x,y,V')
20    xlabel('X');ylabel('Y');zlabel('V(X,Y)');
21  end
```

Program III: Modification of TraLengths to additionally verify the validity of the conditions for a Lyapunov function (decrease of the orbital derivative).

Add the following code to the function TraLengths between lines 19 and 20.

```
1    % verify decrease of orbital derivative
2    hx=(bx-ax)/(xres-1);
3    hy=(by-ay)/(yres-1);
4    As1(1,1)=hx; As1(1,2)=hx; As1(2,1)=0; As1(2,2)=hy;
5    As2(1,1)=0; As2(1,2)=hx; As2(2,1)=hy; As2(2,2)=hy;
6    k=1;
7    for ix=1:xres-1
8      for jy=1:yres-1
9        xm=max(abs(x(ix)),abs(x(ix+1)));
10       ym=max(abs(y(jy)),abs(y(jy+1)));
11       % sigma = ()
12       gV=[(V(ix+1,jy)-V(ix,jy))/hx, ...
13           (V(ix+1,jy+1)-V(ix+1,jy))/hy];
14       gVn=norm(gV,1);
15       err1=0;
16       c1=~(dot(gV,f(0,[x(ix),y(jy)]))+err1*gVn < 0);
17       err2=E(1,xm,ym,As1);
18       c2=~(dot(gV,f(0,[x(ix+1),y(jy)]))+err2*gVn < 0);
```

```
19        err3=E(2,xm,ym,As1);
20        c3=~(dot(gV,f(0,[x(ix+1),y(jy+1)]))+err3*gVn < 0);
21        % sigma = (1 2)
22        gV=[(V(ix+1,jy+1)-V(ix,jy+1))/hx, ...
23            (V(ix,jy+1)-V(ix,jy))/hy];
24        gVn=norm(gV,1);
25        err4=0;
26        c4=~(dot(gV,f(0,[x(ix),y(jy)]))+err4*gVn < 0);
27        err5=E(1,xm,ym,As2);
28        c5=~(dot(gV,f(0,[x(ix),y(jy+1)]))+err5*gVn < 0);
29        err6=E(2,xm,ym,As2);
30        c6=~(dot(gV,f(0,[x(ix+1),y(jy+1)]))+err6*gVn < 0);
31        % check if ~(orbital derivative < 0)
32        if c1 || c2 || c3 || c4 || c5 || c6
33           NotNegx(k)=x(ix)+hx/2;
34           NotNegy(k)=y(jy)+hy/2;
35           k=k+1;
36           continue
37        end
38      end
39   end
40   hold on
41   plot3(NotNegx,NotNegy,zeros(k-1,1),'ro')
```

Further, add the following functions after the code for the function TraLengths

```
1  function Bval=B(r,s,xm,ym)
2    if r==1 && s==1
3      Bval=8*ym;
4    elseif r==2 && s==2
5      Bval=0;
6    else
7      Bval=8*xm;
8    end
9  end
10
11 function Eval=E(i,x,y,A)
12   Eval=0;
13   for r=1:2
14     for s=1:2
15       Eval=Eval+B(r,s,x,y)*A(r,i)*(A(s,i)+A(s,2));
16     end
17   end
18   Eval=0.5*Eval;
19 end
```

References

1. Auslander, J.: Generalized recurrence in dynamical systems. Control Differ. Equ. **3**, 65–74 (1964)
2. Björnsson, J., Giesl, P., Hafstein, S.: Algorithmic verification of approximations to complete Lyapunov functions. In: Proceedings of the 21st International Symposium on Mathematical Theory of Networks and Systems, vol. 0180, pp. 1181–1188, Groningen, The Netherlands (2014)
3. Björnsson, J., Giesl, P., Hafstein, S., Kellett, C., Li, H.: Computation of continuous and piecewise affine Lyapunov functions by numerical approximations of the Massera construction. In: Proceedings of the CDC, 53rd IEEE Conference on Decision and Control, Los Angeles (CA), USA, pp. 5506–5511 (2014)
4. Björnsson, J., Giesl, P., Hafstein, S., Kellett, C., Li, H.: Computation of Lyapunov functions for systems with multiple attractors. Discrete Contin. Dyn. Syst. Ser. A **35**(9), 4019–4039 (2015)
5. Björnsson, J., Hafstein, S.: Efficient Lyapunov function computation for systems with multiple exponentially stable equilibria. Procedia Comput. Sci. **108**, 655–664 (2017). Proceedings of the International Conference on Computational Science (ICCS), Zurich, Switzerland (2017)
6. Conley, C.: Isolated Invariant Sets and the Morse Index. CBMS Regional Conference Series, vol. 38, American Mathematical Society (1978)
7. Doban, A.: Stability domains computation and stabilization of nonlinear systems: implications for biological systems. Ph.D. thesis, Eindhoven University of Technology (2016)
8. Doban, A., Lazar, M.: Computation of Lyapunov functions for nonlinear differential equations via a Yoshizawa-type construction. IFAC-PapersOnLine **49**(18), 29–34 (2016)
9. Doban, A., Lazar, M.: Computation of Lyapunov functions for nonlinear differential equations via a Massera-type construction. IEEE Trans. Autom. Control **63**(5), 1259–1272 (2018)
10. Hafstein, S., Kellett, C., Li, H.: Computing continuous and piecewise affine Lyapunov functions for nonlinear systems. J. Comput. Dyn. **2**(2), 227–246 (2015)
11. Hafstein, S., Valfells, A.: Study of dynamical systems by fast numerical computation of Lyapunov functions. In: Proceedings of the 14th International Conference on Dynamical Systems: Theory and Applications (DSTA). Mathematical and Numerical Aspects of Dynamical System Analysis, pp. 220–240 (2017)
12. Hafstein, S., Valfells, A.: Efficient computation of Lyapunov functions for nonlinear systems by integrating numerical solutions. Nonlinear Dyn. (2019, to be published)
13. Hurley, M.: Chain recurrence and attraction in non-compact spaces. Ergod. Theory Dyn. Syst. **11**, 709–729 (1991)
14. Hurley, M.: Chain recurrence, semiflows, and gradients. J. Dyn. Differ. Equ. **7**(3), 437–456 (1995)
15. Hurley, M.: Lyapunov functions and attractors in arbitrary metric spaces. Proc. Am. Math. Soc. **126**, 245–256 (1998)
16. Khalil, H.: Nonlinear Systems, 3rd edn. Pearson (2002)
17. Li, H., Hafstein, S., Kellett, C.: Computation of continuous and piecewise affine Lyapunov functions for discrete-time systems. J. Differ. Equ. Appl. **21**(6), 486–511 (2015)
18. Patrão, M.: Existence of complete Lyapunov functions for semiflows on separable metric spaces. Far East J. Dyn. Syst. **17**(1), 49–54 (2011)

19. Sastry, S.: Nonlinear Systems: Analysis, Stability, and Control. Springer, New York (1999). https://doi.org/10.1007/978-1-4757-3108-8
20. Sauer, T.: Numerical Analysis, 2nd edn. Pearson (2012)
21. Vidyasagar, M.: Nonlinear System Analysis. Classics in applied mathematics, 2nd edn. SIAM (2002)

Poster Track

Mixed Finite Element Solution for the Natural-Gas Dual-Mechanism Model

Mohamed F. El-Amin[1,2]($\boxtimes$), Jisheng Kou[3], Shuyu Sun[4], and Jingfa Li[4]

[1] College of Engineering, Effat University, Jeddah 21478, Kingdom of Saudi Arabia
momousa@effatuniversity.edu.sa
[2] Mathematics Department, Faculty of Science,
Aswan University, Aswan 81528, Egypt
[3] School of Mathematics and Statistics,
Hubei Engineering University, Xiaogan 432000, Hubei, China
[4] King Abdullah University of Science and Technology (KAUST),
Thuwal 23955–6900, Kingdom of Saudi Arabia

Abstract. The present work is dedicated to studying the transfer of natural gas in shale formations. The governing model was developed on the basis of the model of dual-porosity dual-permeability (DPDP). The mixed finite element method (MFEM) is employed to solve the governing equations numerically. Numerical example is presented and results discussed such as production cumulative rate, pressure and apparent permeability.

Keywords: Shale-gas · Mixed finite element · Porous media · IMPES

1 Introduction

Shale–gas resources represent around 30% of the world resources of the natural gas and it has much promise as a relatively new clean energy since its combustion produces much less CO_2. However, the gas production from the shale formation does not obey the conventional ways of gas/oil production from conventional reservoirs. The shale-formations have very-low permeabilities compared to the conventional formations. The production efficiency of the natural gas from shales increases as the efficiency of the fracture network increases.

The model of flow in fractured porous media contains the effect of fluid transfer from matrix blocks to fractures with dual-mechanism such as dual-porosity dual-permeability. In the conventional models of fractured formations, Darcy's law is employed to describe the flow. However, Darcy's law does not give a suitable description of the flow in shales because the gas flow in the nano-sized pores has slippage on the surface. One of the modeling directions for gas transport in shale formations is to adopt the framework of flow in fractured porous media. Using dual-continua mechanisms, an idealized model was built

© Springer Nature Switzerland AG 2019
J. M. F. Rodrigues et al. (Eds.): ICCS 2019, LNCS 11540, pp. 437–444, 2019.
https://doi.org/10.1007/978-3-030-22750-0_35

by Warren and Root [1] to describe flow in fractured media. In this model, the matrix field is divided as uniform cubes separated by fractures. The dual-continuum models have been developed by several researchers (e.g. Bustin et al. [2]; and [3]) to model gas transport in shales. These models incorporated different physics such as inertial, gas slippage, Knudsen diffusion, etc. Recently, El-Amin et al. [4] used the dual-porosity dual- permeability model with geomechanical effect. Ge et al. [5] presented a quantitative evaluation for organic related pores in unconventional reservoir. A comprehensive review has been introduced on gas transport in tight and shale formations was done by Salama et al. [6]. El-Amin and coauthors [7–9] have extended the mathematical model of shale-gas flow and developed some numerical or analytical solutions.

In the current work, we present a mixed finite element method (MFEM) which is a locally conservative method [10] to solve the dual-porosity dual-permeability (DPDP) shale-gas model. We presented some numerical tests to show the efficiency of the numerical scheme.

2   Modeling and Formulation

In this section, we develop a DPDP model to simulate the natural gas flow in porous media. The DPDP model consists of two mass balance equations, one for flow in matrix blocks and another for flow in fractures. The flow is assumed to be a single-phase and isothermal, and the gravity effect was neglected. The mass balance equation has the form,

$$\frac{\partial M}{\partial t} + \nabla \cdot \rho \mathbf{u} = Q, \tag{1}$$

such that t is the time; M is the mass-accumulation; ρ is the gas density; $\mathbf{u}$ is the velocity; Q is the source-term. In the matrix blocks, there exist two mechanisms, namely, free gas and absorbed gas. So, the mass accumulation term may be written as,

$$M_{fr} = \phi\rho, \tag{2}$$

However, the mass accumulation term for the adsorption process on the matrix surface is,

$$M_{ads} = (1 - \phi)q_a, \tag{3}$$

such that q_a is the volume of the adsorbed-gas on the shale-surface. The adsorption is described by the Langmuir isotherm,

$$q_a = \frac{\rho_s M_w V_L p_m}{V_{std}(P_L + p_m)} \tag{4}$$

where V_{std} and V_L are, respectively, mole volume under standard conditions and the Langmuir volume. ρ_s is the rock density and M_w is the molar weight. P_L and p_m are, respectively, Langmuir pressure and matrix pressure.

Using Eqs. (2)–(4), we get,

$$M = \phi\rho + (1 - \phi)q_a, \tag{5}$$

The mass density can be written as, $\rho_{g,d} = \frac{p_d M_w}{ZRT}, d = m, f$. R, T, m, V and Z are the universal gas constant, temperature, mass, volume, and compressibility factor, respectively. Thus, $Z = \frac{pV}{RT}$, such that p is the pressure. The Peng-Robinson equation of state has been used to calculate Z,

$$Z^3 - (1-B)Z^2 + (A - 3B^2 - 2B)Z - (AB - B^2 - B^3) = 0 \qquad (6)$$

where $A = \frac{a_T p}{R^2 T^2}, B = \frac{b_T p}{RT}$. a and b are coefficients depend on the critical properties, i.e., $a_T = 0.45724 \frac{R^2 T_c^2}{p_c}, b_T = 0.0778 \frac{RT_c}{p_c}$. p_c, is the pressure at the critical point, and T_c is the temperature at the critical point.

The slippage effect takes place in the case of gas flow in tight reservoirs, therefore, the permeability developed to its apparent version, thus,

$$\mathbf{K}_{m,app} = K_m\left(1 + \frac{b_m}{p_m}\right), \qquad (7)$$

where K_m is the intrinsic permeability, b_m is the Klinkenberg effect.

From all the above formulations, the DPDP model may be written as,

$$f_1(p_m)\frac{\partial p_m}{\partial t} - \nabla \cdot \left[\frac{\rho_m K_m}{\mu}\left(1 + \frac{b_m}{p_m}\right)\nabla p_m\right] = -S(p_m, p_f), \qquad (8)$$

$$f_2\frac{\partial p_f}{\partial t} - \nabla \cdot \left[\frac{\rho_f K_f}{\mu}\nabla p_f\right] = S(p_m, p_f) - Q(p_f), \qquad (9)$$

where $f_1(p_m) = \frac{M_w \phi_m}{RT} + \frac{M_w V_L \rho_s}{V_{std}}\frac{P_L(1-\phi_m)}{(P_L+p_m)^2}$ and $f_2 = \frac{M_w \phi_f}{RT}$.

The transfer matrix-fracture term is given by,

$$S(p_m, p_f) = \frac{\sigma \rho K_m}{\mu}(p_m - p_f),$$

The Peaceman's model is used to describe the production-source term as [11],

$$Q(p_f) = \frac{\theta K_f \rho [\bar{p}_f - p_{wf}]}{\mu \ln \frac{r_e}{r_w}},$$

where $r_e = r_c\sqrt{(\Delta x)^2 + (\Delta y)^2}$ is the drainage radius, r_c is a constant and r_w is the well radius. If $\theta = 2\pi$, well will be in field center. If $\theta = \pi/2$, the well will be in the corner. σ is the crossflow coefficient and given by, $\sigma = 4\left(\frac{1}{L_x^2} + \frac{1}{L_y^2} + \frac{1}{L_z^2}\right)$. L_x, L_y and L_z are, respectively, fracture spacing of x, y and z. b_m and b_f are constants [3],

$$b_m = \sqrt{\frac{8\pi RT}{M_w}}\frac{1}{r_w}\left(\frac{2}{\alpha} - 0.995\right)\mu, \qquad b_f = \sqrt{\frac{\pi RT \phi_f}{M_w K_f}}\mu$$

The Knudsen diffusion has the form, $D_{kf} = \sqrt{\frac{\pi RT K_f \phi_f}{M_w}}$. K_f, ϕ_f and μ are the fractures permeability, the fractures porosity, and μ is the gas viscosity. α is a constant.

3   Mixed Finite Element Method

Assume that $\Omega_m \subset \mathbf{R}^d, d \in \{1,2,3\}$ is the matrix polygonal/polyhedral matrix Lipschitz domain; and $\Omega_f \subset \mathbf{R}^d, d \in \{1,2,3\}$ is the fracture polygonal/polyhedral Lipschitz domain. Also, assume that, $\mathbf{L}^2(\Omega) \equiv \left(L^2(\Omega)\right)^d$, such that $L^2(\Omega)$ is the standard space with the boundaries, $\partial\Omega_m = \Gamma_D^m \cup \Gamma_N^m$ and $\partial\Omega_f = \Gamma_D^f \cup \Gamma_N^f$.

The governing equations can be rewritten as,

$$f_1(p_m)\frac{\partial p_m}{\partial t} + \nabla \cdot \mathbf{u}_m = -S(p_m, p_f) \quad \text{in} \quad \Omega_m \times (0,T), \tag{10}$$

$$\mathbf{D}_m(p_m)^{-1}\mathbf{u}_m = -\nabla p_m \quad \text{in} \quad \Omega_m \times (0,T), \tag{11}$$

$$f_2\frac{\partial p_f}{\partial t} + \nabla \cdot \mathbf{u}_f = S(p_m, p_f) - Q(p_f) \quad \text{in} \quad \Omega_f \times (0,T), \tag{12}$$

$$\mathbf{D}_f(p_f)^{-1}\mathbf{u}_f = -\nabla p_f \quad \text{in} \quad \Omega_f \times (0,T), \tag{13}$$

In order to avoid discontinuity, both of the functions $\mathbf{D}_m(p_m)^{-1}$ and $\mathbf{D}_f(p_f)^{-1}$ are moved to the left hand side. Similarly, we can rewrite the initial and boundary conditions of the matrix and fracture domains, as follows,

$$p_m(\cdot,0) = p_f(\cdot,0) = p_0 \quad \text{in} \quad \Omega_m \cup \Omega_f, \quad p_f(\cdot,t) = p_w \quad \text{on} \quad \Gamma_D^f \times (0,T), \tag{14}$$

$$\mathbf{u}_m \cdot \mathbf{n} = 0 \quad \text{on} \quad \Gamma_N^m \cup \Gamma_N^f \times (0,T), \quad \mathbf{u}_f \cdot \mathbf{n} = 0 \quad \text{on} \quad \Gamma_N^f \times (0,T), \tag{15}$$

such that, $\mathbf{D}_m(p_m) = \frac{\rho_m \mathbf{K}_m}{\mu}\left(1 + \frac{b_m}{p_m}\right)$, and $\mathbf{D}_f(p_f) = \frac{\rho_f \mathbf{K}_f}{\mu}$.

Now, let us define the two Raviart–Thomas space (RT_r) subspaces on the partition $\mathcal{T}_h$: $V_h \subset H(\Omega; div)$ and $W_h \subset L^2(\Omega)$ such that r-th order $(r \geq 0)$. The MFE weak formulations are:

$$\left(f_1(p_m^h)\frac{\partial p_m^h}{\partial t}, \varphi\right) + (\nabla \cdot \mathbf{u}_m^h, \varphi) + (S(p_m^h, p_f^h), \varphi) = 0 \tag{16}$$

$$(\mathbf{D}_m(p_m^h)^{-1}\mathbf{u}_m^h, \omega) = (p_m^h, \nabla \cdot \omega), \tag{17}$$

$$\left(f_2\frac{\partial p_f^h}{\partial t}, \varphi\right) + (\nabla \cdot \mathbf{u}_f^h, \varphi) - (S(p_m^h, p_f^h), \varphi) = -(Q(p_f^h), \varphi), \tag{18}$$

$$(\mathbf{D}_f(p_f^h)^{-1}\mathbf{u}_f^h, \omega) = (p_f^h, \nabla \cdot \omega) - \langle P_w, \omega\rangle_{\Gamma_f^D}, \tag{19}$$

for any $\varphi \in W_h$ and $\omega \in V_h$. $p_m^h, p_f^h \in W_h$ and $u_m^h, u_f^h \in V_h$.

4   Numerical Algorithm

The MFE approximations with a quadrature rule has been employed to get an explicit flux. The backward Euler method is used to discretize the time-derivative with a number of N_T time-steps, $\Delta t^n = t^{n+1} - t^n$, such that $n+1$ is the current time, and n is the previous one, and the total time-interval is $[0, T]$. The discretized MFEM equations are given as,

$$\left(f_1^{h,n} \frac{p_m^{h,n+1} - p_m^{h,n}}{\Delta t}, \varphi\right) + (\nabla \cdot \mathbf{u}_m^{h,n+1}, \varphi) + (\mathcal{S}(p_m^{h,n+1}, p_f^{h,n}), \varphi) = 0 \qquad (20)$$

$$(\mathbf{D}_m(p_m^{h,n})^{-1} \mathbf{u}_m^{h,n+1}, \omega) = (p_m^{h,n+1}, \nabla \cdot \omega), \qquad (21)$$

$$\left(f_2^{h,n} \frac{p_f^{h,n+1} - p_f^{h,n}}{\Delta t}, \varphi\right) + (\nabla \cdot \mathbf{u}_f^{h,n+1}, \varphi) - (\mathcal{S}(p_m^{h,n+1}, p_f^{h,n+1}), \varphi) = -(Q_f^{h,n+1}, \varphi) \qquad (22)$$

$$(\mathbf{D}_f(p_f^{h,n})^{-1} \mathbf{u}_f^{h,n+1}, \omega) = (p_f^{h,n+1}, \nabla \cdot \omega) - \langle p_w, \omega \rangle_{\Gamma_f^D}, \qquad (23)$$

Table 1. Physical parameters.

Parameter	Value	Unit	Description
K_m	1.00E$-$04	md	Matrix permeability
K_f	10	md	Fracture permeability
ϕ_m	0.05	–	Matrix porosity
ϕ_f	0.001	–	Fracture porosity
R	8.314	m^{3}Pa/mol K	Gas constant
T	373	K	Temperature
Z	1	–	Compressibility factor
p_0	5	MPa	Initial reservoir pressure
p_w	3	MPa	Bottom hole pressure
M_w	0.016	kg/mol	Molecular weight of CH$_4$
V_{std}	0.0224	m^3/mol	Standard gas volume
P_L	6	MPa	Langmuir pressure
V_L	2.83E$-$03	m^3/kg	Langmuir volume
ρ_s	2550	kg/m^3	Shale rock density
μ	1.02E$-$05	Pa s	Initial gas viscosity
r_w	0.1	m	Wellbore radius
L_x, L_y, L_z	0.2	m	Fracture spacing
α	0.8	–	Constant

Give $p_m^{h,n}$ and $p_f^{h,n}$, the following scheme are used to find pressures and velocities as,

1. Calculate the thermodynamic variables explicitly.
2. Find $p_m^{h,n+1}$ and $\mathbf{u}_m^{h,n+1}$ by solving (20)–(21).
3. Find $p_f^{h,n+1}$ and $\mathbf{u}_f^{h,n+1}$ by solving (22)–(23).

5   Results and Discussions

Numerical test has been presented to show the efficiency of the current scheme. Table 1 shows the physical parameters of the problem under consideration. A 20×20 m domain has been used for computations. Figure 1 shows the distributions of matrix/fracture pressures and velocity. Also, the same figure shows the apparent permeability. From this figure, one notice that both matrix and fractures pressures decrease gradually close to the production well. Also, the apparent permeabilities are reduced. Moreover, It is clear that the apparent permeabilities decrease gradually as it goes farther from the well. It also can be observed that the high the apparent permeability the closer to the production-well. This may be due to the reverse relationship with the pressure. The gas production rate and cumulative production at different values of the slippage factor b_m are plotted against the production time in Fig. 2. This figure indicates that the slippage factor has a significant effect on the gas production rate.

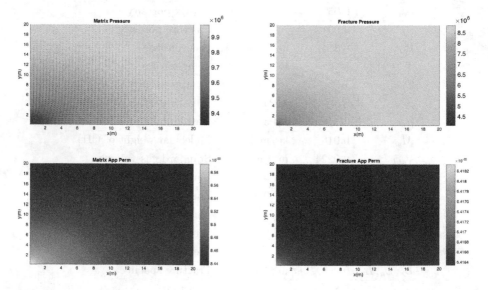

Fig. 1. Distribution of pressure and velocity of matrix and fractures (upper left and right) and apparent permeability of the matrix and fractures (lower left and right).

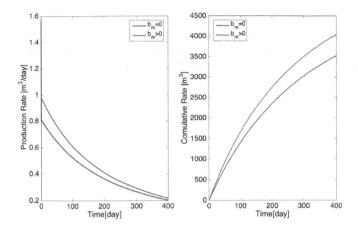

Fig. 2. Gas production rate and cumulative production at different values of b_m.

6   Conclusions

This work presents a mixed finite element method (MFEM) to solve the DPDP model of natural gas transport in shales. Results such as apparent permeability, pressure and production cumulative rate are discussed. In the future work, more realistic scenarios and solution properties of the MFEM scheme will be provided.

References

1. Warren, J.E., Root, P.J.: The behavior of naturally fractured reservoirs. SPE J. **3**, 245–255 (1963)
2. Bustin, A., Bustin, R., Cui, X.: Importance of fabric on the production of gas shales. In: Unconventional Reservoirs Conference in Colorado, USA, 10–12 February 2008. SPE-114167-MS
3. Javadpour, F.: Nanopores and apparent permeability of gas flow in mudrocks (shales and siltstone). J. Can. Pet. Tech. **48**(8), 16–21 (2009)
4. El-Amin, M.F., Kou, J., Sun, S.: Numerical modeling and simulation of shale-gas transport with geomechanical effect. Trans. Porous Media **126**(3), 779–806 (2019)
5. Ge, X., et al.: Investigation of organic related pores in unconventional reservoir and its quantitative evaluation. Energy Fuels **30**(6), 4699–4709 (2016)
6. Salama, A., El-Amin, M.F., Kumar, K., Sun, S.: Flow and transport in tight and shale formations. Geofluids (2017). Article ID 4251209
7. El-Amin, M.F., Amir, S., Salama, A., Urozayev, D., Sun, S.: Comparative study of shale-gas production using single- and dual-continuum approaches. J. Pet. Sci. Eng. **157**, 894–905 (2017)
8. El-Amin, M.F.: Analytical solution of the apparent-permeability gas-transport equation in porous media. Eur. Phys. J. Plus **132**, 129–135 (2017)
9. El-Amin, M.F., Radwan, A., Sun, S.: Analytical solution for fractional derivative gas-flow equation in porous media. Results Phys. **7**, 2432–2438 (2017)

10. Brezzi, F., Douglas, J., Marini, L.D.: Two families of mixed finite elements for second order elliptic problems. Numer. Math. **47**, 217–235 (1985)
11. Peaceman, D.W.: Interpretation of well-block pressures in numerical reservoir simulation. SPE-6893. In: 52nd Annual Fall Technical Conference and Exhibition, Denver (1977)

On the Feasibility of Distributed Process Mining in Healthcare

Roberto Gatta[1], Mauro Vallati[3]([⊠]), Jacopo Lenkowicz[1],
Carlotta Masciocchi[1], Francesco Cellini[2], Luca Boldrini[1],
Carlos Fernandez Llatas[4], Vincenzo Valentini[1,2], and Andrea Damiani[1]

[1] Dipartimento di Diagnostica per Immagini,
Radioterapia Oncologica ed Ematologia,
Istituto di Radiologia Universiá Cattolica del Sacro Cuore, Rome, Italy
roberto.gatta.bs@gmail.com
[2] UOC Radioterapia Oncologica,
Fondazione Policlinico Universitario "A. Gemelli" IRCCS, Rome, Italy
[3] University of Huddersfield, Huddersfield, UK
m.vallati@hud.ac.uk
[4] ITACA-Universitat Politecnica de Valencia, Valencia, Spain

Abstract. Process mining is gaining significant importance in the healthcare domain, where the quality of services depends on the suitable and efficient execution of processes. A pivotal challenge for the application of process mining in the healthcare domain comes from the growing importance of multi-centric studies, where privacy-preserving techniques are strongly needed.

In this paper, building on top of the well-known Alpha algorithm, we introduce a distributed process mining approach, that allows to overcome problems related to privacy and data being spread around. The introduced technique allows to perform process mining without sharing any patients-related information, thus ensuring privacy and maximizing the possibility of cooperation among hospitals.

Keywords: Process mining · Healthcare · Distributed learning

1 Introduction

Process Mining (PM) is an emerging topic aimed at building organizational process representations from real data [3]; it allows, for instance, to identify bottlenecks and undesired or unsuspected processes' paths. Healthcare represents a challenging domain for PM: there is a remarkable cultural gap between clinicians and computer scientists stakeholders; each event can have different meanings depending on the clinical perspective, the specific aim, the period of time or the patient it refers to. Furthermore, the psychology of patients and their response to therapies and drugs can be extremely unpredictable.

The medical sciences generate an ever increasing amount of medical evidence and clinical variables potentially able to play a key role in controlling

© Springer Nature Switzerland AG 2019
J. M. F. Rodrigues et al. (Eds.): ICCS 2019, LNCS 11540, pp. 445–452, 2019.
https://doi.org/10.1007/978-3-030-22750-0_36

diseases or in predicting toxicities. For this reasons, due to the rising number of covariates but the limited number of patients that can be enrolled for each hospital or research center, the trend is to move towards multi-centric clinical trials [6,7]. A PM investigation exploiting multi-centric data sources would be able to extract more evidence and trends, to highlight organizational differences between participating centers, and to show different ways to cope with daily clinical and administrative issues. The perspective of Process Mining applied on different institutional data sources gives rise to problems related to data owner-ship and patients' privacy. The former concerns the need to merge all the data before beginning the investigation: which is something that can, for instance, put a fraudulent data manager in the position to maliciously exploit his or her role; the latter concerns one of the most pivotal constraint in managing clinical data, which is to cope with heterogeneous laws and different, country-dependent requirements. The best way to avoid any privacy-related issue, while still exploit-ing multi-centric data, is to avoid sharing clinical data among different hospitals, and perform analysis and training of algorithms by sharing only few aggregate parameters among hospitals and guaranteeing the global convergence (consen-sus) to an acceptable shared model. This paradigm is commonly termed dis-tributed learning [5,8]. While a number of approaches have been proposed to perform distributed machine learning, there is a lack of approaches for perform-ing distributed multi-centric PM, where patients' privacy and data ownership are ensured and protected.

In this paper, we introduce the idea of distributed process mining, by focus-ing on one of the most well-known algorithms for process discovery: the Alpha algorithm [1]. This algorithm represents a milestone in the history of process discovery algorithms. Due to its simplicity it is widely adopted for describing the basic dynamics of process discovery, and it provides an invaluable ground for describing how distributed process mining can be performed.

2   Alpha Algorithm

The Alpha algorithm (AA) is one of the most well-known algorithms for process discovery [1]. It is widely appreciated for its simplicity, and it provides a very good ground for understanding how to move the first steps in process discovery.

A key concept on which the AA is based is the idea of the Footprint Matrix (FM), which is a squared matrix containing the ordering relations between pairs of events in the event log traces. In order to build this matrix, the following Log-Based ordering relations have to be defined. Given an event log L composed by a set of traces $L = \{\sigma_i\}$, the relation $>$ allows to express if a symbol b follows another symbol a: it is then possible to define that $a > b$ if, in at least one trace in L, in the list of terms t, a appears before b. Formally:

$$a > b \Leftrightarrow \exists \sigma_i = (t_1, t_2, .., t_n) \wedge j \in \{1, .., n-1\} \Rightarrow t_j = a \wedge t_{j+1} = b \quad (1)$$

Using $>$ it is possible to define three other relations: $\rightarrow$, $\sharp$ and $\|$, as:

$$a \rightarrow b \Leftrightarrow a > b \wedge NOT(b > a) \quad (2)$$

Algorithm 1. The Alpha algorithm, given an event log L and a set of traces σ.

1: $T_L = \{t \in T / \exists_{\sigma \in L} t \epsilon \sigma\}$
2: $T_I = \{t \in T / \exists_{\sigma \in L} t = first(\sigma)\}$
3: $T_O = \{t \in T / \exists_{\sigma \in L} t = last(\sigma)\}$
4: $X_L = \{(A,B)/A \subseteq T_L \cap A \neq \emptyset \wedge B \subseteq T_L \wedge A \neq \emptyset \wedge \forall_{a \in A} \forall_{b \in B} a \rightarrow_L$
   $b \wedge \forall_{a1,a2 \in A} a1 \sharp_L a2 \wedge \forall_{b1,b2 \in B} b1 \sharp_L b2\}$
5: $Y_L = \{(A,B) \in X_L / \forall_{(A',B') \in X_L} A \subseteq A' \wedge B \subseteq B' \Rightarrow (A,B) = (A',B')\}$
6: $P_L = \{p_{(A,B)}/(A,B) \in Y_L\} \cup \{i_L, o_L\}$
7: $F_L = \{(a, p_{(A,B)})/A, B \in Y \wedge a \in A\} \cup \{(p_{(A,B)}, b)/(A,B) \in Y \wedge b \in B\} \cup$
   $\{(t, o_L)/t \in T_O\}$
8: $\alpha(L) = (P_L, T_L, F_L)$

$$a \sharp b \Leftrightarrow NOT(a > b) \wedge NOT(b > a) \qquad (3)$$

$$a \| b \Leftrightarrow a > b \wedge b > a \qquad (4)$$

For example, given the event log $L = \{< a,b,c,d >, < a,c,b,d >, < b,c >\}$, the following ordering relations can be specified:

- $a \rightarrow b$, because in the first trace b follows a but the opposite never happens (in the second trace a is not after b).
- $b \| c$, because $b \rightarrow c$ in the first and the third traces but $b \leftarrow c$ in the second.
- $a \sharp d$ because a and d never appear next to each other.

Given the specified ordering relations, the FM can then be built for the considered event log L. For the considered example, the FM would look as follows:

	a	b	c	d
a	$\sharp$	$\rightarrow$	$\rightarrow$	$\sharp$
b	$\leftarrow$	$\sharp$	$\|$	$\rightarrow$
c	$\leftarrow$	$\|$	$\sharp$	$\rightarrow$
d	$\sharp$	$\leftarrow$	$\leftarrow$	$\sharp$

The Alpha algorithm is summarized in eight steps, shown in Algorithm 1. For a detailed description of the algorithm, the interested reader is referred to [2]. For the sake of conciseness, here we provide an overview of the main steps of the Alpha algorithm. In a nutshell, the steps indicated in Algorithm 1 aims: (1) to build the set of symbols adopted in the traces; (2) and (3) to build, respectively, the initial and final transitions; (4) by considering the Footprint Matrix, it is possible to find the groups of symbols which are in a $\sharp$ relation among them and in a $\leftarrow$ relation with all the subsequences (which, again, are in a $\sharp$ relation among them); (5) to reduce the set of the previous set of groups and build the set of states; (6) to define the arcs between couples of items identified in the previous step; (7) to join the arcs with the states; (8) to compose the result and return the structure of the Petri net. The Petri net summarizes the process mined by the algorithm, and can then be exploited for the PM purposes. It should be noted, however, that states in the Alpha algorithm are mapped into nodes of the Petri net. While this can be confusing, this is due to the fact that the notion of state of a Petri net refers to the position of the markings. The interested reader is referred to [9] for an extensive description of Petri nets.

It can be observed that the Alpha algorithm needs only a multi-set of traces comprising exclusively activity labels: the information about users, the data artifacts such as clinical files, or the indications about the premises where processes are carried out, are completely unnecessary to the Alpha algorithm. In such a setting, the reader may object that privacy would not be violated by freely circulating traces (maybe pre-filtered in order to show only literals encoding activity labels) and subsequently join them. At a closer look, however, it becomes apparent that a traditional man-in-the-middle attack can violate the privacy of patients. If the attacker has knowledge (even partial) about the way in which activity labels are encoded, traces can be quickly converted into the corresponding ordered sequence of clinical activities. At this point, patients can be matched with sequences in cases where even a small subsequence of activities is known to the attacker, effectively allowing the attacker to have a complete overview of the health record of the patient.

3   Distributed Alpha Algorithm

It is easy to observe that the Alpha algorithm only requires the FM, T_L, T_I, and T_O sets to perform the process discovery task. Therefore, the event logs do not need to be considered and shared during the computation.

The FM represents an aggregation of the original data, that is generally effectively hiding single patient's pathways. In a canonical non-privacy-preserving multi-centric study, a data manager is required to merge the different event logs L_i collected from each of the participating center i, into a single event log L, and builds a single Footprint Matrix $FM(L)$. The Alpha algorithm can then be applied to compute the global Petri net PN. Here we introduce how to distribute the Alpha algorithm computation.

Each center locally computes its own $FM(L_i)$ and sends it, instead of the event log L_i, to the master node. In this way, the master can only see an aggregation of the data, and has no access to the individual traces. Subsequently, the master *adds* the $FM(L_i)$ by a relation $\overline{\overline{\oplus}}$ in order to obtain:

$$FM(L_1)\overline{\overline{\oplus}}FM(L_2)\overline{\overline{\oplus}}...\overline{\overline{\oplus}}FM(L_n) = FM(L) \tag{5}$$

At this point, having $FM(L)$, the master node can compute the Alpha algorithm following the traditional approach shown in Algorithm 1. It is therefore pivotal to define a relation $\overline{\overline{\oplus}}$ satisfying the property:

$$FM(L_a \cup L_b) = FM(L_a)\overline{\overline{\oplus}}FM(L_b) \tag{6}$$

To understand how this relation should work, in the following we will analyze the behavior of each of the relations involved in the Footprint matrix, with particular regards to matrices composition.

Relation →: The relation $a \rightarrow b$, requires that the symbol a appears at least once immediately before b, and that b never happens immediately before a. It is easy to show that this relation is associative with respect to the composition of event logs. Given two event logs L_j and L_k, if $a \rightarrow_{t1} b$ in L_j and $a \rightarrow_{t2} b$ in L_k, it follows that $a \rightarrow_{t1 \cup t2} b$ in $L_j \bigcup L_k$, because the Condition 2 is preserved. Again, (b) if $a \rightarrow_{t1} b$ and $a \sharp_{t1} b$ the condition holds, because it requires *at least* one occurrence. In the case (c) of $a \rightarrow_{t1} b$ and $a \|_{t1} b$ the situation is different, because $a \|_{t1} b$ introduces an occurrence of $b > a$, so the new relation between a and b becomes $a \|_{t1} b$ (see Eq. 4). The last situation is $a \rightarrow_{t1} b$ and $a \leftarrow_{t2} b$ where evidently the result is $a \|_{t1} b$, because of 4. The case of ← is trivial.

Relation ♯. The relation $a \sharp b$ requires (see Eq. 3) that a and b never occur adjacently in the traces. This means that in composing ♯ with ←, → or ‖, the result becomes respectively ←, → or ‖.

Relation ‖. Once two symbols are in relation of ‖ for L_j, the new evidence of the existence of a ← or → relation in L_k, does not change the more general relation ‖ in $L_j \bigcup L_k$ between the two symbols. The same holds for ♯, so the composition of ‖ and ♯ returns ‖.

Missing Symbols. It is expected that not all the symbols in L_j appear also in L_k, and *vice versa*. In this case the only definition which holds is 2, and the corresponding symbols can be paired with ♯.

The Relations ⊕ and $\overline{\overline{\oplus}}$. We can define the relation ⊕ to join the contribute of different event logs L_j and L_k according to the following rules:

⊕	♯	→	←	‖
♯	♯	→	←	‖
→	→	→	‖	‖
←	←	‖	←	‖
‖	‖	‖	‖	‖

In particular: ♯ is a *neutral element* for the relation ⊕; ‖ is an *absorbing element* for the relation ⊕.

Because the relation ⊕ works only with pairs of symbols, to operate with Footprint matrices we can define $\overline{\overline{\oplus}}$ as:

$$FM(L_T) = FM(L_A) \overline{\overline{\oplus}} FM(L_B)$$

where the i, j element of $FM(L_T)$ is equal to $FM(L_A)_{i,j} \oplus FM(L_B)_{i,j}$ and $FM(L_A)$ and $FM(L_B)$ have the same symbols ordered in the same rows/columns. In other words, $\overline{\overline{\oplus}}$ operates on matrices applying ⊕ to each corresponding element of the two given input matrices. In the case of missing values, for example if $FM(L_B)$ does not have the symbol z which appear in $FM(L_A)$, the column and the row of z should be added in $FM(L_B)$ in order to calculate $FM'(L_B)$ which have, in the same position of $FM(L_A)$, the missing event related to the other with ♯.

Example 1. In order to give a practical example, let us consider the case of two event logs: $L_j = \{< a, b, c, d >, < a, c, b, d >\}$ and $L_k = \{< b, c >\}$. Starting from the two logs, we are interested in generating the FM of $L = L_{j \cup k}$.

By applying the relations defined in Eqs. 2, 3 and 4, for L_j and L_k, the corresponding Footprint matrices $FM(L_j)$ and $FM(L_k)$ would be, respectively:

$FM(L_j)$

	a	b	c	d
a	♯	→	→	♯
b	←	♯	‖	→
c	←	‖	♯	→
d	♯	♯	←	←

$FM(L_k)$

	b	c
b	♯	→
c	←	♯

It is easy to notice that $FM(L_k)$ is smaller then $FM(L_j)$, due to missing symbols in the corresponding event log L_k. In order to obtain matrices of the same shape and size, $FM'(L_k)$ has to be calculated, by filling the rows and columns corresponding to missing symbols with the ♯ relation, following the discussion provided in Sect. 3. As the matrix $FM(L_j)$ already includes all the symbols contained in any of the event logs, there is no need to manipulate it. It is now possible to calculate $FM = FM(L_j) \overline{\overline{\oplus}} FM'(L_k)$. The resulting matrix would then be as follows.

$FM(L_j)$

	a	b	c	d
a	♯	→	→	♯
b	←	♯	‖	→
c	←	‖	♯	→
d	♯	←	←	♯

$\overline{\overline{\oplus}}$ $FM'(L_k)$

	a	b	c	d
a	♯	♯	♯	♯
b	♯	♯	→	♯
c	♯	←	♯	♯
d	♯	♯	♯	♯

$=$ $FM(L)$

	a	b	c	d
a	♯	→	→	♯
b	←	♯	‖	→
c	←	‖	♯	→
d	♯	←	←	♯

∎

Having defined how different FMs can be composed via the appropriate relation, we are now in the position to describe the distributed Alpha algorithm. The distributed Alpha algorithm, inspired by the distributed paradigm proposed by Boyd et al. [4], includes the following steps:

1. Each center calculates the local Footprint matrix FM_{L_i} and the sets T_{L_i}, T_{I_i}, T_{O_i} (as discussed in Sect. 2);
2. the generated structures $< FM_{L_i}, T_{L_i}, T_{I_i}, T_{O_i} >$ are sent to the master;
3. the master node builds the sets:
   - $T_L = \bigcup_{i=1,..,n} T_{L_i}$
   - $T_I = \bigcup_{i=1,..,n} T_{I_i}$
   - $T_O = \bigcup_{i=1,..,n} T_{O_i}$
4. starting from FM_{L_i}, the master node calculates the corresponding FM'_{L_i}. This is the correctly ordered FM, where missing columns/rows –corresponding to missing events in a local event log– are filled with ♯, according to what has been shown in the corresponding section.

5. the overall FM is generated by composing all the FM'_{L_i} via the relation $\overline{\overline{\oplus}}$;
6. the master applies the Alpha algorithm on the structure $< FM, T_L, T_I, T_O >$;

This flow ensures the patient's privacy preservation and allows to apply the Alpha algorithm considering the entire set of available traces.

4 Conclusion

The complex causal relations between involved variables and values, and the objective difficulty of enrolling large cohorts of patients for a single medical center, lead to a growing need to merge data from different centers and perform multi-centric clinical analysis. A PM investigation exploiting multi-centric data sources would be in the best position to highlight organizational differences between different centers, and to show different ways to cope with daily clinical and administrative issues.

In this work, we introduced the first approach to perform multi-centric process mining. Building on top of the well-known Alpha algorithm, we designed a privacy-preserving technique that allows to perform distributed process mining while preserving patients' data privacy. The proposed approach is guaranteed to converge to the same model that would have been generated by merging all the different data sets, and has been empirically tested on a large set of event logs. Future work will focus on extending our approach to different algorithms, and to investigate different aspects of process mining, such as conformance checking and enhancements.

References

1. van der Aalst, W.M.P., Weijters, T., Maruster, L.: Workflow mining: discovering process models from event logs. IEEE Trans. Knowl. Data Eng. **16**, 1128–1142 (2004)
2. van der Aalst, W.: Process Mining: Discovery, Conformance and Enhancement of Business Processes. Springer, Heidelberg (2011). https://doi.org/10.1007/978-3-642-19345-3
3. van der Aalst, W., et al.: Process Mining Manifesto. In: Daniel, F., Barkaoui, K., Dustdar, S. (eds.) BPM 2011, Part I. LNBIP, vol. 99, pp. 169–194. Springer, Heidelberg (2012). https://doi.org/10.1007/978-3-642-28108-2_19
4. Boyd, S., Parikh, N., Chu, E., Peleato, B., Eckstein, J.: Distributed optimization and statistical learning via the alternating direction method of multipliers. Found. Trends Mach. Learn. **3**(1), 1–122 (2011)
5. Damiani, A., et al.: Distributed learning to protect privacy in multi-centric clinical studiest. In: Artificial Intelligence in Medicine (2015)
6. George, M., Selvarajan, S., Dkhar, S., Chandrasekaran, A.: Globalization of clinical trials - where are we heading? Curr. Clin. Pharmacol. **8**(2), 115–123 (2013)
7. Gresham, G., Ehrhardt, S., Meinert, J., Appel, L., Meinert, C.: Characteristics and trends of clinical trials funded by the national institutes of health between 2005 and 2015. Clin. Trials **15**(1), 65–74 (2018)

8. Lindell, Y., Pinkas, B.: Privacy preserving data mining. In: Bellare, M. (ed.) CRYPTO 2000. LNCS, vol. 1880, pp. 36–54. Springer, Heidelberg (2000). https://doi.org/10.1007/3-540-44598-6_3
9. Peterson, J.L.: Petri net theory and the modeling of systems (1981)

How to Plan Roadworks in Urban Regions? A Principled Approach Based on AI Planning

Mauro Vallati[1]([envelope]) [iD], Lukáš Chrpa[2,3] [iD], and Diane Kitchin[1] [iD]

[1] University of Huddersfield, Huddersfield, UK
{m.vallati,d.kitchin}@hud.ac.uk
[2] Artificial Intelligence Center, Czech Technical University in Prague,
Prague, Czech Republic
[3] Faculty of Mathematics and Physics, Charles University in Prague,
Prague, Czech Republic
chrpaluk@fel.cvut.cz

Abstract. Roadworks are required to keep roads in acceptable condition, and to perform maintenance of essential infrastructure. Road agencies are facing the problem of how to effectively plan frequent roadworks. In this paper, we exploit Automated Planning for roadworks planning. We introduce a planning domain model that allows us to plan a set of required roadworks, over a period of time, in a large urban region, by specifying constraints to be satisfied and suitable quality metrics. Our empirical analysis shows the suitability of the proposed approach.

Keywords: AI planning · Roadworks · Urban traffic management

1 Introduction

Despite the pressing need for efficiency, exacerbated by the current growth of urbanisation, there is a lack of approaches designed for supporting agencies in the decision-making process of creating a plan of roadworks for the controlled urban area. Research in the area is mainly focused on providing formal frameworks for the analysis of the impact of works [5], or on the evaluation of the process used by local authorities [6]. For this reason, roadwork plans are usually generated manually, without any explicit notion of "quality", and a limited control on constraints and assessment of the impact of works on the managed network traffic.

Automated planning, which deals with the problem of finding a plan (a sequence of actions) that transforms the environment from an initial state to some desired goal state [7], is an effective tool for decision making. Plans consist of sequences of actions to be performed in order to achieve desired goals. Domain-independent planning requires only to specify a *planning domain model* and a *planning problem description* in a standard language such as PDDL [2], and then

© Springer Nature Switzerland AG 2019
J. M. F. Rodrigues et al. (Eds.): ICCS 2019, LNCS 11540, pp. 453–460, 2019.
https://doi.org/10.1007/978-3-030-22750-0_37

to use one of the available generic planning engines to solve the problem and find a plan. This independence between reasoning and domain knowledge gives engineers a high level of flexibility since planning domain models and planning engines can be used as "black-boxes" that can be easily embedded into larger systems. Examples of embedding planning into larger frameworks include recent works in road traffic accident management [1], and urban traffic control [8].

In this paper, we propose a principled approach to use Automated Planning as an effective Decision Support tool in planning and scheduling roadworks in urban areas. We provide a formalisation of the roadworks planning problem, and we specify and develop a planning domain model that allows the representation of the constraints related to roadworks planning within large urban regions. The domain model is encoded in PDDL [2] that is widely supported by the large number of existing domain-independent planning engines. We empirically validate our approach by considering a urban region of Yorkshire (UK), and by modelling scenarios based on actual roadworks performed in the region.

2 Problem Specification

Here we consider the case in which a number of roadworks need to be performed in a controlled urban region. The controlled region, for which works have to be planned, is represented in terms of roads and areas. Roads r are connected to each other via junctions, and are grouped into areas a. The notion of area allows the effective and efficient encoding of the proximity of roadworks, as discussed later in this section. Each work w is described by a quadruple $\langle R, t_i, t_f, d \rangle$. R indicates the set of roads affected by the work w. The set can include a single road or more, according to the typology and size of w. For the sake of simplicity, here we consider that, if multiple roads are affected by the same work, they are grouped into the same area. d indicates the duration of the work, expressed in time units. Finally, the couple t_i, t_f represents the time window in which the work must be completed such that t_i is the earliest time unit in which the can start, and t_f represents the latest time unit in which the work can be completed, and thus $t_f > t_i$ must hold.

The time window must be at least as long as the duration of the considered work, i.e. $d \leq t_f - t_i$. Time is discretised, following a suitable discretisation for the considered region and type of works. In this paper we consider that each time unit corresponds to one week, but that can be easily adjusted (e.g. to one day) without changes to the model.

The problem is constrained by the maximum number of works that can be planned at the same time in a specific area. More formally, for each time unit t the following must hold:

$$maxW_{a_i} \geq |\{w_j \mid w_j \in W, (active(w_j, t) \wedge in\_area(w_j, a_i))\}| \qquad (1)$$

Where $maxW_{a_i}$ is the maximum number of works that can be active during the same time unit in the a_i area of the controlled region, $active(w_j)$ indicates whether the work w_j is planned or not for the considered time unit, and

$in\_area(w_j, a_i)$ is used to specify if the work w_j has to be performed in the area a_i. A similarly-structured constraint is used to encode cases where, due to the fact a number of roadworks have been assigned to the same company, they can not be executed concurrently due to limited resources:

$$maxPW_{c_i} \geq |\{w_j \mid w_j \in W, (active(w_j, t) \land assigned_{to}(w_j, c_i)\}| \qquad (2)$$

Where $maxPW_{c_i}$ is the maximum number of works that can be performed at the same time by a company c_i, $active(w_j)$ indicates whether the work w_j is planned or not for the considered time unit, and $assigned_{to}(w_j, c_i)$ represents the fact that w_j has been assigned to the company c_i.

The goal is achieved by planning all the required roadworks so that they can be completed within the corresponding defined time windows, while respecting the constraints presented in Eqs. 1 and 2. Planning is done by allocating works to subsequent time units, according to the expected duration.

3   Domain Model Specification

The conceptualisation of the problem requires to specify *object types*, which stand for classes of objects considered in the planning process, *predicates and numeric variables*, which describe the context, and *actions*, which allows the state to be modified. In our model we consider five main object types: *works, roads, time units, companies*, and *areas*. Works consist of the number of roadworks that have to be performed. Roads are the links of the considered urban region. Time units consist of the slots available for planning the roadworks. Companies represent the construction companies involved in the planning task. Areas indicate the various areas in which the controlled region has been divided.

The static part of the model, i.e. the aspects that do not change during the planning process, describes the road network and to some properties of the objects. We consider an abstract road network, where only roads affected by works, and main roads of the network are represented. A predicate *assigned* is used to denote that a work has been assigned to a company. A predicate *inArea* encodes that a road is included in an area of the controlled region. Similarly, a predicate *onRoad* denotes that the corresponding roadwork has to be performed on a road; multiple instances of this predicate can be used to indicate that a roadwork is affecting multiple roads A numeric variable *maxWorks* is used to encode the maximum number of works that can be planned in the same time unit for a given area of the network. Similarly, another numeric variable *maxCompany* is used to encode the maximum number of parallel works a given company can deal with. Note that although the maxWorks and maxCompany values do not change during the planning episode, they might be initially set differently for different planning episodes. Time units objects are ordered using a *subsequent* predicate, which indicates that a time unit $t_i + 1$ is after a corresponding time unit t_i. Each time unit is also assigned a numeric *value*, which corresponds to its position in the problem. For instance, t_1 has a value of 1, and t_2 has a value of 2, etc. Values are exploited for optimisation purposes, and for encoding the time horizon of each roadwork. Finally, numeric variables *startTime* and *deadline* represent the earliest initial and latest final time unit of each work.

The dynamic part of the model encodes the aspects which are under the control of the planning engine. The main task of the planning process is to allocate roadworks to suitable time units so that constraints are satisfied. The *toStart* predicate is used to indicate that the corresponding roadwork has yet to be started, i.e. the planning engine did not start allocating it yet. The numeric variables *duration* denotes the "remaining" duration of a roadwork, in terms of time units. Starting from the actual duration of the roadwork in object, the variable is reduced every time the work is allocated to a time unit. A duration value of 0 indicates that the corresponding work has been completely allocated. The numeric variables *companyWork* and *parallelWork* are used to count, respectively, the number of works assigned to a company and the number of works in the same area, that are planned in the same time unit. Finally, *initiated* is a general numeric variable which is exploited for optimisation purposes. It stores the sum of the differences between the planned starting time and the earliest possible starting time, of all the considered roadworks.

One roadwork cannot be allocated twice to the same time unit; this is a very common constraint for numerical planning and has been encoded using well-known PDDL constructs. It is therefore omitted from the following description of actions.

The actual plan of road works can be generated by a planning engine using the two modelled actions:

- **WorkStart**(w,r,t,a,c) – identifies a time unit t as the starting time of a road work w, that has been assigned to a company c; the work has to be done on road r, which is part of area a. As preconditions, beside requirements on location of roadworks and areas, *toStart* (w) must hold, and the deadline of the work must be satisfied by starting at time unit t. Effects include the fact that the work is considered as started (i.e., the *toStart* predicate is falsified), duration is reduced by one time unit, and the number of works in the area and works performed by company c in time unit t are increased by one.
- **WorkContinue**(w,r,t_b,t_s,a,c) – a work w that has already been allocated to a time unit t_b is allocated to the subsequent time unit t_s (i.e., $t_s = t_b + 1$). The preconditions and effects are the same as the *WorkStart* action except *toStart* (w) is replaced by *onGoingWork*(w, t_b) in the precondition.

The PDDL encoding of two different actions gives a high degree of flexibility to the model, as works with significantly different durations can be considered. Furthermore, the step-by-step approach which is forced by requiring that a different action is used to allocate a work to each time unit guarantees a higher level of control, as at each step all the constraints can be checked.

In addition, to the described actions, two PDDL constraints on the planning trajectory are enforced. Such constraints are used to ensure that the limits on the maximum number of works per area and per company, allocated to one time slot, are respected. The model is available at http://bit.do/roadworkplan.

4   Evaluation of the Approach

The Planning domain model and problems have been encoded in PDDL 2.1. For solving such problems, we selected three planning search techniques implemented by the Metric-FF planner [3]: Fast Forward (FF), Weighted A* (WA*), and Enforced Hill Climbing (EHC). The first technique ignores the quality of the solution found, but focuses on generating a solution as fast as possible, while the considered WA* and EHC approaches do take the quality of plans into account and try to optimise the generated solution. All the techniques are guided by the well-known delete-relaxation heuristic: such heuristic ignores all the negative effects of actions, i.e. it only reasons by considering positive (additive) effects. In this context, quality is measured in terms of the average "delay" of planned starting time of roadworks, with regards to the possible earliest starting time. Better quality corresponds to planes where works, on average, are started as early as possible. This is a key indicator of quality because in a typical urban context, all the roadworks which need to be planned are usually essential, and should be completed as soon as possible.

Experiments have been performed on a system equipped with 2.0 GHz Intel i7-3667U Processors, 8 GB of RAM and Linux operating system. The VAL tool [4] has been used for validating the generated plans, in order to check their correctness with regards to the designed domain model, and also to identify the presence of bugs or flaws in the model.

Our analysis has been focused on the northern part of the Kirklees urban region, that includes 11 different areas.

4.1   Empirical Results

In order to collect informative and realistic data, we analysed the number of road-works being executed in the considered region over a period of time of 2 months, during 2017. According to the observed data, we synthetically generated a set of instances, with the following main characteristics: number of roadworks to be planned ranging from 5 to 20; plan horizon of 6 months, roadworks duration ranging between 1 and 10 weeks. Start time and deadline values have been randomly generated. For each considered number of roadworks to be planned, we randomly generated three instances with different durations and start times. As a general deadline, we required all the works to be completed by the end of the modelled 6-months period. Presented results are averaged on the three instances, in order to take into account noise and variability due to the randomised aspects.

The results of this first set of experiments, designed for testing the fact that the proposed system is effectively able to plan the roadworks required in the controlled region, are presented in Table 1. Unsurprisingly, the use of the FF planning techniques leads to plans of poor quality, with regards to the quality metric considered. This is due to the fact that this approach does not exploit any cost optimisation for generated plans. FF tends to plan roadworks 17 weeks later, on average, than the earliest starting time. By analysing the generated plans, we noticed that the works are frequently scheduled in order to finish exactly by

Table 1. Runtime (CPU-time seconds) needed by FF, WA*, and EHC for providing a plan for an increasing number of roadworks in the considered area, over a 5-month period of time. Quality of generated plans is measured as the average delay (time units) with regards to the first possible start time of the work. Bold indicates best results.

# works	Runtime			Plan quality		
	5	10	20	5	10	20
FF	0.1	**0.3**	1.6	18.3	18.9	19.0
WA*	0.1	0.9	7.8	**1.4**	**1.0**	**1.2**
EHC	0.1	**0.3**	1.5	18.2	18.4	18.5

the deadline, even if it would be possible to complete them before. Surprisingly, plans of poor quality are also generated by the EHC algorithm, which instead exploits cost optimisation. Our intuition is that this behaviour is due to the search space topology, that does not allow the enforced hill climbing approach to identify promising areas to visit in order to increase the quality of an initially identified solution. Instead, the use of a weighted-A* search algorithm allows to find high quality plans: works are started usually at the earliest possible time, except in cases where the constraints on maximum parallel works that can be performed by companies or in specific areas could not be satisfied by starting works earlier.

All plans are generated extremely fast by the considered techniques. WA* is the slowest approach, and requires less than 8 CPU-time seconds to solve the most complex scenario considered in this analysis. The other approaches require less than 2 CPU-time seconds.

Generated plans have been validated using the well-known VAL tool. This guarantees that plans are valid with regards to the exploited domain model. However, this sort of validation does not provide any insights into the actual usefulness of the plans in the real-world applications. In order to check this aspect of the validity of generated plans, they have been visually inspected by transport experts, who confirmed their overall quality and that they look similar to plans one would expect from an expert-generated solution.

4.2 Time Horizons

The size of the time window is an extremely important aspect, as it can significantly affect the shape of the search space: here we assess the impact of time windows on the performance of the proposed planning-based approach. Focusing on the 10 roadworks scenario, we generated four planning problems by varying (i) the duration of roadworks and (ii) the size of the time windows. Works can have a duration of 1 or 10; the time window can be exactly the duration of the work, or can cover the whole modelled period. In other words, in the first problem all the works have duration 1, and the time window is 1 for all of them, in the second problem all the works have duration 10, and the time window is 10 for each of them, etc. Results indicate that the duration of the considered works has no impact on the performance of the considered planning approaches. On the other hand, the size of the time window has a remarkable impact on performance. The more constrained the time windows are, the easier it is for the three considered planning approaches to find a solution: in the more constrained case, all the 3 approaches can find a plan in approximately 0.5 CPU-time seconds. Moreover, the more constrained the problem is, the more similar the quality of the generated solutions is. Conversely, larger time windows lead to a higher computational time (up to 10 CPU-time seconds for the WA* approach), and a remarkable variability in terms of quality of the generated solution plans.

5   Conclusion

In order to assist road agencies and authorities in the critical duty of planning essential roadworks, here we investigated the use of automated planning techniques as an effective decision support tool. The main contributions of this work are: (i) a formalisation of the roadworks planning problem; (ii) a PDDL model encoding of the domain; and (iii) a thorough experimental analysis, that considers realistic scenarios and compared plans generated using three different planning approaches, embodied in the same domain-independent planning engine. The performed experimental analysis demonstrates the extent to which the proposed approach is able to efficiently and effectively create a roadworks plan that satisfies the identified constraints, and follows a notion of quality. Future works will focus on including different degrees of uncertainty in the models, and evaluating the approach in different urban regions.

Acknowledgements. L. Chrpa was partially funded by the Czech Science Foundation (project no. 17-17125Y). M. Vallati was partially supported by the EPSRC grant EP/R51343X/1 (AI4ME).

References

1. Chrpa, L., Vallati, M.: On the exploitation of automated planning for efficient decision making in road traffic accident management. In: Proceedings of CDC (2016)
2. Fox, M., Long, D.: PDDL2. 1: an extension to PDDL for expressing temporal planning domains. J. Artif. Intell. Res. **20**, 61–124 (2003)
3. Hoffmann, J.: The metric-ff planning system: translating "ignoring delete lists" to numeric state variables. J. Artif. Intell. Res. **20**, 291–341 (2003)
4. Howey, R., Long, D., Fox, M.: VAL: automatic plan validation, continuous effects and mixed initiative planning using PDDL. In: Proceedings of ICTAI (2004)
5. Huerne, H.T., Berkum, E.V., Hermelink, W.: A multi objective framework to optimise planning of road works. In: 5th Eurasphalt and Eurobitume Congress (2012)
6. Hussain, R.S., Ruikar, K., Enoch, M.P., Brien, N., Gartside, D.: Process mapping for road works planning and coordination. Built Environ. Proj. Asset Manag. **7**(2), 157–172 (2017)

7. Ghallab, M., Nau, D., Traverso, P.: Automated Planning Theory and Practice. Elsevier Science, Amsterdam (2004)
8. McCluskey, T.L., Vallati, M.: Embedding automated planning within urban traffic management operations. In: Proceedings of ICAPS (2017)

Big Data Approach to Fluid Dynamics Visualization Problem

Vyacheslav Reshetnikov[1], Egor Golubchikov[1], Andrey Pyatlin[1],
Alexey Kuzin[1]([⊠]), Vladislav Kiev[1], Nikolay Shabrov[1],
Alexey Zhuravlev[1,2], and Ekaterina Guseva[1]

[1] Peter the Great St. Petersburg Polytechnic University, St. Petersburg, Russia
kuzin_aleksei@mail.ru
[2] Reutlingen University, Reutlingen, Germany

Abstract. Present work is dedicated to development of the software for interactive visualization of results of simulation of gas dynamics problems on meshes of extra large sizes. Kitware ParaView visualization tool, which is popular among engineers and scientists is used as a frontend. The coupling of client and server instances of ParaView is used in the project. The crucial feature of the work is an application of Apache Hadoop and Apache Spark for distributed retrieving of simulation data from files on hard disk. The data is stored on the cluster in Hadoop Distributed File System (HDFS) managed by Apache Hadoop and is provided to ParaView server by Apache Spark data processing tool.

Keywords: Visualization · Spark · ParaView · Parquet

1 Introduction

The capabilities of modern high performance computers and the level of development of special problem-oriented software packages of predictive modeling allow the user to increase resolution of numerical grids up to the order of billions nodes and more. Files of simulation results on larger meshes are represented by big data arrays, especially in the case of modeling of unsteady processes. Investigations show that there is a growing trend in the size of the data [1]. The size of data retrieved causes a problem of low speed of scientific visualization and analysis of the results.

While software for predictive modeling and hardware available for the wide spectrum of scientists allow to perform fluid dynamics evaluations on meshes up to 10 billions cells, visualization tools do not provide desired efficiency. One of the problems is low visualization speed. Moreover, it is not only linked with the rendering. Childs et al. [2] showed that time of input/output can be two

Supported by Russian Science Foundation (Grant No. 18-11-00245)

The research carried out with the financial support of the grant from the Program Competitiveness Enhancement of Peter the Great St. Petersburg Polytechnic University.

J. M. F. Rodrigues et al. (Eds.): ICCS 2019, LNCS 11540, pp. 461–467, 2019.
https://doi.org/10.1007/978-3-030-22750-0_38

orders higher than rendering and evaluations. One of the ways to reduce the time of input/output is the development of special algorithms for data processing not involving supercomputers technologies. This approach was performed in [3–5]. In this paper authors use another approach which assumes usage of a supercomputer.

There is a lack of good tools which are able to handle extra large grids on the market of visualization systems. The most known scientific visualization brands such as ParaView, TecPlot, COVISE, TechViz do not support information about effective results presentation on the meshes of up to billion nodes large. At the same time, this kind of problem persists for aircrafts designers.

The scientific problem considered in this article is the problem of achieving an effective and rapid interactive visualization and analysis of results of predictive modelling of fluid dynamics problems for modern aircrafts on superlarge meshes. It is assumed that created software can be used by engineers who do not have any special knowledges in IT, that is why it should provide displaying of the results in the most convenient and easy to understand way. The software should give visualization of fields in real or almost real time for comfortable working process.

The key idea underlying the developed software package is the usage of distributed big data analysis tools such as Apache Hadoop in conjunction with Apache Spark [6]. They provide distributed retrieval of the data from cluster nodes that can seriously reduce time of data reading in comparison to traditional sequential approach. Hadoop is mainly used to support Hadoop Distributed File System (HDFS), and the server built on the base of the Spark framework provides distributed processing of queries for retrieving required dataset from a cluster. A plugin to ParaView developed by the authors, plays the role of client of Spark server. It is intended to send queries and is integrated with the server version of ParaView. The user's computer has the client version of ParaView, which receives final results of rendering from server ParaView.

The problem with the Apache Hadoop application for creation of packages for visualization of results of finite element modelling is considered in [7]. The example of an effective usage of HDFS is presented in [8]. An approach similar to the authors one was investigated in [9] with the difference that in that work Apache Hive was used instead of Apache Spark. Also the article [10] should be mentioned in which a hybrid approach is offered that assumes usage of HDFS and Kitware ParaView as a user interface. In the papers [11] and [12] Hadoop and Spark frameworks were applied for analysis and visualization of atmospheric phenomena modeling in Earth science problems. The attention was paid mainly to the analysis tasks.

2   Architecture

In this section the software architecture is described, as well as how particular program parts interact with each other during the typical visualization process.

The software environment is built by "client-server" scheme and has the structure which is shown on the Fig. 1. Basic elements are:

- **Client ParaView.** The client version of the well-known scientific visualization Kitware's ParaView package is installed on the local computer and is intended to provide direct interaction with the user. It visualizes results of rendering obtained from ParaView server. This is the only component of the software, which works on the local machine.
- **Server ParaView.** The server version of ParaView which is installed on the cluster. It provides effective parallel rendering based on the data retrieved with the plugin to ParaView, designed by the authors.
- **Plugin to ParaView.** The plugin is made by the authors and is integrated into the server Paraview and is intended to efficient data reading. The reading is processed through the responds to the client SQL-queries to data server that is also run on the same or another cluster.
- **Data server.** The server is written using Python and uses the Apache Thrift framework. It receives queries from the ParaView plugin and gives data blocks back. The server forwards queries to the Spark system that retrieves data in a distributed manner from Hadoop Distributed File System.

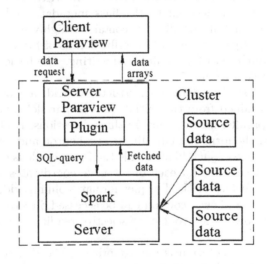

Fig. 1. The scheme of interaction of the software components

The interaction of client and server parts of ParaView is quite traditional approach of usage of ParaView that provides parallel model rendering [13]. Therefore, the development of the ParaView plugin and the data server is the main direction of the authors efforts. From the ParaView point of view the plugin is the regular plugin for the reading of the model data defined on the structured grid. At the present time the plugin gives VTK-object `vtkMultiBlockDataset`.

But instead of direct data reading, plugin forms SQL-query to the data server and retrieves data in response. The server transfers the query to Apache Spark. Spark executes distributed reading, collecting (`collect` operation) and sending data as a response to the SQL-query. The advantage of such a scheme over direct reading from file is that the reading via Spark is performed in parallel on several nodes of the cluster, which can give an increase in speed, especially on large files.

The speed of reading essentially depends on the file format used. Initially, data is presented in text format Tecplot and takes 1.27 GB. It is not suitable format for holding large data so it must be converted into another format. The format to be converted in must meet the following requirements. Firstly, the file size must be as small as possible. Secondly, it must provide relatively fast access to individual blocks of data, which is important in cases there is no need to read the entire file. It must support distributed storage and be readable by Spark. Based on the sum of these requirements, Apache Parquet [14] was chosen as the data format. On the one hand, it meets all requirements above. It is binary format so it has a smaller size in comparison to Tecplot. After conversion one frame of data takes about 400 MB. Besides, it provides fast access, distributed storage and can be read by Spark. Also it has a sufficient flexibility to form the necessary block structure of data storage. The format itself is a set of columns organized into a hierarchical structure under the control of a special scheme. The choice of the scheme is left to the person who writes the file. The scheme is the part of the format written to the file's metadata and can be restored while reading. Thus, a Parquet file can contain a highly complex hierarchical structure. Although Apache does not describe details of format implementation it provides open-source API for reading and writing. The basic ideas of working with Parquet are taken from [15].

As a method of storing data on a structured grid, a block structure was chosen, which is obtained from the initial index of parallelepiped grid elements by dividing it by parallel index planes in three directions. Each block spatially occupies an area in the form of a curvilinear hexahedron and represents a structured grid. The storage of such block inside the Parquet file is organized as a set of separate Parquet columns for each of the node coordinates and field components. Due to the specifics of the Parquet format, column addresses are stored in the file's metadata and each column can be accessed directly without reading the entire file so it solves the problem of selective reading of the necessary grid blocks.

At the moment, the solution is not adaptive to the manner of distribution of simulation output. That is why data must be redistributed in case of change of computer nodes numbers. Solution adaptivity to this kind of changes is the point for further research.

3   Usage Example

An example of visualization of unsteady gas dynamics simulation results on the structured grid of hexahedrons is considered. The source data is written in the form of time layers. Then each layer is initially stored in a separate file in Techplot format. The model contains about $5 \cdot 10^6$ nodes and each Techplot file has size of 1.27 GB.

The processing of separate frame files during direct visualization of these frames sequence on a standalone computer in ParaView takes about 30 s, which is unacceptably slow for interactive mode. Such low speed can be attributed to the fact that Tecplot is not suitable format to hold an information of big data.

The same data is visualized with developed software. All components except of client ParaView are installed on the cluster. The nodes of supercomputer "RSC Tornado" of Saint-Petersburg Polytechnic University Supercomputer Center are used as a cluster for ParaView and data servers. Each node consists of two CPU Intel Xeon E5-2697 v3 (14 cores, 2.6 GHz) and 64 GB RAM DDR4. Simple reading of Tecplot datafile located on one node with Paraview takes about 60 s.

Data files were converted to Parquet format before visualization in the developed environment and the size of one frame was reduced to about 400 MB. There were 20 frames. In total, the size of all frames was about 7.8 GB. The data of such rather small size is used only to illustrate approach. In further research size of each file is proposed to be bigger. During writing to Parquet format the data was transformed as described above: initial index parallelepiped of structured grid was divided by mutually orthogonal index planes into the parallelepipeds of smaller sizes. Inside each box each coordinate and each component field is separate Parquet column. Such a structure is effective on extraction of selected blocks because it does not require reading of the entire file. Each conversion takes about 40–45 s. In total, the whole dataset is converted on 10 MPI threads for 90 s.

Apache Spark is launched on the cluster under control of Slurm system. The reading of Parquet files in Spark is performed by the pyarrow library, as it provides a higher speed and requires significantly less memory than built-in Spark tools for Parquet reading. Data in Spark is represented in non-hierarchical plain structure.

Table 1. Time of reading and displaying of separate frames on 8 nodes of the cluster "RSC Tornado"

Frame	Reading time, s	Total time, s	fps
1	4.655	49.314	0.020
2	2.794	14.142	0.071
3	2.723	11.862	0.084
4	2.790	10.895	0.092
5	2.956	15.596	0.064

The time of displaying of the first five frames using 8 cluster nodes is shown in the Table 1. The second column contains the time of direct reading of the

Parquet file in Spark using pyarrow. The third column is the total time of the frame displaying. The last column is the number of frames per second, i.e. the inverse of the total time. Most of the time is spent on data transfer from the data server to the client ParaView, which indicates the need to use a faster network. The time of reading and transmitting the first frame is longer than the next ones, which is caused first of all by the necessity of reading the grid. The fact is that the grid remains unchanged during the transition from frame to frame, so it is read only once at the first frame. Starting from the second frame the system reads only the values of the displayed field. That is why from the second frame reading time does not change.

The increment of the number of nodes involved does not lead to the expected decrement of reading time in Spark. Finding the cause of this phenomenon is one of the tasks of further research.

4   Conclusions

A software environment for interactive visualization is developed. It provides visualization of simulation results evaluated on large numerical grids. The environment consists of a client ParaView, a server ParaView, a data server that forwards SQL queries to Apache Spark. The latter is used to increase the speed of reading large data by providing distributed access to them.

The experiments have shown the effectiveness of using data in Parquet format. Compared to the Tecplot text format this format provides a smaller file size, is directly readable by Apache Spark, and provides an ability to extract individual data blocks without having to read the entire file.

The experiments also showed the absence of scalability of the data reading speed in Spark with increasing number of nodes and high overhead for data transfer from Spark to the server ParaView. These problems are tasks for further development.

Another challenge is to use more specific SQL queries. These can be requests to get data corresponding to the visible part of the model. In addition, there might be requests to retrieve data distributed across layers. The layer which data should be extracted depends on the camera position. If it corresponds to a higher detalization, the layer must contain more nodes.

References

1. Jin, X., Wah, B.W., Cheng, X., Wang, Y.: Significance and challenges of big data research. Big Data Res. **2**(2), 59–64 (2015)
2. Childs, H., et al.: A contract based system for large data visualization. In: Visualization, VIS 2005, pp. 191–198. IEEE (2005)
3. Belyaev, S., Shubnikov, V., Motornyi, N.: Adaptive screen sampling algorithm acceleration for volume rendering. In: MCCSIS 2018 - Multi Conference on Computer Science and Information Systems; Proceedings of the International Conferences on Interfaces and Human Computer Interaction 2018, Game and Entertainment Technologies 2018 and Computer Graphics, Visualization, Computer Vision and Image Processing 2018, pp. 377–381 (2018)

4. Belyaev, S., Smirnov, P., Shubnikov, V., Smirnova, N.: Adaptive algorithm for accelerating direct isosurface rendering on GPU. J. Electron. Sci. Technol. **16**(3), 222–231 (2018). https://doi.org/10.11989/JEST.1674-862X.71013102

5. Savchuk, D.A., Belyaev, S.Y.: Two-pass real-time direct isosurface rendering algorithm optimization for HTC vive and low performance devices. Paper Presented at the Progress in Biomedical Optics and Imaging - Proceedings of SPIE, vol. 10579 (2018). https://doi.org/10.1117/12.2292183

6. Apache Spark framework. http://spark.apache.org. Apache Spark is developed by Apache company. https://apache.org

7. Lange, B., Nguyen, T.: A Hadoop distribution for engineering simulation. [Research Report] INRIA Grenoble - Rhône-Alpes (2014)

8. Voinov, N., Drobintsev, P., Kotlyarov, V., Nikiforov, I.: Distributed OAIS-based digital preservation system with HDFS technology. In: 2017 20th Conference of Open Innovation Association, (FRUCT), St. Petersburg, pp. 491–497 (2017). https://doi.org/10.23919/FRUCT.2017.8071353

9. Artigues, A., et al.: Scientific big data visualization: a coupled tools approach. Supercomput. Front. Innov. **1**(3), 4–18 (2014)

10. Mitchell, C., Ahrens, J., Wang, J.: VisIO: enabling interactive visualization of ultra-scale, time series data via high-bandwidth distributed I/O systems. In: 2011 IEEE International Parallel & Distributed Processing Symposium (IPDPS), pp. 68–79 (2011)

11. Zhou, S., et al.: A Hadoop-based visualization and diagnosis framework for Earth science data. In: IEEE International Conference on Big Data, pp. 1911–1916 (2015)

12. Zhou, S., Li, X., Matsui, T., Tao, W.: Visualization and diagnosis of earth science data through hadoop and spark. In: IEEE International Conference on Big Data, pp. 2974–2980 (2016)

13. ParaView software. http://www.paraview.org. ParaView is developed by Kitware company. http://www.kitware.com

14. Columnar storage format. http://parquet.apache.org. Parquet is developed by Apache company. https://apache.org

15. Melnik, S., et al.: Dremel: interactive analysis of web-scale datasets. In: Proceedings of the 36th International Conference on Very Large Data Bases, pp. 330–339 (2010)

Dolphin Kick Swimmer Using
the Unstructured Moving Mesh Method

Masashi Yamakawa[1]([⊠]), Norihito Mizuno[1],
and Yongmann M. Chung[2]

[1] Kyoto Institute of Technology, Matsugasaki, Sakyo-ku, Kyoto 6068585, Japan
yamakawa@kit.ac.jp
[2] University of Warwick, Gibbet Hill Road, Coventry CV4 7AL, UK

Abstract. The dolphin kick assumes a vital role in swimming competitions, as it is used after dives and turns in several swimming styles. To improve the swimmer's dolphin kick performance, flows around him were simulated. Using video footage of a male swimmer's joint angles, a 3D model simulation was created. The flows were computed using the unstructured moving grid finite volume method to express the complicated motion of swimmers. The mesh around the swimmer is moved according to his motion. In this method, a geometric conservation law is satisfied as well as a physical one. Furthermore, the moving computational domain method is also adopted for calculation efficiency. The numerical swimmer is finally completed by a coupled computation between motion of human and fluid. In this paper, the simulation results revealed that the influence of the maximum knee oscillation angles affect the speed of the swimmer.

Keywords: Computational fluid dynamics · Unstructured moving mesh ·
Dolphin kick swimming

1 Introduction

The dolphin kick is the swimming style that a swimmer creates wave by wiggling its body like a dolphin in order to propel. As the feature, both arms are fixed on top of its head, and both legs are also drawn up. The surge of its body goes on increasing from fingertip to toe. The style assumes a vital role in swimming competition, because it is adopted after dives and turns in several swimming style. Thus, to clarify the detail mechanism of its movement and to improve are very important.

Against such a background, the measurement of the drag of a swimmer who is pulled under the water with stretching posture was reported [1]. Then, the drag of a swimmer having arm stroke at the crawl swimming style was measured using the MAD system [2] and using the dynamics analysis [3]. However, it would be difficult to produce their detail environment because the experimental apparatus and the measurement method themselves would restrict the movement of a swimmer. The visualization for flows around swimmer were also reported using the approach filming artificial bubbles [4] and using tuft put on skin of a swimmer [5]. But their approaches would affect to its performance of a swimmer. Furthermore, it would be difficult to catch detail phenomena like small vortices around it. While the visualizations for wake of the dolphin kick using PIV were reported [6]. By using PIV, that a couple of vortices around hands generated by arm stroke at the crawl influence the generation of unsteady fluid force were shown [7]. But it was restricted in the two-dimensional measurement.

On the other hand, a lot of results using computational simulation instead of experimentations were reported. The propulsion of the dolphin kick swimmer were estimated [8] using the swimmer model: SWUM [9]. Using the same model, an influence to propulsive speed and efficiency by waviness of swimmer's body was shown [10]. Although this model can capture a tendency of the influence, the human body shape is simplified. Furthermore, the fluid dynamics itself is not calculated in the paper. Therefore, it would be difficult to know a detail force from fluid to human body. While, 3-D computational simulation was carried out for a flow around hand and arm [11], and then it was expended to a flow around a whole body [12]. Comparing the depth of a body from water surface and drag of a swimmer with force during glide swim, the influence of a flow condition around the swimmer by the depth from the surface was reported [13]. The detail human shape and the detail motion of swimming were modeled using 3-D scanner and Autodesk MAYA, then the flow around the model with dolphin kick was calculated using the immersed boundary method [14]. The results showed that the ring vortex created by a motion of kicking lower affects the majority of the propulsion caused by dolphin kick. However, these computations were estimated in constant speed flows only. Since the swimming speed of dolphin kick changes dynamically, it is important to take acceleration and deceleration into consideration to know the detail of the flow mechanism around a swimmer. The flow around a swimmer accelerated and decelerated by fluid force was calculated using the smoothed particle hydrodynamics. The computation [15] showed that an increase of the frequency of dolphin kick cycle boosts the propulsion speed linearly. However, the motion of the human model has not been verified, thus there is doubt about the validity of the result.

In this study, the swimmer shaped model was generated by unstructured mesh. Then, the flows were computed using the unstructured moving grid finite volume method [16, 17] to express the complicated motion of swimmers. This method satisfies both geometric and physical conservation laws using four dimensional space-time unified control volume. Furthermore, the moving computational domain method [18, 19] was adopted to express acceleration and deceleration of the swimmer. The position of the swimmer is decided by a coupled computation between motion of human and fluid. The objective of this paper is to specify the mechanism the flow around the dolphin kick swimmer and to investigate influence which the maximum knee oscillation angles affect the speed of the swimmer.

2 Numerical Approach

2.1 Governing Equations

Governing equations are the continuity equation and the incompressible Navier–Stokes equations. These are written as follows:

$$\nabla \cdot \mathbf{q} = 0, \tag{1}$$

$$\frac{\partial \mathbf{q}}{\partial t} + \frac{\partial \mathbf{E}_a}{\partial x} + \frac{\partial \mathbf{F}_a}{\partial y} + \frac{\partial \mathbf{G}_a}{\partial z} =$$
$$-\left(\frac{\partial \mathbf{E}_p}{\partial x} + \frac{\partial \mathbf{F}_p}{\partial y} + \frac{\partial \mathbf{G}_p}{\partial z}\right) + \frac{1}{Re}\left(\frac{\partial \mathbf{E}_v}{\partial x} + \frac{\partial \mathbf{F}_v}{\partial y} + \frac{\partial \mathbf{G}_v}{\partial z}\right), \tag{2}$$

where q is the velocity vector, $\mathbf{E}_a$, $\mathbf{F}_a$, and $\mathbf{G}_a$ are advection flux vectors in the x, y, and z direction, respectively, $\mathbf{E}_v$, $\mathbf{F}_v$, and $\mathbf{G}_v$ are viscous-flux vectors, and $\mathbf{E}_p$, $\mathbf{F}_p$, and $\mathbf{G}_p$ are pressure terms.

2.2   Numerical Schemes

To express the motion of a swimmer, the unstructured moving grid finite volume method was adopted. The method assures a geometric conservation law as well as a physical conservation law. Then, a control volume in the space-time unified domain (x, y, z, t), which is four-dimensional in the case of three-dimensional flows is used.

To express acceleration and deceleration of a swimmer, moving itself is required instead of general approach which calculates a fixed body in uniform flow. In several moving body approach, the moving computational domain method is suitable for the application. The method which moves the computational domain itself according to motions of a body can remove restrictions of the distance traveled. Furthermore, the method satisfies both geometric and physical conservation laws as it is based on the unstructured moving grid finite volume method. Thus the moving computational domain method is adopted in this paper.

3   Computational Model and Conditions

3.1   Numerical Swimmer Model

To obtain motions to the swimmer model, joint angles data is captured from video footage of a male swimmer as shown in Fig. 1. The average speed of the swimmer in the video is 1.4 m/s. Then the dolphin kick cycle T is 0.65 s.

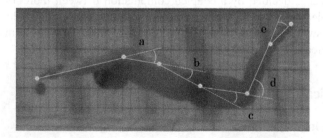

Fig. 1. Video footage of a male swimmer and positions of joint angles (a: shoulder, b: back, c: hip, d: knee, e: ankle)

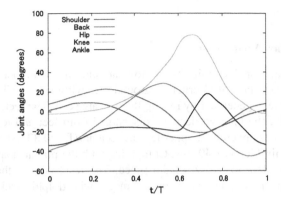

Fig. 2. Changes of five joint angles captured from video footage in a dolphin kick cycle

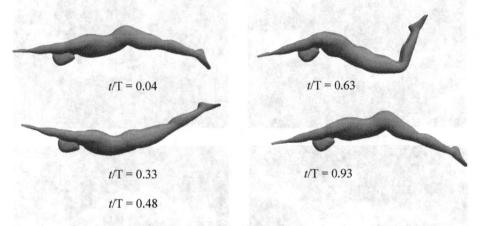

$t/\text{T} = 0.04$

$t/\text{T} = 0.63$

$t/\text{T} = 0.33$

$t/\text{T} = 0.93$

$t/\text{T} = 0.48$

Fig. 3. Dolphin kick movement reproduced

The joint angles of shoulder, back, hip, knee and ankle shown in Fig. 2. These relations of joint angles make a motion of dolphin kick. The sequence of movements for dolphin kick swimmer are shown in Fig. 3.

3.2 Computational Mesh and Conditions

The computational mesh is generated using unstructured type. Here, 1,420,617 tetra-hedron and 726,499 prisms are used. The prism mesh is for boundary layer. Their elements are created by MEGG3D [20].

Dolphin kick motion stars at the speed of the swimmer 1.4 m/s. Then, a coupled computation between motion of human and fluid also starts simultaneously. Reynolds number is 5×10^4. Then number of time steps of a dolphin kick cycle is 17800.

4 Results

4.1 Dolphin Kick Swimming

Under the computational conditions, a flow around the swimmer was calculated. Figure 4 shows isosurface of Q-criterion around the swimmer. At $t/T = 0.33$, vortices around and under legs generated by raising the legs are seen. The feet reach the highest point at $t/T = 0.58$, the long vortices are discharged from feet and toes. In the downward of legs, vortices are created upper legs as seen at $t/T = 0.83$. An average speed in a cycle with dolphin kick is 1.49 m/s. Comparing with the practical speed 1.4 m/s, the difference is equivalent to 6.43% of the speed. The result shows the validity of the computations using this numerical swimmer model with dolphin kick.

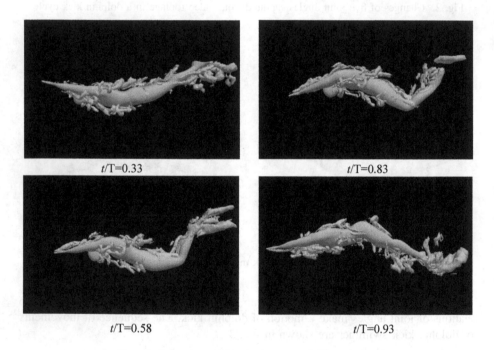

$t/T=0.33$ $t/T=0.83$

$t/T=0.58$ $t/T=0.93$

Fig. 4. Isosurface of Q-criterion around the swimmer

4.2 Influence of Maximum Knee Oscillation Angles

To improve the swimming style with dolphin kick, the influence which the maximum knee oscillation angles affect the speed of the swimmer is investigated. Six different knee oscillation angles (Variant1:80°, V2:70, V3:60, V4:50, V5:40, V6:30) are estimated and their results are compared.

As the results, the average speed of 6 variants are shown in Table 1. The highest speed is taken in Variant 3, then the lowest one is Variant 6. On the other hand, the history of thrust in 6 variants is shown in Fig. 5. So the highest average speed Variant 3 doesn't get the highest thrust. Then the lowest average speed Variant 6 also doesn't take the lowest one. The highest thrust is made by Variant 1 which has the maximum knee angle among the investigation. However Variant 1 also have the lowest thrust because it would have the maximum drag. The drag is made from the resistance of the flow occurred the surface area of the swimmer's body changed by the high angle of knee.

Table 1. Comparison of average speed of 6 variants

Variant number	Average speed
1	1.078
2	1.108
3	1.132
4	1.100
5	1.085
6	1.047

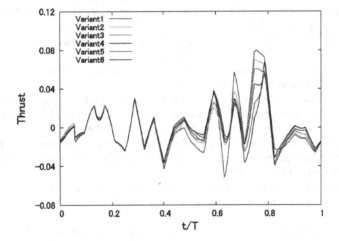

Fig. 5. History of thrust in 6 variants

5 Conclusions

In this paper, flows around the dolphin kick swimmer were computed using unstructured moving mesh method. Furthermore, by combing a moving computational calculating approach and a coupled computation between motion of a swimmer and fluid with the unstructured mesh, the numerical swimmer was structured. Comparing with the practical speed data, the speed of the numerical swimmer made the error less than 7%. Furthermore, around the swimmer's feet, the result could show similar pressure distribution and velocity vectors with the experimentation data. Thus, the validity of the computations using this numerical swimmer model with dolphin kick were shown. To improve the dolphin kick, 6 maximum knee oscillation angles were estimated using the numerical swimmer. In the case of 60° at knee oscillation angle, it was found that the swimmer can take the highest average speed.

Acknowledgments. This publication was subsidized by JKA through its promotion funds from KEIRIN RACE and by JSPS KAKENHI Grant Number 16K06079.

References

1. Lyttle, A., et al.: An instrument for quantifying the hydrodynamic drag of swimmers-a technical note. J. Hum. Mov. Stud. **37**, 261–270 (1999)
2. Toussaint, H., et al.: The determination of drag in front of crawl swimming. J. Biomech. **37**, 1655–1663 (2004)
3. Berger, M., et al.: Determining propulsive force in front crawl swimming: a comparison of two methods. J. Sports Sci. **17**, 97–105 (1999)
4. Arellano, R., et al.: Teaching hydrodynamic concepts related to swimming propulsion using flow visualization techniques in the swimming pool. In: Fifth National Symposium on Teaching Biomechanics in Sports in San Francisco, USA (2001)
5. Toussaint, H., et al.: Pumped-up propulsion during front crawl swimming. Med. Sci. Sports Exerc. **34**, 314–319 (2002)
6. Hochstein, S., et al.: Re-capturing and kinematics in human underwater undulatory swimming. Hum. Mov. Sci. **30**, 998–1007 (2011)
7. Matsuuchi, K., et al.: Unsteady flow field around a human hand and propulsive force in swimming. J. Biomech. **42**, 42–47 (2008)
8. Sugimoto, S., et al.: Simulation analysis for the effects of planter flexion angle increment on the performance during underwater dolphin kick. In: Proceedings of Symposium on Sports Engineering: Symposium on Human Dynamics, pp. 194–199 (2007)
9. Nakashima, M., et al.: Development of swimming human simulation model considering rigid body dynamics and unsteady fluid force for whole body. Trans. Jpn. Soc. Mech. Eng. Ser. B **71**(705), 1361–1369 (2005)
10. Nakashima, M.: Simulation analysis of the effect of trunk undulation on swimming performance in underwater dolphin kick of human. J. Biomech. Sci. Eng. **4**(1), 94–104 (2009)
11. Bixler, B., Riewald, S.: Analysis of a swimmer's hand and arm in steady flow conditions using computational fluid dynamics. J. Biomech. **30**, 713–717 (2002)
12. Bixler, B., et al.: The accuracy of computational fluid dynamics analysis of the passive drag of a male swimmer. Sports Biomech. **6**(1), 81–98 (2007)
13. Sato, Y., et al.: CFD simulation of flows around a swimmer in a prone glide position. Jpn. J. Sci. Swimming Water Exerc. **13**(1), 1–9 (2010)
14. Loebbecke, A., et al.: A computational method for analysis of underwater dolphin kick hydrodynamics in human swimming. Sports Biomech. **8**(1), 60–77 (2009)
15. Cohen, R., et al.: Simulation of dolphin kick swimming using smoothed particle hydrodynamics. Hum. Mov. Sci. **31**, 604–619 (2012)
16. Yamakawa, M., et al.: Numerical simulation for a flow around body ejection using an axisymmetric unstructured moving grid method. Comput. Thermal Sci. **4**(3), 217–223 (2012)

17. Yamakawa, M., et al.: Numerical simulation of rotation of intermeshing rotors using added and eliminated mesh method. Procedia Comput. Sci. **108C**, 1883–1892 (2017)
18. Asao, S., et al.: Simulations of a falling sphere with concentration in an infinite long pipe using a new moving mesh system. Appl. Thermal Eng. **72**, 29–33 (2014)
19. Asao, S., et al.: Parallel computations of incompressible flow around falling spheres in a long pipe using moving computational domain method. Comput. Fluids **88**, 850–856 (2013)
20. Ito, Y.: Challenges in unstructured mesh generation for practical and efficient computational fluid dynamics simulations. Comput. Fluids **85**, 47–52 (2013)

The Performance Prediction
and Improvement of SPH
with the Interaction-List-Sharing Method
on PEZY-SCs

Natsuki Hosono[1,2(✉)] ⓘ and Mikito Furuichi[1]

[1] Japan Agency for Marine-Earth Science and Technology,
3173-25, Showa-machi, Kanazawa-ku, Yokohama, Kanagawa 236-0001, Japan
natsuki.hosono@jamstec.go.jp
[2] RIKEN Center for Computational Science,
7-1-26 Minatojima-minami-machi, Chuo-ku, Kobe, Hyogo 650-0047, Japan

Abstract. The demands for the optimization of particle-based methods with short-range interaction forces such as those in smoothed particle hydrodynamics (SPH) is increasing, especially for many-core architectures. However, because particle-based methods require large amount of memory access, it is challenging to obtain high efficiency for low-byte/FLOP many-core architectures. Hence, an efficient technique, the so-called "multiwalk" method, was developed in an N-body gravitational field. The key of the multiwalk method is in sharing of the interaction lists with multiple particles to offer an efficient use of the cache memory in the double-loops operation for calculating the interactions and reducing the main memory access. However, such performance improvement is not clear for the problems with short-range interaction forces such as those in SPH. In this paper, we proposed a theoretical performance model to examine the tradeoff relations between the memory and the cost of floating point operations to optimise the SPH code. We also validated the model with the wall-clock time spent on the PEZY-SCs (SC1 and SC2).

Keywords: MIMD processors · Smoothed particle hydrodynamics

1 Introduction

Scientific computing is a common technique to solve complex problems which are hard to carry out by laboratory experiments. One important example is that of numerical hydrodynamics for solving fluid-motion problems. The smoothed particle hydrodynamics (SPH) [1,2] is one of the most widely accepted particle-based method which has advantages for problems with large deformations of fluid surface.

One of the disadvantages of particle-based methods compared to mesh-based methods is the calculation cost. When N particles are introduced into a system,

© Springer Nature Switzerland AG 2019
J. M. F. Rodrigues et al. (Eds.): ICCS 2019, LNCS 11540, pp. 476–482, 2019.
https://doi.org/10.1007/978-3-030-22750-0_40

the construction cost of interaction lists for all particles would be $\mathcal{O}(N^2)$ which prevents us from performing large-scale and/or high-resolution numerical simulations. A common technique to speed up calculations is to use a many-core device which is a peripheral unit to CPUs that has a number of microprocessors. The use of many-core devices for SPH simulations poses an interesting challenge in the high-performance computing [3,4]. However, it is not trivial to obtain high efficiency of computation for particle-based methods with recent many-core devices, because a native implementation shows intensive memory access cost compared to that of floating point operations and easily suffers from the memory-bandwidth problem especially in many-core devices with low-byte/FLOP.

One of the clever techniques to address the above-mentioned problem is the "interaction-list-sharing" method that was first developed for the N-body gravitational field problem [5,6]. Let us consider a "group" of particles (i-particles) located close to each other such that they have very similar interaction lists. To calculate the force acting on the i-particles, we need to "sweep" their interaction lists. The simplest implementation of this step is to use a double-loop: the loop sweeping particles in the interaction list inside the loop for the number of i-particles in the group. During this process, the size of the data transfers from the main memory can be reduced, because the data on the cache memory can be reused multiple times for each i-particle in the group.

Because SPH is a particle-based method, the interaction-list-sharing method can be applied. However, gravitational force is a long-range force; any pairs of two particles interact with each other even if they are infinitely distant. On the other hand, SPH involves a short-range force: a particle interacts only with its surrounding particles. Consider N_i particles that share one "shared" interaction list. When we increase N_i, the number of main memory accesses decrease and the size of the shared interaction list increases. However, when we set a large N_i two particles can be assigned to the same group despite them having largely different interaction lists. Such a situation results in an increase of unnecessary computational cost.

These tradeoff relations indicate that there should be a "desirable" value of N_i that minimises the wall-clock time spent on a many-core device which would depend on the choice of the SPH kernel and the computing device we use. Hence, in this paper, we built a model to predict the optimal value of N_i for the efficient computation on the many-core devices. Our performance model depends on the bandwidth and FLOP/s of the device, and arithmetic intensity of the SPH kernel. To validate the proposed model, we performed a benchmark test utilising PEZY Computing's PEZY-SC, which won the 1st place in GREEN500 of November 2018.

This paper is organized as follows. In Sect. 2, we briefly introduce the SPH method and proposed performance-prediction model. In Sect. 2, we also outline the overall procedure of the interaction-list-sharing method for SPH. In Sect. 3, we show the validated result. In Sect. 4, we conclude this study.

2   Methods

In the standard SPH method, the equation of motion is formulated as follows

$$\frac{\Delta v_i}{\Delta t} = -\sum_j m_j \left(\frac{p_i}{\rho_i^2} + \frac{p_j}{\rho_j^2} \right) \nabla_i W(|r_j - r_i|; h_{ij}), \tag{1}$$

where v, t, m, p, ρ, r and h are the velocity, time, the mass, the pressure, the density, the position, and the smoothing length that determines the size of an SPH particle, respectively. The subscript i indicates the label for each particle. h_{ij} is the arithmetic mean of h between particle i and j. Hereafter, we only focus on the performance of Eq. (1) because its computational cost is dominant in the general SPH solver.

The function W is the so-called "kernel function". Herein, we used the Wendland C^2 kernel:

$$W(r, h) = \frac{21}{2\pi} \left(1 - \frac{r}{Hh} \right)_+^4 \left(1 + 4\frac{r}{Hh} \right), \tag{2}$$

where $(\cdot)_+ := \max(\cdot, 0)$. Note that H is a parameter that determines the cutoff size of the kernel function.

Hereafter, a particle that receives moments from its surrounding particles is referred to as i-particle, whereas the particles that give moments to an i-particle are referred to as j-particle. To calculate Eq. (1), an i-particle must have its p_i, ρ_i, h_i and r_i, whereas a j-particle must have its m_j, p_j, ρ_j, h_j and r_j. Assuming double precision, the sizes of an i-particle (S_i) and a j-particle (S_j) were specified as $8 \times 6 = 48$ and $8 \times 7 = 56$ bytes, respectively. The number of arithmetic operations for one interaction (N_{arith}) is 70, assuming that division and square root are 8 times more expensive than the basic arithmetic operations.

For building the interaction lists, we employed a framework named `Framework for Developing Particle Simulator` (FDPS [7]). FDPS takes full responsibility for the construction of interaction lists in parallel. The parts that the users must take care of are the kernel code that works on a many-core device and the communication between the host and the device. FDPS uses the tree method to construct interaction lists, which allows us to reduce the computational cost from $\mathcal{O}(N^2)$ to $\mathcal{O}(N \log_8 N)$ [5,8,9]. After the constructions of interaction lists, FDPS provides the array of i-particles and their interaction lists to the kernel code written by the user. In the interaction-list-sharing method, FDPS makes an aggregation of groups of i-particles and their interaction lists for sending to a single many-core device. Then, the many-core device calculates interactions for particles in several groups simultaneously.

To predict performance improvement using the interaction-list-sharing method, we propose the following theoretical cost model. Consider the double-loop operations to update i-particles with j-particles. A particle in a group of N_i particles shares potentially interacting N_j particles with other particles in the same group. The number N_j depends on N_i and the kernel function's cutoff radius H. The cost for the loop operations (C) contains the memory cost to load

N_j j-particles (C_{load}), to store N_i i-particles (C_{store}), the arithmetic cost for $N_i \times N_j$ pairs of interactions (C_{arith}), and other costs (C_{others}) (e.g., the kernel launching time). Those are summarised as follows:

$$C = C_{\text{load}} + C_{\text{store}} + C_{\text{arith}} + C_{\text{others}}, \tag{3}$$

$$C_{\text{load}} = \frac{N_j S_j}{B}, \quad C_{\text{store}} = \frac{N_i S_i}{B}, \quad C_{\text{arith}} = \frac{N_i N_j N_{\text{arith}}}{F}, \tag{4}$$

where B and F are the memory-bandwidth and the FLOP/s of a device, respectively. Hence, the cost to update a single particle deviated from the result with $N_j = 0$ can be written as

$$\frac{C}{N_i} - \frac{S_i}{B} = \frac{1}{B}\left(\frac{N_j}{N_i}S_j + \frac{B}{F}N_j N_{\text{arith}}\right). \tag{5}$$

Here we assumed that C_{others} is negligible and can be excluded. This cost model shows that the increase of N_i results in the decrease of the memory access $\propto N_j/N_i$ and the increase of the arithmetic costs ($\propto N_j$) to update a single particle. Note that N_j is calculated by counting the number of j-particles inside the spheres centred at each N_i particle position with a radius Hh. Equation (5) indicates that for a given particle-based method, the dependence of the cost to update one particle can be characterized by the inverse of FLOP/byte (B/F). The size of the parameter H is generally chosen between 2.1 and 3.1.

To validate our performance predictions, we tested two PEZY-SC devices, viz., PEZY-SC1 and PEZY-SC2 which are MIMD-type architectures. The bandwidth and FLOP/s for SC1 are $B = 150\,\text{GB/s}$ and $F = 1.50$ TFlops, whereas for SC2 they are $B = 102\,\text{GB/s}$ and $F = 4.09$ TFlops, respectively. Note that PEZY-SC1 does not have floating operation units for division in double precision. Thus, for the PEZY-SC1, the approximate technique are used for square root operations [10] with double precision. For division, we first calculate the division with single precision and then apply Newton-Rhapson method to convert the result to double precision.

3   Results

In this section, we show the results of the validation tests. We put 10^6 SPH particles in cubic lattice with periodic boundary condition so that all particles have the same number of neighbour particles. All calculations are performed with double precision.

Figure 1 shows the predicted wall-clock time by our model and its dependence on N_i. We showed two cases; one is the high $B/F = 0.5$ case (K computer) and another is the low $B/F = 0.025$ case (PEZY-SC2). With the cases of PEZY-SC2 with $H = 2.1$ and 3.1, the curves show a gradual decrease of the wall-clock time with the N_i increases until $N_i \simeq 32$ and then, the wall-clock time increase for $N_i > 64$. This result suggests that the interaction-list-sharing method with $N_i^{\text{opt}} \simeq 32$ is a reasonable choice, where N_i^{opt} is the optimal choice for N_i.

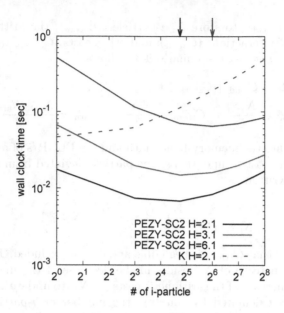

Fig. 1. Theoretical wall-clock time for PEZY-SC2 and K computer. The horizontal axis indicates N_i. The black arrow at the top of the figure indicates N_i to minimise predicted wall-clock time for the cases with $H = 2.1$ and 3.1. The green arrow also indicates optimal N_i, but for $H = 6.1$. The solid black, red, and green curves represent the predicted wall-clock time for the PEZY-SC2 for the cases with $H = 2.1, 3.1,$ and 6.1, respectively. The dashed blue curve represents the predicted result with K computer and $H = 2.1$. (Color figure online)

However, in the case of $H = 6.1$, the optimal choice becomes $N_i^{\mathrm{opt}} \simeq 64$. N_i^{opt} is found to increase with the kernel radius H (see two arrows in Fig. 1). The region $N_i < N_i^{\mathrm{opt}}$ indicates that the efficiency of the device is limited by the memory-bandwidth, whereas in the region $N_i^{\mathrm{opt}} < N_i$, the efficiency is limited by floating-point operations. Hence, N_i^{opt} increases as H increases.

Figure 1 also shows that the shape of the performance curve depends on a computer architecture characterized by B/F. For example, the blue dashed line shows the result with a high $B/F = 0.5$ architecture that predicts the performance in a K computer. The implementation of the interaction-list-sharing method does not result in a performance improvement because the predicted wall-clock time monotonically increases against N_i. In the case of $B/F = 0.5$, the wall-clock time is always limited by floating-point operations which means that the interaction-list-sharing technique is not efficient. Hence, N_i^{opt} would decrease as the B/F of a device increases.

The comparison between the predicted wall-clock time and the measured value are shown below. The results of the benchmark tests for PEZY-SC2 are shown in Fig. 2a. Our cost model characterized by B/F successfully captures the trends of the predicted performance curve and predicts the improvement by

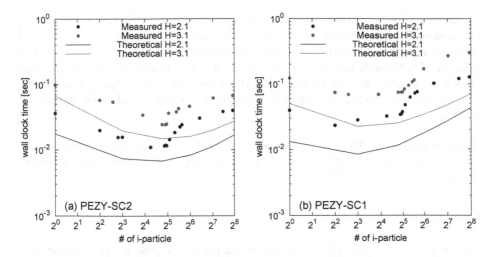

Fig. 2. The theoretical and measured wall-clock time for (a) PEZY-SC2 and (b) PEZY-SC1. The horizontal axis indicates the size of a group N_i. The solid black and red lines indicate the predicted wall-clock times for $H = 2.1$ and 3.1, respectively. The black and red circles indicate the measured wall-clock times for $H = 2.1$ and 3.1, respectively. (Color figure online)

using the interaction-list-sharing method. For example, the numerical experiment with $H = 2.1$ shows the 75% cost reduction when using $N_i = 32$ compared to $N_i = 1$, whereas our model predicts the 61% cost reduction with $N_i = 32$ compared to $N_i = 1$. Conversely, the observed cost in the numerical experiment is approximately 2 times larger than the theoretical cost due to approx. 50% efficiency of computation against theoretical peak performances.

Figure 2b shows a result of performance analysis with PEZY-SC1. The shape of the prediction curve is characterized by B/F; thus, our model predicts a smaller N_i^{opt} than that of PEZY-SC2. The measured wall-clock time shows that N_i^{opt} for PEZY-SC1 is 4, which is consistent with our prediction.

4 Conclusion

This paper examined the impact of the interaction-list-sharing method for SPH simulation. To predict the performance improvement quantitatively, a theoretical cost model was proposed. Through a series of benchmark tests, the proposed model was validated as a scaled performance model successfully suggesting the optimal interaction-list-sharing sizes for problems with a different interaction range H in various many-core architectures, including the MIMD-type architecture. While it appears not to be useful for high B/F machines, the interaction-list-sharing method can be a powerful optimization technique for accelerating particle simulations with short-range interactions especially for low B/F many-core architectures that are widely used in the modern devices.

In this paper, we focus on the optimization of the SPH method; however, the proposed approach to evaluate the performance improvement can be applied to other short-range methods. If we have a typical byte size of a particle, the number of arithmetic operations and the number of j-particles, then the optimal choice of N_i and the computational performance for a given many-core device can be predicted. Hence, we conclude that our prediction model can have a broad utility.

Acknowledgment. We used the super computer Shoubu system B at RIKEN. This work was supported by Post-K Issue 3 "System for Integrated Simulation of Earthquake and Tsunami Hazard and Disaster".

References

1. Gingold, R.A., Monaghan, J.J.: Smoothed particle hydrodynamics - theory and application to non-spherical stars. Mon. Not. R. Astron. Soc. **181**, 375–389 (1977)
2. Lucy, L.B.: A numerical approach to the testing of the fission hypothesis. Astrophys. J. **82**, 1013–1024 (1977)
3. Nishiura, D., Sakaguchi, H.: Parallel-vector algorithms for particle simulations on shared-memory multiprocessors. J. Comput. Phys. **230**(5), 1923–1938 (2011)
4. Nishiura, D., Furuichi, M., Sakaguchi, H.: Computational performance of a smoothed particle hydrodynamics simulation for shared-memory parallel computing. Comput. Phys. Commun. **194**, 18–32 (2015)
5. Barnes, J.E.: A modified tree code: don't laugh; it runs. J. Comput. Phys. **87**, 161–170 (1990)
6. Hamada, T., Narumi, T., Yokota, R., Yasuoka, K., Nitadori, K., Taiji, M.: 42 TFlops hierarchical N-body simulations on GPUs with applications in both astrophysics and turbulence. In: Proceedings of the Conference on High Performance Computing Networking, Storage and Analysis, pp. 1–12, November 2009
7. Iwasawa, M., Tanikawa, A., Hosono, N., Nitadori, K., Muranushi, T., Makino, J.: FDPS: a novel framework for developing high-performance particle simulation codes for distributed-memory systems. In: Proceedings of the 5th International Workshop on Domain-Specific Languages and High-Level Frameworks for High Performance Computing, WOLFHPC 2015, pp. 1:1–1:10. ACM, New York (2015)
8. Barnes, J., Hut, P.: A hierarchical O(N log N) force-calculation algorithm. Nature **324**, 446–449 (1986)
9. Hernquist, L., Katz, N.: TREESPH - a unification of SPH with the hierarchical tree method. Astrophys. J. Suppl. **70**, 419–446 (1989)
10. Kushner, D.: The wizardry of Id [video games]. IEEE Spectr. **39**(8), 42–47 (2002)

Influence of Architectural Features of the SNC-4 Mode of the Intel Xeon Phi KNL on Matrix Multiplication

Ruben Laso$^{(\boxtimes)}$ ⓘ, Francisco F. Rivera ⓘ, and José Carlos Cabaleiro ⓘ

Centro Singular de Investigación en Tecnoloxías da Información (CiTIUS),
Universidade de Santiago de Compostela, Santiago de Compostela, Spain
{r.laso,ff.rivera,jc.cabaleiro}@usc.es

Abstract. The Sub-NUMA Clustering 4 (SNC-4) affinity mode of the Intel Xeon Phi Knights Landing introduces a new environment for parallel applications, that provides a NUMA system in a single chip.

The main target of this work is to characterize the behaviour of this system, focusing in nested parallelization for a well known algorithm, with regular and predictable memory access patterns as the matrix multiplication. It has been studied the effects of thread distribution in the processor on the performance when using SNC-4 affinity mode, the differences between cache and flat modes of the MCDRAM and the improvements due to vectorization in different scenarios in terms of data locality.

Results show that the best thread location is the scatter distribution, using 64 or 128 threads. Differences between cache and flat modes of the MCDRAM are, generally, not significant. The use of optimization techniques as padding to improve locality has a great impact on execution times. Vectorization resulted to be efficient only when the data locality is good, specially when the MCDRAM is used as a cache.

Keywords: Intel Xeon Phi KNL · SNC-4 · MCDRAM · Vectorization

1 Introduction

Manycore architectures as the presented by the Intel Xeon Phi Knights Landing (KNL) provide highly parallel environments with a large number of cores in a single chip, allowing developers to exploit parallelism in their algorithms. In the case of the Intel Xeon Phi KNL, the second generation of the Intel's manycore processors, the most interesting features are the clustering modes and the integrated on-package memory known as MCDRAM (Multi-Channel DRAM).

This work is focused on the SNC-4 (Sub-NUMA Clustering 4) mode, in which the chip is partitioned into four nodes, being considered as a NUMA system. However, this configuration differs from typical NUMA systems in latency and bandwidth. Given that, a characterization of the KNL behaviour in its SNC-4 configuration can be useful for its users. In this work it is also included a brief

© Springer Nature Switzerland AG 2019
J. M. F. Rodrigues et al. (Eds.): ICCS 2019, LNCS 11540, pp. 483–490, 2019.
https://doi.org/10.1007/978-3-030-22750-0_41

comparison between the cache and flat modes of the MCDRAM memory and a study of the performance improvement achieved using vectorization.

To study these elements, different implementations of the classic matrix multiplication have been used, as this algorithm has some interesting properties that fit the goals of this work. It is a well known code and its memory access patterns are predictable, regular and easy to modify by changing the order of the loops, providing different scenarios in terms of data locality. Note that it is not intended to optimize this code, like in [3] or [7], but to use it just as a case study.

Other works have shown studies with very specific benchmarks, testing the MCDRAM in its different modes [12] or obtaining models of its behaviour [10]. Other authors have used benchmarks with dense matrix multiplication of small dimensions [4] or have looked for the roofline model using benchmarks based on sparse matrices algebra [2]. Finally, works like [6] show performance comparison on commonly used software. In contrast, this paper is intended to present a general study with different conditions, as the given when optimizing and parallelizing a code with regular memory patterns as the matrix multiplication.

2   Intel Xeon Phi KNL Architecture and Benchmarks

The Intel Xeon Phi is a manycore processor which bases its architecture in tiles [5]. Each tile has two cores and a 1 MB shared L2 cache memory. Each core has two VPUs (Vector Processing Unit), compatible with the AVX-512 instructions [11], and it is capable of execute up to four threads simultaneously. With a maximum of 36 tiles, the KNL can have up to 72 cores, 144 VPUs, and can execute up to 288 threads concurrently.

This processor has another singular features, like the clustering modes and the MCDRAM memory [13]. Concerning the clustering modes, three main configurations are available: All-to-all, Quadrant and SNC-4. The most interesting one is the SNC-4, where the chip behaves like a singular NUMA system. Additionally, the MCDRAM memory is an integrated high-bandwidth memory (up to 450 GB/s) with 16 GB of capacity that has three different configurations: cache, flat and hybrid modes. Using the cache mode, the MCDRAM behaves as a L3 direct-mapped cache memory. In the flat mode, the MCDRAM is configured as another main memory. In the hybrid mode, the memory is divided 50%–50% or 75%–25% in cache and flat modes, respectively. Note that using SNC-4 and flat modes, there are four NUMA nodes that correspond to the cores and the main memory, and four (of 4 GB each) corresponding to the MCDRAM.

In this study, the parallel (using OpenMP [1]) dense matrix multiplication has been used, $C = A \times B$, $A, B, C \in \mathcal{M}_{n \times n}$. The matrices are located in the MCDRAM when using flat mode, and in DDR4 memory with cache mode. The following nests in the loops has been studied: ijk, ikj, jik and jki. E.g. the ijk is referred as the one which has the i index in the outer-most loop, the index j in the intermediate one, and the k in the inner-most. The matrices elements accessed are always a_{ik}, b_{kj}, c_{ij}. The k loop cannot be parallelized due to race conditions, so nested parallelization has been tested only with ijk and jik nests.

In terms of data locality, the ikj loop should have the best performance because it accesses all the matrices by rows (the best way in C language). The ijk nest accesses the elements of A and C by rows and B by columns, being moderately efficient. The jik nest gets the elements of the matrices B and C by columns, and the elements of A by rows. Finally, the jki order should show the worst performance as all the matrices are accessed by columns.

Given the structure of the code and the organization of the cores in the SNC-4 mode, it is interesting to use a two-level parallelization. In the outer-most loop, iterations are distributed to different thread groups. In the second loop, iterations are shared out to the threads of each group. In the inner-most loop, vectorization can be applied using AVX-512 instructions. The natural distribution, given the architecture of the KNL, would be 4 groups of 64 threads, denoted as 4×64 from now on. This way of sharing out the iterations allows us to consider different locations for the threads of each group, scattering or compacting them across the cores, while the groups will be always scattered.

The basic algorithm where all the data are in contiguous positions, might have performance issues because of the replacements in the caches when using matrices with size $n = a2^b$, $a,b \in \mathbb{N}$, for certain values of b. To solve this problem, another version that uses padding has been implemented, so 64 bytes (the length of a cache line) are added to the end of each row of the matrices.

In the execution of the benchmarks, it has been used the Intel Xeon Phi KNL 7250 using 64 cores and 48 GB of DDR4 memory. The compiler used in the tests was the Intel ICC 18.0.0 with the options -qopenmp, -O2, -xMIC-AVX512 and -vec-threshold0 to ensure that the AVX-512 instructions were used. Other optimization options might transform the code too much, so the memory access patterns might not be the ones written in the source file. Metrics about cache misses and memory performance have been obtained with Intel VTune 2018.

3 Results

This section shows the results of the benchmarks with matrices of dimension $n = 4096$. Other sizes were tested too, achieving similar results.

3.1 Nested Parallelism

A brief summary of the results for the codes using nested parallelization and 4 groups of 16 threads each (4×16) are shown in Tables 1 and 2, where each entry contains the execution times (in seconds) of the sequential and parallel executions and the corresponding speedup.

Padding: The programs that do not use padding show higher execution times than those which implement this technique as it reduces the L2 cache misses up to a 98%. The ijk code shows an improvement of 97% in the sequential execution time, while jik improves up to 98%, showing the relevance of this kind of techniques in the KNL, improving drastically the execution times and changing the behaviour in terms of thread location and vectorization.

Table 1. Summary of the results obtained with the benchmarks without padding.

		Cache, scatter	Cache, compact	Flat, scatter	Flat, compact
ijk	vect.	7586.96 – 238.26 (×31)	7586.93 – 106.29 (×71)	3720.43 – 155.28 (×23)	3720.37 – 144.89 (×25)
	no vect.	3720.42 – 72.90 (×51)	3683.96 – 95.26 (×38)	3720.62 – 154.57 (×24)	3721.08 – 159.03 (×23)
jik	vect.	7601.64 – 1971.87 (×3.8)	7736.51 – 485.35 (×15)	3721.72 – 581.76 (×6.3)	3731.53 – 2158.89 (×1.7)
	no vect.	3729.80 – 531.64 (×7.0)	3965.88 – 302.08 (×13)	3731.37 – 781.83 (×4.7)	3731.61 – 2563.39 (×1.4)

Table 2. Summary of the results obtained with the benchmarks with padding.

		Cache, scatter	Cache, compact	Flat, scatter	Flat, compact
ijk	vect.	247.91 – 6.96 (×35)	242.34 – 91.65 (×2.6)	248.94 – 28.16 (×8.8)	243.80 – 88.76 (×2.7)
	no vect.	404.73 – 7.96 (×50)	400.84 – 83.86 (×4.7)	399.47 – 45.18 (×8.8)	401.81 – 80.67 (×4.9)
jik	vect.	176.25 – 4.93 (×35)	178.17 – 12.58 (×14)	181.18 – 7.08 (×25)	175.72 – 12.17 (×14)
	no vect.	330.00 – 5.28 (×62)	331.51 – 19.24 (×17)	328.87 – 11.42 (×28)	330.05 – 16.91 (×19)

Thread Location: The results of the programs which do not use padding show that the scatter distribution is not always the best. A compact distribution can cause the threads of the same core/tile share cache lines, reducing the traffic in the mesh. This situation also implies a faster communication in comparison with sharing data with external tiles. In addition, a compact distribution divides the resources of each core, probably causing a lower performance. Also, it can saturate the buses of the caches or the interconnection mesh, degrading the execution times, as in the jik nest with the MCDRAM in flat mode.

Vectorization: The codes that do not use padding present poor performance results when using vectorization. However, when padding is applied, improvements up to 54% are obtained, showing that vectorization is only beneficial when the programs have efficient data access patterns.

MCDRAM: Using the MCDRAM in cache mode and having inefficient access patterns causes important performance problems in the vectorized programs, due to the difficulties in feeding the registers. Note that these performance issues do not appear with the flat mode. In contrast, cache mode shows a slightly better performance in parallel executions.

Best Execution: The best execution time obtained using nested parallelism has been 2.66 s (51.66 GFlop/s), achieved using the jik nest, 64 groups of 2 threads with a compact distribution, MCDRAM in cache mode, and using padding and vector instructions.

3.2   One-Level Parallelism

A summary of the results obtained by the codes where only the outer loop is parallelized is shown in Tables 3 and 4, including the ikj and jki nests.

Padding: All loop nests take advantage of this technique because of the significant reduction in the number of cache misses, up to 98%.

Table 3. Summary of the results obtained with the benchmarks without padding.

		Cache, scatter	Cache, compact	Flat, scatter	Flat, compact
ijk	vect.	7586.96 − 292.25 (×25)	7586.93 − 99.18 (×76)	3720.43 − 97.22 (×38)	3720.37 − 110.56 (×33)
	no vect.	3720.42 − 60.37 (×61)	3683.96 − 79.42 (×46)	3720.62 − 97.07 (×38)	3721.08 − 109.36 (×34)
jik	vect.	7601.64 − 177.20 (×42)	7736.51 − 108.61 (×71)	3731.72 − 190.99 (×19)	3731.53 − 189.86 (×19)
	no vect.	3729.80 − 97.26 (×38)	3965.88 − 101.02 (×39)	3731.37 − 188.71 (×19)	3731.61 − 191.46 (×19)
ikj	vect.	79.54 − 9.69 (×8.2)	80.68 − 5.53 (×14)	76.94 − 6.61 (×12)	76.89 − 6.41 (×13)
	no vect.	189.32 − 10.23 (×18)	189.35 − 9.66 (×19)	174.39 − 8.43 (×20)	173.38 − 8.69 (×20)
jki	vect.	8670.13 − 505.25 (×17)	8831.75 − 277.52 (×31)	8624.93 − 516.27 (×16)	8584.76 − 526.59 (×16)
	no vect.	8568.79 − 511.14 (×16)	8497.72 − 245.05 (×34)	8624.20 − 517.28 (×16)	8584.68 − 516.50 (×16)

Table 4. Summary of the results obtained with the benchmarks with padding.

		Cache, scatter	Cache, compact	Flat, scatter	Flat, compact
ijk	vect.	247.91 − 4.59 (×54)	242.34 − 4.39 (×55)	248.94 − 4.56 (×54)	243.80 − 4.75 (×51)
	no vect.	404.73 − 7.33 (×55)	400.84 − 7.65 (×52)	399.47 − 7.20 (×55)	401.81 − 7.21 (×55)
jik	vect.	176.25 − 3.32 (×53)	178.17 − 3.27 (×54)	181.18 − 3.79 (×47)	175.72 − 3.67 (×47)
	no vect.	330.00 − 5.02 (×65)	331.51 − 5.36 (×61)	328.87 − 6.54 (×50)	330.05 − 7.14 (×46)
ikj	vect.	69.75 − 2.61 (×26)	69.54 − 4.58 (×15)	69.61 − 3.67 (×18)	69.42 − 4.63 (×14)
	no vect.	180.51 − 3.92 (×46)	185.23 − 10.80 (×17)	179.36 − 4.08 (×43)	180.27 − 10.77 (×16)
jki	vect.	538.61 − 22.70 (×23)	551.16 − 148.99 (×33)	535.75 − 100.78 (×5.3)	534.93 − 158.36 (×3.3)
	no vect.	550.08 − 22.73 (×24)	545.46 − 147.59 (×26)	535.11 − 98.26 (×5.4)	536.07 − 159.72 (×3.3)

Thread Location: When only the outer most loop is parallelized, the scatter distribution gives the best execution times. The differences are specially noticeable in the ikj and jki cases using padding, improving up to 60% and 85%.

Vectorization: The ikj nest, the best access pattern, takes the maximum benefit of vector instructions, reducing up to 63% the execution time.

MCDRAM: Using one-level parallelism, the cache mode usually achieves a slightly better performance in parallel executions. Even though, differences between flat and cache modes are generally not significant.

Best execution: The best time obtained using one-level parallelism has been of 1.17 s (117.46 GFlop/s), using the ikj order, with padding, 256 threads, MCDRAM in cache mode, and using AVX-512 instructions.

3.3   Comparative with a Real NUMA

A summary of the comparison of the Intel Xeon Phi KNL with a NUMA server that consists of four Intel Xeon E5 4620 is shown in Table 5.

Memory Latencies: To study the effect of non-local accesses in the NUMA server and in the KNL, the numademo command [8] with the sequential STREAM benchmark [9] has been executed in both machines. Results in Table 6 show that the penalties in the NUMA server go from 28% up to 40% while in the KNL these penalties are much lower. With the MCDRAM in cache mode, the differences are between 3.8% and 6.9%. Using the flat mode, both kind of penalties, those related to the DDR memory and the MCDRAM, are around 2% and 3%.

Table 5. Best execution times and speedup of the KNL and NUMA server.

	Sequential		As many threads as cores		All available threads	
	KNL	NUMA	KNL (64 threads)	NUMA (40 threads)	KNL (256 threads)	NUMA (80 threads)
ijk	242.34	545.73	4.39 ($\times$55)	28.76 ($\times$19)	11.55 ($\times$21)	15.04 ($\times$36)
jik	175.72	521.63	3.27 ($\times$53)	33.54 ($\times$15)	2.53 ($\times$69)	16.96 ($\times$30)
ikj	69.42	107.43	2.61 ($\times$26)	4.11 ($\times$26)	1.17 ($\times$59)	2.20 ($\times$48)
jki	535.11	521.71	22.70 ($\times$23)	44.11 ($\times$11)	137.90 ($\times$3.8)	47.92 ($\times$10)

Table 6. Average bandwidth (in MB/s) and penalty percentage given by `numademo` with STREAM copy benchmark.

	NUMA	KNL (cache)	KNL (flat)	
			DDR4	MCDRAM
Local	10,287	9,064	8,266	9,082
Remote	7,409 (28%)	8,712 (3.8%)	8,084 (2.2%)	8,909 (1.9%)
Worst	6,142 (40%)	8,439 (6.9%)	8,066 (2.4%)	8,803 (3.0%)

Padding: This optimization technique has a low impact in the Intel Xeon E5 in comparison with the KNL. In the tests, adding padding to the data has improved the sequential execution times up to 30%.

Vectorization: The Intel Xeon E5 is not compatible with the AVX-512 vector instructions, being only compatible with the AVX2 instruction set. This kind of instructions works with 4 operands with FMA operations, so its performance is lower, improving just up to 9% the execution times.

Thread Location: The performance shows that using more NUMA nodes is not always the best option to improve the performance, like in the Intel Xeon Phi, due to the higher penalties of remote accesses on a real NUMA server.

Best Execution: In the NUMA server the best execution time has been obtained by the ikj nest, computing the matrix multiplication in 2.20 s (62.47 GFlop/s), using 80 threads with a scatter distribution, vectorization and no padding. This performance is noticeably lower than the provided by the KNL.

4   Conclusions

In this work, it has been studied the behaviour of the Intel Xeon Phi KNL in different situations, characterizing its performance in terms of thread location,

MCDRAM mode, vectorization, data locality and, also, comparing the SNC-4 mode of the KNL with a real NUMA system.

Generally, the most efficient thread distribution is the scatter location, using all the NUMA nodes of the KNL, and one or two threads per core. In a real NUMA system, the behaviour is different, depending heavily on the algorithm because of the higher penalties in the communications between nodes.

The way the data is located in memory has a deep impact in the performance in the KNL compared to other processors. In this case, adding padding to the data has produced a reduction of the execution time up to 98%.

Using vector instructions has shown an irregular behaviour. With a low locality, the use of vector instructions had a negative impact. In opposition, with good data locality, their use has improved the execution times up to 60%.

Differences between cache and flat modes of the MCDRAM are, generally, not significant. Flat mode seems to perform better under inefficient data access patterns, but cache mode has usually given better results on parallel codes.

Acknowledgment. This work has received financial support from the Ministerio de Economía, Industria y Competitividad within the project TIN2016-76373-P and network CAPAP-H. It was also funded by the Consellería de Cultura, Educación e Ordenación Universitaria of Xunta de Galicia (accr. 2016–2019, ED431G/08 and reference competitive group 2019–2021, ED431C 2018/19) and network R2016/045.

References

1. Dagum, L., Menon, R.: OpenMP: an industry standard API for shared-memory programming. IEEE Comput. Sci. Eng. **5**(1), 46–55 (1998)
2. Doerfler, D., et al.: Applying the roofline performance model to the Intel Xeon Phi knights landing processor. In: Taufer, M., Mohr, B., Kunkel, J.M. (eds.) ISC High Performance 2016. LNCS, vol. 9945, pp. 339–353. Springer, Cham (2016). https://doi.org/10.1007/978-3-319-46079-6_24
3. Guney, M.E., et al.: Optimizing matrix multiplication on Intel® Xeon Phi TH x200 architecture. In: 2017 IEEE 24th Symposium on Computer Arithmetic (ARITH), July 2017. https://doi.org/10.1109/ARITH.2017.19
4. Heinecke, A., Breuer, A., Bader, M., Dubey, P.: High order seismic simulations on the Intel Xeon Phi processor (knights landing). In: Kunkel, J.M., Balaji, P., Dongarra, J. (eds.) ISC High Performance 2016. LNCS, vol. 9697, pp. 343–362. Springer, Cham (2016). https://doi.org/10.1007/978-3-319-41321-1_18
5. Jeffers, J., Reinders, J., Sodani, A.: Intel Xeon Phi Processor High Performance Programming: Knights Landing Edition. Morgan Kaufmann, San Francisco (2016)
6. Kang, J.H., Kwon, O.K., Ryu, H., Jeong, J., Lim, K.: Performance evaluation of scientific applications on Intel Xeon Phi knights landing clusters. In: 2018 International Conference on High Performance Computing & Simulation (HPCS). IEEE (2018)
7. Kim, R.: Implementing general matrix-matrix multiplication algorithm on the Intel Xeon Phi knights landing processor. Ph.D. thesis, Department of Mathematical Sciences, Seoul National University (2018)
8. Kleen, A.: A NUMA API for Linux. Novel Inc. (2005)

9. McCalpin, J.D., et al.: Memory bandwidth and machine balance in current high performance computers. IEEE Comput. Soc. Tech. Committee Comput. Archit. (TCCA) Newsl. **1995**, 19–25 (1995)
10. Ramos, S., Hoefler, T.: Capability models for many core memory systems: a case-study with Xeon Phi KNL. In: 2017 IEEE International Parallel and Distributed Processing Symposium (IPDPS), pp. 297–306. IEEE (2017)
11. Reinders, J.: AVX-512 instructions. Intel Corporation (2013)
12. Rosales, C., Cazes, J., Milfeld, K., Gómez-Iglesias, A., Koesterke, L., Huang, L., Vienne, J.: A comparative study of application performance and scalability on the Intel knights landing processor. In: Taufer, M., Mohr, B., Kunkel, J.M. (eds.) ISC High Performance 2016. LNCS, vol. 9945, pp. 307–318. Springer, Cham (2016). https://doi.org/10.1007/978-3-319-46079-6_22
13. Sodani, A., et al.: Knights landing: second-generation Intel Xeon Phi product. IEEE Micro **36**(2), 34–46 (2016)

Improving Planning Performance in PDDL+ Domains via Automated Predicate Reformulation

Santiago Franco, Mauro Vallati[(⊠)][ID], Alan Lindsay,
and Thomas Lee McCluskey

School of Computing and Engineering, University of Huddersfield, Huddersfield, UK
{s.franco,m.vallati,a.lindsay,t.l.mcCluskey}@hud.ac.uk

Abstract. In the last decade, planning with domains modelled in the hybrid PDDL+ formalism has been gaining significant research interest. A number of approaches have been proposed that can handle PDDL+, and their exploitation fostered the use of planning in complex scenarios. In this paper we introduce a PDDL+ reformulation method that reduces the size of the grounded problem, by reducing the arity of *sparse* predicates, i.e. predicates with a very large number of possible groundings, out of which very few are actually exploited in the planning problems. We include an empirical evaluation which demonstrates that these methods can substantially improve performance of domain-independent planners on PDDL+ domains.

Keywords: Automated planning · Hybrid reasoning · Reformulation

1 Introduction

Automated planning is one of the most prominent Artificial Intelligence (AI) challenges; it has been studied extensively for several decades and has led to a large number of real-world applications. The growing number of domain-independent PDDL+ planners is fostering the exploitation of planning in complex real-world applications, where notions of continuous processes and discrete events and actions are needed [4]. The use of reformulation and configuration techniques, which can automatically re-represent the planning model in order to increase efficiency and enable a scale up in size of applications that can be handled. In the last decades, research into reformulation techniques has attracted significant attention. Types of reformulation of classical PDDL models include macro-learning [6] and configuration [9].

Hybrid PDDL+ models are amongst the most advanced models of systems and the resulting problems are notoriously difficult for planners to cope with due to non-linear behaviours and immense search spaces. Complexity is exacerbated by the potentially huge size of the fully grounded problems, needed by planners in order to effectively explore the search space, which can make some

© Springer Nature Switzerland AG 2019
J. M. F. Rodrigues et al. (Eds.): ICCS 2019, LNCS 11540, pp. 491–498, 2019.
https://doi.org/10.1007/978-3-030-22750-0_42

Algorithm 1. Reformulation for flattening sparse predicates

Require: D_o, I_o, s^t, a^t
Ensure: D_r, I_r
1:   $SP = statics(D_o); P = predicates(D_o) \cup functions(D_o)$
2:   $D_r = D_o; I_r = I_o$
3:   **for all** p_j in P, **where** $arity(p_j) > 2$ **do**
4:     **if** $sparsity(p_j, I_o) > s^t$ **then**
5:       $p_{stat} = findConstrainingStatic(p_j, SP)$
6:       **if** $p_{stat} \neq None$ **then**
7:         $T_{p_{stat}} = getSparseVariables(p_{stat}, I_o, a^t)$
8:         $C^{new} = makeConstants(T_{p_{stat}}, s_o)$
9:         $D_r = addAsConstants(D_r, C^{new})$
10:        $D_r = updateOpProEv(D_r, T_{p_{stat}}, C^{new})$
11:        $I_r = updatePredsFuncs(D_r, I_r, T_{p_{stat}}, C^{new})$

---

problems impossible to tackle. Particularly, grounding is also strongly affected by predicates' instances that will not be reachable in any state of the problem.

In this paper we introduce a PDDL+ reformulation method that allows to drastically reduce the size of the grounded problem, by reducing the arity of *sparse* predicates, i.e. predicates with a very large number of possible groundings, out of which very few are actually exploited in the planning problems. Arity is reduced by merging suitable objects together, and partially grounding the operators, processes and events in which reformulated predicates are involved. Our experimental analysis, performed on a range of problem from different application domains, shows that the proposed reformulation technique can substantially improve the performance of PDDL+ planning engines, by allowing problems to be grounded and by constraining the search space.

2   The Proposed Reformulation Approach

Our approach relies on identifying sparse predicates that are partially constrained by a static predicate. Through combining the sparsity measure for dynamic predicates with a constraining static predicate, the approach is able to better identify predicates for which the reformulation will have significant impact.

Let us consider an hybrid version of the well-known Rovers domain model, where movements and energy generation via solar are modelled as continuous processes, triggered by actions under the control of a planner, and constrained by appropriate events. In the Rovers domain, rovers are used to make soil and rock samples and to take pictures for various objectives. This requires that the rovers are moved between waypoints in order to establish shots and collect samples from certain positions. The constraints establishing the properties of the rovers and the relationships between waypoints (e.g., that a rover can traverse between waypoints) are encoded as static facts. As with many network based relationships, only a fraction of the potential connections are made available in any particular problem model. Of course, as the number of waypoints (and rovers) grows this fraction will reduce. The problem model reformulation reported in this paper collapses the variables of predicates, creating a model with fewer sparsely instantiated predicates. This procedure can be applied to Rovers, for example, consider replacing the (arity 2) predicate: (`visible` *?waypoint1 ?waypoint2*) with an alternative (arity 1) encoding: (`visible` *?visible_waypoint1_waypoint2*). Whereas in the original version, the domain of possible instantiations is every pair of waypoints; in the second approach, we can reduce the domain by only defining constants for the combinations of waypoints for which the relation holds.

2.1   The Reformulation Algorithm

Algorithm 1 shows how the reformulation of a domain model, D_o, and a problem model, I_o, is performed. Beside the models to be reformulated, the algorithm requires as input a sparsity threshold s^t, which is used to decide whether or not it is useful to perform the reformulation and a parameter, and a^t, which sets the maximum number of variables considered in the reformulation (how these parameters are set is explained below).

In the algorithm (see Algorithm 1) the sparsity of the predicates (Boolean or numeric) with arity greater than 2 are assessed in turn (line 3) to determine if they are suitable for the reformulation step. As a measure of sparsity we compare the set of propositions in the initial state with the possible set of all propositions for the predicate. For example, if we consider a specific example Rovers problem from our benchmark problems, with 4 waypoints and 2 rovers, we can calculate the total set of possible propositions as: $4 \times 4 \times 2 = 32$. In this example, there are 10 instances of `can_traverse` in the initial state and so the sparsity for this predicate is $10/32$.

In the case of a sparse predicate, p_j, the procedure attempts (line 5) to find a static predicate, p_{stat}, such that p_j is only used in transition schemas (that is in the action, process or event schemas) with p_{stat}. We consider predicates as static if instances of the predicate can not be deleted or created during the planning process but, in the case of numeric predicates, their value can be changed. If there is more than one constraining static predicate then one is selected heuristically by selecting the predicate that occurs the most in transition schemas. There are two static facts that constrain the `can_traverse` predicate: `can_traverse` itself and `visible`. The algorithm selects `visible` as it appears in more transition schemas.

2.2   Reformulating the Domain and Problem Models

In the case that p_{stat} exists (e.g., `visible`), a reformulation step is applied using p_{stat} as the basis. In our current system, we have considered subsets of the variables of the static facts and so we add a parameter, a^t, to determine the maximum arity of the reformulation. The best $\max(a^t)$ variables are selected (line 7) using the sparsity of the tuple for p_{stat} in Io. We use T_p to denote a subset of the variables of p. In our example, `visible` has arity 2 and therefore $T_{\texttt{visible}}$

would contain both its variables. The variables in T_{stat} are then combined to form a set of constants, C^{new}, of type, t^{new}, which are added to the domain model. One constant is defined for each distinct combination of these variables for the instances of the predicate in I_o. For example, a new constant is generated for each distinct combination of the waypoints in the instances of the `visible` predicate in the initial state. For instance, (`visible waypoint3 waypoint0`) leads to a new constant `waypoint3_waypoint0` (using the new type).

At this stage (line 10) each of the transition schemas that refer to p_{stat} are reformulated. For example, in Rovers, the transitions with `visible` as a precondition are identified (e.g., `start-navigate`, `communicate_soil_data`). For each predicate (dynamic, static or numeric) in these transitions the algorithm tests to determine if it can be part of this reformulation step. If the predicate is only used in transition schemas with p_{stat}, and $T_{p_{stat}}$ is a subset of the parameters of the predicate then it is selected for reformulation. In the case of `visible`, only the `can_traverse` predicate is constrained by the `visible` predicate and so only these two are selected for reformulation. For each selected predicate, p, (including p_{stat}) a new predicate, p', is made by replacing the variables that are in $T_{p_{stat}}$ with a single variable of type, t^{new} and retaining the other variables (e.g., $arity(p') = arity(p) - |T_{p_{stat}}| + 1$). For example, (`can_traverse ?rover ?waypoint1 ?waypoint2`), is reformulated as (`can_traverse ?rover ?new-type`). The original predicate, p, is omitted from the new model, Dr.

Each transition schema that depends on p_{stat} is partially grounded so that the variables corresponding to those in $T_{p_{stat}}$ are grounded and constants added as necessary (i.e., for referencing the individual objects). This allows the relation between the new constants and the original objects to be maintained. For example, there are new `start-navigate` operators for each of the new constants, e.g., `start-navigate-waypoint3-waypoint2`. In `start-navigate`, each matching of $?waypoint$ is added as constant as it is referred to in other predicates.

Finally, the problem model is reformulated (line 11) by changing those predicates involved in the reformulation to use the constants in C^{new} in the initial state and goal, using a similar approach as described above. Of course, after this step has been applied once, the procedure can be repeated on the reformulated model supporting further combining of variables as appropriate.

3   Experimental Analysis

Four PDDL+ planners at the state of the art are included in the evaluation: ENHSP [8], UPMurphi [3], DINO [7], and SMTPlan [2].

All reported results were achieved by running the planners on a machine equipped with i7-4750HQ CPU, 16 GB of memory, running Ubuntu 16.10 OS. 4 GB of memory were made available for each run, and a 15 CPU-time minutes cut-off time limit was enforced.

The experimental evaluation is performed by considering three benchmark domains, namely Hybrid Rover, Urban Traffic Control, and Baxter.

Table 1. CPU-time seconds needed by the planners to find a satisficing solution. O (R) rows show the results achieved when running the Original (Reformulated) model. X indicates grounded but not solved. "–" means crashed during grounding. NA indicates that the planner is unable to handle the model.

Planner		Baxter					Hybrid rover					UTC				
		1	2	3	4	5	1	2	3	4	5	1	2	3	4	5
ENHSP	O	0.40	X	26.8	13.5	335.7	1.3	18.9	58.54	77.33	–	5.3	10.8	–	–	–
	R	0.45	151.7	23.5	15.2	17.9	1.1	35.0	60.28	65.5	87.5	2.7	3.2	12.5	6.9	30.8
UPMurphi	O	X	X	X	X	X	X	X	X	X	X	7.26	X	–	148.78	X
	R	X	X	X	X	X	X	X	X	X	X	0.62	5.02	29.34	5.2	49.4
DINO	O	12.0	X	X	X	X	116.24	X	X	X	X	X	X	–	X	X
	R	27.6	X	X	X	X	X	X	X	X	X	X	X	X	X	X
SMTPlan	O	0.01	X	X	X	X	0.5	1.8	X	X	X	NA	NA	NA	NA	NA
	R	0.01	X	X	X	X	0.5	1.8	X	X	X	NA	NA	NA	NA	NA

The **Baxter** domain [1] exploits planning for dealing with articulated objects manipulation tasks. The "simplified" domain model has been extended by allowing continuous movements of a joint, modelled via actions and process envelope, on different axis, and by adding events for preventing movements wider than 360°. Problems consider articulated objects composed by 2–5 links. The objects of type `link` has been merged into a new type, and four predicates have then been reformulated: `connected`, `increasing_angle`, `decreasing_angle`, and `use`. Our reformulation approach has been applied following the fact that the `connected` predicate is static, and is exploited in operators and processes to control all the other mentioned predicates. According to the results shown in Table 1 UPMurphi, DINO, and SMTPlan grounded and explored the search space of all the considered problems but only ENHSP solved most of the problems using the original representation. ENHSP solved all but one of the problems using the original models. Remarkably, the use of reformulated models did lead to a significant search speedup, and allows ENHSP to solve all the considered benchmarks. Empirical evidence indicates that the reformulation allows to improve the pruning done by the reachability analysis of ENHSP, leading to a faster expansion and evaluation of the search states. The other planners could only solve the simplest problem in both original and reformulated versions. Interestingly, DINO works better with the original representation. According to our observations, in that specific problem, the DINO heuristic expanded twice the number of states compared to its use with the reformulated model, but to find the same solution. This seems to suggest that, on some simple problem instances, the use of reformulated models can reduce the effectiveness of a domain-independent heuristic.

We extended the well-known **Rover** domain model introduced in IPC-3 by modelling as continuous processes the movements of the rovers, and the energy generation via solar power. Each of the mentioned processes can be controlled by the planner using two actions, and is constrained, where appropriate, via events. The predicate can_traverse has been reformulated by merging the objects of type waypoints, as shown in the previous section. The use of reformulated models allows ENHSP to solve a larger number of problem instances. However, in few cases, the reformulation negatively affected search performance, once the grounding is completed. The larger the problem, the closer the performance between both versions got. According to our observations, the default heuristic exploited by ENHSP is less informative, for this domain, when exploited on reformulated models. However, one large and complex problems, the gap is closed by the fact that the reformulated model allows the planner to explore a significantly larger size of the search space. DINO, running on reformulated models, is able to complete the grounding of all the considered problems, but lacks of the capability of generating plans. SMTPlan is not significantly affected by the different domain models. This seems to be related to the compilation of the PDDL+ model into an SMT encoding that allows to reduce grounding-related issues.

Table 2. Ratio of maximum search space sizes of original vs reformulated representations for the UPMurphi and DINO planners. "–" is used to indicate cases where one of the approaches lead planners to crash during grounding.

	Baxter					Hybrid rover					UTC				
Problem	1	2	3	4	5	1	2	3	4	5	1	2	3	4	5
Ratio	2.42	3.98	3.98	3.13	4.76	1.00	1.05	1.88	2.86	4.18	18.59	37.08	–	18.58	21.05

Finally, the **Urban Traffic Control** (UTC) domain [5] models the use of planning for generating traffic light signal plans, in order to de-congest a urban region. In this analysis we considered the problems introduced by McCluskey and Vallati [5], which involved a network of 10 junctions, and we extended it by considering problems with 20 and 30 junctions, obtained by connecting identical regions. Problems 1–3 have 10, 20 and 30 junctions respectively and only one goal. Problems 4 and 5 have 10 and 20 junctions respectively, both of them have 3 goals. Goals in this domain indicates the requirement of reducing the congestion on a specific link of the network. In this domain, the predicate flowrate has been reformulated by merging the objects of type link into a new type, which represents road links which are connected via a junction. ENHSP and UPMurphi were run using the heuristic proposed by McCluskey and Vallati [5]. Results presented in Table 1 indicate that reformulation has a strong beneficial impact on planners' performance. ENHSP and UPMurphi are able to quickly solve problems involving large networks as they can manage to ground the problem. In ENHSP most of the improvement is due to the faster grounding, and on the largest 30 junction problem 3, to be able to ground it at all. On the contrary, in the case of UPMurphi and DINO, the reformulation boosts also the search performance, as also the size of each state is significantly reduced by the reformulation. Unsurprisingly, the domain-independent heuristic exploited by DINO is not very informative in UTC problems, but the planner is able to ground and to explore a large area of the search space when run on reformulated models. SMTPlan is not able to handle the UTC domain model.

Impact of Reformulation on Search Space Size. Beside the impact on planning performance, it is also important to assess whether the use of the proposed reformulation approach allows to actually reduce the size of a search state, so that the search space can be explored quicker and more efficiently. Table 2 shows how planners DINO and UPMurphi benefit from the reformulation of PDDL+ models, in terms of state size. We consider these planners because they allow to measure effectively and accurately the size of a single search state. Results are presented in terms of ratio of maximum space sizes between original and the corresponding reformulated representation; for instance, a value of 1.0 indicates that there is no difference, while a value of 2.0 means reformulated search can create twice the number of states before running out of memory. In almost every considered instance, reformulation greatly increases the maximum state space, and therefore allows the planners to explore a larger portion of the space, and to generate search states in less time.

4   Conclusion

In this paper, we introduced a reformulation approach that allows to reduce the size of a PDDL+ grounded problem by tackling the arity of sparse predicates. Our experimental analysis showed that the proposed reformulation: (i) effectively reduces the grounding size of hybrid problems, hence allowing planners to deal with them; and (ii) positively affect the size of each search state, leading to a faster and more effective exploration of the search space. Results suggest that PDDL+ reformulation techniques, by allowing larger problems to be reasoned upon by planning engines, can foster the exploitation of planning in real-world applications. Future work is planned on extending the number of objects that can be merged, and to study the importance of the sparsity threshold.

References

1. Capitanelli, A., Maratea, M., Mastrogiovanni, F., Vallati, M.: On the manipulation of articulated objects in human-robot cooperation scenarios. Robot. Auton. Syst. **109**, 139–155 (2018)
2. Cashmore, M., Fox, M., Long, D., Magazzeni, D.: A compilation of the full PDDL+ language into SMT. In: Proceedings of ICAPS (2016)
3. Della Penna, G., Magazzeni, D., Mercorio, F., Intrigila, B.: UPMurphi: a tool for universal planning on PDDL+ problems. In: Proceedings of ICAPS (2009)
4. Fox, M., Long, D.: Modelling mixed discrete-continuous domains for planning. J. Artif. Intell. Res. **27**, 235–297 (2006)
5. McCluskey, T.L., Vallati, M.: Embedding automated planning within urban traffic management operations. In: Proceedings of ICAPS (2017)
6. Newton, M.A.H., Levine, J., Fox, M., Long, D.: Learning macro-actions for arbitrary planners and domains. In: Proceedings of ICAPS (2007)

7. Piotrowski, W.M., Fox, M., Long, D., Magazzeni, D., Mercorio, F.: Heuristic planning for PDDL+ domains. In: Proceedings of IJCAI (2016)
8. Scala, E., Haslum, P., Thiébaux, S., Ramírez, M.: Interval-based relaxation for general numeric planning. In: Proceedings of ECAI (2016)
9. Vallati, M., Hutter, F., Chrpa, L., McCluskey, T.L.: On the effective configuration of planning domain models. In: Proceedings of IJCAI (2015)

The Case of iOS and Android: Applying System Dynamics to Digital Business Platforms

Ektor Arzoglou[✉], Tommi Elo, and Pekka Nikander

Department of Communications and Networking, Aalto University, Helsinki, Finland
{ektor.arzoglou,tommi.elo,pekka.nikander}@aalto.fi

Abstract. Platforms are multi-sided marketplaces that bring together groups of users that would otherwise not have been able to connect or transact. The app markets for Apple iOS and Google Android are examples of such markets. System dynamics is a powerful method to gain useful insight into environments of dynamic complexity and policy resistance. In this paper, we argue that adapted to the context of digital business platforms, the practice of system dynamics facilitates understanding of the role of incentives in such marketplaces for increasing participation, value generation, and market growth. In particular, we describe our efforts to simulate the market competition between iOS and Android in terms of the interacting markets for devices and their apps.

Keywords: Android · iOS · Platform economy · System dynamics

1 Introduction

Digital business platforms, such as Apple App Store, Uber, and AirBnB, are dramatically reducing search and transaction costs. They are multi-sided marketplaces in which two or more user groups benefit from finding each other more easily [1], thus creating *indirect network effects*.[1] For example, in the case of AirBnB[2] and CouchSurfing,[3] the platforms originally allowed people willing to let others use their apartments, to find others who were looking for affordable, cosy accommodation while travelling. In the case of smartphone app stores, the consumers know that they can easily find apps for their smart devices from the store, while the app developers know that the users will look there for apps.

This project has received funding from the European Union's Horizon 2020 research and innovation programme under grant agreement No. 779984.

[1] Direct network effects are positive feedback loops created *within* the same market, meaning that the benefit of a technology to a user depends positively on the number of users of this technology on the *same* side of the market. Indirect network effects are positive feedback loops created *across* the same market, meaning that the benefit of a technology to a user depends positively on the number of users of this technology on *another* side of the market.

[2] https://www.airbnb.com/.

[3] https://www.couchsurfing.com/.

ⓒ Springer Nature Switzerland AG 2019
J. M. F. Rodrigues et al. (Eds.): ICCS 2019, LNCS 11540, pp. 499–506, 2019.
https://doi.org/10.1007/978-3-030-22750-0_43

From the modelling point of view, the market dynamics of digital business platforms has thus far gained relatively little attention (see Sect. 2). Although there have been efforts to model digital business platforms using system dynamics, no studies have yet attempted to model the user and the developer side of the market together in the iOS and Android ecosystems.

In this paper, we model the competition between Apple and Google in the smartphone market using the stocks and flows elements of the *system dynamics modelling* (SDM) methodology. We focus on the dynamic market competition between iOS and Android smartphone platforms in terms of the interplay between the user and developer sides of the market. At this point, a simulation model has been calibrated with publicly available data. The model is able to adequately replicate the historical dynamic interplay between these two competing two-sided markets, based on statistical parameters of *sensitivity* and *threshold*. Our simulation results show that the whole market can be easily captured by an initially inferior player, provided that such a player reaches sufficiently early the exponential feedback loops depicting the direct or indirect network effects.

In particular, we shed light on the factors affecting the competition between Apple and Google in the smartphone industry. With this initial step, we pave the way towards a better understanding of not only digital business platforms in general, but also the reason that digital business platforms tend to show a winner-take-all structure.

To our knowledge, this is the first paper that applies SDM to study the competition between particular, historically recorded digital business platforms. In other words, the prior art on using SDM to model multi-sided markets has focused on abstracting a generic market model, without any reported serious attempts to calibrate the model using historical data.

The rest of the paper is organised as follows. In Sect. 2, we describe the overall background, focusing on previous attempts to capture the dynamics of multi-sided markets and especially the market encompassing Apple and Google. In Sect. 3, we introduce our models and describe the simulation results, which are then discussed in Sect. 4. Section 5 concludes the paper suggesting future work.

2   Background

Before the Internet, many industries were dominated by large search and transaction costs. However, the advent of the Internet and search engines has brought a significant change resulting in companies being able to reach their customers all over the world. This has created a new problem in the form of fake companies and fraudulent services.

Digital business platforms have emerged to solve this second problem by creating an incentive for the suppliers to act in a trustworthy manner, thereby allowing them to enhance their reputation. eBay was perhaps the first platform that managed to capture the required dynamics.

2.1   Related Work

The term "platform", in the meaning of a digital multi-sided market, was coined by Rochet and Tirole in 2003 [6]. Platforms create value by acting as conduits between two (or more) categories of consumers, who would have been unable to connect or transact otherwise [10]. The more consumers enter the platform, the more value they capture as a result of the *indirect network effects* between the user groups. These network effects reflect the exogenous interdependence of demand between consumer groups and shape platform competition [8]. The network effects in such a platform form a self-reinforcing feedback loop, which creates an advantage for early adopters. In addition, as these network effects grow, they act like a barrier to entry for potential competitors [3], under certain conditions, leading to a winner-take-all outcome [4].

Attempts to model digital business platforms using system dynamics are quite rare. Dutta et al. use system dynamics to model the diffusion of iOS and Android based handsets [2]. Ruutu et al. provide a system dynamics simulation model to analyse platform development and platform based competition [7]. Scholten et al. depict network and complementarity effects of a Platform-as-a-Service (PaaS) ecosystem [9]. Von Kutzschenbach and Brønn use a feedback systems approach to illustrate Uber's 'get big fast' (GBF) strategy [11]. Finally, Zolfagharian et al. provide an evidence-based framework that demonstrates why, when, and how system dynamics is combined with other methods [12].

Closest to our present work, Meyer considers path dependency in the context of two-sided markets with indirect network effects which commit users to an inferior technology platform [5]. Meyer uses *agent-based modelling* (ABM) in order to show that third-degree[4] lock-ins exist, but are rather rare. Our work differs from that of Meyer's in that we mainly focus on the historical market share evolution of smartphone platforms, while he has investigated a number of more generic scenarios. Based on these scenarios, we hypothesise that the behaviour of contingent events determining the outcome of market lock-ins could be explained, and possibly driven by, imperfect information and the bounded rationality of the actors.

3   Models

In this paper, we present a high-level and a low-level simulation model for depicting the network effects of users and developers in the two major smartphone platforms, iOS and Android. Our models are based on expert interviews and earlier models presented in the literature. At the time of writing, the models form two architectural layers: a narrow layer that models expert knowledge in a concise *causal loop diagram* (CLD), and an extended layer for simulation purposes.

---

[4] A first-degree lock-in refers to the dominance of a single "best" technology. In contrast, both second- and third-degree lock-ins designate the dominance of an inferior one. A second-degree lock-in describes the dominance of a technology, while a better alternative has since become available. A third-degree lock-in occurs when an inferior technology dominates the market, even though a superior one is available.

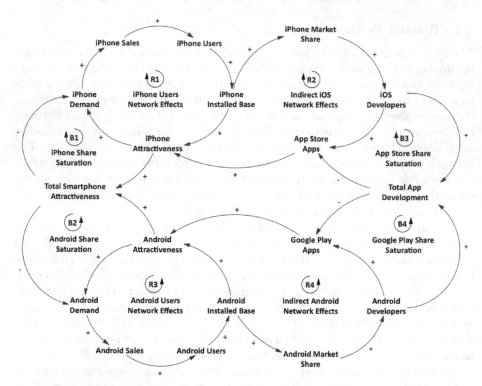

Fig. 1. Smartphone users and developers network effects high-level model

3.1 Expert Models

Figure 1 depicts the high-level model illustrating the overall structure of a multi-sided market. The figure portrays the main network effects of the market: the direct and indirect network effects. The direct network effects are those occurring among the users, who benefit from other users of the respective platform by lowering transaction and learning costs. In contrast, the indirect network effects are those interactions manifested between the users and developers. Furthermore, a negative network effect is postulated among the developers, since the more developers there are on a platform, the fiercer the competition among them. Finally, the phenomena of saturation exemplify the natural limit for smartphone demand and app development. The low-level version can be found in Fig. 3 in the Appendix.

3.2 Simulation Model and Results

Our work focuses on matching historical data with a low-level simulation model. We use publicly available data from Statista[5] for the total smartphone demand, total number of apps, and market shares.

---

[5] https://www.statista.com/.

At the present stage, we have explicit sensitivity and threshold parameters to describe the strength of the direct and indirect network effects. These parameters allow us to easily change the end-states of the simulation. Our goal is to eventually replace these with input variables describing real world effects and the overall ethos in the markets. The manually calibrated models allow us to investigate differences in the effects of static product-specific quality and utility on the lock-ins occurring in the market.

Interestingly, we observed a large number of parameter values forcing the market into a third-degree lock-in. That is, causing an inferior incumbent product to drive out a far "better" product, despite the "better" product being initially able to gain significant market share from the incumbent.

To calculate the effect of the size of the user and developer networks on smartphone attractiveness, we use the following equation:

$$\exp(Sensitivity * \text{Installed Base} \; / \; Threshold)$$

As shown in the equation, attractiveness rises exponentially as the installed base grows relative to the threshold. The sensitivity parameters allow us to vary the strength of the effect of both user and developer network sizes on smartphone attractiveness in sensitivity tests. The threshold parameters are scaling factors representing the users and developers in terms of the size of the installed base and number of apps, respectively. Only above these threshold parameter values do network effects become important. Finally, except for the size of the user and developer networks, the *effect of other factors* parameter aggregates factors, such as the effect of price, features, and availability of the smartphone and the effect of tools and policies for app development.

Figure 2 shows one sample run from our simulation. In this run, iPhone and Android market shares start at the same level, with the remainder of the market being comprised of the shares of other competitors based on historically recorded market data from Statista. Initially, iPhone and Android market shares parallel each other, reaching roughly equal market shares. Thereafter, Android begins to overtake iPhone, whose market share then drops back to 20%. The corresponding sensitivity, threshold, and other factors parameter values for this simple run are shown in Table 1.

4   Discussion

Unlike Meyer's AB model, our SD model integrates phenomena of positive, negative, direct, and indirect network effects with market saturation. At this point, our approach to modelling the difference between iOS and Android is still somewhat ad hoc. Our goal is to eventually replace the current vague sensitivity and threshold parameters with historical market data.

The phenomena related to path dependency, including both positive and negative feedback loops, are of such strength that they will inevitably dominate the simulation results. Meyer describes the market's commitment to an inferior technology platform by investigating first-, second-, and third-degree lock-ins. Although

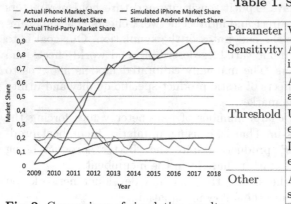

Fig. 2. Comparison of simulation results and real world data

Table 1. Simulation parameter values

Parameter	Variable	iPhone	Android
Sensitivity	Attract. of installed base	1	1
	Attract. of apps	100	100
Threshold	Users network effects	2B	8.5B
	Devs network effects	100k	100k
Other	Attract. of smartphone	1	4.1
	Attract. of app development	1	1

our model is inherently capable of modelling first- and third-degree lock-ins, we are currently unable to investigate second-degree lock-ins. This is due to the ability of our SD model to reflect static quality differences, but inability to yet represent incomplete information, needed by the concept of a second-degree lock-in.

5   Conclusions and Future Work

While platform economy and multi-sided markets are well-established concepts, only a few works have focused on modelling competing platform ecosystems based on historical data. In this paper, we have presented our ongoing work towards understanding the dynamic and multi-dimensional competition between the two major smartphone platforms, iOS and Android. In particular, we have observed that changing generic sensitivity and threshold parameter values, which express the strength of network effects and their relationship to the underlying social penetration factors, can easily lead to major differences in the market share of otherwise similar products. Our model is highly sensitive to changes in these parameter values due to them governing the quantitative exponential effect of the feedback.

In the near future, we will focus on enhancing our model in order to better conform with actual historical data. For this purpose, we plan to add input variables for quantifying the differences between the two companies' strategies, such as the level of product differentiation and resulting price structures. In other words, our goal is to replace the generic feedback loops with more details that describe the manner in which people valuate the differences between products in terms of quality and price.

Acknowledgements. We would like to thank our interviewed domain experts Juri Mattila and Peter Ylén. We also like to thank Santeri Paavolainen and Ali Tabatabaee for their helpful and actionable comments.

Appendix

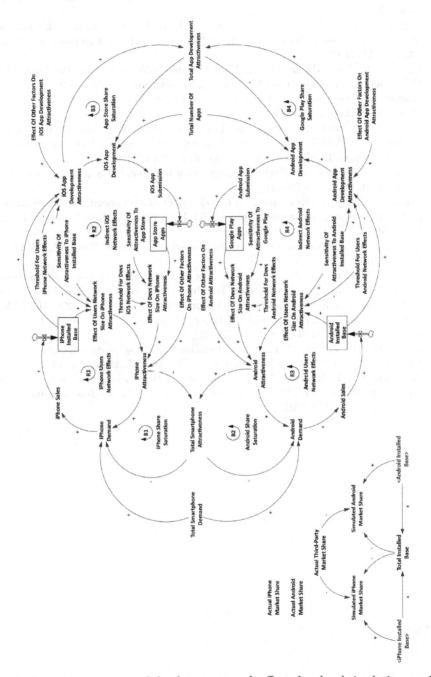

Fig. 3. Smartphone users and developers network effects low-level simulation model

References

1. Boudreau, K., Lakhani, K.: How to manage outside innovation. MIT Sloan Manage. Rev. **50**(4), 69–76 (2009)
2. Dutta, A., Puvvala, A., Roy, R., Seetharaman, P.: Technology diffusion: shift happens—the case of iOS and Android handsets. Technol. Forecast. Soc. Change **118**, 28–43 (2017)
3. Eisenmann, T.R., Parker, G.G., Van Alstyne, M.W.: Opening Platforms: How, When, and Why? Platforms, Markets, and Innovation, pp. 131–162 (2009)
4. Gawer, A., Cusumano, M.A.: Industry platforms and ecosystem innovation. J. Prod. Innov. Manage. **31**(3), 417–433 (2014)
5. Meyer, T.G.: Path Dependence in Two-Sided Markets: A Simulation Study on Technological Path Dependence with an Application to Platform Competition in the Smartphone Industry. Freie Universität Berlin (2012)
6. Rochet, J.C., Tirole, J.: Platform competition in two-sided markets. J. Eur. Econ. Assoc. **1**(4), 990–1029 (2003)
7. Ruutu, S., Casey, T., Kotovirta, V.: Development and competition of digital service platforms: a system dynamics approach. Technol. Forecast. Soc. Change **117**, 119–130 (2017)
8. Rysman, M.: The economics of two-sided markets. J. Econ. Perspect. **23**(3), 125–143 (2009)
9. Scholten, U., Fischer, R., Zirpins, C.: The dynamic network notation: harnessing network effects in PaaS-ecosystems. In: Proceedings of the Fourth Annual Workshop on Simplifying Complex Networks for Practitioners, pp. 25–30. ACM (2012)
10. Van Alstyne, M.W., Parker, G.G.: Platform business: from resources to relationships. GfK Mark. Intell. Rev. **9**(1), 24–29 (2017)
11. Von Kutzschenbach, M., Brønn, C.: Education for managing digital transformation: a feedback systems approach. J. Systemics Cybern. Inf. **15**(2), 14–19 (2017)
12. Zolfagharian, M., Romme, A.G.L., Walrave, B.: Why, when, and how to combine system dynamics with other methods: towards an evidence-based framework. J. Simul. **12**(2), 98–114 (2018)

Sockpuppet Detection in Social Network via Propagation Tree

Jiacheng Li[1,2], Wei Zhou[1(✉)], Jizhong Han[1], and Songlin Hu[1]

[1] Institute of Information Engineering, Chinese Academy of Sciences, Beijing, China
{lijiacheng,zhouwei,hanjizhong,husonglin}@iie.ac.cn
[2] School of Cyber Security, University of Chinese Academy of Sciences,
Beijing, China

Abstract. Sockpuppet detection is a valuable and challenging issue in social network. Current works are continually making efforts to detect sockpuppet based on verbal, non-verbal or network-structure features. However, they do not consider the propagation characteristic and propagation structure of sockpuppet. With our observation, the propagation trees of sockpuppet and ordinary account are different. Sockpuppet' propagation tree is evidently wider and deeper than that of the ordinary one. Based on these observations, we propose a propagation-structure based method to tackle sockpuppet detection problem. The experiment on two real-world datasets of Sina Weibo demonstrates that our method is more robust and outperforms previous methods.

Keywords: Sockpuppet detection · Propagation tree · Social network

1 Introduction

Social networks become a preferential place for information propagation or opinions and to promote ideas [12]. The malicious accounts on social networks lead to serious risks [9]. When the malicious accounts are detected and blocked, they register some new accounts called sockpuppets to continue spreading information. Sockpuppets usually produce malicious and deceptive behavior, such as fraud [11], cyberbullying [2], hate speech [6], and rumors [8]. Therefore, sockpuppet detection is valuable and challenging research issue. We broadly define puppetmaster as an individual that manipulate more than one account.

Prior works on automatic sockpuppet detection have tended to focus on verbal [9], non-verbal [10] and network-structure [7] features. The verbal-based method identify the authorship attribution of sockpuppet [3] by extracting features that capture stylistic, grammatical, and formatting preferences of the authors on 77 groups in Wikipedia and comparing the writing style of account [9]. It assumes that sockpuppets have a similar linguistic preference, such as keywords and topic titles in online discussion forum [15]. [4] is based on byte-level n-grams which are language independent. However, smart puppetmasters would disguise by altering account profile and writing style. Thus non-verbal methods

© Springer Nature Switzerland AG 2019
J. M. F. Rodrigues et al. (Eds.): ICCS 2019, LNCS 11540, pp. 507–513, 2019.
https://doi.org/10.1007/978-3-030-22750-0_44

assume that the non-verbal behavior indicates the intention of puppetmasters, [13] extracts 11 features from contribution's behavior of the accounts, and applies the community detection algorithm to detect sockpuppet group based on the action graph and relationship graph. But most non-verbal features are not fit for different platforms. Existing network structure-based detection methods are subjectively based on user views or emotional similarities. Bu et al. [1] proposed a sockpuppet detection algorithm based on authorship-identification techniques and relationship analysis. The relationships between two accounts are built if they have a similar attitude and similar writing styles. Besides, Kumar et al. [5] constructs the reply network on discussion community and observes that the nodes denoting sockpuppets were more central and highly active. Some community detection based methods have been proposed to leverage the network structure to detect sockpuppet. However, these existing methods almost ignore the propagation characteristic and structure.

In this work, we observe that the differences of propagation trees between sockpuppet and ordinary account which are unusual patterns ignored in previous works. Sockpuppet' propagation tree contains more identical accounts and is unexpectedly wider and deeper than that of the ordinary ones. In addition, the sockpuppet tend to build similar propagation trees. To utilize these patterns of the observations, we construct the propagation tree to detect sockpuppet and extract a set of independent features from propagation tree to detect sockpuppet. To validate the effectiveness, we collect two real-world data sets from Sina Weibo[1]. The experiment demonstrates that our method outperforms previous methods.

2   Problem Formulation

Suppose $G = (V, E)$ be a social network, where V is a set of accounts, $E \in V \times V$ is a set of repost relationship, and $e_{vu}^i \in E$ denotes repost relationship of message i between account v and $u(v, u \in V)$ which reflects propagation of information over G. We formally define the sockpuppet detection problem as: given a set of accounts $U(U \subset V)$, it aims to classify account $u_i(u_i \in U)$ as a sockpuppet account or ordinary account.

3   Observations

We engage in investigation of the difference sockpuppet and ordinary account. (1) Difference between sockpuppet and ordinary account. How difference between sockpuppet and ordinary account on dimensions of propagation tree? The number of identical nicknames. (2) The difference of pairwise accounts. Are the propagation behavior of two individual sockpuppets in the same sockpuppets group more similar than sockpuppet-ordinary account pair?

---

[1] https://weibo.com/.

Difference Between Sockpuppet and Ordinary Account. Combined with Fig. 1b and c, the sockpuppet tend to participate in same discussion of post more than once, in order to maximize the influence of the post. According to structural character, the propagation tree of sockpuppet is deeper and highlights that the message is reposted by sockpuppet will be spread far (1.86 vs 1.75) and wider (4.15 vs 3.51).

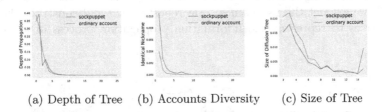

(a) Depth of Tree   (b) Accounts Diversity   (c) Size of Tree

Fig. 1. (a) Shows sockpuppet mainly retweets more than once and the ordinary account tend to do not repost it (2.03 vs 1.86). (b) Demonstrates that sockpuppet is more active than ordinary account (4.60 vs 3.13). (c) Illustrates that sockpuppet tend to participate hot discussion (6.09 vs 5.54).

Difference of Pairwise Accounts. Figure 2 shows the sockpuppets pair is more similar than others through three dimensions: size, depth, and width. It is reasonable that the pairwise sockpuppets behave similarly. It indicates that it is hard for puppetmaster to disguise their identity on propagation behavior.

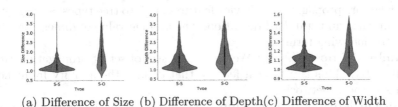

(a) Difference of Size (b) Difference of Depth(c) Difference of Width

Fig. 2. Sockpuppets pair (S-S) refers to two individual sockpuppets that belong to same sockpuppets group, sockpuppet-ordinary account pair (S-O) refers to two accounts that are sockpuppet account and ordinary account separately.

To sum up, we have several discoveries that sockpuppet tend to repost from the other sockpuppet and the message which is reposted by sockpuppet have a wider propagation range than ordinary account. The pairwise sockpuppets tend to behave similarly to each other, in order to enhance the influence of sockpuppets group opinion.

4   Methodology

4.1   Propagation Tree Construction

Similar to Twitter[2], there are two types of posts in Sina Weibo: original posts (tweets) and reposts (retweets). Each reposting log will represents an information propagation process, such as "wow//!B:wonderful//@C:lol". Based on the practice of refereeing to another account in a tweet via "//@username" convention [14], we extract the usernames from reposting log and construct the propagation trees to represent the information propagation process of an account (Fig. 3).

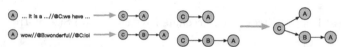

(a) Propagation Flow Extract (b) Propagation Tree construct

Fig. 3. (a) Builds the propagation flow from reposting log. (b) Constructs an propagation tree based on the same root of the propagation flow. We merge the propagation flow of account A which repost from account C. We remove the propagation tree which contains only one node.

4.2   Sockpuppet Account Detection

Given an account u and constructed the propagation trees of account u. Our method capture propagation behavior features fall into tree types: average value, minimum value and standard deviation. The average value of dimension can be seen in the following term:

Number of posts (Np_u): We count the size of set of propagation tree of account $u(D_u)$. This is a typical feature that depicts the activity frequency of accounts in social network.

Average depth of propagation tree (Ad_u): For this feature, we just count maximum depth dp_i of d_i^u. This presents the delay in the message i propagation of account u. $Ad_u = \sum_{i=0}^{Nd_u} \frac{dp_i}{Nd_u}$, where Nd_u is the size of D_u.

Average size of propagation tree (As_u): We count the total number of account (ds_i) of propagation tree of the original message i which account u latest participated (d_i^u). While this feature is trying to capture the coverage of message i which the account u is participated in: $As_u = \sum_{i=0}^{Nd_u} \frac{ds_i}{Nd_u}$

Average number of identical account in tree (Au_u): The goal of this features dn_i which is the number of the same nickname of d_i^u is to model the participation rates of account in the d_i^u. Some accounts prefer to interact with others account by reposting their posts: $Au_u = \sum_{i=0}^{Nd_u} \frac{dn_i}{Nd_u}$

---

[2] https://twitter.com/.

Average maximum depth and width (Ad_u, Aw_u): Maximum depth dd_i is used for presenting one of dimensions of d_i^u: $Ad_u = \sum_{i=0}^{Nd_u} \frac{dd_i}{Nd_u}$. And maximum width dw_i is also used for presenting one of dimensions of d_i^u: $Aw_u = \sum_{i=0}^{Nd_u} \frac{dw_i}{Nd_u}$

Average Depth of only one 1-hop repost of original post (Ah_u): These feature present the depth dh_i of d_i^u with only one child. $Ah_u = \sum_{i=0}^{Nd_u} \frac{dh_i}{Nd_u}$

Average number of children of propagation tree (Ac_u): We take into consideration the number of children dc_i, which represents the diversity of d_i^u. We contain the propagation tree with single child: $Ac_u = \sum_{i=0}^{Nd_u} \frac{dc_i}{Nd_u}$

Average index of type of posts (Pm_u): The type of posts p_t can be divided three types with index of type: posting (1), replying (2) and reposting (3). $Pm_u = \sum_{t=0}^{Np_u} \frac{p_t}{Np_u}$

Average interval between interactions (Pi_u): This is a normalized feature where we compute the time difference between the t-th post p_t and the prior one p_{t-1}. It presents the frequency of which the account u uses the social network: $Pi_u = \sum_{i=0}^{Np_u} \frac{p_t - p_{t-1}}{Np_u}$.

5 Experimental

5.1 Experimental Setup

Datasets. We conduct experiments on two real-world $\mathcal{D}_S$ and $\mathcal{D}_T$ which we crawled tweets from 2017.01 to 2018.10. from Sina Weibo. Accounts are identified as sockpuppets when self-reported sentence pattern such as "This is a sockpuppet of Mix" is matched or other accounts identify them as being controlled by a puppetmaster. Ordinary accounts are randomly selected from the accounts interact with sockpuppets and are not correlated to sockpuppets.

Comparison Method. We consider the following baselines in sockpuppet detection. **Profile Attributes Features**: User profile is the basic information for each account, such as nickname and description. It reflects the lexical preference of puppetmaster. We employ attributes of accounts' homepage and the number of diversity of login device for sockpuppets detection problem. **Verbal Features (Verbal)** [9]: The basis of authorship attributes sockpuppets detection in Wikipedia tries to identify the sockpuppet pair by comparing writing style. It extracts 245 verbal features from each comment of account. **Non-verbal Features (Non-verbal)** [10]: It uses several variables to represent user behavior. Variables of online non-verbal behavior fall under time-independent behavior and time-dependent behavior. For all the methods, 10-fold cross validation is performed and the average results are reported.

5.2 Experimental Result and Discussion

We employ five widely used classification metrics for evaluation: precious (**P**), recall (**R**), F1-score (**F1**) and False Positive Rate (**FPR**). The Table 1 compares several baseline methods and our proposed method over several machine learning algorithms: Logistic regression (LR), Support Vector Machine (SVM), Random Forest (RF), and Adaptive Boosting (ADA). It shows that we obtained the best F1-score using the LR algorithm on different datasets and the LR algorithm appears the most robust among several methods.

Table 1. Sockpuppet accounts detection

Method	Alg	$\mathcal{D}_S$					$\mathcal{D}_T$				
		P	R	F1	ACC	FPR	P	R	F1	ACC	FPR
Profile	SVM	0.644	0.204	0.304	0.246	0.046	0.582	0.764	0.659	0.675	0.391
	RF	0.609	**0.622**	**0.608**	0.775	0.163	0.692	**0.679**	**0.681**	0.740	0.216
	LR	**0.709**	0.526	0.601	0.799	0.090	**0.711**	0.607	0.651	0.728	0.182
	ADA	0.634	0.225	0.324	0.735	0.056	0.621	0.395	0.496	0.630	0.198
Verbal	SVM	0.704	0.154	0.242	0.735	0.027	0.737	0.507	0.587	0.723	0.117
	RF	0.725	**0.578**	0.635	0.809	0.096	0.727	**0.698**	0.708	0.759	0.196
	LR	**0.804**	0.537	**0.635**	0.827	0.054	**0.781**	0.657	**0.710**	0.776	0.167
	ADA	0.735	0.241	0.347	0.750	0.042	0.727	0.545	0.612	0.724	0.144
Non-verbal	SVM	0.630	0.480	0.543	0.765	0.119	0.654	0.456	0.517	0.664	0.168
	RF	0.644	0.491	0.549	0.771	0.115	0.688	0.645	0.660	0.724	0.217
	LR	**0.781**	**0.597**	**0.674**	0.836	0.067	**0.742**	**0.662**	**0.694**	0.757	0.173
	ADA	0.474	0.061	0.103	0.713	0.021	0.575	0.264	0.356	0.618	0.122
Propagation tree	SVM	0.792	0.511	0.618	0.820	0.054	0.743	0.571	0.637	0.734	0.147
	RF	0.771	0.598	0.663	0.828	0.078	0.746	0.657	0.693	0.760	0.165
	LR	**0.840**	**0.633**	**0.719**	0.856	0.052	**0.771**	**0.681**	**0.714**	0.771	0.163
	ADA	0.750	0.511	0.603	0.808	0.071	0.727	0.579	0.637	0.727	0.165

Due to some of the malicious sockpuppets are blocked, we cannot access their profile and some puppetmaster will apply diverse profile information in the same sockpuppets groups, the *Profile Attributes Based* method have the worst performance. *Verbal Based* method identifies sockpuppet through their linguistic traits which assume that sockpuppet have unique linguistic traits, because smart account could apply different writing style to express their idea. *Non-verbal Based* method outperform the Verbal Features method. A plausible explanation is that non-verbal cues are more powerful than verbal cues to characterize account. Our method provides better performance, which achieve the best performance in sockpuppet detection. It indicates that the propagation features based method could capture the sockpuppets' intention.

6   Conclusion

We investigate the difference between the sockpuppet and ordinary account and extract several features from the propagation tree structure to achieve the goal of sockpuppet detection. Then we evaluate the proposed methods on two real-world social network datasets over two subproblems. Compared with several methods, our model shows the best performance.

References

1. Bu, Z., Xia, Z., Wang, J.: A sock puppet detection algorithm on virtual spaces. Knowl.-Based Syst. **37**, 366–377 (2013)
2. Chelmis, C., Zois, D.S., Yao, M.: Mining patterns of cyberbullying on Twitter. In: 2017 IEEE International Conference on Data Mining Workshops (ICDMW), pp. 126–133. IEEE (2017)
3. Hosseinia, M., Mukherjee, A.: Detecting sockpuppets in deceptive opinion spam. In: Gelbukh, A. (ed.) CICLing 2017. LNCS, vol. 10762, pp. 255–272. Springer, Cham (2018). https://doi.org/10.1007/978-3-319-77116-8_19
4. Kešelj, V., Peng, F., Cercone, N., Thomas, C.: N-gram-based author profiles for authorship attribution. In: Proceedings of the Conference Pacific Association for Computational Linguistics, PACLING, vol. 3, pp. 255–264 (2003)
5. Kumar, S., Cheng, J., Leskovec, J., Subrahmanian, V.: An army of me: sockpuppets in online discussion communities. In: Proceedings of the 26th International Conference on World Wide Web, pp. 857–866. International World Wide Web Conferences Steering Committee (2017)
6. Lekea, I.K., Karampelas, P.: Detecting hate speech within the terrorist argument: a Greek case. In: 2018 IEEE/ACM International Conference on Advances in Social Networks Analysis and Mining (ASONAM), pp. 1084–1091. IEEE (2018)
7. Liu, D., Wu, Q., Han, W., Zhou, B.: Sockpuppet gang detection on social media sites. Front. Comput. Sci. **10**(1), 124–135 (2016)
8. Ma, J., Gao, W., Wong, K.F.: Rumor detection on twitter with tree-structured recursive neural networks. In: Proceedings of the 56th Annual Meeting of the Association for Computational Linguistics (Volume 1: Long Papers). vol. 1, pp. 1980–1989 (2018)
9. Solorio, T., Hasan, R., Mizan, M.: A case study of sockpuppet detection in Wikipedia. In: Proceedings of the Workshop on Language Analysis in Social Media, pp. 59–68 (2013)
10. Tsikerdekis, M., Zeadally, S.: Multiple account identity deception detection in social media using nonverbal behavior. IEEE Trans. Inf. Forensics Secur. **9**(8), 1311–1321 (2014)
11. Wang, B., Gong, N.Z., Fu, H.: Gang: detecting fraudulent users in online social networks via guilt-by-association on directed graphs. In: 2017 IEEE International Conference on Data Mining (ICDM), pp. 465–474. IEEE (2017)
12. Yamak, Z., Saunier, J., Vercouter, L.: Detection of multiple identity manipulation in collaborative projects. In: Proceedings of the 25th International Conference Companion on World Wide Web, pp. 955–960. International World Wide Web Conferences Steering Committee (2016)
13. Yamak, Z., Saunier, J., Vercouter, L.: Sockscatch: automatic detection and grouping of sockpuppets in social media. Knowl.-Based Syst. **149**, 124–142 (2018)
14. Yang, J., Counts, S.: Predicting the speed, scale, and range of information diffusion in Twitter. In: ICWSM, vol. 10, PP. 355–358 (2010)
15. Zheng, X., Lai, Y.M., Chow, K.P., Hui, L.C., Yiu, S.M.: Sockpuppet detection in online discussion forums. In: 2011 Seventh International Conference on Intelligent Information Hiding and Multimedia Signal Processing (IIH-MSP), pp. 374–377. IEEE (2011)

Exploring the Performance of Fine-Grained Synchronization and Data Exchange Across Process Boundaries on Modern Multi-core Architectures

Jiri Dokulil$^{(\boxtimes)}$ (iD) and Siegfried Benkner (iD)

Faculty of Computer Science, University of Vienna, Vienna, Austria
{jiri.dokulil,siegfried.benkner}@univie.ac.at

Abstract. Whether to use multiple threads in one process (MPI+X) or multiple processes (pure MPI) has long been an important question in HPC. Techniques like in situ analysis and visualization further complicate matters, as it may be very difficult to couple the different components in a way that would allow them to run in the same process. Combined with the growing interest in task-based programming models, which often rely on fine-grained tasks and synchronization, a question arises: Is it possible to run two tightly coupled task-based applications in two separate processes efficiently or do they have to be combined into one application? Through a range of experiments on the latest Intel Xeon Scalable (Skylake) and AMD EPYC (Zen) many-core architectures, we have compared performance of fine-grained synchronization and data exchange between threads in the same process and threads in two different processes. Our experiments show that although there may be a small price to pay for having two processes, it is still possible to achieve very good performance. The key factors are utilizing shared memory, selecting the right thread affinity, and carefully selecting the way the processes are synchronized.

Keywords: Synchronization · Data movement · Collocated applications

1 Introduction

The increasing heterogeneity of HPC architectures has inspired research into different programming approaches. Task-based runtime systems are one example. By abstracting the work into many small tasks with dependencies, the runtime system is given the ability to better control where and when work is being executed, compared to, for example, MPI where the work to be done by a process running on a specific node is strictly prescribed by the application code.

With data movement being one of the major contributors to runtime and power consumption of modern high performance systems, in situ techniques are

© Springer Nature Switzerland AG 2019
J. M. F. Rodrigues et al. (Eds.): ICCS 2019, LNCS 11540, pp. 514–520, 2019.
https://doi.org/10.1007/978-3-030-22750-0_45

becoming an important way to improve efficiency of HPC systems. By processing data where and when it is generated, we reduce the resources needed to transfer and store the data.

In this paper, we explore how close we can get to this "best case" scenario with two separate processes. We explore the effects of using shared memory (provided by the operating system) and different synchronization mechanisms. It turns out that even with two separate processes, it is possible to get comparable performance to the single-process scenario. By correctly setting thread affinity and synchronizing the two processes, it is possible to even re-use cached data.

Our main contribution is designing, implementing, and running a benchmark aimed specifically at studying data transfers in the case of two collocated applications. Our findings can be used to aid the design of in situ analysis/visualization applications and other systems that require tight coupling of different processes. The experiments were performed on two different machines, using latest many-core architectures from Intel (Skylake) and AMD (Zen). The four Intel Xeon Scalable CPUs in one machine have a total of 80 cores, while the two AMD EPYC processors in the other server have 64 cores together.

2   Local Data Exchange

We will use the terms *producer* and *consumer* to denote the two pieces of code responsible for generating (producer) and reading (consumer) the data. Our main concern is the time it takes the consumer to read the data.

The consumer code might immediately follow the producer code, ensuring it runs right after the consumer on the same thread. Or, the consumer could be in a different thread or even in a different process. Then, some synchronization is required. The consumer needs access to the produced data, which is trivial in a single process, where all threads have access to the same data. If the consumer is in a different process, we can use services provided by the operating system to give the consumer direct access to the producer's memory with the data.

Ideally, we would want a CPU core to finish generating (and writing) the data and then switch immediately to the consumer and start reading the data. One way to achieve this is to set affinity of both the producer thread and the consumer thread to the same CPU core. This can easily be done, even if they belong to different processes. We can block the consumer thread on a synchronization primitive (e.g., a semaphore). When the producer is finished, it unblocks the consumer's thread and suspends itself. This forces the scheduler of the operating system to pick a next thread for execution. As the consumer thread is now active and affinitized to the core, it is likely to be picked.

3   Benchmark Application Design

We have implemented different experiments to try and compare different options for data exchange. They have several things in common. Each experiment uses a single block of memory for the data. The size of the block can be configured. Writing the data is simulated by treating the block as an array of integers and writing 1 to each element. When the data is read, a sum of all elements is produced, to prevent the compiler from optimizing the read away. The sum is performed in a way that makes it easy for the compiler to vectorize it. We measure the time it takes the consumer to read the whole data. Where possible, we also measure the time elapsed between the moment the producer finished writing the data and the moment the consumer starts reading the data. This gives us some estimate of the latency generated by the synchronization. Finally, the experiments can be configured to have the producer and the consumer invalidate their CPU cache before reading the data, to help us check whether the caches are actually being used.

Based on the setup, the producer and consumer might reside in the same thread (then the notification is a NOP and the second for-loop body is moved to the first for-loop), two threads in the same process, or two different processes.

Single Thread. To obtain a baseline, we consider the scenario where a single thread generates the data and then reads it. This is the closest coupling possible and should provide the best performance, but it may be technically challenging or impossible to use in practice.

File. For comparison, we also tested the setup where the producer writes the data to a file which is immediately read by the consumer. To minimize noise, the producer and the consumer are in fact the same thread. The file is written, closed, and immediately opened for reading.

Two Threads, Synchronization via Atomics. The next setup is closer to the intended two-process layout, but it uses two threads in the same process. The two threads are synchronized using atomic operations. When the producer finishes writing the data, it atomically writes a flag (with release semantics) which the consumer keeps checking atomically (with acquire semantics).

Two Processes, Synchronization via Atomics. The setup in this case is the same as in the previous one, except that the threads belong to different processes and use a shared memory region to exchange data and for the atomics-based synchronization.

Two Threads, Synchronization via Promises. To test blocking synchronization, we have used C++11 promises. One promise is used to notify the consumer that the producer is finished and another promise is used to notify the producer that the consumer has finished.

Two Processes, Synchronization via Semaphore. With two processes, the promises cannot be used, so a pair of standard named semaphores is used instead, to achieve the same synchronization pattern.

4   Experimental Evaluation

The experiments were performed on two different servers running Linux. One server contains four Intel Xeon Scalable Gold 6138 processors (Skylake architecture, 20 cores, AVX-512 support, 32 KB L1 data cache per core, 1 MB L2 cache

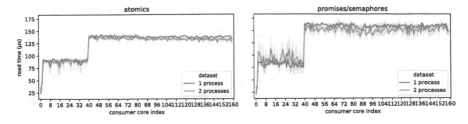

Fig. 1. Read time on the four socket Intel server.

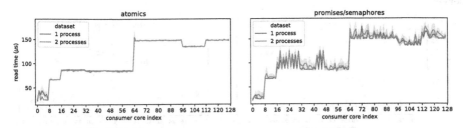

Fig. 2. Read time on the four socket AMD server.

per core and 27.5 MB last level cache in total). The total number of physical cores is 80, but all experiments were executed with Hyper-Threading enabled, the machine supports 160 hardware threads (logical cores).

The other server contains two AMD EPYC 7501 processors (Zen architecture, 32 cores, 32 KB L1 data cache per core, 512 KB L2 cache per core and 64 MB last level cache in total). The total number of physical cores is 64 and there are 128 logical cores.

Unless stated otherwise, the performance figures are read times in microseconds. The number of repetitions (the number of data blocks transferred per one application execution) is 100. Each application configuration is executed 5 times. The graphs show the average value of these 5 executions and 95% confidence intervals. Unless explicitly specified, the size of the data exchanged by the producer and the consumer is 1 MB. The producer is always on core 0.

As a baseline, the read time of the single threaded variant is 20.3 μs on the Intel server and 30.8 μs on the AMD server. If a file is used to exchange data, the read time is 518.0 μs and 365.1 μs on the Intel and AMD servers respectively.

4.1 Core Selection

As we have already explained, both threads used in the experiment are affinitized to a specific core. The producer always uses core 0. In most experiments, the measured performance depends on the core selected for the consumer. In general, there are four different cases: using the same core (0), using the sibling logical core (1), using another core on the same CPU (2–40 or 2–32 for the Intel and AMD machines respectively), and using a core on other CPUs. Figures 1 and 2 show performance for different combinations on the Intel and AMD servers.

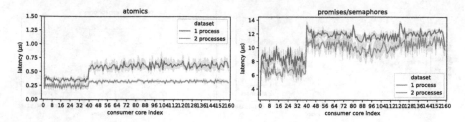

Fig. 3. The latency of synchronization between producer and consumer on the Intel server.

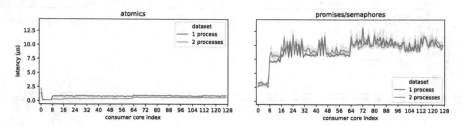

Fig. 4. The latency of synchronization between producer and consumer on the AMD server.

On the AMD CPUs, the performance of 1 process and 2 processes is comparable. When blocking synchronization is used, performance variability increases, but the read performance is very similar. On the Intel CPUs, the results are a bit more interesting, as 2 processes slightly outperform 1 process in many cases, especially when communicating between different CPUs.

Another important factor to consider is latency. In our case, the time elapsed between the producer finishes and the consumer starts. In this time, the producer needs to send a signal to the consumer and the consumer thread needs to start running. We want to keep it as small as possible in order to not waste resources by having the consumer wait unnecessarily long. The measured performance is shown in Figs. 3 and 4.

4.2   Data Size

Figures 5 and 6 show the effect of the different data sizes. We use core 0 for producer and core 2 for consumer. Probably the most interesting observation is that while the performance of using 1 or 2 processes is comparable, the latency increases more when larger data sets are exchanged between two processes. The latency increase is probably due to the fact that the data of the consumer thread is forced out of cache. However, as we use different physical cores, the cache in question is the L3 cache. The latency increases the most as the data size approaches the L3 cache size, but it can be seen earlier.

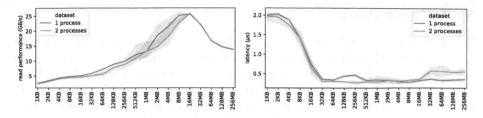

Fig. 5. The effects of data size, on the Intel server.

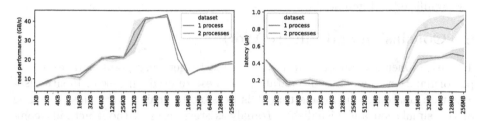

Fig. 6. The effects of data size, on the AMD EPYC server.

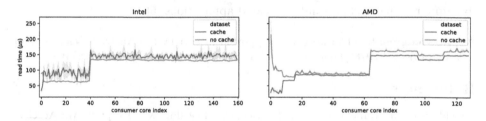

Fig. 7. The effects of cache, on the two architectures.

4.3 Cache Reuse

Figure 7 compares performance with and without data caching. To examine the no-cache behavior, extra work is performed by the producer and consumer to ensure that the caches contain no useful data. A surprising results is that on the Intel architecture, if the producer and consumer do not reside on the same core, the read performance increases and it also becomes much more stable. We believe this is caused by the fact that forcing the data into main memory is beneficial for data pre-fetching. The data is read in a sequential manner from the main memory, which is a best-case scenario for the hardware pre-fetcher. With caching, as the size of the data is the same as the size of the L2 cache, the data is most likely distributed across L2 and L3 caches, which also explains the much larger performance variability between executions, which is clearly visible from the confidence intervals in Fig. 7.

5   Related Work

Benchmark results and even various benchmarking software is widely available [2,4,7]. Many micro-benchmarks are created in an ad-hoc manner as part of development of a larger system. Memory performance is widely studied in HPC [6], including research focused on shared memory performance [3]. Our work focuses on the specific problem of exchanging data between two processes, which are closely coupled on a single node. FastFlow [1] implements a system similar to our proposal for a single process, although the low-level implementation details are significantly different.

6   Conclusion and Future Work

The experiments confirm that it is possible to get very good data exchange performance even when two separate processes are used. In case of in situ visualization/analysis, this means that it is possible to achieve very close coupling of the simulation and visualization/analysis steps, even without actually combining both into a single application. Still, care needs to be taken to achieve good performance. In our future work, we will apply the results to our work on dynamic runtime systems and collocated applications [5].

Acknowledgments. The work was supported in part by the Austrian Science Fund (FWF) project P 29783 Dynamic Runtime System for Future Parallel Architectures.

References

1. Aldinucci, M., Danelutto, M., Kilpatrick, P., Meneghin, M., Torquati, M.: Accelerating code on multi-cores with FastFlow. In: Jeannot, E., Namyst, R., Roman, J. (eds.) Euro-Par 2011. LNCS, vol. 6853, pp. 170–181. Springer, Heidelberg (2011). https://doi.org/10.1007/978-3-642-23397-5_17
2. Bienia, C., Kumar, S., Singh, J.P., Li, K.: The PARSEC benchmark suite: characterization and architectural implications. In: Proceedings of the 17th International Conference on Parallel Architectures and Compilation Techniques, pp. 72–81. PACT 2008, ACM, New York, NY, USA (2008)
3. Bolosky, W.J., Scott, M.L.: False sharing and its effect on shared memory performance. In: Proceedings of the Fourth Symposium on Experiences with Distributed and Multiprocessor Systems (1993)
4. Che, S., Boyer, M., Meng, J., Tarjan, D., Sheaffer, J.W., Lee, S., Skadron, K.: Rodinia: a benchmark suite for heterogeneous computing. In: 2009 IEEE International Symposium on Workload Characterization (IISWC), pp. 44–54, October 2009
5. Dokulil, J., Benkner, S.: Adaptive scheduling of collocated applications using a task-based runtime system. In: 2018 30th International Symposium on Computer Architecture and High Performance Computing (SBAC-PAD), pp. 41–48 (2018)
6. McCalpin, J.D., et al.: Memory bandwidth and machine balance in current high performance computers. IEEE Comput. Soc. Tech. Comm. Comput. Arch. (TCCA) Newsl. **2**, 19–25 (1995)
7. Sakalis, C., Leonardsson, C., Kaxiras, S., Ros, A.: Splash-3: a properly synchronized benchmark suite for contemporary research. In: 2016 IEEE International Symposium on Performance Analysis of Systems and Software (ISPASS) (2016)

Accelerating Wild Fire Simulator Using GPU

C. Carrillo[✉], T. Margalef, A. Espinosa, and A. Cortés

Computer Architecture and Operating Systems Department,
Universitat Autònoma de Barcelona, Barcelona, Spain
{carles.carrillo,tomas.margalef,antoniomiguel.espinosa,
ana.cortes}@uab.cat

Abstract. In the last years, forest fire spread simulators have proven to be very promising tools in the fight against these disasters. Due to the necessity to achieve realistic predictions of the fire behavior in a relatively short time, execution time may be reduced. Moreover, several studies have tried to apply the computational power of GPUs (Graphic Processors Units) to accelerate the simulation of the propagation of fires. Most of these studies use forest fires simulators based on Cellular Automata (CA). CA approaches are fast and its parallelization is relatively easy; conversely, they suffer from precision lack. Elliptical wave propagation is an alternative approach for performing more reliable simulations. Unfortunately, its higher complexity makes their parallelization challenging. Here we explore two different parallel strategies based on Elliptical wave propagation forest fire simulators; the multicore architecture of CPU (Central Processor Unit) and the computational power of GPUs to improve execution times. The aim of this work is to assess the performance of the simulation of the propagation of forest fires on a CPU and a GPU, and finding out when the execution on GPU is more efficient than on CPU. In this study, a fire simulator has been designed based on the basic model for one point evolution in the FARSITE simulator. As study case, a synthetic fire with an initial circular perimeter has been used; the wind, terrain and vegetation conditions have been maintained constant throughout the simulation. Results highlighted that GPUs allow obtaining more accurate results while reducing the execution time of the simulations.

Keywords: Wild fire simulator · Fire front propagation · GPU

1 Introduction

The impact and the damage caused by Forest Fires has been increasing significantly over the last years. In the last decades, several fire spread models have been developed and implemented in computing simulators to help control centers in taking the adequate decisions. However, wildfires are complex systems characterized by a stochastic behaviour, with a large number of involved variables.

© Springer Nature Switzerland AG 2019
J. M. F. Rodrigues et al. (Eds.): ICCS 2019, LNCS 11540, pp. 521–527, 2019.
https://doi.org/10.1007/978-3-030-22750-0_46

Accurate simulators tend to take longer execution times. So, their effectiveness in real-time prediction is reduced. In order to improve the performance of fire spread simulators, several strategies have been developed to reduce the execution time without altering the accuracy of the simulations. In this context, some studies apply multicore architectures by increasing computational power, raising the number of CPUs, [1–3]. At the same time, the increase in the computational power of the Graphical Processing Units (GPUs) has turned them into an ideal tool for the modelling of complex systems. Different works have been carried out to apply the computational capacity of GPUs to accelerate the simulation of forest fire behavior [5,8,9,11,13]. These works have focused on the application on simulators based on Cellular Automata (CA). The main problem is that the simulators based on the CA approach have low intrinsic accuracy. Simulators based on the Huygens principle, or Elliptical Wave Propagation, have higher precision than those based on CA; however, their execution time is higher. In the present work, we focused on FARSITE (*Fire Area Simulator*) [6], which is a forest fire simulator based on the Elliptical Wave Propagation. Two different parallelizations are proposed; on the one hand, we have extracted the FARSITE simulation kernel and implemented it in OpenMP (*Open Multi-Processing*) [4], which is a set of compiler directives, library routines, and environment variables that can be used in any multicore CPU. On the other hand, we used CUDA (*Compute Unified Device Architecture*) to execute the simulation kernel in GPU. The aim of this work is to evaluate the performance of the two parallel strategies and analyse when the execution of one is more efficient than the other. As a first approach to the problem, a synthetic fire is used, which consists of a circular front in flat terrain, with constant wind speed, wind direction and the vegetation conditions throughout the simulation. To be able to compare the different executions (GPU and CPU) the simulations have been performed with different time propagation, in order to analyze in which conditions the execution in the CPU is more efficient than the execution in GPU. This paper is organized as follows. In Sect. 2 the principal characteristics of FARSITE are presented. Section 3 details the methodology used. Section 4 presents the experimental results and, finally, Sect. 5 summarizes the main conclusions and future work.

2   FARSITE Forest Spread Simulator

FARSITE is a simulator which spreads the front of the fire resolving Rothermel's equation [12]. The Rothermel's model is formulated in the following way:

$$R = R_0 \cdot (\overrightarrow{n} + \overrightarrow{\phi}_w + \overrightarrow{\phi}_s) \tag{1}$$

where R_0 represents the rate of spread in a particular point with no wind and no slope, $\overrightarrow{n}$ is the normal direction to the fire perimeter on that particular point, $\overrightarrow{\phi}_w$ is the wind factor and $\overrightarrow{\phi}_s$ the slope factor. In the Elliptical Wave Propagation, the perimeter of the fire is divided into series of points, [10]. To obtain the evolution of the fire perimeter, an ellipse is generated for each point. The shape of the ellipses is determined by the local characteristics at each point.

In this way, the new perimeter is obtained by joining the obtained ellipses, see Fig. 1(a).

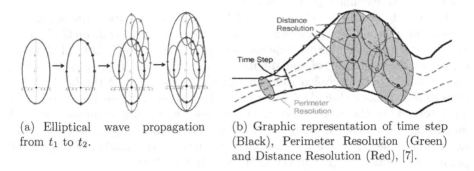

(a) Elliptical wave propagation from t_1 to t_2.

(b) Graphic representation of time step (Black), Perimeter Resolution (Green) and Distance Resolution (Red), [7].

Fig. 1. Forest Fires Spread Simulator. (Color figure online)

In FARSITE there are three different parameters which have a direct impact on the resolution and, therefore, on the execution time [7], see Fig. 1(b):

- **Time Step:** The time step is the maximum amount of time that the conditions at a given point are assumed constant so that the position of the fire front can be projected.
- **Perimeter Resolution:** The perimeter resolution determines the maximum distance between points used to define the fire perimeter. The perimeter resolution controls the ability of a fire perimeter to respond to heterogeneities occurring at a fine scale.
- **Distance Resolution:** The distance resolution is the maximum projected spread distance from any perimeter point. This distance cannot be exceeded in a time step before new local data are used to compute the spread rate.

The precision of FARSITE is directly proportional to the number of points in which the fire front is split. The higher the number of points in the fire perimeter (low *Perimeter Resolution*), the more detail can be reproduced the fire fronts behaviour, consequently, the accuracy of the simulation will be better; therefore, the execution time is longer.

3   Parallelization of Forest Fire Simulator

We extracted the FARSITE simulation kernel and re-implemented in parallel into the FARSITE body. When the fire is propagated in serial, at each time iteration the propagation of the points is done sequentially. Consequently, when the number of points to expand increases, the execution time also increases proportionally. So, simulations with high resolutions provide long execution times, which limits their use in real situations. In the parallel implementations, the point propagation was carried out in parallel. At each time iteration, each thread computes the spread of a single point. When the evolution of all points is finished, the threads are synchronised, and the spread in the next time iteration is performed. All calculations were performed in double precision.

In order to parallelize the code on CPU, we have re-written the simulation kernel code thoroughly with OpenMP. For implementing the code on GPU accelerators, the simulation kernel code has been re-written with CUDA. All data are copied at the beginning of the simulation from the *Host* to *Device*. However, the perimeter data is copied from the *Device* to the *Host* at the final of each time iteration. Each thread only computes the evolution of one single point. We are interested in the evolution of the throughput in a series of simulations with increasing the number of points; we look for the number of points on which the execution in the GPU is faster than the CPU or when the number of propagated points per second is higher in the GPU than in the CPU.

As a first approximation, a synthetic fire was used with an initial circular perimeter. In this particular case, it has been considered a flat terrain, with homogeneous vegetation and constant wind speed and wind direction during the whole simulation.

4   Experimental Study and Results

All calculations reported here were performed using a single GPU and single CPU; we measured the serial CPU performance using a single core, 2, 4 and 6 cores. As execution platform, we have used an Intel(R) Xeon(R) CPU E5-2620 v3 @ 2.40 GHz, with 6 cores and for the GPU simulations, a GeForce RTX 2080 Ti with 4352 CUDA cores was used. The tested propagation times were *1, 2, 5* and *10* h. The *Perimeter Resolution* was modified in each execution by increasing two thousand points at each simulation, so the extreme cases are $2,000$ and $184,000$ points. The higher the number of points, the higher resolution employed for the simulation.

Figure 2 displays the number of points per second of the CPU implementations and the GPU implementation for the different propagation time. In Fig. 2(a) we can see that above $132,000$ perimeter points, the GPU implementation is more efficient than the Serial implementation. However, for this propagation time, all the OpenMP implementations are faster than the GPU application. It can observe that we are in front of a compute-bound problem, so the CPU is quickly saturated in all cases (below $8,000$ points), while the number of propagated points per second grows linearly in the GPU implementation. For 1 h of propagation time, the OpenMP implementations compute more points per second, which means that the OpenMP implementations are more efficient than the GPU implementation. Moreover, we see that the maximum number of propagated points per second decrease faster for all CPU implementations when the propagation time increase than for the GPU. Figure 3(a) shows the maximum propagated points per second for all implementations. It can be seen how the maximum of propagated points per second for each implementation decreases when the time of propagation is increased.

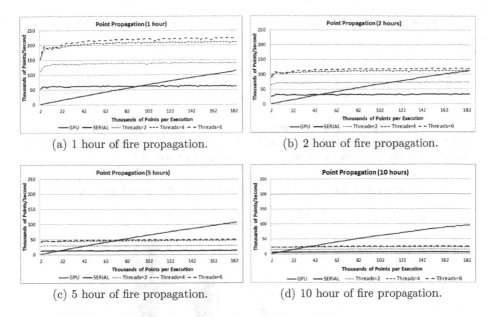

(a) 1 hour of fire propagation.

(b) 2 hour of fire propagation.

(c) 5 hour of fire propagation.

(d) 10 hour of fire propagation.

Fig. 2. Points per second depending on the number of perimeter points by the CPU and GPU.

Table 1. Number of points from which the execution in the GPU is more efficient than the CPU execution in Serial and with 2, 4, and 6 cores.

Propagation time	Number of points			
	Serial	2 Cores	4 Cores	6 Cores
1 h	132, 000			
2 h	66, 000	112, 000		
5 h	28, 000	44, 000	70, 000	74, 000
10 h	14, 000	22, 000	34, 000	38, 000

Figure 3(b) shows the number of points from which the efficiency of the GPU is higher than the efficiency of the CPU. This number depends on the propagation time of the fire. The longer the time propagation of the simulated fire is, the less number of perimeter points is necessary so that the execution in the GPU is faster than the CPU (Table 1).

In Fig. 4 we can see the speed up of the GPU implementation against OpenMP implementation with 6 threads when the fire front is split in 184, 000 perimeter points. We saw that, when we simulated propagation time below two hours, the execution in the CPU is faster when 6 threads are used. Nonetheless, above this propagation time, the execution of the GPU implementation is the fastest one.

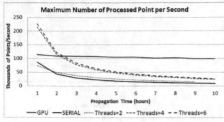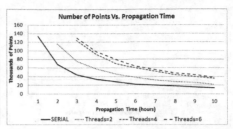

(a) Number maximum of points per second depending on the propagation time.

(b) Number of points from which the GPU is faster than the CPU.

Fig. 3. Execution performance of the different Forest Fire Spread implementations.

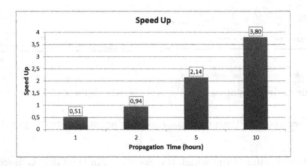

Fig. 4. Speed Up of the GPU implementation versus OpenMP implementation with 6 threads when 184,000 perimeter points are used.

5 Conclusions and Future Work

The computational capabilities of GPUs make them ideal for the simulation of any complex system. In our case, we have focused on the study of forest fire propagation simulators based on the Elliptical Wave Propagation, in particular, FARSITE. In this work, a synthetic fire has been used, with constant wind and vegetation conditions throughout the simulation. The obtained results demonstrated that the use of the GPU open a new way of approaching forest fire spread simulation in the sense that we expect to get more accurate results and, at the same time, faster and, therefore operationally simulation time.

According to the study carried out, for long fire propagation simulations, the GPU implementation is more efficient than the OpenMP implementation. Moreover, GPU is better than CPU when we face compute-bound. We also highlighted that the number of propagated points per second by the GPU is much higher than the number of spread points per second in the CPU in all cases, so the efficiency of the GPU is higher than the CPU.

Future work will be oriented to increase the efficiency of the GPU implementation and use real fires to determine under in which conditions it is better to do the propagation of the fire front in the GPU.

Acknowledgments. This research has been supported by MINECO-Spain under contract TIN2017-84553-C2-1-R and by the Catalan government under grant 2017-SGR-313.

References

1. Artés, T., Cencerrado, A., Cortés, A., Margalef, T.: Core allocation policies on multicore platforms to accelerate forest fire spread predictions. In: Wyrzykowski, R., Dongarra, J., Karczewski, K., Waśniewski, J. (eds.) PPAM 2013. LNCS, vol. 8385, pp. 151–160. Springer, Heidelberg (2014). https://doi.org/10.1007/978-3-642-55195-6_14
2. Brun, C., Margalef, T., Cortés, A., Sikora, A.: Enhancing multi-model forest fire spread prediction by exploiting multi-core parallelism. J. Supercomput. **70**(2), 721–732 (2014). https://doi.org/10.1007/s11227-014-1168-z
3. Cencerrado, A., Artés, T., Cortés, A., Margalef, T.: Relieving uncertainty in forest fire spread prediction by exploiting multicore architectures. In: Proceedings of the International Conference on Computational Science, ICCS 2015, Computational Science at the Gates of Nature, Reykjavík, Iceland, pp. 1752–1761, 1–3 June 2015. https://doi.org/10.1016/j.procs.2015.05.380
4. Dagum, L., Menon, R.: OpenMP: an industry standard api for shared-memory programming. Comput. Sci. Eng. **5**(1), 46–55 (1998)
5. D'Ambrosio, D., Di Gregorio, S., Filippone, G., Rongo, R., Spataro, W., Trunfio, G.A.: A multi-GPU approach to fast wildfire hazard mapping. In: Obaidat, M.S., Filipe, J., Kacprzyk, J., Pina, N. (eds.) Simulation and Modeling Methodologies, Technologies and Applications. AISC, vol. 256, pp. 183–195. Springer, Cham (2014). https://doi.org/10.1007/978-3-319-03581-9_13
6. Finney, M.A.: Farsite: Fire area simulator-model development and evaluation. FResearch Paper RMRS-RP-4 Revised 236, Research Paper RMRS-RP-4 Revised (1998)
7. Farsite tutorial website (2007). http://fire.org/downloads/farsite/WebHelp/using_farsite_help.htm
8. Gregorio, S.D., Filippone, G., Spataro, W., Trunfio, G.A.: Accelerating wildfire susceptibility mapping through GPGPU. J. Parallel Distrib. Comput. **73**(8), 1183–1194 (2013). https://doi.org/10.1016/j.jpdc.2013.03.014
9. Hoang, R.V.: Wildfire Simulation on the GPU. Ph.D. thesis, university of Nevada (2008)
10. Knight, I., Coleman, J.: A fire perimeter expansion algorithm-based on huygens wavelet propagation. Int. J. Wildland Fire **3**, 73–84 (1993)
11. Ntinas, V.G., Moutafis, B.E., Trunfio, G.A., Sirakoulis, G.C.: Parallel fuzzy cellular automata for data-driven simulation of wildfire spreading. J. Comput. Sci. **21**, 469–485 (2017). https://doi.org/10.1016/j.jocs.2016.08.003
12. Rothermel, R.: A mathematical model for predicting fire spread in wildland fuels. Technical Report INT-GTR-115. (Ogden, UT) (1972)
13. Sousa, F.A., dos Reis, R.J.N., Pereira, J.C.F.: Simulation of surface fire fronts using fireLib and GPUs. Environ. Model. Softw. **38**, 167–177 (2012). https://doi.org/10.1016/j.envsoft.2012.06.006

Augmented Reality for Real-Time Navigation Assistance to Wheelchair Users with Obstacles' Management

Sawssen Ben Abdallah[1], Faiza Ajmi[1], Sarah Ben Othman[1(✉)],
Sébastien Vermandel[2], and Slim Hammadi[1]

[1] CRIStAL Laboratory UMR 9198, Ecole Centrale of Lille, Cité Scientifique,
59651 Villeneuve-d'Ascq, France
Sara.ben-othman@centralelille.fr
[2] APF France Handicap C-RNT, 25 rue Corneille Roubaix,
59100 Roubaix, France

Abstract. Despite a rapid technological evolution in the field of technical assistance for people with motor disabilities, their ability to move independently in a wheelchair is still limited. New information and communication technologies (NICT) such as augmented reality (AR) are a real opportunity to integrate people with disabilities into their everyday life and work. AR can afford real-time information about buildings and locations' accessibility through mobile applications that allow the user to have a clear view of the building details. By interacting with augmented environments that appear in the real world using a smart device, users with disabilities have more control of their environment. In this paper, we propose a decision support system using AR for motor disabled people navigation assistance. We describe a real-time wheelchair navigation system equipped with geological mapping that indicates access path to a desired location, the shortest route towards it and identifies obstacles to avoid. The prototyped wheelchair navigation system was developed for use within the University of Lille campus.

Keywords: Geological mapping · Augmented reality · Wheelchair users · Short path · Inclusion · Smart glasses · Accessibility

1 Introduction

Currently, around 1% of the general population uses wheelchairs. For 850,000 of these people, motor deficiency is predominant. Despite the existence of several New information and communication technologies (NICT) which can be used to help and assist them, most of the research done concerns sensory disabilities, namely deafness and blindness. However, living conditions improvement for people with motor disabilities could represent an immense hope for them. This improvement includes mobility assistance, which consists of guiding them during their travels. This assistance covers the journey from a source point to a desired destination. It concerns public transport, information on places accessible to people with a wheelchair, tourist information, etc. Each person has some specific criteria for choosing a path while moving from a place

© Springer Nature Switzerland AG 2019
J. M. F. Rodrigues et al. (Eds.): ICCS 2019, LNCS 11540, pp. 528–534, 2019.
https://doi.org/10.1007/978-3-030-22750-0_47

to another. Normal people usually choose the shortest path, but disabled people may choose a longer route that does not include an uphill for example. They do not have the ability to make their own choices for their movements. They usually try to memorize routes and accessibility information for every environment they visit which can be problematic in future times if building architecture has been changed, in case of an under-construction road, or the presence of a dynamic or static obstacle that could vary from his previous experience. Motor disabled people cannot even easily try a new route because they do not have the information about the surroundings and feel disposed to various kinds of risks such as being blocked in front of a building with no elevator. Many wheelchair users balk to visit a foreign place, as they have no information about the new environment and its accessibility conditions [1]. Some wheelchair users may address these obstacles and others may not. Therefore, obstacle detection becomes an important point while dealing with wheelchair users. Obstacles include narrow aisles, lighting, bad weather, sidewalk width, door handles and/or door pressure, gravel surfaces, etc. [2].

Navigation assistance becomes a crucial point for people with motor deficiency. NICT are a suitable option to improve motor disabled people inclusion in their environment while moving. Among various technologies that exist nowadays, AR technology is suited to make motor disabled people life easier and more comfortable. This technology mixes real and virtual objects using the 3D interaction devices of virtual reality in real space.

2   Related Work

In the literature, various studies have been conducted to detach wheelchair users from dependency. They have been working on social inclusion improvements of motor disabled people. In [3], authors have classified motor disabled people into different categories based on their hand stability, and have evaluated various types of interfaces. In [4], a smart wheelchair that can be controlled by gestures has been presented.

As well [5] submitted a smart and powered wheelchair based on pattern recognition that could help users become more independent without much physical effort. According to many studies, the number of motor disabled people has been increasing worldwide and there should be solutions able to provide autonomy for wheelchair users. In [6, 7] authors have developed a robotic arm, allowing wheelchair users to individually collect an object from a shelf. [8] has presented an algorithm which returns the optimum path from a source to a destination, and re-orients the user if an obstacle is detected along the path. In fact, navigation path has been studied in robotic movement context [9–11], where the robot studies all environments and proposes a path to reach the destination while averting any obstacle that can be found in the user environment. Navigation path is based on several approaches such as metaheuristic techniques: genetic algorithm, Dijkstra algorithm, graph theory, etc. These algorithms are designed to provide the optimal path. In [3] authors described an application based on AR and radio frequency identification (RFID) technologies allowing wheelchair users to do shopping and browsing independently with the help of a smartphone or a tablet. Besides, the smart glasses "Glasschair" are used to control the electric wheelchairs with the help head movements[1]. The meaning of the word AR is vast; some of the expert uses a general meaning for it, although others mean something very definite. AR, as a technological facilitator, is becoming on top of charts in different fields such as health, education, science, navigation, etc. [12].

In the literature, a small number of studies have been dealing with the wheelchair user's requirement using AR. In the previous studies, numerous methods with very different approaches for motor disabled people movement assistance have been proposed. However, none of these systems provides real-time navigation assistance with obstacles management for wheelchairs users.

3 Solution Architecture

3.1 Presentation

With the help of ORA-2 [1] a smart glass that features mobile computing and AR, the ARSAWP (augmented reality system for the assistance of wheelchair people) gives an augmented touch to accessibility information. It is an innovative way to interact between users and accessibility information in the real world; it gives a secure guidance from a place to another. ARSAWP ensures real-time navigation, provides the shortest path to the destination and redirects the user to an emergency route in case of obstacles are encountered.

3.2 Solution Features

Figure 1 represents the global architecture of ARSAWP system, which is composed of AR and Android technology combined. Wheelchair user connects via his profile to the ARSAWP system. All of these flows and connections are managed by a connection management system.

Wearing the AR smart glasses, wheelchair user is able to choose a place to go to among several destinations according to his current location. But before that, ARSAWP gives him a glimpse of accessibility information concerning the desired destination whether it is accessible, accessible with assistance or not accessible at all. Once the user has chosen his destination, ARSAWP will proceed to creating the accessibility graph towards the desired location in real-time according to the user's GPS coordinates and the area cartography. We propose an innovative method based on Dijkstra algorithm to calculate the shortest path. In case there's an obstacle blocking the user's way, he can report it. The ARSAWP can then proceed to rerouting to another obstacle-free path. If obstacles are already reported by other users, then obstacles management is handled in terms of notifying wheelchair users of potential obstacles reported that is when a redirection process towards an emergency route is applied by modifying the proposed algorithm based on Dijkstra algorithm. ARSAWP handles a real-time display of obstacles and paths on a map using google maps API.

---

[1] https://www.rehacare.com/cgi-bin/md_rehacare/lib/pub/tt.cgi/GLASSCHAIR_Controlling_
wheelchairs_with_smart_glasses.html?oid=46573&lang=2&ticket=g_u_e_s_t.

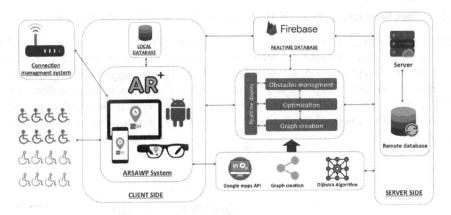

Fig. 1. Global architecture

This operation is managed by applying the proposed method based on Dijkstra algorithm alongside accessibility graphs and Google maps API. All these data are stored in a local database and a remote database hosted in a cloud server. Each reported obstacle is recorded in a Firebase, which is a real time database, so that ARSAWP can instantaneously display the obstacles on the map.

3.3 Methodology Implementation

Figure 2 shows the proposed algorithm flowchart combined with the real-time behavior of our system. The algorithm ShortPath() returns the optimum path from source to destination. It uses the following arrays:

Visited[i]=1; if the shortest is found from src to i	Visited[i]=0; if not
d[i] denotes the optimum distance from 'src' to 'i'	parent[i] denotes the parent of node 'i' in the shortest path

4 Simulation and Results

Various tests were performed for different scenarios (location not accessible, a route with obstacles and accessible with assistance) in Lille University in the north of France. We considered students using wheelchair to test the developed prototype.

4.1 User Scenario 1: Navigation with Obstacles

In this use case scenario, whenever obstacles have been reported in the path selected by the user, ARSAWP handles obstacles management by notifying him and redirects him towards an alternative route proposed by our proposed algorithms as shown in Fig. 2. At the same time, ARSAWP displays in real-time obstacles and paths on the map that are saved in Firebase, which is a real time database.

4.2   User Scenario 2: Desired Destination Accessible with Assistance

If the user has to choose a place to go to according to his current location, ARSAWP notifies him that this location is accessible with assistance. For instance, if the entrance door that is not wide enough as shown in Fig. 3, our system suggests finding assistance for him. It can also notify him in real time whether the toilets in the first floor are in service or not. Once arrived at destination, ARSAWP checks if the user has arrived at destination. If this is the case, a gauge of satisfaction is proposed to the user as shown in Fig. 3.

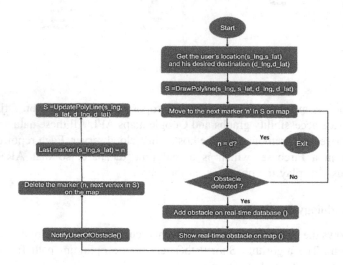

Fig. 2. Algorithm flowchart

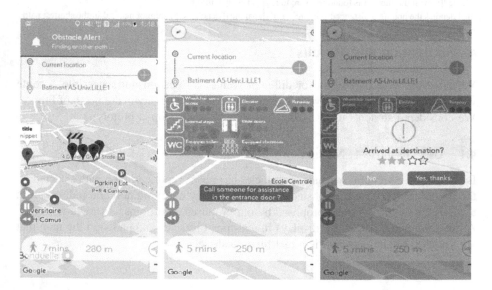

Fig. 3 Navigation system and accessibility information

5   Discussion

ARSAWP system is primarily aimed at providing access information to wheelchair users to avoid creating this wall of not getting the right and useful information. Our solution meets the need of comfort and well being of people with physical and mobility impairments. In fact, after a long and an interesting discussion made with them, they clearly expressed their expectations from this kind of system. Afterwards, we collected all these data and were able to specify the main features of our system which uses AR to increase the accessibility of buildings and locations through a mobile application that allows the user to have a clear view of the building details such as the existence of an elevator, an access ramp in the entrance and if they need assistance in the entrance door. AR makes it possible for users to overcome limitations and difficulties of the environment and allows new ways of understanding it so that wheelchair users will feel at ease. By interacting with augmented environments that appear in the real world using a smart device, users with disabilities have more control of their environment and interact with several information that they could not access before. The proposed algorithm validates the specifications and criteria that a system must provide to wheelchair users in order to ensure a proper route and a smooth navigation from a source to a destination by guiding them through the shortest path in real-time. Besides, using AR and the proposed algorithm we provide the necessary accessibility information about the destination and can notify the users of a blocked path and redirect them to an emergency route for a better guidance. Most importantly, we need to ensure that the next step would be to enhance the developed system to meet the expectations of wheelchair users whom we interviewed.

6   Conclusion and Perspectives

ARSAWP is able to help many wheelchair users providing accessibility information and allowing them to navigate freely and surely from a source to a destination in real-time. We aim to make this system available not only for wheelchair users but for all people who present other kind of disabilities, the system will provide guidance according to the user's profile and to the type of disability. The next step will be to make the experience more accurate. In fact, an obstacle verification service has to be included to be certain of the integrity of the obstacle being reported. The best route for people with physical and mobility impairments is not only the shortest distance route from a source to a destination but also the route with minimum of efforts and energy spent that is why distance is negligible compared with time because the more wheelchair users spend time to get to their desired destination the more effort they provide. So taking time into consideration is a must. In addition, we plan to make ARSAWP system more comfortable and easy to use to meet the users' needs by adding vocal guidance as a feature.

References

1. Thapar, N., et al.: A pilot study of functional access to public buildings and facilities for persons with impairments. Disabil. Rehabil. **26**(5), 280–289 (2004)
2. Neis, P., Zielstra, D.: Generation of a tailored routing network for disabled people based on collaboratively collected geodata. Appl. Geogr. **47**, 70–77 (2014)
3. Rashid, Z., Melià-Seguí, J., Pous, R., Peig, E.: Using augmented reality and internet of things to improve accessibility of people with motor disabilities in the context of smart cities. Future Gener. Comput. Syst. **76**, 248–261 (2017)
4. Kuno, Y., Murashima, T., Shimada, N., Shirai, Y.: Interactive gesture interface for intelligent wheelchairs. In: 2000 IEEE International Conference on Multimedia and Expo. ICME 2000. Proceedings. Latest Advances in the Fast Changing World of Multimedia (Cat. No.00TH8532), vol. 2, pp. 789–792 (2000)
5. Leaman, J., La, H.M.: A comprehensive review of smart wheelchairs: past, present, and future. IEEE Trans. Hum. Mach. Syst. **47**(4), 486–499 (2017)
6. Tsui, K., Kim, D., Behal, A Kontak, D., Yanco, H.A.: "I want that": human-in-the-loop control of a wheelchair-mounted robotic arm. Appl. Bionics Biomech. **8** (2011)
7. Kumar, S., Rajasekar, P., Mandharasalam, T., Vignesh, S.: Handicapped assisting robot. In: 2013 International Conference on Current Trends in Engineering and Technology (ICCTET), pp. 88–91 (2013)
8. Nandini, D., Seeja, K.R.: A novel path planning algorithm for visually impaired people. J. King Saud Univ. Comput. Inf. Sci. (2017)
9. Minguez, J., Montano, L.: Sensor-based robot motion generation in unknown, dynamic and troublesome scenarios. Robot. Auton. Syst. **52**(4), 290–311 (2005)
10. Yared, R., Défago, X., Iguchi-Cartigny, J., Wiesmann, M.: Collision prevention platform for a dynamic group of asynchronous cooperative mobile robots. JNW **2**, 28–39 (2007)
11. Khatib, O.: Real-time obstacle avoidance for manipulators and mobile robots. In: Cox, I.J., Wilfong, G.T. (eds.) Autonomous Robot Vehicles, pp. 396–404. Springer, New York (1990). https://doi.org/10.1007/978-1-4613-8997-2_29
12. Olmedo, H.: Virtuality continuum's state of the art. Procedia Comput. Sci. **25**, 261–270 (2013)

p3Enum: A New Parameterizable and Shared-Memory Parallelized Shortest Vector Problem Solver

Michael Burger[(✉)], Christian Bischof, and Juliane Krämer

Fachbereich Informatik, Technische Universität Darmstadt, Hochschulstraße, 10, 64289 Darmstadt, Germany
{michael.burger,christian.bischof}@sc.tu-darmstadt.de, jkraemer@cdc.informatik.tu-darmstadt.de

Abstract. Due to the advent of quantum computers, quantum-safe cryptographic alternatives are required. Promising candidates are based on lattices. The hardness of the underlying problems must also be assessed on classical hardware. In this paper, we present the open source framework p3Enum for solving the important lattice problem of finding the shortest non-zero vector in a lattice, based on enumeration with extreme pruning. Our parallelized enumeration routine scales very well on SMP systems with an extremely high parallel efficiency up to 0.91 with 60 threads on a single node. A novel parameter ν within the pruning function increases the probability of success and the workload of the enumeration. This enables p3Enum to achieve runtimes for parallel enumerations which are comparable to single-threaded cases but with higher success rate. We compare the performance of p3Enum to publicly available libraries and results in the literature. For lattice dimensions 66 to 88, p3Enum performs the best which makes it a good building block in lattice reduction frameworks.

1 Introduction

We need cryptography in our daily lives to secure the applications like social media and online banking. However, we know that all public-key cryptography based on prime factorization and elliptic curves in use today will be broken once large-scale quantum computers exist. Therefore, a vivid field of research is post quantum cryptography, where cryptographic algorithms that withstand attacks with quantum computers are developed. Algorithms based on lattices are promising since they are versatile and efficient. One of the frequently studied algorithmic problems in lattice cryptography is the shortest vector problem (SVP) which cannot be solved by quantum computers efficiently. However, to be practical, lattice-based cryptography does not only have to withstand attacks with quantum computers, but the exact hardness in relation to classical computers and HPC systems has also to be determined so that secure parameters, e.g., key sizes, can be chosen. Regarding the SVP, the most promising techniques

© Springer Nature Switzerland AG 2019
J. M. F. Rodrigues et al. (Eds.): ICCS 2019, LNCS 11540, pp. 535–542, 2019.
https://doi.org/10.1007/978-3-030-22750-0_48

to solve it are extreme pruning [2] and sieving [4]. To understand their full potential, they have to be analyzed on parallel systems. To that end, we present p3Enum which is a parallelized, parameterizable open source framework based on extreme pruning.

2   Preliminaries

We denote vectors with bold lower case letters, e.g., $\mathbf{u}$, matrices with bold upper case letters, e.g., $\mathbf{B}$, and scalars with normal lower case letters, e.g., β. $\mathbf{B}^{m \times n}$ stands for an $m \times n$ matrix. If the dimensions are clear from the context, we simply write $\mathbf{B}$. Integers are denoted by $\mathbb{Z}$ and the real numbers by $\mathbb{R}$. The standard inner product is denoted by $\langle \cdot, \cdot \rangle$ and the Euclidean norm by $\|\cdot\|$.

A lattice of dimension d is a discrete additive subgroup of $\mathbb{R}^d$. Every lattice $\Lambda \subset \mathbb{R}^d$ can be represented by a basis, i.e., a set of $\mathbb{R}$-linearly independent vectors $\mathbf{B} = \{\mathbf{b}_1, \ldots, \mathbf{b}_n\} \subset \mathbb{R}^d$ such that $\Lambda = \Lambda(\mathbf{B}) = \mathbb{Z}\mathbf{b}_1 + \cdots + \mathbb{Z}\mathbf{b}_n$. We identify lattice bases with matrices whose columns represent the basis vectors. In this case, d is called the dimension of the lattice and $n \leq d$ is called its rank. If $n = d$, the lattice is called a full-rank lattice. All lattices within this work are full-rank. The Gram-Schmidt (GS) basis (obtained by GS orthogonalization) of a basis $\mathbf{B}$ is denoted by $\mathbf{B}^* = \{\mathbf{b}_1^*, \ldots, \mathbf{b}_n^*\} \subset \mathbb{R}^d$, the respective GS-lengths by $\|\mathbf{b}_1^*\|^2, \ldots, \|\mathbf{b}_n^*\|^2$, and the GS-coefficients by $\mu_{i,j}$ with $1 \leq j < i \leq n$.

The quality of a lattice basis $\mathbf{B}$ can, e.g., be measured by the decrease of the series $\|\mathbf{b}_1^*\|^2, \ldots, \|\mathbf{b}_n^*\|^2$, or by the value of $\|\mathbf{b}_1^*\|^2$. Improving the quality of a basis is called *basis reduction*. Geometrically, basis reduction means, in particular, to make the basis vectors shorter and more orthogonal. The most commonly used basis reduction algorithm is BKZ 2.0 [2]. BKZ 2.0 works on local blocks of lattices of dimension $\beta < n$ and optimizes the basis by sliding over all basis vectors in contiguous blocks. A basis processed by BKZ with block size β is called BKZ-β reduced. Solutions for the SVP or approximately good solutions, delivered by a so-called SVP-oracle, are required within each local block. The two most common types of SVP-oracles employ enumeration [2,3,5] with extreme pruning or sieving algorithms [1,4]. p3Enum uses enumeration and hence searches for coefficient vectors $\mathbf{u}$ fulfilling $\|\mathbf{u} * \mathbf{B}\| < \overline{A}$ in a heuristically pruned search tree. The extreme pruning function $\mathbf{A} = (A_1, \ldots, A_n)$ with $A_1 \leq A_2 \leq \cdots \leq A_n \leq \overline{A}$ determines the maximal costs A_i for a partial solution vector with length i. Large parts of the search tree are cut off and in general the search has to be repeated several times on randomized input bases to succeed [2].

3   Related Work

The single-threaded, template-based fplll C++-library [3] implements important algorithms from the lattice domain like LLL, BKZ, or enumeration with extreme pruning. Pruned enumeration is possible for bases with $n \leq 90$.

Kuo et al. [5] presented an implementation of extreme pruning on GPUs. The enumeration tree is split into starting vectors generated on the CPU which are completed on GPUs. The pruning function **A** is a scaled polynomial of degree eight.

Aono et al. [2] developed progressive BKZ which avoids a predefined BKZ-strategy but starts with a small β and iteratively increases it in appropriate steps. The pruning functions are based on the so-called full enumeration cost (FEC) which results from benchmarks, heuristics, and optimized estimates for $\overline{A}$.

Concerning sieving, SubSieve [4] employs progressive sieving [6] which works on sublattices instead of directly solving the SVP on the whole lattice basis. It also takes advantage of the fact that the output of sieving a list of short vectors. This allows to solve the n-dimensional SVP with sieving calls on $(n-\delta)$-dimensional sublattices, where δ is heuristically determined.

Albrecht et al. [1] combine the principles of SubSieve with further algorithmic improvements into the General Sieve Kernel (G6K). G6K processes the basis in non-contiguous blocks and its parallelized C++-implementation holds the record in the Darmstadt SVP challenge[1] (D-SVPC), where researchers are invited to search short vectors within provided random lattices.

4   Implementation of p3Enum

p3Enum is implemented in C++11[2]. The randomized bases are reduced by two different BKZ calls to the fplll library. First, we execute a classical BKZ without pruned enumeration calls on a relatively small block size of pre-$\beta \in \{2, \ldots, 36\}$. It is followed by a call with $\beta \in \{2, \ldots, 54\}$ and pruning and heuristical early abortion enabled so that BKZ 2.0 terminates when the heuristic detects no further considerable improvements in the basis quality. The values of pre-β and β depend on the dimension of the lattice and we try to weight the required runtimes in a relation of $2:1$ since our experiments showed the best performance for this combination.

For the extreme pruning function, we employ the polynomial of degree eight from [5] (cf. Sect. 3) scaled to the respective dimension of the lattice. We evaluate the polynomial at each position $l \in \{1, \ldots, n\}$ and multiply the result $\in \ ]0, \ldots, 1.0]$ with $\overline{A}$. This value is assigned to the respective entry A_l of the pruning function vector **A**.

4.1   Parallelization Strategy

Our parallelization strategy is twofold. First, the enumeration itself is parallelized and second, multiple instances of basis reduction are executed in parallel.

The parallelization of the enumeration is strongly related to the approaches of [2,5]. Within the OpenMP parallel region, the first thread arriving starts at

---

[1] https://www.latticechallenge.org/svp-challenge/.

[2] https://github.com/MiBu84/p3enum.

a pre-defined depth $\eta = 10$ which is chosen based on experiments. This thread enumerates vectors from η to the root which are smaller than the corresponding A_i's, called *candidates*. They are inserted in a thread-safe, shared ring-queue, developed by us. The size can be reconfigured at runtime and the fixed size allows to calculate its memory requirements. If the queue is filled above a given threshold, all threads except the one filling the queue start processing the partial tree at level $\eta + 1$.

Experiments with the fplll library showed that running multiple instances on the same compute node does not have a considerable negative effect on the runtime compared to running one instance. Hence, p3Enum performs multiple instances on different randomized bases in parallel to make efficient use of the compute capabilities of modern computers. This also considerably reduces, if not even removes, the drawback that serial BKZ-implementations prevent efficient parallelization on an SMP-system, as mentioned in [5]. In that way, we create a bunch of randomized, reduced bases for processing in about the same time as a single basis otherwise. One drawback of this approach is that the time to reduce a basis varies for randomized instances. Consequently, some threads finish faster than others. Empirical experiments show a difference of about 2 in the runtime between the fastest and the slowest thread.

4.2   Parameterized Workload

Performing experiments with the progressive BKZ library [2] show a conspicuous behavior. The library internally decides whether to execute the pruned enumeration in the OpenMP-parallelized or the single-threaded version. For many dimensions <85, the heuristic chooses the single-threaded version since the workload is too small and the predicted single-threaded execution time far below 1 s. Comparably fast runtimes can be achieved with p3Enum: When directly running pruned enumeration on a random lattice of dimension 80, which is BKZ-30 reduced and a tight bound $\overline{A}$ for the shortest vector is known (as described below), the single-threaded execution time is below 0.01 s and a parallelization will not pay off. Hence, we introduce a new pruning-parameter ν which works in the following way. We first evaluate the polynomial of [5] for the considered entry $l \in \{1, \ldots, n\}$, resulting in a value $\alpha \in]0, \ldots, 1.0]$. Now, we update $\alpha = \min(\alpha + \nu, 1.0)$ and finally calculate the value of the pruning function $\mathbf{A}$ at entry l by $A_l = \alpha \cdot \overline{A}$. Mathematically, we shift the graph of the pruning function along the positive y-axis.

In that way, we increase the probability of keeping the shortest vector in the pruned tree and increase the workload so that a parallelization pays off. The goal is to set ν such that it enables p3Enum to efficiently run parallel enumerations in the time of single-threaded enumerations, but with increased success rate.

4.3   Heuristics to Improve the Performance

To reduce the number of randomized bases to be processed, the bases reduced in parallel are processed in ascending order of their $||\mathbf{b}_1^*||^2$ values. Mainly for smaller dimensions <80 this reduces the number of processed bases at no additional cost.

To cope with the different runtimes of parallel BKZ-instances, the program measures the runtimes during the first round of reductions and, based on the observed timing variance between threads (cf. Sect. 4.1), sets a time limit for the BKZ-calls in the following rounds to 2.1 times of the fastest reduction in the first round. If the $||\mathbf{b}_1^*||^2$ value of the BKZ calls terminated because of this time limit lies in the range of the $||\mathbf{b}_1^*||^2$'s of the other bases, the basis is processed normally, otherwise it is discarded.

5   Experiments and Results

5.1   Methodology

system1 nodes are dual socket Intel E5-2680 v3 CPUs (24 cores) with 64 GB of RAM. *system2* nodes are quad socket Intel E7-4890 v2 CPUs (60 cores) with 1024 GB of RAM. For p3Enum and fplll we use gcc 8.2.0 and for SubSieve gcc 4.9.4. All random lattices are in Goldstein-Mayer form of the D-SVPC with seeds (0, 237, 6880, 97575, 98937). To have an upper bound for the length of the vectors, we performed, like [4], several runs of fplll's pruned SVP-routine with a target success probability of 99%.

5.2   Performance Analysis

The parameterization for p3Enum ($\nu \in [0.03, \ldots, 0.3]$, $\beta \in \{2, \ldots, 58\}$, pre-$\beta \in \{2, \ldots, 38\}$) was based on empirical test runs and constant for a lattice dimension d. It is available on p3Enum's github. Figure 1 summarizes the runtimes and splits the data into three diagrams to refine the logarithmic scale on the y-axis.

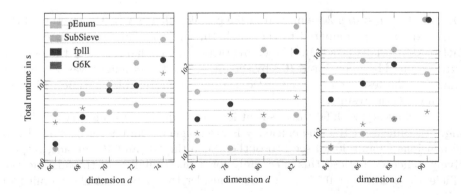

Fig. 1. Performance comparison for dimensions 76-90 on *system1* nodes.

Table 1. Speedup of p3Enum when compared to other solutions.

Solver	n												
	66	68	70	72	74	76	78	80	82	84	86	88	90
fplll [3]	1.5	1.4	1.9	1.8	2.8	1.7	3.3	3.8	5.4	4.1	3.7	4.9	4.8
GPUEnum [5]	-	-	-	-	-	2.4	4.8	4.5	4.2	3.8	2.9	3.6	1.7
ProSieve [6]	15.7	21.9	29.3	42.9	58.7	50.3	113	109	148	-	-	-	-
SubSieve [4]	1.6	2.3	2.2	3.4	5.0	3.6	7.2	7.5	10.5	7.7	7.2	7.5	4.9
G6K [1]	1.5	1.5	1.9	1.8	1.9	1.2	2.5	1.3	1.6	1.1	1.1	1.0	0.34

p3Enum, SubSieve, and fplll are visualized as dots which are created by the average of at least 75 measurement points (at least 15 per seed). Finally, the stars indicate the average runtime of G6K whose runtime are taken from [1].

Figure 1 shows that p3Enum delivers faster runtimes than fplll and SubSieve for all tested dimensions. Table 1 summarizes the speedup of p3Enum compared to other solutions.

We calculated the speedup when employing p3Enum compared to the implementation in the first column. The values in italics result from least squares fitting and extrapolation of the other data points given in the literature. p3Enum outperforms the GPU-enumeration from [5] in all dimensions considered, although [5] employs a system with eight GPUs, providing more theoretical FLOPs and consuming much more energy than *system1*. Progressive sieving [6] is also slower in all dimensions. G6K, however, is faster than p3Enum in dimension 90. This seems to underpin the assumption of [4] that a good implementation of the novel sieving algorithms will outperform all pruned enumeration solvers somewhere around dimension 90. Since p3Enum delivers the best performance in $n = \{66, \ldots 88\}$ it is a very suitable SVP-oracle within basis reduction for that range.

5.3   Scaling Behavior and Efficiency

On our 60-core *system2* we measured the parallel efficiency by the fraction of the achieved speedup compared to the single-threaded baseline with four different lattice dimensions (seed 0). The enumeration was configured such the 60 core runs required between 5 and 10 s by setting ν. The target length was set shorter than the shortest vector in the lattices to have reproducible runtimes. Figure 2 summarizes the results.

The efficiency with 60 threads is at least 0.85, demonstrating very good scaling. For dimension 100, the efficiency is even higher than 0.9 for 60 threads. In dimensions 80 and 90 we see two outliers at 5 threads and 10 threads, respectively. Since we noticed this effect only for a small number of threads and the efficiency is still over 0.75, we do not consider this a relevant shortcoming of p3Enum.

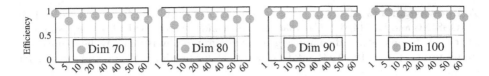

Fig. 2. Parallel efficiency on *system2*. x-axis: # threads employed.

5.4 New Shortest Vectors in Darmstadt SVP Challenge

We found shorter solutions with a different seed in higher, already solved dimensions for in the D-SVPC. Table 2 compares the fastest succeeding trial of p3Enum (row 1) and SubSieve (row 3), the runtimes of former solutions in the D-SVPC (row 2), and the average time of G6K from [1]. The underlying parameterization for p3Enum varies and may not be the ideal one (cf. D-SVPC entries for details).

Table 2. Comparison of runtime in s for higher dimensions.

Solver	n									
	91	92	93	94	95	96	97	98	99	100
p3Enum	25	397	233	279	76	190	65	107	897	4000
D-SVPC	$4.4 \cdot 10^5$	$3.7 \cdot 10^5$		3600	1800	5400			7200	27360
SubSieve	553	823	1642	1015	2083	1447	2872	1734	3467	3363
G6K	-	312	-	375	-	815	-	995	-	1964

Table 2 shows that higher dimensions can be solved with a competitive runtime. p3Enum's time for dimensions 91, 95, and 97 is smaller than the shortest run of fplll and SubSieve achieved on the 90-dimensional bases. Additionally, the results highlight the randomness in the runtime of SubSieve in higher dimensions. Although we performed five runs, the fastest runtime in dimension 94 is considerably lower than in dimension 93. The seeds for G6K were different and no direct comparison is possible.

6 Conclusion

We introduced the open source framework p3Enum for solving the SVP with its additional parameter ν enabling a parallel efficiency rate of more than 0.9 on a 60-core system by adjusting the workload and the success probability. p3Enum is the fastest solver in dimensions 66–88 compared to available SVP solutions. Hence, p3Enum can be employed as a building block in lattice reduction frameworks. To further increase p3Enum's performance we will extend it with MPI and implement a search for the pruning function as realized in [2,3].

Acknowledgments. This work has been co-funded by the DFG through CRC 1119 CROSSING and BI 714/6-1. Calculations were conducted on the Lichtenberg computer of the TU Darmstadt, and computing resources granted by RWTH Aachen University under project prep0016. We thank L. Ducas et al. for providing the preliminary version of [1].

References

1. Albrecht, M., Ducas, L., Herold, G., Kirshanova, E., Postlethwaite, E.W., Stevens, M.: The general sieve kernel and new records in lattice reduction. Cryptology ePrint Archive, Report 2019/089 (2019)
2. Aono, Y., Wang, Y., Hayashi, T., Takagi, T.: Improved progressive BKZ algorithms and their precise cost estimation by sharp simulator. In: Fischlin, M., Coron, J.-S. (eds.) EUROCRYPT 2016. LNCS, vol. 9665, pp. 789–819. Springer, Heidelberg (2016). https://doi.org/10.1007/978-3-662-49890-3_30
3. The fplll development team. fplll, a lattice reduction library (2016). https://github.com/fplll/fplll
4. Ducas, L.: Shortest vector from lattice sieving: a few dimensions for free. In: Nielsen, J.B., Rijmen, V. (eds.) EUROCRYPT 2018. LNCS, vol. 10820, pp. 125–145. Springer, Cham (2018). https://doi.org/10.1007/978-3-319-78381-9_5
5. Kuo, P.-C., et al.: Extreme enumeration on GPU and in clouds. In: Preneel, B., Takagi, T. (eds.) CHES 2011. LNCS, vol. 6917, pp. 176–191. Springer, Heidelberg (2011). https://doi.org/10.1007/978-3-642-23951-9_12
6. Laarhoven, T., Mariano, A.: Progressive lattice sieving. In: Lange, T., Steinwandt, R. (eds.) PQCrypto 2018. LNCS, vol. 10786, pp. 292–311. Springer, Cham (2018). https://doi.org/10.1007/978-3-319-79063-3_14

Rendering Non-Euclidean Space in Real-Time Using Spherical and Hyperbolic Trigonometry

Daniil Osudin[✉], Chris Child, and Yang-Hui He

City, University of London, London, UK
daniil.osudin.1@city.ac.uk

Abstract. We introduce a method of calculating and rendering shapes in a non-Euclidean 2D space in real-time using hyperbolic and spherical trigonometry. We record the objects' parameters in a polar coordinate system and use azimuthal equidistant projection to render the space onto the screen. We discuss the complexity of this method, renderings produced, limitations and possible applications of the created software as well as potential future developments.

Keywords: non-Euclidean geometry · Spherical trigonometry ·
Hyperbolic trigonometry · Azimuthal equidistant projection ·
Polar coordinate system · Real-time

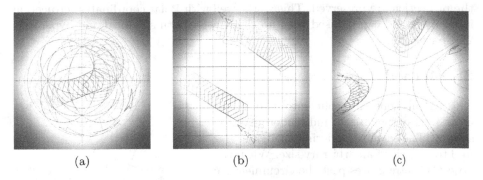

(a)   (b)   (c)

Fig. 1. Time-lapse images of multiple objects moving through spherical (a), planar (b) and hyperbolic (c) 2D space calculated and rendered by the described software

© Springer Nature Switzerland AG 2019
J. M. F. Rodrigues et al. (Eds.): ICCS 2019, LNCS 11540, pp. 543–550, 2019.
https://doi.org/10.1007/978-3-030-22750-0_49

1   Introduction

Non-Euclidean geometry is a field that studies any space that arises from changing Euclid's fifth postulate [1] or changing the metric requirement. In spherical geometry, Fig. 2(a), all geodesics (shortest paths in a non-planar space) intersect:

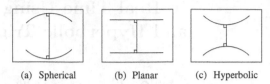

(a) Spherical   (b) Planar   (c) Hyperbolic

Fig. 2. Comparison of parallel lines in the 2D spaces

don't preserve the distance and appear to 'bend' towards each other. In Hyperbolic geometry, Fig. 2(c), each line has an infinite number of parallel lines, as they appear to 'bend' away from each other.

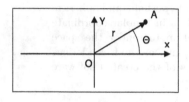

Fig. 3. Point A with polar coordinates r and θ

We present a method for calculating the object's position and its vertices in polar coordinates using spherical [2] or hyperbolic trigonometry [3,4]. A polar coordinate system of the form (r, θ) is used in this model for all calculations instead of Cartesian coordinates. The centre of the of the screen is taken as a reference point $O(0,0)$ for the distance coordinate, r, while eastbound is the reference direction for the bearing coordinate, θ.

This allows the same coordinates to be used irrespective of the correct curvature. In order to render the curved space onto a flat 2D screen, we are using azimuthal equidistant projection. By definition, distances and bearing from the centre of the projection are preserved. This works well with Polar coordinates, projection is intuitive and can be used with no changes for both spherical and hyperbolic 2D spaces (Fig. 3).

2   Method

The calculations are split into two parts: movement of the objects and rendering of the shapes. The screen (rendering space) is limited to a circle of an arbitrary size. When the object's centre moves past the circumference of the circle, it is repositioned to the antipodal point on the circle with the velocity preserved. This is implemented in order to keep the objects in the visible area on the screen.

Shape has a list of position vectors for each vertex in local coordinates with object's position being the reference point and reference direction is taken as the reverse of its position vector (Fig. 4).

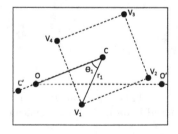

Fig. 4. O $(0,0)$, reference point; C (r_c, θ_c), position and local reference point; V_x (r_x, θ_x), vertices; OO', reference direction; CC', local reference direction

2.1   Rendering the Shape

Let $K \in [-1, 1] \subset \Re$ s.t. $K = 0 \Rightarrow$ Euclidean Geometry;

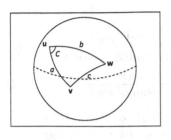

Fig. 5. Spherical triangle

$$K > 0 \Rightarrow \text{Spherical Geometry}, r = \frac{1}{\sqrt{K}}$$

Theorem 1. *For a sphere of radius r and hence Gaussian curvature $K = \frac{1}{r^2}$ and a spherical triangle on its surface described by points u, v and w, connected by great circles that form the edges a, b, c (interpreted as subtended angles) and an angle C (Fig. 5), the spherical law of cosines [5] states:*

$$\cos \frac{c}{r} = \cos \frac{a}{r} \cos \frac{b}{r} + \sin \frac{a}{r} \sin \frac{b}{r} \cos C \qquad (1)$$

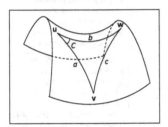

Fig. 6. Hyperbolic triangle

$$K < 0 \Rightarrow \text{Hyperbolic Geometry}, k = -\frac{1}{\sqrt{K}}$$

Theorem 2. *For a hyperbolic plane with Gaussian Curvature $K = -\frac{1}{k^2}$ and a hyperbolic triangle on its surface described by points u, v and w, connected by geodesics that form the edges a, b and c, as well as an angle C (Fig. 6), the hyperbolic law of cosines [6] states:*

$$\cosh \frac{c}{k} = \cosh \frac{a}{k} \cosh \frac{b}{k} - \sinh \frac{a}{k} \sinh \frac{b}{k} \cos C \qquad (2)$$

Note: To simplify the equations below, all lengths are assumed to have been divided by r or k depending on the value of K.

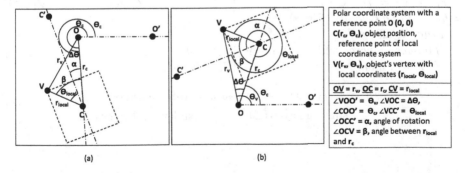

Fig. 7. Finding the θ and r coordinates of an object's vertices through a hyperbolic/spherical triangle OCV; Case (a): $\theta_{local} + \alpha < \pi$; case (b): $\theta_{local} + \alpha > \pi$

Corollary 1. *Given:* $O(0,0)$, $C(r_c, \theta_c)$, $V(r_v, \theta_v)$, $\underline{OC} = r_c$, $\underline{CV} = r_{local}$, $\angle COO' = \theta_c$, $\angle OCC' = \alpha$, $\angle VCC' = \theta_{local}$
    Find: r_v, $\theta_v = ?$
    If $K > 0$, then: *If $K < 0$, then:*

$$\cos r_v = \cos r_c \cos r_{local} + \qquad \cosh r_v = \cosh r_c \cosh r_{local} -$$
$$\sin r_c \sin r_{local} \cos \beta \qquad \sinh r_c \sinh r_{local} \cos \beta \qquad (3)$$

$$\cos \Delta \theta_v = \frac{\cos r_{local} - \cos r_c \cos r_v}{\sin r_c \sin r_v} \qquad \cos \Delta \theta_v = \frac{\cosh r_c \cosh r_v - \cosh r_{local}}{\sinh r_c \sinh r_v} \qquad (4)$$

In order to find r_v, first find $\beta = \alpha + \theta_{local}$; if $\Pi < \beta < 2\Pi$, use the explementary angle instead to determine to which side of $\underline{OC}$ the triangle lies. Depending on that $\Delta\theta$ is then added to or subtracted from θ_c to find θ_v (Fig. 7).

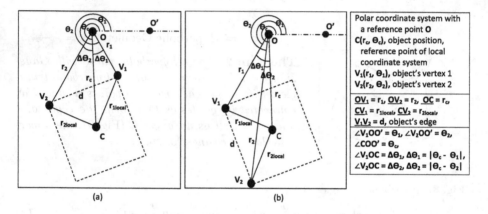

Fig. 8. Finding the length of edge d and the angle $\Delta\theta$. Case (a), $\Delta\theta_1$ and $\Delta\theta_2$ diverge, so $\Delta\theta$ is the sum; case (b), angles converge, so $\Delta\theta$ is the absolute value of the difference.

Corollary 2. *Given:* $O(0,0)$, $C(r_c, \theta_c)$, $V_1(r_1, \theta_1)$, $V_2(r_2, \theta_2)$, $\underline{OC} = r_c$, $\underline{OV_1} = r_1$, $\underline{OV_2} = r_2$, $\underline{CV_1} = r_{1local}$, $\underline{CV_2} = r_{2local}$, $\angle COO' = \theta_c$, $\angle V_1OO' = \theta_1$, $\angle V_2OO' = \theta_2$ (Fig. 8)
    Find: d, $\Delta\theta = ?$
    If angles converge, $\Delta\theta = \|\Delta\theta_1 - \Delta\theta_2\|$; if angles diverge, $\Delta\theta = \|\Delta\theta_1\| + \|\Delta\theta_2\|$
    If $K > 0$, then: *If $K < 0$, then:*

$$\cos d = \cos r_1 \cos r_2 + \qquad \cosh d = \cosh r_1 \cosh r_2 -$$
$$\sin r_1 \sin r_2 \cos \Delta\theta \qquad \sinh r_1 \sinh r_2 \cos \Delta\theta \qquad (5)$$

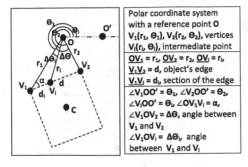

| Polar coordinate system |
| with a reference point O |
| $V_1(r_1, \theta_1)$, $V_2(r_2, \theta_2)$, vertices |
| $V_i(r_i, \theta_i)$, intermediate point |
| $OV_1 = r_1$, $OV_2 = r_2$, $OV_i = r_i$, |
| $V_1V_2 = d$, object's edge |
| $V_1V_i = d_i$, section of the edge |
| $\angle V_1OO' = \theta_1$, $\angle V_2OO' = \theta_2$, |
| $\angle V_iOO' = \theta_i$, $\angle OV_1V_1 = \alpha$, |
| $\angle V_1OV_2 = \Delta\theta$, angle between |
| V_1 and V_2 |
| $\angle V_1OV_i = \Delta\theta_i$, angle |
| between V_1 and V_i |

Fig. 9. Finding intermediate points in order to render the edge.

Note: distance d is divided into a number of equal parts in order to find the distance d_i for each of the points on the edge V_1V_2. The number of segments depends on the object tesselation variable.

Corollary 3. *Given:* $O(0,0)$, $V_1(r_1, \theta_1)$, $V_2(r_2, \theta_2)$, $V_i(r_i, \theta_i)$, $OV_1 = r_1$, $OV_2 = r_2$, $V_1V_2 = d$, $V_1V_i = d_i$, $\angle V_1OO' = \theta_1$, $\angle V_2OO' = \theta_2$, $\angle V_1OV_2 = \Delta\theta$
Find: r_i, $\theta_i = ?$

If $K > 0$, then:

$$\cos\alpha = \frac{\cos r_2 - \cos r_1 \cos d}{\sin r_1 \sin d}$$

If $K < 0$, then:

$$\cos\alpha = \frac{\cosh r_1 \cosh d - \cosh r_2}{\sinh r_1 \sinh d} \quad (6)$$

$$\cos r_i = \cos r_1 \cos d_i + \sin r_1 \sin d_i \cos\alpha$$

$$\cosh r_i = \cosh r_1 \cosh d_i - \sinh r_1 \sinh d_i \cos\alpha \quad (7)$$

$$\cos\Delta\theta_i = \frac{\cos d_i - \cos r_1 \cos r_i}{\sin r_1 \sin r_i}$$

$$\cos\Delta\theta_i = \frac{\cosh r_1 \cosh r_i - \cosh d_i}{\sinh r_1 \sinh r_i} \quad (8)$$

α *is calculated to find the angle opposite r_i. Then r_i and subsequently $\Delta\theta_i$ can be found using the cosine rule (illustrated on Fig. 9). Then to find actual coordinates of the point V_i, r_i should be multiplied by r or k depending on the value of K; $\Delta\theta_i$ should be added to or subtracted from angle $\theta 1$, depending on the direction of the edge d, determined previously.*

2.2   Updating Object Position

Corollary 4. *Given:* $O(0,0)$, $C_{t0}(r_{t0}, \theta_{t0})$, $C_{t1}(r_{t1}, \theta_{t1})$, $OC_{t0} = r_{t0}$, $C_{t0}C_{t1} = r_p$, $\angle C_{t0}OO' = \theta_{t0}$, $\angle OC_{t0}C'' = \gamma_{t0}$, $\angle OC_{t0}C'_{t0} = \beta_{t0}$
Find: r_{t1}, θ_{t1}, γ_{t1}, $\beta_{t1} = ?$
γ_{t0} *should be 0 to π, if calculated value is $\gamma_{t0} > \pi$, take the explemntary angle. This indicates the movement direction with respect to the reference point* (Fig. 10).
Let $\angle OC_{t1}C_{t0} = \gamma'_{t1}$
If $K > 0$, then: If $K < 0$, then:

$$\cos r_{t1} = \cos r_{t0} \cos r_p + \sin r_{t0} \sin r_p \cos\alpha$$

$$\cosh r_{t1} = \cosh r_{t0} \cosh r_p + \sinh r_{t0} \sinh r_p \cos\alpha \quad (9)$$

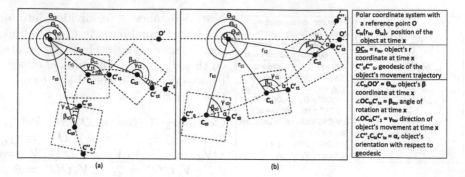

Fig. 10. Movement of the object along a hyperbolic in Spherical (a) and Hyperbolic (b) space. Orientation with respect to the geodesic is kept the same (angle α is constant) if the object is not rotating.

$$\cos \Delta\theta = \frac{\cos r_p - \cos r_{t0} \cos r_{t1}}{\sin r_{t0} \sin r_{t1}} \qquad \cos \Delta\theta = \frac{\cosh r_{t0} \cosh r_{t1} - \cosh r_p}{\sinh r_{t0} \sinh r_{t1}} \qquad (10)$$

$$\cos \gamma'_{t1} = \frac{\cos r_{t0} - \cos r_p \cos r_{t1}}{\sin r_p \sin r_{t1}} \qquad \cos \gamma'_{t1} = \frac{\cosh r_p \cosh r_{t1} - \cosh r_{t0}}{\sinh r_p \sinh r_{t1}} \qquad (11)$$

$\alpha = \beta_{t0} - \gamma_{t0}$. α *is the difference between rotation direction and the geodesic of movement* $(C"_0 C"_1)$, *it does not change if the object is not rotating. Hence,* $\beta_{t1} = \gamma_{t1} + \alpha$. *Because* γ'_{t1} *and* γ_{t1} *are supplementary angles,* $\gamma_{t1} = \Pi - \gamma'_{t1}$.

To find the θ *coordinate, either subtract or add* $\Delta\theta$ *to the* θ_c *depending on whether the angle* α *or its explementary angle is used for this calculation.*

3   Results

3.1   Implementation

Using the method described above and OpenGL, we created a software capable of calculating the objects and rendering the vector graphics in a non-Euclidean space with constant curvature in the range of $-1 \leq K \leq 1$. Figure 1 shows the time-lapses of multiple objects in spherical (a), planar (b) and hyperbolic (c) geometries. They show movement through different geodesics at $K = 1$, $K = 0$ and $K = -1$ respectively. Starting positions as well as shape definitions of each object are the same across all time-lapses (grid-lines have been created and rendered as separate objects). The software can calculate the object moving in arbitrary direction with arbitrary speed as well as starting from arbitrary position in the space.

Curvature of the world can be modified in real-time using keyboard inputs in a similar manner to controlling the object's acceleration and orientation. Another feature is the cut-off of the world at a distance of N pixels. This can be seen in the hyperbolic and planar time-lapse images. While these spaces should be infinite, we chose to limit them in order to keep objects within the boundaries of the screen (non-shaded area). We created a video [7] displaying the implementation.

3.2 Complexity Analysis

Positions of each vertex need to be calculated, requiring $O(v)$ time, where v is the number of vertices. Subsequently, intermediate points have to be computed, requiring $O(i)$ time to find all of the points on a single edge, where i is the level of tessellation. Complexity to render the world with s number of shapes is therefore $O(s*v*i)$. The best case would be equal to $O(n)$ complexity, if two of the terms are negligibly small. The worst case can be approximated to $O(n^3)$ if all terms were comparably large. Spatial complexity for shape rendering is only $O(v*i)$ as previous shape's data is rewritten to store the next shape's data. So either $O(n)$ in the best case or $O(n^2)$ in the worst case.

Only one movement calculation per object is required and the previous position record is overwritten, both spatial and time complexity is $O(n)$, where n is the number of objects in the world.

Trigonometric and hyperbolic functions in the calculations are slower to compute than simpler operations, hence additional cost (implementation dependent). For example the AGM iteration [8] method is faster than the previously common Taylor series method.

4 Discussion

Implementation does not affect performance up to a certain number of objects or tessellation amount. The focus was on implementing the method correctly and having it work continuously under any curvature in the range $-1 \le K \le 1$.

The next step in the project's development is improving the execution time using parallelised calculations. Subsequent calculation of the points creates a bottleneck, which can be solved by performing some calculations directly on the GPU. Other approaches are considered as well, including lookup tables to speed up trigonometric calculations, for example, Frank Rochet's implementation [9]; or finding intermediate points from a geodesic equation.

Potential applications for the software include education about non-Euclidean geometry (more intuitive than standard projections: Poincare disk and Upper Half-Plane models); cartography [10] (the engine could be modified to efficiently convert data into different projections); ecology [11] and climatology [12] (modelling dynamic systems); Astrophysics (modelling systems of cosmological objects and gravitational fields) and video games (game engine for a real-time continuous non Euclidean space, unlike HyperRogue [13], which uses step by step implementation).

References

1. Heath, T.L.: Euclid's Elements. Dover, New York (1956). (translated)
2. Todhunter, I.: Spherical Trigonometry For the use of colleges and schools. Project Gutenberg License (1886). (republished 12 November 2006)
3. Carslaw, H.S.: The Elements of Non-Euclidean Plane Geometry and Trigonometry. Longmans, Green and Co., London (1916)
4. Traver, T.: Trigonometry in the hyperbolic plane (2014). Accessed December 2017, Manuscript
5. Gellert, W., Gottwald, S., Hellwich, M., Kästner, H., Küstner, H.: The VNR Concise Encyclopedia of Mathematics, 2nd edn. Van Nostrand Reinhold, New York (1989). ch. 12
6. Gray, J.: Non-Euclidean geometry–a re-interpretation. Historia Mathematica **6**, 236–258 (1979)
7. Osudin, D., Child, C., He, Y.-H.: Rendering non-Euclidean space in realtime using spherical and hyperbolic trigonometry (2019). https://youtu.be/A1ZCFh5qfNg
8. Brent, R.P.: Multiple-precision zero-finding methods and the complexity of elementary function evaluation (2010). http://arxiv.org/abs/1004.3412v2. Accessed 26 Aug 2018
9. Rochet, F.: Fast trigonometry functions using lookup tables (2004). http://www.flipcode.com/archives/Fast_Trigonometry_Functions_Using_Lookup_Tables.shtml. Accessed 30 Aug 2018
10. Gartner, G., Huang, H.: Recent research developments in modern cartography in Europe, Issue 1: EuroCarto 2015 (2015)
11. Sutherland, C., Fuller, A.K., Royle, J.A.: Modelling non-Euclidean movement and landscape connectivity in highly structured ecological networks. Methods Ecol. Evol. **6**, 169–177 (2015)
12. Frei, C.: Interpolation of temperature in a mountainous region using nonlinear profiles and non-Euclidean distances. Int. J. Climatol. **34**, 1585–1605 (2013). https://doi.org/10.1002/joc.3786
13. Zeno Rogue Games: Hyperrogue (2019). http://roguetemple.com/z/hyper/

Improving Academic Homepage Identification from the Web Using Neural Networks

Jiapeng Zhao[1,2], Tingwen Liu[1,2(✉)], and Jinqiao Shi[1,3]

[1] Institute of Information Engineering, Chinese Academy of Sciences, Beijing, China
{zhaojiapeng,liutingwen,shijinqiao}@iie.ac.cn
[2] School of Cyber Security, University of Chinese Academy of Sciences,
Beijing, China
[3] Key Lab of Trustworthy Distributed Computing and Service (BUPT),
Ministry of Education, Beijing, China

Abstract. Identifying academic homepages is a fundamental work of many tasks, such as expert finding, researcher profile extraction and homonym researcher disambiguation. Many works have been proposed to obtain researcher homepages using search engines. These methods only extract features at the lexical-level from each single retrieval result, which is not enough to identify homepage from retrieval results with high similarity. To address this problem, we first make deep-insight improvements on three aspects. (1) Fine-gained features are designed to efficiently detect whether the researcher's name appears in retrieval results; (2) Establishing correlation of multiple retrieval results for the same researcher; (3) Obtaining semantic information involved in URL, title and snippet of each retrieval result by recurrent neural networks. Afterwards, we employ a joint neural network framework which is able to make comprehensive use of these informative information. In comparison with previous work, our approach gives a substantial increase of 10%–11% accuracy on a real-world dataset provided by AMiner. Experimental results demonstrate the effectiveness of our method.

Keywords: Academic homepage identification · Retrieval results ·
Semantic information representation · Joint neural network

1 Introduction

The academic homepage of a researcher usually contains lots of profile information and descriptions, such as employment status, research interests, contact information and publications. These are essential resources for the digital library access portals [8] to collect the researcher's metadata. In general, there are two frequently-used ways to collect academic homepages. One is to monitor the official websites of known research institutes and make a binary classification on each crawled webpage to determine whether it is an academic homepage. The other

© Springer Nature Switzerland AG 2019
J. M. F. Rodrigues et al. (Eds.): ICCS 2019, LNCS 11540, pp. 551–558, 2019.
https://doi.org/10.1007/978-3-030-22750-0_50

is to use researcher names and some additional information (such as researcher's affiliation and research interests) as search engine queries to retrieve related webpages from the web, and choose one retrieval result as the academic homepage. In this paper, we focus on the second way to collect academic homepages massively only with retrieval results of search engines including URLs, titles and snippets, because it can be deployed as an API service and respond rapidly with low resource cost. However, building an accurate module that automatically identifies researcher homepages from retrieval results is not easy, owing to the following technical challenges. First, one researcher may have multiple webpages associated with him/her. Second, search engines may split the query into multiple fragments to obtain more retrieve results, but introduce more noise at the same time. Then one's homepage may rank very low in retrieval results.

In this paper, we focus on the academic homepage identification. In comparison with previous work that only extracted some simple statistical features from each retrieval result, Our key contributions are as follows: (1) We proposed a novel solution to identify researcher homepages via search engines, and demonstrated the effectiveness of our approach on a publicly-available dataset. It not only obtains remarkable improvements with respect to the accuracy, but also performs more stable through computing precision and recall by selecting different proportions of test results. (2) We designed four types of novel features to help identify homepage from high similarity retrieval results. (3) We presented a joint neural network model, which allows different kinds of neural networks being trained synchronously, and thus makes full use of hand-crafted features information and sequence information.

2   Related Work

Relevant work on academic homepage identification using retrieval results of search engines first appeared in TREC's track [1]. It's an entity-oriented web search task. The task aims at finding homepages for four types of entities: organization, location, person, and product. To identify an academic homepage, many query-dependent features can be effectively utilized. Tang et al. [10] used researcher's name and affiliation name as queries of search engines and selected the best retrieval results as researcher homepage, but only hand-craft features from URL are used in their work.

In the perspective of feature extraction, there are three shortcomings in previous work. First, whether researcher's name appears in URL, title or snippet is a critical factor for homepage identification. It can't be judged by simply string matching. Second, relevance between retrieval results has not been explored, while previous work only considers a single retrieval result. Third, semantic information involved in URL, title or snippet have not been utilized.

3 Academic Homepage Identification

Given a researcher and his affiliation, the query statement will be "researcher name + affiliation", such as "Clark T. C. Nguyen, UC Berkeley Engineering", defined as Q. Each query Q has N retrieval results, named as QR pair.

3.1 Feature Analysis of Academic Homepages

Through the analysis of academic homepages, query statements and retrieval results, we find some typical features and summarize them into four types. These four types of features are described in Table 1.

Query-dependent (QD) Feature. The QR pair's order in the retrieval results is positively related to being a homepage. Thus we extract the QR pair's order as a feature. Other features are from previous work [3], which can be divided into two parts. First, number of researcher's name and researcher's affiliation fragments in URL, title or snippet. Second, keywords related to homepage.

Table 1. Summary of features designed for each QR pair

Type	Function	Feature description	Dim[a]
QD	Order()	QR pair's order in retrieval results of a researcher	1
	Length(U/T/S)	Length of URL/Title/Snippet/RN/RI	5
	Exist(U, special_char)[b]	Each special char(/,=?&-_%~) exist in URL or not	9
	Exist(U, num_frag)	Pure digital fragments in URL or not	1
	Score(U/T/S, RN_frag)[c]	Score of RN fragments in URL/Title/Snippet	1
	Score(U/T/S, RI_frag)	Score of RI fragments in URL/Title/Snippet	1
	Exist(U, domain)	Each domain name in URL or not	29
	Exist(T/S, keyword)	Each keyword exists in Title/Snippet or not	254
ES	Exist(U/T/S, RN)	Researcher's Name (RN) in URL/Title/Snippet	3
	Exist(U/T/S, RI)	Researcher's Institute (RI) in URL/Title/Snippet	3
LC	Rank(feature_value)	Each feature rank value of a researcher	304
	Norm(feature_value)	Each feature normalized value of a researcher	304
SE	Embed(U/T/S)[d]	Semantic embedding learned from URL/Title/Snippet	256

[a] Dim: Dimension of features.
[b] Exist(s, t): Whether string fragment t exists in s.
[c] Score(s, t): Number of fragments t exists in s divided by the number of fragments.
[d] Embed(seq): Learning semantic embedding from sequence seq.

The keywords contain homepage related topic words such as "students", "member", and "committee" extracted from the web-contents of homepages using the Latent Dirichlet Allocation (LDA) model and some high frequency words.

Entity-saliency (ES) Feature. Whether researcher's name and researcher's affiliation appear in the retrieval result is quite important for the identification of the homepage. However, researcher's name has various forms. ES feature aims to make entities more easy to be detected. Hence, some heuristic rules are set to splice name fragments, like from "Jiawei Han" to "JiaweiHan" or from "Charu C. Aggarwal" to "CharuAggarwal".

Local-contextual (LC) Feature. Existing works only consider a single QR pair, while ignoring the relationship between QR pairs of the same researcher. LC Feature able to establish relations between QR pairs of the same researcher, which contains: Rank: ranks of specific QR pair; Mean: the mean of feature values of top N QR pairs; Variance: the variance of feature values of top N QR pairs; Normalized feature: normalizing feature values by standard deviation.

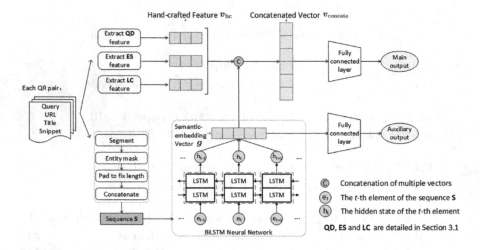

Fig. 1. Architecture of our joint model.

Semantic-embedding (SE) Feature. We first perform segmentation and entity masking operation to URL, title and snippet. Entity masking aims to encode name and institute in a unified way. Then researcher's name fragment, researcher's institute fragment and numerical fragment appear could be encoded to three fixed numbers. Sequences of URL, title and snippet are padded to fixed length and concatenated to a single sequence S. They were fed into a Bi-directional LSTMs (BiLSTM) [5] model to obtain the semantic information.

3.2 Joint Model in Our Work

In order to integrate hand-crafted features and semantic-embedding feature, we employ a joint neural network and adopt a joint training mode. The joint neural network model, as shown in Fig. 1, aims to improve the identification ability of the model by combining hand-crafted features and semantic information.

The joint neural network model contains two inputs: the first part takes hand-crafted features as input, denoted by v_{hc}; the second part takes sequence information as input, denoted by v_{seq}. Let v_{seq} input to one layer BiLSTM, the output of BiLSTM layer is a global sentence-level hidden vector g which detailed in Sect. 3.1. The joint neural network model contains two outputs, namely main output and auxiliary output. At the training stage, these two outputs share the same label. At the testing stage, they will output a probability value range from 0 to 1 represent the score of QR pair. The value of main output is regarded as the final score of QR pair.

The joint training model has two advantages. One is to make the BiLSTM and embedding layer being trained smoothly, even if the joint loss value is very high. The other is to make use of semantic information involved in sequences. For a single QR pair i, the loss value can be calculated by formula Eq. (1) and the batch random gradient descent as Eq. (2), m is the batch size. We set a tunable parameter λ to control the joint loss function Eq. (3). The goal of parameter estimation is to find the optimal θ^* to minimize the joint loss function $\mathcal{L}_{joint}$, y represent the label of the data, $h_\theta(x)$ represent a series of linear or nonlinear transformations.

$$\mathcal{L}(h_\theta(x_i), y_i) = -y_i log(h_\theta(x)) - (1 - y_i)log(1 - h_\theta(x)) \tag{1}$$

$$\mathcal{L}(x, y) = \sum_{i=1}^{m} \mathcal{L}(h_\theta(x_i), y_i) \tag{2}$$

$$\mathcal{L}_{joint}(x, y) = \lambda \mathcal{L}_{seq}(x, y) + (1 - \lambda)\mathcal{L}_{hc}(x, y) \tag{3}$$

4 Experiments

4.1 Experimental Setup

We conduct our experiments on a real-world dataset provided by AMiner[1]. It contains 20,445 researchers and 203,019 corresponding retrieval results where each researcher has 8 to 11 retrieval results. The dataset falls into three parts including a training set (6000 researchers, 59675 QR pairs, 5677 homepages), a validation set (2435 researchers, 24187 QR pairs, 2267 homepages) and a test set (12010 researchers, 119157 QR pairs, 11364 homepages). In order to present impact of our approaches, we set the following 6 groups comparison experiments.

(1) Baseline (BL_SVM/BL_RSVM). This experiment uses features from previous work [3,10], as described in the query-dependent feature part of Sect. 3.1.

---

[1] https://biendata.com/competition/scholar/.

The SVM [2] and RankSVM [6] models are utilized. (2) Hand-crafted features with SVM (HF_SVM). The SVM model which uses the hand-crafted features proposed in this paper as input. (3) Hand-crafted features with deep neural network (HF_DNN). A deep neural network model which only uses the hand-crafted features to classify the homepages. It could be seen as the baseline of the Joint Neural Network. (4) Semantic-embedding feature with biLSTM (SF_BiLSTM). It's an auxiliary classifier trained to identify the homepages. (5) Combined features with joint neural network (CF_JNN). This experiment combines hand-crafted features and semantic-embedding feature together and adopts joint training mode, detailed in Sect. 3.2.

To better evaluate these approaches, we set three types of evaluation criterion. Accuracy (only if the identified page equals to the labeled homepage, the page is considered to be correct.) precision recall curves [4] and the mean reciprocal rank [9]. For the network configuration, the parameters of the dropout probability are tuned to 0.25, the layer of deep neural network is set to 3, the semantic embedding vector size is set to 256, the batch size is 300 and the optimizer is Adam [7] with a learning rate of 0.001. We implement neural models based on Keras[2] and directly use its default parameter initialization strategy. Since academy homepages are only one tenth of retrieval results, the weight proportion of the positive and negative data is set to 9:1 heuristically to overcome the problem of data imbalance.

Table 2. Accuracies (%) of different approaches.

Method	BL_SVM	BL_RSVM	HF_SVM	HF_DNN	SF_BiLSTM	CF_JNN
Accuracy	59.03	60.17	61.91	67.02	62.96	**71.04**

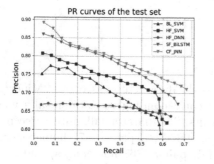
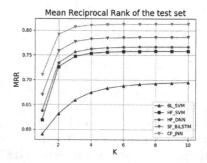

Fig. 2. P-R curves and MRR values of different approaches on test set.

---

[2] https://keras.io/.

4.2 Experimental Results and Discussion

We first report the accuracy of 6 groups of different experiments in Table 2. Our CF_JNN approach achieves best results. The accuracy is 71.12% in the validation set and 71.04% in the test set. In comparison with baseline approach BL_SVM and BL_RSVM, which accuracy are 60.31% and 60.17%, our approach performs 10–11% better than the baseline. There are two main reasons: one is that our more effective features, which could be proved from the comparison between BL_SVM and HF_SVM; the other is that our joint model and it could be seen from the comparison between HF_DNN and CF_JNN.

From the accuracy of HF_DNN, SF_BiLSTM and CF_JNN, we observe that the joint neural network significantly outperform both SF_BiLSTM and HF_DNN, which means the joint model is effective. According to PR curves in Fig. 2, our improved features and joint neural network are more stable than previous work. From the PR curves of BL_SVM and HF_SVM, although they have similar accuracy, our features have higher F1 values and perform more stable in most cases. The result of the RankSVM model unable to draw the P-R curves, since it's a comparison between retrieval results of the same author instead of giving a global score. These demonstrate the effectiveness of our approach.

5   Conclusion and Further Work

In this paper, we study the problem of academic homepage identification using retrieval results from the search engine. To fully leverage both structural and content information in retrieval results, we propose a joint neural network model to identify academic homepage using both carefully designed features and semantic embeddings. We conduct experiments on a real-world dataset and the experimental results demonstrate the effectiveness of our approach. Our future directions is to investigate the performance of our approach for identifying the related webpages of other entities, such as institute, medicine and weapon.

Acknowledgments. This work was supported in part by the National Key Research and Development Program of China under Grant No. 2016YFB0801003.

References

1. Balog, K., Serdyukov, P., De Vries, A.P.: Overview of the TREC 2010 entity track. Norwegian University of Science and Technology Trondheim, Technical report (2010)
2. Chang, C.C., Lin, C.J.: LIBSVM: a library for support vector machines. ACM Trans. Intell. Syst. Technol. (TIST) **2**(3), 27 (2011)
3. Das, S., Mitra, P., Giles, C.: Learning to rank homepages for researcher-name queries. In: The International Workshop on Entity-Oriented Search, SIGIR, pp. 53–58. Citeseer (2011)
4. Davis, J., Goadrich, M.: The relationship between precision-recall and ROC curves. In: Proceedings of the 23rd International Conference on Machine Learning, pp. 233–240. ACM (2006)

5. Graves, A., Jaitly, N., Mohamed, A.r.: Hybrid speech recognition with deep bidirectional LSTM. In: 2013 IEEE Workshop on Automatic Speech Recognition and Understanding (ASRU), pp. 273–278. IEEE (2013)
6. Joachims, T.: Optimizing search engines using clickthrough data. In: ACM Conference on Knowledge Discovery and Data Mining (2002)
7. Kingma, D.P., Ba, J.: Adam: a method for stochastic optimization. arXiv preprint arXiv:1412.6980 (2014)
8. Li, H., et al.: CiteSeer χ: a scalable autonomous scientific digital library. In: Proceedings of the 1st International Conference on Scalable Information Systems, p. 18. ACM (2006)
9. Radev, D.R., Qi, H., Wu, H., Fan, W.: Evaluating web-based question answering systems. In: LREC (2002)
10. Tang, J., Zhang, J., Yao, L., Li, J., Zhang, L., Su, Z.: ArnetMiner: extraction and mining of academic social networks. In: Proceedings of the 14th ACM SIGKDD International Conference on Knowledge Discovery and Data Mining, pp. 990–998. ACM (2008)

Combining Fuzzy Logic and CEP Technology to Improve Air Quality in Cities

Hermenegilda Macià[1]([✉]) [ID], Gregorio Díaz[1] [ID], Juan Boubeta-Puig[2] [ID],
Edelmira Valero[3] [ID], and Valentín Valero[1] [ID]

[1] School of Computer Science, University of Castilla-La Mancha, Albacete, Spain
{hermenegilda.macia,gregorio.diaz,valentin.valero}@uclm.es
[2] Department of Computer Science and Engineering, University of Cádiz,
Puerto Real, Cádiz, Spain
juan.boubeta@uca.es
[3] Higher Technical School of Industrial Engineering,
University of Castilla-La Mancha, Albacete, Spain
edelmira.valero@uclm.es

Abstract. Road traffic has become a main source of air pollution in urban areas. For this reason, governments are applying traffic regulations trying to fulfill the recommendations of Air Quality (AQ) standards in order to reduce the pollution level. In this paper, we present a novel proposal to improve AQ in cities by combining fuzzy logic and Complex Event Processing (CEP) technology. In particular, we propose a flexible fuzzy inference system to improve the decision-making process by recommending the actions to be carried out on each pollution scenario. This fuzzy inference system is fed with pollution data obtained by a CEP engine and weather forecast from domain experts.

Keywords: Complex Event Processing · Fuzzy logic · Air Quality

1 Motivation

Nowadays, Air Quality (AQ) in cities is becoming a great environmental problem, posing multiple challenges in terms of management and mitigation of harmful pollutants. A poor air quality can cause health problems and reduce the quality of life, with children, elderly people and those with existing heart and lung conditions begin the most affected. The World Health Organization (WHO) [13] has estimated that atmospheric pollution caused around 4.2 million premature

This study was funded in part by the Spanish Ministry of Science and Innovation and the European Union FEDER Funds under Grants TIN2015-65845-C3-2-R, TIN2015-65845-C3-3-R and TIN2016-81978-REDT, and also by the JCCM regional projects SBPLY/17/180501/000276/1 and SBPLY/17/180501/000276/2, both of them co-financed by the European Union FEDER Funds.

J. M. F. Rodrigues et al. (Eds.): ICCS 2019, LNCS 11540, pp. 559–565, 2019.
https://doi.org/10.1007/978-3-030-22750-0_51

deaths worldwide in 2016. As an illustration of the public awareness importance of this problem, different organizations worldwide report daily air quality on the basis of major pollutants such as EPA (Environmental Protection Agency) [2] in the USA and EEA (European Environment Agency) [5] in the EU.

Nevertheless, data reported by air quality monitoring stations indicate that very often air pollutants levels exceed the limits and guidelines reported by these international agencies, which have harmful effects on human health and/or the environment as a whole. Clearing up the air in big cities is not an easy challenge because it necessarily involves taking concrete actions, even if they are not popular. Therefore governments and local authorities face to a complex make-decision process.

In an increasingly interconnected world, the role of computational science with novel models, algorithms and tools driving efficient application in environmental systems can be crucial to find proposals to reduce air pollution in large urban centers implementing new policies. In the present paper, our aim is to propose the combined use of Complex Event Processing (CEP) [6] –a technology that allows us to process and correlate large volumes of data by matching event patterns- and Fuzzy Inference System (FIS) –a mechanism based on fuzzy logic [14], which provides a formal methodology for representing, manipulating, and implementing a heuristic knowledge about how to control a system allowing to model complex systems in a more intuitive manner. In this way, our proposal presents an unprecedented work combining fuzzy logic and CEP to improve the decision-making process by recommending the actions to be carried out on each pollution scenario.

In particular, this approach allows us to make decisions about traffic regulations on a daily basis, considering the levels of environmental pollution according to the air quality levels accepted by the international recommendations. For this purpose, both the automatically monitored pollution levels for the previous days and the weather forecast for the following day provided by the domain expert are used to make a decision about the specific type of traffic regulation to be applied, if necessary.

The rest of the paper is organized as follows. Section 2 describes the proposal based on CEP and fuzzy logic to improve AQ using traffic regulation, then Sect. 3 tested it on a real AQ dataset. Finally, Sect. 4 concludes the paper and provides possible lines of future work.

2   An Approach to Improve AQ in Cities

We focus on the NO_2 pollutant, since this is possibly the most significant pollutant in the case of large cities, due to the high values it reaches with respect to the AQ standards under the Air Quality Directive from the European Environment Agency (EEA), and WHO air quality guidelines. This pollutant is mainly produced by combustion engines, so traffic restrictions are applied during episodes of high pollution, for instance, banning vehicles from driving [8]. However, this methodology could be applied over any other pollutants or a combination of them. Specifically, we focus on the limit value of $40\,\mu g/m^3$ annual average, and our objective is to address long-term improvements to AQ in order not to exceed this limit value. Other policies and proposals for AQ improvement could also be applied in combination with our proposal, such as investing in public transport, cycling lanes, etc.

2.1 Methodology

Our main objective is to outline an easy and readable methodology combining CEP technology and fuzzy logic in order to establish an action plan to reduce NO_2 emissions of road traffic in each possible scenario.

A Fuzzy Inference System (FIS) is a system that uses fuzzy set theory to map inputs (features in the case of fuzzy classification) to outputs (classes in the case of fuzzy classification). This mapping provides a basis from which decisions can be made. One of the most commonly FIS used is Mamdani's fuzzy inference method [9], proposed by Ebrahim Mamdani in 1975. This mechanism involves the following steps: (1) Definition of the fuzzy sets and fuzzification of each input, (2) Definition of fuzzy if-then rules, (3) Aggregation of the consequent by applying the rules, and (4) Defuzzification of the output. In this way, we consider as an input of our FIS three parameters: (α, β, γ), where α is the last day average of NO_2; β is a weighted average of NO_2 for the two previous days, and γ is the weather forecast for tomorrow. In this way, α and β are data obtained for the monitored places by using a CEP engine and γ is a parameter provided from a domain expert. The output of the FIS is the μ parameter, which is a natural value ranging from 0 to 10 that determines the action plan. We will use the fuzzy logic designer of the fuzzy logic MATLAB toolbox [10] for our proposed FIS.

Our methodology to improve the AQ in cities consists of several steps, which are introduced as follows:

- **Step 1.** The implementation of the event patterns to compute the α and β FIS inputs. These patterns are encoded in Esper EPL [3,4], by using the *AirMeasurement* schema defined below, which has three attributes: the time at which the measure was taken, the station identifier, and the NO_2 value. For each station, the *NO2_Daily_Avg* pattern computes its daily average of NO_2 (i.e. α), while the *NO2_Historical_Avg* pattern calculates the (0.3 and 0.7) weight averages of NO_2 (i.e. β) for the day before yesterday and yesterday, respectively. Note that other periods of time and weights could also be considered.

```
create schema AirMeasurement (timestamp long, stationId string,
    no2 float);

/* NO2 PATTERNS */

@Name("NO2_Daily_Avg")
insert into NO2_Daily_Avg
select current_timestamp() as timestamp,
    a1.stationId as stationId,
    avg(a1.no2) as value
```

```
from pattern [(every a1 = AirMeasurement)].win:time_batch(1 days)
group by a1.stationId;

@Name("NO2_Historical_Avg")
insert into NO2_Historical_Avg
select current_timestamp() as timestamp,
  a1.stationId as stationId,
  ((a1.value * 0.3) + (a2.value * 0.7)) as value
from pattern [((every a1 = NO2_Daily_Avg) ->
  a2 = NO2_Daily_Avg(a2.stationId = a1.stationId) ->
  a3 = NO2_Daily_Avg(a3.stationId = a2.stationId))].win:time(3 days);
```

– **Step 2.** Define the corresponding classification for (α, β, γ) inputs and the μ output, establishing different rating levels for these inputs and outputs. To not exceed the limit value of $40\,\mu g/m^3$ annual average of NO_2, we propose the classification for α and β shown in Table 1, but other possibilities can be studied.

Table 1. NO_2 levels on a 24 h average basis.

Rating	Label	$\mu g/m^3$
Level A	Very Good	0–40
Level B	Good	>40–50
Level C	Moderate	>50–60
Level D	Bad	>60–70
Level E	Very Bad	>70

The γ parameter is provided by a domain expert according to Pasquill Stability Classes [12], with values ranging from 0 to 8 (Table 2). Nevertheless, and for greater simplicity only 3 levels will be considered (but any other number of levels would be possible). Then, ratings A and B are considered as *Favourable*, C, D and E as *Neutral*, and F and G *Unfavourable*. Furthermore, five action plans are proposed depending on the μ value (see Table 3).

Table 2. Pasquill stability classes

Rating	Pasquill stability classes	Scoring
A	Extremely unstable conditions	<2
B	Moderately unstable conditions	2–3
C	Slightly unstable conditions	3–4
D	Neutral conditions	4–5
E	Slightly stable conditions	5–6
F	Moderately stable conditions	6–7
G	Extremely stable	> 7

– **Step 3.** Definition of the fuzzy sets for α, β, γ inputs and the μ output. We consider the Gaussian combination membership function (*gauss2mf*) for α and β (Fig. 1) and the Triangular-shaped membership function (*trimf*) for γ and μ (the details in each case can be found in [7]). Note that it is possible to change membership functions (parameters and type) according to domain expert's recommendations based on previous experiences.

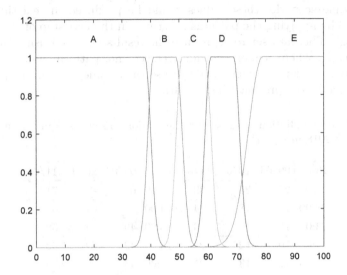

Fig. 1. Fuzzy sets for α and β

Table 3. Action plans

Label	Action plan	Scoring
AP0	No Measures	0–2
AP1	Informative actions/Warning	2–4
AP2	To Reduce maximum velocity	4–6
AP3	To Reduce the number of vehicles	6–8
AP4	Only are allowed some special vehicles	8–10

– **Step 4.** Definition of the rules for the fuzzy inference process. The whole proposal [7] includes 17 fuzzy if-then- rules (all of them with the same priority). Some of these rules are the following:

```
1. If (alpha is A) then (mu is AP0)
7. If (alpha is C) and (gamma is Favourable) then (mu is AP1)
14. If (alpha is D) and (beta is D) and (gamma is Unfavourable) then
    (mu is AP4)
```

3   Discussion

We have tested our methodology on an AQ dataset of Madrid city, available at [8]. As an example, we have considered the data set obtained from the *Aguirre's school* station on April 09th, 10th and 11th, 2018 (shown in Table 4), obtaining the values $\alpha = 61.75$ and $\beta = 50.12$. Furthermore, the weather forecast was obtained from the weather archive available at [11], obtaining the value $\gamma = 3$. In a real time scenario, these values would be produced in real time by the CEP engine by applying the patterns described in the first step of the proposed methodology. The data set to reproduce the results obtained using this data is available at [7], where readers can find the schema, patterns and data used in this scenario. Readers are committed to use the on-line software https://www. esperonline.net/ to reproduce this scenario.

Table 4. Madrid pollution data set for the station "Escuelas Aguirre on April 9th, 10th and 11th 2018 in $\mu g/m^3$

Hour	09/04	10/04	11/04	Hour	09/04	10/04	11/04
01:00	25	25	39	13:00	61	-	71
02:00	16	18	34	14:00	-	61	65
03:00	11	14	23	15:00	49	64	58
04:00	10	12	22	16:00	42	69	62
05:00	12	13	21	17:00	46	66	56
06:00	17	15	29	18:00	52	73	66
07:00	35	27	58	19:00	52	72	70
08:00	59	54	98	20:00	52	73	85
09:00	77	66	115	21:00	55	67	76
10:00	80	77	96	22:00	47	100	67
11:00	72	73	86	23:00	42	82	59
12:00	62	67	72	24:00	29	54	52

Using the FIS defined in the previous section we obtain that $\mu = 5.53$ and then both action plans AP2 and AP3 could be considered for application, but the membership grade for $\mu = 5.53$ is higher for the fuzzy set corresponding to the *AP2* (0.735) action plan than that of the *AP3* plan (0.265), so the *AP2* action plan will be finally recommended.

4   Conclusions and Future Work

In this paper, a methodology combining CEP and fuzzy logic has been proposed for improving the AQ in large cities. Thanks to the use of CEP technology, some of the input parameters required by the proposed FIS are automatically calculated. Additionally, fuzzy logic improves the decision-making process by recommending the actions to be carried out on each scenario.

One of the main advantages of our proposal is its flexibility and feasibility to be adapted to new scenario requirements. For instance, new input variables and fuzzy rules can easily be included in the FIS model as well as new modifications of the action plan in order to reduce more the air pollution.

As future work, we plan to extend this work by including other relevant pollutants (and the interactions between them) specified in the AQ international standards, for instance PM and/or O_3, as well as considering other objectives for NO_2 pollution reduction such as not exceeding the value of $200\,\mu g/m^3$ of the NO_2 hourly average more than 18 times in a year. Moreover, more experiments would be included together with a deep study of the optimality of the proposal.

Furthermore, this approach could be integrated with our ITS proposed in [1], by including an appropriate Fuzzy Inference System following this methodology. This would allow us to improve the traffic regulation by means of a more realistic approach.

References

1. Díaz, G., Macià, H., Valero, V., Boubeta-Puig, J., Cuartero, F.: An intelligent transportation system to control air pollution and road traffic in cities integrating CEP and colored petri nets. Neural Comput. Appl. (2018). https://doi.org/10. 1007/s00521-018-3850-1
2. Environmental Protection Agency: Air Pollution. https://airnow.gov/. Accessed 23 Jan 2019
3. EsperTech: Esper EPL Online. http://esper-epl-tryout.appspot.com/epltryout/ index.html. Accessed 10 Jan 2019
4. EsperTech: Esper reference documentation. http://www.espertech.com/esper/ esper-documentation/. Accessed 10 Jan 2019
5. European Environmental Agency: Air Pollution. https://www.eea.europa.eu/ themes/air/air-quality-index. Accessed 23 Jan 2019
6. Luckham, D.: Event Processing for Business: Organizing the Real-Time Enterprise. Wiley, Hoboken (2012)
7. Macià, H., Díaz, G., Boubeta-Puig, J., Valero, E., Valero, V.: A dataset for the paper "Combining fuzzy logic and CEP technology to improve air quality in cities". https://dx.doi.org/10.17632/4njddyw68t.1. Accessed 3 Apr 2019
8. Madrid City Council: Portal web de Calidad del Aire del Ayuntamiento de Madrid. http://www.mambiente.munimadrid.es/sica/scripts/. Accessed 23 Jan 2019
9. Mamdani, E., Assilian, S.: An experiment in linguistic synthesis with a fuzzy logic controller. Int. J. Man-Mach. Stud. **7**(1), 1–13 (1975). https://doi.org/10.1016/ S0020-7373(75)80002-2
10. MathWorks: MATLAB. http://www.mathworks.com. Accessed 10 Jan 2019
11. meteoblue: Archivo meteorológico Madrid. https://www.meteoblue.com/es/ tiempo/pronostico/archive/madrid_espa%C3%B1a_3117735. Accessed 23 Jan 2019
12. Pasquill, F., Michael, P.: Atmospheric Diffusion, 2nd edition. Phys. Today **30**, 55 (1977). https://doi.org/10.1063/1.3037599
13. World Health Organization: Air Pollution. https://www.who.int/airpollution/ ambient/health-impacts/en/. Accessed 23 Jan 2019
14. Zadeh, L.: Fuzzy sets. Inf. Control **8**(3), 338–353 (1965). https://doi.org/10.1016/ S0019-9958(65)90241-X

Parallel Parametric Linear Programming Solving, and Application to Polyhedral Computations

Camille Coti[1], David Monniaux[2(✉)], and Hang Yu[2]

[1] LIPN, CNRS UMR 7030, Université Paris 13, Sorbonne Paris Cité,
99, avenue Jean-Baptiste Clément, 93430 Villetaneuse, France
Camille.Coti@lipn.univ-paris13.fr
[2] Univ. Grenoble Alpes, CNRS, Grenoble INP, 38000 Grenoble, France
{david.monniaux,hang.yu}@univ-grenoble-alpes.fr

Abstract. Parametric linear programming is central in polyhedral computations and in certain control applications. We propose a task-based scheme for parallelizing it, with quasi-linear speedup over large problems.

1   Introduction

A *convex polyhedron*, or *polyhedron* for short here, in dimension n is the solution set over $\mathbb{Q}^n$ (or, equivalently, $\mathbb{R}^n$) of a system of inequalities (with integer or rational coefficients). Polyhedra in higher dimension are typically used to enclose the reachable states of systems whose state can be expressed, at least partially, as a vector of reals or rationals; e.g. hybrid systems or software [3].

The conventional approaches for polyhedral computations are the *dual description* (using both vertices and faces) and *Fourier-Motzkin elimination*. They both suffer from high complexity on relevant cases. We instead express image, projection, convex hull etc. as solutions to *parametric linear programmings*, where parameters occur linearly within the objective. A solution to such a program is a quasi-partition of the space of parameters into polyhedra, with one optimum associated to each polyhedron. The issue is how to compute this solution efficiently. In this article, we describe how we parallelized our algorithm.

2   Sequential Algorithms

Here we are leaving out how polyhedral computations such as projection and convex hull can be reduced to parametric linear programming—this is covered in the literature [4,6]—and focus on solving the parametric linear programs.

Grenoble INP—Institute of Engineering Univ. Grenoble Alpes.

© Springer Nature Switzerland AG 2019
J. M. F. Rodrigues et al. (Eds.): ICCS 2019, LNCS 11540, pp. 566–572, 2019.
https://doi.org/10.1007/978-3-030-22750-0_52

2.1 Non-parametric Linear Programming (LP)

A linear program with n unknowns is defined by a system of equations $AX = B$, where A is an $m \times n$ matrix; a solution is a vector X such that $X \geq 0$ on all coordinates and $AX = B$. The program is said to be *feasible* if it has at least one solution, *infeasible* otherwise. In a non-parametric linear program one considers an objective C: one wants the solution that maximizes $C^T X$. The program is deemed *unbounded* if it is feasible yet it has no such optimal solution.

Example 1. Consider the polygon P defined by $x_1 \geq 0$, $x_2 \geq 0$, $3x_1 - x_2 \leq 6$, $-x_1 + 3x_2 \leq 6$. Define $x_3 = 6 - 3x_1 + x_2$ and $x_4 = 6 + x_1 - 3x_2$. Let $X = (x_1, x_2, x_3, x_4)$, and then P is the projection onto the first two coordinates of the solution set of $AX = B \wedge X \geq 0$ where $A = \begin{bmatrix} 1 & -3 & 0 & -1 \\ -3 & 1 & -1 & 0 \end{bmatrix}$ and $B = \begin{bmatrix} 6 \\ 6 \end{bmatrix}$.

An LP solver takes as input (A, B, C) and outputs "infeasible", "unbounded" or an optimal solution. Most solvers work with floating-point numbers and their final answer may be incorrect: they may answer "infeasible" whereas the problem is feasible, or give "optimal solutions" that are not solutions, or not optimal.

In addition to a solution X^*, solvers also provide the associated *basis*: X^* is defined by setting $n - m$ of its coordinates to 0 (known as *nonbasic variables*) and solving for the other coordinates (known as *basic variables*) using $AX^* = B$, and the solver provides the partition into basic and nonbasic variables it used. If a floating-point solver is used, it is possible to reconstruct an exact rational point X^* using that information and a library for solving linear systems in rational arithmetic. One then checks whether it is truly a solution by checking $X^* \geq 0$.

The optimal basis also contains a proof of optimality of the solution. We compute the objective function $C^T X$ as $\sum_{i \in N} \alpha_i X_i + c$ where N is the set of indices of the nonbasic variables and c is a constant, and conclude that the solution obtained by setting these nonbasic variables to 0 is maximal because all the α_i are nonpositive. If X^* is not a solution of the problem ($X^* \geq 0$ fails) or is not optimal, then we fall back to an exact implementation of the simplex algorithm.

Example 1 (continued). Assume the objective is $C = \begin{bmatrix} 1 & 1 & 0 & 0 \end{bmatrix}$, that is, $C^T X = x_1 + x_2$. From $AX = B$ we deduce $x_1 = 3 - \frac{3}{8}x_3 - \frac{1}{8}x_4$ and $x_2 = 3 - \frac{1}{8}x_3 - \frac{3}{8}x_4$. Thus $x_1 + x_2 = 6 - \frac{1}{2}x_3 - \frac{1}{2}x_4$.

Assume x_3 and x_4 are nonbasic variables and thus set to 0, then $X^* = (x_1, x_2, x_3, x_4) = (3, 3, 0, 0)$. It is impossible to improve upon this solution: as $X \geq 0$, changing the values of x_3 and x_4 can only decrease the objective $o = 6 - \frac{1}{2}x_3 - \frac{1}{2}x_4$. This expression of o from the nonbasic variables can be obtained by linear algebra once the partition into basic and nonbasic variables is known.

While the optimal value $C^T X^*$, if it exists, is unique for a given (A, B, C), there may exist several X^* for it, a situation known as *dual degeneracy*. The same X^* may be described by different bases, a situation known as *primal degeneracy*, happening when more than $n - m$ coordinates of X^* are zero, and thus some basic variables could be used as nonbasic and the converse.

2.2   Parametric Linear Programming (PLP)

For a *parametric* linear program, we replace the constant vector C by $C_0 + \sum_{i=1}^{k} \mu_i C_i$ where the μ_i are parameters.[1] When the μ_i change, the optimum X^* changes. Assume temporarily that there is no degeneracy. Then, for given values of the μ_i, the problem is either unbounded, or there is one single optimal solution X^*. It can be shown that the region of the $(\mu_1, \ldots, \mu_k)$ associated to a given optimum X^* is a convex polyhedron (if $C_0 = 0$, a convex polyhedral cone), and that these regions form a quasi partition of the space of parameters (two regions may overlap at their boundary, but not in their interior) [4–6]. The output of the parametric linear programming solver is this quasi-partition, and the associated optima—in our applications, the problem is always bounded in the optimization directions, so we do not deal with the unbounded case.

Let us see in more details about how to compute these regions. We wish to attach to each basis (at least, each basis that is optimal for at least one vector of parameters) the region of parameters for which it is optimal.

Example 1 (continued). Instead of $C = \begin{bmatrix} 1\,1\,0\,0 \end{bmatrix}$ we consider $C = \begin{bmatrix} \mu_1\ \mu_2\ 0\ 0 \end{bmatrix}$. Let us now express $o = C^T X$ as a function of the nonbasic variables x_3 and x_4:

$$o = (3\mu_1 + 3\mu_2) + \left(-\tfrac{3}{8}\mu_1 - \tfrac{1}{8}\mu_2\right) x_3 + \left(-\tfrac{1}{8}\mu_1 - \tfrac{3}{8}\mu_2\right) x_4 \qquad (1)$$

The coefficients of x_3 and x_4 are nonpositive if and only if $3\mu_1 + \mu_2 \geq 0$ and $\mu_1 + 3\mu_2 \geq 0$, which define the cone of optimality associated to that basis and to the optimum $X^* = (3, 3, 0, 0)$.

The description of the optimality polyhedron by the constraints obtained from the sign conditions in the objective function may be redundant: containing constraints that can be removed without changing the polyhedron. Our procedure [7] for removing redundant constraints from the description of a region R_1 also provides a set of vectors outside of R_1, a feature that will be useful.

Assume now we have solved the optimization problem for a vector of parameters D_1, and obtained a region R_1 in the parameters (of course, $D_1 \in R_1$). We store the set of vectors outside of R_1 provided by the redundancy elimination procedure into a "working set" W to be processed, choose D_2 in it. We compute the region R_2 associated to D_2. Assume that R_2 and R_1 are adjacent, meaning that they have a common boundary. We get vectors outside of R_2 and add them to W. We pick D_3 in W, check that it is not covered by R_1 or R_2, and, if it is not, compute R_3, etc. The algorithm terminates when W becomes empty, meaning the $R_1, \ldots$ produced form the sought quasi-partition.

This simplistic algorithm can fail to work because it assumes that it is discovering the adjacency relation of the graph. The problem is that, if we move from a region R_i to a vector $D_j \notin R_i$, it is not certain that the region R_j generated from D_j is adjacent to R_i—we could miss some intermediate region. We modify our traversal algorithm as follows. The working set contains pairs (R, D') where

---

[1] There exists another flavor of PLP with parameters in the right-hand sides of the constraints.

Algorithm 1. Concurrent push on the shared region structure.

procedure PUSH_REGION(R)
  atomic ($i \leftarrow n_{\text{fill}}$; $n_{\text{fill}} \leftarrow n_{\text{fill}} + 1$)
  $regions[i] \leftarrow R$
  while $n_{\text{ready}} < i$ **do**

    possibly use a condition variable
    instead of spinning
  end while   ▷ $n_{\text{ready}} = i$
  atomic $n_{\text{ready}} \leftarrow i + 1$
end procedure

Algorithm 2. Task for parallel linear programming solver.

push_tasks adds new tasks to be processed (different under TBB and OpenMP).
test_and_insert(T, x) checks whether x already belongs to the hash table T, in which case it returns true; otherwise it adds it and returns false. This operation is atomic.

procedure PROCESS_TASK((R_{from}, D))
  $R_{\text{cov}} \leftarrow is\_covered(D, regions)$
  if $R_{\text{cov}} ==$ none **then**
    $basis \leftarrow float\_lp(A, B, C(D))$
    if $\neg test\_and\_insert(bases, basis)$
  then
      $X^* \leftarrow exact\_point(basis)$
      $o \leftarrow exact\_objective(basis)$
      if $\neg(X^* \geq 0 \wedge o \leq 0)$ **then**
      $(basis, X^*) \leftarrow exact\_lp(A, B, C(D))$
      end if
      $S \leftarrow sign\_conditions(basis)$
      $R \leftarrow eliminate\_redundancy(S)$
      for each constraint i in R **do**
        $D_{\text{next}} \leftarrow compute\_next(R, i)$
        $push\_tasks(D_{\text{next}})$
      end for
      $push\_region(R, X^*)$
      $R_{\text{cov}} \leftarrow R$

  end if
  end if
  if $\neg are\_adjacent(R_{\text{from}}, R_{\text{cov}})$ **then**
    $D' \leftarrow midpoint(R_{\text{from}}, R_{\text{cov}}, D)$
    $W \leftarrow W \cup \{(R_{\text{from}}, D')\}$
  end if
end procedure

procedure $is\_covered(D, regions)$
  for $i \in 0 \ldots n_{\text{ready}} - 1$ **do** ▷ n_{ready} to be read at every loop iteration
    $(R, X^*) \leftarrow regions[i]$
    if D covered by R **then**
      return(R)
    end if
  end for
  return(none)
end procedure

R is a region and $D' \notin R$ a vector (there is a special value none for R). The region R' corresponding to D' is computed. If R and R' are not adjacent, then a vector D'' in between R and R' is computed, and (R, D'') added to the working set. This ensures that we obtain a quasi-partition in the end. Additionally, we obtain a spanning tree of the region graph, with edges from R to R'.

The last difficulty is degeneracy. We have so far assumed that each optimization direction corresponds to exactly one basis. In general this is not the case, and the interiors of the optimality regions may overlap. This hinders performance. The final result is no longer a quasi-partition, but instead just a covering of the parameter space—enough for projection, convex hull etc. being correct.

3   Parallel Parametric Linear Programming

Our algorithms are designed in a task-based execution model. The sequential algorithm executes tasks taken from a working set, which can themselves spawn new tasks. In addition, it maintains the set *regions* of regions already seen, used:

(i) for checking if a vector D belongs to a region already covered (*is_covered*); (ii) for checking adjacency of regions; (iii) for adding new regions found. Therefore, in a parallel task model, this algorithm is straightforwardly parallel. The regions are inserted into a concurrent array. We investigated two task scheduling strategies. A *static* approach starts all the available tasks, waits for them to complete and collects all the new tasks (R, D) into the working set, until no new task is created and the working set is empty. A *dynamic* approach allows adding new tasks to the working set dynamically and runs the tasks until that set is empty.

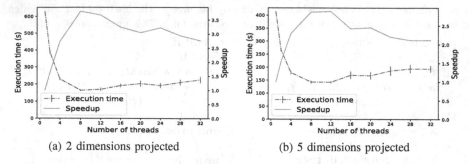

(a) 2 dimensions projected   (b) 5 dimensions projected

Fig. 1. 9 constraints, no redundant ones, 16 variables, 2–36 regions, OpenMP.

The number of tasks running to completion (not aborted early due to a test) is the same as the number of generated regions. The *is_covered*(D, *regions*) loop can be easily parallelized as well. We opted against it as it would introduce a difficult-to-tune second level of parallelism.

We implemented these algorithms using Intel's Thread Building Blocks (TBB [8]) and OpenMP tasks [1], both providing a task-based parallelism model with different features.

The dynamic task queue can be implemented using TBB's `tbb::parallel_do`, which dynamically schedules tasks from the working set on a number of threads. The static scheduling approach can simply be implemented by a task synchronization barrier (such as OpenMP's barrier).

That first implementation of the dynamic task scheduling approach was slow. The working set often contained tasks such that the regions generated from them were the same, leading to redundant computations. The workaround was to add a hash table storing the set of bases (each being identified by the ordered set of its basic variables) that have been or are currently being processed. A task will abort after solving the floating-point linear program if it finds that its basis is already in the table.

4   Performance Evaluation

We implemented our parallel algorithms in C++, with three alternate schemes selectable at compile-time: no parallelism, OpenMP parallelism or TBB.

All benchmarks were run on the Paranoia cluster of Grid'5000 [2] and on a server called Pressembois. Paranoia has 8 nodes, each with 2 Intel® Xeon®

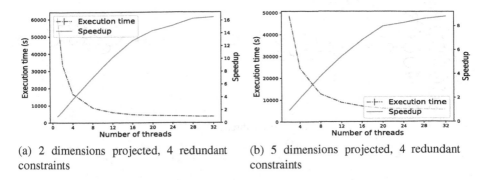

(a) 2 dimensions projected, 4 redundant constraints

(b) 5 dimensions projected, 4 redundant constraints

Fig. 2. 24 constraints, 10 variables, 8–764 regions, OpenMP.

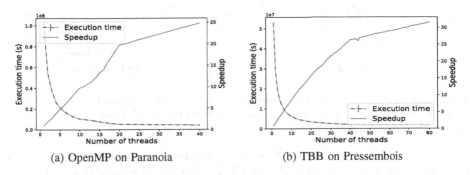

(a) OpenMP on Paranoia

(b) TBB on Pressembois

Fig. 3. 120 constraints, 50 variables, 1 dimension projected, 3459–3718 regions.

E5-2660v2 CPUs (10 cores, 20 threads/CPU) and 128 GiB of RAM. Code was compiled using GCC 6.3.1 and OpenMP 4.5 (201511). The nodes run Linux Debian Stretch with a 4.9.0 kernel. Pressembois has 2 Intel Xeon Gold 6138 CPU (20 cores/CPU, 40 threads/CPU) and 192 GiB of RAM. It runs a 4.9 Linux kernel, and we used GCC 6.3. Every experiment was run 10 times. The plots presented in this section provide the average and standard deviation. Paranoia was used for the OpenMP experiments, whereas Pressembois was used for TBB.

We evaluated our parallel parametric linear programming implementation by using it to project polyhedra, a very fundamental operation. We used a set of typical polyhedra, with different characteristics: numbers of dimensions, of dimensions to be projected and of constraints, sparsity. Here we present a subset of these benchmarks, each comprising 50 to 100 polyhedra.

On problems that have only few regions, not enough parallelism can be extracted to exploit all the cores of the machine. For instance, Fig. 1 presents two experiments on 2 to 36 regions using the OpenMP version. It gives an acceptable speed-up on a few cores (up to 10), then the computation does not generate enough tasks to keep the additional cores busy. As expected, when the solution has many regions, computation scales better. Figure 2 presents the performance obtained on polyhedra made of 24 constraints, involving 8 to 764 regions, using the OpenMP version. The speed-up is sublinear, especially beyond 20 cores.

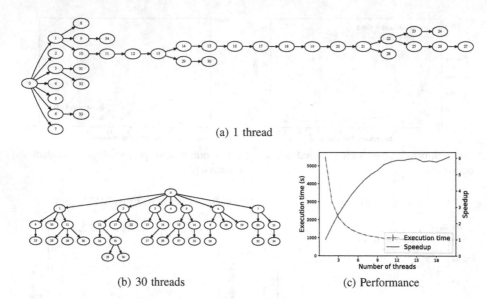

(a) 1 thread

(b) 30 threads   (c) Performance

Fig. 4. Generation graph of the regions from one polyhedron, computed with 1 and 30 threads. The region graphs, depending on overlaps etc., are different; the numbers in both trees have no relationship.

On larger polyhedra, with 120 constraints and 50 variables, the speedup is close to linear with both OpenMP and TBB (Fig. 3). The parallelism extracted from the computation is illustrated by Fig. 4, on a polyhedron involving 29 constraints and 16 variables. Figure 4b shows the number of parallel tasks.

References

1. OpenMP Application Programming Interface, 4.5 edn. (2015)
2. Cappello, F., et al: Grid'5000: a large scale and highly reconfigurable grid experimental testbed. In: International Workshop on Grid Computing. IEEE/ACM (2005)
3. Cousot, P., Halbwachs, N.: Automatic discovery of linear restraints among variables of a program. In: POPL, pp. 84–96. ACM Press (1978)
4. Jones, C., et al.: On polyhedral projections and parametric programming. J. Optim. Theory Appl. **138**(2), 207–220 (2008)
5. Jones, C.N., Kerrigan, E.C., Maciejowski, J.M.: Lexicographic perturbation for multiparametric linear programming with applications to control. Automatica **43**, 1808 (2007)
6. Maréchal, A., Monniaux, D., Périn, M.: Scalable minimizing-operators on polyhedra via parametric linear programming. In: Ranzato, F. (ed.) SAS 2017. LNCS, vol. 10422, pp. 212–231. Springer, Cham (2017). https://doi.org/10.1007/978-3-319-66706-5_11
7. Maréchal, A., Périn, M.: Efficient elimination of redundancies in polyhedra by ray-tracing. In: Bouajjani, A., Monniaux, D. (eds.) VMCAI 2017. LNCS, vol. 10145, pp. 367–385. Springer, Cham (2017). https://doi.org/10.1007/978-3-319-52234-0_20
8. Reinders, J.: Intel Threading Building Blocks: Outfitting C++ for Multi-core Processor Parallelism. O'Reilly Media Inc., Sebastopol (2007)

Automating the Generation of Comparison Weights for Enhancing the AHP Decision-Making Process

Karim Zarour[1,2(✉)], Djamel Benmerzoug[1], Nawal Guermouche[2], and Khalil Drira[2]

[1] LIRE Laboratory, University of Constantine 2 - Abdelhamid Mehri, Constantine, Algeria
{zarour.karim,djamel.benmerzoug}@univ-constantine2.dz
[2] LAAS-CNRS, University of Toulouse, Toulouse, France
{nguermou,khalil}@laas.fr

Abstract. The Analytic Hierarchy Process (AHP) method is widely used to deal with multi-criteria decision-making problems thanks to its simplicity and flexibility. However, it is often criticized for subjectivity and inconsistency in assigning the comparison weights that are based on expert judgments. In order to remedy these shortcomings, we propose in this paper an algorithm that automatically generates the pairwise comparison weights of alternatives according to each considered criterion. In addition, we demonstrate through an example that the judgment matrices constructed by the algorithm are very consistent.

Keywords: Analytic Hierarchy Process · Multi-criteria decision-making · Automatic generation of comparison weights · Consistency ratio · Subjectivity

1 Introduction

Real-world decision-making problems are becoming increasingly complex due to the large number of alternatives, heightened uncertainty, shorter deadlines, greater pressure, environment dynamicity, etc. [1]. Several multi-criteria decision-making (MCDM) methods have been proposed such as TOPSIS, and ELECTRE but the most used and popular one is the Analytic Hierarchy Process (AHP) method that attracts decision-makers by its simplicity and flexibility [4, 6, 9]. Despite its advantages, the AHP method is often criticized for subjectivity and inconsistency in assigning comparison weights. Indeed, the assigned weights must be readjusted until the Consistency Ratio (CR) defined by [7] is equal to or less than 10%. In addition, experts may be asked several times for pairwise comparisons, which is not practical, time-consuming, and very annoying when the number of criteria or alternative is high.

© Springer Nature Switzerland AG 2019
J. M. F. Rodrigues et al. (Eds.): ICCS 2019, LNCS 11540, pp. 573–580, 2019.
https://doi.org/10.1007/978-3-030-22750-0_53

In this paper, we propose an algorithm that automatically generates the pairwise comparison weights of alternatives and fills the corresponding judgment matrices by taking as input the real values of alternatives according to a set of criteria. The main objective of the algorithm is to dispense with the tedious task of assigning weights, which is usually done manually and therefore the decision-making process time could be significantly reduced. In addition to the automation, we demonstrate through an example on hosting offer selection that the proposed algorithm constructs very consistent judgment matrices.

The remainder of this paper is organized as follows. Section 2 overviews the AHP method and its limitations. Some related work are addressed in Sect. 3. Section 4 presents and explains the proposed algorithm. Section 5 demonstrates the usefulness of the algorithm through an illustrative example. Finally, Sect. 6 concludes this paper and provides directions for future work.

2   AHP Limitations

The AHP is a method developed by Saaty [8] for resolving decision problem with multiple criteria through four major steps [7, 8]. Beyond its several advantages like the hierarchical structuring of decision problems, the AHP method relies on the experience and judgments of experts who give weights value directly. In this assessment way, the weights may be attributed with prejudices and the results may have subjectivity. Moreover, the influence of human factor in AHP method can lead to wrong decisions. Another issue may arise when experts need to be asked several times for pairwise comparisons. For instance, a decision-making problem of 4 criteria and 3 alternatives requires to ask experts 18 times. We find that this is not practical and may become very annoying when the number of criteria or alternative is high. Furthermore, the AHP method is supported by the Expert Choice tool but the comparison weights are filled manually.

3   Related Work

There exists many work in the literature that improve the AHP method but most of them focus on the consistency of judgment matrices on the grounds that it is the main weakness of the AHP method. For example, Lin et al. [6] develop an adaptive AHP approach that uses a genetic algorithm to recover the relative importance weights of the considered criteria. Benitez et al. [1] provide an optimization method for improving consistency based on the minimization of the distance between each two judgment matrices. Khatwani and Kar [5] propose an algorithm that can adjust the entries of judgment matrices iteratively until reaching a desired level of consistency.

Certain researchers also aim to improve AHP in group decision-making. For instance, two consensus models have been defined in [2] for group decision-making by using the row geometric mean prioritization method. Huang *et al.* [4] demonstrate that the efficiency of AHP can be significantly enhanced via an optimal expert allocation. However, although consulting several experts for evaluating criteria and alternatives may reduce the bias of personal subjectivity, it requires further calculations and a longer decision process. Furthermore, few work have been realized for improving the AHP method in other aspects. For example, Xiulin and Dawei [9] aim to simplify the calculations needed to construct the judgment matrices by adopting a scale of only three values instead of nine values. In [3], the authors present and test a model based on Multi-layer Perceptron (MLP) neural networks that is capable of completing missing values in AHP judgment matrices. However, to the best of our knowledge, there is no work in the literature that automates the generation of pairwise comparison weights in the AHP method.

4   An Automatic Comparison of Alternatives

To overcome the AHP limitations tackled previously, we propose an algorithm (Algorithm 1) that automatically generates the pairwise comparison weights and fills the judgment matrix corresponding to each criterion. The algorithm takes as input a matrix 'TabVal' containing the real values of alternatives according to a set of criteria. The structure of input and output matrices is presented in Fig. 1.

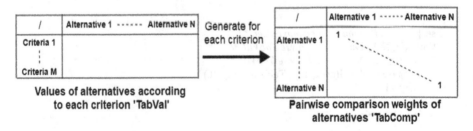

Fig. 1. Automatic generation of pairwise comparison matrices

Algorithm 1: Automatic filling of alternative comparison matrices.
Declaration
TabComp : a square matrix $[(N+1) \times (N+1)]$ of real numbers; /* *an empty comparison matrix, N is the number of alternatives to compare* */
AltComp : { TabComp }; /* *a set of comparison matrices (a TabComp for each criterion$_k$)* */
TabVal : a matrix $[(M+1) \times (N+1)]$ of real numbers; /* *TabVal is a matrix containing the values of alternatives according to each criterion* */
/* *TabComp and TabVal have the same number of columns N (the number of compared alternatives), M is the number of considered criteria.* */
i, j, k : integer;
Begin
  Scalling (TabVal); /* *Scaling the real values of alternatives* */
  AltComp ← { };
  For (k=1 ; k ≤ M; k++) Do { /* *for each criterion$_k$* */
  TabComp$_k$ = New TabCom ();   /* *a new comparison matrix for each criterion$_k$* */
  For (i=1 ; i ≤ N ; i++) Do {
    TabComp$_k$(alt$_i$, alt$_i$) ← 1; /* *fill in the matrix diagonal with 1* */
  }
  i ← 1; j ← 2;
  If (Criterion$_k$ IS Criterion to be minimized) Then {
    While (i ≤ N-1) Do {   /* *browse TabVal* */
      While (j ≤ N) Do {   /* *browse TabVal* */
      Compare (TabVal(crit$_k$, alt$_i$), TabVal(crit$_k$, alt$_j$));   /* *this procedure compares 2 alternatives and inserts the comparison weight into TabComp$_k$* */
        j ← j +1;
      }
      i ← i +1; j ← i +1;
    }
  }
  Else {   /* *Criterion$_k$ is to be maximized* */
    While (i ≤ N-1) Do {   /* *browse TabVal* */
      While (j ≤ N) Do {   /* *browse TabVal* */
      Compare (TabVal(crit$_k$, alt$_j$), TabVal(crit$_k$, alt$_i$));
        j ← j +1;
      }
      i ← i +1; j ← i +1;
    }
  }
  AltComp ← AltComp ∪ { TabComp$_k$ };   /* *add to the result the comparison matrix of alternatives for the criterion$_k$* */
  }
  Return (AltComp);
End ;

Besides the automation, we will also show through the example of the next section that the proposed algorithm generates consistent judgment matrices (i.e. the consistency ratio is less than or equal to 10%). Indeed, judgment matrix consistency is one of the most challenging problems when using AHP [2, 9].

In order to respect the Saaty's scale [8], we have integrated in the beginning of the algorithm a procedure called 'Scalling', which allows to put the real values on a scale of numbers ranging from 0 to 8. After creating an alternative comparison table and filling its diagonal with '1', the 'Compare' procedure is iteratively called within two nested loops. This procedure takes as parameter two values characterizing two different alternatives according to a given criterion. Next, it generates the two comparison weights (one weight to compare Alternative i and Alternative j and another for the opposite) and inserts them in the 'TabCom' comparison matrix corresponding to the criterion in question. The instructions of 'Scaling' and 'Compare' procedures are described in the following:

Procedure Scaling (TabVal: Matrix [(M+1) × (N+1)] of real numbers)
Declaration MaxValue: float; i, k: integer;
Begin
  For (k=1 ; k ≤ M; k++) Do { /* for each criterion$_k$ */
    MaxValue ← TabVal(crit$_k$, alt$_1$);
    For (i=2; i ≤ N; i++){   /* search for the maximum value of alternatives according to criterion$_k$ */
      If (TabVal(crit$_k$, alt$_i$) > MaxValue) Then {
        MaxValue ← TabVal(crit$_k$, alt$_i$);
      }
    }
    For (i=1 ; i ≤ N; i++) {
      TabVal(crit k, alt i) ← $\frac{TabVal(crit\,k, alt\,i) * 8}{MaxValue}$; /* scaling each value */
    }
  }
End;

Procedure Compare (X: float, Y: float)
Declaration difference : float;
Begin
  difference ← X - Y;
  If (difference ≥ 0) Then {
    TabComp(alt$_i$, alt$_j$) ← $\frac{1}{difference + 1}$; TabComp(alt$_j$, alt$_i$) ← $difference + 1$;
  }
  Else {
    TabComp(alt$_i$, alt$_j$) ← $|difference| + 1$; TabComp(alt$_j$, alt$_i$) ← $\frac{1}{|difference| + 1}$;
  }
End;

5   Illustrative Example

In order to illustrate the application of our algorithm, we suppose a company that wants to host its online booking platform and it hesitates between seven dedicated servers. Hence, the company decided to apply the AHP method in order to choose the hosting offer that best meets its requirements. The characteristics of the dedicated servers proposed by each online hosting offer that the company faces are represented in Table 1 (values that are at the top of cells).

Table 1. Values of alternatives according to each criterion

Offer name/Criteria	Magic Epsilone	OVH EG-32	Magic Delta	Bluehost Prenium	eStruxture Intel Xeon	Serverroom Intel E5-2630L	Hivelocity Skylake
CPU frequency (Ghz)	2.6	3.8	3.06	2.5	3.3	2.0	3.4
	5.5	8	6.4	5.2	6.9	4.2	7.2
	0.08	0.3	0.13	0.07	0.17	0.05	0.2
CPU cores	8	4	12	4	4	6	4
	5.3	2.7	8	2.7	2.7	4	2.7
	0.2	0.06	0.44	0.06	0.06	0.12	0.06
RAM (GB)	128	32	96	16	32	32	64
	8	2	6	1	2	2	4
	0.43	0.05	0.24	0.03	0.06	0.06	0.13
Storage (TB)	2	8	2	1	1	0.48	0.96
	2	8	2	1	1	0.5	1
	0.11	0.53	0.11	0.06	0.06	0.07	0.06
Bandwidth (Mbps)	150	500	150	200	100	300	1024
	1.2	3.9	1.2	1.6	0.8	2.3	8
	0.06	0.18	0.06	0.07	0.04	0.1	0.49
Price (€/month)	199	99.99	129	89	137	129	145
	8	4	5.2	3.6	5.6	5.2	5.8
	0.03	0.24	0.13	0.29	0.1	0.13	0.08

In what follows, we describe how our algorithm intervenes in the AHP method for solving this MCDM problem after structuring it hierarchically.

5.1 Comparison of Alternatives

Before comparing alternatives, the algorithm scales their real values and the values of alternatives after scaling are represented in the middle in Table 1. Once the real values are scaled, the algorithm generates a comparison matrix of alternatives according to each criterion. Table 2 represents the comparison matrices generated in our example for the criterion of CPU frequency (the absolute weights are represented without parentheses and the relative weights in parentheses) and it is the same principle for the other five criteria.

Table 2. The pairwise comparisons generated according to the CPU frequency

	Magic Epsilone	OVH EG-32	Magic Delta	Bluehost	eStruxture	Serverroom	Hivelocity	Average
Magic Epsilone	1 (0.08)	1/3.5 (0.09)	1/1.9 (0.06)	1.3 (0.09)	1/2.4 (0.07)	2.3 (0.11)	1/2.7 (0.07)	8%
OVH EG-32	3.5 (0.28)	1 (0.32)	2.6 (0.32)	3.8 (0.26)	2.1 (0.34)	4.8 (0.23)	1.8 (0.35)	30%
Magic Delta	1.9 (0.15)	1/2.6 (0.12)	1 (0.12)	2.2 (0.15)	1/1.5 (0.11)	3.2 (0.15)	1/1.8 (0.11)	13%
Bluehost	1/1.3 (0.06)	1/3.8 (0.08)	1/2.2 (0.06)	1 (0.07)	1/2.7 (0.06)	2 (0.10)	1/3 (0.07)	7%
eStruxture	2.4 (0.19)	1/2.1 (0.15)	1.5 (0.18)	2.7 (0.19)	1 (0.16)	3.7 (0.18)	1/1.3 (0.15)	17%
Serverroom	1/2.3 (0.03)	1/4.8 (0.07)	1/3.2 (0.04)	½ (0.03)	1/3.7 (0.04)	1 (0.05)	¼ (0.05)	5%
Hivelocity	2.7 (0.21)	1/1.8 (0.18)	1.8 (0.22)	3 (0.21)	1.3 (0.21)	4 (0.19)	1 (0.20)	20%

5.2 Verification of the Comparison Consistency

We have used the Expert Choice tool to check the consistency of the pairwise comparisons and we observe in Fig. 2 that excepted the storage criterion, the consistency ratio (CR) is fairly stable and very far from the acceptable limit of 10%. This proves that our algorithm generates very consistent comparison matrices. Furthermore, we have verified through tests that the algorithm generates consistent comparison matrices (the CR varies between 1% and 3%) whatever the number of alternatives.

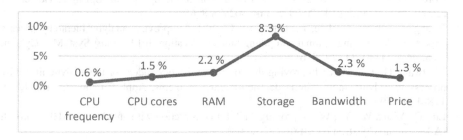

Fig. 2. The consistency ratio corresponding to each alternative comparison matrix

The last step consists calculating the final score of each alternative by multiplying the final values of alternatives (shown at the bottom of cells in Table 1) by the final weights of criteria that depend company's requirements.

6 Conclusion and Future Work

In order to enhance the AHP method, we have proposed in this paper an algorithm that automatically generates the pairwise comparison weights of alternatives and construct a judgment matrix for each considered criterion. The algorithm takes as input the real values of alternatives according to a set of criteria and puts them on a scale of eight values so that the generated weights respect the Saaty's scale. The applicability and usefulness of our algorithm has been demonstrated through an example on hosting offer selection. In addition to the automation, the illustrative example has also shown that the judgment matrices generated by our algorithm are very consistent.

For future work, we envisage to integrate into the proposed algorithm a procedure that quantify qualitative criteria. We also want to incorporate new methods or reuse existing ones that allow readjusting the comparison weights of judgment matrices in order to further reduce the consistency ratio. Moreover, it would also be interesting to compare the obtained consistency ratios with those of similar work. Finally, we intend to extend the Expert Choice software with the proposed algorithm through a plugin.

References

1. Benitez, J., Delgado-Galvan, X., Izquierdo, J., Pérez-Garcia, R.: Improving consistency in AHP decision-making processes. Appl. Math. Comput. **219**(5), 2432–2441 (2012)
2. Dong, Y., Zhang, G., Hong, W., Xu, Y.: Consensus models for AHP group decision making under row geometric mean prioritization method. Decis. Support Syst. **49**(3), 281–289 (2010)
3. Gomez-Ruiz, J.A., Karanik, M., Peláez, J.I.: Improving the consistency of AHP matrices using a multi-layer perceptron-based model. In: Cabestany, J., Sandoval, F., Prieto, A., Corchado, J.M. (eds.) IWANN 2009. LNCS, vol. 5517, pp. 41–48. Springer, Heidelberg (2009). https://doi.org/10.1007/978-3-642-02478-8_6
4. Huang, E., Zhang, S., Lee, L., Chew, E., Chen, C.: Improving analytic hierarchy process expert allocation using optimal computing budget allocation. IEEE Trans. Syst. Man Cybern. Syst. **46**(8), 1140–1147 (2016)
5. Khatwani, G., Kar, A.K.: Improving the cosine consistency index for the analytic hierarchy process for solving multi-criteria decision-making problems. Appl. Comput. Inform. **13**(2), 118–129 (2017)
6. Lin, C., Wang, W., Yu, W.: Improving AHP for construction with an adaptive AHP approach (A3). Autom. Constr. **17**(2), 180–187 (2008)
7. Saaty, T.L.: Decision making with the analytic hierarchy process. Int. J. Serv. Sci. **1**(1), 83–98 (2008)
8. Saaty, T.L.: The Analytic Hierarchy Process. McGraw-Hill, New York (1980)
9. Xiulin, S., Dawei, L.: An improvement analytic hierarchy process and its application in teacher evaluation. In: Proceedings of the 5th International Conference on Intelligent Systems Design and Engineering Applications, pp. 169–172. IEEE (2014)

Parallel Algorithm Based on Singular Value Decomposition for High Performance Training of Neural Networks

Gabriele Maria Lozito[1], Valentina Lucaferri[1(✉)], Mauro Parodi[2],
Martina Radicioni[1], Francesco Riganti Fulginei[1], and Alessandro Salvini[1]

[1] Roma Tre University, Rome, Italy
valentina.lucaferri@uniroma3.it
[2] University of Genoa, Genoa, Italy

Abstract. Neural Networks (NNs) are frequently applied to Multi Input Multi Output (MIMO) problems, where the amount of data to manage is extremely high and, hence, the computational time required for the training process is too large. Therefore, MIMO problems are often split into Multi Input Single Output (MISO) problems; MISOs are further decomposed into several Single Input Single Output (SISO) problems. The aim of this paper is to present an optimized approach for NNs training based on properties of Singular Value Decomposition (SVD), allowing to decompose the MISO NN into a collection of SISO NNs. The decomposition provides a two-fold advantage: firstly, each SISO NN can be trained by using a one-dimensional function, namely a limited dataset, and then a parallel architecture can be implemented on a PC-cluster, decreasing the computational cost. The parallel algorithm performance are validated by using magnetic hysteresis dataset with the aim to prove the computational speed up by preserving the accuracy.

Keywords: Neural Networks · Parallel computing ·
Magnetic hysteresis

1 Introduction

Neural Networks (NNs) for Multi-Input- Multi-Output (MIMO) problems have gained wide attention in many scientific fields [3,6], especially regarding complex non-linear systems without a closed-form solution. In these applications, a NN, able to learn non-linear relationship between quantities by means of a dataset, represents a reasonable alternative to the use of mathematical models [17,19]. The main drawback of any neural model is, however, the necessity of a large amount of data/measurements for its set up; in fact, a NN with m inputs and n outputs can be directly implemented to approximate a non-linear function, provided the amount of data to manage is not too large; moreover the high computational cost, needed for the wide size of training patterns and for complex NN architecture, makes the direct approach inefficient.

© Springer Nature Switzerland AG 2019
J. M. F. Rodrigues et al. (Eds.): ICCS 2019, LNCS 11540, pp. 581–587, 2019.
https://doi.org/10.1007/978-3-030-22750-0_54

Therefore, several authors make the problem easier by dividing MIMO NN into a collection of n Multi-Input-Single-Output (MISO) NNs [8,9], further, split into several Single-Input-Single-Output (SISO) NNs; in this way, their structure appears to be considerably simplified and the time required by the whole learning process results to be lower. Singular Value Decomposition (SVD) turns out to be a powerful instrument for switching between MISO NNs and a group of SISO NNs, since it is defined for any array of data and it provides a low-rank matrix, which is a good approximation to the original one [1,7]; the efficiency of this approach has already demonstrated in different works [12,16]. The presented paper concerns the implementation of parallel algorithm on a PC-cluster that is able to address the SISO NN training process to a cluster node, separately, decreasing strongly the computational cost. Moreover, the parallel architecture obtained from SVD-based approximation is applied to magnetic hysteresis data and the study of its ability to model the magnetic hysteresis constitutes the main issue of this work. Indeed, although the magnetic hysteresis problem has been already solved by the authors through a suitably trained NN [2,10,11,13], it is still a difficult task [4,5,14], especially when dealing with materials having large variance in characteristics and NN must be trained with a dataset describing these conditions. Obtaining such data involves a whole set of measurements that can be very difficult to perform; thus, to solve this problem in a feasible way, the parallel SVD-based algorithm is applied to a 2D-array, which represents the magnetic permeability, μ, obtained starting from experimental values of magnetic field, H, and flux density, B [15].

This paper is organized as follows: Sect. 2 summarizes some of the main properties of the SVD technique, emphasizing the decomposition of bidimensional functions. Section 3 focuses on the parallel algorithm implemented for training each SISO NN: firstly, MISO-SISO decomposition is introduced, pointing out the training of each SISO NN, then, PC-cluster characteristics are introduced. In Sect. 4 the performances of parallel algorithm are evaluated by using magnetic hysteresis data and, finally, in Sect. 5, the conclusions follow.

2   The Approximation Method: SVD

Among the methods used to simplify a MISO problem into a collection of SISO, the SVD represents a valid tool, that allows to approximate a multivariate functions without losing accuracy.

The SVD of a rectangular matrix $\mathbf{A} \in R^{m \times n}$ is the factorization of $\mathbf{A}$ into the product of three matrices: $\mathbf{A} = \mathbf{U}\mathbf{\Sigma}\mathbf{V}^{\mathbf{T}}$ where $\mathbf{U} = (u_1 \ldots u_m) \in R^{m \times m}$, $\mathbf{V} = (v_1 \ldots v_n) \in R^{n \times n}$ and $\mathbf{\Sigma} \in R^{m \times n}$ is a diagonal one having its non-zero diagonal entries equal to the singular values $\sigma_{k,k}$ with $k = 1, ..., p$ and $p = \min(m,n)$, written in descending order $\sigma_{1,1} > \sigma_{2,2} > ... > \sigma_{p,p} > 0$. Thus:

$$\mathbf{A} = \sum_{k=1}^{p} \sigma_{k,k} \mathbf{u}^{\langle k \rangle} (\mathbf{v}^T)^{\langle k \rangle} \tag{1}$$

where $\mathbf{u}^{\langle k \rangle} \subset R^m$ and $\mathbf{v}^{\langle k \rangle} \subset R^n$ are orthonormal vectors coincident with the column vectors of matrices $\mathbf{U}$ and $\mathbf{V}^T$, respectively. Usually, the smallest singular values are neglected and an approximated version of (1) is considered:

$$\hat{\mathbf{A}} = \sum_{k=1}^{\hat{p}} \sigma_{k,k} \mathbf{u}^{\langle k \rangle} (\mathbf{v}^T)^{\langle k \rangle} \tag{2}$$

with $\sigma_{1,1} > \sigma_{2,2} > ... > \sigma_{\hat{p},\hat{p}} > 0$ and $\hat{p} < min(m,n)$, whereas $\sigma_{j,j}$ is forced to be 0, for $\hat{p} < j < p$.

Let us assume that the $\mathbf{A}$ entries are coming from the sampling of a bivariate function $f(x_1, x_2)$, so $A_{i,j} = f(x_{1,i}, x_{2,j})$ with $i = 1 \ldots m$ and $j = 1 \ldots n$. Correspondingly, the columns of the matrices $\mathbf{U}$ and V^T are arising from the sampling of unknown univariate functions, $u_{k,i} = \psi_k(x_{1,i})$ and $v_{k,j}^T = \eta_k(x_{2,j})$. Once obtained $\hat{p}$ with a fixed accuracy ε, the approximation can be written as follows:

$$\hat{A}_{i,j} = \hat{f}(x_{1,i}, x_{2,j}) = \sum_{s=1}^{\hat{p}} \sigma_s \hat{\psi}_s(x_{1,i}) \hat{\eta}_s(x_{2,j}) \tag{3}$$

The different unknown univariate functions in (3) can be evaluated by appropriate curve fitting techniques, such as NNs.

3   Parallel Algorithm for SISO NNs Training

3.1   MISO NN Decomposition

The SVD-based approximation of bivariate functions described in the previous section can be exploited for optimizing the feed-forward MISO NNs learning process. Thus, the idea is to split the MISO NN into several SISO NN by applying the SVD-based approximation of multivariate functions, to reduce the time of learning process and to simplify the NN architecture.

Let us consider an array of data $\mathbf{A}$, obtained by sampling a 2D-function $f(x_1, x_2)$, and define a decomposition operator, D, which indicates the SVD of function $f(x_1, x_2)$, $D\{f(x_1, x_2)\} = \sum_{i=1}^{\hat{p}} \sigma_i \psi_i(x_1) \eta_i(x_2)$.

When the operator is applied, it generates $2 \times \hat{p}$ univariate functions and each of them can be approximated by a SISO NN, as shown in Fig. 3.

Every SISO NN presents a simple feed-forward architecture, composed by an input neuron, one hidden layer and an output neuron. The main advantage is that each NN is independent from the other, so its training can be implemented as a stand-alone process being part of a larger parallel architecture.

3.2   Parallel Computing in a PC-Cluster System

The principal aim of using a PC-cluster is to improve the performance for large computational task, such as MISO NN training, that can be divided into smaller tasks distributed around the nodes. The SISO NN training is implemented on

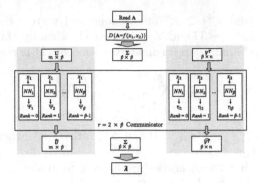

Fig. 1. Scheme of parallel algorithm implemented for SISO NNs training.

a high performance computing, HPC, cluster, that consists of 4 compute nodes, each one composed of 24 CPUs. A Single Program Multiple Data, SPMD, parallel model is performed; thus, the master node sends the program, involving training set and algorithm, to the others computational nodes and it commands every one to execute the learning process of a SISO NN. The communication is realized through the Message Passing Interface standard, MPI, which allows the programmer to easily pass the message from one of computers to the others by calling the function offered from the libraries. Generally, a parallel process requires a communicator, containing a group of processes, whose number defines its size, r; each process has a *rank* inside the communicator, namely a number that permits to recognize it. So, the implementation of parallel NNs training involves a communicator having $n \times \hat{p}$ size, where n indicates the dimension of the initial array of data; especially, $r = 2 \times \hat{p}$ for 2-D array, $n = 2$. As shown in Fig. 2, each process returns a SISO NN, able to approximate the one-dimensional function, representing the $\mathbf{U}$ matrix columns or $\mathbf{V^T}$ matrix rows. Once, the $\mathbf{U}$ and $\mathbf{V}$ matrices have been reconstructed, the approximation of original array can be obtained, by using Eq. 2.

4   Validation of SVD-Based Parallel Algorithm: Magnetic Hysteresis Data

With the aim to validate the presented approach, the parallel algorithm is used to train a neural network implemented for modelling magnetic hysteresis. A collection of asymmetric loops has been generated for building the data set for NN training by employing Preisach model in Matlab environment [18]. Each loop has been sampled in n points and for every couple of coordinates $[H_k, B_k]$ the corresponding differential magnetic permeability, μ_k has been computed. As a result, an array of data $\mu(B_n, H_n)$ function of H and B variables is obtained to test the SVD-based algorithm.

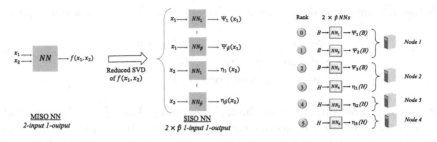

Fig. 2. Parallelization of training process. From left to right: transformation of MISO NN into several SISO NNs. Each NN training represents a process of rank k, with $k = 1...\hat{p}$, executed by a cluster node.

Thus, assuming $x_1 = B_1, ..., B_m$, $x_2 = H_1, ..., H_n$ and $f(x_1, x_2) = \mu(H, B)$, the SVD applied to bivariate function returns

$$\mu_{i,j} = \mu(B_i, H_j) = \sum_{s=1}^{p} \sigma_s \psi_s(B_i) \eta_s(H_j) \tag{4}$$

To reduce the total number of NNs required for reconstructing the original function, a threshold value is fixed to $\epsilon = 0.001$ in order to obtain a reduced SVD form with $\hat{p} = 3$.

$$\hat{\mu}_{i,j} = \sum_{s=1}^{3} \sigma_s \hat{\psi}_s(B_i) \hat{\eta}_s(H_j) \tag{5}$$

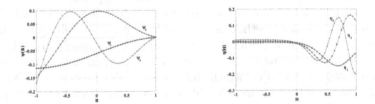

Fig. 3. Results on validation test of NNs: left $\psi_s(B)$; right $\eta_s(H)$.

After decomposition, the total number of SISO NNs results to be 6:3 univariate functions of B, $\psi_1(B), \psi_2(B), \psi_3(B)$; 3 univariate functions of H, $\eta_1(H), \eta_2(H), \eta_3(H)$. Each SISO NN has a feed-forward architecture and it was trained by using Levenberg Marquardt algorithm. The NN architecture consists of a input neuron, representing the B or H variable sampled in n points, the hidden layer is composed by 8 neurons and the output neuron returns the one dimensional function $\psi(B)$ or $\eta(H)$, respectively. The training set for each NN is made by 70 of 100 sampled points belonging to the univariate function, which is intended to approximate. So, for computing the parallel learning process, the master node

Table 1. Computational time results achieved for the bivariate function, $\mu(B, H)$.

System	Training time (s)
Serial implementation	30.81
Parallel implementation	7.24

has to send a program, including the training set and Levenberg Marquardt algorithm, to the others one. Hence, a communicator, involving 6 processes distributed between the 4 available compute nodes, is set up as shown in Fig. 1, with the purpose to minimize the computational effort of a single node. The performance of NN is compared with the points excluding the ones in the training, especially, test set is composed by 30 samples. The NN shows a good accuracy in modelling, Fig. 3 shows that each NN is able to reproduce the univariate function trend, with a $MSE \simeq 10^{-8}$. Finally, the NN outputs are exploited to reconstruct the μ array, following the Eq. 5. Table 1 shows a comparison between the time required by whole learning process with and without parallel architecture implementation; a strong reduction of computational time is achieved.

5   Conclusions

An optimization algorithm able to perform parallel learning process of SISO NNs, by exploiting the SVD properties has been presented. A well-known challenging topic is constituted by the modelling of magnetic hysteresis, a typical MISO problem. Thus, a procedure for magnetic hysteresis loops identification has been proposed. The method is based on the parallelization of NNs training processes, that allows to accurately model the hysteresis problem achieving a reduction of computational cost. The implemented SISO-based approach allows to provide a solution, even when the conventional use of a MISO strategy fails. Results show that the proposed method is a suitable tool for modelling the magnetic hysteresis, it is capable of reducing the processing time strongly and, at the same time, preserving the accuracy of solution. The presented technique constitutes an effective solution based on which the more complex problem of 3D magnetic hysteresis can be solved, hence, should be considered a valid starting point for future developments. In particular, it is worth noticing that the method can, in general, suit any multidimensional problem.

References

1. Bizzarri, F., Parodi, M., Storace, M.: SVD-based approximations of bivariate functions. In: IEEE International Symposium on Circuits and Systems, vol. 5, p. 4915. IEEE 1999 (2005)
2. Cardelli, E., Faba, A., Laudani, A., Riganti Fulginei, F., Salvini, A.: A neural approach for the numerical modeling of two-dimensional magnetic hysteresis. J. Appl. Phys. **117**(17), 17D129 (2015)
3. Chen, M., Ge, S.S., How, B.V.E.: Robust adaptive neural network control for a class of uncertain mimo nonlinear systems with input nonlinearities. IEEE Trans. Neural Netw. **21**(5), 796–812 (2010)

4. Duan, N., Xu, W., Wang, S., Zhu, J., Guo, Y.: Hysteresis modeling of high-temperature superconductor using simplified Preisach model. IEEE Trans. Magn. **51**(3), 1–4 (2015)
5. Handgruber, P., Stermecki, A., Biro, O., Goričan, V., Dlala, E., Ofner, G.: Anisotropic generalization of vector Preisach hysteresis models for nonoriented steels. IEEE Trans. Magn. **51**(3), 1–4 (2015)
6. Hovakimyan, N., Calise, A.J., Kim, N.: Adaptive output feedback control of a class of multi-input multi-output systems using neural networks. Int. J. Control **77**(15), 1318–1329 (2004)
7. Huynh, H.T., Won, Y.: Training single hidden layer feedforward neural networks by singular value decomposition. In: Fourth International Conference on 2009 Computer Sciences and Convergence Information Technology, ICCIT 2009, pp. 1300–1304. IEEE (2009)
8. Jianye, L., Yongchun, L., Jianpeng, B., Xiaoyun, S., Aihua, L.: Flaw identification based on layered multi-subnet neural networks. In: 2009 Second International Conference on Intelligent Networks and Intelligent Systems, pp. 118–121. IEEE (2009)
9. Kabir, H., Wang, Y., Yu, M., Zhang, Q.J.: High-dimensional neural-network technique and applications to microwave filter modeling. IEEE Trans. Microw. Theory Tech. **58**(1), 145–156 (2010)
10. Laudani, A., Lozito, G.M., Riganti Fulginei, F.: Dynamic hysteresis modelling of magnetic materials by using a neural network approach. In: 2014 AEIT Annual Conference-From Research to Industry: The Need for a More Effective Technology Transfer (AEIT), pp. 1–6. IEEE (2014)
11. Laudani, A., Lozito, G.M., Riganti Fulginei, F., Salvini, A.: Modeling dynamic hysteresis through fully connected cascade neural networks. In: 2016 IEEE 2nd International Forum on Research and Technologies for Society and Industry Leveraging a better tomorrow (RTSI), pp. 1–5. IEEE (2016)
12. Laudani, A., Salvini, A., Parodi, M., Riganti Fulginei, F.: Automatic and parallel optimized learning for neural networks performing mimo applications. Adv. Electr. Comput. Eng. **13**(1), 3–12 (2013)
13. Makaveev, D., Dupré, L., De Wulf, M., Melkebeek, J.: Modeling of quasistatic magnetic hysteresis with feed-forward neural networks. J. Appl. Phys. **89**(11), 6737–6739 (2001)
14. Rasilo, P., et al.: Modeling of hysteresis losses in ferromagnetic laminations under mechanical stress. IEEE Trans. Magn. **52**(3), 1–4 (2016)
15. Riganti Fulginei, F., Salvini, A.: Neural network approach for modelling hysteretic magnetic materials under distorted excitations. IEEE Trans. Magn. **48**(2), 307–310 (2012)
16. Riganti Fulginei, F., Salvini, A., Parodi, M.: Learning optimization of neural networks used for MIMO applications based on multivariate functions decomposition. Inverse Prob. Sci. Eng. **20**(1), 29–39 (2012)
17. Salvini, A., Riganti Fulginei, F.: Genetic algorithms and neural networks generalizing the Jiles-Atherton model of static hysteresis for dynamic loops. IEEE Trans. Magn. **38**(2), 873–876 (2002)
18. Salvini, A., Riganti Fulginei, F., Pucacco, G.: Generalization of the static Preisach model for dynamic hysteresis by a genetic approach. IEEE Trans. Magn. **39**(3), 1353–1356 (2003)
19. Turk, C., Aradag, S., Kakac, S.: Experimental analysis of a mixed-plate gasketed plate heat exchanger and artificial neural net estimations of the performance as an alternative to classical correlations. Int. J. Therm. Sci. **109**, 263–269 (2016)

In-Situ Visualization with Membrane Layer for Movie-Based Visualization

Kohei Yamamoto and Akira Kageyama(✉)

Department of Computational Science, Kobe University, Kobe 657-8501, Japan
kage@port.kobe-u.ac.jp

Abstract. We propose a movie-based visualization for High Performance Computing (HPC) visualization. In this method, a viewer interactively explores a movie database with a specially designed application program called a movie data browser. The database is a collection of movie files that are tied with the spatial coordinates of their viewpoints. One can walk through the simulation's data space by extracting a sequence of image files from the database with the browser. In this method, it is important to scatter as many viewpoints as possible for smooth display. Since proposing the movie-based visualization method, we have been developing some critical tools for it. In this paper, we report the latest development for supercomputers to apply many in-situ visualizations with different viewpoints in a Multiple Program Multiple Data (MPMD) framework. A key feature of this framework is to place a membrane-like layer between the simulation program and the visualization program. Hidden behind the membrane layer, the simulation program is not affected by the visualization program even if the number of scattered viewpoints is large.

Keywords: Visualization · Movie-based visualization · HPC

1 Introduction

The imbalance between processor speed and network bandwidth in High Performance Computing (HPC) systems leads in-situ visualization to provide hope for the future [4,5,10–12]. In general, in-situ visualization deprives users of interactive control of visualization parameters. We cannot change the visualization-related variables after a simulation, unless we adopt special methods for rendering [7,15]. These visualization-related variables include the applied visualization algorithms, their parameters, and the camera settings. It is not reasonable to resubmit a simulation job when, for example, we just want to observe a phenomenon from a different view position.

We have proposed a way to realize an interactive viewing of in-situ visualization that can be applied with standard rendering [6]. The key point of this method is to apply multiple in-situ visualizations from different viewpoints.

© Springer Nature Switzerland AG 2019
J. M. F. Rodrigues et al. (Eds.): ICCS 2019, LNCS 11540, pp. 588–594, 2019.
https://doi.org/10.1007/978-3-030-22750-0_55

As in the bullet-time method used in the movie industry, a number of visualization cameras are placed in the simulation space. Each camera takes a sequence of visualization images as the simulation goes on. In contrast to the steering approach [2,9], the positions of the cameras are predefined before the simulation. After the simulation, images from each camera are combined into a movie file. Therefore, the output of this in-situ visualization method is a collection of movie files, rather than numerical data files. Labeled with the cameras' locations, the collection of movie files composes a movie database. Here we refer to this approach to HPC visualization as "movie-based visualization", since it is a generalization of image-based visualization [1].

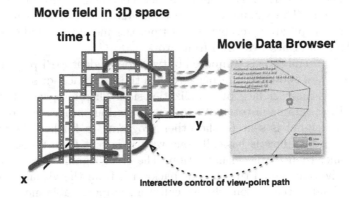

Fig. 1. Interactive exploration of movie database.

After we proposed the idea of movie-based visualization, we developed a specially designed PC application that extracts a sequence of still images from the movie database and shows the sequence as an animation in a PC window. The application, the movie data browser, enables its user to interactively change the viewpoint by dragging the PC's mouse or by keyboard input while a movie is playing forward or backward in time; see Fig. 1.

A key point of this method is to place multiple cameras as densely as possible in the simulation space. In contrast to the bullet-time method, our visualization cameras have no mutual occlusion, regardless of how dense they are, but applying many in-situ visualizations for a simulation is a challenging task.

2   Membrane Layer Method

Major (post hoc) visualization applications are now provided with in-situ libraries or tools, such as Catalyst [3] for ParaView and libsim [18] for VisIt. Various in-situ visualization infrastructures have also been developed for HPC, such as Embree [17] and OSPray [16]. More general framework for high-performance I/O middleware ADIOS [8] is also used in in-situ visualizations. We are also developing an in-situ visualization tool that is based on KVS [14], a general-purpose visualization software tool. We use the off-screen rendering mode of KVS for visualizations on CPU cores of supercomputer systems.

In order to make the movie-based visualization method a practical tool for simulations, it is necessary to avoid the speed deterioration of the simulation by the in-situ generation of many movie files. The key idea proposed in this paper is to separate the simulation from the visualization by placing a semipermeable membrane-like layer between them. The two programs are "invisible" to each other (Fig. 2).

Both the simulation and the visualization are parallelized. They are basically independent MPI programs with their own `MPI_Init` and `MPI_Finalize` functions. The semipermeable membrane between them is also an independent parallel program. The simulation program sends data to the membrane, and the visualization program receives them from the membrane under a Multiple Program Multiple Data (MPMD) framework.

In fact, the membrane is composed of two independent MPI programs. They correspond to the two sides of a plane; one is the front face, or simulation side, and the other is the back face, or visualization side. Therefore, we have four independent MPI programs in total. When one of the four programs stops because of an error, a signal is sent to the other programs to finalize the whole job. The numerical data flow is basically one way; it is sent from the simulation to the membrane, then to the visualization. The error signal of the visualization program is the only signal that is sent backward, from the visualization to the simulation. That is the reason why we call it a semipermeable membrane.

Owing to the MPMD framework, we can apply multiple in-situ visualizations in an asynchronous way. In contrast to the synchronous cases, we apply multiple in-situ visualizations on other processor nodes of the supercomputer system while the simulation is running. For a larger computational load for the visualization, or for a larger number of viewpoints of the in-situ visualization, we just allocate more computer nodes for additional MPI processes for the visualization program.

Since the membrane's front face is devoted only to receiving data from the simulation, the simulation program can assume that each data transfer is completed without delay. The simulation program does not wait for completion of the visualization for each item of data.

The membrane's back face stores the latest simulation data and passes them to the visualization program when they are requested. If new data are sent from the simulation before completing the visualization of previous data, the stored data in the membrane is overwritten. In other words, it is possible that some image frames are missing in the final product of the in-situ visualization movies when it takes an unusually long time to render an image. We accept it as unavoidable in this framework at present but it could be improved in the future.

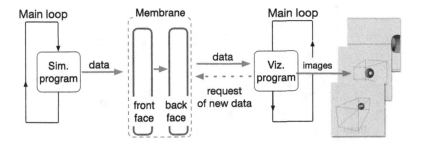

Fig. 2. The membrane layer between the simulation and visualization programs.

3 Experiments

To demonstrate the applicability of the membrane method to in-situ visualization of practical simulations on supercomputers, we have performed a simple computational fluid dynamics simulation on a supercomputer system, Oakforest-PACS (Fujitsu PRIMERGY CX600 M1, 8208 nodes of Intel Xeon Phi) for the well-known smoke-ring formation [13]. The simulation region is a rectangular box of size $L_x \times L_y \times L_z = 30\,\text{m} \times 10\,\text{m} \times 10\,\text{m}$. Periodic boundary conditions are assumed in all $(x, y,$ and $z)$ directions. Compressible Navier-Stokes equations for an ideal gas are solved with a second-order central finite difference method for the spatial derivatives and an explicit fourth-order Runge-Kutta method for time integration. The code is parallelized based on a three-dimensional decomposition with $24\,(= 6 \times 2 \times 2)$ domains. An MPI process is assigned to each domain.

In this experiment, we visualize three scalar fields, pressure p, mass density ρ, and enstrophy density $q = |\nabla \times \boldsymbol{v}|^2$, where $\boldsymbol{v}$ is the flow velocity. The front face of the membrane consists of three (same as the number of fields) MPI processes, and the back face of it has the same number of MPI processes. A process of the front face receives one of the scalar fields from the simulation and its counterpart process of the back face passes the scalar data to the visualization program on request. As for the visualization program, we allocate 27 cameras in this experiment, with one MPI process for each camera. In total, we invoke 57 MPI processes (57 processor nodes): 24 (for simulation) + 3 (for front face of the membrane) + 3 (for back face of the membrane) + 27 (for visualization).

Figure 3 shows visualizations by an isosurface of q by four cameras out of 27 cameras in total. Other visualizations for different physical variables lead to basically the same images. The resulting movie files have sufficiently high temporal resolution (high number of frames per second (FPS)) to analyze the fluid dynamics in detail.

We next confirm the effect of the membrane as a barrier against increased load for the visualization program. We compare two different styles of in-situ visualization: one with the membrane-based asynchronous visualization method and the other a fully synchronous visualization (i.e., in-situ visualization embedded into the simulation code).

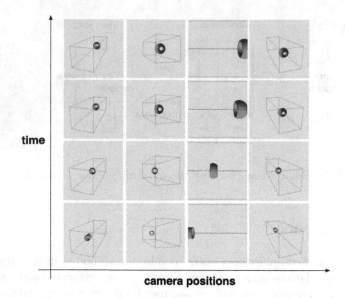

Fig. 3. Snapshot sequences of the propagation of a vortex ring, visualized from different viewpoints. It shows four of the 27 viewpoints. The vortex ring is visualized by an isosurface of the enstrophy density.

We perform asynchronous and synchronous in-situ visualizations on the same supercomputer system with the same simulation set-ups and physical parameters as in the previous experiment. We measure execution time for the simulations with different numbers of in-situ visualization cameras. As in the previous experiment, we solve the vortex ring formation with 24 MPI processes, applying in-situ visualizations every 100 simulation loops.

The numbers of cameras considered are 8, 16, 32, 64, and 128. We compare the elapsed time in each case for 10,000 simulation steps. In the embedded synchronous in-situ case, the total number of MPI processes is fixed, while in the membrane layer method, the number of MPI processes varies.

The result of the comparison is shown in Fig. 4. The elapsed time presented in this figure is defined as the average of five runs. This figure shows that the membrane method is always faster than the synchronous method and, more importantly, it keeps the execution time for the simulation constant even if we increase the number of in-situ visualization cameras.

4   Summary

We have proposed a movie-based visualization method for HPC [6]. In this method, the viewer interactively explores a database of movie files, rather than a numerical dataset, as in a standard post hoc visualization. The movie database is a collection of movie files that are labeled with the spatial coordinates of the viewpoint of each movie.

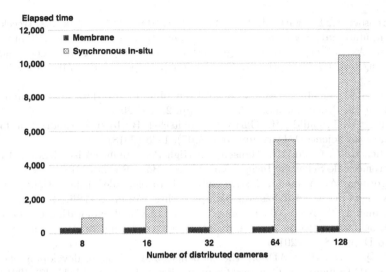

Fig. 4. Elapsed time for the vortex ring simulation with and without the membrane layer.

In this paper, we have proposed a framework for in-situ visualization on HPC systems that enables simulation researchers to apply a large number of in-situ visualizations with different viewpoints without slowing down the simulation. The key idea is to place a semipermeable membrane layer between the simulation and visualization programs. The membrane is implemented by two MPI programs that correspond to two faces, the front face and the back face of the membrane.

To confirm the effect of the membrane layer, we have compared two in-situ visualizations, with and without the membrane, for a computational fluid dynamics simulation. Without the membrane, the execution time grows as the number of cameras increases. In contrast, with the membrane, the execution time stays almost constant. This means that the increased computational load for a larger number of cameras is safely absorbed by an increased number of MPI processes for visualization, without affecting the simulation program. The membrane method described in this paper could be regarded as a promising approach to parallel in-situ visualizations, not only for movie-based visualization, but also to in-situ visualizations in HPC in general.

References

1. Ahrens, J., Jourdain, S., O'Leary, P., Patchett, J., Rogers, D.H., Petersen, M.: An image-based approach to extreme scale in situ visualization and analysis. In: International Conference for High Performance Computing, Networking, Storage and Analysis, SC, pp. 424–434 (2014)

2. Atanasov, A., Bungartz, H.J., Mehl, M., Mundani, R.P., Rank, E., Treeck, C.V.: Computational steering of complex flow simulations. In: Wagner, S., Steinmetz, M., Bode, A., Müller, M. (eds.) High Performance Computing in Science and Engineering, pp. 63–74. Springer, Berlin (2009). https://doi.org/10.1007/978-3-642-13872-0_6

3. Ayachit, U., et al.: ParaView catalyst: enabling in situ data analysis and visualization. In: Proceedings of ISAV 2015, pp. 25–29 (2015)

4. Bennett, J.C., Childs, H., Garth, C., Hentschel, B.: In situ visualization for computational ScienceD. Dagstuhl Rep. **8**(07), 1–43 (2018)

5. Bethel, E.W., Childs, H., Hansen, C.: High Performance Visualization: Enabling Extreme-scale Scientific Insight. CRC Press, Boca Raton (2013)

6. Kageyama, A., Yamada, T.: An approach to exascale visualization: interactive viewing of in-situ visualization. Comput. Phys. Commun. **185**, 79–85 (2014)

7. Kawamura, T., Noda, T., Idomura, Y.: In-situ visual exploration of multivariate volume data based on particle based volume rendering. In: Proceedings of ISAV 2016 D, pp. 18–22 (2017)

8. Liu, Q., et al.: Hello ADIOS: the challenges and lessons of developing leadership class I/O frameworks. Concurr. Comput.: Pract. Exp. **26**, 1453–1473 (2014)

9. Long, L.N., Plassmann, P.E., Sezer-Uzol, N., Jindal, S.: Real-time visualization and steering of large-scale parallel simulations. In: 11th International Symposium on Flow Visualization, pp. 1–12 (2004)

10. Ma, K.L.: In situ visualization at extreme scale: challenges and opportunities. IEEE Comput. Graph. Appl. **29**, 14–19 (2009)

11. Ma, K.L., Wang, C., Yu, H., Tikhonova, A.: In-situ processing and visualization for ultrascale simulations. J. Phys.: Conf. Ser. **78**(1), 1–10 (2007)

12. Ross, R.B., et al.: Visualization and parallel I/O at extreme scale. J. Phys.: Conf. Ser. **125** (2008)

13. Saffman, P.G.: Vortex Dynamics. Cambridge University Press, Cambridge (1992)

14. Sakamoto, N., Koyamada, K.: KVS: a simple and effective framework for scientific visualization. J. Adv. Simul. Sci. Eng. **2**(1), 76–95 (2015)

15. Tikhonova, A., Correa, C.D., Kwan-Liu, M.: Explorable images for visualizing volume data. In: IEEE Pacific Visualization Symposium 2010, PacificVis 2010 - Proceedings D (VIDi), pp. 177–184 (2010)

16. Wald, I., et al.: OSPRay - a CPU ray tracing framework for scientific visualization. IEEE Trans. Vis. Comput. Graph. **23**(1), 931–940 (2017)

17. Wald, I., Woop, S., Benthin, C., Johnson, G.S., Ernst, M.: Embree: a kernel framework for efficient CPU ray tracing. ACM Trans. Graph. **33**(4), 8 (2014)

18. Whitlock, B., Favre, J.M., Meredith, J.S.: Parallel in situ coupling of simulation with a fully featured visualization system. In: Eurographics Symposium on Parallel Graphics and Visualization, pp. 101–109 (2011)

Genetic Algorithm based EV Scheduling for On-Demand Public Transit System

Thilina Perera[✉], Alok Prakash, and Thambipillai Srikanthan

Nanyang Technological University, 50 Nanyang Avenue, Singapore 639798, Singapore
pere0004@e.ntu.edu.sg, {alok,astsrikan}@ntu.edu.sg

Abstract. The popularity of real-time on-demand transit as a fast evolving mobility service has paved the way to explore novel solutions for point-to-point transit requests. In addition, strict government regulations on greenhouse gas emission calls for energy efficient transit solutions. To this end, we propose an on-demand public transit system using a fleet of heterogeneous electric vehicles, which provides real-time service to passengers by linking a zone to a predetermined rapid transit node. Subsequently, we model the problem using a Genetic Algorithm, which generates routes and schedules in real-time while minimizing passenger travel time. Experiments performed using a real map show that the proposed algorithm not only generates near-optimal results but also advances the state-of-the-art at a marginal cost of computation time.

Keywords: Evolutionary computation · On-demand transit

1 Introduction

On-demand transit is a mode of transportation, where vehicles operate with flexible routes and schedules within a zone to serve passenger ride requests [5]. Static on-demand transit, where all ride requests are known a-priori, has been used in transportation of elderly and disabled. In contrast, dynamic on-demand transit, where ride requests are received in real-time has gained traction lately due to technological developments in GPS-based location tracking, mobile communication and wireless payment. Consequently, many dynamic on-demand transit services (e.g. Uber, Lyft, Grab) linking drivers with riders using a smartphone application have been introduced recently [10]. However, on-demand public transit, where a fleet of small/medium size buses deployed to provide shared transit services to commuters is still an emerging area. For example, in Singapore, the first phase of on-demand transit service was contracted only in 2018 [6]. In an on-demand public transit service, buses respond to real-time demand by dynamically routing to pick-up and drop-off passengers. The quality of service (QoS)

This research project is partially funded by the National Research Foundation Singapore under its Campus for Research Excellence and Technological Enterprise (CREATE) programme with the Technical University of Munich at TUMCREATE.

J. M. F. Rodrigues et al. (Eds.): ICCS 2019, LNCS 11540, pp. 595–603, 2019.
https://doi.org/10.1007/978-3-030-22750-0_56

of on-demand public transit services is measured by factors such as maximum waiting-/riding- time etc. Hence, it imposes additional constraints on the operator of the on-demand public transit service (in contrast to a fixed route service) to intelligently manage the fleet such that profitability is maintained while assuring high QoS. This is further exacerbated with strict government regulations [11] to curb the usage of petroleum-based fuels, which necessitates the use of alternatives such as electric vehicles (EV). However, usage of EVs, even for fixed-route public transit, is at an early stage even in developed countries [8].

To this end, we proposed an on-demand public transit service using EVs in [9]. However, the use of EVs for public transit necessitates novel algorithms for optimal, real-time scheduling and routing to satisfy multiple constraints such as EV capacity, driving range and passenger travel time. Existing works by Uchimura et al. [12], Kawamura and Mukai [4], Tsubouchi et al. [2], Markovíc et al. [7], Uehara et al. [3], Archetti et al. [1], Zhu et al. [14] and Perera et al. [9] propose on-demand bus services with flexible routes serving demand originating in a zone. However, some of the works have been tested only for a limited number of buses [4,12], scenarios with small number of passengers [2] and for relatively low rate of arrival of passenger requests [7]. Hence, the scalability of the proposed algorithms for relatively large number of passengers or vehicles and higher passenger incoming rates has not been validated. In some works, the proposed systems require advance reservations due to the excessive computation time of the algorithms and hence, is not suitable for real-time on-demand transit solutions [3]. Also, authors in [1,4,12] do not consider the limited driving range of EVs. Zhu et al. [14] proposed a dynamic path planning strategy based on a greedy algorithm for a peer-to-peer ride-sharing service. In the experiments, authors use characteristics of a Tesla model S car with five seats and supercharger facilities. Thus, the algorithm needs to be further verified with EV characteristics used for public transit. Similarly, our previous work in [9] proposed a hybrid GA to model an on-demand first mile transit service. However, we assume that at the scheduling instance, all EV have the same driving range and capacity, which in a practical scenario is limited to an instance such as service initialization. Thus, we advance the state-of-the-art by proposing a genetic algorithm (GA), which generates near-optimal results for a heterogeneous fleet of EV in real-time.

2   Proposed On-Demand Public Transit System

The proposed system consists of a heterogeneous fleet of EV (different seating capacities and driving ranges) dispersed in a zone, which respond to the ride-requests of passengers by picking them up from their origin and dropping them off at their destination. EV have a designated service zone, a maximum capacity and a driving range. Further, we assume that passengers request for the on-demand service near to their departure time using a smartphone. Thus, passenger origins, destinations and real-time traffic data can be saved in the database for computations. However, in this work, we assume that all passengers travel to a common destination such as the nearest transit hub. The objective is to devise a near-optimal set of routes and schedules in real-time for the fleet of EV such that total passenger travel time (waiting time + riding time) is minimized within constraints. We assume that the proposed GA is executed periodically and passenger requests received during that time as the demand and also that at execution time all EV are empty. Also, compared to [9], which guarantees a fixed maximum travel time constraint for each passenger, we introduce a location-based maximum travel time constraint. This modification ensures ride-sharing to be more practical and user-friendly for passengers nearer to the destination.

3 Methodology

Block diagram of the proposed GA is given in Fig. 1. First, the initial population is generated using random assignment and local search strategies. Next, the fitness of each chromosome is evaluated. Thereafter, parents are selected to perform genetic operations. After that, offspring are validated for the constraints followed by fitness evaluation. Finally, the population for the next generation is selected based on the fitness of individual chromosomes. At the end of population selection, if the termination criteria is satisfied the process outputs the schedule of the EV of the fittest chromosome, failing which the process is repeated.

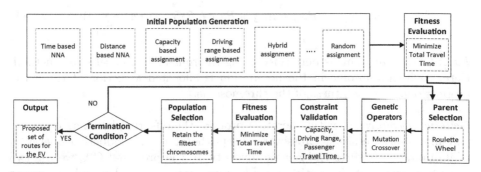

Fig. 1. Block Diagram of the Proposed GA

Initial Population Generation: Initial population is a set of feasible solutions (chromosomes), which satisfy all the constraints. Each chromosome represents the route of all EV. The route shows the sequence in which passengers (genes) are picked-up by an EV. Thus, we propose to use the path-based encoding scheme to represent a chromosome. Hence, each passenger is represented using an integer. Figure 2 shows an example of 10 passengers, 3 EV and a chromosome. In a traditional GA for the VRP, initial population is generated using random assignment of passengers to each vehicle. However, as shown by Xiang et al. [13] good initial solutions expedite the convergence of GA. Hence, in addition to random assignment, we use nearest neighbor assignment (NNA), priority based assignment and hybrid assignment strategies to allocate passengers to EV. In the NNA, each passenger is assigned to the nearest EV in terms of time or distance.

In the priority based assignment, priority is given to EV based on selected features such as EV capacity and driving range. For example, a chromosome can be generated by assigning passengers to the EV with the least remaining capacity. This strategy yields better results for cases with heterogeneous fleets. In the hybrid method, we use multiple combinations of the above strategies. However, in all chromosomes, EV capacity, driving range and passenger travel time constraints are checked to validate feasible solutions.

Fig. 2. Path-based Encoding of Chromosomes

Fitness Evaluation: Fitness of a chromosome is calculated by summing the fitness of each EV. Equation 1 shows the fitness of EV_1 of the chromosome in Fig. 2. Here, $R_{x,y}$ is the riding time from location x to y, and W_j is the waiting time for passenger j at the time of scheduling. However, based on the objective of the study, the reciprocal is used for evaluation of fitness.

$$F_{V_1} = (R_{V_1,8} + R_{8,5} + R_{5,3} + R_{3,2} + R_{2,Dest}) * 4 + \sum_{j=1}^{4} W_j; \tag{1}$$

Parent Selection: This is performed using the roulette wheel selection method. Here, each chromosome in the population is allocated a section of the wheel proportionate to the fitness of the chromosome.

Genetic Operators: Genetic operators produce the offspring by using crossover and mutation operators. Crossover reflects the actual reproduction operation while mutation ensures the genetic diversity. In the proposed GA, we use two crossover operators, heuristic and adoption crossover. In heuristic crossover shown in Fig. 3a, Clarke-Wright savings heuristic is used to find the gene for crossover. In adoption crossover shown in Fig. 3b, a single parent will transfer a gene to the other parent. Here also, Clarke-Wright savings heuristic is used to find the gene. Further, during adoption, we use the 2-opt local search to find the best position to insert the gene. For mutation, we use three operators, displacement, insertion and exchange mutation. In displacement mutation shown in Fig. 4a, the position of a randomly selected sub-tour of a parent is exchanged to optimize the route. We use the 2-opt local search heuristic to find the optimal position to exchange the sub-tour. In insertion mutation shown in Fig. 4b, a randomly selected gene is inserted into a different position in the route using 2-opt local search heuristic. In exchange mutation shown in Fig. 4c, two randomly selected genes from a parent are exchanged to produce a new offspring. This operator is used to preserve randomness of the population.

(a) heuristic   (b) adoption

Fig. 3. Crossover Operators

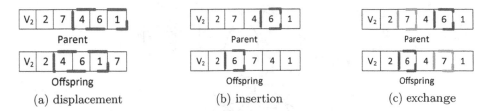

(a) displacement   (b) insertion   (c) exchange

Fig. 4. Mutation Operators

Constraint Validation: The offspring are validated for EV capacity and driving range violations and maximum passenger travel time violations. At this stage, an offspring which violates the constraints is removed from the population.

Population Selection: The proposed GA uses a constant population size. Hence, only the fittest chromosomes are retained in each generation. Thus, some parents and offspring are discarded from the population in each generation.

Termination Condition: The GA is terminated if we observe five consecutive generations without an improvement of the fittest chromosome. When it terminates, the GA outputs the best route that exist for the EV.

4   Results

GA proposed in Sect. 3 is implemented in C++. Baseline results are generated by modifying the optimal mathematical model presented in [9]. It is implemented using IBM ILOG CPLEX optimization studio 12.7.1 and solved using the in-built constraint programming solver. Runtime is measured on a PC with 16 GB RAM, running Windows 10 on an Intel Xeon E5-1650V2 CPU at 3.50 GHz. We generate test data using the real map shown in Fig. 5 in a locality surrounding a university. A university often provides fixed route shuttle services to the nearest rapid transit node, which further affirms the suitability of the selected locality. In this zone, demand can originate from any location. However, for clarity, we have limited them to 40 existing bus-stops. Figure 5 shows 5 bus-stops and the destination. In each experiment, origins of demand and supply are distributed randomly. Further, real-time traffic data is used to determine the travel time of passengers to the destination using existing public transit (Transit time) and private vehicles (Single Occupancy Vehicle (SOV) time). These values are used as upper and lower bounds of the travel time respectively for comparison. All parameters used in the experiments are given in Table 1. Here, a QoS of $2x$ implies that each passenger is guaranteed that travel time (waiting time + riding time) of the on-demand service will be less than double the time of a private vehicle ride (under prevailing traffic) to the destination. Further, we classify EV utilization as low (25%–49%), medium (50%–79%) and high (80%–100%).

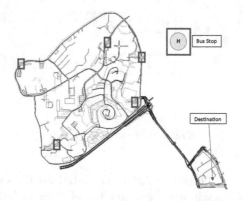

Fig. 5. Locality for Experimental Data Generation

Table 1. Parameter Values

Range	No. of EV	Capacity of EV	Driving range of EV	No. of passengers	QoS
Minimum	1	8	25 km	1	$2x$
Maximum	30	12	35 km	200	$4x$

Computation Time: Figure 6a and 6b show the variation of the computation time and average passenger travel time of the proposed GA in comparison to Perera et al. [9] (henceforth referred to as state-of-the-art (SoA)). Experiments are performed with 25 EV with a driving range of 30 km, seating capacity of 8 and QoS of $4x$. Even though computation time increases with demand for both cases, it is marginally higher of the proposed GA. This is due the increase in initial population and the number of genetic operators. However, as given in Fig. 6b, the benefit of the proposed GA is significant compared to the SoA. Figure 6b also shows the average SOV- and transit- times. Results show that the average travel time obtained from the proposed GA is closer to the SOV time when utilization is low. Also, when utilization is high the proposed GA performs significantly better compared to the SoA.

Performance: In all experiments, parameters are randomly selected from Table 1 except for a fixed QoS of $4x$. Figures 7a & 7b and Fig. 7c & 7d show the results for small and large test cases respectively. In general, the proposed

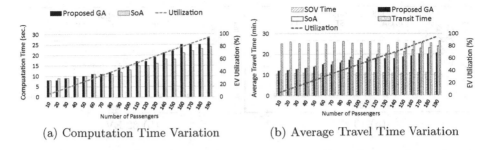

(a) Computation Time Variation   (b) Average Travel Time Variation

Fig. 6. Homogeneous Fleet

GA gives near optimal results. On average, we observe a deviation of **4.1%** of the travel time of the proposed GA compared to the baseline. Baseline results are not computed for large test cases due to the exponential time complexity. However, we observe that the travel time of the proposed GA is within the bounds except in Fig. 7c with 7 EV, where the travel time of the proposed GA is marginally higher than the upper-bound due to the high utilization. However, passengers still receive a door-to-door service compared to fixed route transit.

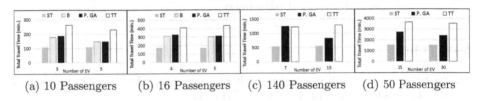

(a) 10 Passengers   (b) 16 Passengers   (c) 140 Passengers   (d) 50 Passengers

Fig. 7. Heterogeneous Fleet. ST: SOV Time, B: Baseline, P.GA: Proposed GA, TT: Transit Time

QoS and EV Utilization: Figure 8a and 8b show the variation of total travel time for different QoS and EV utilization levels for a case with 80 passengers. In all experiments, values of the parameters are randomly selected as given in Table 1 except for a fixed capacity of 10. Here, we observe the improved performance of the system (low QoS) at low utilization. This is due to the low ride-sharing ratio. Further, we observe that when utilization is high (Fig. 8a), the total travel time of the on-demand system marginally exceeding the transit time. However, passengers are still guaranteed a seat. Also, when both EV utilization and QoS values are high (80%, 2x, 20 min.) the on-demand system is unable to find a schedule which meets all the constraints. Further, in instances such as Fig. 8a, we observe that the proposed location-based QoS method generating a route in comparison to the fixed QoS method affirming that passengers get a better service from the location-based QoS method.

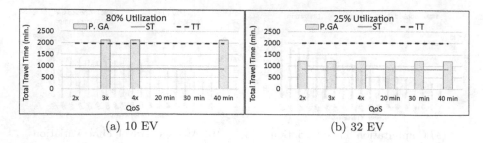

(a) 10 EV   (b) 32 EV

Fig. 8. Total Travel Time Variation with Location-based and Fixed QoS. P.GA: Proposed GA, ST: SOV Time, TT: Transit Time

5   Conclusion

This work proposes an on-demand public transit system using EV, which satisfies the point-to-point transit requests of passengers and a GA to solve the problem. Experiments performed using a real map show that the proposed GA not only generates near-optimal routes and schedules, but also improves the SoA at a marginal cost of computation time. In future, we plan to consider scheduling of EV in the presence of allocated passengers.

References

1. Archetti, C., et al.: A simulation study of an on-demand transportation system. Int. Trans. Oper. Res. **25**, 1137–1161 (2018)
2. Tsubouchi, K., et al.: Innovative on-demand bus system in Japan. IET Intell. Transp. Syst. **4**(4), 270–279 (2010)
3. Uehara, K., et al.: Evaluation of a hierarchical cooperative transport system using demand responsive bus on a dynamic simulation. IEICE Trans. Fundam. **E99–A**(1), 310–318 (2016)
4. Kawamura, K., Mukai, N.: Optimization of transport plan for on-demand bus system using electrical vehicles. In: Velásquez, J.D., Ríos, S.A., Howlett, R.J., Jain, L.C. (eds.) KES 2009. LNCS (LNAI), vol. 5712, pp. 656–663. Springer, Heidelberg (2009). https://doi.org/10.1007/978-3-642-04592-9_81
5. Koffman, D.: Operational experiences with flexible transit services. TCRP Synthesis of Transit Practice (2004)
6. Land Transport Authority, Singapore: LTA awards first phase of tender for on-demand bus service trial (2018). https://bit.ly/2GUD2EG
7. Marković, N., et al.: Optimizing dial-a-ride services in Maryland: benefits of computerized routing and scheduling. Transp. Res. C Emer. Tech. **55**, 156–165 (2015)
8. Nichola Groom, Reuters: U.S. transit agencies cautious on electric buses despite bold forecasts (2017). https://reut.rs/2nSSZq5
9. Perera, T., Prakash, A., Gamage, C.N., Srikanthan, T.: Hybrid genetic algorithm for an on-demand first mile transit system using electric vehicles. In: Shi, Y., et al. (eds.) ICCS 2018. LNCS, vol. 10860, pp. 98–113. Springer, Cham (2018). https://doi.org/10.1007/978-3-319-93698-7_8

10. Shaheen, S., et al.: Shared Mobility: Current Practices and Guiding Principles (2016). https://bit.ly/2v5LZtx
11. The Guardian: France to ban sales of petrol and diesel cars by 2040 (2017). https://bit.ly/2utuWNF
12. Uchimura, K., et al.: Demand responsive services in hierarchical public transportation system. IEEE TVT **51**(4), 760–766 (2002)
13. Xiang, Z., et al.: A fast heuristic for solving a large-scale static dial-a-ride problem under complex constraints. Eur. J. Oper. Res. **174**(2), 1117–1139 (2006)
14. Zhu, M., et al.: An online ride-sharing path-planning strategy for public vehicle systems. IEEE T-ITS (2018)

Short-Term Irradiance Forecasting on the Basis of Spatially Distributed Measurements

Antonino Laudani, Gabriele Maria Lozito, Valentina Lucaferri, and Martina Radicioni[(✉)]

Roma Tre University, Rome, Italy
martina.radicioni@uniroma3.it

Abstract. The output power of photovoltaic (PV) systems is heavily influenced by mismatching conditions that can drastically reduce the power produced by PV arrays. The mismatching power losses in PV systems are mainly related to partial or full shading conditions, i.e. non-uniform irradiation of the array. An essential point is the detection of the irradiance level in the whole PV plant. The use of irradiance sensors is generally avoided because of their cost and necessity for periodic calibration. In this work, an Artificial Neural Network (ANN) based method is proposed to forecast the irradiance value of each panel constituting the PV module, starting from a number of spatially distributed analytical irradiance computations on the array. A 2D random and cloudy 12 h irradiance profile is generated considering wind action; the results show that the implemented system is able to provide an accurate temporal prevision of the PV plant irradiance distribution during the day.

Keywords: Photovoltaic · Mismatch · Irradiance · Neural networks

1 Introduction

The technological progress of the last decades together with the demographic increase have shown the need of finding renewable energy sources able to stem global warming and pollution produced by fossil fuels and to meet the global energy demand [6]. These features have led to a rapid expansion of PV technologies in the world market. However, although several advances have been made in PV technology, some factors still persist that severely lower the efficiency of PV devices. Differences in the principal parameters of the modules of a PV system may produce discrepancies in the modules behavior; this problem is addressed as mismatching [17]. Among mismatch effects, shading is the principal concern as it is the main aspect that decreases the output power [1,5]. Shading is also responsible for overheating and aging of the cells or modules [15]: the reduction in current production affecting a shaded PV cell influences the whole PV device behavior since hot spots may appear causing irreversible damages. A common practice aimed at limiting the effects coming from these power losses is the use

© Springer Nature Switzerland AG 2019
J. M. F. Rodrigues et al. (Eds.): ICCS 2019, LNCS 11540, pp. 604–611, 2019.
https://doi.org/10.1007/978-3-030-22750-0_57

of bypass diodes that are generally connected in parallel to each cell or group of cells [2,18]; however, this strategy makes multiple peaks arise in the P-V characteristic of the device and this heavily affects the efficiency of the system showing the need of Maximum Power Point Tracking (MPPT) methods [7,9]. As a matter of facts, the presence of multiple peaks makes the detection of the Global MPP (GMPP) hard since, despite the accuracy of traditional MPPT methods, they may remain trapped in Local MPP (LMPP). Hence, in this situation, more sophisticated MPP tracking strategies are needed. It is evident the impact that the identification of a shading scenario has on the choice and use of the afore mentioned strategies. The particular shading pattern on a device affects the nature of the procedure to be applied. As a result, a tool able to provide a time prediction of the shadow distribution is an essential requirement. In this paper a method is proposed to forecast the trend of the shadows on a PV array in order to obtain a satisfying shading prevision in a 12 h period. The approach makes use of an ANN based method and does not need any irradiance sensors since the measurement of such a quantity is performed by an analytical strategy based on the well-known five-parameters model for PV systems. As a matter of facts, it is better to avoid the use of pyranometers because of their cost and tricky setup; to face this problem several alternative strategies have been developed to assess the value of irradiance without making use of sensors [3,4,14]. The cost reduction and the simplicity of using this analytical method make devices based on it integrable in large-scale PV systems. Moreover, the ANN choosen ensures accuracy and speed, despite of its simple architecture [12]. The paper is organized as follows: in Sect. 2 the five-parameters model, also known as one-diode model, is explained and in the second the closed-form formulation the analytical extraction of irradiance is shown; the ANN based method developed is presented in Sect. 3; in Sect. 4 the conclusions will follow.

2   Analytical Irradiance Sensor

2.1   Circuital Model for PV Devices

Several mathematical models for PV cells and modules are available in literature; among them the most used is the one-diode model also known as five-parameters model. The model, shown in Fig. 1 allows to express the current I of the device as a function of the five parameters and of the number of cells connected in series, N_s:

$$I = I_{irr} - I_D - \left(\frac{V + IR_S N_s}{R_P N_s}\right) \tag{1}$$

with the diode current, I_D, equivalent to:

$$I_D = I_0 exp(\frac{V + IR_S N_s}{(\frac{nKT}{q})N_s}) \tag{2}$$

The parameters appearing in the afore mentioned formulas are here explained: I_{irr} is the photo-current, $q = 1.602 \times 10^{-9}C$ is the electron charge,

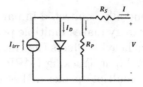

Fig. 1. One-diode model for a PV device.

$k = 1.38 \times 10^{-23} \frac{J}{K}$ the Boltzmann constant, n the diode ideality factor, R_s and R_p the series and shunt resistances and I_0 the diode reverse saturation current. The parameters appearing in (1) are, in turn, dependent on the values of irradiance and temperature. Such dependencies are shown in the following part. It has to be mentioned that SRC, i.e. *Standard Reference Condition*, is used to refer to parameters measured in the condition of $T = 25\ C°$ and $G = 1000\ \frac{W}{m^2}$:

$$I_{irr} = I_{irr(SRC)} + K(T - T_{SRC}))\frac{G}{G_{SRC}} \tag{3}$$

$$I_0 = I_{0,SRC}\left(\frac{T}{T_{SRC}}\right)^3 e^{\left[\frac{E_{g,ref}}{kT_{SRC}} - \frac{E_g}{kT}\right]} \tag{4}$$

$$R_p = R_{p,SRC}\frac{G_{SRC}}{G} \tag{5}$$

$$R_s = R_{s,SRC} \tag{6}$$

$$n = n_{SRC} \tag{7}$$

$$E_g = E_{g,SRC}(1 - 0.0002677(T - T_{SRC})) \tag{8}$$

The extraction of the five parameters [10,11,13] is a key point since allows the application of the analytical formula for the calculation of irradiance through the PV plant. This paper makes use of the procedure proposed in [11] where an efficient algorithm ensuring global convergence; this strategy allows to provide feasible solution to the system of non-linear equations.

2.2   Closed-Form Formulation for Irradiance

The topic of irradiance sensing is of central importance for PV applications since needs to be measured on the surface of the device under exam. As a matter of fact, while measuring temperature is trivial, knowing the value of irradiance is quite hard. Such a quantity is highly dependent on the inclination of the panel with respect to sunrays, therefore, to exactly estimate the value of irradiance shades have to be carefully taken into account. As a result of this, a standard practice consists in placing sensors close to the device. This leads to some relevant issues: firstly, measurements based on the use of pyranometers are generally expensive because of the cost of the device itself and the need of periodic calibrations, then, the measurement system is quite difficult to be set up since the pyranometer should be hold perfectly parallel to the PV plant and, lastly, the PV device may have a different spectral response from that of the sensor. Hence, the analytical calculation of irradiance on the PV array is a suitable option allowing to avoid the mentioned drawbacks. In [3] the authors propose a closed-form expression for solar irradiance based on the knowledge of the five parameter of the circuital model of the PV cell:

$$\frac{G}{G_{STC}}\left(N_p I_{irr,STC} + N_p \alpha_T \left(T - T_{STC}\right) - \frac{V + I N_s R_{s,STC}/N_p}{N_p N_s R_{p,STC}/N_p}\right) =$$
$$I + N_p I_0 \left[e^{\left(\frac{V + I N_s R_{s,STC}/N_p}{N_S n k T}\right)} - 1\right] \tag{9}$$

Given these information together with the value of temperature and the measurement of operating voltage and current, the formula of Eq. (9) is able to accurately estimate irradiance in any point of the PV array.

3 Irradiance Prediction System

3.1 Irradiance Calculation Based on Spatial Distributed Measurements

The proposed approach is based on a set of distributed measurements of solar irradiance on a number of modules of the array, that are collected at a certain timestep t. Let us focalize on a set of such modules, for example 5 elements arranged to form a cross shape; at each timestep, t, the triplet $V(k, t)$, $I(k, t)$ and $T(k, t)$ of each kth device is measured. In order to use expression (9), the five parameter of the model have to be extracted. Following the method shown in [11], a unique and reasonable solution for the five parameters identifying the one-diode model is found. Such a procedure is based on some algebraic manipulations of (1), allowing to establish that, among the five parameters, R_s and n are the only independent variables of the problem: thus the search space is a two-dimensional one and it is possible to use very fast and accurate algorithms for searching the solution. Then, the operating point of each module and its temperature are collected through measurements on the device; these information, togheter with the values of the five parameters are injected into (9) to obtain the values of irradiance on the five cross-shape arranged modules.

3.2 Forecasting an Irradiance Profile by Using Artificial Neural Networks

The mapping capabilities of an ANN depend on its non-linear nature that can be attributed to the non-linear activation function. In absence of this latest element the ANN would be equivalent to a linear interpolator. The implemented ANN is able to forecast the future irradiance on any module of the cross at time $(t + 1)$ by knowing the past values at instant t on the whole cross, that we suppose constituted by N modules. Firstly, a random 12 h (from 6 a.m. to 6 p.m.) 2D

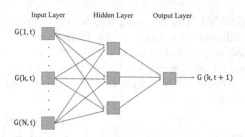

Fig. 2. Architecture of the proposed FFNN.

cloudy irradiance profile is generated. In particular, each PV module of the array is supposed to be irradiated by a waveform coming from the overlapping of a set of randomly generated spikes to a Gaussian waveform with a peak value of $1000\ W/m^2$. The array is a square matrix with 100 points per side: each couple of coordinates identifies a single PV module of the array. The implemented ANN is a multi-input and single-output feed-forward neural network (FFNN) with a hidden layer constituted by 3 neurons, Fig. 2. 10000 timesteps have been considered: 1000 have been used for the training set and the other 9000 constitute the validation one; the training samples are choosen randomly on the whole set in order not to have them accumulated in a specific area. Each of the inputs of the ANN is a row with the values of irradiance of one of the N modules of the array calculated for each timestep, while the target is a row containing the values of irradiance on the kth module, with k to be choosen from the N panels set, with an assigned time shift. The greater the time shift, the greater the forecast time gap. The prevision is efficient even when increasing the elements of the cross. The presented ANN achieves satisfactory performances, with a Mean Squared Error below 10% on the whole dataset, even if it does not involve a recurrent architecture (RNN). As a matter of facts, the choice of an FFNN allows to avoid problems that generally affect RNNs, i.e. the issue of vanishing or exploding gradients [8,16], that make the training of the neural network hard, especially when dealing with large datasets. Moreover, the use of RNNs requires a very careful initialization of the neural network parameters. In addition, the training of an RNN is performed through complex and computationally demanding algorithms, unlike an FFNN that can be efficiently trained by using simply implementable on-line algorithms; this aspect is relevant since makes the neural network suitable for embedded solutions, able to provide a real-time training of the FFNN. The proposed method is described in the flowchart of Fig. 3. The results of the presented solution are shown in Fig. 4 for a PV cross of 5 modules. It can be seen that the FFNN is able to correctly fit the irradiance trend on the whole time frame.

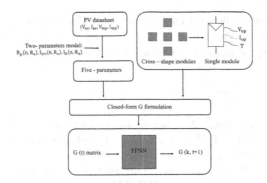

Fig. 3. Flowchart of the implemented algorithm.

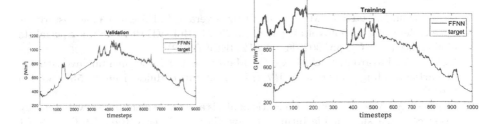

Fig. 4. Training and validation results of the FFNN-based solution.

4   Conclusions

This work has presented an ANN based procedure able to predict the time trend of solar irradiance on a PV array. A 12 h random cloudy profile has been generated and its development over time has been simulated by using a shift function. A preliminary part has seen the extraction of irradiance values over the PV plant. A performing analytical tool has been used able to accurately assess irradiance on each panel avoiding to embed costly pyranometers into the PV set-up. The strength of the closed-form irradiance computation consists in the requirement of simply obtainable information, i.e. operating voltage and current and cell or module temperature. Thus, the formulation allows an easy mapping of irradiance along the whole PV structure. The prediction part of the paper, has been based on the use of FFNNs that have been proved to be powerful tools with accurate forecasting capabilities, despite their static nature. The error, both on training and on test, is of about 10%, a satisfying percentage when dealing with time previsions. Moreover, the simplicity of the feed-forward architecture, consisting in a unique hidden layer, ensures fast training of the network: the time needed to train a single FFNN does not exceed half a second. Definitely, the developed method provides a fast and accurate way to forecast irradiance; as a matter of facts, the prediction of such a quantity is a central issue when addressing the problem of partial shading since irradiance knowledge is a critical factor for the adoption of strategies aimed at retrieving the power lost, thus allowing to increase the efficiency of any PV plant.

References

1. Alonso-Garcia, M., Ruiz, J., Chenlo, F.: Experimental study of mismatch and shading effects in the I–V characteristic of a photovoltaic module. Sol. Energy Mater. Sol. Cells **90**(3), 329–340 (2006)
2. Bauwens, P., Doutreloigne, J.: Reducing partial shading power loss with an integrated smart bypass. Sol. Energy **103**, 134–142 (2014)
3. Carrasco, M., Laudani, A., Lozito, G.M., Mancilla-David, F., Riganti Fulginei, F., Salvini, A.: Low-cost solar irradiance sensing for PV systems. Energies **10**(7), 998 (2017)
4. Cruz-Colon, J., Martinez-Mitjans, L., Ortiz-Rivera, E.I.: Design of a low cost irradiance meter using a photovoltaic panel. In: 2012 38th IEEE Photovoltaic Specialists Conference (PVSC), pp. 002911–002912. IEEE (2012)
5. Dolara, A., Lazaroiu, G.C., Leva, S., Manzolini, G.: Experimental investigation of partial shading scenarios on PV (photovoltaic) modules. Energy **55**, 466–475 (2013)
6. Hepbasli, A.: A key review on exergetic analysis and assessment of renewable energy resources for a sustainable future. Renew. Sustain. Energy Rev. **12**(3), 593–661 (2008)
7. Ishaque, K., Salam, Z.: A review of maximum power point tracking techniques of pv system for uniform insolation and partial shading condition. Renew. Sustain. Energy Rev. **19**, 475–488 (2013)
8. Jozefowicz, R., Zaremba, W., Sutskever, I.: An empirical exploration of recurrent network architectures. In: International Conference on Machine Learning, pp. 2342–2350 (2015)
9. Koutroulis, E., Blaabjerg, F.: A new technique for tracking the global maximum power point of PV arrays operating under partial-shading conditions. IEEE J. Photovoltaics **2**(2), 184–190 (2012)
10. Laudani, A., Fulginei, F.R., Salvini, A.: High performing extraction procedure for the one-diode model of a photovoltaic panel from experimental I–V curves by using reduced forms. Sol. Energy **103**, 316–326 (2014)
11. Laudani, A., Fulginei, F.R., Salvini, A.: Identification of the one-diode model for photovoltaic modules from datasheet values. Sol. Energy **108**, 432–446 (2014)
12. Laudani, A., Lozito, G.M., Fulginei, F.R., Salvini, A.: On training efficiency and computational costs of a feed forward neural network: a review. Comput. Intell. Neurosci. **2015**, 83 (2015)
13. Laudani, A., Riganti Fulginei, F., Salvini, A., Lozito, G.M., Coco, S.: Very fast and accurate procedure for the characterization of photovoltaic panels from datasheet information. Int. J. Photoenergy **2014** (2014)
14. Mancilla-David, F., Fulginei, F.R., Laudani, A., Salvini, A.: A neural network-based low-cost solar irradiance sensor. IEEE Trans. Instrum. Meas. **63**(3), 583–591 (2014)
15. Manganiello, P., Balato, M., Vitelli, M.: A survey on mismatching and aging of PV modules: the closed loop. IEEE Trans. Ind. Electron. **62**(11), 7276–7286 (2015)

16. Pascanu, R., Mikolov, T., Bengio, Y.: On the difficulty of training recurrent neural networks. In: International Conference on Machine Learning, pp. 1310–1318 (2013)
17. Petrone, G., Ramos-Paja, C.: Modeling of photovoltaic fields in mismatched conditions for energy yield evaluations. Electr. Power Syst. Res. **81**(4), 1003–1013 (2011)
18. Silvestre, S., Boronat, A., Chouder, A.: Study of bypass diodes configuration on PV modules. Appl. Energy **86**(9), 1632–1640 (2009)

Multi-GPU Acceleration of the iPIC3D Implicit Particle-in-Cell Code

Chaitanya Prasad Sishtla[✉], Steven W. D. Chien, Vyacheslav Olshevsky,
Erwin Laure, and Stefano Markidis

KTH Royal Institute of Technology, Stockholm, Sweden
{sishtla,wdchien,slavik,erwinl,markidis}@kth.se

Abstract. iPIC3D is a widely used massively parallel Particle-in-Cell
code for the simulation of space plasmas. However, its current implementation does not support execution on multiple GPUs. In this paper, we
describe the porting of iPIC3D particle mover to GPUs and the optimization steps to increase the performance and parallel scaling on multiple
GPUs. We analyze the strong scaling of the mover on two GPU clusters and evaluate its performance and acceleration. The optimized GPU
version which uses pinned memory and asynchronous data prefetching
outperform their corresponding CPU versions by $5-10\times$ on two different
systems equipped with NVIDIA K80 and V100 GPUs.

Keywords: GPU porting and optimization · iPIC3D · Particle-in-Cell

1 Introduction

The advent of large supercomputers with multiple accelerators per computational node is impacting the development of large scientific applications. The
two current largest supercomputers in November 2018 Top500 list, *Summit* and
Sierra, feature six and four V100 NVIDIA GPUs per node respectively providing
a theoretical peak performance of 750 and 500 Tops/s (operations in mixed precision) [7] per node. Hence, it is important to exploit the computational power
from GPUs on supercomputers. The PIC method is one of the main tools for the
simulation of plasmas [1]. The method was initially developed in the late Fifties
and early Sixties and then further improved by using more sophisticated numerical schemes, such as semi-implicit and fully-implicit schemes [8], and combining
fluid and kinetic equations for plasmas [11].

iPIC3D has been designed for the kinetic simulation of space plasmas on large
supercomputers [9]. Its main application is the study of magnetic reconnection in
Earth's magnetotail [14] and dayside magnetopause, kinetic turbulence [13] and
interaction of solar wind with Earth's magnetosphere [15,16] and comets [17]. It
works as stand-alone code and as part of a multi-physics framework, called Space
Weather Modeling Framework (SWMF), for the simulation for space weather
[3]. During the last decades the code has been improved by using advanced

© Springer Nature Switzerland AG 2019
J. M. F. Rodrigues et al. (Eds.): ICCS 2019, LNCS 11540, pp. 612–618, 2019.
https://doi.org/10.1007/978-3-030-22750-0_58

parallelization strategies, optimized I/O that have been developed during European-Commission funded EPiGRAM [10] and SAGE projects [12]. It is written in C++ and uses MPI for internode communication. The code has shown scalability up to 80% parallel efficiency on one million MPI processes [10]. However, iPIC3D does not currently support execution on GPUs which limits its usage on supercomputers with GPUs. Since 2008 several studies focused on PIC code porting to efficiently use GPU systems. The first seminal work on this topic is by Stantchev et al. [18]: it presents a PIC porting to GPUs focusing on the optimization of the interpolation step. Optimization of the data layout in PIC codes are presented in Refs. [5,6]. However, all these previous works use simple formulation of PIC algorithm, and porting of a semi-implicit PIC method does not exist in the literature. Our work aims to fill this gap and present the porting of a semi-implicit PIC method to GPUs. In particular, we focus on describing the steps for porting the iPIC3D application on multi-GPU systems: Tegner and Kebnekaise [4] featuring NVIDIA K80 and Tesla V100 GPUs.

2 Methodology

The iPIC3D simulation is initialized first by setting particle positions and velocities and assigning electric and magnetic field values on grid points (**Initialization**). After the initialization, a computational cycle is repeated until the end of simulation. Each computational cycle consists of three basic steps: **Fields solver** - where the electric and magnetic fields are calculated from the semi-implicit formulation of Maxwell's equations on a grid by solving a linear system, **Particle Mover** - where new particle positions and velocities are computed using the electric and magnetic field values on the grid points and interpolating them at the particle positions, and **Moments Calculation** - where particle moments of the distribution function, such as density, current and pressure are calculated on the grid by interpolation.

We performed a dedicated profiling of the problem that is considered in this work, the so-called GEM (Geospace Environmental Modelling) challenge [2] as shown in Table 1. Particle mover is clearly the most expensive operation and so we focus on the porting and optimization of the iPIC3D particle mover while calculation of fields and moments still remain on CPU. Detailed descriptions about the particle mover can be found in [8,11,17]. We present the performance of the multi-GPU iPIC3D by reporting the harmonic mean of our main figure of merit - Millions of Particles Advanced per second (MPA/s) obtained by dividing the total number of particles in the simulation by the average time spent in the particle mover per computational cycle. The standard deviation is plotted as an error bar and shows minimum variability between different simulation runs.

Each numerical experiment has a computational grid consisting of $64 \times 64 \times 32$ cells, with 4 particle species. We perform a maximum of 3 predictor-corrector iterations in the mover. Unless otherwise is specified, 216 particles per cell is used for each species and each simulation is repeated six times with the first being a warmup run. For simplicity, when we refer to one K80 GPU in subsequent text, we refer to one GK210 GPU engine.

Table 1. Percentage of execution time per cycle for the three PIC steps in a typical iPIC3D benchmark run on a CPU, varying the number of particles per cell (ppc).

Part of code	% time spent				
	27ppc	64ppc	125ppc	216ppc	343ppc
Fields solver	6.12	2.79	1.47	0.87	0.55
Particle mover	68.81	71.42	72.23	72.77	73.14
Moments calculation	25.04	25.76	26.31	26.35	26.29

2.1 Porting to Multi-GPU Systems

We use NVIDA CUDA for porting iPIC3D to GPU and associate each MPI process with one GPU device which is allocated according to the rank of the MPI process. Since the particle mover is responsible for updating the new position of particles, it requires the following information: Grid (geometry) information about the particle's neighbour nodes which remains unchanged throughout the simulation, values of the electromagnetic field on these grid nodes which is updated every computational cycle, and current position and velocity of the particle to be updated. The number of particles in the simulation may vary due to open boundary conditions and injection of particles from the simulation boundary.

For each particle species, GPU kernel of the particle mover is launched such that each thread is responsible for updating one particle. We thus enable more particles per cell while having limited GPU memory. Double precision is used in the entire computation process. We improve the particle mover in three stages, each being an optimization of the former.

Simple Synchronous Implementation. We first allocate memory for grid and field data on both the host and device as the sizes are known beforehand. Second, the grid data is copied during the initialization, and it will remain on the device for the entire course of the simulation. We allocate all the remaining available device memory for particles. To avoid future resizing, we allocate the same amount of memory for the particle data on the host. In each cycle the field data is copied to device memory, the CUDA kernel is launched and the new particle positions and velocities are copied back to host memory. The above process is repeated for each particle species and is performed in a blocking, synchronous manner.

Host Memory Pinning. CUDA performs Direct Memory Access (DMA) through PCI-E to move data between the device and host. However, since the operating system allocates virtual memory in a pageable fashion, data in this address space must first be copied to a staging area in memory before DMA can be performed. We implement host memory pinning by replacing the allocation of host memory for field and particles with `cudaMallocHost()` or `cudaHostAlloc()`, the APIs provided by CUDA to perform page-locked memory allocation. In this prototype allocation, we uniformly allocate 3GB as the maximum possible size for each particle species.

Data Prefetching to GPU Memory. The CUDA API supports asynchronous memory transfers by means of `cudaMemcpyAsync()` which can be effectively used to overlap CUDA calls with CPU computation. We use CUDA stream to ensure the order of kernel execution and data movement that are performed asynchronously to the host. The implementation of data prefetching is summarized in Fig. 1. Steps 1 and 3 are asynchronous data transfers which are respectively called before and after the fields solver. In Step 5 the mover kernel is followed by `cudaStreamSynchronize()`, and after synchronization, the updated data for the first species is copied to the host, and the asynchronous copying of the particle data of the next species to the GPU is initiated.

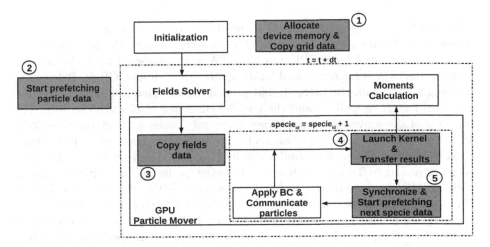

Fig. 1. The flowchart of the iPIC3D code with the GPU particle mover using data prefetching is shown. The white blocks correspond to instructions executed on the CPU while the grey blocks correspond to CUDA code. Dashed lines indicate where in the host the relevant CUDA code is called.

3   Results

Even a simple synchronous ('naive') porting of the particle mover gives a substantial performance benefit on one GPU as demonstrated by the profiling results presented in Table 2. The acceleration, computed as the ratio of the mover execution times $A = T_{CPU}/T_{GPU}$, ranges from 4–4.5 on K80 to 8.7 on V100. Each improvement in the mover, use of pinned memory and prefetching, makes execution faster. The prefetch mover gives 30% better acceleration than the simple mover on K80. The striking $A = 25$ acceleration of the prefetch mover on V100 is affected by the slow execution on the corresponding CPU.

To investigate the parallel performance of the three movers we did a strong scaling study. The same experiment (with the same initial configuration and number of particles) was repeated employing 2, 4, and 8 MPI processes with 1 GPU per MPI process on Tegner Haswell+K80 nodes. As a reference, a purely

Table 2. The average time spent in the particle mover over 10 cycles using a single GPU (one MPI process) in different testing environments.

Type of node	Particle mover execution times (in seconds)			
	CPU	Naive	Pinned	Prefetch
Tegner (Haswell+K80)	15.33	3.28	3.05	2.44
Kebnekaise (Broadwell+K80)	15.20	3.84	3.44	2.87
Kebnekaise (Skylake+V100)	36.82	4.20	2.02	1.43

CPU study was also performed. The results of the scaling study for Tegner (Haswell+K80) are shown in Fig. 2a. The prefetch mover exhibits a peak performance (measured for 8 GPUs) of 243MPA/s, as compared to 206MPA/s for pinned, 146MPA/s for naive, and 52MPA/s for the run at 8 CPUs. Relative to the single-GPU run, the prefetch and pinned movers give the parallel speedup of $S = 6.1$ at $N = 8$ GPUs and the parallel efficiency of $E = S/N = 76\%$. The speedup of the naive mover implementation only reaches 4.7 giving $E = 59\%$. The same parallel scaling experiments for up to 16 MPI processes on Kebnekaise Broadwell+K80 nodes show similar results (Fig. 2d). The parallel speedup on 8 nodes is similar to Tegner's K80 nodes, however the figures are somewhat lower, with 222MPA/s given by the prefetch mover on 8 GPUs. The parallel efficiency of the pinned and prefetch movers at 16 GPUs is 73%, while for the naive mover it is only 44%. Therefore, asynchronous prefetching of particle data to GPUs is essential for the parallel performance of the mover. Scaling performance exhibited by the experiments on the Kebnekaise V100 nodes is shown in Fig. 2c. The results of the scaling study on V100 GPU show that the movers perform significantly better than on K80 both in terms of speedup and absolute figures. The prefetch mover performs the best with a peak performance of 622MPA/s, as compared to 437MPA/s for pinned and 204MPA/s for the naive mover using 8 GPUs. We get a nearly optimal scaling for the prefetched mover which exhibits a parallel speedup of $S = 7.9$ at $N = 8$ GPUs resulting in a parallel efficiency of 98.8%. The parallel efficiency for the pinned and naive movers are 97.7% and 94.6% respectively.

We ran a series of experiments on Tegner's Haswell+K80 nodes in the same setup as above, with prefetch mover, varying the number of particles in the system and the number of GPUs in order to study the impact of number of particles on parallel performance. The results are summarized in Fig. 2b. The parallel scaling appears very similar for different number of particles. However, there is a clear tendency of the simulation with 125 particles/cell to outperform others. The peak performance for 125 particles/cell at 8 GPUs is 246MPA/s, while for 27 particles/cell it is lower, 212MPA/s. The degrading in performance for the higher number of particles is not so significant, with 240MPA/s for the run with 343 particles/cell.

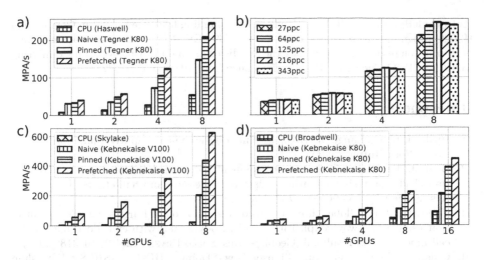

Fig. 2. (a) The performance of the GPU porting schemes compared on Tegner (using the Haswell CPU and the K80 node). (b) The performance of the prefetched GPU porting schemes compared on the K80 nodes of Tegner by varying the number of particles per cell in the simulation. The performance of the GPU porting schemes compared for Kebnekaise. (c) nodes with Broadwell CPUs and K80 GPUs; (d) nodes with Skylake CPUs and V100 GPUs.

4 Discussion and Conclusion

We have designed and implemented porting of the particle mover in the semi-implicit PIC code to GPUs. Numerical experiments using a typical space plasma physics simulations have shown that GPU movers clearly outperform the purely CPU implementation, being 5–10 MPA/s faster. The experiments on K80 and V100 GPUs have shown that memory pinning and prefetching is essential to reach a good parallel performance. The best performance and scaling efficiency is exhibited by the prefetch mover. Its parallel efficiency reaches 73% on 16 K80 GPUs, while the naive implementation of the mover results in the parallel efficiency of only 44%. Finally, the prefetch mover does not exhibit a substantial dependency of its performance and scaling on the number of particles in the simulation. The question, whether the performance of a GPU-ported particle mover depends on the number of particles, or some other parameters of the system, such as the number of predictor-corrector iterations, is worth further investigation. Further work should also include porting of the particle distribution function moment calculation to GPUs, or its possible merging with the mover phase of the computational cycle. Implementation of the efficient particle mover on GPUs is the most essential feature required for adapting iPIC3D to modern and forthcoming HPC architectures, and optimizing performance of the large-scale kinetic plasma simulations.

Acknowledgements. This work has received funding from the European Commission H2020 program, Grant Agreement No. 801039 (EPiGRAM-HS). Experiments were performed on resources provided by the Swedish National Infrastructure for Computing (SNIC) at PDC Center for High Performance Computing and HPC2N.

References

1. Birdsall, C.K., Langdon, A.B.: Plasma Physics via Computer Simulation. CRC Press, Boca Raton (2004)
2. Birn, J., Hesse, M.: Geospace Environment Modeling (GEM) magnetic reconnection challenge: resistive tearing, anisotropic pressure and hall effects. J. Geophys. Res. Space Phys. **106**(A3), 3737–3750 (2001)
3. Chen, Y., et al.: Global three-dimensional simulation of Earth's dayside reconnection using a two-way coupled magnetohydrodynamics with embedded particle-in-cell model: initial results. J. Geophys. Res. Space Phys. **122**(10), 10–318 (2017)
4. Chien, S.W.D., et al.: TensorFlow Doing HPC. ASHES Workshop arXiv:1903.04364, March 2019
5. Decyk, V.K., Singh, T.V.: Adaptable particle-in-cell algorithms for graphical processing units. Comput. Phys. Commun. **182**(3), 641–648 (2011)
6. Decyk, V.K., Singh, T.V.: Particle-in-cell algorithms for emerging computer architectures. Comput. Phys. Commun. **185**(3), 708–719 (2014)
7. Markidis, S., et al.: NVIDIA tensor core programmability, performance & precision. In: 2018 IEEE IPDPSW (2018)
8. Markidis, S., Lapenta, G.: The energy conserving particle-in-cell method. J. Comput. Phys. **230**(18), 7037–7052 (2011)
9. Markidis, S., et al.: Multi-scale simulations of plasma with iPIC3D. Math. Comput. Simul. **80**(7), 1509–1519 (2010)
10. Markidis, S., et al.: The EPiGRAM project: preparing parallel programming models for exascale. In: Taufer, M., Mohr, B., Kunkel, J.M. (eds.) ISC High Performance 2016. LNCS, vol. 9945, pp. 56–68. Springer, Cham (2016). https://doi.org/10.1007/978-3-319-46079-6_5
11. Markidis, S., et al.: PolyPIC: the polymorphic-particle-in-cell method for fluid-kinetic coupling. Front. Phys. **6**, 100 (2018)
12. Narasimhamurthy, S., et al.: The SAGE project: a storage centric approach for exascale computing. In: Proceedings of the 15th ACM International Conference on Computing Frontiers, pp. 287–292. ACM (2018)
13. Olshevsky, V., et al.: Energetics of kinetic reconnection in a three-dimensional null-point cluster. Phys. Rev. Lett. **111**(4), 045002 (2013)
14. Peng, I.B., et al.: Energetic particles in magnetotail reconnection. J. Plasma Phys. **81**(2) (2015)
15. Peng, I.B., et al.: Kinetic structures of quasi-perpendicular shocks in global particle-in-cell simulations. Phys. Plasmas **22**(9), 092109 (2015)
16. Peng, I.B., et al.: The formation of a magnetosphere with implicit particle-in-cell simulations. In: ICCS, 01–03 June 2015, Reykjavik University, Reykjavik, Iceland, pp. 1178–1187 (2015)
17. Sishtla, C.P., et al.: Particle-in-cell simulations of plasma dynamics in cometary environment. arXiv preprint arXiv:1901.09638 (2019)
18. Stantchev, G., et al.: Fast parallel particle-to-grid interpolation for plasma PIC simulations on the GPU. J. Parallel Distrib. Comput. **68**(10), 1339–1349 (2008)

Reducing Symbol Search Overhead on Stream-Based Lossless Data Compression

Shinichi Yamagiwa[1(✉)], Ryuta Morita[2], and Koichi Marumo[2]

[1] Faculty of Engineering, Information and Systems, University of Tsukuba,
1-1-1 Tennodai, Tsukuba, Ibaraki, Japan
yamagiwa@cs.tsukuba.ac.jp
[2] Department of Computer Science, University of Tsukuba,
1-1-1 Tennodai, Tsukuba, Ibaraki, Japan
{morita,marumo}@padc.cs.tsukuba.ac.jp

Abstract. Lossless data compression is emerged to utilize in the Big-Data applications in the recent days. The conventional algorithms mainly generate a symbol lookup table to replace a frequent data pattern in the inputted data to a symbol, and then compresses the information. This kind of the dictionary-based compression mechanism potentially has an overhead problem regarding the number of symbol matchings in the table. This paper focuses on a novel method to reduce the number of searches in the table using a bank separation technique. This paper reports design and implementation of the bank select method on the LCT-DLT, and shows the performance evaluations to validate the effects of the method.

Keywords: Lossless data compression ·
Stream-based data compression · Dictionary-based compression ·
Interconnection

1 Introduction

Increasing the demands for handling BigData applications, it becomes one of the important techniques to processing a huge size data in computer systems. Because it is impossible to reduce the fast generation rate of the BigData itself, one of the ideal solutions is to minimize the data by applying the lossless data compression algorithm.

Lossless data compression algorithm has been studied for these three decades regarding mainly the *dictionary-based* compression. The LZW [3] is one of the well-known algorithms. Those algorithms exploit the frequent data patterns and create a dictionary that contains replacement rules to smaller data. Using the rules, the compressor generates a compressed data sequence. Decompressor decodes the sequence using the dictionary. However, it has fatal disadvantage

© Springer Nature Switzerland AG 2019
J. M. F. Rodrigues et al. (Eds.): ICCS 2019, LNCS 11540, pp. 619–626, 2019.
https://doi.org/10.1007/978-3-030-22750-0_59

due to the processing style. When we consider to apply the algorithm to a data stream such as continuous sensor data, we need to prepare a data buffer for the dictionary. The size of the dictionary is not deterministic and decompression is blocked until the dictionary is prepared fully after the compression processes entire input data. Moreover, the data stream must be terminated in chunks to generate the dictionary. Finally, it can not continuously compress and decompress in pipeline manner. Thus, the conventional dictionary-based algorithms are not suitable for realtime compression for the continuous data stream.

We have developed a new compression algorithm called *LCA-DLT* [2]. It is able to compress data stream without buffering and blocking inputted data to the compressor and the decompressor. The compression algorithm applies an idea that any data stream can be expressed by a binary tree, and compresses a data pair to a compressed symbol. Cascading modules of the two-to-one compression, it compresses long data patterns. Using a dynamic histogram management for the symbol lookup table under the limited number of entries in the table, it can compress data stream in a pipeline manner and hides information to reproduce its symbol lookup table in compressed data. In decompressor side, when it receives any compressed data, it reproduces a histogram of the symbol lookup table and decompresses the data to originals. This mechanism does not need any buffer for creating entire symbol lookup table due to the hided reproducible information for the histogram. Moreover, it can compress a continuous data without scattering original data stream. This mechanism is suitable for hardware implementation due to availability of parallel pattern matching in the symbol lookup table. However, when we consider software implementation, it needs to search compressing or decompressing symbol pairs from the symbol lookup table sequentially. When the number of patterns in the table is N, it never avoids the overhead of the search operations from $O(N)$. When the number of cascaded modules increases, it linearly increases. Thus, using LCA-DLT algorithm, this paper will try to reduce the searching overhead in the dictionary-based lossless compression mechanism.

2   Stream-Based Lossless Data Compression Algorithm: LCA-DLT

The table of LCA-DLT has any number N of entries and the i-th entry E_i includes a pair of the original symbols $(s0_i, s1_i)$, a compressed symbol S_i, and frequent counter $count_i$. The compressor side uses the following rules: (1) reading two symbols $(s0, s1)$ from the input data stream and if the symbols match to $s0_i$ and $s1_i$ in a table entry E_i, after counting up the $count_i$, it outputs S_i as the compressed data, (2) if the symbols do not match to any entry in the table, it outputs $(s0, s1)$ and register an entry $(s0_k, s1_k, S_k, count_k = 1)$ where S_k is the index number of the entry, and (3) if all entries in the table are used, all $count_i$ where $0 \leq i < N$ are decremented until any count(s) become zero and the corresponding entries are deleted from the table.

In the decompressor side, assume that a compressed data S is transmitted from the compressor, the subsequent steps are equivalent to the compressor's, but the symbol matching is performed based on S_k in an entry. If the compressed symbol S matches to S_k in a table entry, it outputs $(s0_k, s1_k)$. If not, it reads another symbol S' from the compressed data stream and outputs a pair of (S, S') and then the pair is registered in the table. When the table entry is full, the same operation as the compressor side is performed. These operations provide a reproducible dynamic histogram on a limited number of lookup table entries.

Here, let us focus on overhead of the symbol matching operation in the lookup table when we implement LCA-DLT in software. LCA-DLT supports two-to-one comparisons. It allows a fixed number of matching operations in the table entries. However, the number of entry search operations in the symbol lookup table is not predictable because it is inevitably performed by sequential search from the top of the table. Unfortunately, it occupies a large part of the entire compression and decompression operations.

When we consider to divide the symbol lookup table into multiple groups, we can expect to reduce the number of search operations in the table. However, it is not so simple in the case of the conventional dictionary-based compression algorithms because the matching complexity of the search operations is expected in the worst case as $O(NM)$ where the N is the number of entries in the table and the M is the length of matching symbols in an entry. Both of M and N are very variable according to the inputted data. Besides, the number of symbol matching is fix to two in LCA-DLT. Therefore, the complexity of the search operation $O(2N)$ where N is the number of the table entries and also is defined as a fixed value. This means that we can expect to reduce the entire compression and decompression processing times drastically if we divide the number of the table entries by the number of groups (i.e. N/k where the k is the number of groups). In this paper we will propose the technique to divide the symbol lookup table into multiple groups and the technique will effectively reduce the number of search operations. We will call the groups the *banks of symbol lookup table*, and will propose the speedup technique using the LCA-DLT based on the bank select method for the symbol lookup table.

3   Bank Select Method for Stream-Based Lossless Data Compression

Now, let us explain the bank select method using LCA-DLT. The technique is quite simple. The symbol lookup table is divided into N_b banks. When a data pair arrives, it is associated to one of the banks in the table. Then LCA-DLT algorithm is applied to the bank. When the number of the table entries is N, the number of entries related to the compression is N/N_b. For example, when the table has 256 entries and the number of banks is 16, LCA-DLT will manage the dynamic histogram operation in a bank with 16 entries. Thus, in the case of LCA-DLT, the complexity of the number of search operations becomes $O(N/N_b)$. The decompressor has the same organization in symbol lookup table.

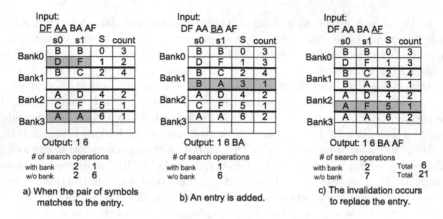

Fig. 1. Compression example of LCA-DLT with the bank select method.

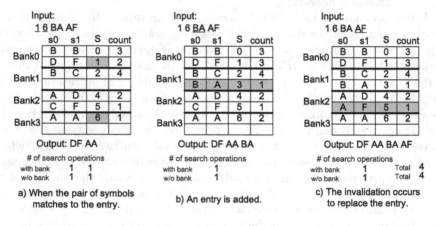

Fig. 2. Decompression example of LCA-DLT with the bank select method.

Let us consider how a bank is selected. Here, we employ two simple selection methods; order major and data major selection. One is the sequence index i of the inputted data stream. In this case the *hash* function becomes an equation with i, for a simple example, such as the round-robin selection $n_b = i \bmod N_b$. This hash function has the characteristics that there is no relationship to the inputted data itself. Therefore, the bank indices are totally selected among the entire table entries even if the entropy of the data is dynamically changing. This causes multiple assignments of the table entries of the same data pair among multiple banks. Another function uses a part of the inputted data itself. In this case, the selected bank indices are spread among the entire table entries according to the entropy of the inputted data stream. Let us introduce a simple example to keep fast hash calculation such as $n_b = \sum_{k=0}^{K/2-1} 2^k \cdot s_{0,k} + \sum_{k=K/2}^{K-1} 2^k \cdot s_{1,k-K/2}$ where $K = \log_2 N_b$ and $s_{i,j}$ is a bit of s_i. n_b is combined by $\log_2 K$ bits made from the LSBs of inputted symbols (s_0, s_1). Here, the combined bits can be generated by picking up the least significant $K/2$ bits from s_0 and s_1 respectively. In this case, the bank selection will be depending on the entropy of the entire data elements. This would cause data concentration in several banks if the number of data combinations is few (i.e. the entropy of the data is low). We will evaluate the effects of the hash methods in the performance evaluation.

We shall show examples of the compression and the decompression flows. Figures 1 and 2 illustrate examples under the conditions: N and N_b are 8 and 4 respectively, and the hash function is the n_b of the data major approach. We assume that the inputted data stream consists of ASCII characters in 8bits. The combined K bits are organized by the least significant bit of each ASCII code of the input data pair. In this case, the least significant bit of each symbol in the pair is combined and finally the two-bit bank number n_b is decided.

Regarding the compression operation, we assume that after several data has been processed and then the part of the stream "DFAABAAF" is inputted to the compressor. Figure 1(a) shows the case when the first two pairs match to the table entries. Here, we count the numbers of search operations with/without the bank select method. In the case of "DF", the number of search operations is two in both cases. However, in the case of "AA", because the bank number is 3 decided by the combination of the LSB of "A"s (0x41). Therefore, the bank select method finds the entry from the first search in the bank. However, in the case without banks, it searches the entry from the top of the table. It needs six search operations. Figure 1(b) shows the case when the pair is inserted to a bank. In order to know that the inputted data pair must be inserted to an empty entry, the case without bank needs to search all the entries in the table, and then it inserts a new entry. Therefore, it needs 6 search operations. Besides, the case with bank performs just a search operation in a bank. Finally, Fig. 1(c) shows the case of invalidation operation to replace the entry. It is the same as the case of the insertion operation. It needs to count maximally the number of entries in a bank. However, without the bank, it needs to search all the entries in the table and finds an entry to replace it. According to the total number of the search operations in the compression example, the one without banks becomes almost four times more search operations than the one with banks. Thus, the complexity of the search operation becomes approximately $O(N/N_b)$.

On the other hand, regarding the decompression operation, the decompressor reads an inputted symbol and searches the entry associated by the symbol value as the index of the table. In the case when the compressed symbols match to the entries as shown in Fig. 2(a), the numbers of search operations for '1' and '6' are both one time. This is the same when the case without bank. The new entry insertion is performed for the table such as the case of "BA" as shown in Fig. 2(b), after the decompressor searches the index 'B' in the table. However, 'B' as the index of the table is more than the number of entries. It knows that there is not any matched entries and then insert it to a new entry. Therefore, the number of search operations is always one. Finally, in the case of invalidation of the entry as shown in Fig. 2(c), the number of search operations is the same as the insertion. Thus, the numbers of search operations with/without the bank select method are totally the same.

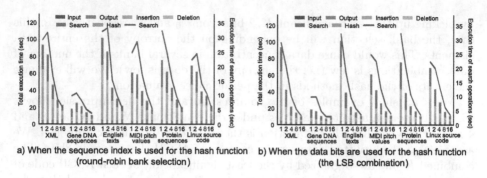

a) When the sequence index is used for the hash function
   (round-robin bank selection)

b) When the data bits are used for the hash function
   (the LSB combination)

Fig. 3. Comparisons of compression times of LCA-DLT with the bank select method.

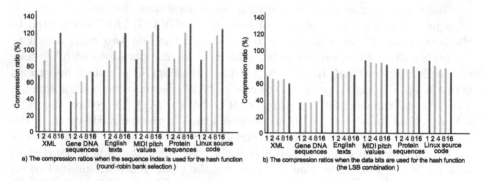

a) The compression ratios when the sequence index is used for the hash function
   (round-robin bank selection)

b) The compression ratios when the data bits are used for the hash function
   (the LSB combination)

Fig. 4. Comparisons of compression ratios of LCA-DLT with/without the bank select method among different hash functions.

As we have seen in the examples above, we have confirmed that the bank select method has the effective speedup technique for the compression. Even if the processing speed is fast, it is not a novel technique if the effective compression ratio degrades than the compressor without the bank select method. In the next section, we will discuss the actual computing speed and the compression ratio to validate the proposed technique.

4 Performance Evaluation

We used several patterns of 10 Mbyte benchmark data of a text collection available from [1]. We have implemented compressor/decompressor with/without the bank select method using C#. The execution time measurement has been performed in a 64 bit Windows 8 PC with Intel Core i7 2.7 GHz and 8 GByte memory.

First, we measured the elapsed times of individual operations in the compression process: *Insertion* is the time for inserting a new symbol pair entry to the table, *Deletion* is the one for deleting an entry(s) from the table by the invalidation operation, and *Search* is the one for searching operation for the input pair, as shown in Fig. 3 when four compressor modules are cascaded. The bars show the execution times of the individual operations with varying the number of banks from 1 to 16. The total number of entries in the table is configured to 256. In the case when the number of banks is 1 as shown in the graph, it is precisely equal to the case without banks. The line shows the speedup ratio where the execution time of search operations is divided by the one without banks. We have evaluated the speedup applying different kinds of the hash function: one is the hash function based on sequence index, and another is based on a bit combination of inputted data.

According to the graphs, in both cases of the different hash functions, we have confirmed that the search operation occupies a large part of the total execution. Therefore, when the number of banks is increased, the time of the search operations is linearly decreased for two times when it is changed to the twice number of the banks. The total execution times of Fig. 3(a) are larger than the ones of Fig. 3(b). This means that insertion and deletion times are larger because the compression ratios become worse as we will discuss in the next evaluation.

Next, Fig. 4 shows the compression ratios. The compression ratios in the case when the hash function uses the sequence index become worse when the number of banks is increased. The hash function selects all the banks fairly. Because the invalidation operations occur frequently among all the banks, the table search often does not match the input pair to the corresponding one in the table. On the other hand, in the case when the hash function uses a bit combination of symbol pair. It results better compression ratios than the round-robin case. When the hash function uses the data itself, the bank selection is affected directly by data entropy. Therefore, the corresponding symbol pair is always mapped to the same bank. This causes high probability to match the inputted symbol pair to the entry in the bank.

5 Conclusion

We have proposed a novel method to speedup lossless compression operation called the bank select method. It divides the symbol lookup table to multiple banks. The banks are selected by the hash function. We applied it to the stream-based lossless data compression algorithm called LCA-DLT. The evaluation has shown the drastic improvement of execution time. We have confirmed that the method is effective to reduce the search operations in compression.

Acknowledgement. This work is partially supported by JSPS KAKENHI Grant Number 15H02674 and JST CREST Grant Number JPMJCR1402.

References

1. http://pizzachili.dcc.uchile.cl/
2. Yamagiwa, S., Marumo, K., Sakamoto, H.: Stream-based lossless data compression hardware using adaptive frequency table management. In: Zhan, J., Han, R., Zicari, R.V. (eds.) BPOE 2015. LNCS, vol. 9495, pp. 133–146. Springer, Cham (2016). https://doi.org/10.1007/978-3-319-29006-5_11
3. Ziv, J., Lempel, A.: A universal algorithm for sequential data compression. IEEE Trans. Inf. Theor. **23**(3), 337–343 (1977). https://doi.org/10.1109/TIT.1977. 1055714

Stabilized Variational Formulation for Solving Cell Response to Applied Electric Field

Cesar Augusto Conopoima[1], Bernardo Martins Rocha[1], Iury Igreja[1(✉)] (iD),
Rodrigo Weber Dos Santos[1], and Abimael Fernando Dourado Loula[2]

[1] Universidade Federal de Juiz de Fora, Juiz de Fora, MG, Brazil
iuryigreja@ice.ufjf.br
[2] Laboratório Nacional de Computação Científica, Petropolis, RJ, Brazil

Abstract. In this work a stabilized variational formulation is proposed to solve the interface problem describing the electric response of cells to an applied electric field. The proposed stabilized formulation is attractive since the discrete operator resulting from finite element discretization generates a definite linear system for which efficient iterative solvers can be applied. The interface problem describing the cell response is solved with a primal variational formulation and the proposed stabilized formulation. Both methods are compared in terms of the approximation properties of the primal and the Lagrange multiplier variable. The computational performance of the methods are also compared in terms of the mean number of iterations needed to solve one time step during the polarization process of an isolated square cell. Moreover, numerical experiments are performed to validate the convergence properties of the methods.

Keywords: Cell interface problem · Primal varitional formulation · Stabilized variational formulation

1 Introduction

The numerical study of the electrical activity of biological cells in conductive medium subject to applied electric field is of interest to the medical community [1,2]. The development of a general tool to numerically investigate the electric field distribution in biological cells has been studied by [3–5]. In particular, employing finite element methods, numerical simulations of individual cells and cluster of cells when subjected to an applied electric field were developed by [3] using a primal hybrid variational formulation, introduced by Raviart-Thomas in [6].

Recently in [5], a framework to solve the electric field distribution at the tissue scale based in an EMI cell model has been proposed. The primal hybrid formulation with dual Lagrange multiplier was also used and a mixed formulation with $H(div)$ approximation spaces was developed with demonstrated optimal rates in convergence tests. These works use dual space as proposed in [7] to approximate

© Springer Nature Switzerland AG 2019
J. M. F. Rodrigues et al. (Eds.): ICCS 2019, LNCS 11540, pp. 627–634, 2019.
https://doi.org/10.1007/978-3-030-22750-0_60

the Lagrange multiplier used to enforce the interface condition associated to the primal variable. However, since the linear system resulting from the discrete operator of this formulation is positive indefinite, iterative methods based on Krylov subspaces usually present poor performance for large linear systems.

Another approach for the hybrid primal formulation was firstly proposed and analyzed by [8], where a stabilization term related to the Lagrange multiplier definition are included in order to circumvent the compatibility conditions between spaces necessary for stability of the Lagrange multiplier of the primal hybrid formulation. This methodology generate positive definite linear systems and consequently improves the computational solution.

In this context, this work presents a variational formulation to solve the problem of the response of a square cell to an applied electric field, using a stabilization technique based on the inclusion of terms related to Lagrange multiplier definition as proposed in [8]. This approach is interesting since when compared to the primal hybrid formulation a positive definite linear system associated to the discrete operator of finite elements is obtained. The capabilities of the proposed approach are demonstrated through numerical studies that present optimal rates of convergence for the primal variable and the Lagrange multiplier and lower computational cost when compared with primal hybrid method.

2   The Model Problem

Consider a square cell with conductivity given by the second order tensor κ_i, in a bounded domain Ω_e with electrical conductivity given by the second order tensor κ_e subjected to an externally applied electric field $\mathbf{E}$ as depicted in Fig. 1. The system of equations of electric current conservation for this system without source current, can be written as a function of the electric potential inside and outside the cell (u_i, u_e), as:

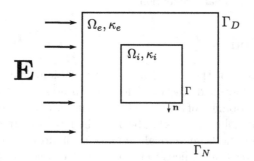

Fig. 1. A square cell in a conductive medium

$$-\text{div}(\boldsymbol{\kappa}_e \nabla u_e) = 0, \quad \text{in } \Omega_e, \tag{1}$$
$$-\text{div}(\boldsymbol{\kappa}_i \nabla u_i) = 0, \quad \text{in } \Omega_i,$$
$$-\boldsymbol{\kappa}_i \nabla u_i \cdot \mathbf{n} = -\boldsymbol{\kappa}_e \nabla u_e \cdot \mathbf{n} = I_m, \quad \text{on } \Gamma,$$
$$u_e = \hat{u}, \quad \text{on } \Gamma_D,$$
$$\boldsymbol{\kappa}_e \nabla u_e \cdot \mathbf{n}_e = 0, \quad \text{on } \Gamma_N,$$

Where $\hat{u}$ is the electric potential applied on Γ_D. In this system of equations, Γ denotes the cell membrane, $\mathbf{n}$ is the unitary normal vector pointing outward of Γ and I_m [A $\cdot$ cm^{-2}] is the transmembrane current which depends on the transmembranic potential, that is defined as $V_m = u_i - u_e$ on Γ.

The transmembrane current I_m is written as the contribution of two main currents, the capacitive and the resistive or ionic current, and is given by $I_m = C_m \frac{\partial V_m}{\partial t} + I_{ion}(V_m)$ on Γ.

2.1 Variational Formulation

The variational formulation of the conservation of electric current of the unicellular system (1) is: *find* $u = (u_i, u_e) \in X$ *with* $X = H^1(\Omega_i) \times H^1(\Omega_e)$ *such that*

$$\int_\Omega \boldsymbol{\kappa} \nabla u \cdot \nabla v \, dx - \int_\Gamma \boldsymbol{\kappa} \nabla u \cdot \mathbf{n}[v] ds = 0, \quad \forall v \in X, \tag{2}$$
$$-\boldsymbol{\kappa}_i \nabla u_i \cdot \mathbf{n} = -\boldsymbol{\kappa}_e \nabla u_e \cdot \mathbf{n} = I_m, \quad \text{on } \Gamma,$$
$$[u] = V_m, \quad \text{on } \Gamma,$$

where $[u] = u_i|_\Gamma - u_e|_\Gamma$ is the scalar jump of electric potential on the membrane, $\boldsymbol{\kappa} = (\boldsymbol{\kappa}_i, \boldsymbol{\kappa}_e)$ denotes the electric conductivity inside and outside the cell, $H^1(\Omega_i)$ and $H^1(\Omega_e)$ are the Sobolev space of functions with first-order derivative square-integrable on the bounded domains Ω_i and Ω_e respectively and $X = H^1(\Omega_i) \times H^1(\Omega_e)$ is the product space.

Primal Hybrid Variational Formulation. An additional Hilbert space defined on Γ is included to impose the interface condition ($[u] = V_m$ on Γ) of the primal variable in a weak sense via the use of a Lagrange multiplier. Given K_i and K_e a non-overlapping domain triangulation for Ω_i and Ω_e with linear finite elements, and a polynomial base of degree at most k denoted by $P_k(K)$ defined on each triangulation, it is possible to write the discrete version of the primal variational formulation as: *find* $u_h = (u_{i_h}, u_{e_h}) \in X_h$ *and* $\lambda_h \in \mathcal{M}_h(\Gamma_{ie})$ *such that*

$$\int_\Omega \boldsymbol{\kappa} \nabla u_h \cdot \nabla v_h dx + \int_\Gamma \lambda_h[v_h] ds = 0, \quad \forall v_h \in X_h, \tag{3}$$
$$\int_\Gamma [u_h] \mu_h ds = \int_\Gamma V_m \mu_h ds, \quad \forall \mu_h \in \mathcal{M}_h(\Gamma_{ie}),$$

with,

$$X_h = \{(v_{i_h}, v_{e_h}) \in (H^1(K_i) \times H^1(K_e)) : v_{h_i}|_{K_i} \in P_k(K_i); v_{h_e}|_{K_e} \in P_k(K_e)\}, \quad (4)$$

$$\mathcal{M}_h(\Gamma_{ie}) = \{\mu_h \in C^0(\Gamma_{ie}); \mu_h|_{\Gamma_{ie}} \in p_k(\Gamma_{ie})\}, \quad (5)$$

where Γ_{ie} denotes a domain triangulation for $\Gamma = \Omega_i \cap \Omega_e$, such that Γ_{ie} is uniquely discretized by linear finite elements of dimension $(d-1)$ with $d = 2$ inherited from K_i or K_e, and $p_k(\Gamma_{ie})$ is the space of polynomials of degree at most k on each edge Γ_{ie} on the interface Γ.

Stabilized Primal Variational Formulation. Based on the discrete variational formulation (3), it is possible to include stabilization terms as proposed and analyzed in [8], to have the following equivalent stabilized primal variational formulation: *find* $u_h = (u_{i_h}, u_{e_h}) \in \mathcal{W}_h$ *and* $\lambda_h \in \mathcal{M}_h(\Gamma_{ie})$, *such that*

$$\int_\Omega \kappa \nabla u_h \cdot \nabla v_h dx + \int_\Gamma \lambda_h [v_h] ds + \alpha h \int_\Gamma (\lambda_h + \kappa \nabla u_h \cdot \mathbf{n})(\mu_h + \kappa \nabla v_h \cdot \mathbf{n}) = 0,$$
$$(6)$$

$$\int_\Gamma [u_h] \mu_h ds = \int_\Gamma V_m \mu_h ds,$$

for all $v_h \in \mathcal{W}_h$ *and* $\mu_h \in \mathcal{M}_h(\Gamma_{ie})$, *with:*

$$\mathcal{W}_h = \{(v_{i_h}, v_{e_h}) \in (H^{3/2}(K_i) \times H^{3/2}(K_e)) : v_{h_i}|_{K_i} \in P_k(K_i); v_{h_e}|_{K_e} \in P_k(K_e)\}.$$
$$(7)$$

Here, $\alpha \in \mathbb{R}$ is an arbitrary real constant and h is the characteristic length of the finite element triangulation. Note that for $\alpha = 0$ the primal hybrid form is recovered. In order to evidence an important relation, we can rewrite Eq. (6) as:

$$\int_\Omega \kappa \nabla u_h \cdot \nabla v_h dx + \int_\Gamma \lambda_h [v_h] ds + \alpha h \int_\Gamma (\lambda_h + \kappa \nabla u_h \cdot \mathbf{n})(\kappa \nabla v_h \cdot \mathbf{n}) = 0, \quad (8)$$

$$\int_\Gamma ([u_h] + 2\alpha h(\lambda_h + \{\kappa \nabla u_h \cdot \mathbf{n}\})) \mu_h ds = \int_\Gamma V_m \mu_h ds, \quad (9)$$

where $\{\kappa \nabla u_h \cdot \mathbf{n}\} = \frac{\kappa_i \nabla u_{i_h} \cdot \mathbf{n} + \kappa_e \nabla u_{e_h} \cdot \mathbf{n}}{2}$, denotes an average quantity. Then, Eq. (9) can be rearranged in the following form:

$$\int_\Gamma [u_h] \mu_h ds = \int_\Gamma V_m \mu_h ds - 2\alpha h \int_\Gamma \mu_h (\lambda_h + \{\kappa \nabla u_h \cdot \mathbf{n}\}) ds. \quad (10)$$

To show that on the approximated case by finite elements method, the interface condition over Γ of the transmembrane potential will be imposed with an error associated to the following relation: $e_{\lambda_h} = \lambda_h + \{\kappa \nabla u_h \cdot \mathbf{n}\}$. Note that, since $\alpha \in \mathbb{R}$ the influence of the term e_{λ_h} may be reduced by choosing small values for α. Additionally, note that from Eq. (10) a pair (u_h, λ_h) solving problem (8) also satisfies the following relation for the Lagrange multiplier λ_h:

$$\lambda_h = \frac{V_m - [u_h]}{2\alpha h} - \{\kappa\nabla u_h \cdot \mathbf{n}\}, \quad \text{on } \Gamma. \tag{11}$$

3   Numerical Results

In this section the spatial convergence of the stabilized hybrid and primal hybrid formulations are evaluated for an idealized problem of the polarization of a square cell. The computational perfomance of both methods, is also compared in terms of the mean number of iterations needed to solve one time step during 0.5 [μs] of the polarization process for different iteratives methods and preconditioning techniques.

3.1   Convergence of the Method

One interesting property of the proposed method is the possibility to approximate discontinuous solutions on the interface of two non-overlapping domains.

Exact solution for the model problem (1) on the unicellular cell system depicted in Fig. 1 is obtained in [5] using the method of manufactured solution. The exact solution is then obtained for $f_i = -8.0\pi^2 \sin(2\pi x)\sin(2\pi y)(1 + e^{-t})$, $f_e = -8.0\pi^2 \sin(2\pi x)\sin(2\pi y)$ on the bounded domain $\Omega = \Omega_i \cup \Omega_e$ with $\Omega_i = [0.25, 0.75] \times [0.25, 0.75]$ and $\Omega_e = [0.0, 1.0] \times [0.0, 1.0] \setminus \Omega_i$.

In order to test the h-convergence of the methods an arbitrary fixed time is chosen and the forcing terms on each subdomain (f_i, f_e) are defined. The jump on the cell interface Γ is computed as $V_m = e^{-t}\sin(2\pi x)\sin(2\pi y)$. In this scenario the convergence test of the primal hybrid and the stabilized formulations are presented in Table 1 below.

Table 1. Errors and order of convergence for the approximations of u_h and λ_h obtained by the finite elements discretization of the primal hybrid variational formulation and stabilized variational formulation with $\alpha = 0.01$. The $\|\cdot\|$ norm denotes the $L^2(\Omega)$ norm used to measure the error of the primal variable, for the Lagrange multiplier $\|\cdot\|$ denotes a mesh dependant norm defined in [7].

n	Primal hybrid				Stabilized			
	$\|u - u_h\|$	Order	$\|\lambda - \lambda_h\|$	Order	$\|u - u_h\|$	Order	$\|\lambda - \lambda_h\|$	Order
8	5.22e-2	–	6.76e-1	–	5.10e-2	–	6.31e-1	–
16	1.46e-2	1.84	1.71e-1	1.98	1.45e-2	1.56	1.61e-1	1.91
32	3.77e-3	1.90	4.26e-2	1.99	3.76e-3	1.85	4.01e-2	1.99
64	9.51e-4	1.93	1.06e-2	2.00	9.49e-4	1.96	9.96e-3	2.00
128	2.38e-4	1.95	2.63e-3	2.00	2.38e-4	1.99	2.48e-3	2.00
256	6.12e-5	1.96	6.60e-4	2.00	5.95e-5	2.00	6.22e-4	2.00

3.2 Computational Aspects

In this section the mean number of iterations to solve a time step on the polarization process of an isolated square cell by an explicit time discretization with $\delta t = 0.01$ [μs], is used to compare the performance of the primal hybrid and the stabilized formulations. To study the performance of the formulations we solve the conservation of electric current given by the system of equations (1) with $\kappa_i = 5.0$ [S · cm^{-1}], $\kappa_e = 20.0$ [S · cm^{-1}] and an applied electric potential difference between right and left of $0.1[V]$ (see Fig. 2).

Since the linear system associated to the primal hybrid formulation is positive indefinite, the Minimal Residual Method (MINRES) with a Jacobi preconditioner was used and considered as a reference for further comparisons against the stabilized formulation with other solvers. For the stabilized formulation the associated linear system is definite and, therefore, the GMRES and Stabilized Biconjugate Gradient (BICGSTAB) method preconditioned by the incomplete LU factorization ILU(0) are successful in solving it in a significantly lower number of iterations. This indicates an improvement in terms of computational performance of the proposed stabilized method when compared to the primal variational form.

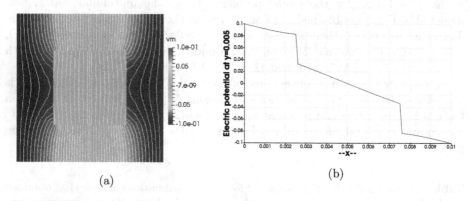

(a)   (b)

Fig. 2. (a) Electric potential and isocontour of an square cell in a conductive medium under an applied electric field after 0.5 [μs] of polarization; (b) electric potential magnitude at $y = 0.005$.

Table 2. Average number of iterations and standard deviation for different iterative methods to solve one time step for the distribution of electric potential on the unicellular system after 0.5 [μs] of polarization (50 time steps). The absolute tolerance for the iterative solver was set to $1e - 10$. All the linear systems were solved with the high-performance scientific library PETSc [9].

Preconditioner	Jacobi	ILU(0)	ILU(0)
Iterative method	MINRES	GMRES	BICGSTAB
Primal Hybrid	928 ± 114	-	-
Stabilized $\alpha = 1e - 5$	-	188 ± 13	92 ± 7

Table 2 shows the average number of iterations needed to solve 0.5 [μs] of cell polarization for both formulations using different combination of iterative solvers and preconditioners for the stabilized formulation. The empty entries in the table, correspond to cases where the preconditioner could not be constructed due to the matrix structure or the iterative solver was not able to solve the problem for the given tolerance.

4 Conclusions

In this work a stabilized formulation for an interface problem was introduced in the context of the response of a isolated square cell to an externally applied electric field. Through numerical experiments it was demonstrated that the proposed stabilized method converges with optimal convergence order for both primal and Lagrange multiplier variables.

The discrete operator from the stabilized formulation results in a definite matrix that allows the use of effective iterative solvers and preconditioning techniques that significantly improves the convergence properties of the iterative solution. Even though, the time consumed by the application phase of the ILU(0) preconditioner is bigger than the application phase of the jacobi preconditioner, the global perfomance of the ILU(0) preconditioned GMRES solver or the ILU(0) preconditioned BICGSTAB solver outperfom the Jacobi preconditioned MINRES solver due to the reduced number of iterations needed to achieve the demanded tolerance.

The convergence rate of the Lagrange multiplier for the presented stabilized variational formulation is optimal for straight interfaces, this meant that when solving irregular shape cells, the convergence rate of the Lagrange multiplier is degradated and the optimal convergence for the primal variable is conserved. Concerning the Krylov iterative methods, it is possible to exploit the block structure of the discrete operator in order to implement more effective preconditioning techniques, this characteristic could be explored in future works.

Acknowledgements. This study was financed in part by the Coordenação de Aperfeiçoamento de Pessoal de Nível Superior - Brasil (CAPES) - Finance code 001.

References

1. Chen, C., Smye, S.W., Robinson, M.P., Evans, J.A.: Membrane electroporation theories: a review. Med. Biol. Eng. Comput. **44**, 5–14 (2006)
2. Orlowski, S., Mir, L.M.: Cell electropermeabilization: a new tool for biochemical and pharmacological studies. Biochimica et Biophysica Acta (BBA)-Rev. Biomembr. **1154**, 51–63 (1993)
3. Ying, W., Henriquez, C.S.: Hybrid finite element method for describing the electrical response of biological cells to applied fields. IEEE Trans. Biomed. Eng. **54**, 611–620 (2007)

4. Agudelo-Toro, A., Neef, A.: Computationally efficient simulation of electrical activity at cell membranes interacting with self-generated and externally imposed electric fields. J. Neural Eng. **10**, 26019 (2013)
5. Tveito, A., Jæger, K.H., Kuchta, M., Mardal, K.-A., Rognes, M.E.: A cell-based framework for numerical modeling of electrical conduction in cardiac tissue. Front. Phys. **5**, 48 (2017)
6. Raviart, P.-A., Thomas, J.M.: Primal hybrid finite element methods for 2nd order elliptic equations. Mathe. Comput. **31**, 391–413 (1977)
7. Lamichhane, B.P.: Higher Order Mortar Finite Elements with Dual Lagrange Multiplier Spaces and Applications. 1st edn. University of Stuttgar (2006)
8. Ewing, R.E., Wang, J., Yang, Y.: A stabilized discontinuous finite element method for elliptic problems. Numer. Linear Algebra Appl. **10**, 83–104 (2003)
9. Balay, S., Gropp, W.D., McInnes, L.C., Smith, B.F.: Modern Software Tools in Scientific Computing, 1st edn. Birkhäuser Press, Boston (1997)

Data-Driven Partial Derivative Equations Discovery with Evolutionary Approach

Mikhail Maslyaev, Alexander Hvatov$^{(\boxtimes)}$, and Anna Kalyuzhnaya

ITMO University, Kronsersky pr. 49, 197101 St. Petersburg, Russia
`alex_hvatov@corp.ifmo.ru`

Abstract. The data-driven models are able to study the model structure in cases when a priori information is not sufficient to build other types of models. The possible way to obtain physical interpretation is the data-driven differential equation discovery techniques. The existing methods of PDE (partial derivative equations) discovery are bound with the sparse regression. However, sparse regression is restricting the resulting model form, since the terms for PDE are defined before regression. The evolutionary approach, described in the article, has a symbolic regression as the background instead and thus has fewer restrictions on the PDE form. The evolutionary method of PDE discovery (EPDE) is tested on several canonical PDEs. The question of robustness is examined on a noised data example.

Keywords: Data-driven model · PDE discovery ·
Evolutionary algorithms · Symbolic regression

1 Introduction

Data-driven algorithms are usually considered as the source of models when the connection between the data samples is not known a priori. There are various data-driven models. As an example, deep neural networks models, regression, combined evolutionary-based models [5] and other models and their combinations. However, most of the existing models are unsuitable for interpretation. Therefore, for cases, when the researcher is interested in the process of the model's decision making, other methods should be applied.

Data-driven algorithms are a solution for cases of systems, that we lack knowledge about. Nevertheless, in most cases raw observational data are available. The data-driven algorithms bring the ability to build the model for dynamical systems from time-series of data, received from in-field or laboratory observations. The development of the data-driven methodology of partial differential equations (PDE) derivation, combined with recent advances in technologies of measurements and probing, brings new opportunities for studying of metocean dynamic systems.

Sparse regression is considered to be the main tool for selection of the leading terms of the differential equations [7]. The applied regularization is based on the

© Springer Nature Switzerland AG 2019
J. M. F. Rodrigues et al. (Eds.): ICCS 2019, LNCS 11540, pp. 635–641, 2019.
https://doi.org/10.1007/978-3-030-22750-0_61

addition of the L1 norm of the calculated weights to the least-square expression. One of the most popular methods used in PDE discovery is the least absolute shrinkage and selection operator (LASSO). The main feature of LASSO is the ability to mutate the loss function. Zero weights are chosen for terms, that poorly fit the input data, and, therefore, identify the structure of the PDE.

Previously, the problem of the discovery of the differential equation structure has been developed in a number of papers. From derivation of systems of equations, defining physical laws, by means of symbolic regression [3,9] to study dynamic systems, that are represented by a system of partial differential equations [2,8]. Also, in the last years, the deep learning methods are becoming popular [1,6].

The methods of PDE derivation, used in previous papers, usually utilize regression over the set of the pre-determined terms, that are usually comprised of different polynomial combinations of derivatives and functions. This limitation provides only the discovery of equations, that have a corresponding structure. The method, presented in this paper is referred below as EPDE. It is based on a combination of sparse regression and an evolutionary algorithm. The proposed way of calculation of terms' weights values includes the application of linear regression over the non-normalized data for selected terms.

The paper is organized as follows, Sect. 2 describes the problem of data-driven PDE discovery in details. Also, in Sect. 2 dataset for experiments is described. Section 3 describes the data-driven PDE discovery algorithm based on evolutionary optimization. Section 4 is dedicated to the analysis of algorithm precision, stability, and robustness. Section 5 concludes the paper.

2 Problem Statement and Data Acquisition

The developed EPDE algorithm is aimed at the derivation of the dynamic systems governing equation by time series, containing information about the studied variable (temperature, velocity, etc.). At first, the approach must be applied for test cases, including artificially created data, acquired from numerically solved equations to check the algorithm convergence. For further tests, a noise of selected magnitude can be added to data to investigate the reaction of the algorithm to it.

In this work, the algorithm was tested on the wave equation, Burgers and Korteweg-de Vries equations. They were solved numerically with the application of a finite-difference scheme to approximate time and spatial derivatives. For instance, the Crank-Nicolson method was utilized to solve the Burgers equation.

From the acquired field of equation solution, its time and spatial derivatives are calculated in order to be utilized further in regression. These derivatives are calculated by the finite-difference method, or from polynomial interpolation function depending on the presence of noise in the data.

After derivatives are obtained, it is possible to create vectors of spatial data for a specific time point. In the same time normalization of each of these time frames should be held. It can be done with the highest variable value for that time point, or by time frame's L2-norm. Finally, data vectors are created by compositions of all time frames for the studied period.

Finally, the feature vectors $F(j)$ are formed by structures, such as the product shown in Eq. 1:

$$F(j) = \begin{bmatrix} (u'(t_1, x_0) * u_t(t_1, x_0))^N \\ \vdots \\ (u'(t_m, x_n) * u_t(t_m, x_n))^N \end{bmatrix} = U_x * U_t \tag{1}$$

On a balance, the data preparation step consists of representing data and it's spatial and time derivatives in vectors. After these steps, features are collected in forms, similar to Eq. 1, in order to perform the optimization procedure.

3   Algorithm Description

The proposed algorithm includes two parts: the evolutionary algorithm that generates a small group of terms that are called individuals and sparse regression that allows choosing significant terms in the set of individuals.

To find values of the weights α, that is representing the system's PDE, it is possible to define the loss function (Eq. 2) in the following way, using the defined set of features and target vectors, created in the previous section:

$$\min_{\alpha} \left(\sum_{k=0}^{p} \|F_k \alpha - F_{target,k}\|_2^2 + \lambda \|\alpha\|_1 \right) \tag{2}$$

Where p is the number of features selected for the regression algorithm and λ represents a regularization parameter. This application of the regularized regression is not able to discover the true values of the weights due to the fact, that it uses normalized vectors of target and features. However, it is able to select leading ones with their sign. Due to the addition of L1-norm, the loss functions must be minimized, using optimization algorithms, that are able to work with non-differentiable functions, such as the subgradient method.

After the structure is found, the coefficients are defined with non-normalized data, i.e. features are constructed from their initial form and regression is used to find the final values coefficients. Usually, in regression all possible combinations [8] of the feature vectors Eq. 1 are chosen for minimization problem Eq. 2. Thus, the optimization problem complexity grows exponentially as the maximal order of the derivative increases. With the evolutionary algorithm, described below, one can use multiple reduced optimization problems instead of full regression on a complete terms library.

The second element of the EPDE method is the evolutionary algorithm. By its iterations, the evolutionary algorithm should be able to select and preserve the most appropriate elements of the resulting equation. Therefore, the sparse regression is done on every iteration of the evolutionary algorithm for every candidate in the population with a random selection of target among the set of terms.

To initiate the method, it is required to create a population of individuals, represented by chromosomes, where each gene represents a combination of functions and their derivatives. An evolutionary algorithm is able to vary the chromosomes in two ways: crossover, that represents the exchange of corresponding genes between two individuals, and mutation, which involves random alteration of chromosome's genes. In the examined case, the mutation is held by the conversion of one term to the other randomly generated one.

Due to the specification of the task, every individual represents a specific case of the equation, having its own features matrix and the target vector. Vectors F(i), that compose the columns of the feature matrix S (Eq. 3), are created as a product of a randomly selected number of feature factors Eq. 1:

$$
S = \left[\; F(1) \; F(2) \; F(3) \; ... \; \right] \tag{3}
$$

It should be emphasized, that the number of feature vectors in Eq. 3 is the parameter of the evolutionary algorithm. The second remark is that, in contrast to the existing algorithms [4, 7, 8], the target feature is chosen randomly, whereas in the sparse-regression only cases time-derivative is used.

While mutation is usually applied to all individuals of the population, crossover occurs only between the most eligible of them. To select candidates for crossover, the fitness function should be introduced. For the task of partial differential equation derivation, it can be introduced by a norm of the difference between the target term and the expression with other ones i.e. regression error, calculated for all of the training data:

$$
f_{fitness} = \frac{1}{\|F \cdot \alpha - F_{target}\|_2} \tag{4}
$$

A manner of the population's participation in crossover should be defined before the initiation of an algorithm. The crossover procedure is schematically shown in Fig. 1.

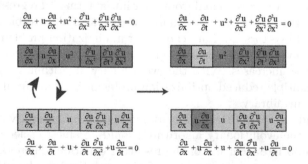

Fig. 1. An example of implemented crossover between two chromosomes, where each of them represent PDE

After the sparse regression application, one more regression step is required. It is initialized over the set of terms, selected by non-zero weights in the previous step. In this step, non-normalized fields of variables are used as a feature and target vectors. This approach is uncommon in general machine learning due to its limitations on variables of different scale, where the algorithm is not able to properly generalize data and discover a contribution of each feature. However, in this particular case, the structure of the an equation, represented by weights of features, is already known, and these variables must be evaluated according to their scale.

The described algorithm allows one to reduce regression space. Additionally, it allows to theoretically find ordinary differential equation instead of the PDE since target feature is not restricted by the highest time-derivative. This is required for potential one-dimensional static problems ODE discovery.

4   Validation

To analyze the algorithm performance, it is necessary to make sure, that it has the following qualities: stability, approximation, and convergence. These qualities are dependent on each other, and to prove them, it is enough to check, if any two of them are fulfilled. Due to the reasons of convenience, in the research, stability, and convergence of the algorithm are studied. Convergence of the PDE deriving algorithm manifests in the improvement of the quality of the algorithm with the reduction of a step of the grid, from that is adopts data. Stability can be proved by addition of the noise to the input PDE solution and test, how this corruption affects the structure of the resulting equation.

The algorithm has proved to be capable of discovering partial differential equations structure and calculating the values of weights for the selected terms for all of the studied equations.

The selected part of the solution matrix has influence over the results of regressions and, therefore, defines the equation's structure. The results were tested on the parts of the matrix from 1.0 to 0.1 of its size. On the lesser sizes of the selected matrix part, especially for cases, when the selected part contains an only small part of the solution non-zero values, the algorithm can have difficulties, deriving wrong structures. The results of the matrix division are presented in Table 1.

We note that for Table 1 different number of points was taken in order to check the performance of the algorithm. For Burger's equation, the 256×256 grid was taken whereas for the Korteweg-de Vries equation - 1024×1024 points.

The previously mention effect remained in this scenario: the algorithm only had issues in discovering the structure of the governing equation. For cases, when it succeeded, the true values of the weights were calculated correctly, even on minor parts of the equation's solution matrix.

To check the evolutionary algorithm stability, the noise is added to the entire solution's field. It is added from a normally distributed random variable with

Table 1. Discovered structure of the equations for different input matrix section.

Data part	Burger's correct	Burger's wrong	KdV correct	KdV wrong
0.9	$\frac{\partial u}{\partial t}, \frac{\partial^2 u}{\partial x^2}, u\frac{\partial^2 u}{\partial x}$	-	$\frac{\partial u}{\partial t}, \frac{\partial^3 u}{\partial x^3}, u\frac{\partial^2 u}{\partial x}$	-
0.7	$\frac{\partial u}{\partial t}, \frac{\partial^2 u}{\partial x^2}, u\frac{\partial^2 u}{\partial x}$	-	$\frac{\partial u}{\partial t}, \frac{\partial^3 u}{\partial x^3}, u\frac{\partial^2 u}{\partial x}$	-
0.5	$\frac{\partial^2 u}{\partial x^2}, \frac{\partial u}{\partial t}$	$\frac{\partial^2 u}{\partial x^2}\frac{\partial u}{\partial t}$	$\frac{\partial u}{\partial t}, \frac{\partial^3 u}{\partial x^3}, u\frac{\partial^2 u}{\partial x}$	-
0.4	$\frac{\partial u}{\partial t}, \frac{\partial^2 u}{\partial x^2}, u\frac{\partial^2 u}{\partial x}$	-	$\frac{\partial u}{\partial t}, \frac{\partial^3 u}{\partial x^3}, u\frac{\partial^2 u}{\partial x}$	-
0.1	$\frac{\partial^2 u}{\partial x^2}, \frac{\partial u}{\partial t}$	$\frac{\partial^2 u}{\partial x^2}\frac{\partial u}{\partial t}$	-	$\frac{\partial^2 u}{\partial x^2}, \frac{\partial^2 u}{\partial t^2}$

zero mean value and dispersion taken as the fraction of maximal value. As the invariant noise measure, Eq. 5 is used.

$$Q_{noise} = \frac{\|w_0 - \widetilde{w}\|_2}{\|w_0\|_2} * 100 \tag{5}$$

With w_0 in Eq. 5 the initial (clean) solution field is designated, $\widetilde{w}$ is the field with noise added, $\| \cdot \|_2$ is the matrix's Frobenius norm.

For comparison, we take the latest supplementary code for the article [7] from GitHub repository. Same Burger's equation solution field and same noise procedure implementation were taken. It should be noted, that we compare "basic" versions of the algorithms. For the sparse regression more sophisticated derivative procedure and meta-parameter optimization for the regression algorithm could be implemented, which, definitely, increases the quality of both algorithms.

Polynomial derivatives procedure was utilized, also for the sparse regression improved ridge regression with $\alpha = 10^{-6}$ was taken.

For the Bruger's equation, after certain noise level limit $Q_{noise} \approx 0.11$ the classical algorithm loses the ability to discover the term $\frac{\partial^2 u}{\partial x^2}$ without an additional regression tuning. However, it is still able to catch the leading term. The EPDE, however, is able to find the right structure up to the $Q_{noise} \approx 0.11$.

As seen the evolutionary approach allows one to extend the noise level which is allowed for all terms of the initial equation discovery. The term coefficients discovery precision is increased, which leads to more stable equation discovery and allows one to discover the equations in a more robust way.

5   Conclusions and Discussion

In the paper evolutionary approach for PDE discovery is described. In contrast to the existing algorithms based on the regression on a complete terms library it has the following advantages:

- Regression is done on a reduced space, i.e. only a small amount of features is taken for the regression;
- More flexible features choice allows to obtain wider space of possible differential operators;

- No restriction on the target function is allowing to obtain more sophisticated forms of differential operators including ODEs;

The possible disadvantages could be:

- Possible extended computation time due to the stochastic process of the initial population initialization, population crossover and mutation;
- Additional procedures are required in order to maintain the robustness of the algorithm, i.e. in order to obtain the same model for the data of the same origin;

The proposed method can be considered as a base point for the data-driven PDE discovery with an evolutionary approach. In the article, the main stages of the methods are shown. Every stage could be improved, for example, a more sophisticated grid function differentiation method could be taken to increase precision and stability. Also, more advanced evolution methods could be used in order to increase computation efficiency and stability.

Acknowledgement. This research is financially supported by The Russian Science Foundation, Agreement No. 17-71-30029 with co-financing of Bank Saint Petersburg.

References

1. Berg, J., Nyström, K.: Data-driven discovery of pdes in complex datasets. J. Comput. Phys. **384**, 239–252 (2019)
2. Bongard, J., Lipson, H.: Automated reverse engineering of nonlinear dynamical systems. Proc. Nat. Acad. Sci. **104**(24), 9943–9948 (2007)
3. Gray, G.J., Murray-Smith, D.J., Li, Y., Sharman, K.C., Weinbrenner, T.: Nonlinear model structure identification using genetic programming. Control Eng. Pract. **6**(11), 1341–1352 (1998)
4. Kondrashov, D., Chekroun, M.D., Ghil, M.: Data-driven non-markovian closure models. Physica D **297**, 33–55 (2015)
5. Kovalchuk, S.V., et al.: A conceptual approach to complex model management with generalized modelling patterns and evolutionary identification. Complexity **2018** (2018)
6. Raissi, M., Perdikaris, P., Karniadakis, G.E.: Physics informed deep learning (part ii): data-driven discovery of nonlinear partial differential equations. arXiv preprint arXiv:1711.10566 (2017)
7. Rudy, S.H., Brunton, S.L., Proctor, J.L., Kutz, J.N.: Data-driven discovery of partial differential equations. Sci. Adv. **3**(4), e1602614 (2017)
8. Schaeffer, H.: Learning partial differential equations via data discovery and sparse optimization. Proc. R. Soc. A **473**(2197), 20160446 (2017)
9. Winkler, S., Affenzeller, M., Wagner, S.: New methods for the identification of nonlinear model structures based upon genetic programming techniques. Syst. Sci. Wroclaw **31**(1), 5 (2005)

Predicting Cervical Cancer
with Metaheuristic Optimizers
for Training LSTM

Andre Quintiliano Bezerra Silva$^{(\boxtimes)}$ (iD)

Instituto Federal de Mato Grosso do Sul, Jardim, Brazil
andre.bezerra@ifms.edu.br
http://www.ifms.edu.br

Abstract. Disease prediction can be extremely helpful in saving people, especially when we are diagnosed with cancer. Cervical cancer, also known as uterine cancer, is the fourth most frequent cancer in women with an estimated 570,000 new cases in 2018 representing 6.6% of all female cancers. In accordance with World Health Organization (WHO), the mortality rate for cervical cancer reaches 90% in the underdeveloped nations and that the high mortality rate found in it could suffer a substantial reduction if there were: prevention, effective screening, treatment programs and early diagnosis. Artificial Neural Networks (ANN) has been helping to provide predictions in healthcare for several decades. Most research works utilize neural classifiers trained with backpropagation (BP) learning algorithm to achieve cancer diagnosis. the traditional BP algorithm has some significant disadvantages, such as training too slowly, easiness to fall into local minima, and sensitivity of the initial weights and bias. In this work, we use a type of Recurrent Neural Network (RNN), known as Long Short-Term Memory (LSTM), whose main characteristic is the ability to store information in a series of temporal data. Instead of training the network with the backpropagation, the LSTM network was trained using five different metaheuristic algorithms: Cuckoo Search (CS), Genetic Algorithm (GA), Gravitational Search (GS), Gray Wolf Optimizer (GWO) and Particle Swarm Optimization (PSO). From results obtained can be observed that metaheuristic algorithms had performances above 96%.

Keywords: Machine learning · Cervical cancer ·
Metaheuristic algorithms · Long Short-Term Memory (LSTM)

1 Introduction

Cancer of the uterus is a type of pathology that develops in the lining of the uterus (endometrium). This type of cancer affects the female reproductive organs, located in the lower part of the uterus, near the vaginal canal [1]. According to a survey conducted by WHO [2], Cervical cancer is the fourth most frequent cancer in women with an estimated 570,000 new cases in 2018 representing

© Springer Nature Switzerland AG 2019
J. M. F. Rodrigues et al. (Eds.): ICCS 2019, LNCS 11540, pp. 642–655, 2019.
https://doi.org/10.1007/978-3-030-22750-0_62

6.6% of all female cancers. Approximately 90% of deaths from cervical cancer occurred in low and middle-income countries. The proliferation of cancer cells occurs by the non-treatment of lesions that are mostly caused by the transmissible HPV virus [1].

The development of cervical cancer is usually slow and preceded by abnormalities in the cervix. However, the absence of early stage symptoms might cause carelessness in prevention. Additionally, a lack of effective screening programs aimed at detecting and treating precancerous conditions is a key reason for the much higher cervical cancer incidence in developing countries.

Several studies have been working to detect this type of cancer using different types of machines learning classifiers such as Decision Tree [3], Support Vector Machines (SVM) [4], Gaussian Naive Bayes [5] and Artificial Neural Network [6]. RNN is one of the underlying network architectures used to develop other deep learning architectures. However, RNNs can suffer from two problems: vanishing gradients or exploding gradients. The gradients carry information used in the RNN parameter update, thus, when the gradient becomes too small, the parameter updates become insignificant which means in real learning is done. The LSTM is a special type of recurring networks and was created by [7] to solved the above problems of RNN. Its popularity has grown in recent years as an RNN architecture for various applications, such as: text compression in natural language [8,9], text recognition [10,11], speech recognition [12–14], gesture recognition [15–17].

The traditional method for performing recurrent network training is Back Propagation. BP is primarily relying on the approach of gradient-descent for shrinking the calculated squared error. However, traditional training algorithms have some drawbacks such as slow speed of convergence, easy to fall into local minimum and the training process takes longtime [18].

This work intends to design an adaptation of the LSTM network in order to apply these metaheuristic algorithms to searching for optimal values for its bias and weights. In addition, the study can be used in the Oncological area, aiming to subsidize the clinical decision in patients with suspected cervical cancer.

The rest of the article is organized as follows: after this introduction, we explain the essential RNN and LSTM fundamentals. In Sect. 3, metaheuristic optimizers are explained in brief. Section 4, presented details about the process of data collection. Section 5, provides the results of the comparison followed by final remarks and future development in Sect. 6.

2   Neural Networks

2.1   RNN

Recurrent Neural Networks have the ability to receive signals from both the input layer and the hidden layer at the previous time iteration [19]. In some ways, the hidden layer simulates the operation of a memory. In addition, Recurrent Neural Networks work with sequential data on both the input layer and the output layer, fitting perfectly in the context of time series.

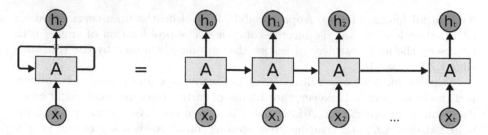

Fig. 1. Neuron of the Hidden layer of the Recurrent Neural Network [Olah, 2014].

In [20] explains that the hidden state of the Recurrent Neural Network receives information from the independent variables, as well as from its own processing result from the previous iteration, as shown in (Fig. 1). In this example, X_t is represented by the input data at time iteration t, which will be processed by the neuron A. h_t is represented by the output of neuron A in time iteration t, which can be used in the next iterations by the same neuron, as behavior. The practical effect of this is similar to the behavior of a short-term memory and is very useful in the context of time-series prediction, given that the data of the input and output layers are sequential data [21].

2.2 Long Short-Term Memory (LSTM)

Long Short-Term Memory networks are a recurrent and deep model of neural networks. It is a typical neural network architectures based on neurons and introduced the concept of memory cell. The memory cell can keep its value for a short or long time as a function of its inputs, which allows the cell to remember what is important and not only of its last computed value, that is, it can capture long dependencies range and nonlinear dynamics.

This type of network has been widely used and been able to achieve some of the best results when placed compared to other methods [23]. This fact is especially observed in the field of Natural Language Processing, and in calligraphy recognition is considered the state-of-the-art [24]. The architecture of LSTM cell is displayed in (Fig. 2).

Each LSTM block consists of a forget gate, input gate and an output gate. In Fig. 2, a basic LSTM cell with a step wise explanation of the gates is shown and on the top an other illustration of the cell connected into a network is shown. In the first step the forget gate looks at h_{t-1} and x_t to compute the output f_t (Eq. 1) which is a number between 0 and 1. This is multiplied by the cell state C_{t-1} and yield the cell to either forget everything or keep the information. In the next step the input gate is computing the update for the cell by first multiplying the outputs i_t and $\tilde{C}_t$ (Eqs. 2 and 3) and then adding this output to the input $C_{t-1} * f_t$, which was computed in the step before. Finally the output value has to be computed, which is done by multiplying o_t with the $tanh$ of the result of the previous step, which can be seen in (Eqs. 5 and 6).

$$f_t = \sigma(W_f \cdot [h_{t-1}, x_t] + b_f) \tag{1}$$

$$i_t = \sigma(W_i \cdot [h_{t-1}, x_t] + b_i) \tag{2}$$

$$\tilde{C}_t = tanh(W_C \cdot [h_{t-1}, x_t] + b_C) \tag{3}$$

$$C_t = f_t * C_{t-1} + i_t * \tilde{C}_t \tag{4}$$

$$o_t = \sigma(W_o \cdot x_t[h_{t-1}, x_t] + b_o) \tag{5}$$

$$h_t = o_t * tanh(C_t) \tag{6}$$

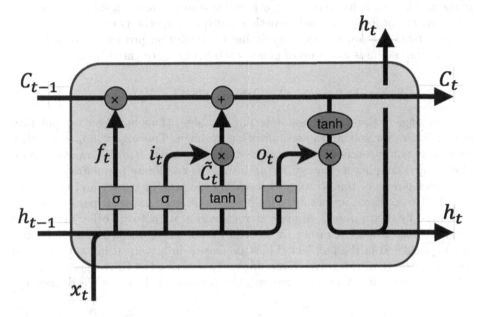

Fig. 2. Basic architectures of LSTM cells.

3 Metaheuristic Optimizers

3.1 Cuckoo Search Algorithm - CSA

The Cuckoo Search Algorithm is an evolutionary metaheuristic algorithm developed by [25], based on the behavior of the reproduction of some cuckoo bird species. In the process, these species lay their eggs in the nests of other birds. Some of these eggs, which are very similar to the eggs of the host bird, have the opportunity to grow and become adult cuckoos. This technique has been widely applied in global optimization problems, due to its simplicity and effectiveness, as well as its rapid convergence and the ability to avoid local minimums [25, 26]. CS can be described using following three idealized rules: 1. Each cuckoo lays one egg at a time and dumps it in a randomly chosen nest. 2. The best nests with the high quality of eggs will carry to the next generations. 3. The number of available host nest is fixed and if a host bird identifies the cuckoo egg with the probability of pa $\in [0, 1]$ then the host bird can either throw them away or abandon them and build a new nest.

3.2   Genetic Algorithm - GA

The Genetic Algorithm is one of the most consolidated methods of computational intelligence [27]. It is inspired by the theory of evolution of species. In this case, the best solution for a given problem is found from the combination of possible solutions that are improved at each iteration. At each iteration or generation, a new population of possible solutions or individuals is created from the genetic information of the best individuals of the previous generation population. The algorithm represents a possible solution using a simple structure, called a chromosome, to which genetically inspired operators of selection, crossing and mutation are applied, simulating the evolution process of the solution. Each chromosome is made up of genes, each being represented by bits.

3.3   Gravitational Search Algorithm - GSA

GSA belongs to the class of non-deterministic algorithms based on a population of candidates for solving the optimization problem. This algorithm, as in other evolutionary approaches, is characterized by generating an initial (random) population and defining a strategy for updating the candidate population. In GSA the development of this strategy is based on the laws of gravity and movement [28]. In this context, agents are considered as objects and their performance is measured by their masses. All mass attracts and is attracted to other masses due to gravitational force, causing the overall movement of the whole set (population) considered in the problem. Thus the masses intertwine using a direct form of communication, the gravitational force. To describe the GSA algorithm, consider a system with N agents (masses), the position of the agent i is defined by:

$$X_i = (X_i^1, \cdots, X_i^d, \cdots, X_i^n), \qquad for \quad i = 1, 2, \cdots, N \qquad (7)$$

where x_i^d presents the position of the agent i in the dimension d and n is the search space dimension. After evaluating the current population fitness, the mass of each agent is calculated as follows:

$$M_i(t) = \frac{m_i(t)}{\sum_{j=1}^{N} m_j(t)} \qquad (8)$$

where

$$m_i(t) = \frac{fit_i(t) - worst(t)}{best(t) - worst(t)} \qquad (9)$$

where $fit_i(t)$ represent the fitness value of the agent i at time t. The $best(t)$ and $worst(t)$ are the best and worst fitness of all agents, respectively and defined as follows:

$$best(t) = min_{j \in (1, \cdots N)} fit_j(t), \quad worst(t) = max_{j \in (1, \cdots N)} fit_j(t) \qquad (10)$$

To evaluate the acceleration of an agent, total forces from a set of heavier masses applied on it should be considered based on a combination of the law of gravity according to:

$$F_i^d(t) = \sum_{j \in kbest, j \neq i} rand_j G(t) \frac{M_j(t) \times M_i(t)}{R_{i,j}(t) + \epsilon} (x_j^d(t) - x_i^d(t)) \qquad (11)$$

where $rand_j$ is a random number in the interval $[0, 1]$, $G(t)$ is the gravitational constant at time t, M_i and M_j are masses of agents i and j, ϵ is a small value and $R_{i,j}(t)$ is the Euclidean distance between two agents, i and j. $kbest$ is the set of first K agents with the best fitness value and biggest mass, which is a function of time, initialized to K_0 at the beginning and decreased with time. Here K_0 is set to N (total number of agents) and is decreased linearly to 1. By the law of motion, the acceleration of the agent i at time t, and in direction d, $a_i^d(t)$, is given as follows:

$$a_i^d(t) = \frac{F_i^d(t)}{M_i^d(t)} = \sum_{j \in kbest, j \neq i} rand_j G(t) \frac{M_j(t)}{||X_i(t), X_j(t)||_2 + \epsilon} (x_j^d(t) - x_i^d(t)) \qquad (12)$$

Finally, the searching strategy on this concept can be described by following equations:

$$v_i^d(t + 1) = rand_i \times v_i^d(t) + a_i^d(t) \qquad (13)$$
$$x_i^d(t + 1) = x_i^d(t) + v_i^d(t + 1) \qquad (14)$$

where x_i^d, v_i^d and a_i^d represents the position, velocity and acceleration of ith agent in dth dimension, respectively. $rand_i$ is a uniform random variable in the interval $[0, 1]$. This random number is applied to give a randomized characteristic to the search. It must be pointed out that the gravitational constant $G(t)$ is important in determining the performance of GSA and is defined as a function of time t:

$$G(t) = G_0 \times exp\left(- \beta \times \frac{t}{t_{max}}\right) \qquad (15)$$

3.4   Gray Wolf Algorithm - GWO

The Gray Wolf Algorithm (GWO) is a computational optimization technique created by [29] based on the hunting behavior of gray wolves (Canis lupus). This species usually lives in packs of 5 to 12 individuals. These adopt a well-defined and narrow hierarchy. The leader of the wolves is called the Alpha (α), who is responsible for making decisions related to hunting, time, place of rest, etc. The second level is called Beta (β), which supports the Alphas making decisions. These are also strong candidates to take the lead in losing an Alpha. The lowest level of the hierarchy is occupied by Omega (ω), who play the role of scapegoat and must satisfy the whole group. The third level is occupied by the Delta (δ), responsible for the safety of the pack [30]. The grey wolves encircling behavior to hunt for a prey can be expressed as:

$$\boldsymbol{D} = \left|\boldsymbol{C} \cdot \boldsymbol{X}_p(t) - \boldsymbol{X}_{(t)}\right| \qquad (16)$$

$$X(t+1) = X_p(t) - A \cdot D \tag{17}$$

where t indicates the current iteration, $A = 2a \cdot r_1 - a, C = 2 \cdot r_2$, where components of a are linearly decreased from 2 to 0 over the course of iterations and r_1, r_2 are random vectors in $[0, 1]$, X_p is the prey's position vector, and X indicates the position vector of a grey wolf. The second main phase is the hunting phase and it can be modeled as:

$$D_\alpha = \left| C_1 \cdot X_\alpha - X \right|, D_\beta = \left| C_2 \cdot X_\beta - X \right|, D_\beta = \left| C_3 \cdot X_\delta - X \right| \tag{18}$$

$$X_1 = X_\alpha - A_1 \cdot (D_\alpha), X_2 = X_\beta - A_2 \cdot (D_\beta), X_3 = X_\delta - A_3 \cdot (D_\delta) \tag{19}$$

$$X(t+1) = \frac{X_1 + X_2 + X_3}{3} \tag{20}$$

3.5 Particle Swarm Optimizatio - PSO

The PSO algorithm came from the observation of the social behavior, birds in a flock and fish shoals [31]. The problem is expressed through the objective function. The quality of the solution represented by a particle is the value of the objective function at the position of this particle. The term particle is used, in analogy to physics, to have well defined position and velocity vector but has no mass or volume. Already the term swarm, represents a set of possible solutions. The velocity which takes the particle position close to pbest (own best position) and gbest (overall best position among the particles) is given by:

$$v_{id}^{k+1} = w_{id}^k + c_1 \times rand \times (pbest_{id} - s_{id}^k) + c_2 \times rand \times (gbest_{id} - s_{id}^k) \tag{21}$$

The current searching position of the particle can be modified by:

$$s_{id}^{k+1} = s_{id}^k + v_{id}^{k+1} \tag{22}$$

where s_{id}^k is the current searching point, c_1 and c_2 are learning factors w_{id}^k is weight function for velocity which is given by:

$$w_i = w_{max} - \frac{w_{max} - w_{min}}{k_{max}} \cdot k \tag{23}$$

where w_{max} and w_{min} are the maximum and minimum weights respectively, k_{max} and k are the maximum and current iteration number [32].

4   Data Description

The dataset was collected at Hospital Universitario de Caracas in Caracas, Venezuela and it was published in 2017 by University of California, Irvine. According to the study [33], most of the patients belong to the lowest socioeconomic status with low income and educational level, being the population with the highest risk. The age of the patients spans between 13 and 84 years old.

All patients are sexually active and 98% of them have been pregnant at least once. The dataset comprises demographic information, habits, and historic medical records of 858 patients and 32 features as well as four targets (Hinselmann, Schiller, Cytology and Biopsy). This paper focuses on studying the Biopsy target as it recommended by the literature review. The attributes in the dataset have been presented in the Table 1.

Table 1. Inputs with their respective type and valid amount of data.

Attributes	Types	Valids	Attributes	Types	Valids
Age	int	858	STDs: genital herpes	int	753
First sexual intercourse	int	851	AIDS	bool	753
Number of pregnancies	int	802	STDs: molluscum contagiosum	bool	753
Smokes	bool	845	HIV	bool	753
Smokes (years)	int	845	Hepatitis B	bool	753
Smokes (packs/year)	int	845	HPV	bool	753
Hormonal Contraceptives	bool	750	Number of diagnosis	int	858
Hormonal Contraceptives (years)	bool	750	STDs: syphilis	bool	753
STDs: cervical condylomatosis	bool	753	STDs: vulvo-perineal condylomatosis	bool	753
Number of sexual partners	int	832	Pelvic inflammatory disease	bool	753
STDs: vaginal condylomatosis	bool	753	Dx: Cancer	bool	858
STDs	bool	753	Dx: CIN	bool	858
STDs (number)	int	753	Dx: HPV	bool	858
STDs: condylomatosis	bool	753	Dx	bool	858
IUD	bool	741	IUD (years)	int	741

4.1 Preprocessing

Real-world information mostly tend to contain low quality data which could not be used directly in mining process without pre-processing. Numerous aspects may influence the performance of a learning system due to data quality. There ate two steps in preprocessing:

Data Cleaning. The cervical dataset suffers from a vast number of missing cells due to the lack of information. Removes all data for an observation that has one or more missing values is a quick solution and typically is preferred in cases where the percentage of missing values is relatively low. Because the amount of data lost is very small relatively to the size of the data set, the omission of the few samples with missing features was the best strategy for not influencing the analysis. The final data set was 668 instances and 34 attributes.

Normalization. Since the data of cancer cervical has features with different ranges, normalization was necessary. The goal of normalization is to change the values of numeric columns in the dataset to a common scale, without distorting differences in the ranges of values. Thus, all the features in our experiments using [0,1] normalization.

5   Simulation and Experiments

This section demonstrates the main test scenarios that have been conducted. All neural networks models used in this article were built in framework, called Keras [?]. Keras is a framework for building deep neural networks with Python and enables us to build state-of-the-art, deep learning systems. All computations are performed using a PC/Intel Core i7-5775R processor with 32 GB DDR4 RAM, GPU GeForce GTX 1080 and 1 hard drive SSD of 300 GB SATA 5 Gb/s.

The data set was used to feed the models: LSTM-CS, LSTM-GA, LSTM-GSA, LSTM-GWO and LSTM-PSO, then weights and biases are updated to get better accuracy results and less error rate. The main objective of these experiments is to compare all models of LSTM using different parameters, such as, numbers of hidden neurons, population and iteration for each model are chosen (Table 2).

The training and testing on the same set, the mean accuracy is determined using five-fold cross-validation.

Figures 3 show the accuracy of the classifiers with different numbers of neurons in hidden layer, but without changing the number of iterations and populations, remained fixed (total of 40) for all architectures presented. LSTM-PSO

Table 2. Parameters that will be modified in the scale of 4 *to* 40. In each test, one parameter changes while the others remain fixed.

Trained networks	Iteration	Population	Number of hidden neurons
LSTM-CS	4-40	4-40	4-40
LSTM-GA	4-40	4-40	4-40
LSTM-GSA	4-40	4-40	4-40
LSTM-GWO	4-40	4-40	4-40
LSTM-PSO	4-40	4-40	4-40

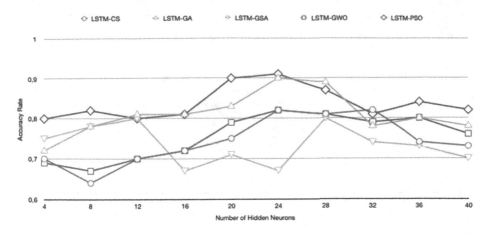

Fig. 3. Variations of the LSTM networks with the metaheuristic algorithms and their respective accuracy rate, with different number of neurons.

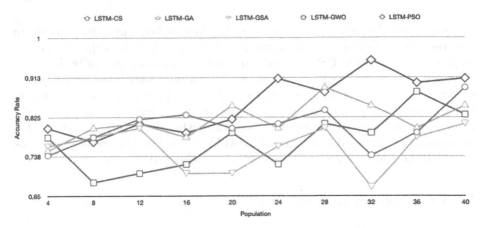

Fig. 4. Variations of the LSTM networks with the metaheuristic algorithms and their respective accuracy rate, with different population size.

and LSTM-GA arrived close to their accuracy, reaching 92% and 91% success, respectively, while LSTM-GSA had the worst result. In this test, the increase in the number of neurons did not increase the accuracy of the models. The hidden neuron can influence the error on the nodes to which their output is connected. Thus, too much hidden neurons cause too much flexibility and this leads over-fitting, and therefore, the neural networks have overestimate the complexity of the target problem.

Figure 4 shows the accuracy of neural models increasing the population size. Each LSTM was tested with different population numbers to find the ideal solution, increasing the population at each step. For the classification procedures consistently different population sizes (from 5 to 40) were used. The number of

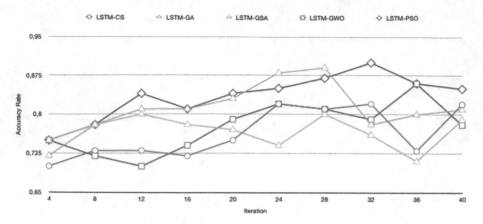

Fig. 5. Variations of the LSTM networks with the metaheuristic algorithms and their respective accuracy rate, with different number of iterations.

iterations and the number of neurons in the hidden layer were fixed to 40. In this test, we show that increasing the population size increases the accuracy of all models. The LSTM-PSO network, achieved an accuracy rate of 96%.

Figure 5 shows increase the number of iterations, while the other parameters remained fixed at 40. Despite the increase in the accuracy of the models, this improvement was quite small. A high number of iterations allows that the algorithm converges to the optimal solution, although in this case study, the performance of all models except LSTM-GSA is roughly equivalent. Again, the LSTM-PSO model obtained the best result, reaching 91% accuracy.

6   Conclusions and Future Work

The mortality rate from cervical cancer worldwide has increased significantly in recent years. The cervical tumor, threatens not only the motherhood of women, but especially their lives. This paper has proposed a LSTM network, as an alternative to deal with slow convergence and convergence to local minimum in neural networks trained with backpropagation. The paper used five different optimizers based on Nature Inspired Algorithms, such as: Cuckoo Search, Genetic Algorithm, Gravitational Search, Gray Wolf Optimizer and Particle Swarm Optimization for training LSTM neural networks and apply all models in a cervical cancer dataset. The tests were performed on a dataset containing data from Venezuelan women and they were performed by changing three parameters: the number of neurons, the number of iterations and the population size. The developed neural networks were applied to diagnose cancer patients based on a number of content related features. The LSTM-BPTT approach was evaluated and compared with other neural networks trained by five different optimizers. After that, the five optimizers were compared based on their accuracy results. The experiments showed that LSTM with PSO performed better than LSTM with the other algorithms in the term of optimizers. The accuracy for LSTM with PSO was 0.92 by varying the number of neurons, 0.96 when there was modification in the population and 0.91 changing the number of iterations.

In addition, other observations could be made according to the experiments performed. The increase in the number of neurons was not interesting for the experiment. The accuracy declines with increasing the number of neurons in the hidden layer, and this might be due to the increased complexity in the network's structure, which requires more training time to converge. On the other hand, the increase of the population was beneficial for the performance of models. Meanwhile, the increase in iterations did not show significant improvements.

Finally, this paper has provided a method for detecting women with cervical cancer with higher accuracy resulting by applying LSTM to overcome the back-propagation problems. The future work for this paper will include using other ways to address the missing data problem, verification of the optimization algorithms used in other types of neural networks and investigating the effectiveness of LSTM and algorithms with larger datasets. Prediction of different diseases with different datasets using LSTM neural networks could also be considered.

Acknowledgments.. This research is supported by Instituto Federal de Mato Grosso do Sul.

References

1. National Cancer Institute Homepage. https://www.inca.gov.br/. Accessed 15 Dec 2018
2. World Health Organization Homepage. https://www.who.int/cancer/prevention/diagnosis-screeningcervical-cancer/. Accessed 12 Dec 2018
3. Sharma, S.: Cervical cancer stage prediction using decision tree approach of machine learning. Int. J. Adv. Res. Comput. Commun. Eng. **5**(4) (2016)
4. Zhang, J., Liu, Y.: Cervical cancer detection using SVM based feature screening. In: Barillot, C., Haynor, D.R., Hellier, P. (eds.) MICCAI 2004. LNCS, vol. 3217, pp. 873–880. Springer, Heidelberg (2004). https://doi.org/10.1007/978-3-540-30136-3_106
5. Alwesabi, Y., Choudhury, A., Won, D.: Classification of cervical cancer dataset. In: IISE Annual Conference (2018)
6. Devi, M.A., Ravi, S., Vaishnavi, J., Punitha, S.: Classification of cervical cancer using artificial neural networks. Procedia Comput. Sci. **89**, 465–472 (2016)
7. Hochreiterm, S., Schmidhuber, J.: Long short-term memory. Neural Comput. **9**(8), 1735–1780 (1997). https://doi.org/10.1162/neco.1997.9.8.1735
8. Sakti, S., Ilham, F., Neubig, G., Toda, T., Purwarianti, A., Nakamura, S.: Incremental sentence compression using LSTM recurrent networks. In: IEEE Workshop on Automatic Speech Recognition and Understanding (ASRU), Scottsdale, AZ, pp. 252–258 (2015)
9. Gogate, M., Adeel, A., Hussain, A.: A novel brain-inspired compression-based optimised multimodal fusion for emotion recognition. In: IEEE Symposium Series on Computational Intelligence (SSCI), Honolulu, HI, pp. 1–7 (2017)

10. Breuel, T.M.: High performance text recognition using a hybrid convolutional-LSTM implementation. In: 14th IAPR International Conference on Document Analysis and Recognition (ICDAR), Kyoto, pp. 11–16 (2017)
11. Su, M., Wu, C., Huang, K., Hong, Q.: LSTM-based text emotion recognition using semantic and emotional word vectors. In: First Asian Conference on Affective Computing and Intelligent Interaction (ACII Asia), Beijing, pp. 1–6 (2018)
12. Sarma, K., Sarma, M.: Acoustic modelling of speech signal using artificial neural network: a review of techniques and current trends, pp. 287–303. IGI Global (2015). https://doi.org/10.4018/978-1-4666-8493-5.ch012
13. Sak, H., Senior, A., Beaufays, F.: Long short-term memory based recurrent neural network architectures for large vocabulary speech recognition. In: Interspeech, pp. 338–342 (2014)
14. Billa, J.: Dropout approaches for LSTM based speech recognition systems. In: IEEE International Conference on Acoustics. Speech and Signal Processing (ICASSP), Calgary, AB, pp. 5879–5883 (2018)
15. Lin, C., Wan, J., Liang, Y., Li, S.Z.: Large-scale isolated gesture recognition using a refined fused model based on masked res-C3D network and skeleton LSTM. In: 13th IEEE International Conference on Automatic Face Gesture Recognition, pp. 52–58 (2018)
16. Zhu, G., Zhang, L., Shen, P., Song, J.: Multimodal gesture recognition using 3-D convolution and convolutional LSTM. IEEE Access **5**, 4517–4524 (2017)
17. Xu, S., Xue, Y.: A long term memory recognition framework on multi-complexity motion gestures. In: 14th IAPR International Conference on Document Analysis and Recognition (ICDAR), Kyoto, pp. 201–205 (2017)
18. Park, H.: Part 2: multilayer perceptron and natural gradient learning. New Gener. Comput., 79–95 (2006)
19. Sundermeyer, M., Ney, H., Schluter, R.: From feedforward to recurrent lstm neural networks for language modeling. IEEE/ACM Trans. Audio Speech Lang. Process. **23**(3), 517–529 (2015)
20. Barreto, J.M.: Introdução as redes neurais artificiais. In: V Escola Regional de Informáttica. Sociedade Brasileira de Computação, Regional Sul, Santa Maria, Florianópolis, Maringá, pp. 5–10 (2002)
21. Goodfellow, I., Bengio, Y., Courville, A.: Deep Learning, 1st edn. Massachusetts Institute of Technology, Cambrigde (2016)
22. Schmidhuber J.: Deep learning in neural networks: an overview. In: Neural Networks, pp. 85–117 (2015)
23. Graves, A.: Supervised Sequence Labelling with Recurrent Neural Networks. Studies in Computational Intelligence, vol. 385. Springer, Heidelberg (2012). https://doi.org/10.1007/978-3-642-24797-2
24. Graves, A., Liwicki, M., Fernández, S., Bertolami, R., Bunke, H., Schmidhuber, J.: A novel connectionist system for unconstrained handwriting recognition. IEEE Trans. Pattern Anal. Mach. Intell., 855–868 (2009). https://doi.org/10.1109/TPAMI.2008.137
25. Yang, X.S., Deb, S.: Cuckoo search: recent advances and applications. Neural Comput. Appl. **24**(1), 169–174 (2014)
26. Biswas, S., Kundu, S., Das, S.: Inducing niching behavior in differential evolution through local information sharing. IEEE Trans. Evol. Comput. **19**(2), 246–263 (2015)
27. Holland, J.H.: Adaptation in Natural and Artificial Systems. The University of Michigan Press, Ann Arbor (1975)

28. Rashedi, E.: Gravitational Search Algorithm. M.Sc. thesis, Shahid Bahonar University of Kerman, Kerman, Iran (2007)
29. Mirjalili, S., Mohammad, S., Lewis, A.: Grey wolf optimizer. Adv. Eng. Softw. **69**, 46–61 (2014)
30. Long, W., Songjin, X.: A novel grey wolf optimizer for global optimization. In: Advanced Information Management, Communicates, Electronic and Automation Control Conference (IMCEC), pp. 1266–1270. IEEE (2016)
31. Kennedy, J., Eberhart, R.: Particle swarm optimization. In: IEEE International Conference on Neural Network, Perth, Australia (1995)
32. Yang, X.S., Deb, S.: Engineering optimisation by cuckoo search. J. Math. Model. Numer. Optim. **1**(4), 330–343 (2010)
33. Fernandes, K., Cardoso, J.S., Fernandes, J.: Transfer learning with partial observability applied to cervical cancer screening. In: Alexandre, L.A., Salvador Sánchez, J., Rodrigues, J.M.F. (eds.) IbPRIA 2017. LNCS, vol. 10255, pp. 243–250. Springer, Cham (2017). https://doi.org/10.1007/978-3-319-58838-4_27

Top k 2-Clubs in a Network: A Genetic Algorithm

Mauro Castelli[1], Riccardo Dondi[2], Sara Manzoni[3], Giancarlo Mauri[3], and Italo Zoppis[3(✉)]

[1] NOVA Information Management School (NOVA IMS),
Universidade Nova de Lisboa, Campus de Campolide, 1070-312 Lisboa, Portugal
mcastelli@novaims.unl.pt
[2] Università degli Studi di Bergamo, Bergamo, Italy
riccardo.dondi@unibg.it
[3] Università degli Studi di Milano-Bicocca, Milano, Italy
{sara.manzoni,giancarlo.mauri,italo.zoppis}@unimib.it

Abstract. The identification of cohesive communities (dense subgraphs) is a typical task applied to the analysis of social and biological networks. Different definitions of communities have been adopted for particular occurrences. One of these, the *2-club* (dense subgraphs with diameter value at most of length 2) has been revealed of interest for applications and theoretical studies. Unfortunately, the identification of *2-clubs* is a computationally intractable problem, and the search of approximate solutions (at a reasonable time) is therefore fundamental in many practical areas. In this article, we present a genetic algorithm based heuristic to compute a collection of *Top k 2-clubs*, i.e., a set composed by the largest k 2-clubs which cover an input graph. In particular, we discuss some preliminary results for synthetic data obtained by sampling Erdös-Rényi random graphs.

Keywords: Community optimization · 2-club maximization ·
Genetic algorithms · Graphs

1 Introduction

The identification of communities within a network is a typical task that has been widely applied in different contexts. In particular, dense subgraphs (i.e., cohesive communities) have perceived the attention of the scientific literature oriented to the analysis of social [1,15,17,18] and biological networks [3,19]. A standard approach to compute dense subgraphs is focused on the identification of structures known as *cliques*: complete subgraphs whose vertices are pairwise connected by edges. However, the use of a clique is too binding for specific applications. For example, a critical issue arises when missing data are persistent in the considered analysis. In this case, the missing information does not allow to represent all links of a dense community, thus requiring to search alternative

© Springer Nature Switzerland AG 2019
J. M. F. Rodrigues et al. (Eds.): ICCS 2019, LNCS 11540, pp. 656–663, 2019.
https://doi.org/10.1007/978-3-030-22750-0_63

notions of dense subgraphs. For this reason, different definitions of community have been introduced in literature "by relaxing" to some extent the concept of clique (see, e.g., [13] for the concept of *relaxed clique*).

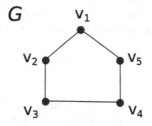

Fig. 1. The figure shows a 2-club of 5 vertices. Notice that, each subgraph of 4 vertices is not a 2-club.

In this paper, we focus on a distance-based relaxation of the clique model. In other words, instead of seeking structures where distances between pairs of vertices are equal to 1 (i.e., cliques), we will consider dense subgraphs where the distance between vertices can be at most s. Such a structure is generally known as *s-club*. In particular, due to the importance of this problem for social [1,15,17,18] and biological networks [3,19], we will consider the case of $s = 2$.

From the computational point of view, the literature concerning s-clubs has mainly focused on the identification of s-clubs of maximum size (a problem known as *Max-s-club*), and its complexity has been extensively studied. Although the results have shown the NP-hardness of the problem for each $s \geq 1$ [5], polynomial-time approximation algorithms, with factor $|V|^{1/2}$, are available for every $s \geq 2$ [2]. Recently, the problem has been investigated for restricted graph classes [11,12], and even for its parameterized complexity. The problem has been shown to be fixed-parameter tractable, when the parameter is the size of the sought s-club [6,14,20].

In many real applications, the objective is to find a set of cohesive subgraphs of the original input graph (rather than a single subgraph covering the input). Following the approach proposed in [10], this paper considers the problem of computing the set of largest k 2-clubs, with $k \geq 1$. We will denote this problem as *Top k-2-clubs*. Notice that other problems that seek for 2-clubs have been considered recently. In [9], it is considered the problem of finding a maximum set of disjoint s-clubs of at least a given size, while in [7] it is considered the problem of finding a minimum set of s-clubs that covers the input graph.

The identification of *Top-k-2-clubs* turns to be NP-hard (as Max-2-clubs is NP-hard), for this reason we design a genetic algorithm based heuristic by defining: first, a specific set of search operators for obtaining GA's approximate solutions, then a greedy approach to extrapolate the k top different approximations. While GA optimization is not new in literature, the interest on designing new heuristics and special operators for the applied models is still required to deal with the intractability of computational problems, with new applications in different contexts [8,22].

The paper is organized as follows. In Sect. 2 we provide the definitions and we introduce the problem we are interested in. In Sect. 3, we discuss the GA-based approach to seek approximate solutions for the *Top-k-2-clubs*. In Sect. 4, we report numerical evaluations based on Erdös-Rényi random graphs. Section 5 discusses the preliminary results and the future development of our research.

2   Preliminaries

Let $G = (V, E)$ be a graph, and $V' \subseteq V$ a subset of the input vertices V. Denote by $G[V']$ the subgraph of G induced by V', and $d_G(u, v)$ the *distance* (i.e., length of a shortest path) between two vertices u, v. The *diameter* of $G = (V, E)$ is defined as $\max_{u,v \in V} d_G(u, v)$, i.e., the maximum distance between any two vertices in V.

Given a graph $G = (V, E)$, a 2-club in G is a subgraph $G[W]$, with $W \subseteq V$, that has diameter of at most 2. The property of being a 2-club is not *hereditary*[1]. This means that if a graph G is a 2-club, then a subgraph of G is not necessarily a 2-club (see Fig. 1).

We introduce now the formal definition of the problem, denoted as *Top-k-2-clubs*. The *Top-k-2-clubs* maximizes an objective function that considers the size of the 2-clubs in the solution (since we want to compute large 2-clubs) and a distance function *dist* to provide graphs that are significantly different. The parameter λ allows us to define how much relevance has the distance with respect to the size of the 2-clubs.

Problem 1 *Top-k-2-club*
Input: *a graph* $G = (V, E)$, *a value* $\lambda > 0$.
Output: *a set* $\mathcal{W} = \{G[W_1], \ldots, G[W_k]\}$ *of k 2-clubs, with* $1 \leq k < |V|$ *and* $W_i \subseteq V$, *that maximizes the following value*

$$r(\mathcal{W}) = \sum_{i=1}^{k} |W_i| + \lambda \sum_{i=1}^{k-1} \sum_{j=i+1}^{k} dist(G[W_i], G[W_j])$$

where

$$dist(G[W_i], G[W_j]) = \begin{cases} 2 - \frac{|W_i \cap W_j|^2}{|W_i||W_j|} & \text{if } W_i \neq W_j, \\ 0 & \text{else.} \end{cases}$$

Notice that *Top-k-2-clubs*, when $k = 1$, is exactly Max-2-club. Since Max-2-club on an input graph $G = (V, E)$ is NP-hard [3] and not approximable within factor $|V|^{1/2-\varepsilon}$, for each $\varepsilon > 0$ [2], the same properties hold for *Top-k-2-clubs*.

---

[1] This property can be extended to each $s \geq 2$.

3   A Genetic Algorithm for the Top-k-s-club Problem

As reported above, the *Top-k-2-clubs* is NP-hard, thus making optimization potentially impracticable. Our approach here is to provide approximate solutions by designing dedicated genetic operators. Let $G[V']$ be a 2-club of the input graph $G = (V, E)$, for some set of vertices $V' \subseteq V$. We represent GA's solutions as binary chromosomes c, of size $|V|$, such that for each $v_i \in V'$, $c[i] = 1$ ($c[i] = 0$ if $v_i \in V \setminus V'$). With a slight abuse of notation, we denote by $G[c]$ the subgraph of G induced by the representation of chromosome c. Furthermore, $V[c]$ and $E[c]$ represent the set of vertices, V', and edges of $G[c] = G[V']$. With the given representation, a set of k chromosomes, which are involved in the offspring generation, is interpreted as a set of hypothesis of feasible solutions (i.e., hypotheses of potential 2-clubs). To quantify the validity of such (assertions) hypothesis, chromosomes will be then evaluated by the fitness function.

3.1   Fitness Function

We design the fitness function to promote new offspring in such a way that "feasible" chromosomes, able to properly express 2-clubs (i.e., with a correct diameter value ≤ 2) will gradually adapt through specific mutation and crossover operators.

Consider a chromosome c, and the graph $G = (V, E)$. In order to apply a fitness function which promotes the representation of large subgraphs $G[c]$, when $0 \leq \text{diam} \leq 2$, we consider the following.

$$f(n_v, \text{diam}; S) = \begin{cases} n_v & \text{if } 0 \leq \text{diam} \leq 2; \\ \frac{1}{n_v} & \text{if } 2 < \text{diam}, \end{cases} \tag{1}$$

where n_v is the number of vertices of $G[c]$.

Notice that, when diam > 2 (i.e., unfeasible solutions) we obtain fitness values which decrease asymptotically for large subgraphs with size n_v, thus penalizing the corresponding chromosomes.

3.2   Mutation

The following types of mutations are considered with equal probability.

- Mutation 1. In this case, mutation is applied to correct hypotheses sparingly and consistently. For this, consider the set $V_+ = \{v_i : c[i] = 1\}$ and the associated graph $G[V_+]$, which should correspond to some feasible 2-club. In order to check such a feasibility, we randomly sample a vertex $v' \in V_+$, and for each pair $(v_i, v'), v_i \in V_+ \setminus \{v'\}$ we verify whether the minimum length between v_i and v' is of at most 2. If this is not the case, $c[i]$ is flipped to 0.
- Mutation 2. Similarly to the previous case this operator has the objective to sparingly increment the size of a solution. We consider now $V_- = \{v_j : c[j] = 0\}$ and the current subgraph $G[V_+]$ induced by c. In order to consistently add

vertices to V_+, we sample some v' from V_- (equivalently, we will have some j' such that $v' = v_{j'} \in V_- : c[j'] = 0$) to check whether the shortest distance of v' from vertices in V_+ is not larger than 2. In this case, the corresponding bit $c[j']$ is flipped to 1.

– Standard Mutation. This is a standard mutation procedure where bits of the selected chromosome are randomly switched on or off.

3.3   Cross-Over

The cross-over operations operations are selected with equal probability.

– Standard cross-over. Parts of parents' chromosomes are copied and mixed in new offspring with standard one-point crossover.
– Logical AND/OR cross-over. New offspring are generated by applying logical AND and logical OR operations between parents.

Table 1. Performances: Models; Av. Fitness (Fit); Av. Inp. Diameter (InD); Av. Out. Diameter (OutD); Av. Jaccard (AvJ); Av. Covering (AvC); Av. User (T1) and System Time (T2)

Models	AvFit	InD	OutD	AvJ	AvC	T1	T2
ER(100,0.1)	12.4	4	2	0.31	0.13	98.4	1.30
ER(150,0.1)	16.8	3	2	0.47	0.12	154	1.86
ER(200,0.1)	27.1	3	2	0.69	0.14	233	2.49
ER(300,0.1)	35.1	3	2	0.74	0.12	370	3.02
ER(400,0.1)	41.1	3	2	0.78	0.11	453	3.90

3.4   Selection of Top k 2-Clubs

The final step of the algorithm selects k 2-clubs (chromosomes) from the population. Denote by $\mathcal{W}$ the solution of *Top-k-2-clubs* we are computing. We apply a greedy procedure consisting of k iterations. In the first iteration, a 2-club of the population (denoted by $G[W_1]$) having maximum cardinality is added to $\mathcal{W}$.

Consider iteration t, with $2 \leq t \leq k$, of the greedy procedure. Assume that $\mathcal{W} = \{G[W_1], \ldots, G[W_{t-1}]\}$, iteration t adds to $\mathcal{W}$ a 2-club $G[W_t]$ of the population that maximizes the following value

$$|W_t| + \lambda \sum_{i=1}^{t-1} \sum_{j=i+1}^{t} dist(G[W_i], G[W_j]).$$

4   Numerical Experiments

The goal of our experiments was to check the capability of GAs to provide feasible solutions in reasonable computational time. The whole procedure described above was coded in R using the "GA" package [21]. In this work we use synthetic data by sampling Erdos-Renyi (ER) random graphs, $ER(n, p = 0.1)$, with different number of vertices, n [4].

To ensure both the correctness of the Genetic Algorithm and the computational tractability of our approach, we followed a standard practice of the evolutionary methods; that is, maintaining the tractability of genetic operators while promoting, at the same time, new evaluable offspring, which in our case provide feasible diameter values for the obtained solutions.

Similarly, the termination of the genetic algorithm is guaranteed by standard criteria. We have set both the revaluation number of the fitness with respect to new populations (equivalently, the number of GA iterations) and the number of new consecutive generation without fitness improvement. Finally, to get a more robust performance evaluation, each Erdos graph was repeatedly sampled 3 times. Performances are reported in Table 1. The following comments summarize our results.

1. All models are correct, being (2-clubs) with diameter ≤ 2.
2. Since the considered problem is computationally intractable, it is not possible to compare the optimal solution with those provided by the described approach. In order to give, at least, a qualitative idea of the validity of the obtained approximations, we considered both the average covering (output / input vertices) of the input graph (obtained through the returned solutions, i.e., 10 largest solutions) and a measure of how such solutions differ from each other. In this way, the higher the number of covering ratio (and the more dissimilar are the solutions), the more input graph cover is qualitatively effective. To this aim, the Jaccard index was applied. Specifically, we remind that the smaller this index, the more dissimilar are the solutions. Observing Table 1 more convincing solutions seem to be those obtained for a small number of input vertices. Moreover, the coverage performances do not decrease in an "obvious" way in the considered models.
3. Covering performances are reported. In this case, we get, on average, a covering value at least of 10%.
4. A reasonable system time is observed after execution (T2 ≤ 4 s).

5   Conclusion

Identifying cohesive subgraphs within a network is a typical task with many applications in different important fields. In this paper, we reported our work in progress by considering the case where a collection of *Top-k-2-clubs* (i.e., largest different cohesive subgraphs) is maximized, providing large communities of a network covering. The computational hardness of the problem makes it impracticable to get optimal solutions. Here, we designed a set of dedicated GA operators to return approximate solutions at reasonable costs.

The preliminary results reported in this paper show we can get correct solutions in a reasonable time. Although compelling solutions seem to be provided for small graphs only ($n \leq 150$), the 10% of the input graph size is covered.

Some aspects of this research can certainly be considered in a future extension. For example, a detailed analysis should be applied to optimize the parameters of the applied models. Although different *igraph* (R package) default values (i.e., probabilities of the search operators) have been used, accurate tuning should be properly considered to evaluate performance improvement. In this case, for example, the *irace* R package [16] can be easily applied for this purpose. Furthermore, only *s-clubs* with $s = 2$ were considered. Our approach can further be scaled up to any value $s \geq 2$ and k (number of 2-clubs). Finally, real case analysis cannot be neglected in this research, and this will be another extension for a future version of the work.

Acknowledgments. This work was partially supported by national funds through FCT (Fundação para a Ciência e a Tecnologia) under project DSAIPA/DS/0022/2018 (GADgET).

References

1. Alba, R.D.: A graph-theoretic definition of a sociometric clique. J. Math. Sociol. **3**, 113–126 (1973)
2. Asahiro, Y., Doi, Y., Miyano, E., Samizo, K., Shimizu, H.: Optimal Approximation Algorithms for Maximum Distance-Bounded Subgraph Problems. Algorithmica (2017)
3. Balasundaram, B., Butenko, S., Trukhanov, S.: Novel approaches for analyzing biological networks. J. Comb. Optim. **10**(1), 23–39 (2005)
4. Bollobas, B.: Random Graphs. Cambridge University Press (2001)
5. Bourjolly, J., Laporte, G., Pesant, G.: An exact algorithm for the maximum k-club problem in an undirected graph. Eur. J. Oper. Res. **138**(1), 21–28 (2002)
6. Chang, M., Hung, L., Lin, C., Su, P.: Finding large k-clubs in undirected graphs. Computing **95**(9), 739–758 (2013)
7. Dondi, R., Mauri, G., Sikora, F., Zoppis, I.: Covering with clubs: complexity and approximability. In: Iliopoulos, C., Leong, H.W., Sung, W.-K. (eds.) IWOCA 2018. LNCS, vol. 10979, pp. 153–164. Springer, Cham (2018). https://doi.org/10.1007/978-3-319-94667-2_13
8. Dondi, R., Mauri, G., Zoppis, I.: Orthology correction for gene tree reconstruction: theoretical and experimental results. Procedia Comput. Sci. **108**, 1115–1124 (2017)
9. Dondi, R., Mauri, G., Zoppis, I.: On the tractability of finding disjoint clubs in a network. Theor. Comput. Sci. (2019). https://doi.org/10.1016/j.tcs.2019.03.045
10. Galbrun, E., Gionis, A., Tatti, N.: Top-k overlapping densest subgraphs. Data Min. Knowl. Discov. **30**(5), 1134–1165 (2016). https://doi.org/10.1007/s10618-016-0464-z
11. Golovach, P.A., Heggernes, P., Kratsch, D., Rafiey, A.: Finding clubs in graph classes. Discrete Appl. Math. **174**, 57–65 (2014)

12. Hartung, S., Komusiewicz, C., Nichterlein, A.: Parameterized algorithmics and computational experiments for finding 2-clubs. J. Graph Algorithms Appl. **19**(1), 155–190 (2015)
13. Komusiewicz, C.: Multivariate algorithmics for finding cohesive subnetworks. Algorithms **9**(1), 21 (2016)
14. Komusiewicz, C., Sorge, M.: An algorithmic framework for fixed-cardinality optimization in sparse graphs applied to dense subgraph problems. Discrete Appl. Math. **193**, 145–161 (2015)
15. Laan, S., Marx, M., Mokken, R..J.: Close communities in social networks: boroughs and 2-clubs. Social Netw. Anal. Min. **6**(1), 20:1–20:16 (2016)
16. López-Ibáñez, M., Dubois-Lacoste, J., Pérez Cáceres, L., Stützle, T., Birattari, M.: The irace package: iterated racing for automatic algorithm configuration. Oper. Res. Persp. **3**, 43–58 (2016). https://doi.org/10.1016/j.orp.2016.09.002
17. Mokken, R.: Cliques, clubs and clans. Qual. Quant. Int. J. Methodol. **13**(2), 161–173 (1979)
18. Mokken, R.J., Heemskerk, E.M., Laan, S.: Close communication and 2-clubs in corporate networks: Europe 2010. Soc. Netw. Anal. Min. **6**(1), 40:1–40:19 (2016)
19. Pasupuleti, S.: Detection of protein complexes in protein interaction networks using n-clubs. In: Marchiori, E., Moore, J.H. (eds.) EvoBIO 2008. LNCS, vol. 4973, pp. 153–164. Springer, Heidelberg (2008). https://doi.org/10.1007/978-3-540-78757-0_14
20. Schäfer, A., Komusiewicz, C., Moser, H., Niedermeier, R.: Parameterized computational complexity of finding small-diameter subgraphs. Optim. Lett. **6**(5), 883–891 (2012)
21. Scrucca, L.: GA: a package for genetic algorithms in R. J. Stat. Softw. **53**(4), 1–37 (2013)
22. Weise, T., Zapf, M., Chiong, R., Nebro, A.J.: Why is optimization difficult? In: Chiong, R. (ed.) Nature-Inspired Algorithms for Optimisation, vol. 193, pp. 1–50. Springer, Heidelberg (2009). https://doi.org/10.1007/978-3-642-00267-0_1

CA-RPT: Context-Aware Road Passage Time Estimation for Urban Traffic

Ying Liu[1,2]([✉]), Zhenyu Cui[1,2][iD], Tianlin Zhang[1,2,3,4], Jiaxu Leng[1,2,3,4],
Weihong Xie[3], and Liang Zhang[4]

[1] School of Computer Science and Technology,
University of Chinese Academy of Sciences, Beijing 100190, China
`yingliu@ucas.ac.cn`
[2] Key Lab of Big Data Mining and Knowledge Management,
Chinese Academy of Sciences, Beijing 100190, China
[3] School of Economics and Commerce, Guangdong University of Technology,
Guangzhou 510006, China
[4] School of Applied Mathematics, Guangdong University of Technology,
Guangzhou 510006, China

Abstract. Road passage time is an important measure of urban traffic. Accurate estimation of road passage time contributes to the route programming and the urban traffic planning. Currently, the estimation of road passage time for a particular road is usually based on its historical data which is simple to express the general law of road traffic. However, with the increase of the number of roads in the urban area, the connection between roads becomes more complex. The existing methods fail to make use of the connection between different roads and the road passage time, merely based on its own historical data. In this paper, we propose a road passage time estimating model, called "CA-RPT", which utilizes the contextual information between road connections as well as the date and time period. We evaluate our method based on a real geolocation information data set collected by mobile APP anonymously. The results demonstrate that our method is more accurate than the state-of-the-art methods.

Keywords: Road passage time · Contextual information ·
Road connection

1 Introduction

In recent years, the introduction of smart transportation has gradually changed the manner people travel. One of the widely used applications, in modern urban, is the estimation of road passage time. Accurately estimating road passage time is an important measure in guiding people's travel and avoiding the potential high blockage.

Road passage time estimation, which can reveals the congestion, is an important, but challenging task in smart traffic and has been widely used in travel time

© Springer Nature Switzerland AG 2019
J. M. F. Rodrigues et al. (Eds.): ICCS 2019, LNCS 11540, pp. 664–673, 2019.
https://doi.org/10.1007/978-3-030-22750-0_64

estimation [1,2] which is of great importance for real-time traffic monitoring [3], driving directions choice [4] and transportation resource scheduling [5]. Existing solutions, e.g., based on support vector regression [6] or spare flow capacity on the concerned road links [7], focus on the information of the road itself including the historical data rather than the contextual information from the connections between different roads.

Road passage time is affected by factors, such as the length, width, and the grade of the road, etc. In general, it takes longer time to travel on a longer, narrower, and lower-grade road under the same natural conditions and human activities.

However, all of the above are the objective characteristics of the road itself. Due to human activities, the judgment of the passage time of a road is inseparable from the time slots and date. Most of the previous studies used the historical data of a road to predict its future behavior, which uses a model to fit the recent or the historical data of a single node, and applies the given model to predict the upcoming road passage time.

In modern urban transportation, the connections between roads are more and more complicated and multiple factors may impact the road patency. The upstream and downstream, the number of adjacent roads, the congestion level of a road may affect the passage time estimation more or less. Therefore, the passage time estimation of a road is not a simple prediction with its own historical data or objective property. However, existing methods failed to utilize this part of the information, but by fitting historical data on a single road to fit the law of road passage time. The generalization ability of this approach is limited. Because it can only predict according to a general law, and does not have the ability to estimate the passage time based on temporary accidents, congestion, etc. which is occurring on some roads. Considering the influence of the following factors, we would like to use the road contextual information to finish the estimation.

Based on previous research, contextual information has been proven to be important in the field of psychology [8–10] and computer vision [11–13], which has also been well recognized for years that the contextual information is helpful in object recognition [14–16]. Inspired by the use of contextual information, this paper would like to propose an estimation model creatively, called CA-RPT which combines the road information with the contextual information into a novel structure thus to improve the prediction accuracy. The rest of the paper is organized as follow: Sect. 2 explain our approach; Sect. 3 is the experimental result and evaluation; Sect. 4 is the conclusion of this paper.

2 Our Approach

In this section, we will present our approach, including the overall architecture and the modeling process.

2.1 Overview

The architecture of the proposed CA-RPT is presented in Fig. 1.

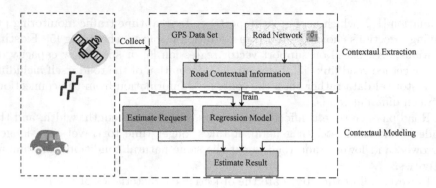

Fig. 1. The architecture of CA-RPT.

CA-RPT is consist of two parts. The first part, called "Contextual Extraction", using GPS data, which collected in real time and processed into the passage time of each road, and road network to extract contextual information. The second part, which called "Contextual Modeling", use the former extracted contextual information and original GPS data set to fit a model, which utilizes the historical road passage time to estimate the one in the next few moments. The process of "Contextual Modeling" could be considered as a regression model, which use the road passage time for a period of time before a certain time point to predict the possible road passage time in the future. Since the process of contextual extraction does not add new dimensions, we could take the existing appropriate regression method for prediction.

In the following, we will explain the two parts of CA-RPT in turn.

2.2 Contextual Information

Inspired by contextual information for image recognition, we believe that the same approach can be used in passage time estimation. Before contextual extraction, we should firstly define some context concepts in traffic to be mentioned as follows. Similarly, we first define the context of the road as shown in Eq. 1.

r_i: a road wait to be estimated
r_i': a road adjacent to r_i.
$F(r_i)$: the sum of the contextual information of the road r_i.
$\psi(r_i')$: the road passage time of r_i'.
λ_i: the impact scale factor ($\lambda_i \in (0,1)$).

$$F(r_i) = \sum_{r_i'(road\ adjacent\ to\ r_i)} \lambda_i * (\psi(r_i') + F(r_i'))$$ (1)

The equation indicates that all the information available to a road consists of two parts, the road passage time and the contextual information obtained from its neighboring roads. Compared with existing methods, which only use the historical data of one road, our approach adds the road's own historical time to the adjacent road multiply by the coefficient λ. This formula is obviously a recursive formula. In order to use the context information, we must first find a simple way of obtaining contextual information. Hence we get partial derivative of Eq. 1 with respect to variable r'_i and obtain the Formula 2, which shows that the contextual information of each road can be obtained by multiplying the passage time of each adjacent road by the influence factor separately.

$$\frac{\partial F(r_i)}{\partial r'_i} = \lambda_i * \psi'(r'_i) + \lambda_i * F'(r'_i) \tag{2}$$

Based on the above facts, the contextual information of different roads can be added independently. Thus, for each r'_i in the Eq. 1, we can put them into the same equation recursively. Then we can recursively get the Eq. 3, where $\omega_{r'_i}$ represents the adjacent trajectory between r_i and r'_i.

$$F(r_i) = \sum_{r'_i \in (roads)} \psi(r'_i) * \prod_{r'_j \in \omega_{r'_i}} \lambda_j \tag{3}$$

According to Eq. 3, the contextual information of each road is the accumulation of the road passage time obtained by a series of roads adjacent to it. The contextual information acquired by each street is multiplied by the impact factor on its corresponding adjacent trajectory, which are obviously exponentially related, to form the contribution of the road to be estimated. In this way, we get complete context information for a road, which includes contextual information of all the other roads.

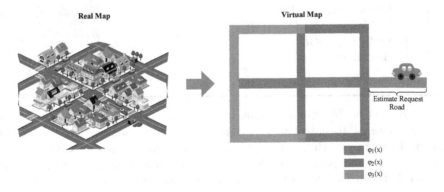

Fig. 2. Contextual information based on distance.

For an intuitive view, as shown in Fig. 2, $\phi_i(x)$ represents the sum of contextual information of road x, and i represent there are i roads belong to the trajectory from each road to road x.

2.3   Contextual Extraction

Based on the conclusions of the previous section, we perform contextual extraction to obtain contextual information for a road to assist us in making road passage time estimation. Obviously, for each road, the cumulative impact of other roads on it is unique and independent. Therefore, the core of contextual extraction is to calculate the sum of contextual information for each road. For a road to be estimated, we design an algorithm to mine the contextual information on its adjacent roads, called "Contextual Information Extraction in Traffic Network" (CIETN), as shown in Algorithm 1. Here, an additional parameter must be used, called Dep, to limit the number of ϕ_i we calculate.

Algorithm 1. CIETN

Input: The index of road to be estimated(i);The depth of context-aware
         information extraction(Dep).
Output: The contextual information extracted by CIETN(Σ).
1   **Function** *CIETN(Index i, Deep Dep)*
2 | $List_A \leftarrow Push(i, \lambda = 1)$;
3 | $Visit_{r_i} = True$;
4 | **while** $Dep \geq 0$ **do**
5 | | Dep = Dep-1;
6 | | **for** *each (i, λ) in $List_A$* **do**
7 | | | $\phi_i \leftarrow \lambda * RoadPassageTime\ in\ History(r_i)$;
8 | | | $\Sigma \leftarrow \Sigma + \phi_i$;
9 | | | **for** $r_j adjacent to r_i$ **do**
10 | | | | **if** (r_j) *and* $(r_i\ to\ r_j)$ *is Not Visited* **then**
11 | | | | | $List_B \leftarrow Push(j, \lambda * \lambda_j)$;
12 | | $Visit_{each(r_i)} = True$;
13 | | $Visit_{each(r_i\ to\ r_j)} = True$;
14 | | $List_A \leftarrow List_B$;
15 | | clear($List_B$);
16 | return Σ;

Algorithm 1 is the core algorithm of contextual extraction, which uses hierarchical traversal to calculate the contextual information on the road to be extracted. There are three points that need special explanation. For the first one, the additional parameter, Dep, is necessary based on the fact that if the distance between two roads is far enough, the mutual impact can be ignored. In other words, we must only extract contextual information within a certain distance from the road to be extracted. Secondly, the parameter λ is calculated based on the basic properties of the road, including the length, width and level of the road. That is, the λ of a road is fixed and does not change over time. Last but not least, when we get the passage time of a road, we get historical data for a period of time. These fixed-length historical data are processed by the same sampling process and become a fixed-length one-dimensional tensor, which makes the contextual information Σ returned by the algorithm also a one-dimensional tensor. And this tensor is consistent with the length of the original data.

The complexity of this algorithm is $O(N^{Dep})$, where N represent the connected roads and Dep represent the depth of extraction, but it is far from this level in practical applications.

Considering the crossing of modern cities, there are often only four roads at the junction. As a result, the N in the above paragraph is small. Moreover, we often set the Dep to a number within 5, which ensures that the time cost of the algorithm is small.

After extraction algorithm CIETN, contextual information for each road, which will be used in contextual modeling along with the original road historical passage time.

2.4   Contextual Modeling

The task of contextual modeling is to train a predictor to predicts future passage time of a road. According to experience, road passage time is likely to show a trend in different time periods and different dates. In order to verify our conjecture, we first randomly selected several roads with different timing characteristics for analysis.

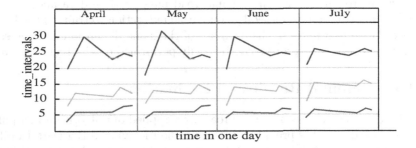

Fig. 3. Contextual passage time statistics on partial roads in different months in average.

As shown in Fig. 3, the ordinate represents the contextual passage time of a road in history, and its unit is 10 s while the abscissa represents the time in one day from day to night.

After cluster analysis, we found that the contextual passage time of most roads show a similar trend within single hour, however, what is not obvious is the information to show trend of the roads between the months. So we consider further subdividing the data to mine more potential features that can be distinguished.

Then, in each week, we clustered each of the randomly selected streets, as shown in Fig. 4. The results show that the contextual passage time of the same

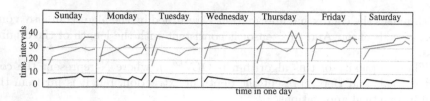

Fig. 4. Contextual passage time statistics on partial roads in different months.

road is relatively obvious due to the factors of the week. Therefore, different days of a week should be considered separately when establishing a regression model during the contextual modeling phase.

What's more, during the cluster analysis experiment, we also found that the contextual passage time of the road in one week day are often grouped into one class, which shows a strong correlation between road transit time and time. Therefore, we consider using the LSTM(Long Short Term Memory Network) model [17] to solve this regression problem, which is applied in multiple time-related regression problems successfully [18–20]. LSTM changes the weight of the self-loop by increasing the input threshold, forgetting the threshold and outputting the threshold. In this way, when the model parameters are fixed, the integral scale at different times can be dynamically changed, thereby avoiding the problem of gradient disappearance or gradient expansion.

In our approach, we use the original road passage time data and the road contextual information extracted by contextual extraction to train on the LSTM network. These two parts of the data are of the same dimension and the same length, so our modeling process can be easily performed on the LSTM network.

3   Experimental Result and Evaluation

To verify the effectiveness of our approach, we applied our algorithm in a public data set. This data set is provided by the Guizhou Provincial Big Data Development Authority which is composed of the urban road network information and the geographical location information of users, which collected anonymously by the mobile APP in real time. These raw data are processed and merged to form traffic information for the city at full time without blind spots. It is composed of three data sets, which described by three tables: Tables 1, 2, 3.

Table 1. The description of data set.1.

Attribute name	Type	Tip
link_ID	string	Unique identifier for each road segment (link)
length	double	Length of the link (m)
width	double	Width of the link (m)
link_class	int	The level of link, e.g., number 1 represents main road

Table 2. The description of data set.2.

Attribute name	Type	Tip
link_ID	string	Unique identifier for each road segment (link)
in_links	string	Direct upstream link of the link, link_IDs were split by #
out_links	string	Direct downstream link of the link, link_IDs were split by #

Table 3. The description of data set.3.

Attribute name	Type	Tip
link_ID	string	Unique identifier for each road segment (link)
date_time	date	Date, e.g., '2015-10-01'
time_interval	string	Time_interval, e.g., [2015-09-01 00:00:00, 2015-09-01 00:00:10]
travel_time	double	Average passage time of the vehicles on the road(s)

Table 1 is the attribute data set of road (link), which is a data set describing the connection of major roads in the city. Wherein, each traffic direction of each road is composed of a plurality of road segments (links), and the data set provides a unique identifier, length, width, and road type of each link. Table 2 is a link upstream and downstream connection data set, which is a data set describing the connection between links. The links includes the upstream and downstream connection according to the direction in which the vehicle is allowed to pass. The data set includes the direct upstream link and the direct downstream link of each link. Table 3 is a link historical passage time data set. The data set records the average passage time on each link in different time periods of the history (2 min is a time period). The average passage time of each time period is determined by the time period when the vehicle entering the link.

The data set we use provides a total of 132 links of static information, as well as upstream and downstream connections between these links which also includes the average passage time of each road on each time period from March 2016 to May 2016, March 2017 to June 2017 and July 2016.

The evaluation index we use is MAPE (mean absolute percent error). Generally, the lower the MAPE value, the higher the accuracy of the model, such as the Formula 4, where ttp means the estimate value, ttr means the real value, N means the number of estimate value, T_i means the number of estimate time intervals in link i.

$$MAPE = \frac{sum_{i=1}^{N} sum_{j=1}^{T_i} |ttp_{i,j} - ttr_{i,j}|}{sum_{i=1}^{N} T_i} \tag{4}$$

We randomly selected the time period to make predictions and ensure that there is sufficient data for training. In the experiment, we used some common methods to make estimation and use the model(LSTM) which has the best performance in estimation accuracy as our regression model for contextual modeling. The experimental results are shown in Table 4.

Table 4. The MAPEs of different methods.

Method	MAPE
Linear regression	0.5451
Logistic regression	0.3324
Xgboost	0.3159
LSTM	0.2881
Context-aware LSTM	0.2762

The experiment result shows that the MAPE value of our approach decreased by 0.0119 compared with the most accurate existing method, which shows that our context-aware modeling method can effectively improve the accuracy of estimation which is better than any existing method.

4   Conclusion

In this article, inspired by the successful application of contextual mining in multiple domains, we proposed a road passage time estimation framework(CA-RPT) whose process is divided into two parts, contextual extraction and contextual modeling. The former part use historical data of roads and road network information to extract contextual information in traffic. Based on these information, the later one make estimation by context-aware modeling. Experiments show that the road passage time estimated by CA-RPT is more accurate than existing methods, by obtaining and effectively utilizing road contextual information. What is more, in the experiment, the λ and the depth of contextual extraction(Dep) we used is an empirically based constant, which does not necessarily produce the best results in our modeling process. It can be expected that if these constant can be trainable, the estimation accuracy might be further improved.

Acknowledgement. This project was partially supported by Grants from Natural Science Foundation of China 71671178, 9154620, and also supported by Guangdong Provincial Science and Technology Project 2016B010127004.

References

1. Wang, Y., Zheng, Y., Xue, Y.: Travel time estimation of a path using sparse trajectories (2014)
2. Wang, Z., Fu, K., Ye, J.: Learning to estimate the travel time. In: Proceedings of the 24rd ACM SIGKDD International Conference on Knowledge Discovery and Data Mining. ACM (2018)
3. Chawla, S., Zheng, Y., Hu, J.: Inferring the root cause in road traffic anomalies. In: IEEE International Conference on Data Mining. IEEE (2013)
4. Jing, Y., Yu, Z., Zhang, C., et al.: T-drive: driving directions based on taxi trajectories. In: Sigspatial International Conference on Advances in Geographic Information Systems (2010)
5. Yuan, N.J., Zheng, Y., Zhang, L., et al.: T-finder: a recommender system for finding passengers and vacant taxis. IEEE Trans. Knowl. Data Eng. **25**(10), 2390–2403 (2013)
6. Wu, C.H., Ho, J.M., Lee, D.T.: Travel-time prediction with support vector regression. IEEE Trans. Intell. Transp. Syst. **5**(4), 276–281 (2004)
7. Liang, Z., Wakahara, Y.: Real-time urban traffic amount prediction models for dynamic route guidance systems. EURASIP J. Wirel. Commun. Netw. **2014**(1), 85 (2014)
8. Bar, M.: Visual objects in context. Nat. Rev. Neurosci. **5**(8), 617–629 (2004)
9. Oliva, A., Torralba, A.: The role of context in object recognition. Trends Cogn. Sci. **11**(12), 520–527 (2007)
10. Palmer, E.: The effects of contextual scenes on the identification of objects. Mem. Cogn. **3**, 519–526 (1975)
11. Carbonetto, P., De Freitas, N., Barnard, K.: A statistical model for general contextual object recognition. Lect. Notes Comput. Sci. **3021**, 350–362 (2004)
12. Chen, L.C., Papandreou, G., Kokkinos, I., et al.: DeepLab: semantic image segmentation with deep convolutional nets, atrous convolution, and fully connected CRFs. IEEE Trans. Pattern Anal. Mach. Intell. **40**(4), 834–848 (2016)
13. Galleguillos, C., Belongie, S.: Context based object categorization: a critical survey. Comput. Vis. Image Underst. **114**(6), 712–722 (2010)
14. Divvala, S.K., Hoiem, D., Hays, J.H., et al.: An empirical study of context in object detection (2009)
15. Galleguillos, C., Rabinovich, A., Belongie, S.: Object categorization using co-occurrence, location and appearance. In: IEEE Conference on Computer Vision and Pattern Recognition. IEEE (2008)
16. Leng, J. , Liu, Y.: An enhanced SSD with feature fusion and visual reasoning for object detection. Neural Comput. Appl. (2018)
17. Gers, F.A., Schmidhuber, J., et al.: Learning to forget: continual prediction with LSTM. Neural Comput. **12**(10), 2451–2471 (2000)
18. Duan, Y., Lv, Y., Wang, F.Y.: Travel time prediction with LSTM neural network. In: IEEE International Conference on Intelligent Transportation Systems (2016)
19. Selvin, S., Vinayakumar, R., Gopalakrishnan, E.A., et al.: Stock price prediction using LSTM, RNN and CNN-sliding window model. In: International Conference on Advances in Computing (2017)
20. Marchi, E., Ferroni, G., Eyben, F., et al.: Multi-resolution linear prediction based features for audio onset detection with bidirectional LSTM neural networks. In: IEEE International Conference on Acoustics (2014)

Modelling and Analysis of Complex Patient-Treatment Process Using GraphMiner Toolbox

Oleg Metsker[1], Sergey Kesarev[1(✉)], Ekaterina Bolgova[1],
Kirill Golubev[1], Andrey Karsakov[1], Alexey Yakovlev[1,2],
and Sergey Kovalchuk[1]

[1] ITMO University, Saint Petersburg, Russia
olegmetsker@gmail.com, kesarevs@gmail.com,
{ekaterina_bolgova, golubev1251, karsakov,
kovalchuk}@corp.ifmo.ru
[2] Almazov National Medical Research Centre, Saint Petersburg, Russia
yakovlev_an@almazovcentre.ru

Abstract. This article describes the results of multidisciplinary research in the areas of analysis and modeling of complex processes of treatment on the example of patients with cardiovascular diseases. The aim of this study is to develop tools and methods for the analysis of highly variable processes. In the course of the study, methods and algorithms for processing large volumes of various and semi-structured series data of medical information systems were developed. Moreover, the method for predicting treatment events has been developed. Treatment graph and algorithms of community detection and machine learning method are applied. The use of graphs and machine learning methods has expanded the capabilities of process mining for a better understanding of the complex process of medical care. Moreover, the algorithms for parallel computing using CUDA for graph calculation is developed. The improved methods and algorithms are considered in the corresponding developed visualization tool for complex treatment processes analysis.

Keywords: Graph mining · Process mining · Community detection ·
Process modeling · Cardiology · Complex process analysis

1 Introduction

Data modeling is traditionally the way to understand better the processes, which provides excellent opportunities to foresee changes. The data of medical information systems, describing the processes of health care, contains empirical information about the treatment of patients, on the basis of which it is possible to develop models of these processes [1]. Electronic medical records (EHR) describing the process of providing medical care consist of many different types of process elements. Information in the EHR is contained both in a structured form and semi-structured form (for example, protocols of operations, descriptions of diagnoses, anamnesis of the patient's diseases,

J. M. F. Rodrigues et al. (Eds.): ICCS 2019, LNCS 11540, pp. 674–680, 2019.
https://doi.org/10.1007/978-3-030-22750-0_65

diaries of observation in resuscitation, conclusions about examinations, and other medical records).

The aim of this study is to develop tools and methods for the analysis of complex processes of medical care based on poorly structured data of medical information systems using high-performance algorithms. During solving this scientific problem, methods of adaptation of process mining technology have been developed to identify a complex process of providing medical care, taking into account the personal characteristics of the patient. Appointment of treatment procedures with the individual characteristics of the patient in mind contributes to modern approaches to the organization of health care (value-based health care, P4 medicine). The presence of predictive models based on empirical information is a qualitative characteristic for the health system in the management of the quality of treatment.

2 Conceptual Approaches to Process Space Analysis

The health care processes for different categories of patients may vary significantly. For patients in the same homogeneous group, the providing medical care process can be practically typical. However, in complex cases with concomitant pathological processes on the results of treatment of patients with cardiovascular disease can affect a significant number of factors.

2.1 Phase Space Analysis and Variability Reduction

The first stage of development of the patient-treatment model: reducing the variability of the phase space of the rendering process of patients with cardiovascular diseases. Process mining methods use discrete data about events of treatment episodes to identify typical ways of providing medical care, patterns identification, and deviations from the standard ways. For presentation and primary analysis of the processes of medical care in the field of the basic relational structure of the undirected graph is used, namely the identification of community structure, relationships, and relationships, or abstracted links between the instances of the processes. Also, it is possible to quantify the network function of the spread of diseases, based on random local interactions [2]. At this stage, the model is often determined as a complete systematized graph or a random graph as Erdos-Renyi or Poisson graphs [3, 4]. These models are used to quantify network characteristics such as connectivity, the existence, and size of a giant component, the distribution, and extent of elements, the degree of connectivity, and the determination of node parameters, including the clustering factor. In graphical models, the data are considered as a set of random variables, indexed nodes of the graph, where probabilistic dependences between elements are fixed. For example, directed graphs [2, 5] in the form of Bayesian networks, where each random variable is independent of others. Undirected graphical models, also called Markov random fields [6, 7], describe processes, where variables defined on two sets of nodes, are statistically independent. A key tool in working with graphical models is the Hammersley–Clifford theorem [6–9], with the corresponding positivity of the conditions, the factors of the joint distribution of the graphical model are equal to the product of potentials.

2.2   Graph-Based Process Space Representation

It is proposed to use the graph representation of space (GPS) process model states $M : G^{(M)} = \langle V^{(M)}, E^{(M)} \rangle$. GPS can be used to study the functional and operational characteristics of the existing base of precedents. Each vertex of the graph corresponds to a subset of admissible model realizations of the object of research, which are considered to be identical within the framework of this interpretation (in the limiting case, the subset consists of a single instance): $v_i^{(M)} \in V^{(M)} : v_i^{(M)} = \left\{ v_{i,j}^{(M)} \right\} \subset M, \forall v_{i,j_1}^{(M)}, v_{i,j_2}^{(M)} \in v_i^{(M)} : \delta\left(v_{i,j_1}^{(M)}, v_{i,j_2}^{(M)} \right) = 0$, where δ – a specified proximity measure, usually defined as $\delta : M \times M \to \mathbb{R}$. The edges of the graph correspond to the level of proximity not lower than some boundary $e = \left\langle v_{i_1}^{(M)}, v_{i_2}^{(M)} \right\rangle \in E^{(M)} : \forall v_{i_1,j_1}^{(M)} \in v_{i_1}^{(M)}, v_{i_2,j_2}^{(M)} \in v_{i_2}^{(M)} : \delta\left(v_{i_1,j_1}^{(M)}, v_{i_2,j_2}^{(M)} \right) < \delta_0$ and can have weights varying depending on the actual proximity of the elements in the composition of the vertices according to a given measure δ. The topology and properties of the graph allow us to analyze the features of the functional and operational characteristics specified in the space M: to identify the cluster, to simplify the assessment of the proximity measures for the partially observed instances of the studied objects, to assess possible alternatives to the development of situations, etc. As a result, the most significant effect of this approach is observed within the framework of the Big Data concept and the corresponding models.

At the first stage of the study, the journal of events on the treatment of cardiovascular planned patients with stenting and ACS (acute coronary syndrome) patients (12900 patients) of intensive cardiac care unit (ICCU) and cardiac departments were analyzed using traditional process mining tools[1]. The results were poorly interpreted because the process map was characterized by variability and high complexity. As a result, we developed our solution for visualization of the phase space of the treatment process.

The edges of the graph denote the value of the symmetric difference $\left(\delta\left(v_1^{(M)}, v_2^{(M)} \right) = v_1^{(M)} \Delta v_2^{(M)} \right)$ between ordered sets, the vertices denote the sets of processed events.

The primary purpose of this phase of the experiment is automatic detection the communities of the process graph. Graph describing the process of medical care and the differences in care for different groups of patients with cardiovascular diseases in the intensive care unit was developed. Community detection is a way to highlight the structure of the phase space. The algorithm of community detection is proposed in Fig. 1.

The quality of partitions proposed by the community detection algorithm is estimated by the metric called modularity. The algorithm proposed for iterative optimization of the modularity score [10]. It is designed to work with weighted undirected graph structures.

---

[1] http://www.fluxicon.com/disco/.

Input: Graph (V, E), where V is the set of all graph's vertices and E – the set of all graph's edges

```
1  let V' = {}, E' = {} ;
2  repeat
3  |   if V', E' are not empty then
4  |   |   V = V', E = E';
5  |   let C = {{v} for v in V};
6  |   for i in V do
7  |   |   let c_i = community from C to which i belongs;
8  |   |   for j in neighbors(i) do
9  |   |   |   let c_j = community from C to which j belongs;
10 |   |   |   modularity gain(i, j) =
   |   |   |     modularity({c_i \ {i}, c_j ∪ {i}, all other communities from C}) − modularity(C);
11 |   |   k = arg max_j (modularity gain(i, j));
12 |   |   if modularity gain(i, k) > 0 then
13 |   |   |   C = {c_i \ {i}, c_k ∪ {i}, all other communities from C};
14 |   for c in C do
15 |   |   add new vertex v_c to V';
16 |   for (i, j) in E do
17 |   |   let c_i = community from C to which i belongs;
18 |   |   let c_j = community from C to which j belongs;
19 |   |   if (v_{c_i}, v_{c_j}) not in E' then
20 |   |   |   add (v_{c_i}, v_{c_j}) to E';
21 |   |   |   weight(v_{c_i}, v_{c_j}) = weight(i, j);
22 |   |   else
23 |   |   |   weight(v_{c_i}, v_{c_j}) + = weight(i, j);
24 until V = V' and E = E';
```

Fig. 1. The algorithm of community detection of close cases of treatment

2.3   Communities Interpretation and Modelling

The second stage of development of the patient-treatment model: analysis of the treatment process of patients in particular groups. In the second stage, it is possible to analyze and simulate processes within each individual community. At this scale of treatment process detailing it is possible to calculate the probability of events and treatment pathways with the least variability. For example, the most frequent event of patients with the hypertensive disease is "primary examination consultation of a cardiologist" (61,507 times in 40,537 episodes). Most patients who have passed the necessary examinations at the stage of applying for qualified help or at the primary level are sent to the initial consultation of a specialist. Some of the patients who applied need echocardiography, if this study was not carried out, more than six months have passed since the previous study or a protocol of low-quality study is presented.

Graphs are a more general class of structures than sets, sequences, and data trees. Graph mining is used to analyze repetitive patterns and perform specificity, discrimination, classification, and cluster analysis of large data sets. Using medical data, it is possible a specification of the clusters of the graph events of treatment for patient data in the cluster (community). Classification and cluster analysis of graph datasets can be studied by integrating them with the process of identifying frequent graph patterns. Analysis of similarity measures of events sets to analyze the nodes proximity of the graph of treatment of patients with cardiovascular diseases can be used to predict and rank new treatment processes, and the study of empirical data already accumulated in the medical information system (MIS).

Most methods of data mining of graph clusters suggest that the results do not depend on the personal characteristics of the patient. There may be relationships between individual characteristics and patients treatment graph clusters (communities). Identification of links between objects in such networks provides an assessment of the significance of different links concerning data on the treatment process and the patient, which may be the basis for requesting this information in the next stages of treatment. For the clinical process, this can quantitatively justify the priority of diagnostic procedures. In terms of treatment quality - to define quality indicators and priority of their implementation.

3   The Model Development

Treatment processes are directly related to the processes occurring in the human body and vice versa. The analysis of complex treatment processes in general involves a system of two main types of models: patient models (models of pathophysiological processes, calculation of the clinical events probability, including the probability of death and the risks of complications, etc.); models of medical care processes (treatment processes, administrative processes, logistics processes, the effects of therapy, the appropriateness of the procedure).

3.1   Prediction of Processes' Dynamics

In the next experiment on the analysis of data from the MIS determined the content and structure describing the treatment process in the form of a graph. Data on the treatment process are discrete and represent a variety of events (for example, admission to the department, echocardiography, heart surgery, transfer to another department, general blood test, etc.). The symmetric difference between ordered sets of patient treatment events is a method of calculating the weight of the graph edge, expressed in the number of different elements. The algorithm identified communities described above. Further calculations for several graphs at various stages of treatment are made: one day, three days, five days, ten days, all events (Fig. 2)

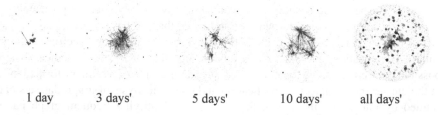

| 1 day | 3 days' | 5 days' | 10 days' | all days' |

Fig. 2. Representation of the phase space of treatment processes at different stages.

It was possible to identify patients in terms of movement in these communities find similar trajectories and classify them by machine learning methods for trajectory prediction.

In the obtained graphs, some vertex communities are founded, which at different stages of treatment have different compositions (eczema-pliers of patient treatment processes). It was possible to identify the pattern of treatment processes in terms of movement in these communities at various stages using machine learning methods to predict the path in the areas of phase space.

Then the problem of classification into a certain class of patients is solved. Machine learning methods (SVC, Random Forest, KNeigbor, Logistic Regression, Naïve bayes, GB - gradient boosting, ensemble V1 = RandomForest + KNeigbor, ensemble V2 − RandomForest + LogisticRegression) were used to train the model. The following predictors were used: minimum hemoglobin level, maximum troponin, maximum ALT, maximum AST, maximum creatinine level, maximum PLT level, maximum glucose level, age, gender). Ensemble V2 showed coating ROC 88%.

4   Experimental Setting with GraphMiner Toolbox

Interactive toolbox developed for the analysis of complex processes taking into account all the methods described above. In this study, we conduct an experiment using this tool to analyze complex patient-treatment processes. The main elements of the software package:

- Knowledge base with data analysis (Data Mining/Process Mining/Text Mining).
- Computing core. The experimental study calculations were implemented using the GPU.
- Interactive Visualizer [11]. Supported processor architecture: x86, x86-64, supported platform: .NET 4.0 Programming language: C#.

The visualization of a resulting graph with a force-directed layout algorithm allows the user to analyze patients' communities further, find misbehaviors in an algorithm and detect a different clustering or higher-level similarities between the communities. The overview of the visualization tool is demonstrated in the video available on YouTube[2]. This tool visualizes a three-dimensional non-oriented force-directed graph layout, where nodes are patients and links between them depending on their likelihood, calculated in a previous chapter. The user interface allows the user to further explore and analyze the graph with different actions.

---

[2] https://youtu.be/EH74f1w6EeY.

5   Conclusion and Future Work

The relevance of the development of custom tools of intellectual analysis, taking into account the specifics of the subject area of data and the processes described by them is not in doubt. This practice integrates domain-specific knowledge with data analysis techniques and provides data mining solutions for specific areas. Graph mining visualization integrates visual elements into data mining, process mining to discover implicit and useful knowledge from large sets of medical data. Parallel computing technologies provide a reasonable response in the calculations and interactions. For the analysis, MIS data offered several well-proven methods: classification, clustering, analysis of the significance of predictors, correlation analysis, etc. Trends in data mining and the methods considered in this study (improved scalable integration of data mining with data storage and interactive knowledge bases).

Further work is possible in the direction of studying various graphs. The issues of identifying the community, improving the performance of calculations, machine learning methods, visualization of results are also relevant in further research. It is also possible to scale these methods and toolbox for other process areas.

Acknowledgments. This work financially supported by the Ministry of Education and Science of the Russian Federation, Agreement #14.575.21.0161 (26/09/2017). Unique Identification RFMEFI57517X0161.

References

1. Yang, L., Zhang, J.: Automatic transfer learning for short text mining. EURASIP J. Wirel. Commun. Netw. **2017**(1), 42 (2017)
2. Chamley, C.: Rational Herds: Economic Models of Social Learning. Cambridge University Press, Cambridge (2004)
3. Jackson, M.: Social and Economic Networks. Princeton University, Princeton (2008)
4. Newman, M.: Networks: An Introduction. Oxford University Press, Oxford (2010)
5. Edwards, D.: Introduction to Graphical Modelling. Springer, Heidelberg (2000). https://doi.org/10.1007/978-1-4612-0493-0
6. Kindermann, R., Snell, J.L.: Markov random fields and their applications. Am. Math. Soc. **12** (1980)
7. Willsky, A.S.: Multiresolution Markov models for signal and image processing. Proc. IEEE **90**(8), 1396–1458 (2002)
8. Besag, J.: Spatial interaction and the statistical analysis of lattice systems. J. Roy. Stat. Soc. **1974**, 192–236 (1974)
9. Hammersley, J.M., Handscomb, D.C.: Monte Carlo Methods. Chapman Hall, Boca Raton (1964)
10. Blondel, V., et al.: Fast unfolding of communities in large networks. iopscience.iop.org
11. Karsakov, A., et al.: Toolbox for visual explorative analysis of complex temporal multiscale contact networks dynamics in healthcare. core.ac.uk

Combining Algorithmic Rethinking and AVX-512 Intrinsics for Efficient Simulation of Subcellular Calcium Signaling

Chad Jarvis[1], Glenn Terje Lines[1], Johannes Langguth[1], Kengo Nakajima[2], and Xing Cai[1,3(✉)]

[1] Simula Research Laboratory, P.O. Box 134, 1325 Lysaker, Norway
xingca@simula.no
[2] Information Technology Center, The University of Tokyo, Tokyo, Japan
[3] Department of Informatics, University of Oslo, Oslo, Norway

Abstract. Calcium signaling is vital for the contraction of the heart. Physiologically realistic simulation of this subcellular process requires nanometer resolutions and a complicated mathematical model of differential equations. Since the subcellular space is composed of several irregularly-shaped and intricately-connected physiological domains with distinct properties, one particular challenge is to correctly compute the diffusion-induced calcium fluxes between the physiological domains. The common approach is to pre-calculate the effective diffusion coefficients between all pairs of neighboring computational voxels, and store them in large arrays. Such a strategy avoids complicated if-tests when looping through the computational mesh, but suffers from substantial memory overhead. In this paper, we adopt a memory-efficient strategy that uses a small lookup table of diffusion coefficients. The memory footprint and traffic are both drastically reduced, while also avoiding the if-tests. However, the new strategy induces more instructions on the processor level. To offset this potential performance pitfall, we use AVX-512 intrinsics to effectively vectorize the code. Performance measurements on a Knights Landing processor and a quad-socket Skylake server show a clear performance advantage of the manually vectorized implementation that uses lookup tables, over the counterpart using coefficient arrays.

Keywords: Subcellular calcium dynamics ·
Piecewise constant coefficients · AVX-512 ·
Xeon Phi Knights Landing · Xeon Skylake

1 Introduction

The heart needs to be electrically stimulated so that it can contract during every heartbeat. The calcium ion is particularly important for the muscle contraction, and subcellular calcium signaling involves fine-scale physiological details. On the surface of sarcoplasmic reticulum (SR), i.e., each cell's internal calcium storage,

© Springer Nature Switzerland AG 2019
J. M. F. Rodrigues et al. (Eds.): ICCS 2019, LNCS 11540, pp. 681–687, 2019.
https://doi.org/10.1007/978-3-030-22750-0_66

there are calcium-sensitive channels called ryanodine receptors (RyRs). They open up when being attached by calcium ions. The typical distance between the cell membrane and the SR surface is 10–20 nm, and this narrow gap is referred to as the cleft space. There are normally between 10 and 100 RyRs inside such a cleft, and together they form a calcium release unit (CRU).

Computer simulations are impor-
tant for studying subcellular calcium signaling, where the subcellular space is composed of several *physiological domains* (see Fig. 1). These domains, each with distinct properties, are irreg-ularly shaped and elaborately con-nected. A resulting challenge is to effi-ciently compute the diffusion-induced, inter-domain calcium fluxes with phys-iological realism. The main focus of this paper is on utilizing the 512-bit vector length on the Xeon Phi Knights Landing (KNL) and Xeon Skylake pro-cessor architectures. In the context of simulating calcium signaling, we will show that manual vectorization needs to be combined with some algorithmic

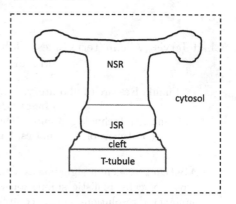

Fig. 1. A 2D schematic of the actual 3D subcellular space when divided into five physiological domains.

rethinking to fully release the hardware performance potential.

2   Computing Diffusion-Induced Calcium Fluxes

The whole 3D spatial domain is covered by a uniform mesh of box-shaped com-putational voxels. Approximated concentrations of the various calcium species are sought at the center of each voxel (i, j, k). The irregular 3D interior geome-tries are imbedded into the mesh, such that each voxel belongs *uniquely* to one of the physiological domains. If $u_s(x, y, z, t)$ denotes the concentration of a specific calcium species, then the diffusion-induced increment of u_s over a time step Δt is calculated by a second-order finite volume/difference method as

$$\Delta t \left(\frac{\sigma_s(x_{i+\frac{1}{2}}, y_j, z_k)(u_{s,i+1,j,k} - u_{s,i,j,k}) + \sigma_s(x_{i-\frac{1}{2}}, y_j, z_k)(u_{s,i-1,j,k} - u_{s,i,j,k})}{h^2} \right.$$
$$+ \frac{\sigma_s(x_i, y_{j+\frac{1}{2}}, z_k)(u_{s,i,j+1,k} - u_{s,i,j,k}) + \sigma_s(x_i, y_{j-\frac{1}{2}}, z_k)(u_{s,i,j-1,k} - u_{s,i,j,k})}{h^2} \tag{1}$$
$$\left. + \frac{\sigma_s(x_i, y_j, z_{k+\frac{1}{2}})(u_{s,i,j,k+1} - u_{s,i,j,k}) + \sigma_s(x_i, y_j, z_{k-\frac{1}{2}})(u_{s,i,j,k-1} - u_{s,i,j,k})}{h^2} \right).$$

To ensure accuracy, the diffusion coefficient needs to be evaluated at the boundary between two neighboring voxels. For example, $\sigma_s(x_{i+\frac{1}{2}}, y_j, z_k)$ denotes the *effective diffusion coefficient* between voxels (i, j, k) and $(i+1, j, k)$. If the two voxels are inside the same physiological domain d, then $\sigma_s(x_{i+\frac{1}{2}}, y_j, z_k)$ naturally attains the constant diffusion value for species u_s in domain d, denoted by σ_s^d. Care is needed when voxel (i, j, k) is inside domain d whereas voxel $(i + 1, j, k)$ is inside another domain e. In such a case, the two voxels are on the border between two physiological domains. To faithfully represent the physiology, we adopt the following formula for evaluating the effective diffusion coefficient:

$$\sigma_s(x_{i+\frac{1}{2}}, y_j, z_k) = \begin{cases} \sigma_s^d & \text{if voxels } (i, j, k) \text{ \& } (i + 1, j, k) \text{ both in domain } d; \\ \frac{\sigma_s^d + \sigma_s^e}{2} & \text{if voxel } (i, j, k) \text{ in domain } d, \text{ voxel } (i + 1, j, k) \text{ in } e, \\ & \text{and the two domains allow flux in-between;} \\ 0 & \text{if voxels } (i, j, k) \text{ \& } (i + 1, j, k) \text{ in two domains} \\ & \text{that do not allow flux in-between.} \end{cases} \tag{2}$$

3 Implementation

The diffusion-related computation, of the form (1), is the most time-consuming part of a 3D subcellular simulation. Since the effective diffusion coefficient between a pair of neighboring voxels may invoke a complex formula of the form (2), care must be taken to ensure the computational efficiency.

3.1 The Coefficient-Array Approach

A commonly used approach is to *pre-calculate* all the effective diffusion coefficients once and for all. This requires three logically 3D arrays to be prepared: alpha_x, alpha_y and alpha_z. For example, alpha_x contains values of $\frac{\Delta t}{h^2} \cdot \sigma_s(x_{i+\frac{1}{2}}, y_j, z_k)$, whereas the arrays alpha_y and alpha_z correspond to the y and z-directions. The diffusion computation can be implemented as below.

```
const int x_offset = ny*nz; const int y_offset = nz;
int xi, yi, zi, pos; double u_value;

for (xi=1; xi<nx-1; xi++) {
  for (yi=1; yi<ny-1; yi++) {
    #pragma ivdep
    for (zi=1; zi<nz-1; zi++) {
      pos = xi*x_offset + yi*y_offset + zi;
      u_value = u[pos];
      // x-direction contribution
      du[pos] += alpha_x[pos]*(u[pos+x_offset]-u_value)
                +alpha_x[pos-x_offset]*(u[pos-x_offset]-u_value);
      // y-direction contribution
      du[pos] += alpha_y[pos]*(u[pos+y_offset]-u_value)
                +alpha_y[pos-y_offset]*(u[pos-y_offset]-u_value);
      // z-direction contribution
      du[pos] += alpha_z[pos]*(u[pos+1]-u_value)
                +alpha_z[pos-1]*(u[pos-1]-u_value);
    }
  }
}
```

Listing 1. The coefficient-array approach to implementing the diffusion computation.

The coefficient-array implementation is fully justified if the effective diffusion coefficients indeed vary everywhere in space. For the example of five physiological domains, however, there are only $5 \times 5 = 25$ combinations of the effective diffusion constant between any pair of neighboring voxels. An enormous memory footprint overhead is thus associated with the three arrays alpha_x, alpha_y and alpha_z, needed per calcium species. Moreover, a considerable amount of memory traffic arises from (repeatedly) reading these coefficient arrays.

3.2   The Lookup-Table Approach

A memory-friendly approach is thus to pre-calculate for each calcium species a small lookup table (e.g. named coef_table) of length num_domains*num_domains. It stores the different combinations of the effective coefficient constant. A logically 3D array of char values, named dm_ids, is assumed to store the physiological domain type info for all the computational voxels. Listing 2 shows this "lookup-table" approach to implementing the diffusion computation.

```
const int x_offset = ny*nz; const int y_offset = nz;
int xi, yi, zi, pos; double u_value, *coef;
char di_m, di_p;

for (xi=1; xi<nx-1; xi++) {
  for (yi=1; yi<ny-1; yi++) {
    #pragma ivdep
    for (zi=1; zi<nz-1; zi++) {
      pos = xi*x_offset + yi*y_offset + zi;
      u_value = u[pos];
      // Focusing on the corresponding row in the lookup table
      coef = coef_table + (dm_ids[pos]*num_domains);
      // x-direction contribution
      di_m = dm_ids[pos-x_offset]; di_p = dm_ids[pos+x_offset];
      du[pos] += coef[di_p]*(u[pos+x_offset]-u_value)
                +coef[di_m]*(u[pos-x_offset]-u_value);
      // y-direction contribution
      di_m = dm_ids[pos-y_offset]; di_p = dm_ids[pos+y_offset];
      du[pos] += coef[di_p]*(u[pos+y_offset]-u_value)
                +coef[di_m]*(u[pos-y_offset]-u_value);
      // z-direction contribution
      di_m = dm_ids[pos-1]; di_p = dm_ids[pos+1];
      du[pos] += coef[di_p]*(u[pos+1]-u_value)
                +coef[di_m]*(u[pos-1]-u_value);
    }
  }
}
```

Listing 2. The lookup-table approach to implementing the diffusion computation.

Code vectorization is needed on processers with SIMD capability to get good performance. On the Xeon Phi Knights Landing and Xeon Skylake architectures, manual vectorization through AVX-512 intrinsics can be essential. Due to space limits, we cannot show the detailed code using AVX-512 intrinsics. It suffices to say that the mask variants of AVX-512 intrinsics allow a much more elegant manual vectorization than the previous generations of AVX intrinsics.

4   Performance Measurement and Comparison

Two hardware testbeds have been used. The first testbed is one node on the Oakforest-PACS system [8], i.e., a 68-core Xeon Phi KNL processor of model

7250 that has in total 272 hardware threads. The other testbed is a 4-socket server with Xeon Skylake Gold 16-core CPUs of model 6142, i.e., in total 64 CPU cores and 128 hardware threads. (Due to a suboptimal configuration, however, only four of the six memory channels are occupied per CPU.) The cores of both systems are capable of two 512-bit wide FMA operations per clock cycle. The C compilers used on the two systems are ICC v18 on the KNL node and GCC v8.2 on the Skylake server.

Table 1 lists the time measurements obtained on the two hardware testbeds. Specifically, we have employed two real-world simulations of subcellular calcium signaling, one using a small-scale global mesh of $168 \times 168 \times 168$ voxels, the other using a medium-scale $672 \times 672 \times 168$ mesh. The geometries are based on images obtained from confocal microscopy of rat ventricular myocytes, see [6] for details. In total four implementations (parallelized with OpenMP) have been tested. Two of them correspond to the "plain" coefficient-array and lookup-table approaches, i.e., Listings 1 and 2. These are denoted as CA_auto and LUT_auto because partial vectorization is automatically enabled by the compilers. In comparison, CA_man and LUT_man denote the implementations that explicitly use AVX-512 intrinsics.

In almost all cases manual vectorization (CA_man or LUT_man) gives better performance than compiler auto-vectorization (CA_auto or LUT_auto). On the KNL node, the CA_auto version performs better than the LUT_auto version using up to 32 OpenMP threads. This seems to suggest that ICC does a better job

Table 1. Time measurements of four implementations for the diffusion computation.

Code version	CA_auto		LUT_auto		CA_man		LUT_man	
Testbed	KNL	Skylake	KNL	Skylake	KNL	Skylake	KNL	Skylake
Global computational mesh: $168 \times 168 \times 168$, time steps: 16000								
Serial performance	2006.7	963.0	3368.6	846.1	1675.7	861.2	1301.7	463.8
2 OpenMP threads	1146.2	459.4	1767.7	415.6	1003.6	415.3	714.7	210.2
4 OpenMP threads	580.2	187.0	891.7	178.8	510.3	176.2	360.3	89.9
8 OpenMP threads	295.7	101.9	450.4	94.0	258.9	95.6	185.8	48.7
16 OpenMP threads	155.4	58.4	230.0	53.1	136.1	55.7	97.4	28.3
32 OpenMP threads	89.6	43.2	113.0	30.8	79.2	43.2	50.5	21.4
64 OpenMP threads	66.5	40.9	60.1	21.4	62.8	40.9	30.4	18.9
128 OpenMP threads	50.2	45.9	46.2	23.0	54.8	46.8	27.0	22.9
256 OpenMP threads	52.4	N/A	44.0	N/A	54.4	N/A	27.6	N/A
Global computational mesh: $672 \times 672 \times 168$, time steps: 1000								
Serial performance	1905.5	885.7	3381.9	851.1	1606.2	779.2	1284.6	472.7
2 OpenMP threads	1096.9	455.5	1757.9	438.4	989.7	405.6	695.8	252.2
4 OpenMP threads	551.6	236.0	880.8	222.7	498.8	212.2	347.7	130.7
8 OpenMP threads	277.9	134.7	442.2	118.6	252.8	118.4	175.8	68.2
16 OpenMP threads	141.7	69.4	221.3	62.0	128.5	64.7	89.0	36.9
32 OpenMP threads	80.6	47.8	106.9	33.2	72.9	47.8	43.7	26.1
64 OpenMP threads	60.3	58.2	54.4	36.6	57.9	56.6	23.7	30.5
128 OpenMP threads	43.8	67.2	39.3	49.1	49.8	64.3	21.8	42.9
256 OpenMP threads	49.4	N/A	35.7	N/A	46.5	N/A	22.4	N/A

with auto vectorization for CA_auto. On the Skylake server LUT_auto always outperforms CA_auto. This is likely due to the much smaller memory footprint of LUT_auto. Comparing the KNL node with the Skylake server, it is clear that a single Skylake core is much more powerful than a single KNL core. However, the performance advantage of the Skylake server decreases with an increasing number of OpenMP threads used. Figure 2 plots all the time usages related to the medium-scale simulation (672 × 672 × 168 voxels).

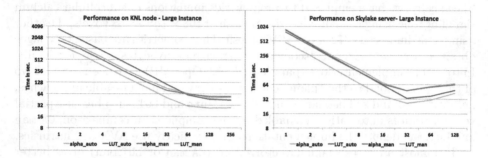

Fig. 2. Comparing the time usage of the four implementations on diffusion computation for the medium-scale simulation (672 × 672 × 168 voxels).

5   Related Work and Concluding Remarks

There exist many mathematical models of calcium signaling, see [4] for a review. The model adopted by the present paper differs from most of the existing work in that the individual RyRs are accurately resolved with realistic positions and geometries. Moreover, the shape and location of the various physiological domains are reproduced from medical imaging data. This represents a major improvement of the modeling strategy used in [1], where the different domains are simplified to co-exist "on top of each other" throughout the subcellular space. The computational capability of the KNL architecture has been studied in previous publications, such as [2,5,7], using both real-world simulators and well-known benchmarks. However, we are not aware of a detailed study of applying the new AVX-512 intrinsics on the KNL architecture. Regarding the Xeon Skylake server architecture, the existing work such as [3] does not seem to have specifically studied the applicability and performance of AVX-512 intrinsics either.

The work presented in this paper is only a first step towards physiologically realistic simulations of subcellular calcium signaling. Such simulations will eventually require using a large number of KNL nodes or high-end Skylake server nodes. We have shown that manual code vectorization through AVX-512 intrinsics is necessary for ensuring good single-node performance. Lookup tables are known to be notoriously difficult to vectorize for a performance advantage. However, with AVX-512 we have seen a clear benefit of the manually vectorized lookup-table implementation, compared with a plain code intended for compiler auto-vectorization.

We want to stress that the strategy of using a lookup table is not restricted to simulations of subcellular calcium signaling. It is applicable to any diffusion-similar calculation where the diffusion coefficient is patch-wise constant. As long as the number of different "patch" types is small, all the different combinations of inter-patch diffusion coefficients can be pre-calculated in a small lookup table. The AVX-512 intrinsics, with the `mask` variants, allow a much more elegant vectorization of this optimization strategy than the previous AVX intrinsics, while giving an additional performance boost due to SIMD.

Acknowledgements. The work is partially supported by the Research Council of Norway (grant No. 251186/F20) and a JHPCN grant (No. JHPCN-jh180024) from Japan.

References

1. Chai, J., et al.: Towards simulation of subcellular calcium dynamics at nanometre resolution. Int. J. High Perform. Comput. Appl. **29**(1), 51–63 (2015)
2. Doerfler, D., et al.: Applying the roofline performance model to the Intel Xeon Phi Knights landing processor. In: Taufer, M., Mohr, B., Kunkel, J.M. (eds.) ISC High Performance 2016. LNCS, vol. 9945, pp. 339–353. Springer, Cham (2016). https://doi.org/10.1007/978-3-319-46079-6_24
3. Hammond, S., Vaughan, C., Hughes, C.: Evaluating the Intel Skylake Xeon processor for HPC workloads. In: 2018 International Conference on High Performance Computing and Simulation (HPCS), pp. 342–349. IEEE (2018)
4. Izu, L.T., Xie, Y., Sato, D., Bányász, T., Chen-Izu, Y.: Ca^{2+} waves in the heart. J. Mol. Cell Cardiol. **58**, 118–124 (2013)
5. Kang, J.H., Kwon, O.K., Jeong, J., Lim, K., Ryu, H.: Performance evaluation of scientific applications on Intel Xeon Phi Knights Landing clusters. In: 2018 International Conference on High Performance Computing and Simulation (HPCS), pp. 338–341. IEEE (2018)
6. Kolstad, T.R., et al.: Ryanodine receptor dispersion disrupts Ca^{2+} release in failing cardiac myocytes. eLife **7**, e39427 (2018). https://doi.org/10.7554/eLife.39427
7. Langguth, J., Jarvis, C., Cai, X.: Porting tissue-scale cardiac simulations to the Knights Landing platform. In: Kunkel, J.M., Yokota, R., Taufer, M., Shalf, J. (eds.) ISC High Performance 2017. LNCS, vol. 10524, pp. 376–388. Springer, Cham (2017). https://doi.org/10.1007/978-3-319-67630-2_28
8. Homepage of Oakforest-PACS at JCAHPC. http://jcahpc.jp/eng/ofp_intro.html

Ocean Circulation Hindcast at the Brazilian Equatorial Margin

Luiz Paulo de Freitas Assad[2]([⊠]), Raquel Toste[1],
Carina Stefoni Böck[1], Dyellen Soares Queiroz[3], Anne Goni Guedes[3],
Maria Eduarda Pessoa[3], and Luiz Landau[1]

[1] Laboratório de Métodos Computacionais em Engenharia,
Rua Athos da Silveira Ramos, 274, Bloco G1, Cidade Universitária,
Rio de Janeiro, RJ 21941-916, Brazil
[2] Departamento de Meteorologia, LAMCE/COPPE/UFRJ,
Av. Athos da Silveira Ramos, 149, Centro de Tecnologia – Bloco I – Sala 214,
Cidade Universitária, Rio de Janeiro, RJ 21941-999, Brazil
lpaulo@lamce.coppe.ufrj.br
[3] Enauta (Queiroz Galvão Exploração e Produção S.A.),
Avenida Almirante Barroso, 52, Centro, Rio de Janeiro, RJ, Brazil

Abstract. The growth of the activities of the Petroleum Industry in the Brazilian Equatorial Margin, reinforces the need for the environmental knowledge of the region, which will be potentially exposed to risks related to such activities. The environmental importance of this region evidences the need to deepen and systematize not only the knowledge about the environmental sensitivity of the region, but also about the characteristics that will exert influence over it. The Costa Norte Project can be identified with one of these initiatives. The project has as one of the main objectives to evaluate the efficiency of the use of marine hydrodynamic environmental computational modeling methods to represent the marine dynamics over that region. In this paper a regional ocean computational model was used to produce an inedited ten year hindcast simulation in order to represent the main aspects associated with mesoscale climatological ocean circulation at the Brazilian equatorial margin. This article aims to present the methodology and the results analysis and evaluation associated to the cited hydrodynamic computational simulation. The obtained results clearly demonstrated the ocean model potential to represent the most important ocean variables space and time distribution over the studied region. Comparative analysis with observed data demonstrated good agreement with temperature, salinity and sea surface height fields generated by the implemented model. The Costa Norte Project is carrying out under the Brazilian National Petroleum Agency (ANP) R&D levy as "Investment Commitment to Research and Development" and is financially supported by Enauta O&G company.

Keywords: Ocean modelling · Regional ocean circulation ·
Climatological hindcast · Brazilian equatorial margin · Costa norte project

© Springer Nature Switzerland AG 2019
J. M. F. Rodrigues et al. (Eds.): ICCS 2019, LNCS 11540, pp. 688–701, 2019.
https://doi.org/10.1007/978-3-030-22750-0_67

1 Introduction

The growth of the activities of the Petroleum Industry in the Brazilian Equatorial Margin, reinforces the need for ocean circulation knowledge of the region, which will be exposed to greater risks of environmental impacts related to such activities. Potential oil spill accidents associated with these activities would bring several environmental damages for that region. The nature of this vast region in terms of the magnitude of energy forcing (currents, tides, waves and winds) coupled with the high local biological wealth, both in terms of species and ecosystems makes several sectors of society turn their attention to potential activities to be implemented in this area. It stands out, from the environmental point of view, the presence of what are considered the largest areas of mangroves on the planet (Filho et al. 2005).

All this environmental particularity evidences the need to deepen and systematize not only the knowledge about the environmental sensitivity of the region, but also about the characteristics that will exert influence in the operations of the oil industry and contingency actions in case of eventual accidents. The Costa Norte Project can be identified with one of these initiatives. Financially supported by Enauta O&G company, the project has as one of the main objectives to evaluate the efficiency of the use of marine hydrodynamic environmental computational modeling methods to represent the marine dynamics, particularly in the region that includes some marine sedimentary basins the estuaries of Amazon and Pará rivers. It should be noted that the aforementioned project is carry out under the Brazilian National Petroleum Agency (ANP) R&D levy as "Investment Commitment to Research and Development".

In this paper a regional ocean computational model was used to proceed with an inedited ten year hindcast simulation in order to represent the main aspects associated with mesoscalc ocean circulation at the Brazilian equatorial margin. This article aims to present the methodology and the results analysis associated to the cited hydrodynamic computational simulation. An evaluation of the results obtained with the simulation will also be presented based on the comparisons of some ocean variables produced by the model and remote and in situ data sets available for the study region. It is important to emphasize that the presented development corresponds to a first step towards the implementation of an operational regional ocean model for the Equatorial Brazilian Margin.

2 Study Area

In this section it will be described a brief overview about the main aspects associated ocean dynamics near the Brazilian equatorial margin. There are, four main ocean dynamics forcing's at the studied region: The North Brazil Current flow, tides, amazon river discharge and trade winds (Fig. 1).

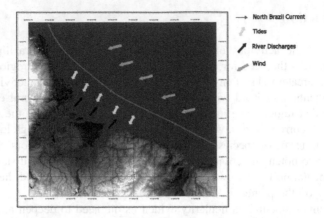

Fig. 1. Schematic representation of the main environmental forcing to Amazon Continental Shelf ocean dynamics.

The atmospheric circulation in this region is dominated by trade winds, with the strongest season from December to April and winds from the Northeast direction (Geyer et al. 1996). Ocean dynamics has strong annual and interannual variability related to local and remote influence of atmospheric processes with distinct spatial scales. The North Brazil Current (NBC) represents the main influence for the marine circulation in a regional scale. The NBC is a strong western boundary ocean current originated from the South Equatorial Current (SEC) (Geyer et al. 1996; Nittrouer and DeMaster 1996), and influences Amazon River plume transport (Limburner et al. 1995). Also, NBC closes the equatorial ocean circulation gyre generated by the wind. It also feeds the North Equatorial Counter Current (NECC) (Johns et al. 1998). The Amazon River discharge is also an important contributor to Amazon Continental Shelf hydrodynamics. Its average discharge is 180,000 m³/s which represents around 16% of the global river discharge into the oceans and its seasonality is associated with Andes freezing processes with minimum discharge in November and maximum in May (Geyer et al. 1996). The continental shelf has its circulation strongly influenced by the tides. The region is dominated by a macrotidal regime with variations of up to 8 m. The tides have semidiurnal periodicity. It is important to emphasize the all of these environmental forcing act in different time and spatial scales at the equatorial margin ocean dynamics.

3   Methodology

In this section, information will be presented regarding the computational hydrodynamic model, the boundary and initial conditions used as well as details regarding the methodology of analysis and evaluation of the obtained results.

3.1   The Hydrodynamic Model

The numerical model to be used is the Regional Ocean Modeling System (ROMS) (Shchepetkin and Mcwilliams 2005). The ROMS is a free surface model that solves the primitive equations of the oceans in a Arakawa C-grid, through a system of curvilinear horizontal coordinates. This system allows the grid to be defined along the coastline, which allows different resolutions in the contours of the grid and favors the observation of flows close to the coastal contours (Hedstrom 2012).

The Numerical grid

The generated numerical grid is comprised between latitudes 11.95 °N and 5.45 °S and between longitudes 55.95 °W and 35.54 °W. It is composed of 246 × 210 cells, which represents the horizontal resolution of 1/12° (Fig. 2). The numerical grid has 30 vertical levels. In Fig. 2, it is represented the numerical grid domain and the bathymetry field used by the model.

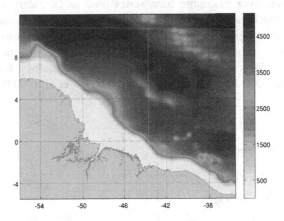

Fig. 2. Numerical grid domain and the bathymetry field used by the model simulation.

The model was forced (used as boundary conditions) by three of the main ocean dynamic forcing described on Sect. 2 that are winds, river discharge and the NBC flow. The model also used as initial conditions the mass field of the study area represented by the temperature and salinity fields.

3.2   Initial and Boundary Ocean Conditions

It was acquired daily mean ocean velocity (meridional and zonal), mass (temperature and salinity) and sea surface height fields from the Copernicus Marine Environment Monitoring Service (CMEMS) with 1/12° horizontal resolution with 40 vertical levels for the simulation period. This data set was processed and used as initial and boundary ocean conditions for the implemented model. In Fig. 3 it is represented the current field over the sea surface temperature field (Fig. 3) and the sea surface velocity field (Fig. 3) used as initial conditions for the hindcast simulation.

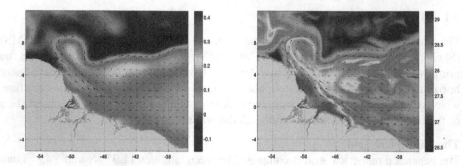

Fig. 3. Sea surface current field over the velocity distribution (left) and sea surface temperature field (right) used as initial conditions. The data were acquired from CMEMS.

River Discharge

The climatological river discharge time series used as boundary conditions by the ocean regional model were obtained for three rivers: Amazon, Tocantins and Pindaré (Fig. 4). The cliamtologies for Amazon and Pindaré river were estimated from river discharge time series acquired from HYBAM and international research laboratory and Brazilian National Water Agency respectively. The climatology for Tocantins river discharge was the same that one presented by Nikiema et al. (2007).

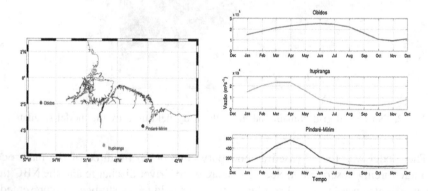

Fig. 4. River discharge station sites (left) and river discharge time series plot for each site.

3.3   Atmospheric Boundary Conditions

It was acquired the datasets from Reanalysis 2 project from National Centers of Environmental Prediction (Kalnay et al. 1996; Kanamitsu et al. 2002), with space resolution of 2° and time resolution of 6 h for the entire simulation period. The acquired data correspond to the zonal and meridional wind stress field (Fig. 5). It was also acquired the heat and mass fluxes over the sea surface (not shown).

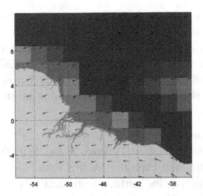 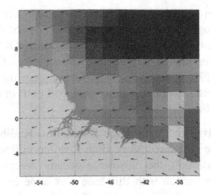

Fig. 5. Average wind stress field (N.m^{-2}) for the year of 2007. On left it is represented the zonal component of the wind stress field and on the right the meridional component of the wind stress field. The vectors indicate the resultant direction of the wind.

3.4 Model Evaluation

The results of the simulations were compared with remote and *in situ* data observations, considering the variables sea surface temperature, salinity and sea surface height. These comparisons were made in order to evaluate the ocean model skill in obtain a consistent climatological regional ocean representation of the study region. The databases acquired and used in each of the analyzes will be described below. The conducted analysis will be also described at the end of this section.

PIRATA Project

The Prediction and Research Moored Array in the Tropical Atlantic (PIRATA) project is a multinational cooperation program between Brazil, France and the United States, initiated in 1977, in which the three countries share the tasks of implementing and maintaining the observation network (Servain et al. 1998; GOOS 2017). The project

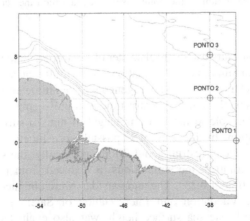

Fig. 6. Location of the PIRATA buoys used for evaluation analysis superimposed on the bathymetry field of the study area.

corresponds to an *in situ* observation network composed of meteoceanographic buoys designed to monitor several variables of the ocean-atmosphere interaction processes in the tropical Atlantic Ocean region (GOOS 2017). Among others, the measured variables include temperature and salinity of the sea surface (Servain et al. 1998; GOOS 2017), which were used for comparison with the simulations. Only three buoys are located within the numerical grid used in the simulations and their locations are shown in Fig. 6.

Advanced Very High Resolution Radiometer

The Advanced Very High Resolution Radiometer (AVHRR) (Price 1984). With 0.05° resolution, the mesoscale oceanic features are well captured by this data base. The comparison with this base was done by analyzing the surface fields and time series extracted at four points of the numerical grid as shown in Fig. 7. These points were chosen in order to obtain a representative analysis of the different areas in the numerical grid. This evaluation is important in order to observe not only the temporal but also the spatial patterns of temperature variation.

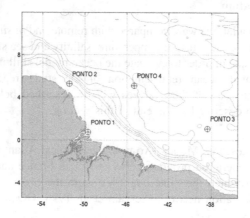

Fig. 7. Localization of the points used to extract the time series of SST.

AVISO Project

Passive microwave sensors use the backscatter of electromagnetic waves on the surface of the sea and thus obtain remote information even with the presence of clouds, which makes the mesh of estimates more homogeneous. The altimeter is an example of this type of passive sensor; through the return time of a pulse emitted vertically from the location, the sensor is able to measure the height of the sea surface in relation to its already known position. In this way, it is possible to estimate the height of the sea surface in relation to the terrestrial geoid. More specifically, the sea surface anomaly, which was used to compare the simulations, is deduced from the height of the sea surface with a resolution of 0.25°. The anomaly data were obtained through the French database Archiving, Validation and Interpretation of Satellite Oceanographic data (AVISO). As for SST, the sea surface height was also evaluated in the four points shown in Fig. 7.

3.5   Evaluation Analysis

The analysis of the results of the climatological round was performed using the bases described in the previous item, for sea surface temperature, sea surface height and salinity. Basically, the analyzes were performed considering the bias and the mean square error (RMSE), calculated according to Eqs. 1 and 2.

$$bias = y_t^{mod} - y_t^{obs} \tag{1}$$

$$RMSE = \sqrt{\frac{1}{N}\sum_{t=1}^{N}(y_t^{obs} - y_t^{mod})^2} \tag{2}$$

where t is time, N is the number of observations, y^{obs} is the observation e y^{mod} is the model result.

4   Results

4.1   The Hindcast Simulation

After the preprocessing of the initial and boundary conditions the model was integrated for a long period of ten years from December 27, 2006 to June 30, 2017. The choice of this period is justified because it corresponded to the period in which all bases used as boundary conditions were available. The simulation was performed for the mentioned period, considering an internal time step of 600 s and an external time step of 20 s. The results for the period prior to January 2007 were discarded. In Fig. 8 the kinetic energy time series plot is presented, and it is possible to verify that the model adjustment is found quickly achieved, at the beginning of the simulation. This behavior is consistent since the fields used as initial and boundary conditions have the same spatial resolution as the regional grid used in this step.

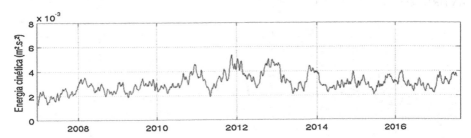

Fig. 8. Average kinetic energy time series plot integrated for the entire grid region for the period from December 27, 2006 to June 30, 2017.

4.2   Analysis

Sea Surface Temperature - AVHRR

For the evaluation of the SST generated by the model the equations presented in Sect. 3 were applied with the data of the AVHRR sensor for the points showed in Fig. 7. Figure 9 shows the time series plot of the model's SST and the AVHRR data (top panel) and the bias between the two series (lower panel) referring to Point 1. It is observed that the model overestimates the temperature values at different times in the series, cyclically, with anomalies above 2 °C and in some instants reaching 4 °C. Meanwhile, the average of the bias is approximately 1.0 °C. It is important to note that Point 1 is located in the region of the mouth of the Amazon River, where there is intense variability, which may contribute to the greatest differences.

Point 2 (Fig. 10) does not exhibit the same behavior as Point 1 and the model and data series tend to exhibit the same behavior. Thus, the bias presents smaller and average values of approximately - 0.2 °C, that is, in general, there is an underestimation in the SST. In contrast to what happens in Point 1, Point 2 is located in a region whose dynamics is most behaved and dominated by the Northern Current of Brazil (Johns et al. 1990).

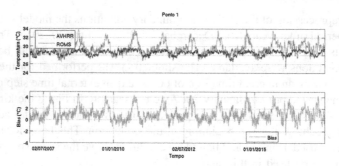

Fig. 9. Time series plot of SST (upper panel) and bias (lower panel) between model and AVHRR sensor data in Point 1.

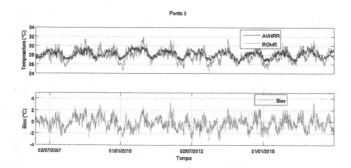

Fig. 10. Time series plot of SST (top panel) and bias (bottom panel) between model and AVHRR sensor data in Point 2.

In Point 3 (not shown), little variability is observed in the SST from the AVHRR data, with the temperature oscillating around 28 °C. The model, however, captured greater variability. In general, there was an underestimation of SST by the average model of −0.9 °C. In Point 4 (not shown) the two series exhibit similar cyclic behavior, which indicates that the climatology is satisfactorily representing the features of that region. Despite the differences between the two series over time, especially in the representation of the minimum temperatures, the average value of the bias was −1.2 °C, which still shows a good correlation between the two series.

Sea Surface Temperature - PIRATA

The same evaluation conducted with AVHRR data was cone with the PIRATA data. The surface temperature data of the three buoys located in the study area (Fig. 6) were compared with the time series generated by the model at the nearest grid points. Figure 11 shows the time series plots of SST of the climatological model and PIRATA in the upper panel, and the bias between the two series, in the lower panel, obtained for Point 1. Several intervals of time are observed with no data recorded on this PIRATA buoy, and the same occurs on the other buoys located in Points 2 and 3. A good agreement between the modeled values and the PIRATA data is observed, which can also be verified by the low values of bias throughout the series.

At point 2 (mot shown), which has larger sample gaps, larger differences are observed between the modeled and observed values than at Point 1. Despite this, the bias does not exceed mean values of −0.7 °C, indicating a good correlation between they. The time series of Point 3 (not shown) is, among others, the one with the greatest differences, and for the most part, the climatological model underestimates the maximum temperatures and overestimates the minimum ones. Nevertheless, as in Point 2, on average, the difference between the two series is −0.97 °C, also indicating a good correlation between them.

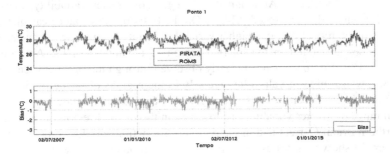

Fig. 11. Time series plots of the surface temperature (top panel) and bias (bottom panel) between the model results and the PIRATA Project data in Point 1, in degrees Celsius.

Salinity

For the analyzed period, as well as for temperature, gaps were observed in the measurement of salinity in the three PIRATA buoys. In spite of this, it is possible to verify that the model adequately represented surface salinity, especially in the float located in

Point 1 (Fig. 12). In Point 2 and in Point 3 (not shown), errors were observed in the representation of the peaks of lower salinities. However, overall, there is a good representation of seasonality in Point 3.

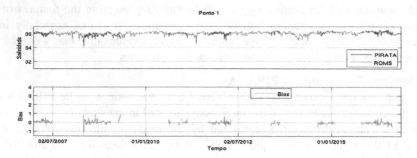

Fig. 12. Time series of salinity on the surface (top panel) and bias (bottom panel) between the model result and the data from the Point 1 PIRATA Project

Sea Surface Height

The comparison of the free surface elevation between the climatology generated by ROMS and the AVISO database was performed in two stages. The first one consisted in quantifying the time variance of the sea level anomaly at each point of the numerical grid, representing the turbulent variability. The second stage consisted in extracting time series from the sea surface anomaly at the sampling points shown in Fig. 3, located in the region of interest.

Figure 13a represents the horizontal variance field of the sea surface anomaly estimated from the AVISO data. There is little variability, with the exception of the coastal region of Amapá and part of the Amazon River mouth region, which present higher values of variance. Although it is a region of greater variability, the highest values of variance do not exceed 0.14 m. On the basis of the variance calculated from the results of the model run, this higher variability is not observed near Amapá, and the highest values are found in the internal region of the mouth of the Amazon River (Fig. 13b). This difference is most clearly observed through the bias between the fields (Fig. 13c), whose highest values lie precisely in the regions of greater variance of the AVISO data base. Nevertheless, similar features occur in the oceanic region (Fig. 13), where a range of greater variability is observed in the internal region of retroflexion of the Northern Brazil Current and in the region of formation and detachment of vortices.

Figure 14 shows the series extracted from the climatological results generated by the ROMS and the AVISO fields in Point 1. The greatest variability at this point is observed for the AVISO when compared to the model results, which is reflected in higher values of bias, which for the most part ranges from −0.5 to 1.0 m. It should be noted that Point 1 is very close to the region of greatest variability found in Fig. 14a, which also explains the differences found.

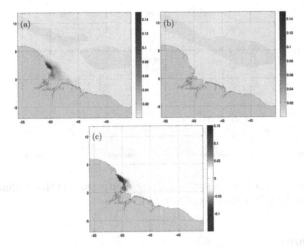

Fig. 13. Mean variance field of the sea surface anomaly, from 2007 to 2016, referring to the database of AVISO (a), the climatology generated by ROMS (b) and the bias between the two fields (c), in m².

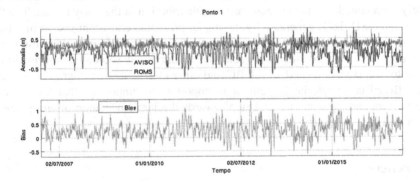

Fig. 14. SSH anomaly mean variance time series plot (top panel) and bias (bottom panel) between the model and the AVISO data in Point 1, in meters.

For point 2 (Fig. 15), the values of bias are very close to zero and the variability of the series is expressively smaller. By verifying the location of these points (Fig. 7) in relation to the variability fields shown in Figs. 14a and b, it is noted that these are located in regions of low variability, both when considering the modeled results and the AVISO observation data. In spite of this low variability, the climatology generated by the ROMS presents a very similar behavior to the data of the base AVISO in these points, where it is possible to observe that the model captured the cyclic oscillation behavior of the sea level, evidenced mainly through the time series in Point 3 (not shown).

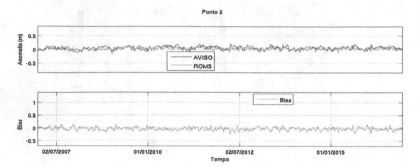

Fig. 15. Time series plot of the sea surface anomaly (top panel) and bias (bottom panel) between the model and the AVISO data in Point 2, in meters.

5   Conclusions

A regional ocean computational model was implemented and executed in order to obtain an inedited ten-year climatological ocean hindcast for the Brazilian equatorial margin. From the conducted evaluation analysis, it was verified that the model adequately represented the main ocean variables distribution in the study region. The SST, salinity and SSH fields generated by the model presented consistent results in offshore regions of the study area and less consistent results in coastal areas. The results indicated that the produced hindcast could be an important source of information for climatological studies related with NBC and other mesoscale oceanographic processes in the Brazilian Equatorial Margin. It is important to emphasize that the presented development corresponds to a first step towards the implementation of an operational regional ocean model using data assimilation techniques for the study region.

References

Geyer, W.R., et al.: Physical oceanography of the Amazon shelf. Cont. Shelf Res. **16**(5/6), 575–616 (1996)

GOOS: GOOS-Brasil. Aliança Regional para a Oceanografia no Atlântico Sudoeste Superior e Tropical – OCEATLAN (2017)

Hedstrom, K.S.: Draft Technical Manual for a Coupled Sea-Ice/Ocean Circulation Model (version 4). Relatório Técnico OCS Study BOEM 2012-0xx, US Department of Interior, Bureau of Ocean Energy Management (2012)

Johns, W.E., Lee, T.N., Schott, F.A., Zantopp, R.J., Evans, R.H.: The North Brazil current retroflection: seasonal structure and eddy variability. J. Geophys. Res. **95**(C12), 2156–2202 (1990)

Johns, W.E., Lee, T.N., Beardsley, R.C., Candela, J., Limeburner, R., Castro, B.M.: Annual cycle and variability of the North Brazil current. J. Geophys. Res. **28**, 103–128 (1998)

Kalnay, E., et al.: The NCEP/NCAR 40-year reanalysis project. Bull. Amer. Meteor. Soc. **77**(1), 437–471 (1996)

Kanamitsu, M., Ebisuzaki, W., Woollen, J., Yang, S.-K., Hnilo, J.J., Fiorino, M., Potter, G.L.: NCEP-DOE AMIP-II reanalysis (R-2). Bull. Am. Meteor. Soc. **83**(1), 1631–1643 (2002)

Limeburner, R., Beardsley, R.C., Soares, I.D., Lentz, S.J., Candela, J.: Lagragian flow observations of the Amazon River discharge in the North Atlantic. J. Geophys. Res. **100**(C2), 2401–2415 (1995)

Nikiema, O., Devenona, J., Bakloutib, M.: Numerical modeling of the Amazon River plume. Cont. Shelf Res. **27**(1), 873–899 (2007)

Nittrouer, C.A., DeMaster, D.J.: The Amazon shelf setting: tropical, energetic, and influenced by a large river. Cont. Shelf Res. **16**(5/6), 553–573 (1996)

Price, J.C.: Land surface temperature measurements from the split window channels of the NOAA 7 advanced very high resolution radiometer. J. Geophys. Res. **89**, 7231–7237 (1984)

Shchepetkin, A.F., McWikkiams, J.C.: The regional oceanic modeling system (ROMS): a split-explicit, free-surface, topography-following-coordinate oceanic model. Ocean Model. **9**, 347–404 (2005)

Servain, J., et al.: A pilot research moored array in the tropical atlantic (PIRATA). Bull. Amer. Meteor. Soc. **79**, 2019–2031 (1998)

Filho, S., Martins, P.W.: Costa de manguezais de macromaré da Amazônia: cenários morfológicos, mapeamento e quantificação de áreas usando dados de sensores remotos. Rev. Bras. de Geofís. **23**(4), 427–435 (2005)

A Matrix-Free Eigenvalue Solver for the Multigroup Neutron Diffusion Equation

Amanda Carreño[1(✉)], Antoni Vidal-Ferràndiz[1], Damian Ginestar[2], and Gumersindo Verdú[1]

[1] Instituto Universitario de Seguridad Industrial, Radiofísica y Medioambiental, Universitat Politècnica de València, València, Spain
{amcarsan,gverdu}@iqn.upv.es, anvifer2@upv.es
[2] Instituto Universitario de Matemática Multidisciplinar, Universitat Politècnica de València, València, Spain
dginesta@mat.upv.es

Abstract. The stationary neutron transport equation describes the neutron population and thus, the generated heat, inside a nuclear reactor core. Obtaining the solution of this equation requires to solve a generalized eigenvalue problem efficiently. The majority of the eigenvalue solvers use the factorization of the system matrices to construct preconditioners, such as the ILU decomposition or the ICC decomposition, to speed up the convergence of the methods. The storage of the involved matrices and incomplete factorization demands high quantities of computational memory although a the sparse format is used. This makes the computational memory the limiting factor for this kind of calculations in some personal computers. In this work, we propose a matrix-free preconditioned eigenvalue solver that does not need to have the matrices allocated in memory explicitly. This method is based on the block inverse-free preconditioned Arnoldi method (BIFPAM) with the innovation that uses a preconditioner that is applied from matrix-vector operations. As well as reducing enormously the computational memory, this methodology removes the time to assembly the sparse matrices involved in the system. A two-dimensional and three-dimensional benchmarks are used to study the performance of the methodology proposed.

Keywords: Neutron diffusion · Eigenvalue problem · Lambda modes · Matrix free · Block method

1 Introduction

The simulation of the reactor kinetics is a fundamental objective to ensure safe operation of nuclear reactors. The steady-state neutron transport equation [10] is the equation that describes the neutron flux and then, the generated heat power in every region of the reactor in steady-state.

© Springer Nature Switzerland AG 2019
J. M. F. Rodrigues et al. (Eds.): ICCS 2019, LNCS 11540, pp. 702–709, 2019.
https://doi.org/10.1007/978-3-030-22750-0_68

Different equations have been successfully used to approximate the neutron transport equation. Usually, all eliminate the energy dependence of the equations by means of a multi-group approximation. The dependence on the direction of the neutrons depends on the selected method. In this work, the multigroup neutron diffusion equation is chosen. This equation is analogous to the Fick's law for the diffusion of species and the Fourier equations in heat transfer. This equation states, for each energy group $g = 1, \ldots, G$, as

$$- \nabla \cdot (D_g \nabla \phi_g) + \Sigma_{tg} \phi_g(\boldsymbol{r}) = \sum_{\substack{g'=1 \\ g' \neq g}}^{G} \Sigma_{s, g' \to g} \phi'_g(\boldsymbol{r}) + \frac{1}{\lambda} \sum_{g'=1}^{G} \chi_g \nu_g \Sigma_{fg'} \phi_{g'} \quad (1)$$

where ϕ_g denotes the neutron flux of the energy group g. The coefficients, D_g, Σ_{tg}, $\Sigma_{fg'}$, χ_g and $\Sigma_{s, g' \to g}$ depend of the energy group g and group g'.

The spatial discretization scheme chosen for the diffusion equations is a continuous Galerkin finite element method to obtain an algebraic generalized eigenvalue problem, where the largest eigenvalue λ, also known as k-effective (k_{eff}), shows the criticality of the reactor and its corresponding eigenvector the distribution of the flux in the reactor core. Moreover, it is interesting computing several modes to develop modal methods that allows integrating the time-dependent equation.

Most eigenvalue problems, that arise from the different approximations to deterministic neutron transport equations, have been classically solved with the power iteration method. However, recently the Krylov-Schur method [12], the Generalized Davidson [5] or the block inverse-free preconditioner Arnoldi method (BIFPAM) [2], are becoming increasingly popular for this type of computations. These methods permit to solve the eigenvalue problem faster than the power iteration when the spectral distribution of the eigenvalues is very clustered. The bottleneck of all these methods is the preconditioner used. The eigenvalue solvers can use the factorization of the system matrices to construct preconditioners, such as the ILU decomposition or the ICC decomposition, to speed up the convergence of the methods. However, this type of factorizations demands a high level of computational memory to assemble the matrices and the preconditioners. Henceforth, preconditioners based on matrix-vector product are needed. In this work, we use the BIFPAM with a matrix-free preconditioner based on the block Gauss-Seidel method and the Chebyshev polynomial that does not need to have the matrices constructed explicitly eliminating setup costs for the matrix assembly and reducing storage requirements. Other matrix-free approaches have been studied in [3,4,7].

2   Algebraic Eigenvalue Problem

The problem (1) is spatially discretized by means of a high-order continuous Galerkin Finite Element Method (FEM), implemented with the help of library deal.II [1]. The discretization transforms the differential problem into an algebraic generalized eigenvalue problem of the form

$$Ax = \lambda Bx. \quad (2)$$

More details about the finite element spatial discretization are explained in [12].

As we are solving an energy multigroup diffusion equations problem, we can take advantage of the block structure of the matrices A and B, where each block is symmetric and positive definite. Most of the blocks in the lower part of matrix A and far from the diagonal in B are zero. For example, the matrices of the C5G7 benchmark studied in Sect. 5 have the following block structure,

$$
\begin{pmatrix}
A_{11} & \cdots & A_{17} \\
A_{21} & \cdots & A_{27} \\
A_{31} & \cdots & A_{37} \\
A_{41} & \cdots & A_{47} \\
0 & \cdots & 0 \\
0 & \cdots & 0 \\
0 & \cdots & 0
\end{pmatrix}
\begin{pmatrix}
x_1 \\ x_2 \\ x_3 \\ x_4 \\ x_5 \\ x_6 \\ x_7
\end{pmatrix}
= \lambda
\begin{pmatrix}
B_{11} & 0 & 0 & 0 & 0 & 0 & 0 \\
B_{21} & B_{22} & 0 & 0 & 0 & 0 & 0 \\
B_{31} & B_{32} & B_{33} & 0 & 0 & 0 & 0 \\
B_{41} & B_{42} & B_{43} & B_{44} & B_{45} & 0 & 0 \\
0 & 0 & 0 & B_{54} & B_{55} & B_{56} & 0 \\
0 & 0 & 0 & 0 & B_{65} & B_{66} & B_{67} \\
0 & 0 & 0 & 0 & 0 & B_{76} & B_{77}
\end{pmatrix}
\begin{pmatrix}
x_1 \\ x_2 \\ x_3 \\ x_4 \\ x_5 \\ x_6 \\ x_7
\end{pmatrix}.
\tag{3}
$$

3   Matrix-Free Strategy

This strategy computes the matrix-vector products on the fly in a cell-based interface. For instance, we can consider that a finite element Galerkin approximation that leads to the block matrix $A_{b,b}$ takes a vector u as input and computes the integrals of the operator multiplied by trial functions, and the output vector is v. The operation can be expressed as a sum of K cell-based operations,

$$
v = A_{b,b}u = \sum_{k=1}^{K} P_k^T A_{b,b}^k P_k u
$$

where P_k denotes the matrix that defines the location of cell-related degrees of freedom in the global vector and $A_{b,b}^k$ denotes the submatrix of $A_{b,b}$ on cell k. This sum is optimized through *sum-factorization*. Details about the implementation are explained in [6]. The main difficult of this strategy is to obtain efficient algebraic solvers that only use matrix-vector multiplications.

In this work, three matrix storage schemes are used. The first one, allocated all the block matrices in a compressed sparse row *CRS* way. The second one stores the diagonal block matrices of B in a sparse way to permit the computation of an incomplete LU factorization of these blocks. The rest of the blocks are implemented with the matrix-free operator (*non-diagonal*). Finally, all block matrices are implemented with the matrix-free technique in the *full matrix-free* scheme.

4   Eigenvalues Solver

This section is devoted to present the preconditioned block algorithm based on the Block inverse-free preconditioned Arnoldi method (BIFPAM) for finding the q largest eigenvalues in magnitude and their corresponding eigenvectors of the generalized eigenvalue problem (2).

Given problem (2) and an initial block approximation

$$([\lambda_{0,1}, \lambda_{0,1}, \ldots, \lambda_{0,q}], [x_{0,1}, x_{0,1}, \ldots, x_{0,q}]),$$

one could obtain a new approximation $[x_{1,1}, x_{1,1}, \ldots, x_{1,q}]$ from the union of the p bases of the Krylov subspaces $K_m(A - \lambda_{0,i}B, x_{0,i})$, $1 \leq i \leq q$ of order m, where

$$K_m(A - \lambda_{0,i}B, x_{0,i}) := \text{span}\{x_{0,i}, (A - \lambda_{0,i}B)x_{0,i}, \ldots, (A - \lambda_{0,i}B)^m x_{0,i}\}.$$

Then, the original generalized eigenvalue problem is projected onto these bases. Arnoldi method is used to construct each basis K_m. This process is repeated to obtain the next iterations.

The convergence of this method improves with the application of a preconditioner of the matrix $C_{k,i} = A - \lambda_{k,i}B$. One can use preconditioners that come from the factorization of the matrices involved in the system, but it implies to assemble these matrices. In this work, we use an approximation of the matrix B^{-1} as preconditioner that it is shown in other works that it is more efficient than using a preconditioner of $C_{k,i}$ or $C_{1,1}$. This permits using a block preconditioner (by using the block structure of the matrix B), without assembling any additional matrix, with the advantage that the blocks of this matrix are symmetric and positive definite. This causes an improvement in the implementation of the Algorithm. In particular, we choose the block Gauss-Seidel preconditioner [9] as shown in Algorithm 1.

Algorithm 1. Block Gauss-Seidel preconditioner.

Input: Matrix B and vector $x = [x_1; \ldots; x_{Bl}]$.
Output: Vector $y = [y_1; \ldots; y_{Bl}]$, result of applying the preconditioner of B to x.

1: **for** $b = 1$ to $nblocks$ **do**
2:    $t = x_b$
3:    **for** $c = 1$ to b **do**
4:       Compute $t = t - B_{b,c}y_c$
5:    **end for**
6:    Solve $B_{b,b}t = y_b$
7: **end for**

The previous algorithm only applies matrix-vector multiplications except in line 6, where some linear systems related with the block diagonal matrices are needed to be solved. In this work, the conjugate gradient (CG) is applied to solve these linear systems preconditioned with a Chebyshev preconditioner [11], provided by the library `deal.II` [1]. The degree of the Chebyshev polynomial has been set to 3. These auxiliary systems of the preconditioner are solved with a low maximum number of iterations (50).

5   Numerical Results

The performance of the proposed matrix-free method is tested with the C5G7 benchmark introduced by the Nuclear Energy Agency (NEA) [8]. In particular, we test the 2D configuration version and the 3D configuration version. It consists of a nuclear reactor core with MOX and UO_2 square fuel assemblies surrounded by a moderator region.

First, we start with the 2D configuration benchmark. Figure 1 shows the proposed mesh used to spatially discretize the square regions. The area in gray color represents the neutron fuel region. The degree of the polynomial for the finite element method has been set to 2. The computation using the mesh of Fig. 1 requires 28 900 finite element cells and 812 063 degrees of freedom. The number of eigenvalues computed is 4.

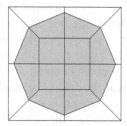

Fig. 1. Mesh considered in the FEM for the fuel pin in the C5G7 benchmark.

We compare the performance of the matrix-free preconditioner with the incomplete LU preconditioner. This last strategy is called 'CG-ILU'. This application needs the assembly of the block matrices B_{gg} to make ILU factorizations. Then, we propose to use the conjugate gradient method with the same number of maximum iterations, but in this case, the Chebyshev preconditioner is applied to solve linear systems. We denote this implementation as 'CG-CHEB'. For this last methodology, we compare three types of matrix-free implementations described in Sect. 3.

Table 1 displays the computational memory required by the matrix and preconditioner operators, the CPU time to set the matrices and the preconditioners and finally, the total CPU time to reach a residual error in the generalized eigenvalue problem less than 10^{-7} for each methodology. Table 1 shows that the CSR strategy is outperformed by the rest of the methodologies in terms of memory consumption. We can observe that the CG-ILU methodology solves the problem in the fastest way but this implementation does not allow to reduce the computational memory. If the 'full matrix-free' type is considered the computational memory is greatly reduced but the computational time is increased by a factor of 6 respect to 'CG-ILU' and a factor of 2 respect to the 'Non-Diag.' implementation.

Table 1. Computational results for the 2D-C5G7.

Methodology	Matrix-free type	Mat. memory	Prec. memory	Mat. + Prec. CPU time	Total CPU time
CG-ILU	Non-Diag.	193 Mb	160 Mb	1 s	184 s
CG-CHEB	Non-Diag.	193 Mb	-	0.8 s	570 s
CG-CHEB	Full matrix-free	20 Mb	-	0.1 s	1110 s
CG-CHEB	CSR	2375 Mb	-	4.6 s	573 s

In the next, we consider the 3D version of the C5G7 benchmark problem. This problem has the same radial configuration as the two dimensional version, and then, the discretization is used in this direction has been the same than the 2D version (Fig. 1). The axial discretization is done by extruding the two dimensional mesh by axial plane. The mesh used has $264\,992$ finite element cells and $2\,343\,865$ degrees of freedom. The configuration of the finite element method to compute the solution for this case is the same than in the previous case. In this case the number of eigenvalues requested is 1.

Now, we compare the matrix-free strategy presented in this work with the other methodologies described. Table 2 displays the CPU memory and the CPU time required by the matrix and preconditioner to set the matrices and the preconditioners and the total CPU time to reach a residual error in the generalized eigenvalue problem less than 10^{-7}. The same pattern of results described above can be observed in this Table. However, for the 3D case the reduction of the memory is considerable. In this way, if the required memory for the problem is low the authors recommends to use the 'CG-ILU' preconditioner. In this case, the CSR strategy has not been computed due to its high memory demands. However, for personal computers with low memory resources the full matrix-free strategy with the 'CG-CHEB' preconditioner is a reasonable option to solve the problem.

Table 2. Computational results for the 3D-C5G7 with $r = 1$.

Methodology	Matrix-free type	Mat. memory	Prec. memory	Mat. + Prec. CPU time	Total CPU time
CG-ILU	Non-Diag.	13415 Mb	10668 Mb	143 s	608 s
CG-CHEB	Non-Diag.	13415 Mb	-	62 s	3534 s
CG-CHEB	Full matrix-free	2410 Mb	-	4 s	7224 s

6 Conclusions

The multigroup neutron diffusion equation has been selected to approximate the neutron transport equation. A finite element method is used to discretize the problem obtaining an algebraic generalized eigenvalue problem. In this work, we propose a matrix-free preconditioned eigenvalue solver that does not need to have the matrices allocated in memory explicitly. This method is based on the BIFPAM and uses as preconditioner the Chebyshev process. Moreover, different types of matrix-free implementation for the problem are proposed. The performance of these schemes are tested by using the C5G7 benchmark.

Numerical results concludes that the ILU preconditioner is more efficient than the Chebyshev preconditioner in terms of CPU time, but needs to access to assembled block diagonal matrix elements. However, the Chebyshev preconditioner, although is less efficient, can be implemented without any block matrix assembled and removes the time to assembly the sparse matrices involved in the system. The reduction of the memory in the full matrix-free strategy is greater as the size of the problem increases. It means, if the size of the problem is low, the authors recommends to use the 'CG-ILU' preconditioner. On the other hand, for the personal computers (with low memory resources) the full matrix-free strategy with the 'CG-CHEB' preconditioner is a reasonable option to solve the problem.

Adknowledgements. This work has been partially supported by the Ministerio de Economía y Competitividad under projects ENE2014-59442-P, MTM2014-58159-P and BES-2015-072901 and the Universitat Politècnica de València under project PAID-06-18.

References

1. Bangerth, W., Hartmann, R., Kanschat, G.: Deal. II - a general-purpose object-oriented finite element library. ACM Trans. Math. Softw. **33**(4), 24 (2007)
2. Carreño, A., Vidal-Ferràndiz, A., Ginestar, D., Verdú, G.: Block hybrid multilevel method to compute the dominant λ-modes of the neutron diffusion equation. Ann. Nuclear Energy **121**, 513–524 (2018)
3. De Simone, V., di Serafino, D.: A matrix-free approach to build band preconditioners for large-scale bound-constrained optimization. J. Comput. Appl. Math. **268**, 82–92 (2014)
4. Gondzio, J.: Matrix-free interior point method. Comput. Optim. Appl. **51**(2), 457–480 (2012)
5. Hamilton, S.P., Evans, T.M.: Efficient solution of the simplified pn equations. J. Comput. Phys. **284**, 155–170 (2015)
6. Kronbichler, M., Kormann, K.: A generic interface for parallel cell-based finite element operator application. Comput. Fluids **63**, 135–147 (2012)
7. Kronbichler, M., Wall, W.A.: A performance comparison of continuous and discontinuous galerkin methods with fast multigrid solvers. SIAM J. Sci. Comput. **40**(5), A3423–A3448 (2018)
8. OECD/NEA: Benchmark on deterministic transport calculations without spatial homogenisation. Technical Report NEA/NSC/DOC(2003) 16, NEA (2003)

9. Saad, Y.: Iterative Methods for Sparse Linear Systems, 2nd edn. Society for Industrial and Applied Mathematics, Philadelphia (2003)
10. Stacey, W.M.: Nuclear Reactor Physics. Wiley, Hoboken (2007)
11. Varga, R.S.: Matrix Iterative Analysis, 2nd edn. Springer, Berlin (2009)
12. Vidal-Ferràndiz, A., Fayez, R., Ginestar, D., Verdú, G.: Solution of the lambda modes problem of a nuclear power reactor using an h–p finite element method. Ann. Nucl. Energy **72**, 338–349 (2014)

Path-Dependent Interest Rate Option Pricing with Jumps and Stochastic Intensities

Allan Jonathan da Silva[1,2]([⊠]), Jack Baczynski[1],
and João Felipe da Silva Bragança[3]

[1] National Laboratory for Scientific Computing, LNCC,
Petrópolis, RJ 25651-075, Brazil
{allanjs,jack}@lncc.br
[2] Federal Center for Technological Education, CEFET/RJ,
Itaguaí, RJ 20271-110, Brazil
[3] Federal University of the State of Rio de Janeiro - UNIRIO,
Rio de Janeiro, RJ 20070-030, Brazil
joaofelipee@gmail.com,
http://www.lncc.br, http://www.cefet-rj.br,
http://www.unirio.br

Abstract. We derive numerical series representations for option prices on interest rate index for affine jump-diffusion models in a stochastic jump intensity framework with an adaptation of the Fourier-cosine series expansions method, focusing on the European vanilla derivatives. We give the price for nine different Ornstein-Uhlenbeck models enhanced with different jump size distributions. The option prices are accurately and efficiently approximated by solving the corresponding set ordinary differential equations and parsimoniously truncating the Fourier series.

Keywords: Interest rate · Option pricing · COS method · AJD models

1 Introduction

The main interest rate option traded in B3 is the IDI (interfinancial deposit index) Option, which is of European type and cash settled at maturity. The IDI is an index that accumulates from an initial value according to the daily calculated effective market rate DI. Analytical solutions for pricing IDI options can be found for the short rate given by the Vasicek model [14] and the CIR model [3]. [1,12] developed the closed-form with the known Hull-White model [11]. [4] implemented the HJM model to price IDI options. The problem was numerically solved via a finite difference method in [13] and via an alternative generic method in [2]. Another result was developed by [10], where the model is sensitive to changes in monetary policy. A discussion about this type of path-dependent option is found in two books [6,7].

© Springer Nature Switzerland AG 2019
J. M. F. Rodrigues et al. (Eds.): ICCS 2019, LNCS 11540, pp. 710–716, 2019.
https://doi.org/10.1007/978-3-030-22750-0_69

The COS method is a Fourier inversion method introduced by [9]. It is a procedure to calculate probability density functions and expectations via cosine series. It originally relies on the explicit formula for the characteristic function of the state variable. This paper accomplishes an original contribution for the financial literature via devising a way to apply the COS method to affine jump-diffusion (AJD) models which do not pursue analytical solutions for the Riccati equations. More specifically, we develop option prices for AJD models with stochastic jump intensities, which can capture different market stress scenarios.

The paper is organized as follows: in Sect. 2 we review the Fourier-cosine expansion method to recover density functions and calculate the price of financial derivatives. In Sect. 3 we present the pricing problem and show the ordinary differential equations which give the solution of the AJD characteristic functions of the interest rate processes. In Sect. 4 we develop the coefficients related to the payoff functions. In Sect. 5 we present the probability distributions of the models and the convergence of the prices.

2   Fourier Series Method

An interesting, fast and accurate pricing method based on Fourier series expansions was recently proposed by [9] for options on stocks, with good potential to be shaped for other derivatives. Let $f : [0, \pi] \longrightarrow \mathbb{R}$ be an integrable function. A change of variable $\xi = \pi \frac{x-a}{b-a}$ is considered in order to have an even function. Then, the Fourier-cosine series expansion of f in the interval $[a, b]$ is

$$f(x) = \frac{a_0}{2} + \sum_{j=1}^{\infty} a_j \cos\left(j\pi \frac{x-a}{b-a} \right),$$ (1)

where

$$a_j = \frac{2}{b-a} \int_a^b f(x) \cos\left(j\pi \frac{x-a}{b-a} \right) dx, \quad j \geq 0.$$ (2)

Let us assume that $f \in L^1(\mathbb{R})$ and that we explicitly know the Fourier transform of f. The approximation in the interval $[a, b]$ of the coefficients of the Fourier-cosine expansion of f is

$$a_j = \frac{2}{b-a} \int_a^b f(\xi) \Re\left(e^{ij\pi \frac{\xi-a}{b-a}} \right) d\xi \approx \frac{2}{b-a} \Re\left(e^{-ij\pi \frac{a}{b-a}} \hat{f}\left(\frac{j\pi}{b-a} \right) \right) \triangleq A_j.$$ (3)

The approximation of f is given by the following Fourier-cosine series:

$$f(x) \approx \frac{A_0}{2} + \sum_{j=1}^{n} A_j \cos\left(j\pi \frac{x-a}{b-a} \right), \quad x \in [a, b],$$ (4)

for an appropriated chosen n.

Let X_T be a random variable with probability density function f_{X_T} with known characteristic function $\hat{f}(\omega) = \int_{\mathbb{R}} e^{ix\omega} f(x)dx \approx \int_a^b e^{ix\omega} f(x)dx$. The price of a European call option $C(t,T)$ with payoff function g and strike K is

$$C(t,T) \approx \frac{A_0}{2} \int_a^b g(x)dx + \sum_{j=1}^n A_j \int_a^b g(x)cos\left(j\pi\frac{x-a}{b-a}\right)dx. \qquad (5)$$

Hence, the series approximation of the option price is given by

$$C(t,T) = \mathbb{E}[g(x)|X_t] \approx \frac{A_0 B_0}{2} + \sum_{j=1}^n A_j B_j, \qquad (6)$$

where the A_j coefficients are given by (3) and

$$B_j = \int_a^b g(x)cos\left(j\pi\frac{x-a}{b-a}\right)dx, \quad \text{for } j = 0, 1, ..., n. \qquad (7)$$

The choices of the integration limits (a, b) for the approximation were proposed in [9]. The scenario we are dealing with is that where no closed-form exists for the characteristic function. Then, the limits are achieved by numerically differentiate the cumulant function of the model.

3 IDI Options with Affine Jump-Diffusion Models

We assume an interest rate market with underlying probability space $(\Omega, \mathbb{F}, \mathbb{P})$ equipped with a filtration $\mathbb{F} = (\mathcal{F}_t)_{t\in[0,T]}$ where $\mathbb{P}$ is the risk neutral measure.

Let r_t be the spot continuously compounding interest rate given by

$$dr_t = \mu(r_t, t)dt + \sigma(r_t, t)dB_t + JdN(\lambda t), \qquad (8)$$

where $\mu(r_t, t)$ is the mean, $\sigma(r_t, t)$ is the volatility and B_t the standard Wiener process. N is a pure jump process with positive intensity $\lambda = \lambda_0 + \lambda_1 r_t$ and jump amplitudes J, which are i.i.d. and independent of B_t.

According to the B3 protocols, the DI rate is the average of the interbank rate of a one-day-period, calculated daily and expressed as the effective rate per annum. So, the ID index accumulates discretely, according to

$$y_t = y_0 \prod_{k=1}^t (1 + DI_k)^{\frac{1}{252}}, \qquad (9)$$

where k denotes the end of day and DI_k assigns the corresponding DI rate. More details about the DI index is found in [7].

If $r_t = \ln(1 + DI_t)$, the index can be represented by the following continuous compounding expression $y_t = y_0 e^{\int_0^t r_s ds}$, where r_t is given by (8). The price for the option with strike K and maturity in t is given by

$$C_0 = \mathbb{E}\left[\max\left(y_0 - Ke^{-\int_0^t r_s ds}, 0\right)\Big|\mathcal{F}_0\right]. \qquad (10)$$

From now on we benefit from the procedure found in [8] which focus in obtaining characteristic functions of affine jump-diffusion (AJD) models. We find a function of the solution of the AJD model from which a characteristic function should be obtained.

Theorem 1. *The characteristic function of the integrated process* $x_t = \int_0^t r_s ds$ *where* r_s *is given by an AJD model of the form* (8) *is*

$$\hat{f}(x_t, iu) = \mathbb{E}\left[\exp\left(iux_t\right)|r_0\right] = \exp(\alpha(t) + \beta(t)r_0), \tag{11}$$

where

$$\alpha'(s) = \beta(s)\kappa + \lambda_0 \left[\mathbb{E}\left(e^{\beta(s)J}\right) - 1\right], \tag{12}$$

$$\beta'(s) = \beta(s)\theta + \frac{1}{2}\beta(s)^2\sigma^2 - iu + \lambda_1 \left[\mathbb{E}\left(e^{\beta(s)J}\right) - 1\right], \tag{13}$$

with boundary conditions $\alpha(0) = \beta(0) = 0$.

Proof. See [8].

The variable λ_0 multiplying the expectation in (12) gives the constant intensity of the jumps and the variable λ_1 multiplying the expectation in (13) gives the slope when the intensity of the jumps is stochastic. Closed-form solutions for $\left[\mathbb{E}\left(e^{\beta(s)J_i}\right) - 1\right]$ are found in [5] for exponential, normal and gamma jumps.

When dealing with stochastic intensity for the jumps, closed-form solutions for the Riccati equations do not exist. Solving (12) and (13) numerically with the Runge-Kutta algorithm gives the characteristic function we use in A_j coefficients of the series (6). Finite difference method is then used to differentiate the cumulant-generating function in order to prescribe the integration limits.

In this paper, we deal with a variety of Ornstein-Uhlenbeck processes. The Vasicek model without jumps, the Vasicek model with exponential positive and negative jumps, the Vasicek model with stochastic positive and negative exponential jumps, the Vasicek model with normal and stochastic normal jumps and the Vasicek model with gamma and stochastic gamma jumps.

4   IDI Option Pricing with the COS Method

The characteristic function of the random variable $\int_0^t r_s ds$ enters in the A_j coefficients in Eq. (6). Whence, this suffices for calculating the A_j. So we only have to calculate the corresponding B_j coefficients in order to price the IDI option. We consider from now on the vanilla call option case as shown in Eq. (10). To the best of authors' knowledge, this is the first paper to study this class of models for path-dependent interest rate derivatives.

Theorem 2. *The B_j coefficients for vanilla IDI call options are given by*

$$B_0 = \int_{-\ln\left(\frac{y_0}{k}\right)}^{b} y_0 - ke^{-x}dx = y_0 \left(\ln\left(\frac{y_0}{k}\right) + b - 1\right) + e^{-b}k, \qquad (14)$$

and

$$B_j = \int_{-\ln\left(\frac{y_0}{k}\right)}^{b} \left(y_0 - ke^{-x}\right) \cos\left(\frac{\pi j\,(x - a)}{b - a}\right) dx \qquad (15)$$

Proof. The vanilla IDI call option is given by (10). Integrating it according to Eq. (7) gives Theorem 2.

5   Numerical Results

In the following panels we exhibit the probability density functions of the integrated process of each model under study and the corresponding convergence analysis of the IDI call option prices. In Fig. 1 we show the Vasicek model with exponential jumps. The model is enhanced with positive and negative jumps. In Fig. 2 we show the Vasicek model with normal jumps. In Fig. 3 we show the Vasicek model with gamma jumps. All above models were enhanced with constant and stochastic intensities. The base parameters for the figures are: $r_0 = 0.1$, $\kappa = 0.25$, $\theta = 0.1$, $\sigma = 0.04$, $\lambda_0 = 1$, $\lambda_1 = 10$, $\eta = 0.01$, $\mu = 0$, $\Sigma = 0.015$, $p = 1.5$, $T = 5$, $y_0 = 100000$ and $K = 165000$.

We highlight that the numerical COS method converges with few terms in the Fourier series, namely around fifteen terms. It takes a half of second to calculate the price performing the Runge-Kutta method inside each term of the series and achieving an error of the order of 10^{-3}. The computer used for all experiments has an Intel Core i5 CPU, 2.53 GHz. The code was written in MATLAB 7.8.

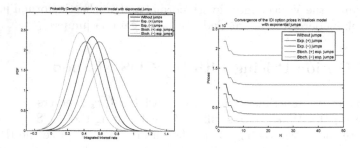

Fig. 1. (Left) Probability density functions for Vasicek model with exponential jumps. (Right) Convergence analysis for Vasicek model with exponential jumps.

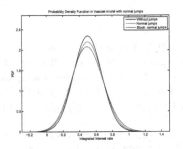
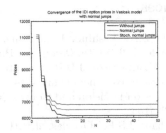

Fig. 2. (Left) Probability density functions for Vasicek model with normal jumps. (Right) Convergence analysis for Vasicek model with normal jumps.

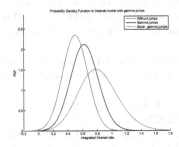
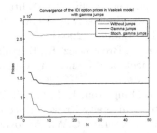

Fig. 3. (Left) Probability density functions for Vasicek model with gamma jumps. (Right) Convergence analysis for Vasicek model with gamma jumps.

6   Conclusion

We extended the range of application of the COS method to interest rate derivatives contracts. We benefited from the procedure found in [8] to obtain characteristic functions related to affine jump-diffusion (AJD) models. In this paper, we are not restricted to explicit solutions of the characteristic function. We provided the probability densities for interest rate models with stochastic intensities and the corresponding prices for a financial product found in the Brazilian market, the IDI option. We devised a path-dependent function that corresponds to the integral of the interest rate process, from which the numerical values for the associated characteristic function was calculated via Runge-Kutta method. We show that the prices converge fastly regarding the number of terms of the Fourier series. To the best of our knowledge, this is the fastest method to numerically calculate the price of a path-dependent interest rate option.

Acknowledgments. This work was partially supported by CNPq under the grant 140840/2015-0.

References

1. Almeida, L.A., Yoshino, J., Schirmer, P.P.S.: Derivativos de renda-fixa no Brasil: Modelo de Hull-White. Pesquisa e Planejamento Econômico **33**, 299–333 (2003)
2. Baczynski, J., Otazú, J.B.R., Vicente, J.V.: A new method for pricing interest-rate derivatives in fixed income markets. In: Proceedings of the 2017 IEEE 56th Annual Conference on Decision and Control (CDC) (2017). https://doi.org/10.1109/CDC.2017.8264105
3. Barbachan, J.S.F., Ornelas, J.R.H.: Apreçamento de opções de IDI usando o modelo CIR. Estudos Econômicos **33**(2), 287–323 (2003). https://doi.org/10.1590/S0101-41612003000200003
4. Barbedo, C.H., Vicente, J.V., Lion, O.B.: Pricing Asian interest rate options with a three-factor HJM model. Revista Brasileira de Finanças **8**(1), 9–23 (2010)
5. Bouziane, M.: Pricing Interest-Rate Derivatives: A Fourier-Transform Based Approach. Springer, Berlin (2008). https://doi.org/10.1007/978-3-540-77066-4
6. Brace, A.: Engineering BGM. Financial Mathematics Series. Chapman & Hall/CRC, London (2008)
7. Carreira, M., Brostowicz, R.: Brazilian Derivatives and Securities: Pricing and Risk Management of FX and Interest-Rate Portfolios for Local and Global Markets. Palgrave Macmillan UK (2016)
8. Duffie, D., Singleton, K.J.: Credit Risk: Pricing, Measurement, and Management. Princeton University Press, Princeton (2003)
9. Fang, F., Oosterlee, C.W.: A novel pricing method for European options based on Fourier-cosine series expansions. SIAM J. Sci. Comput. **31**(2), 826–848 (2008). https://doi.org/10.1137/080718061
10. Genaro, A.D., Avellaneda, M.: Pricing interest rate derivatives under monetary policy changes. Int. J. Theor. Appl. Financ. **21**(6) (2018). https://doi.org/10.1142/S0219024918500371
11. Hull, J., White, A.: One-factor interest rate models and the valuation of interest-rate derivatives securities. J. Finan. Quant. Anal. **28**(2), 235–253 (1993). https://doi.org/10.2307/2331288
12. Junior, A.F., Grecco, F., Lauro, C., Francisco, G., Rosenfeld, R., Oliveira, R.: Application of Hull-White model to Brazilian IDI options. In: Annals of Brazilian Finance Meeting (2003)
13. da Silva, A.J., Baczynski, J., Vicente, J.V.: A new finite difference method for pricing and hedging fixed income derivatives: comparative analysis and the case of an Asian option. J. Comput. Appl. Math. **297**, 98–116 (2016). https://doi.org/10.1016/j.cam.2015.10.025
14. Vieira, C., Pereira, P.: Closed form formula for the price of the options on the 1 day Brazilian interfinancial deposits index IDI1. In: Annals of the XXII Meeting of the Brazilian Econometric Society, vol. 2. Campinas, Brazil (2000)

Composite Data Types in Dynamic Dataflow Languages as Copyless Memory Sharing Mechanism

Aurelien Bloch[1]([✉])[ID], Endri Bezati[2][ID], and Marco Mattavelli[1][ID]

[1] EPFL SCI STI MM, École Polytechnique Fédérale de Lausanne,
Lausanne, Switzerland
{aurelien.bloch,marco.mattavelli}@epfl.ch
[2] EPFL VLSC, École Polytechnique Fédérale de Lausanne, Lausanne, Switzerland
endri.bezati@epfl.ch

Abstract. This paper presents new optimization approaches aiming at reducing the impact of memory accesses on the performance of dataflow programs. The approach is based on introducing a high level management of composite data types in dynamic dataflow programming language for the memory processing of data tokens. It does not require essential changes to the model of computation (MOC) or to the dataflow program itself. The objective of the approach is to remove the unnecessary constraints of memory isolations without introducing limitations to the scalability and composability properties of the dataflow paradigm. Thus the identified optimizations allow to keep the same design and programming philosophy of dataflow, whereas aiming at improving the performance of the specific configuration implementation. The different optimizations can be integrated into the current RVC-CAL design flows and synthesis tools and can be applied to different sub-networks partitions of the dataflow program. The paper introduces the context, the definition of the optimization problem and describes how it can be applied to dataflow designs. Some examples of the optimizations are provided.

Keywords: Dynamic dataflow programs · RVC-CAL ·
Shared memory · Composite data types

1 Introduction

In recent years the difficulties of CMOS technologies to scale-up by increasing the processors frequency, led the processor research and industry to investigate the scale-out by increasing the number of processing units using multi-core, many-core architecture combined with different memory architectures and possibly programmable HW elements building heterogeneous platform. However, these new platforms require software developments to be adapted to the specific platform architecture to take full advantage of the potential hardware parallelism. Such constrains introduce new challenges to software design such as the

© Springer Nature Switzerland AG 2019
J. M. F. Rodrigues et al. (Eds.): ICCS 2019, LNCS 11540, pp. 717–724, 2019.
https://doi.org/10.1007/978-3-030-22750-0_70

portability of applications across platforms or the ability for the programmer to properly abstract and correctly design algorithms using imperative programming languages that take advantage of the processing power available in terms of massive parallelism.

High-level dataflow programming languages are well recognized to be able to overcome those issues [7]. They are used in several fields for modeling data-driven algorithms and in many application areas such as video and audio processing, bioinformatics, financial trading and packet switching. Their essential feature is to be able to abstract parallelism regardless of the targeted hardware platform [4].

The nice properties of dataflow MoC are valid for any type of memory architectures, ranging from the most restricted architectures, for which each memory component is only accessible by a single computational unit, to full shared memory architectures, for which memory is freely accessible by any processing element. However, the performance of dataflow programs on less constrained platforms using the same implementation assumptions of more constrained platforms may result sub-optimal.

Indeed to guarantee the absence of data races in highly parallel platforms, dataflow programs relies on the concept of a full memory isolation for each computational kernel called *actors*. This assumption which provides the guarantees of correctness of the executions for any mapping of the network of actors on any platform, may lead to memory inefficiencies when some actor partitions (i.e. dataflow network partitions) share some or all memory elements.

The paper presents a new approach for data sharing between actors of a dataflow network that reduces the amount of data transfers without changing the model of computation or the semantic of a given application, but only changing the data transfers implementation. The cases for which the communication buffers can be implemented more efficiently, by removing memory isolation constraints for specific partitioning of the dataflow network, are first identified and then three different optimized implementation solutions are defined.

The paper is structured as follows: Sects. 2 and 3 present the context of the dataflow model of computation and compiler. Section 4 provide an overview of the related work. Section 5 presents the design proposition of this new approach, discusses application cases and present different implementations. Finally, Sect. 6 concludes the paper and outlines other directions for further investigations and more effective optimizations.

2   Dataflow Model of Computations

A dataflow program is composed of a (hierarchical) directed graph called, *network*, where each node is an actor and each directed edge is a lossless and order preserving communication channel called *buffer*. These buffers are used to asynchronously transmit atomic data packets called *tokens* between actors. Different dataflow MoC have been defined. A common characteristic is the fact that actors do not have access to a shared memory allowing parallel executions without data race.

Dynamic Process Network (DPN) is one of the most expressive MoC where the actors consumption and production of tokens can vary according to the nature of the available inputs and their internal states [8]. This flexibility is well suited for designing, real-world and complex algorithms at the cost of facing more challenging analysis and optimization problems.

In this work RVC-CAL, a dataflow programming language standardized by the MPEG committee, which fully captures the behavioral features of the DPN model of computation [6] is used. In RVC-CAL, each actor can contain a set of atomic firing functions, called *actions* and internal state variables. When an actor is executed, only a single action can fire at a time. The firing of an action depends on the input availability and values of its tokens, the output available spaces, and the internal state of the actor.

3   Dataflow Compiler

The Open RVC-CAL Compiler (Orcc) is an open-source Integrated Development Environment (IDE) based on Eclipse and dedicated to dataflow programming [1]. It is the compiler used in this work and is mainly a source-to-source compiler that translates the RVC-CAL application into another programming language depending on the backend selected during compilation.

In this work, the Xronos backend [3] is used. It generates from an RVC-CAL description, a C++ implementation with all the necessary library dependencies. The objective is to improve the quality of the generated code by minimizing the overall amount of memory copies by providing when compatible with the dataflow MoC, specific memory sharing mechanism across actors. The introduction of such optimizations can improve the performance of implementations for specific partitioning and scheduling configurations. It can also provide an extension of the design exploration space and yield new scheduling, partitioning and buffer size design points for the design space exploration framework, TURNUS [2].

4   Related Work

A first approach for solving this problem of memory sharing across actors has been presented in the same design context [9]. it is proposed to have actual shared variables, breaking the encapsulation of actors. To do so, internal variables that are shared among multiple actors are tagged with @shared. In addition, a Shared Memory Controller (SMC) along with a specific protocol were designed for access synchronization. The solution has shown the benefice of relevant performance gains due to the instant access to the shared memory once granted access to it and low overhead of the synchronization protocol. The drawback of the solution is that designers have to modify the model of computation and break the principle of memory encapsulation of actors to allow to share their internal states. This means than the compiler cannot guarantee that the generated code is free of data races and that the validity solution has to rely on the designer knowledge. This brings the solution closer to what it is obtained in more traditional settings in which parallelism is obtained by introducing additional synchronization barriers to general purpose imperative languages.

There also exists in the literature some work done such as [5] but they mostly target Synchronous Dataflow (SDF) MoC.

5   Design Proposition

The proposed solution aim at optimizing the performance of the implementation of the generated code for data communication between actors, when targeting a hardware platform where actors partitions are mapped to processing units having access to a common physical memory. In doing so, the model of computation remains unchanged keeping all the properties guaranteed by construction by applying code synthesis and compilation.

5.1   Composite Data Types

The current implementation of the code generation for the standard version of RVC-CAL, processes inter-actor communications by instantiating buffers using primitive data types. A consequence of this relatively low granularity of data transfer may impact the performance of the application compared to design using other models of computation. Furthermore depending on the application some amounts of data might need to be copied over different buffers through the dataflow network even if not all data is relevant for the processing of an actor internal algorithm. To illustrate this fact an example can be useful. Consider an action that produce five tokens to its output buffer at each firing. Figure 1a represents the content of such a buffer after two firings.

Currently, when synthesizing code for such simple program, the Orcc compiler generates a loop, that copy the tokens from the internal memory of the actor to the memory of the output buffer. This makes the amount of copy to the same physical memory for each firing linearly proportional to the production of tokens.

The approach described in this paper is to introduce composite data types as objects manipulated with pointers, which would allow fewer data copies. Lists (arrays) are considered here an example. Figure 1b shows how the status of the buffer might look like, when list to represent data in the same example of two firings of five tokens each are used. It can be observed that instead of having a loop copying the ten primitive values, it is only necessary to copy the two pointers to the corresponding memory chunks. This makes the amount of data copy proportional to the number of moved chunks instead of the number of tokens. This approach keeps the same philosophy for avoiding data races like what Orcc is currently implementing. In fact, it does not propose to modify the model by offering shared memories between actors and still can rely on buffers to synchronize communication between them.

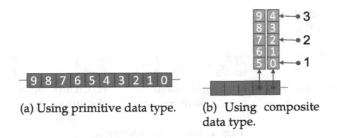

(a) Using primitive data type.   (b) Using composite data type.

Fig. 1. Buffer filled with two firings of five primitive tokens each.

5.2   Buffer Identification

In this section the different cases where this optimization can be applied are identified. First of all, this proposition can only be beneficial for actions that produce multiple tokens in a single firing. This is the case for example, when actions uses a *repeat* expression. In addition to that, three different configuration cases can be identified.

The first one is when a buffer has multiple fan-out as shown in Fig. 2a. In this case, it is necessary to duplicate the composite data to avoid the data race. This is due to the fact that pointers are used, to be copied at the place of the data. If the data is not duplicated, each actor has a reference to the same piece of data, which might yield data races problems. This configuration should only improve performances to a fraction inversely proportional to the fan-out numbers, as only the first actor will have access to the original data whereas the others need a copied version.

The second case is when a list is transmitted only between two actors as shown in Fig. 2b. In this case the proposition will not result in any performance improvement as it is already optimized by the current implementation of the Orcc compiler. Indeed, instead of generating the tokens to a local array and then copying this data to the output buffer memory, the compiler generated code uses a pointer to the buffer memory and store the tokens directly there when they are available. In the same way, the consumer actor (actor B in the schema) use a pointer to the buffer memory to directly process the data read, instead of first copying them locally and then processing them. This optimization prevent the introduction of list to bring performance improvement in this particular case.

The third case is the one where the use of composite data types can provide the higher performance gain. It can be identified when multiple actors process the same composite structure of data. An illustration of this case is depicted in Fig. 2c. The achievable performance improvement is proportional to the size of the chunk of data and to the number of stages the same data is processed by a different actor.

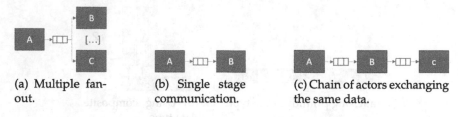

(a) Multiple fan-out.

(b) Single stage communication.

(c) Chain of actors exchanging the same data.

Fig. 2. Buffer identification.

5.3   Implementations Discussion

In this section the implementation challenges that need to be addressed so that the Orcc compiler is able to generate valid C++ code implementing the proposed optimization is discussed.

Fully Dynamic Solution. In this case the consumer can read list of token of any sizes regardless of the size of the emitted list. This solution is the most general. It means that actors can read chunks of data smaller than the actual list and even read chunks that span across two physical data allocations. An illustration of this implementation is depicted in Fig. 1b, where the pointers returned to the reader at each firing are the red arrows. In this example, the read size is 2 and it can be observed that the third read (composing or the numbers 4 and 5) span across the two continuous memory allocation. This fully flexible settings raises two implementation challenges.

One is that since a single continuous allocation can be linked to from different actors during the runtime of the application, it can be difficult to pinpoint when this memory chunk can be released especially in environment with no native garbage collection like in C++. For that, we used *std::shared_ptr* that offers a kind of autorelease mechanism once a memory chunk is no longer referenced by any actors or buffers.

Another technicality to be solved is the need to transparently handle reads that can reach multiple memory chunks. For this purpose a custom proxy class has been developed that is returned, instead of a direct pointer to a memory block, that act as a middle man and handles reads through an indirection, which can affect performance and prevent processor vectorization. This side-effect might be somewhat mitigated if the proxy is used only in the corner cases where it is necessary and if the memory allocator used is tuned so that most consecutive chunks would be allocated consecutively in memory removing the need also in these cases. A custom allocator that would be used explicitly and provided with network specific information to make the use of consecutive allocation more frequent could also be considered.

Semi Dynamic Solution. This case is a constrained version of the previous more general case, where the consumer can only read at each firing a number of tokens that is a divider of the size of the produced list. This constraint is equivalent to impose that a read would never reach across two different memory allocation, which removes the need for any proxy or special allocator, while keeping some amount of flexibility. The difficulty here is to be able to guarantee that this property is always satisfied to allow the safe usage of this implementation solution.

Static Solution. In the third case, the reader has always to consume an entire chunk of data seen in this case as an object. This is the most restrictive configuration, but also the simpler to implement. The need to use memory releasing mechanisms can be avoided as only a single actor or buffer can reference an object at any time allowing for an explicit freeing of memory from the actor itself when the chunk is no longer needed.

6   Conclusions

This paper presents a new approach for the synthesis of efficient implementations of data sharing between actors of a dynamic dataflow networks in the context of the RVC-CAL programing language. The approach introduces composite data types as a ways to avoid data copies whenever possible. It shows for which buffer configuration the optimization solutions can be beneficial and specifies the different ways of implementing them in C++ depending on the flexibility, given to the rate at which an actor can consume data.

Future work considers automatizing the selection by the Orcc compiler of the generated solution depending on the configuration (i.e. Buffer configuration, Network partition, targeted platform). Moreover, this would enable more design point to be considered by the TURNUS framework.

References

1. Orcc. http://github.com/orcc/orcc. Accessed Apr 2019
2. Casale-Brunet, S.: Analysis and optimization of dynamic dataflow programs. Technical report EPFL (2015)
3. Casale-Brunet, S., Bezati, E., Mattavelli, M.: Programming models and methods for heterogeneous parallel embedded systems. In: 2016 IEEE 10th International Symposium on Embedded Multicore/Many-core Systems-on-Chip (MCSoC), pp. 289–296. IEEE (2016)
4. Castrillon, J., Leupers, R.: Programming Heterogeneous MPSoCs. Tool Flows to Close the Software Productivity Gap. Springer, Switzerland (2013). https://doi.org/10.1007/978-3-319-00675-8
5. Desnos, K., Pelcat, M., Nezan, J.F., Aridhi, S.: Distributed memory allocation technique for synchronous dataflow graphs. In: 2016 IEEE International Workshop on Signal Processing Systems (SiPS), pp. 45–50. IEEE (2016)
6. Eker, J., Janneck, J.: CAL language report: Specification of the CAL Actor Language. Technical Memo UCB/ERL M03/48, Electronics Research Laboratory, University of California at Berkeley, December 2003

724   A. Bloch et al.

7. Kahn, G.: The semantics of a simple language for parallel programming. In: Rosenfeld, J.L. (ed.) Information Processing, pp. 471–475. North Holland, Amsterdam, Stockholm, Sweden, August 1974
8. Lee, E., Parks, T.: Dataflow process networks. In: Proceedings of the IEEE, pp. 773–799 (1995)
9. Modas, A., Casale-Brunet, S., Stewart, R., Bezati, E., Ahmad, J., Mattavelli, M.: Shared-variable synchronization approaches for dynamic data flow programs. In: 2018 IEEE International Workshop on Signal Processing Systems (SiPS), pp. 263–268. IEEE (2018)

A Coupled Food Security and Refugee Movement Model for the South Sudan Conflict

Christian Vanhille Campos[1,2], Diana Suleimenova[3], and Derek Groen[3(✉)]

[1] Universidad Complutense de Madrid, Madrid, Spain
[2] Université Paris Diderot, Paris, France
[3] Department of Computer Science, Brunel University London,
London, UK
Derek.Groen@brunel.ac.uk

Abstract. We investigate, through data sets correlation analysis, how relevant to the simulation of refugee dynamics the food situation is. Armed conflicts often imply difficult food access conditions for the population, which can have a great impact on the behaviour of the refugees, as is the case in South Sudan. To test our approach, we adopt the Flee agent-based simulation code, combining it with a data-driven food security model to enhance the rule set for determining refugee movements. We test two different approaches for South Sudan and find promising yet negative results. While our first approach to modelling refugees response to food insecurity do not improve the error of the simulation development approach, we show that this behaviour is highly non-trivial and properly understanding it could determine the development of reliable models of refugee dynamics.

Keywords: Multiscale modelling · Agent-based modelling ·
Forced displacement · Data-driven simulation

1 Introduction

Forced displacement has reached record levels in 2018, and so has the number of people facing severe food insecurity. As of June 2018, 68.5 million people are forcibly displaced worldwide, 25.4 million of which are refugees[1]. According to World Food Programme (WFP)[2], 124 million people are currently facing *crisis* (IPC phase 3) food insecurity conditions or worse, which corresponds to a 15% increase with respect to last year's figures. Despite the current efforts by the global community, not enough is being done to improve the situation. Simulations could prove to be key in trying to understand what drives such crises and predict how they might evolve. This precisely is what a generalized

---

[1] https://www.unhcr.org/figures-at-a-glance.html.
[2] https://sway.office.com.

© Springer Nature Switzerland AG 2019
J. M. F. Rodrigues et al. (Eds.): ICCS 2019, LNCS 11540, pp. 725–732, 2019.
https://doi.org/10.1007/978-3-030-22750-0_71

simulation development approach (SDA) and Flee code proposed by Suleimenova et al. [1] intends to do, attempting to establish accurate and reliable models for refugee destination predictions for a variety of conflicts over different periods of time. Indeed, achieving such a feat could be determinant in helping NGOs and governments alike make better-informed decisions regarding humanitarian support for refugee crises.

In this paper, we focus on the situation in South Sudan, which was established in 2011. Important tensions and violence in South Sudan has had tremendous consequences for the civilian where there are now 7 million people starving and 4.3 million people forcibly displaced (2.5 million of which are refugees)[3]. Given the conflict situation and the famine in South Sudan, its population is enduring and it is fair to wonder what role food security plays here in the movement of people. Several reports stress that many people are fleeing their homes due to starvation[4]. However, how the food conditions precisely relate to the conflicts remains unclear. Both NGOs and news companies report that the famine, while being strongly linked to the climate, is also heavily influenced by the war[5]. Therefore, to assess how relevant to the modelling of food security is in South Sudan, we conducted a correlation analysis to see how it relates to conflict occurrence. We conclude that while there exists a positive correlation between the two phenomena, food security conditions may still need to be accounted to construct a faithful model. In addition, we have coupled the Flee code with a data-driven food security model to create a multiscale model using a generalized SDA and enhanced the rule set that guides agents' behaviour. We compare our results to those using Flee alone and to observations by international agencies.

2   Related Works

Understanding and forecasting forced displacement movements have become the main interest for international agencies and governments, thus motivating further scientific research in this field [2]. Many different techniques are gaining importance, such as machine learning [3], simulation-based approaches and agent-based modelling (ABM) [4]. In ABM, a complex system is modelled as a set of autonomous decision-making agents that behave accordingly with their environment based on a set of rules [5]. This has proven to be especially useful when modelling how individuals and their environment interact and evolve over time, as is the case in migration and forced displacement movements. This technique is thus being used extensively to model armed conflicts response [6], to predict humanitarian assistance need [7], to predict refugee routes [8] or to understand climate effects on migration [9]. However, these current models do not incorporate the effects of famine and starvation on refugee movements.

With regards to refugee arrival predictions, we rely on a generalized SDA, which is based on an ABM and is used to forecast refugee distributions across

---

[3] https://www.wfp.org/Situation-Reports/South-Sudan.

[4] http://www.unhcr.org/news/briefing.

[5] http://reporting.unhcr.org/sites/default/files/UNHCR.

camps (see Fig. 1). A full explanation of the SDA, which is beyond the scope of this short paper, is provided by Suleimenova et al. [1]. We run our simulations using a modified version of the Flee code[6], obtaining results for the refugee distribution in camps, both without and with a model to reflect the food security aspects. We validate simulations against official UNHCR refugee registration counts. The SDA is also optimized to allow for easy adaptation to different conflicts, policy choices and assumptions, using the FabFlee automation toolkit [10]. This adaptability, however, should not become a limitation in terms of accuracy while being generalized to any given conflict.

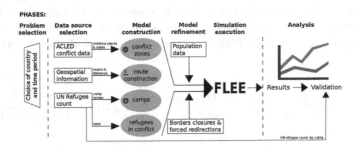

Fig. 1. Schematic representation of a generalized SDA and the Flee code   (source: Suleimenova et al. [1])

3   Input Data Analysis

To understand the effects of the food security situation on the refugee dynamics in South Sudan, we assess its relation to the conflict. We use data sources for conflict (or battle) occurrence from ACLED[7] and famine-like stress classifications of food insecurity indexes from the Integrated Food Security Phase Classification (IPC) project[8]. IPC distinguishes five different food security phases ranging from *Minimal stress* to *Famine* and classifies the population of different regions accordingly.

We look for linear correlations in time and space using two different sets of variables, namely the temporal and spatial analyses, which differ in the normalization. We normalize the fraction of the population classified as IPC phase 3 or worse[9] and number of conflicts by their maximum over time, as well as by their maximum for each state[10] (see Eqs. 1 and 2). We then create scatter

---

[6] Available at: https://github.com/djgroen/flee-release.

[7] Available at: https://www.acleddata.com.

[8] Available at: https://www.ipcinfo.org.

[9] This percentage is often considered to represent the population at risk, and is widely referenced as a good estimator of the food security of a certain region.

[10] By defining variables in this way we guarantee that they all take values ranging from 0 to 1 and that they are coherently normalized, making it easier to compare the data sets.

plots of these variables with linear least-square fits, first for each different region $(F_{space}(X,t)$ vs $C_{space}(X,t)$ with X a fixed state of South Sudan) and second for each month of the conflict $(F_{time}(x,T)$ vs $C_{time}(x,T)$ with T a fixed month).

Considering that the variables of interest take values ranging from 0 to 1, in most cases the correlation is actually quite weak, with average values of the mean square error (MSE) above 0.05 both for the temporal and geographical analyses, and very few cases where a lower MSE than that is observed (see Table 1). Moreover, we observe an important variability in the parameters of the fits (i.e. the slope of correlations), with standard deviations of the same order as or even higher than the average values. Overall, conflict emergence cannot simply portray food security conditions in a model as not only the quality but also the precise form of the correlations between the two data sets is highly variable, making it very difficult to express the effects of one in terms of the other.

$$F_{time}(x,t) = \frac{N^P_{IPC \geq 3}(x,t)}{\max_x[N^P_{IPC \geq 3}(x,t)]} \qquad C_{time}(x,t) = \frac{N^C(x,t)}{\max_x[N^C(x,t)]} \qquad (1)$$

$$F_{space}(x,t) = \frac{N^P_{IPC \geq 3}(x,t)}{\max_t[N^P_{IPC \geq 3}(x,t)]} \qquad C_{space}(x,t) = \frac{N^C(x,t)}{\max_t[N^C(x,t)]} \qquad (2)$$

The temporal correlations are largely consistent regarding the sign of the slope, with more than 96% of the cases having a positive relationship as compared to only 27% for spatial correlations. This result shows that famine is not local to conflict zones but is still frequently higher when more conflict events occur. This suggests that starving tactics may indeed have been implemented in this struggle, as indicated in the literature.

Table 1. The percentage of data sets with a MSE below 0.05 provides a comparative measure of the quality of the spatial and temporal correlations respectively. Such a value of the MSE corresponds to an error of the order of 20% for a set of variables, which is already considerably high. The average slope, its standard deviation and the percentage of data sets with same-sign (positive) correlations provide a measure of the consistency. Full results are available at DOI: https://10.17633/rd.brunel.8053340.

	States	Months
$MSE \leq 0.05$	36.4%	11.5%
Positive slope	26.9%	96.2%
Average slope	0.130	0.524
Std. dev. of the slope	0.259	0.267

4 Implementation

As we have presented so far, the food conditions appear relevant to the model, at least in South Sudan. We now present a simple tentative way to include these features in the model that will also allow for it to be extrapolated to any other conflict to which the model might be applicable. We confront the results obtained by implementing these modifications with the official UNHCR refugee counts and compare the error with the results of the original model offers [1].

Modification 1: We make movechances in other locations dependent on the IPC index of each region at every time-step using the following function: $IPC_{x,t} + 0.3(1 - IPC_{x,t})$, where $IPC_{x,t}$ is the fraction of the population estimated by IPC to be in a *stress* situation in the region of the location at a given time. With this definition, we guarantee that, on average, the fraction of the population estimated to be affected by food insecurity conditions leaves such locations, in an attempt to recreate real refugee behaviour.

Modification 2: We assume that food insecurity leads to refugees departing exclusively for that reason. We do this by expanding the list of spawn locations to include both conflict locations and locations with IPC indexes greater than 0. The weighted probability of agents spawning in each location is set as follows: (a) for conflict zones, equal to the full population according to the last census. (b) for IPC locations that are not conflict zones, equal to the population times $IPC_{x,t}/100$.

5 Simulation Setup

To implement these changes in the code, we use FabFlee, which is an automation tool facilitating the study of the effects of policy decisions on the refugee situation and relies on Flee for underlying simulations. FabFlee allows for easy changes in simulation settings, such as border closures, forced redirection, speed changes, etc. The basic workflow of the tool consists of (i) loading the conflict we want to work on, including locations, routes and conflicts information, (ii) performing desired changes to the simulation settings and (iii) instantiating the modified version of the conflict and run it to obtain the results. The advantage of performing multiple different simulations for the same conflict without having to manually edit every part of the code each time makes FabFlee well suited for studying the effects of food security on refugee migrations (instead of having to construct a complete simulation from scratch, just add few modifying modules). This allows ensuring that both simulations run under the same general conditions.

We also incorporate a separate submodel in the original Flee code [1], which either updates the movechances of neutral locations or modifies the spawning probabilities of refugees over time. The data-driven food security submodel updates the affected locations' parameters when a change occurs in the IPC index of their region as the simulation evolves. We include new commands in FabFlee: one to run the simulation including the food security aspects, and one to generate side-by-side comparison graphs of key metrics between both types of simulations.

6   Results

In Fig. 2, we present the time evolution of the total error we obtain with the two proposed modifications together with the total error of the original Flee implementation under the same conditions for comparison. To calculate this error we use the following equation:

$$E(t) = \frac{\sum_{x \in S}(|n_{sim,x,t} - n_{data,x,t}|)}{N_{data,all}} \tag{3}$$

where $n_{sim,x,t}$ gives the number of refugees in camp x at time t according to the simulation and $n_{data,x,t}$ according to the UNHCR data. S is the set of all camps and therefore $N_{data,all} = \sum_{x \in S} n_{data,x,t}$.

In Fig. 2a, we compare the movechance modification of Flee with the original implementation, correcting the original such that the average movechance across locations is identical in both simulations. We perform this correction because the validation error is known to be sensitive to this parameter. The error behaviour of both runs are highly similar, with the error of the modified implementation being $\approx -1.79\%$ worse on average. These observations imply that our first proposed implementation of the food considerations doesn't translate to a more accurate prediction of the refugee distribution across camps in South Sudan.

In Fig. 2b, we compare the error of the simulation with modified spawning behaviour with the original Flee run. Here, the movechance is identical in both runs, so no corrections are required. We observe a higher error in the modified run which lasts for 517 days, upon which it drops and reaches up to an $\approx 21.95\%$ improvement on the final day of simulation. The average improvement (fraction of decrease in the error) over time is negative ($\approx -9.49\%$), meaning that this

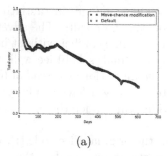
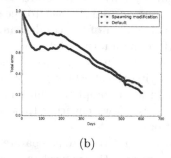

(a)   (b)

Fig. 2. Total error of the two modified implementations of FLEE compared to the original rule set when using (a) modified movechance calculations and (b) modified spawn location algorithm. The consistency of these results has been tested by running multiple replicas of the same simulation, obtaining very low variability.

implementation of the food considerations also does not lead to a better refugee distribution forecast in this conflict.

7 Discussion

In this paper, we have explored the importance of food security for refugee movements in the current South Sudan conflict and propose a computational model based on automated SDA. Although we find that food security conditions are important in refugee dynamics in South Sudan through our data analysis, it has not yet become clear to us in what way the two phenomena exactly interrelate. We investigated two hypotheses in a very tentative fashion as a demonstrator. First, food security influences the likelihood of travelling refugees to depart and second, food insecure locations may act as a cause for refugees choosing to depart in the first place. When implemented in our coupled simulation, neither scenario leads to a lower error, but one of the hypotheses does affect a sensitive parameter in the simulation (movechance).

To understand the relation clearly more in-depth investigations are required. These include bolstering the simulations against the sensitivity of movechance (e.g., by estimating refugee departure dates in the validation set), testing a wider range of hypotheses, and to validate across a much wider range of conflicts. We expect that extensive automation and more sophisticated coupling (as discussed in [11]) will be essential in making this possible. Lastly, we need to do extensive data analysis to improve the model's accuracy. Examples include studies on the origins of the refugees (ethnic, religious, etc.), group dynamics, other environmental conditions and average trajectories. A better understanding of these aspects should then lead to a significant improvement in the model's predictive capacities.

Acknowledgements. This work was supported by the VECMA and HiDALGO projects, which has received funding from the European Union Horizon 2020 research and innovation programme under grant agreement No. 800925 and 824115.

References

1. Suleimenova, D., Bell, D., Groen, D.: A generalized simulation development approach for predicting refugee destinations. Sci. Rep. **7**, 13377 (2017)
2. Ahmed, M.N., et al.: A multi-scale approach to data-driven mass migration analysis. In: CEUR Workshop Proceedings, vol. 1831 (2016)
3. Sfyridis, A., Cheng, T., Vespe, M.: Detecting vessels carrying migrants using machine learning. In: ISPRS Annals of Photogrammetry, Remote Sensing and Spatial Information Sciences. Volume IV-4/W2, pp. 53–60, October 2017
4. Sokolowski, J.A., Banks, C.M.: A methodology for environment and agent development to model population displacement. In: Proceedings of the 2014 Symposium on Agent Directed Simulation, Tampa, Florida, Society for Computer Simulation International (2014)

5. Bonabeau, E.: Agent-based modeling: methods and techniques for simulating human systems. Proc. Nat. Acad. Sci. **99**(suppl 3), 7280–7287 (2002)
6. Williams, N.E., O'Brien, M.L., Yao, X.: Using survey data for agent-based modeling: design and challenges in a model of armed conflict and population change. In: Grow, A., Van Bavel, J. (eds.) Agent-Based Modelling in Population Studies. TSSDMPA, vol. 41, pp. 159–184. Springer, Cham (2017). https://doi.org/10.1007/978-3-319-32283-4_6
7. Crooks, A.T., Wise, S.: GIS and agent-based models for humanitarian assistance. Comput. Environ. Urban Syst. **41**, 100–111 (2013)
8. Hébert, G.A., Perez, L., Harati, S.: An agent-based model to identify migration pathways of refugees: the case of Syria. In: Perez, L., Kim, E.-K., Sengupta, R. (eds.) Agent-Based Models and Complexity Science in the Age of Geospatial Big Data. AGIS, pp. 45–58. Springer, Cham (2018). https://doi.org/10.1007/978-3-319-65993-0_4
9. Entwisle, B., et al.: Climate shocks and migration: an agent-based modeling approach. Popul. Environ. **38**(1), 47–71 (2016)
10. Suleimenova, D., Bell, D., Groen, D.: Towards an automated framework for agent-based simulation of refugee movements. In: Winter Simulation Conference (WSC) 2017, pp. 1240–1251. IEEE (2017)
11. Groen, D., Knap, J., Neumann, P., Suleimenova, D., Veen, L., Leiter, K.: Mastering the scales: a survey on the benefits of multiscale computing software. Phil. Trans. R. Soc. A **377**, 20180147 (2018)

A Proposal to Model Ancient Silk Weaving Techniques and Extracting Information from Digital Imagery - Ongoing Results of the SILKNOW Project

Cristina Portalés[1]([⊠]) , Javier Sevilla[1] , Manolo Pérez[1] ,
and Arabella León[2]

[1] Universitat de València, Blasco Ibáñez 13, 46013 València, Spain
cristina.portales@uv.es
[2] Garín 1820 S.A., Ramon Villarroya, 15, 46113 Moncada, València, Spain

Abstract. Three dimensional (3D) virtual representations of the internal structure of textiles are of interest for a variety of purposes related to fashion, industry, education or other areas. The modeling of ancient weaving techniques is relevant to understand and preserve our heritage, both tangible and intangible. However, ancient techniques cannot be reproduced with standard approaches, which usually are aligned with the characteristics of modern, mechanical looms. The aim of this paper is to propose a mathematical modelling of ancient weaving techniques by means of matrices in order to be easily mapped to a virtual 3D representation. The work focuses on ancient silk textiles, ranging from the 15th to the 19th centuries. We also propose a computer vision-based strategy to extract relevant information from digital imagery, by considering different types of images (textiles, technical drawings and macro images). The work here presented has been carried out in the scope of the SILKNOW project, which has received funding from the European Union's Horizon 2020 research and innovation program under grant agreement No. 769504.

Keywords: Silk · Weaving · Modelling · 3D · Image processing

1 Introduction

Knowing the internal structure of textiles is of relevance for a variety of areas, such as fashion, industry, education or even cultural heritage. Indeed, the modeling of ancient weaving techniques such as *brocato*, *damask* or *espolín* can be of relevance to understand and preserve our heritage, both tangible and intangible.

In the field of research, different authors have dealt with the 3D representation textiles and/or clothes [1–4], although not all of them deal with the internal structure of textiles at the yarn level. A good example is the work presented in [3], where a representation of knitted cloth at the yarn level is proposed, which treats yarn-yarn contacts as persistent, avoiding expensive contact handling altogether. On the other hand, there also exist a variety of software for the 3D virtual representation of textiles, such as WiseTex [5, 6] or TexGen [7]. However, many of the ancient silk weaving

© Springer Nature Switzerland AG 2019
J. M. F. Rodrigues et al. (Eds.): ICCS 2019, LNCS 11540, pp. 733–740, 2019.
https://doi.org/10.1007/978-3-030-22750-0_72

techniques cannot be reproduced with standard software solutions, which are usually aligned with the characteristics of modern, mechanical looms. For example, the spines that weave the handlooms cannot be reproduced on modern looms. This is because the wefts representing pictorial parts in these ancient fabrics are centered in certain places (prepared by the weaver), while the mechanical looms go automatic and pass their plots from side to side of the fabric (Fig. 1). Because of that, a new mathematical representation embedding these techniques is needed, that allows an easy transfer to 3D drawing tools.

Fig. 1. Making a Valencian *espolín* with a Jacquard loom. It can be seen how the golden threads do not go from side to side of the fabric, but remain concentrated around the pictorial motifs.

The aim of this paper is to propose a basic mathematical representation of ancient weaving techniques, focusing on ancient silk textiles, ranging from the 15th to the 19th centuries. We also propose a computer vision-based strategy to extract relevant information from digital imagery, by considering different types of images. The work here presented has been carried out in the scope of SILKNOW [8], an ongoing three-year project (2018–2021) that aims at preserving silk digital heritage. As part of this project, a module called Virtual Loom will be designed and implemented. This module is meant to preserve the fragile ancient art of weaving by directly "cloning" the way it was woven. By means of this tool, high resolution 3D models of the textiles will be produced, based on mathematical models of weaving techniques and the pictorial forms embedded in images. Having information about the weaving technique (to which mathematical models will be defined) and the image itself (mainly, technical drawings and textiles images), virtual models of the fabrics will be automatically produced with a virtual reality motor generation-based software, such as Unity 3D [9], that additionally allows the generation of multi-platform solutions.

2 Representation of Ancient Weaving Techniques

2.1 Single Matrix Representation

In Fig. 2, the representation of a basic weave is given, following the usual convention. As it can be seen, this example can be fully represented with only one matrix. The given matrix indicates the level of the warp, with level 0 indicating the surface of the textile, and the level 1 indicating the level below the textile. As the threads interact one with other, the wefts lay just in the other level: when a warp is in level 0, the weft is in level 1 and the other way round. The example shown in Fig. 2 correspond to a plain weave or tabby. Other basic weaves such as twill and satin, can be also represented in this way.

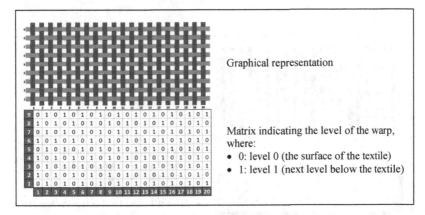

Graphical representation

Matrix indicating the level of the warp, where:

- 0: level 0 (the surface of the textile)
- 1: level 1 (next level below the textile)

Fig. 2. Representation of a fabric with a basic weave (plain weave or tabby). The warp (vertical yarns) are shown in blue color; the weft (horizontal yarns) are represented in red color. No pictorial parts are considered in this example. (Color figure online)

2.2   Multiple Matrix Representation

Matricial forms can also be used to represent complex weavings involving different layers, as suggested by [10, 11], where the topological coding of a multilayered weave is based on warp yarns paths. However, for the representation of ancient silk textiles more than one matrix is needed. This is because the yarns representing pictorial parts in these ancient fabrics are centered in certain places (prepared by the weaver), so they do not go from side to side of the fabric. As they are horizontally weaved, they are aligned with the weft yarns, but constitute an additional layer that is not homogeneous alongside with the textile.

Therefore, pictorial parts involving other threads need to be represented with additional matrices, such as the one we propose in Fig. 3. Representing the threads of pictorial parts independently and considering both sides of the textile, allows accounting from possible overlays on the backside. This could be done simple looking where there are "ones" for a same cell in matrices representing threads of different colors. For the 3D modelling, a logical order needs to be established to see in what level representing each thread for each cell. As all the matrices have the same size (rows, columns), the complete representation of a textile can be in form of a hypercube with different dimensions (Di), which will be ordered in the following way:

- D_0: the matrix embedding the pattern or background
- D_1: the matrix of the pictorial part of yarn 1, for the front side of the textile
- D_2: the matrix of the pictorial part of yarn 1, for the backside of the textile
- ...
- D_{n-1}: the matrix of the pictorial part of yarn n/2, for the front side of the textile
- D_n: the matrix of the pictorial part of yarn n/2, for the backside of the textile

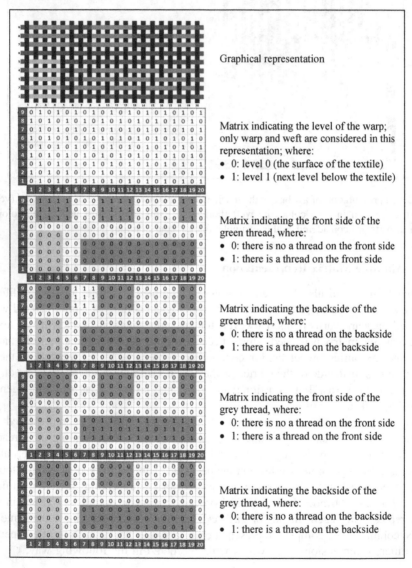

Fig. 3. Representation of a textile with pictorial parts. For the sake of brevity, the two matrices corresponding to the orange yarn have been omitted.

3   Extracting Information from Imagery

3.1   Types of Images

Most of the images belonging to the digital collections correspond to photographs of ancient textiles taken with an RGB camera (Fig. 4a). Therefore, this is the main type of information that will be considered for the Virtual Loom. Other types of images can be also used in order to extract more information on the textiles or to complement the RGB images, such as technical drawings (Fig. 4b), macro images (i.e. close-up photography) (Fig. 4c), images in other spectral bands, etc.

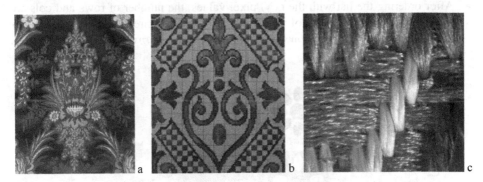

Fig. 4. Examples of images, where: (a) image of a textile (detail); (b) image of a technical drawing (detail); (c) macro image of a textile.

3.2   Methodology

Strategy 1: given an image of a textile. The way of proceed in this case is the following (Fig. 5):

- The user will manually select the area interactively.
- If not available from the text, the user will specify the number of different yarns (i.e., the number of colors).
- The image will be posterized to get an image with so many plain colors as number of different yarns.
- The image will be processed in order to extract the background and pictures with different colors. To that end, users will interactively click in one part of the image that contains the background.
- The user will specify, for the background and each of the pictorial part, a weaving technique. For instance, for the background he/she will be able to choose between tabby, twill or satin with different variations.
- If not available from the text, the user will manually specify the number of weft yarns and warp yarns.
- A matrix will be built based on the aforementioned values, following the conventions explained in Sect. 2; rows and columns of the matrix will be build according the number of weft yarns and warp yarns.

Strategy 2: given an image of a technical drawing of a textile. To extract the number of weft and warp yarns the following procedure is followed (Fig. 6):

- To begin, the original image is converted to greyscale values.
- Then, a local threshold is applied in order to highlight borders. In this way, the original grid is depicted, as schematized in the "thresholded image".
- A morphological operator is used to find crosses in the image. The resulting image will be composed of crosses, although it might happen that some of them are missing if the grid was not fully reproduced in the previous step.
- In order to retrieve pixel coordinates of the centers of the crosses, blobs are computed, which contain such information.
- After ordering the list with the *(x,y)* pixel values, the number of rows and cols are retrieved. Then, the number of weft and warp yarns are directly computed: *number of weft yarns = number of rows + 1*; *number of warp yarns = number of cols + 1*.

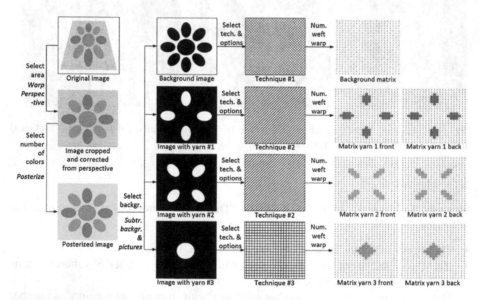

Fig. 5. Methodology used in strategy 1 to derive the matrices that will be used for the Virtual Loom.

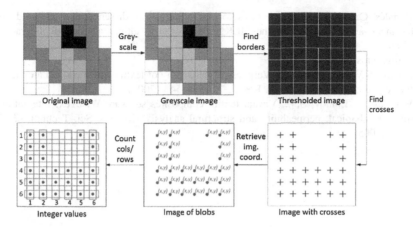

Fig. 6. Methodology used in strategy 2 to derive the number of weft and warp yarns.

4 Conclusion and Future Work

Ancient silk textiles are an endangered heritage, not only the textiles themselves, but also their weaving techniques. In order to preserve this legacy, in this paper we show a proposal to represent such techniques with matrices, and to extract information from digital imagery. This information is key to build 3D models of their internal structure, what constitute the next step in our research.

Acknowledgments. The research leading to these results is in the frame of the "SILKNOW. Silk heritage in the Knowledge Society: from punched cards to big data, deep learning and visual/tangible simulations" project, which has received funding from the European Union's Horizon 2020 research and innovation program under grant agreement No. 769504.

References

1. Etzmuss, O., Keckeisen, M., Strasser, W.: A fast finite element solution for cloth modelling. In: Proceedings of 11th Pacific Conference on Computer Graphics and Applications 200, pp. 244–251 (2003)
2. Jiang, Y., Chen, X.: Geometric and algebraic algorithms for modelling yarn in woven fabrics. J. Text. Inst. **96**, 237–245 (2005)
3. Cirio, G., Lopez-Moreno, J., Otaduy, M.A.: Efficient simulation of knitted cloth using persistent contacts. In: Proceedings of the 14th ACM SIGGRAPH/Eurographics Symposium on Computer Animation - SCA 2015, pp. 55–61. ACM Press, Los Angeles (2015)
4. Kaldor, J.M., James, D.L., Marschner, S.: Simulating knitted cloth at the yarn level. In: ACM SIGGRAPH 2008 Papers, pp. 65:1–65:9. ACM, New York (2008)
5. WiseTex Suite. https://www.mtm.kuleuven.be/Onderzoek/Composites/software/wisetex
6. Lomov, S.V.: WiseTex software suite (2005). https://www.mtm.kuleuven.be/Onderzoek/Composites/software/downloads/wisetex-overview.pdf
7. TexGen. http://texgen.sourceforge.net/index.php/Main_Page

8. Portalés, C., et al.: Interactive tools for the preservation, dissemination and study of silk heritage—an introduction to the SILKNOW project. Multimodal Technol. Interact. **2**, 28 (2018)
9. Unity. https://unity3d.com
10. Lomov, S.V., Huysmans, G., Verpoest, I.: Hierarchy of textile structures and architecture of fabric geometric models. Text. Res. J. **71**, 534–543 (2001)
11. Verpoest, I., Lomov, S.V.: Virtual textile composites software WiseTex: integration with micro-mechanical, permeability and structural analysis. Compos. Sci. Technol. **65**, 2563–2574 (2005)

A Comparison of Selected Variable Ordering Methods for NFA Induction

Tomasz Jastrząb$^{(\boxtimes)}$ (iD)

Institute of Informatics, Silesian University of Technology, Gliwice, Poland
Tomasz.Jastrzab@polsl.pl

Abstract. In the paper, we study one of the fundamental problems of grammatical inference, namely the induction of nondeterministic finite automata (NFA). We consider the induction of NFA consistent with the given sets of examples and counterexamples. We transform the induction problem into a constraint satisfaction problem and propose two variable ordering methods to solve it. We evaluate experimentally the proposed variable ordering methods and compare them with a state-of-the-art method. Additionally, through the experiments, we assess the impact of sample sets sizes on the performance of the induction algorithm using the respective variable ordering methods.

Keywords: Grammatical inference ·
Nondeterministic finite automata · Variable ordering ·
Constraint satisfaction

1 Introduction

In the paper, we deal with automata, which are important for numerous practical applications, such as compiler design, bioinformatics [20], and grammatical inference [19]. These automata are *finite*, *nondeterministic* and *minimal* in terms of the number of states. They are given by $A = (Q, \Sigma, \delta, q_0, Q_F)$, where Q is the finite set of states of the automaton, Σ is the input alphabet, $\delta : Q \times \Sigma \to 2^Q$ is the transition function, $q_0 \in Q$ is the initial state and $Q_F \subseteq Q$ is the set of final states [11]. The automata are also *consistent* with the given sample $S = (S_+, S_-)$, in which S_+ denotes the examples, and S_- denotes the counterexamples. The automaton A is said to be consistent with the sample S if for each word $w \in S_+$ there exists a sequence of transitions between state q_0 and at least one state $q \in Q_F$ and for each word $w \in S_-$ the condition does not hold.

In order to find a minimal nondeterministic finite automaton (NFA) consistent with the given sample S, we transform the induction problem into a constraint satisfaction problem (CSP) as discussed in [12,15,18]. Using the CSP

The preliminary version of this paper was presented at ICGI 2018 [13]. In the current paper, we substantially extended the algorithmic descriptions and made the experimental part much more comprehensive.

© Springer Nature Switzerland AG 2019
J. M. F. Rodrigues et al. (Eds.): ICCS 2019, LNCS 11540, pp. 741–748, 2019.
https://doi.org/10.1007/978-3-030-22750-0_73

formulation we ask whether for the given sample S and a given positive integer k there exists a k-state automaton consistent with S. By taking $k = 1, 2, \ldots$, we ensure that we find the *minimal* NFA for the given sample.

The main contributions of the paper are as follows. Firstly, we discuss two variable ordering methods devised specifically to solve the given CSP formulation of the problem. And secondly, we perform comprehensive experiments on input samples built of amino acid sequences from WaltzDB database [2]. To assess how the sizes of sets S_+ and S_- affect the performance of the induction algorithms, we consider samples for which $|S_+| \ll |S_-|$, $|S_+| = |S_-|$, and $|S_+| \gg |S_-|$ hold.

The paper is organized into 6 sections. In Sect. 2 we present a brief literature review. In Sect. 3 we recall the formulation of minimal NFA induction problem as a CSP. In Sect. 4 we present the parallel induction algorithm and the two proposed variable ordering methods. We present the results of conducted experiments in Sect. 5. Finally, Sect. 6 contains the conclusions.

2   State of the Art

The problem of finding a minimal NFA consistent with the given sample is hard. In particular, it cannot be done from polynomial time and data [6]. It was also shown that inducing a minimal NFA given a deterministic finite automaton (DFA) is PSPACE-complete [16].

The existing NFA induction methods focus around state merging algorithms, designed initially for DFAs. The merging is performed over a prefix tree acceptor constructed for the sample S. The algorithms remove redundant states keeping the consistency with the sample. Among the existing algorithms there are *DeLeTe2* [8], NRPNI [1], and the state merging methods proposed in [5,9]. Some algorithms transform the induction problem to CSPs [12,14,15,18].

The variable ordering methods are either *static* – defined before the algorithm begins, or *dynamic* – changing as the algorithm proceeds. Examples of static methods include *deg* [7] and *ddeg* heuristics based on the initial and current number of constraints involving the variable (degree). Dynamic methods include domain-size based method *dom* [10], as well as *dom-deg* [3], *dom-ddeg* [17] or *dom-futdeg* [4] methods, which follow the *dom* heuristic but in case of ties use the initial, current or future degree of the variables, respectively.

3   Problem Formulation

Since we assume the fixed number of states k, to find the NFA we only need to find the transition function δ and the set of final states Q_F. As the result, the following statements about the given CSP always hold [12,15,18]:

1. Let $l = |\Sigma|$ be the number of symbols in the alphabet. Then we need exactly $n = k^2 l + k$ binary variables x and y.
2. For each pair of states $p, q \in Q$ and each symbol $a \in \Sigma$, if $q \in \delta(p, a)$, then $x_i = 1$, for some $i \in [1, k^2 l]$. Likewise, $x_i = 0$ means that $q \notin \delta(p, a)$ holds.

3. For each state $p \in Q_F$, $y_i = 1$ holds for some $i \in [1, k]$. Otherwise, $y_i = 0$.
4. Let i, j, $i \neq j$ be the indices of x variables. Let $a, b \in \Sigma$, and $p, q, r \in Q$, $p \neq r, q \neq r$. Then the x variables are ordered as follows:
   (a) If $a < b$ in lexicographical order, then $i < j$ holds.
   (b) Given the transitions $p \xrightarrow{a} r$ and $q \xrightarrow{a} r$, if p precedes q in Q, then $i < j$.
   (c) Given the transitions $r \xrightarrow{a} p$ and $r \xrightarrow{a} q$, if p precedes q in Q, then $i < j$.
5. If p precedes q in Q, then $i < j$, where i, j are the y variables indices.

Example 1. Let the set of states be $Q = \{q_0, q_1\}$, and so $k = 2$, and let $\Sigma = \{a, b\}$. Then variables y_1 and y_2 correspond to the possible final states q_0 and q_1. Since $a < b$ lexicographically, and $q_0 < q_1$ in set Q, then the following relations hold: variable x_1 corresponds to the transition $q_0 \xrightarrow{a} q_0$, variable x_2 corresponds to the transition $q_0 \xrightarrow{a} q_1$, ..., variable x_8 corresponds to the transition $q_1 \xrightarrow{b} q_1$.

Given the above considerations, a *path* is a product of x variables followed by a y variable. Using Example 1, the product $x_1 x_6 x_8 y_1$ corresponds to the transitions $q_0 \xrightarrow{a} q_0$, $q_0 \xrightarrow{b} q_1$, and $q_1 \xrightarrow{b} q_1$, i.e., a word abb ending in state q_1.

Let us note that for the given number of states k and for each word w, there are exactly $k^{|w|}$ paths on which this word can be spelled out. There are also $k^{|w|}$ products of length $|w| + 1$ corresponding to these paths. In the sequel, we use P_{mi}^w to denote the m-th product of x variables corresponding to word w, and ending in variable y_i. Hence an NFA is consistent with the sample S when:

1. for all words $w \in S_+ \setminus \{\lambda\}$ it holds that (examples acceptance):

$$\sum_{i=1..k} \sum_{m=1..k^{|w|-1}} P_{mi}^w y_i = 1, \tag{1}$$

2. for all words $w \in S_- \setminus \{\lambda\}$ it holds that (counterexamples rejection):

$$\sum_{i=1..k} \sum_{m=1..k^{|w|-1}} P_{mi}^w y_i = 0. \tag{2}$$

Note that the summation is performed according to the rules of Boolean algebra. Note that (1) is satisfied iff there exists a path for word w that ends in a final state. Note also that (2) is satisfied iff either there is no path for word w, or the path ends in a non-final state. Finally, if $\lambda \in (S_+ \cup S_-)$ holds, then we set:

$$y_1 = \begin{cases} 1, \text{for } \lambda \in S_+, \\ 0, \text{for } \lambda \in S_-. \end{cases} \tag{3}$$

The above condition allows to reduce the solution space to be searched [14].

4   Algorithms and Methods

Hereinafter, let C be the set of constraints, $c_+ \in C$ be a constraint given by (1), and $c_- \in C$ be a constraint given by (2). Let $|c|$ be the number of *active* (non-zero) products P in (1) or (2), and $d(c, x)$ be the number of active products in c

▷ **input:** i, v – index and value of the most recently set variable, initially empty
▷ **global:** f – flag showing that a solution was found, initially **false**
1: **procedure** INDUCENFA(i, v)
2:     **if** $i \neq \emptyset$ **and** $v \neq \emptyset$ **then** $f, p \leftarrow$ EVALUATE(i, v)
3:     **if** $f =$ **true then** print solution; **return**   ▷ solution found
4:     **if** $p =$ **false then return**   ▷ contradiction found – backtracking needed
5:     $i, v \leftarrow$ REORDER()
6:     INDUCENFA(i, v)
7:     **if** $f =$ **false then** INDUCENFA($i, !v$)   ▷ assigns opposite value to x_i
8:     **if** $f =$ **false then** unset x_i
9: **end procedure**

Fig. 1. The induction algorithm

▷ **input:** i, v – index and value of the most recently set variable
1: **procedure** EVALUATE(i, v)
2:     **if** $v = 0$ **then**
3:         **for all** $c_+ : c_+$ is not satisifed **and** $d(c_+, x_i) > 0$ **do**
4:             $|c_+| \leftarrow |c_+| - d(c_+, x_i); \forall x_j : x_i, x_j \in P \wedge P \in c_+$, update $d(c_+, x_j)$
5:         **end for**
6:         **if** $\exists c_+ : |c_+| = 0$ **then** $p \leftarrow$ **false else** $p \leftarrow$ **true**
7:     **else**
8:         **if** $\exists c_- : d(c_-, x_i) > 0 \wedge \exists P \in c_- : P = 1$ **then** $p \leftarrow$ **false; return** f, p
9:         $\forall c_+ : d(c_+, x_i) > 0 \wedge \exists P \in c_+ : P = 1$, mark c_+ as satisfied
10:        **if** all c_+ are satisfied **then** $f \leftarrow$ **true;** $\forall x_j : x_j = \emptyset, x_j \leftarrow 0$ **else** $p \leftarrow$ **true**
11:    **end if**
12:    **return** f, p
13: **end procedure**

Fig. 2. The EVALUATE procedure for *min-max-ex* variable ordering method

containing variable x. The variable ordering methods pertain to the x variables. The y variables are set first to produce independent instances of the CSP, solved in parallel [12, 14]. It also applies to the *deg* method used in the experiments.

The induction procedure INDUCENFA, executed for each independent instance of the CSP, is given in Fig. 1. The procedure consists of the evaluation phase (represented by the EVALUATE procedure in line 2), and the ordering phase (represented by the REORDER procedure in line 5). They both differ depending on the selected variable ordering method.

The *min-max-ex* Method. The EVALUATE procedure (Fig. 2) implements a 'fail-fast' behavior, by checking first the constraints that may result in a contradiction (lines 6 and 8). It aims at explicitly satisfying constraints $c_+ \in C$ (line 9), setting the other x_i variables to zeros if all c_+ are satisfied (line 10). It updates $|c_+|$ and $d(c_+, x_j)$ using the techniques described in [12] (lines 3–5).

The REORDER procedure (Fig. 3) selects a constraint $c'_+ \in C$, for which $|c'_+|$ is minimal (line 2) and the index of the most frequent variable $x_i \in c'_+$ (line 3). It sets v to zero (line 4). The ordering makes the evaluation shorter, by choosing constraint with fewer products and enforcing the 'fail-fast' behavior.

```
1: procedure REORDER
2:     c'_+ ← argmin_{c_+ ∈ C} |c_+|          ▷ considers only not-yet-satisifed c_+
3:     i ← argmax_{x_i ∈ c'_+} d(c'_+, x_i)   ▷ considers only variables x_i : x_i = ∅
4:     return i, 0
5: end procedure
```

Fig. 3. The REORDER procedure for *min-max-ex* variable ordering method

The *min-max-cex* Method. The second variable ordering method, called *min-max-cex*, differs from *min-max-ex* in that, that in the EVALUATE procedure, for $v = 1$ it only checks for a contradiction in any $c_- \in C$. For $v = 0$ it checks for a contradiction in any $c_+ \in C$, it updates the $|c_-|$ and $d(c_-, x_j)$ values as in lines 3–5, and checks for satisfied constraints c_-. If all c_- are satisfied it sets all unset x_j variables to 1. In the REORDER procedure, in line 2 we look for a constraint $c_- \in C$, such that $|c_-|$ is minimal, and we propose that $v \leftarrow 1$ in line 4.

The *deg* Method. In the EVALUATE procedure the *deg* method first checks for any contradictions, either in c_+ or in c_- (for $v = 0$ or $v = 1$, respectively), and then it checks for satisfied constraints. We require that all constraints are explicitly satisfied. Since the method is static, the order of variables is established before the induction begins. So the REORDER procedure merely returns the next variable according to the already known order. It sets $v \leftarrow 0$ in line 4 of Fig. 3.

Example 2. Due to space limitations, we provide an example trace of the induction algorithm at https://github.com/tjastrzab/iccs.

5   Experimental Evaluation

We generated 450 samples based on the sets of amino acid sequences [2]. The number of sequences in set S_+ (resp. S_-) was 5 (resp. 45), 25 (resp. 25), and 45 (resp. 5), for the samples, for which $|S_+| \ll |S_-|$, $|S_+| = |S_-|$, and $|S_+| \gg |S_-|$ hold. The algorithms were implemented in Java and run on an Intel Xeon E5-2640 2.60 GHz processor (16 cores, 8 GB RAM). The time limit was 3 hours. We considered two scenarios. In *Exp. 1*, we sought the first consistent k-state automaton. In *Exp. 2*, the configuration of the final states was also given to see how the algorithms perform when searching for a specific NFA. All minimal automata had 2 states.

The distribution of the run times, given by the minimum, maximum and average values, is shown in Table 1. Note that in all cases, the average times for the best- and worst-performing methods, differ by an order of magnitude or more. Moreover, the balanced samples are the hardest to solve, since the average times are the longest for both *Exp. 1* and *Exp. 2*. The *min-max-ex* method prevails when $|S_+| \ll |S_-|$ holds, while *min-max-cex* is the fastest for the cases in which $|S_+| \gg |S_-|$ is true. This is expected since the time spent in the REORDER procedure is then much shorter (as the ordering is based on S_+ or S_- only). The *min-max-cex* method is also the fastest for the hardest sample type. This proves that the new methods are competitive with respect to the *deg* method.

Table 1. Minimum, maximum and average run times (in seconds)

Sample type	min-max-ex		deg		min-max-cex	
	Exp. 1	Exp. 2	Exp. 1	Exp. 2	Exp. 1	Exp. 2
Minimum and maximum run times						
$|S_+| \ll |S_-|$	0.0, 0.2	0.0, 10.8	0.0, 32.9	0.0, 5752.7	0.0, 5795.5	0.0, 5678.8
$|S_+| = |S_-|$	0.0, 8083.2	0.0, 8058.2	0.2, 4681.9	0.3, 4612.5	0.1, 25.5	0.1, 118.8
$|S_+| \gg |S_-|$	0.0, 7327.1	0.0, 9085.7	0.0, 2652.3	0.0, 8618.0	0.0, 1.0	0.0, 1.8
Average run times						
$|S_+| \ll |S_-|$	0.0	0.1	0.5	58.7	79.2	229.9
$|S_+| = |S_-|$	460.6	647.7	176.4	279.9	4.6	7.6
$|S_+| \gg |S_-|$	129.6	471.8	49.6	269.4	0.2	0.2

To explain the differences in the run times, we counted the number of calls of the INDUCENFA procedure, i.e., the number of visited solution tree nodes. We observed that the number of calls differed by two to four orders of magnitude.

Finally, to observe the relation between the run time performance and the automaton size, we counted the number of transitions in each NFA. We observed that *min-max-ex* and *deg* methods produce NFAs of similar size, while the *min-max-cex* method generates on average 17–30 transitions more. It is so because *min-max-ex* and *deg* methods produce mostly the transitions necessary to accept the examples. The *min-max-cex* method satisfies all counterexamples and makes the remaining variables equal to one, which generates more transitions.

6   Conclusions

In the paper we have investigated the problem of nondeterministic finite automata induction. The experiments have shown that the proposed variable ordering methods perform better than the state-of-the-art one, especially in case of imbalanced sizes of sets S_+ and S_-. The result is important since it is not uncommon that, for instance, we know just a few factors causing a disease (set S_+) and much more factors that are not responsible for this particular disease (set S_-). Hence, being able to classify these factors efficiently and correctly using the induced NFAs, can be of help in some bioinformatics tasks.

Acknowledgment. The research was supported by National Science Centre Poland (NCN), project registration no. 2016/21/B/ST6/02158.

References

1. Alvarez, G., Ruiz, J., Cano, A., García, P.: Nondeterministic Regular Positive Negative Inference NRPNI. In: Proceedings of CLEI 2005, pp. 239–249 (2005)
2. Beerten, J., Van Durme, J., Rousseau, F., Schymkowitz, J.: WALTZ-DB. Database of amyloid forming peptides (2014). http://waltzdb.switchlab.org/
3. Bessière, C., Régin, J.-C.: MAC and combined heuristics: two reasons to forsake FC (and CBJ?) on hard problems. In: Freuder, E.C. (ed.) CP 1996. LNCS, vol. 1118, pp. 61–75. Springer, Heidelberg (1996). https://doi.org/10.1007/3-540-61551-2_66
4. Brélaz, D.: New methods to color the vertices of a graph. Commun. ACM **22**(4), 251–256 (1979)
5. Coste, F., Fredouille, D.: Unambiguous automata inference by means of state-merging methods. In: Lavrač, N., Gamberger, D., Blockeel, H., Todorovski, L. (eds.) ECML 2003. LNCS (LNAI), vol. 2837, pp. 60–71. Springer, Heidelberg (2003). https://doi.org/10.1007/978-3-540-39857-8_8
6. de la Higuera, C.: Characteristic sets for polynomial grammatical inference. Mach. Learn. **27**, 125–138 (1997)
7. Dechter, R., Meiri, I.: Experimental evaluation of preprocessing techniques in constraint satisfaction problems. In: Proceedings of IJCAI 1989, pp. 271–277. Morgan Kaufmann Publishers Inc., San Francisco (1989)
8. Denis, F., Lemay, A., Terlutte, A.: Learning regular languages using RFSAs. Theor. Comput. Sci. **313**(2), 267–294 (2004)
9. García, P., Vázquez Parga, M., Alvarez, G., Ruiz, J.: Universal automata and NFA learning. Theor. Comput. Sci. **407**(1–3), 192–202 (2008)
10. Harallick, R.M., Elliot, G.L.: Increasing tree search efficiency for constraint satisfaction problems. Artif. Intell. **14**, 263–313 (1980)
11. Hopcroft, J., Ullman, J.: Introduction to Automata Theory, Languages, and Computation. Addison-Wesley Publishing Company (1979)
12. Jastrząb, T.: On parallel induction of nondeterministic finite automata. Procedia Comput. Sci. **80**, 257–268 (2016)
13. Jastrząb, T.: Performance evaluation of selected variable ordering methods for NFA induction (2018). http://icgi2018.pwr.edu.pl/public/ex-abstracts/jastrzab18.pdf
14. Jastrząb, T.: Two parallelization schemes for the induction of nondeterministic finite automata on PCs. In: Wyrzykowski, R., Dongarra, J., Deelman, E., Karczewski, K. (eds.) PPAM 2017. LNCS, vol. 10777, pp. 279–289. Springer, Cham (2018). https://doi.org/10.1007/978-3-319-78024-5_25
15. Jastrząb, T., Czech, Z.J., Wieczorek, W.: Parallel induction of nondeterministic finite automata. In: Wyrzykowski, R., Deelman, E., Dongarra, J., Karczewski, K., Kitowski, J., Wiatr, K. (eds.) PPAM 2015. LNCS, vol. 9573, pp. 248–257. Springer, Cham (2016). https://doi.org/10.1007/978-3-319-32149-3_24
16. Jiang, T., Ravikumar, B.: Minimal NFA problems are hard. SIAM J. Comput. **22**(6), 1117–1141 (1993)
17. Smith, B.M., Grant, S.A.: Trying harder to fail fast. In: Proceedings of ECAI 1998, pp. 249–253. Wiley (1997)
18. Wieczorek, W.: Induction of non-deterministic finite automata on supercomputers. In: Proceedings of ICGI 2012, vol. 21, pp. 237–242 (2012)

19. Wieczorek, W.: Grammatical Inference. SCI, vol. 673. Springer, Cham (2017). https://doi.org/10.1007/978-3-319-46801-3
20. Wieczorek, W., Unold, O.: Use of a novel grammatical inference approach in classification of amyloidogenic hexapeptides. Comput. Math. Methods Med. **2016**, 8 (2016). article ID 1782732

Traffic3D: A Rich 3D-Traffic Environment to Train Intelligent Agents

Deepeka Garg$^{(\boxtimes)}$, Maria Chli, and George Vogiatzis

Aston University, Birmingham B4 7ET, UK
{gargd,m.chli,g.vogiatzis}@aston.ac.uk

Abstract. The last few years marked a substantial development in the domain of Deep Reinforcement Learning. However, a crucial and not yet fully achieved objective is to devise intelligent agents which can be successfully taken out of the laboratory and employed in the real world. Intelligent agents that are successfully deployable in true physical settings, require substantial prior exposure to their intended environments. When this is not practical or possible, the agents benefit from being trained and tested on powerful test-beds, effectively replicating the real world. To achieve traffic management at an unprecedented level of efficiency, in this paper, we introduce a significantly richer new traffic simulation environment; Traffic3D. Traffic3D is a unique platform built to effectively simulate and evaluate a variety of 3D-road traffic scenarios, closely mimicking real-world traffic characteristics including faithful simulation of individual vehicle behavior, precise physics of movement and photo-realism. We discuss the merits of Traffic3D in comparison to state-of-the-art traffic-based simulators. Along with deep reinforcement learning, Traffic3D facilitates research across various domains such as object detection and segmentation, unsupervised representation learning, visual question answering, procedural generation, imitation learning and learning by interaction.

Keywords: Virtual reality 3D traffic simulator ·
Intelligent transportation systems · Machine learning · Deep learning

1 Introduction

Training agents to autonomously act in real-world settings involves challenges that transcend beyond the competency of commonly-used annotated data in a typical supervised learning setting. In control optimization problems such as traffic infrastructure optimization, the control strategies that lead to the optimal solution are not obvious at the outset. Equipping an agent with the ability to learn and discover effective solutions, enables it to be adaptable to a variety of settings. However, learning often depends on aggressive exploration of unknown space; as to generalize well in an unseen 3D environment, an agent is required to demonstrate robustness to different variations of the environment it is to be

© Springer Nature Switzerland AG 2019
J. M. F. Rodrigues et al. (Eds.): ICCS 2019, LNCS 11540, pp. 749–755, 2019.
https://doi.org/10.1007/978-3-030-22750-0_74

exposed to. Consequently, to be able to perform well within a dynamic physical environment, an agent is required to observe and learn from a considerable set of environment states, where the agent is also able to perceive the outcome of its behavior by receiving feedback from the environment it is interacting with. Real-world physical environments fulfill these requirements, but they are expensive and unsafe to train in and are hard to scale. Especially in the traffic optimization case, among others, safety and economic cost concerns in obtaining sufficing amount of diversity (different traffic conditions and different seasons) in the training data preclude this vital training from being carried out in situ.

(a)

(b)

(c)

(d)

Fig. 1. Different views of a junction in Traffic3D. (a) Top-down view. (b) A night scene. (c) A rainy day. (d) A snowy day.

In addition, deep learning solutions (such as deep neural networks), the state-of-the-art paradigm used to effectively train agents to autonomously accomplish tasks (such as autonomous traffic infrastructure optimization) are known to depend on a large amount of training data to achieve peak performance, which further makes training an autonomous agent in the physical world impractical. An alternative is to train agents in simulations, providing a viable environment for protocol development.

Agents trained in more realistic settings are known to learn features which are generalizable to their corresponding real-world environments [1]. However, the most-widely used state-of-the-art traffic simulators fail to deliver crucial functionalities that are indispensable for authentic traffic simulation [2]. They lack in realism, diversity, stochasticity and perception challenges of the real-world.

To bridge the gap between simulations and real-world traffic specifications, we created a traffic micro-simulation tool; Traffic3D. Traffic3D is an interactive 3D-environment designed to precisely simulate traffic entities, their interactions and their emergent physical and visual properties.

Table 1. Comparison between different traffic-based and deep learning-based simulation environments.

Environment	Suitable for traffic simulation	Photo-realistic	3D	Physics	Customizable
SUMO [2]	Yes	No	No	Yes (with restrictions^a)	Yes (with restrictions^b)
VISSIM [2]	Yes	Yes	Yes	Yes (with restrictions^c)	Yes (with restrictions^d)
TORCS [3]	Yes	Yes	Yes	Yes (with restrictions^e)	Yes (with restrictions^f)
Virtual KITTI [1]	Yes	Yes	Yes	No^g	Yes
CHALET [4]	No	Yes	Yes	Yes	Yes
AI2-THOR [5]	No	Yes	Yes	Yes	Yes
ATARI [6]	No	No	No	No	No
DeepMind Lab [7]	No	No	Yes	No	Yes
Traffic3D	**Yes**	**Yes**	**Yes**	**Yes (with high degree of realsim)**	**Yes (fully)**

^{a} no proper reactive control to random incidents like collisions between vehicles.
^{b} does not support simulation of autonomous vehicles and does not prioritize public transport.
^{c} unrealistic lane-closing behavior.
^{d} restrictions in customizing delay.
^{e} limited sensor suite.
^{f} does not support road intersection simulation.
^{g} information not available.

2   Related Work

Traffic optimization being an established field, a number of traffic simulators have been developed. Also, a handful of deep learning platforms are available for training and benchmarking purposes of deep learning agents, mostly built around computer game environments. Pell et al. [2] intensively reviewed a comprehensive set of traffic simulation platforms. The review acknowledges that none of the currently-used traffic simulation tools are able to adequately deliver important functionalities that are fundamental to faithful traffic simulation. The existing traffic models lack in detail and flexibility, and we believe that a detailed network model with efficient real-time traffic information collection capabilities is necessary to simulate heterogeneous transportation networks. In Table 1, we summarize the capabilities of a few widely-used traffic and deep learning-based simulation environments over important simulation characteristics; photo-realistic graphical rendering, 3D nature of objects, simulation physics and flexibility to customize the simulation environment according to the requirements of the application. It is evident that no single simulator supports comprehensive traffic-based research and analysis. In contrast, Traffic3D provides a flexible platform to facilitate research for novel traffic optimization protocol development and analysis in an environment that is as close as possible to a physical one.

3   Our Simulation Environment: Traffic3D

The main contribution of this paper is our gamified simulator; Traffic3D. Traffic3D consists of a diverse range of traffic scenes including a variety of photo-realistic vehicles (such as emergency vehicles, personal and public transport vehicles) and street furniture (sidewalks and traffic lights etc.). Scenes include, from the clear view of a sunny day, to the blurred scene of a dimly-lit night, a rainy and a snowy day (as shown in Fig. 1). We use real-world traffic images as a reference to create 3D-traffic scenes with photo-realistic lighting and texture. To facilitate control optimization, we developed a framework to enable an agent perform actions in a 3D-traffic environment and subsequently, perceive the outcomes of its actions. Our framework includes integration of Unity engine with deep learning python support. Figure 1 shows different views Traffic3D's junction's graphical display. In the next couple of paragraphs we discuss architectural design and performance capabilities of Traffic3D.

3.1   Traffic3D's Design and Architecture

Traffic3D is built of two elements; (1) the traffic environment built within the Unity Game Development Platform[1] and (2) a light-weight python API that supports online learning. We designed this architecture to facilitate implementation of different control algorithms for various traffic-based and general applications. The traffic simulation environment and the learning agent (implemented in python) form a client-server system. The traffic environment acts as a client while the learning agent acts as server. The learning agent listens for http requests from the traffic environment and responds immediately as soon as a request is made. Figure 2 illustrates Traffic3D's traffic environment and the learning agent's bilateral interaction pipeline.

---

[1] https://unity3d.com/.

3.2   Traffic3D's Performance

The socket data exchange mechanism we implemented to facilitate interaction between the simulation and the deep learning python module works accurately and seamlessly. However, the deep learning training is typically long. In order to accelerate training, Traffic3D allows control of the frame rate and speed of the simulation without affecting its quality. Collecting data using an accurate simulator is still faster and safer to than doing so directly from the physical world [8]. At the same time, to boost Traffic3D's performance in computationally-intensive machine learning tasks, we are currently exploring an alternative architecture version for Traffic3D. Under this architecture, the simulation engine will be running as a powerful server application, interacting with another server which is responsible for the machine learning. Visualization software, such as Unity, will act as a client to the simulation server.

4   Traffic3D's Properties

Here, we discuss some of the key simulation properties of Traffic3D. This section highlights Traffic3D's potential in creating natural-looking and realistically-operating traffic environment with believable visuals and physics.

4.1   Traffic3D's Physical Properties

To ensure high-precision simulation, Traffic3D allows effective modelling of complex physical interactions between the transportation entities based on mass, friction and other forces such as gravity. Within the Traffic3D environment, vehicles are independently tuned to react appropriately accordingly to their input parameters and also to the presence of other vehicles and traffic infrastructure (such as their progressive slow-down and acceleration). Vehicles in Traffic3D are well-equipped to behave realistically. They have appropriate sensors (i.e. ray-cast sensors which can be customized to use as an equivalent to Lidar and microwaves) and programmed behaviour to avoid collisions, overtake and give way to emergency vehicles.

4.2   Traffic3D's Visual Properties

Unlike most of the state-of-the-art traffic simulators (such as SUMO), Traffic3D provides high-definition graphical rendering facilitating transfer of developed models to the real-world with no or minimum pre-training. In Traffic3D, light, material, texture and scale work in synergy to make digital content look as close to a real scene as possible. The lighting techniques used in Traffic3D create immersive scene lighting, mimicking the real-world effects. Objects' shadows can be dynamically cast on the scene, adding further realism to the environment.

4.3   Traffic3D's Diversity of Parameters

To ensure stability and generalizability of agents trained using Traffic3D to different variations of the environment, Traffic3D supports plug-n-play architecture such that distinct scenes can be reliably incorporated. Essentially, Traffic3D facilitates seamless creation and switching between different types of scenes during runtime without interrupting the simulation or causing frame-rate hiccups. For example, traffic density can be dynamically varied independently, or around different times of the day (rush and quiet hours), as well as different weather conditions (e.g. rain, snow) during run-time, to study quantitatively the impact of different times and weather conditions on the learning agent's behavior. The current level of diversity offered by Traffic3D includes: (1) vehicle number, types, models, colors and sizes, (2) vehicles' trajectory configuration and speed, (3) different road layouts and surface textures, (4) variety of street furniture, (5) different lighting and weather conditions and (6) different camera position and orientation to comprehensively capture a wide range of possible aspects of a traffic scene.

4.4   Traffic3D's Sensory Data Collection Capabilities

Traffic3D facilitates fast, inexpensive, limitless, diverse and photo-realistic collection of traffic data for various research purposes such as online reinforcement learning and supervised learning. Cameras in Traffic3D replicate the operation of real-world cameras; supporting capture of photo-realistic images and videos with proper field-of-view and depth-of-view. To study the effects of occlusions in the traffic environment, which is common in real-world traffic scenarios; vehicles and street furniture that are not currently being seen by the camera can have their rendering disabled. Multiple cameras can be deployed at the same time within a scene to perceive different aspects of the scene and their views can be combined in different ways. For example, in addition to the main view of the complete scene, when simulating a car, camera output to show a rear view mirror footage can also be simulated separately, if needed.

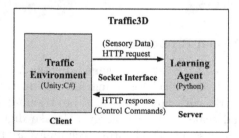

Fig. 2. The design architecture of Traffic3D encompassing the simulation environment and a python deep learning agent.

4.5   Traffic3D's Reusability

Traffic3D consists of a library of pre-created traffic environments. In addition, it offers complete flexibility over the deployment of new simulation scenarios; allowing complete customization over vehicle spawning, vehicle routing, road layouts and street furniture placement. Apart from creating simulation scenes manually, which can be tedious and less intuitive, Traffic3D also allows its users to programmatically generate scenes, even during run-time. Any of the traffic elements (such as vehicles) available within the project can be programmatically placed at any position and their physical (such a vehicle's size, mass and its speed) and visual behavior (such as a vehicle's texture) can also be programmatically configured.

5   Conclusion

This paper introduced our novel traffic simulation tool; Traffic3D. Traffic3D is created to build visually and physically intelligent traffic simulation models to build and test new technology with the goal to its eventual deployment in the real-world. It supports unique simulation features such as complex physical phenomenon, creation of relevant content such as traffic objects with appropriate background and photo-realism, comprehensibility, robustness, adaptability, partial observability challenges and inexpensive collection of diverse training data. Going forward, to further improve realism and preserve objects' geometry and other aspects, we intend to incorporate photogrammetry to enable automatic creation of traffic environments within Traffic3D from real-world traffic images and videos.

References

1. Gaidon, A., Wang, Q., Cabon, Y., Vig, E.: Virtual worlds as proxy for multi-object tracking analysis. In: Proceedings of the IEEE Conference on Computer Vision and Pattern Recognition, pp. 4340–4349 (2016)
2. Pell, A., Meingast, A., Schauer, O.: Trends in real-time traffic simulation. Transp. Res. Procedia **25**, 1477–1484 (2017)
3. Wymann, B., Espié, E., Guionneau, C., Dimitrakakis, C., Coulom, R., Sumner, A.: TORCS, the open racing car simulator. Software, vol. 4, no. 6 (2000). http://torcs.sourceforge.net
4. Yan, C., Misra, D., Bennnett, A., Walsman, A., Bisk, Y., Artzi, Y.: CHALET: cornell house agent learning environment. arXiv preprint arXiv:1801.07357 (2018)
5. Kolve, E., Mottaghi, R., Gordon, D., Zhu, Y., Gupta, A., Farhadi, A.: AI2-THOR: an interactive 3D environment for visual AI. arXiv preprint arXiv:1712.05474 (2017)
6. Bellemare, M.G., Naddaf, Y., Veness, J., Bowling, M.: An evaluation platform for general agents: the arcade learning environment. J. Artif. Intell. Res. **47**, 253–279 (2013)
7. Beattie, C., et al.: Deepmind lab. arXiv preprint arXiv:1612.03801 (2016)
8. Zhu, Y., et al.: Target-driven visual navigation in indoor scenes using deep reinforcement learning. In: 2017 IEEE International Conference on Robotics and Automation (ICRA), pp. 3357–3364. IEEE (2017)

Energy Efficiency Evaluation
of Distributed Systems

James Phung[1(✉)], Young Choon Lee[2], and Albert Y. Zomaya[1]

[1] School of Computer Science, The University of Sydney, Sydney, Australia
{james.phung,albert.zomaya}@sydney.edu.au
[2] Department of Computing, Macquarie University, Sydney, Australia
young.lee@mq.edu.au

Abstract. Rapid growth in Big Data and Cloud technologies has fueled rising energy demands in large server systems such as data centers, leading to a need for effective power management. In this paper, we investigate the energy consumption characteristics of data-intensive distributed applications in terms of the CPU and memory subsystem. To this end, we develop *PowerSave* as a lightweight software framework that enables dynamic reconfiguration of power limits. PowerSave uses Running Average Power Limit (RAPL) to impose power limits. Our evaluation study, conducted on three different real systems, demonstrates that for workloads typical of servers used in data centers, higher power caps correlate with higher overall CPU energy use.

Keywords: Big Data · Cloud · Energy monitoring ·
Power management · RAPL · Virtualization

1 Introduction

The rise of Cloud technologies and Big Data has led to rapid growth in storage and computational power requirements, and consequently, energy demands. An example of this phenomenon is the popularity of virtual computing services provided by Cloud service providers such as Amazon Web Services. These services range from database services to bare metal processing power. Likewise, scientific and engineering applications such as the Montage astronomy application [6] impose enormous processing and data demands, requiring that large amounts of real-time data (often on a terabyte scale) be collected and processed in a short period of time.

In the interests of economic and environmental sustainability, it is desirable to design large-scale server systems with the goal of optimizing energy use. To this end, various energy-aware scheduling algorithms [4,9], power models [3,7,11] and techniques to reduce power consumption of data centers [2,10,12] have been developed.

James Phung's work was supported by an Australian Government Research Training Program (RTP) Scholarship.

In this paper, we investigate the impact of CPU power limiting on CPU energy consumption and performance for a Macbook Pro laptop and two different types of Amazon EC2 virtualized instances. We use the RAPL feature of modern Intel CPUs to implement power limiting and monitor real-time energy use.

In order to streamline the process of setting power limits, we have developed PowerSave, a power management framework that provides a user-friendly interface for setting or locking various RAPL power limits. Also, we use an RAPL energy tracker that has been developed in-house in order to monitor the RAPL counters in real-time.

Specific contributions of this paper are as follows:

1. We develop PowerSave and an RAPL energy tracker to streamline the energy monitoring procedures; and
2. We evaluate the impact of CPU power limiting on CPU energy consumption and performance on a laptop, a virtual machine and a cluster of bare metal instances using a variety of big data benchmarks (HiBench benchmarks [5]).

We use HiBench benchmarks to investigate the impacts of CPU power limiting. We use various combinations of CPU power limits, turbo setting and benchmarks.

Our evaluation shows that there is a correlation between CPU power limits and CPU energy efficiency when CPU utilization is high. Energy efficiency rapidly decreases if the CPU consumes power beyond the thermal design power for sustained periods.

The remainder of this paper is organized as follows. Section 2 discusses related work. Section 3 describes the basic functionality of our power management framework *PowerSave*. Section 4 describes the implementation details of *PowerSave*. Section 5 describes our experimental setup, evaluation methodology and results for our experiments. We draw our conclusions in Sect. 6.

2   Related Work

The RAPL feature of Intel CPUs has been applied by various researchers in a range of energy efficiency and monitoring applications. In [13], the problem of achieving proportionality in energy usage relative to workload is explored. The author proposes "a methodology to efficiently cap the power consumption of applications while meeting strict performance targets". To confirm the validity of the power-capping scheme, the author runs the SPECweb benchmark for various workloads. During the benchmark, a Watts Up power meter is used to measure the energy usage of the entire system while subsystem energy use is measured using the RAPL interface's energy meters.

In [8], Khan et al. model the system-level power consumption of two different computers using RAPL counters: one contains an Intel Haswell CPU and the other contains an Intel Skylake CPU. They establish that there is a strong correlation between system power usage and CPU power usage. Also, they investigate important issues such as performance overhead, sampling rate, and potential register overflows. Furthermore, they conduct their investigations on some non-bare metal Amazon AWS EC2 instances.

In [2], the researchers develop a new MapReduce workload manager to address the problem of processing batch MapReduce workloads efficiently in situations where some MapReduce jobs have to be performed on demand with little tolerance for latency such as an ad-hoc database query. To validate the workload manager's performance, the researchers performed simulations involving Facebook MapReduce traces. In addition, a live experiment was performed on Amazon EC2.

Energy-aware resource allocation is also studied in [1]. Unlike [2], the authors of [1], consider the problem of allocating virtual machines to servers and virtual machine migration in order to minimize energy use. New algorithms are designed and tested by simulation.

3   PowerSave

PowerSave[1] is a lightweight software framework that allows the user to specify power limits for the CPU package, PP0, PP1 and DRAM using a text interface. Two CPU package power limits can be set: a short-term power limit and long-term power limit. Each of these limits is defined in terms of maximum average power consumption over a time window; both the average power value and time window are set by the user. It may be used in physical systems that feature an RAPL enabled CPU; also, it may be used in bare metal and virtualized instances if RAPL support is available. Thus, PowerSave enables rapid testing of the impact that different CPU power limits have on energy consumption and application performance in distributed systems.

4   Implementation

In this section, we describe the implementation details of our PowerSave framework and RAPL energy tracker.

4.1   PowerSave

Our PowerSave framework consists of the RAPLCap library and a C program PowerSet that provides a user-friendly interface. The framework can operate in a variety of UNIX environments.

The C program uses the RAPLCap library to read and write to RAPL registers that control the power limiting behaviors of the CPU and memory subsystem. When the user runs the C code by invoking it on the UNIX command line, it checks to see if the CPU supports RAPL. If RAPL support is not found, the program aborts with an error message; otherwise, the user is presented with the

---

[1] https://github.com/SydneyDSysGuy/PowerSave/blob/master/README.md.

current RAPL power limits for the CPU package, PP0, PP1 and DRAM. Also, the user is notified as to whether the power limits are locked or not. If the power limits are unlocked, the user is prompted to enter new power limits if they wish to change them; also, the user is offered the choice of locking the power limits (an operation that cannot be reversed without restarting the system). Whenever the user requests a power limit change, the appropriate RAPL register is updated immediately.

4.2   RAPL Energy Tracker

Our RAPL energy tracker is a C program that can be run in a variety of UNIX environments. The CPU's RAPL counters are accessed via an interface provided by the UNIX *msr* (model specific registers) driver that provides raw access to RAPL counters. RAPL counters are sampled at regular intervals with a nominal sampling rate of one sample per second. RAPL energy readings are exported to a specified text file in real-time. The user is notified of the energy usage in real-time via the console.

5   Experiments

In this section, we describe our experimental setup, evaluation of our hypothesis, and results.

5.1   Experimental Setup

For our experiments, we use three different systems: a 2017 model Macbook Pro containing an Intel Mobile Core i5 Kaby Lake 2.3 GHz CPU (hereafter called "Macbook Pro"), an Amazon AWS EC2 i3.metal instance (hereafter called "AWS i3.metal") and a cluster of 11 Amazon AWS EC2 m4.large instances (hereafter called "AWS m4.large"). For the Macbook Pro machine, we run our experiments directly on the machine; we do not use any virtualization technique. For the Amazon AWS instances, we run them on an *on-demand* basis; we do not use reserved spot instances or dedicated server reservations.

We use the RAPL energy tracker as described in Subsect. 4.2 to capture RAPL energy values in real-time. A sampling rate of one measurement per second is used. The readings are saved to comma separated values (CSV) files for later offline analysis and evaluation.

The PowerSet utility is used to set the CPU RAPL power caps to the required values. The *cpufreq-set* utility is used to adjust the CPU frequency to the appropriate value and toggle the turbo boost feature as required. Idle time is allowed between experiments to permit the CPU (and system) to cool to its normal idle temperature.

5.2   Energy Efficiency Evaluation

To verify our hypothesis that there is an optimum RAPL power limit for minimizing energy use, we use a series of real-world MapReduce workloads. These workloads are representative of real-world data analytics jobs. They are drawn from the HiBench benchmark suite [5]. They are Aggregate, Join, PageRank, Scan, Sort, TeraSort and WordCount.

We run our experiments for a range of RAPL CPU package power limits in the case of the Macbook Pro and AWS i3.metal instance, or for various numbers of Hadoop DataNodes in the case of the AWS m4.large cluster. We run them without turbo boost when a power limit has been set; otherwise, we run them with turbo boost. The workload is allowed to run to completion.

5.3   Results

Figures 1, 2 and 3 show the average RAPL counter readings for CPU package and DRAM for various workloads, RAPL CPU package power limit and turbo settings running on the Macbook Pro, AWS i3.metal instance and AWS m4.large virtual cluster. For the Macbook Pro, average RAPL counter readings for PP0 are also shown.

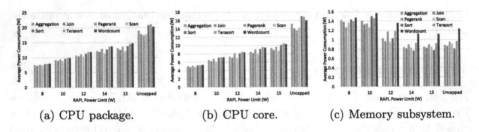

(a) CPU package.   (b) CPU core.   (c) Memory subsystem.

Fig. 1. Average power use for CPU package, CPU core and memory subsystem for various workloads on Macbook Pro.

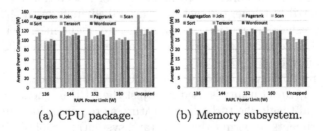

(a) CPU package.   (b) Memory subsystem.

Fig. 2. Average power use for CPU package and memory subsystem for various workloads on AWS i3.metal instance.

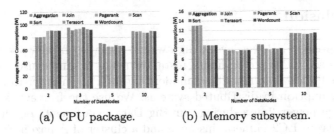

(a) CPU package.   (b) Memory subsystem.

Fig. 3. Average power use for CPU package and memory subsystem for various workloads on AWS m4.large virtual cluster.

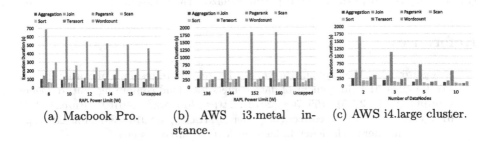

(a) Macbook Pro.   (b) AWS i3.metal instance.   (c) AWS i4.large cluster.

Fig. 4. Total execution durations for various workloads on various systems.

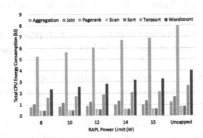

Fig. 5. Total CPU Energy Consumption for various workloads on Macbook Pro.

Figures 4(a) and (b) show the total execution durations for various workloads, RAPL CPU package power limit and turbo settings running on the Macbook Pro and AWS i3.metal instance.

Figure 4(c) shows the total execution durations for various workloads running on the AWS m4.large virtual cluster for various numbers of DataNodes.

Figure 5 shows the CPU energy consumption for various workloads, RAPL CPU package power limit and turbo settings running on the Macbook Pro.

6　Conclusion

We have developed PowerSave, a lightweight software framework that allows the user to set CPU power limits instantly using a simple user interface. It exploits the RAPL functionality of modern Intel CPUs in order to reduce CPU energy consumption in distributed systems. We use the framework to dynamically change power limits during our Big Data experiments on the Macbook Pro, an Amazon EC2 i3.metal instance and a cluster of Amazon EC2 m4.large instances.

We have shown that for heavy workloads, there exists an optimum power limit that minimizes energy use per unit of work. We have demonstrated that this is the case on both the laptop and Amazon EC2 instances.

References

1. Beloglazov, A., Abawajy, J., Buyya, R.: Energy-aware resource allocation heuristics for efficient management of data centers for cloud computing. Future Gener. Comput. Syst. **28**(5), 755–768 (2012). https://doi.org/10.1016/j.future.2011.04. 017. http://www.sciencedirect.com/science/article/pii/S0167739X11000689. special section: energy efficiency in large-scale distributed systems
2. Chen, Y., Alspaugh, S., Borthakur, D., Katz, R.: Energy efficiency for large-scale mapreduce workloads with significant interactive analysis. In: Proceedings of the 7th ACM European Conference on Computer Systems, EuroSys 2012, pp. 43–56. ACM, New York (2012). https://doi.org/10.1145/2168836.2168842. http://doi.acm.org/10.1145/2168836.2168842
3. Dargie, W.: A stochastic model for estimating the power consumption of a processor. IEEE Trans. Comput. **64**(5), 1311–1322 (2015). https://doi.org/10.1109/TC. 2014.2315629
4. Hoffmann, H.: JouleGuard: energy guarantees for approximate applications. In: Proceedings of the 2015 Symposium on Operating Systems (SOSP 2015), ACM (2015)
5. Intel: HiBench Suite (2018). https://github.com/intel-hadoop/HiBench
6. Jacob, J.C., et al.: Montage; a grid portal and software toolkit for science grade astronomical image mosaicking. Int. J. Comput. Sci. Eng. **4**(2), 73–87 (2009)
7. Kansal, A., Zhao, F., Liu, J., Kothari, N., Bhattacharya, A.A.: Virtual machine power metering and provisioning. In: Proceedings of the 1st Symposium on Cloud Computing, ACM (2010)
8. Khan, K.N., Hirki, M., Niemi, T., Nurminen, J.K., Ou, Z.: RAPL in action: experiences in using RAPL for power measurements. ACM Trans. Model. Perform. Eval. Comput. Syst. **3**(2), 9–26 (2018). https://doi.org/10.1145/3177754. http://doi.acm.org/10.1145/3177754
9. Lee, Y.C., Zomaya, A.Y.: Energy conscious scheduling for distributed computing systems under different operating conditions. IEEE Trans. Parallel Distrib. Syst. **22**(8), 1374–1381 (2011). https://doi.org/10.1109/TPDS.2010.208
10. Lee, Y.C., Zomaya, A.Y.: Energy efficient utilization of resources in cloud computing systems. J. Supercomputing **60**(2), 268–280 (2012). https://doi.org/10.1007/s11227-010-0421-3. http://dx.doi.org/10.1007/s11227-010-0421-3

11. Möbius, C., Dargie, W., Schill, A.: Power consumption estimation models for processors, virtual machines, and servers. IEEE Trans. Parallel Distrib. Syst. **25**(6), 1600–1614 (2014)
12. Outin, E., Dartois, J.E., Barais, O., Pazat, J.L.: Seeking for the optimal energy modelisation accuracy to allow efficient datacenter optimizations. In: Proceedings of the 16th IEEE/ACM Symposium on Cluster, Cloud and Grid Computing (CCGrid), IEEE (2016)
13. Subramaniam, B.: Metrics, models and methodologies for energy-proportional computing. In: 2014 14th IEEE/ACM International Symposium on Cluster, Cloud and Grid Computing, pp. 575–578, May 2014. https://doi.org/10.1109/CCGrid.2014.119

Support for High-Level Quantum Bayesian Inference

Marcin Przewięźlikowski[1], Michał Grabowski[1], Dariusz Kurzyk[2],
and Katarzyna Rycerz[1(✉)]

[1] Institute of Computer Science, AGH, al. Mickiewicza 30, 30-059 Kraków, Poland
{przewiez,mgrabow}@student.agh.edu.pl, kzajac@agh.edu.pl
[2] Institute of Theoretical and Applied Informatics, Polish Academy of Sciences,
Bałtycka 5, 44-100 Gliwice, Poland
dkurzyk@iitis.pl

Abstract. In this paper, we present `AcausalNets.jl` - a library supporting inference in a quantum generalization of Bayesian networks and their application to quantum games. The proposed solution is based on modern approach to numerical computing provided by Julia language. The library provides a high-level functions for Bayesian inference that can be applied to both classical and quantum Bayesian networks.

Keywords: Quantum Bayesian networks · Quantum games ·
Julia language

1 Introduction

Bayesian networks [10] are probabilistic models which, among their numerous use cases, allow to model complex systems of interconnected random events in games of chance and their influence on each other. There are several approaches to generalizing Bayesian probability theory into the quantum realm [7,8,10].

Introducing quantum probability into game theory opens up new opportunities for finding optimal strategies in various games of chance. Apart from well known work on representing quantum strategies as unitary gates [9] or applying quantum entanglement to find optimal strategies [2], there is also relatively new approach to apply quantum Bayesian networks for that purpose [5]. This new approach has not yet been fully explored, therefore we focus on methods, algorithms and numerical support for researchers working this topic.

Proper numerical tools are required due to high complexity of computations essential to perform such experiments. In particular, performing Bayesian inference in Acausal networks, a quantum generalization of Bayesian networks, is not yet fully supported among numerical libraries.

In this paper we present `AcausalNets.jl` - a library providing a high-level functions for Bayesian inference that can be applied to both classical and quantum Bayesian networks. The library takes advantage of the new approach to numerical computing offered by Julia language [1].

© Springer Nature Switzerland AG 2019
J. M. F. Rodrigues et al. (Eds.): ICCS 2019, LNCS 11540, pp. 764–770, 2019.
https://doi.org/10.1007/978-3-030-22750-0_76

Organization of the Paper. The paper is organized as follows: in Sect. 2 we summarize relevant related work which provided basis and inspiration for this paper. In Sect. 3 we describe Bayesian networks and their usage in higher detail. We also briefly delve into the inference algorithms implemented in `AcausalNets.jl`. In Sect. 4 we provide in brief detail the principles we followed when implementing the library. Section 5 sums up our results when recreating and expanding experiments first conducted and described in [5]. In Sect. 6 we provide a brief overview of our ideas for further improvements and enhancements of `AcausalNets.jl`.

2   Related Work

Belief Propagation algorithms that we adapt for quantum Bayes networks in `AcausalNets.jl` library are covered to a fuller extent in [4] and [10]. The adaptation was done using theoretical foundations for generalizing Bayesian probability theory into the quantum realm described in [7,8]. To our best knowledge, there is no numerical support for quantum version of that algorithm.

Additionally, our work was inspired by `BayesNets.jl`[1] - a Julia library designed for high-level operations on classical Bayesian Networks.

In general, there is a huge variety of software related to quantum information implemented in numerous programming languages, most notably `QuantumInformation.jl` [3] implemented in `Julia`. The set of functionalities these libraries provide, while wide, is focused mainly on low-level optimized matrix operations. `AcausalNets.jl` makes use of matrix operations implemented in `QuantumInformation.jl` in its implementation of Belief Propagation algorithms.

3   Classical and Quantum Bayesian Networks

Bayesian Networks are probabilistic graphical models used for describing systems of random variables and correlations between their probability distributions. Those networks take a form of a directed acyclic graph, where vertices denote the variables and edges - correlations between them. Bayesian network is usually represented using set of multivariate distributions, which accounts for both variables distributions and their correlations. Typical applications of Bayesian networks include: calculating probability of a given variable values configuration or inferring probability distributions of given variables based on known states of other variables in the network. The variable connection types differ depending on their relation is classical or quantum.

Classical Version - Conditional Dependence. In classical version, as in Fig. 1, the variables in a Bayesian Network may be causally dependent on one other, which means their distributions are conditional. If distribution of variable

---

[1] https://github.com/sisl/BayesNets.jl.

V_2 is conditionally dependent on the distribution of variable V_1, a *happens-before* relationship between V_1 and V_2 is implied.

Quantum Version - Acausal Relationships. When generalized to the quantum domain, Bayesian networks can also describe acausal relationships between variables, which may be interpreted as a quantum entanglement between them. As opposed to a causal relationship, such a system of variables is not described as one system at two times, but rather as two systems in a single time frame. An example of such a network is shown in Fig. 2.

Fig. 1. A Bayesian network where there is a dependency of C on A and B (arrows). Source: [5]

Fig. 2. States of systems A and B are entangled (zigzag line), and there is classic dependence of C on A and B (arrows). Source: [5]

Inference Algorithms. Inference in Bayesian networks - calculating the probability distribution of a subsystem of variables based on already known variables - is an NP-hard problem [6]. However, there are algorithms which successfully approximate such computations with considerably lower computational complexity. AcausalNets.jl provides implementations of two algorithms for performing inference in Bayesian networks: a non-optimal naive algorithm, as well as the Belief Propagation algorithm [4,10]. Moreover, the second algorithm has been generalized to the quantum domain based on [7].

4   Design of the Library

AcausalNets.jl API has been designed with simplicity of computations regarding discrete quantum probability systems in mind. It allows the user to perform inference in Bayesian Networks described in Sect. 3.

Defining Bayesian Networks. To define a Bayesian network, one must first define systems of random variables which make up the network. Then a graph is built of those variables by inserting them in topological order. This requirement is essential, since Bayesian network is a directed acyclic graph. An example of the code which defines a network is shown in Fig. 3.

Performing Inference. Inference can be performed with or without evidence - the known state of some of the random variables in the network. Known states are represented by a system with appropriate variables and their known distribution. Next, function *infer(network, variables, evidence, strategy)* is used for conducting inference for variables and evidence passed as arguments. By default, naive

```
# 1.Defining variables
var_a = Variable(:a, 3) # location of the prize
var_b = Variable(:b, 3) # player's first choice
var_c = Variable(:c, 3) # host's choice

# 2.Defining system distribution for variable C
roCwAB = QuantumDistribution(diagm(0 =>[
0,1/2,1/2,  0,0,1,  0,1,0,  0,0,1,
1/2,0,1/2,  1,0,0,  1/2,1/2,0]))

# 3.Defining other systems using previously declared
# distribution and variables
sys_c_ab = DiscreteQuantumSystem( [var_a, var_b],[var_c],
    roCwAB)
sys_b= ...
sys_a=...
.....
# 4.Defining a Bayesian network
network = AcausalNet()
push!(network, sys_a)
push!(network, sys_b)
push!(network, sys_c_ab)
```

Fig. 3. Using `AcasualNets.jl` API to build a bayesian network analogous to Fig. 1

```
# 5.Declaring evidence
ev = Evidence{QuantumDistribution}[
    Evidence{QuantumDistribution}(
        [var_b],  QuantumDistribution(ketbra(1,1,3))),
    Evidence{QuantumDistribution}(
        [var_c], QuantumDistribution(ketbra(3,3,3)))
    ]
# 6.Performing inference on created network
inferred = infer(network, [var_a], evidence)
```

Fig. 4. Using `AcasualNets.jl` API to perform Baycsian inference.

inference algorithm is used, but it can be changed in library's *Inference* module. The example code for Monty Hall game use case [5] is shown in the Fig. 4.

Usage of the Julia Language. The software benefits strongly from Julia type-system [1], which allows to implement general operation on Bayesian Networks based on the type of the distributions the network is dealing with. It is possible because of Julia's type parametrization properties. More specifically, as shown in the Fig. 5, `BayesNet` type is parametrized with the type of `DiscreteSystem`, which specifies the math operations to be perform on probability distributions of the variables. For example, in case of `DiscreteQuantumSystem` such operations are performed in accordance with definitions of quantum conditional operators [8].

`AcausalNets.jl` has been implemented in version 1.0 of Julia language [1]. Source code is publicly available on[2]. The repository also contains more example use cases[3] in a convenient form of interactive Jupyter Notebooks[4].

---

[2] https://github.com/mikegpl/AcausalNets.jl.

[3] https://github.com/mikegpl/AcausalNets.jl/tree/master/notebooks.

[4] https://jupyter.org.

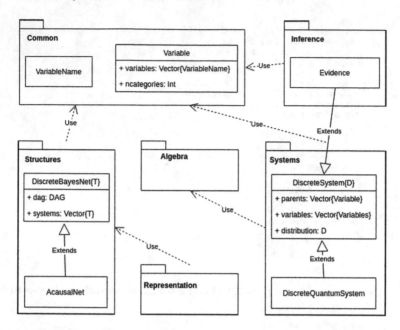

Fig. 5. The structure of `AcausalNets.jl` modules and types.

5   Monty Hall Game Example Use Case

As a use case we use results of quantum Monty Hall game from [5] modeled as Bayesian network shown in Fig. 2. We consider a case where there occur quantum effects between the event A - the placement of the prize and B - the initial choice of the player. We choose two example quantum probability distributions. Equation (1) models the situation where A and B are entangled in the way that due to quantum effects the placement of the prize always turns out to be the same as the initial choice of the player.

$$\rho_{AB\_same} = \frac{1}{3}(|00\rangle + |11\rangle + |22\rangle)(\langle 00| + \langle 11| + \langle 22|) \tag{1}$$

Equation (2) models the situation where A and B are entangled in the way that due to quantum effects placement of the prize always turns out to be different that the initial choice of the player.

$$\rho_{AB\_diff} = \frac{1}{6}(|01\rangle + |10\rangle)(\langle 01| + \langle 10|)$$
$$+\frac{1}{6}(|02\rangle + |20\rangle)(\langle 02| + \langle 20|) + \frac{1}{6}(|12\rangle + |21\rangle)(\langle 12| + \langle 21|) \tag{2}$$

Next, we construct linear combination of these two situations for $\lambda \in (0,1)$:

$$\rho_{AB} = \lambda\rho_{AB\_same} + (1-\lambda)\rho_{AB\_diff} \tag{3}$$

We aim to find how the probability of the player wining the game by staying with his initial choice depends on λ. We construct the Bayesian network as on Fig. 2 with an appropriate ρ_{AB} and perform inference to obtain the probabilities. We show our results in Fig. 6.We find that for $\lambda = 0.6$, the game is fair.

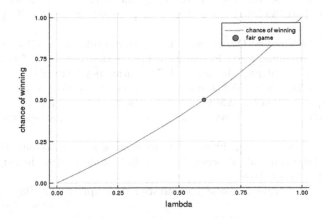

Fig. 6. Probability of winning the prize, when player does not change the door for the initial state given by $\rho_{AB} = \lambda \rho_{AB\_same} + (1 - \lambda)\rho_{AB\_diff}$

The above results show that using high level API of `AcausalNets.jl` library gives the same results as direct calculations presented in [5].

6   Conclusions and Future Work

High-level tools enable researchers to better organize, perform and document their experiments, although sometimes for the price of their flexibility. `AcausalNets.jl` has been created with aim to abstract out tedious calculations, provide a high-level API for quantum Bayesian inference and leverage the numerical potential of Julia language. We have successfully used our library to reproduce the results of experiments on the quantum version of Monty Hall originally presented in [5]. Moreover, through automating a lot of necessary computations, `AcausalNets.jl` helps experiment with more complex Bayesian networks. In the future we plan to fully implement the quantum version of Belief Propagation algorithm and test its efficiency on bigger networks. This may lead to interesting new research results in quantum information as well as machine learning in the future.

Acknowledgements. This research was partly supported by the National Science Centre, Poland – project number 2014/15/B/ST6/05204.

References

1. Bezanson, J., Edelman, A., Karpinski, S., Shah, V.: Julia: a fresh approach to numerical computing. SIAM Rev. **59**(1), 65–98 (2017). https://doi.org/10.1137/141000671
2. Eisert, J., Wilkens, M., Lewenstein, M.: Quantum games and quantum strategies. Phys. Rev. Lett. **83**, 3077–3080 (1999). https://doi.org/10.1103/PhysRevLett.83.3077
3. Gawron, P., Kurzyk, D., Pawela, Ł.: QuantumInformation.jl–a Julia package for numerical computation in quantum information theory. PLOS ONE **13**(12), e0209358 (2018). https://doi.org/10.1371/journal.pone.0209358
4. Huang, C., Darwiche, A.: Inference in belief networks: a procedural guide. Int. J. Approx. Reason. **15**(3), 225–263 (1996). https://doi.org/10.1016/S0888-613X(96)00069-2. http://www.sciencedirect.com/science/article/pii/S0888613X-96000692
5. Kurzyk, D., Glos, A.: Quantum inferring acausal structures and the monty hall problem. Quantum Inf. Process. **15**(12), 4927–4937 (2016). https://doi.org/10.1007/s11128-016-1431-8
6. Kwisthout, J.: Lecture notes: computational complexity of Bayesian networks (2015)
7. Leifer, M.S., Spekkens, R.W.: Towards a formulation of quantum theory as a causally neutral theory of Bayesian inference. Phys. Rev. A **88**, 052130 (2013). https://doi.org/10.1103/PhysRevA.88.052130
8. Leifer, M., Poulin, D.: Quantum graphical models and belief propagation. Annals Phys. **323**(8), 1899–1946 (2008). https://doi.org/10.1016/j.aop.2007.10.001
9. Meyer, D.A.: Quantum strategies. Phys. Rev. Lett. **82**, 1052–1055 (1999). https://doi.org/10.1103/PhysRevLett.82.1052
10. Yedidia, J.S., Freeman, W.T., Weiss, Y.: Understanding belief propagation and its generalizations. In: Exploring Artificial Intelligence in the New Millennium, pp. 239–269. Morgan Kaufmann Publishers Inc., San Francisco (2003). http://dl.acm.org/citation.cfm?id=779343.779352

Financial Time Series: Motif Discovery and Analysis Using VALMOD

Eoin Cartwright[(✉)], Martin Crane, and Heather J. Ruskin

Centre for Advanced Research Computing & Complex Systems Modelling
(ARC-SYM), School of Computing, Dublin City University, Dublin 9, Ireland
eoin.cartwright3@mail.dcu.ie
https://www.dcu.ie/arcsym/index.shtml

Abstract. Motif discovery and analysis in time series data-sets have a wide-range of applications from genomics to finance. In consequence, development and critical evaluation of these algorithms is required with the focus not just detection but rather evaluation and interpretation of overall significance. Our focus here is the specific algorithm, *VALMOD*, but algorithms in wide use for motif discovery are summarised and briefly compared, as well as typical evaluation methods with strengths. Additionally, Taxonomy diagrams for motif discovery and evaluation techniques are constructed to illustrate the relationship between different approaches as well as inter-dependencies. Finally evaluation measures based upon results obtained from *VALMOD* analysis of a GBP-USD foreign exchange (F/X) rate data-set are presented, in illustration.

Keywords: Motif analysis · Motif evaluation · *VALMOD*

1   Introduction

Sequential data are found in many applications ranging from Healthcare [1] to Seismology [2], Machine Learning [3] and Finance [4]. Recurrent patterns (*motifs*) are common and can occur both within and between individual time series [5]. Motif identification can help pre-processing in other high level data mining tasks e.g. time series clustering and classification, rule discovery and summarisation [6]. Similarly, evaluating motif content can aid better understanding of the underlying processes generating data in a given domain.

A brief summary of state-of-the art in motif discovery and evaluation follows, with strengths and limitations indicated. Analysis of motif contributions is nontrivial, and combined analyses are usually required, depending on data features. In prior work [7], we compared two popular methods (*MrMotif* and *VALMOD*), the latter particularly suited to variable motif length analysis. Interest here is on an early evaluation of motif results from *VALMOD* for a GBP-USD F/X rate data-set.

2   Motif Discovery Techniques: Summary

Although a suite of motif discovery techniques are available (illustrated, Fig. 1), two principal approaches form the basis for many applications in the literature. These

© Springer Nature Switzerland AG 2019
J. M. F. Rodrigues et al. (Eds.): ICCS 2019, LNCS 11540, pp. 771–778, 2019.
https://doi.org/10.1007/978-3-030-22750-0_77

are, (i) The *CK* algorithm [8] which uses Random Projection (*RP*) to create a collision matrix and (ii) Symbolic Aggregate Approximation (*SAX*) which utilises Piecewise Aggregate Approximation (*PAA*) with breakpoints and symbolisation to discretize sequential data into a symbolic string [9].

A Taxonomy diagram of type and inter-relationships is provided, Fig. 1. Following the recent survey paper by [10], this is designed to aid analysis and interpretation. The majority of techniques apply to the time domain, and rely on *approximation* of the time-series to provide motif candidates within a suitable timeframe. In comparison, relatively few frequency-based approaches to motif discovery appear in the literature, with one notable exception, the *SIMD* [11] algorithm, which uses a *Wavelet-based* approach.

Given issues such as an initial word (or motif) target length requirement for a Brute Force (*BF*) approach and the computational expense involved, series approximation using *SAX* was considered initially. This offers significant advantages as it allows utilisation of string analysis techniques for motif detection, borrowed from the study of DNA sequences. Notable algorithms in this domain are *MrMotif* [12], an algorithm which examines *SAX* at increasing resolutions and *SEQUITUR* [13] which implements a grammar-based approach on these symbolic strings.

Initial limitations for the exact approach were overcome by use of *Brute Force* (*BF*) in combination with early abandonment, allowing motif identification in a linear timeframe (*SBF*). In consequence, the *MK* algorithm [5] has underpinned many extensions including *top-k* [14] and Variable Length Motif Discovery (*VLMD*), [15]. A twofold improvement in performance compared to *SBF* was offered by *Quick-Motif* [16] with preference shifting towards a deterministic approach to motif discovery. More recently still, performance improvements and increased scalability have been achieved through a series of algorithms based on approximation for the Matrix Profile technique: (examples include *STAMP* [24], *STOMP* [25] & *VALMOD* [26]).

A summary table of techniques and algorithms from the literature is given, Table 1. A similar split between exact and approximate methods, as noted already for motif discovery, (Fig. 1), is evident here also, with most algorithms reliant on some form of data discretisation, (notably *iSAX*).

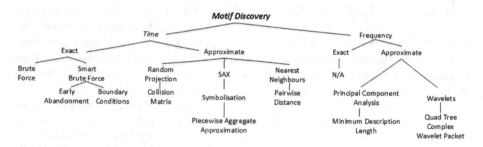

Fig. 1. Motif Discovery Technique Taxonomy: principal techniques and their interdependencies are shown here.

Table 1. Named motif discovery algorithm & model techniques: main algorithm techniques & features are illustrated. Attributes & applications are also given.

Year	Time/Frequency	Exact/Approx	Algorithm Named	Full Name	Technique Used	Algorithm Characteristics
2002	T	Approx	EMMA	Enumeration of Motif through Matrix Approximation	Piecewise AggregateApprox (PAA) & Breakpoints to create SAX	First to address problem of repeated patterns in time series. [9]
2002	T	Approx	PERUSE	Pattern Extraction from Real-valued sequences Using Expectation maximization	Expectation Maximization algorithm	Looks for a known pattern. [17]
2003	T	Approx	PROJECTION/CK	Chiu & Keogh Algorithm	Random Projection Collision Matrix	Built on Projection Alg. Can have "Don't Care" sections for noisy data. [8]
2009	T	Exact	MK	Mueen & Keogh Algorithm	Brute Force (BF) with early abandoning	First with exact solution in reasonable time vis-à-vis BF [5]
2009	T	Exact	DAME	Disk Aware Motif Enumeration	Bottom-up search on memory blocks of an "order line". While using a linear bound to prune off distance computations	"order line": created by computing distance of each point from random point & sort all such in ascending order. Applicable to large data sets.[18]
2010	T	Approx	MrMotif		Symbolic Aggregate Approximation (SAX)	Uses iSAX & Space Saving (SS) algorithm. Starting from low iSAX resolution & refined to higher. [12]
2010	T	Approx	AMG	Adaptive Motif Generation	Vector Suffix Tree Time Lag Matrix	Candidate motifs are generated then refined to detect desired patterns using a Time lag Matrix.[19]
2011	T	Exact	k-Motif		Generalized MK Algorithm	MK corresponds to case of finding a single motif (i.e. $k = 1$). [14]
2011	T	Approx	VLMD	Variable Length Motif Discovery Algorithm	SAX MK Sliding Window	Obviates predefined window length value. Returns exact solution to variable-length motif discovery problem. [15]
2012	T	Approx	kBMD	K-Best Motif Discovery	Extension of VLMD Algorithm	k-Best Motif Discovery (kBMD). Extension of VMLD above. [20]
2013	T	Exact	MOEN	Motif Enumeration	Smart Brute Force (SBF)	Adjustment of boundary conditions. [21]
2013	T	Approx	PLMD	Proper Length Motif Discovery	kBMD VMLD	Continuation of kBMD above. More efficient by early termination. [22]
2015	T	Approx	MDLats	Motif Discovery method for Large-scale Time Series	Random Projection Euclidian distance DTW	Like other discovery methods' structure but uses Hadoop for scalability. [1]
2015	F	Approx	SIMD	Shift Invariant Feature Extraction for Motif Discovery	Quad Tree-Complex Wavelet Packet (QT-CWP)	Applied to time & scale shifted data. [11]
2015	T	Exact	Quick-Motif		SBF MK Hilbert R-tree Min bounding rectangle (MBR) pairs	Minimum 2x faster than MK. Also scalable. [16]
2016	T	Approx	MDM	Multi-Dimensional Motif	SAX SEQUITUR	Applied to multidimensional data. [23]
2016	T	Approx	STAMP	Scalable Time series Anytime Matrix Profile	Matrix Profile Matrix Profile Index	Creates two meta time series, the matrix profile and the matrix profile index. These two data objects explicitly or implicitly contain the answers to many data mining tasks. [24]
2017	T	Approx	STOMP	Scalable Time series Ordered-search Matrix Profile	STAMP	Extension of STAMP to large datasets. STAMP evaluates distance profiles in a random order while STOMP performs an ordered search. GPU unit added to improve performance [25]
2018	T	Approx	VALMOD	Variable Length Motif Discovery	Matrix Profile Lower Bound Distance Profile	Solution returning all motifs within a given range of lengths. [26]

3   Motif Evaluation Techniques Summary

In assessing importance of a given motif (or motif set) some measures are calculated exclusively based on the pattern's information-content, while others are based on how these relate to the underlying data within which they appear. Three main approaches to motif evaluation were proposed in [27]. These are: *Class-Based(CBM)*, *Theoretical Information(TIM)* and *Mixed Measures(MM)*.

CBM measures do not rely upon motif structure, but on the number of occurrences in a given category. Hence, they are applicable to any deterministic motif, whereas *TIM* measures have a probabilistic basis and *MM* a combination of both. We show a Taxonomy for motif evaluation based on examples in the literature, (Fig. 2).

To date *CBM* and *TIM* measures predominate, typically rated on the basis of *Discrimination Power* and explained variability (*F-Ratio*). Achieving a meaningful evaluation, of motif occurrence and importance, generally requires statistical inference from more than one complementary technique as well as flexible treatment of mis-(or partial) matches and identification.

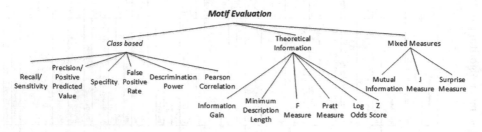

Fig. 2. Motif Evaluation Technique Taxonomy: principal techniques and their group-dependencies.

The following sub-sections outline main motif evaluation approaches with an initial application given in Sect. 4.

Class-Based Measures (*CBM*)

The ideal (or *signature*) motif [28] matches all sequences within a target family and does not overlap with any sequences outside it. As the ideal occurs rarely however the motif *quality* is illustrated by other measures: e.g. for *CBM*, these are usually *Sensitivity*, *Specificity* and *Positive Predicted Value(PPV)*, based on comparison possibilities for a given sequence and target family, Table 2.

Sensitivity (*S*) is the proportion of the target family within a data-set correctly (i.e. exactly) matched by a motif. Specificity (S_p or *Recall*) indicates non-matches while Positive Predicted Value (*PPV* or *Precision*) is the percentage of data correctly matched by a motif and also belonging to the target family: formulae see [27].

Table 2. Class-based measures (*CBM*) motif comparison possibilities.

	Target family	Not target family
Matches motif	True positive (T_p)	False positive (F_p)
Does not match motif	False negative (F_n)	True negative (T_n)

A signature motif requires both *Sensitivity* and *Positive Predicted Value* = 100%. Other notable measures include the *F-Ratio* for overall quality of match and *Discrimination Power* which provides an indication of the rarity of a given pattern.

Theoretical Information Measures (*TIM*)
These measures analyse the degree and nature of information encoded in a motif. The principle of *Minimum Description Length* (*MDL*) is used to rank motifs, assuming the best is the minimum length possible (thus *reducing* the overall series length when re-encoded). *MDL* can also be used in the detection of motifs with 'wobble' (or inexact match).

Common statistical techniques such as the *Z-Score*, (based on Gaussian assumptions), can be used to identify functionally important regions within a data set and as an initial pruning mechanism before other significance measures are calculated. In determining incidence of unexpected motifs, the *Log-Odds* calculates the probability of occurrence in relation to a given distribution, e.g. *Binomial, Uniform* or other. Commonly, either *Bernoulli* or *Markov* models are used for motif symbol counts, depending on whether those symbols occurring within a sequence are independent or conditional.

Another useful *TIM* is the *Pratt* measure, used to rank motifs when 'flexible gaps' are permitted in symbol content. A two-step approach applies, whereby information is first encoded by the motif, then a penalty factor is introduced when gaps occur.

Hybrid (or Mixed) Measures (*MM*)
For *MM*, *Class-Based* and *Theoretical Information* measures are combined to gain a better appreciation of a motif's functional significance within a given data-set. Numerous occurrences within a data-set of a given motif do not necessarily imply importance, while a functionally significant motif may occur infrequently but still contain valuable information. *MM* techniques include *Mutual Information*, the *J-Measure* and the *Surprise (or S-)measure* [27].

4   Motif Evaluation Examples

We briefly illustrate points from Sects. 2 and 3 with reference to Financial data from [29]. A GBP vs USD daily F/X series provides input for *Mr Motif, SBF, Mueen Keogh* and *VALMOD* algorithms with motif data set location for the same motif length criteria the objective (Table 3). Similar motif locations are returned (even for small sample size) so that algorithm features best suited to the application can guide tool choice.

Table 3. *MrMotif, Quick Motif, Smart Brute Force, Mueen Keogh & VALMOD* Compared

Motif length		100		150		200	
FX series	Algorithm	Execution time (s)	Dataset location	Execution time (s)	Dataset location	Execution time (s)	Dataset location
GBP V USD	MrMotif	0.123	41, 1921, 2161, 2461, 2601, 4181 2461, 2601, 4181	0.125	1081, 1591, 2251	0.122	161, 201, 601
	Quick Motif	0.111	1751, 2584	0.093	3039, 1701	0.05	1045, 244
	Smart Brute Force	0.247	1751, 2584	0.243	3039, 1701	0.269	244, 1045
	Mueen Keogh	0.171	1751, 2584	0.163	1701, 3039	0.154	244, 1045
	VALMOD (single length)	0.484	2585, 1752	0.468	3040, 1702	0.437	1046, 245

The *VALMOD* algorithm was chosen for further tests due to its ability to parse a *user-provided* range of lengths. *VALMOD* source code was amended to return a complete set of candidate motifs for given length, serving as input for a bespoke application written in C#. The original series can be displayed, *VALMOD* criteria chosen and motif evaluation measures applied to the discovered motif set, as shown, (Fig. 3).

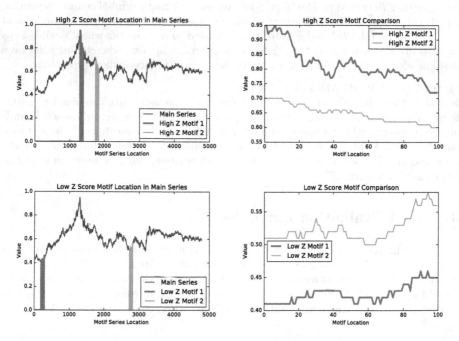

Fig. 3. Sample *VALMOD* motif results analysis of GBP vs USD FX dataset (Motif Length 100, Z-Score technique)

The user can choose a target family, based on an area of interest in the series, allowing T_p, F_p, T_n & F_n values with corresponding formulae to be calculated. Similarly motif locations can be shown within the data set and ordered by *Z-Score*, (Fig. 3).

5   Conclusions and Future Work

The growing importance of identifying repeated sub-sections or motifs in sequential data is outlined. Taxonomy diagrams illustrating motif discovery and evaluation techniques are provided while state of the art discovery algorithms are listed and characterised. The *VALMOD* algorithm is found to provide a sound basis for evaluation of motif occurrence in a financial data-set and examples are provided, indicative of its potential as an investigative tool in this context.

Clearly desirable for the future however, is an implementation of *SAX*, permitting refinement of the MDL principle applied to motif ranking and analysis, (with 'wobble' or less precise matching), as well as to discovery of common motifs over multiple series.

References

1. Liu, B., Li, J., Chen, C., Tan, W., Chen, Q., Zhou, M.: Efficient motif discovery for large-scale time series in healthcare. IEEE Trans. Ind. Inform. **11**, 583–590 (2015)
2. Cassisi, C., et al.: Motif discovery on seismic amplitude time series: the case study of Mt Etna 2011 eruptive activity. Pure Appl. Geophys. **170**, 529–545 (2012)
3. Lecun, Y., Bengio, Y., Hinton, G.: Deep learning. Nature **521**, 436–444 (2015)
4. Guan, Q., An, H., Liu, N., An, F., Jiang, M.: Information connections among multiple investors: evolutionary local patterns revealed by motifs. Sci. Rep. **7**, 14034 (2017)
5. Mueen, A., Keogh, E., Zhu, Q., Cash, S., Westover, B.: Exact discovery of time series motifs. In: Proceedings of 2009 SIAM International Conference on Data Mining (2009)
6. Son, N., Anh, D.: Discovery of time series k-motifs based on multidimensional index. Knowl. Inf. Syst. **46**, 59–86 (2015)
7. Cartwright, E., Crane, M., Ruskin, H.J.: Abstract: Motif Discovery & Evaluation Focus on Finance. https://sites.google.com/view/econophysics-colloquium-2018
8. Chiu, B., Keogh, E., Lonardi, S.: Probabilistic discovery of time series motifs. In: Proceedings of 9th ACM SIGKDD International Conference on Knowledge Discovery and Data Mining (2003)
9. Lin, J., Keogh, E., Lonardi, S., Patel, P.: Finding motifs in time series. In: Proceedings of 2nd Workshop on Temporal Data Mining, pp. 53–68 (2002)
10. Torkamani, S., Lohweg, V.: Survey on time series motif discovery. Wiley Interdiscip. Rev.: Data Min. Knowl. Disc. **7**, e1199 (2017)
11. Torkamani, S., Lohweg, V.: Shift-Invariant Feature Extraction for Time-Series Motif Discovery (2015)
12. Castro, N., Azevedo, P.: Multiresolution motif discovery in time series. In: Proceedings of 2010 SIAM International Conference on Data Mining (2010)
13. Nevill-Manning, C., Witten, I.: Identifying hierarchical structure in sequences: a linear-time algorithm. J. Artif. Intell. Resol. **7**, 67–82 (1997)
14. Lam, H., Pham, N., Calders, T.: Online discovery of top-k similar motifs in time series data. In: Proceedings of 2011 SIAM International Conference on Data Mining (2011)

15. Nunthanid, P., Niennattrakul, V., Ratanamahatana, C.: Discovery of variable length time series motif. In: The 8th (ECTI) Association of Thailand (2011)
16. Li, Y., U, L., Yiu, M., Gong, Z.: Quick-motif: an efficient and scalable framework for exact motif discovery. In: 2015 IEEE Proceedings of International Conference on Data (2015)
17. Oates, T.: PERUSE: an unsupervised algorithm for finding recurring patterns in time series. In: 2002 IEEE Data Mining, Proceedings (2002)
18. Mueen, A., Keogh, E., Bigdely-Shamlo, N.: Finding time series motifs in disk-resident data. In: 2009 IEEE Data Mining (2009)
19. Wang, L., Chng, E., Li, H.: A tree-construction search approach for multivariate time series motifs discovery. Pattern Recogn. Lett. **31**, 869–875 (2010)
20. Nunthanid, P., Niennattrakul, V., Ratanamahatana, C.: Parameter-free motif discovery for time series data. In: 9th (ECTI) Association of Thailand (2012)
21. Mueen, A., Chavoshi, N.: Enumeration of time series motifs of all lengths. Knowl. Inf. Syst. **45**, 105–132 (2014)
22. Yingchareonthawornchai, S., Sivaraks, H., Rakthanmanon, T., Ratanamahatana, C.: Efficient proper length time series motif discovery. In: IEEE Data Mining (2013)
23. Balasubramanian, A., Wang, J., Prabhakaran, B.: Discovering multidimensional motifs in physiological signals for personalized healthcare. IEEE J-STSP **10**(5), 832–841 (2016)
24. Yeh, C., et al.: Time series joins, motifs, discords and shapelets: a unifying view that exploits the matrix profile. Data Min. Knowl. Disc. **32**, 83–123 (2017)
25. Zhu, Y., et al.: Exploiting a novel algorithm and GPUs to break the ten quadrillion pairwise comparisons barrier for time series motifs and joins. Knowl. Inf. Syst. **54**, 203–236 (2017)
26. Linardi, M., Zhu, Y., Palpanas, T., Keogh, E.: Matrix profile X. In: Proceedings of 2018 International Conference on Management of Data - SIGMOD 2018 (2018)
27. Ferreira, P., Azevedo, P.: Evaluating deterministic motif significance measures in protein databases. Algorithms Mol. Biol. **2**, 1–20 (2007)
28. Jonassen, I., Collins, J., Higgins, D.: Finding flexible patterns in unaligned protein sequences. Protein Sci. **4**, 1587–1595 (1995)
29. Time Series Data Library - Data provider – DataMarket. https://datamarket.com/data/list/?q=provider:tsdl

Profiling of Household Residents' Electricity Consumption Behavior Using Clustering Analysis

Christian Nordahl[✉], Veselka Boeva, Håkan Grahn, and Marie Persson Netz

Blekinge Institute of Technology, 371 79 Karlskrona, Sweden
{christian.nordahl,veselka.boeva,hakan.grahn,marie.netz}@bth.se

Abstract. In this study we apply clustering techniques for analyzing and understanding households' electricity consumption data. The knowledge extracted by this analysis is used to create a model of normal electricity consumption behavior for each particular household. Initially, the household's electricity consumption data are partitioned into a number of clusters with similar daily electricity consumption profiles. The centroids of the generated clusters can be considered as representative signatures of a household's electricity consumption behavior. The proposed approach is evaluated by conducting a number of experiments on electricity consumption data of ten selected households. The obtained results show that the proposed approach is suitable for data organizing and understanding, and can be applied for modeling electricity consumption behavior on a household level.

Keywords: Ambient Assisted Living ·
Non-intrusive remote monitoring

1 Introduction

The world's population is getting older. By 2050, projections state that the number of individuals over 60 will be around 2.1 billion. Keeping individuals in their own homes, and delaying their entrance to the health and elderly care systems, can help to offload costs and work from the already strained health care systems. Likewise, the elderly population often want to keep living independently at home, but they also want a sense of safety without any intrusion in their lives [13]. Remote monitoring, and assistance, is one way to provide the safety of the residents. Traditionally, remote monitoring has been performed with video surveillance. However, with the introduction of smart homes, i.e. houses with built in sensors and actuators, Ambient Assisted Living (AAL) has emerged and allows for monitoring and assistance without the use of cameras and with less intrusion of the residents' privacy.

With the adoption of smart meters in the electrical power grids, we have the opportunity to collect high resolution electricity consumption data remotely on

© Springer Nature Switzerland AG 2019
J. M. F. Rodrigues et al. (Eds.): ICCS 2019, LNCS 11540, pp. 779–786, 2019.
https://doi.org/10.1007/978-3-030-22750-0_78

a household level. This type of data can be used to get insight into the residents' habits and activities, with low impact and intrusion of the residents' privacy. We may detect abnormalities and changes of residents' behavior through analyzing their daily household electricity consumption. For example, dementia, and other neurodegenerative diseases, cause changes in the behavior of the individual in different ways, e.g., they can provoke insomnia, apathy, restlessness etc. [8]. We believe that changes like these, in the individual's daily behavior, can be caught by his/her electricity consumption activities.

Most current research related to household electricity consumption has mainly revolved around creating consumer profiles by clustering households together [3] and comparing different households to determine and predict abnormal consumption patterns [2]. But, there has been some research as well on household electricity consumption. For example, Zhang et al. [14] analyze energy consumption data on a household level to identify days when the residents have gone on vacation. Further, we have previously studied the use of prediction models for electricity consumption behavior [10].

In this paper, we present and evaluate a cluster analysis approach for organizing, understanding, and modeling household electricity consumption data, a continuation of our work in [9]. Our aim is to study the possibility of using the knowledge discovered by such analysis for creating consumption behavior signatures on a household level. The long-term goal is to investigate whether the created signatures can be used for identifying abnormal behavior in daily life and apply this outlier detection model in health care applications, e.g., for monitoring early stages of dementia or other neurodegenerative diseases. The developed consumption signatures can be considered as predefined activities and can be used for detecting abnormal consumption patterns in order to notify the environment (relatives and health care professionals) if early signs of dementia occurs repeatedly at home.

2   Clustering Analysis Approach

Data Pre-processing. The electricity consumption data collected in this study is gathered with a one-minute resolution and is measured in kWh (kilowatt hours). To be able to determine and profile a behavior of the household, we divide the time series into 24 h profiles. This is an intuitive division of the data, as it allows us to capture a daily behavior which we then can analyze and use to model a routine daily behavior.

We set a maximum limit of 10% of the entire day or 20 consecutive minutes of missing data to remove that day from the data set. For the remaining profiles, we impute missing values by using linear interpolation. We then aggregate the electricity consumption data into a one hour resolution due to that resolution being more common for today's smart meters. Finally, we standardize the time series profiles using z-standardization, or Z-score, because we are more interested in the general shapes of the time series and not the actual amplitudes.

Clustering Algorithms and Validation Measures. Three partitioning algorithms are commonly used for data analysis to divide the data objects into k disjoint clusters [7]: k-means, k-medians, and k-medoids clustering. The three partitioning methods differ in how the cluster center is defined. In k-means clustering, the cluster center is defined as the mean data vector averaged over all objects in the cluster. In k-medians, the median is calculated for each dimension in the data vector to create the centroid. Finally, in k-medoids clustering, which is a robust version of the k-means, the cluster center is defined as the object with the smallest sum of distances to all other objects in the cluster, i.e., the most centrally located point in a given cluster. We have used k-medoids, since having an actual consumption profile as the cluster's centroid (medoid) is more representative of the consumption behavior compared to creating a synthetic centroid.

There are two major categories in which we can divide cluster validation measure to: *external* and *internal*. External measures are used when you have prior knowledge of the data and validate according to the ground truth and internal measures validate based on the data and clusters themselves [5]. We use three internal validation measures for analyzing the data and to select the optimal clustering scheme. We have selected one validation measure for assessing compactness and separation - *Silhouette Index* [11], one for assessing connectedness - *Connectivity* [6], and one for assessing tightness and dealing with arbitrary shaped clusters - *IC-av* [1].

Distance Measures. The simplest and most widely used approach to measure the distance, or dissimilarity, between two data points in an n-dimensional space (in our case two time series) is to calculate the Euclidean Distance (ED). ED calculates the distance between the two time series by aligning the ith point of one time series with the ith point of the other. ED is fast but it is sensitive to outliers and it cannot identify similarities between two segments if they are shifted out of phase [4]. Therefore, we also investigate the use of Dynamic Time Warping (DTW).

DTW measures the dissimilarity of two time series in a similar way, but instead of strictly calculating point by point it allows for some elasticity. One point in one of the time series can be aligned against one or more points in the other [12], which allows the identification of similar shapes even though they are out of phase.

3   Experiments and Results

3.1   Data

We have gathered electricity consumption data from 9909 anonymous households, collected with a 1-minute interval. Initially, the household data have gone through the pre-processing stage, as explained in Sect. 2. Then we have selected

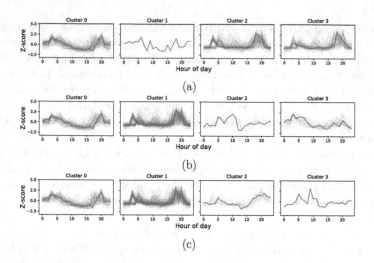

Fig. 1. Clustering solution generated by k-medoids for $k = 4$, using DTW as a dissimilarity measure and supported to be the best by; (a) IC-av, (b) SI, and (c) Connectivity.

the 10 households with the largest number of daily profiles. 3 of these 10 households contained a few clearly abnormal profiles which we excluded from further analysis. At the final stage, the 10 selected households contain between 345 and 353 daily profiles. We then selected 1 of the 10 studied households as the representative and we discuss and interpret the results obtained on its data for the rest of our study.

3.2   Results and Analysis

Estimation of the Number of Clusters. We run the k-medoids clustering algorithm using the two distance metrics (ED and DTW) for all values of k between 2 and 9. The clustering algorithm is run 100 times for each k and with the cluster medoids randomly initialized. All clustering solutions are then evaluated using the cluster validation measures mentioned in Sect. 2. We then look upon each measure individually and in combination to determine which k is the appropriate. Based on the scores generated by the validation measures, we decide upon $k = 5$ for ED and $k = 4$ for DTW. However, the scores we evaluated are the best ones generated from each individual measure, i.e., the scores are not necessarily generated from the same clustering solutions.

Clustering Analysis. In Fig. 1 we show the clustering solutions produced by k-medoids using DTW as a distance metric. All three validation measures support different clustering solutions. Therefore, we compare the three solutions and choose one of them that will be used to analyze the produced household consumption signatures. SI and Connectivity do however, share two clusters with the same signature and both of them contain mostly the same profiles.

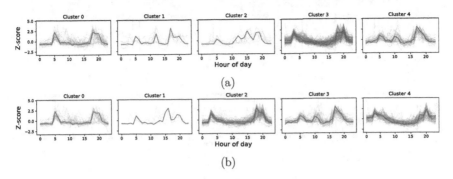

Fig. 2. Clustering solution generated by k-medoids for $k = 5$, using ED as a dissimilarity measure and supported to be the best by; (a) IC-av and Connectivity, and (b) SI.

Both these solutions have two major and two smaller clusters. IC-av, on the other hand, generates a solution with three major clusters and only one small cluster. The solutions chosen by SI and Connectivity are fairly similar, with only a few profiles difference between them. Therefore, we have chosen the clustering solution supported by SI (Fig. 1(b)) as its smallest clusters contain more profiles compared to their counterparts in the solution preferred by Connectivity.

In Fig. 2 we present the clustering solutions produced by ED. In this scenario, both IC-av and Connectivity support the same clustering solution, while SI has a solution of its own. However, the differences between the two solutions are not as distinct as in the case of DTW. We notice that they share the same signature for two of the clusters, and one additional cluster has a very similar signature. The solution promoted by SI does not have an equally as large cluster as the other solution has. In addition, in the solution proposed by IC-av and Connectivity we have two small clusters. In case of ED, we use the majority rule for choosing which clustering solution is the best and thereby it is the one proposed by IC-av and Connectivity (Fig. 2(a)).

Consumption Behavior Signatures. The produced cluster centroids, which can be seen as the signatures of the electricity consumption habits of the household, are shown in Fig. 3. We can see that the two different distance measures support different consumption signatures. For instance, DTW (Fig. 3(a)) focuses more on the general shape of the electricity consumption profiles, while ED (Fig. 3(b)) favours more the exact time of the days when the consumption peaks are happening. This supports our expectations, since DTW is an elastic measure that stretches the time series in the time axis to find an optimal alignment. Evidently, the different distance measures favour different electricity consumption profiles and logically, this will affect the intended analysis and built signatures. For example, we notice clear morning and evening consumption peaks recognized by Cluster 1 of the DTW solution (see Fig. 3(a)) and Clusters 0 and 3 of the ED signatures (see Fig. 3(b)), respectively. However, the additional consumption

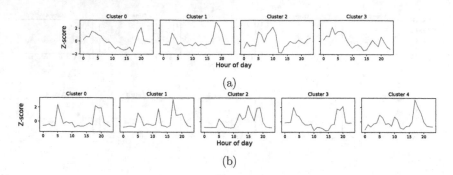

(a)

(b)

Fig. 3. Electricity consumption signatures created from the cluster medoids from the chosen clustering solutions generated by k-medoids with (a) $k = 4$ (DTW) and (b) $k = 5$ (ED).

peak in the middle of the day seen in Cluster 1 of the ED solution is not clearly presented in any of the four signatures supported by DTW.

Context Based Signatures. In order to investigate further the electricity consumption behavior of households, we create context based signatures that represent how the weekdays are distributed between each cluster. These signatures are shown in Fig. 4. They can be used to further improve and refine the consumption behavior model and allow us to gain additional knowledge about the household's behavior.

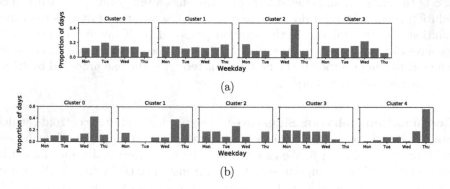

(a)

(b)

Fig. 4. Context based signature of the weekdays from the clustering solution generated by k-medoids with (a) $k = 4$ (DTW) and (b) $k = 5$ (ED).

Comparing the distribution of weekdays produced by DTW (see Fig. 4(a)) and ED (see Fig. 4(b)), it is apparent that DTW divides the days more evenly than ED. The signatures of the three larger clusters, 0, 1, and 3, are fairly monotonic. ED, on the other hand, has a more distinct separation between working days and weekends. Clusters 0, 1, and 4 all contain more weekend days, while Cluster 3 has almost only working days. Cluster 2, which is the smallest cluster only containing 11 signatures, has a more diverse spread of its days. It is further interesting to notice that the signatures of Clusters 0 and 3 (see Fig. 3(b)) are very similar, which can be an indication that these could be merged into a single cluster. However, this is not strongly supported by the context presented in Fig. 4(b), since the first cluster presents a typical working day consumption behavior while the second one is more representative for the weekends.

3.3 Discussion

The produced electricity consumption signatures are representatives of the *current* electricity consumption behavior of the residents. To detect changes in the residents' behavior, we can apply our approach on a new portion of data presenting the electricity consumption for the next time period. If new signatures are created through the clustering process, this might be a sign of a new behavior of the resident.

The proposed method can also be used to produce better prediction models. The generated clusters give a clearer distinction between normality and abnormality of the electricity consumption profiles. For example, in Fig. 2(a) it is clear that clusters 0, 3, and 4, contain a majority of the electricity consumption profiles. Training the prediction models only on the contents of these clusters would give a more accurate model.

As mentioned before, it is beneficial to use some elasticity when comparing the electricity consumption profiles. Using DTW, we allow for changes in time for the individual electricity consumption profiles. However, we may cluster some signatures which probably should not be regarded as similar, e.g., if a resident is sick and stays in bed for a few extra hours and then performs his/her daily routines. We would like to identify this as a behavioral change, but DTW identifies this as normal. Introducing a time window for DTW to warp would remedy this.

4 Conclusions and Future Work

In this paper we propose a clustering analysis approach for profiling a households electricity consumption behavior. The proposed approach is evaluated on real electricity consumption data from 10 anonymous households. The results show that we can create electricity consumption signatures that model electricity consumption behaviors of the household residents. Further, we have found that Euclidean distance (ED) produce clusters that are more focused on the exact times of consumption peaks, Dynamic Time Warping (DTW) was better at identifying the shapes of the consumption peaks. We have also identified that ED has a clear distinction between working days and weekend days between the clusters, while DTW has a more monotonic distribution of days.

We are currently in the final stages of collecting both electricity and water consumption from a set of elderly residents with the help of our local elderly care system. The subjects will be continuously interviewed and monitored during the study to be able to accurately label the data, i.e. changes in their behavior. The collected data will be used for further evaluation and validation of the approach proposed in this study.

References

1. Baya, A.E., Granitto, P.M.: How many clusters: a validation index for arbitrary-shaped clusters. IEEE/ACM Trans. Comput. Biol. Bioinform. **10**(2), 401–414 (2013)
2. Chalmers, C., Hurst, W., Mackay, M., Fergus, P.: Profiling users in the smart grid. In: The Seventh International Conference on Emerging Networks and Systems Intelligence (2015)
3. Chen, T., Mutanen, A., Järventausta, P., Koivisto, H.: Change detection of electric customer behavior based on AMR measurements. In: 2015 IEEE Eindhoven PowerTech, pp. 1–6. IEEE (2015)
4. Ding, H., Trajcevski, G., Scheuermann, P., Wang, X., Keogh, E.: Querying and mining of time series data: experimental comparison of representations and distance measures. Proc. VLDB Endow. **1**(2), 1542–1552 (2008)
5. Halkidi, M., Batistakis, Y., Vazirgiannis, M.: On clustering validation techniques. J. Intell. Inf. Syst. **17**(2–3), 107–145 (2001)
6. Handl, J., Knowles, J., Kell, D.B.: Computational cluster validation in post-genomic data analysis. Bioinformatics **21**(15), 3201–3212 (2005)
7. MacQueen, J., et al.: Some methods for classification and analysis of multivariate observations. In: 5th Berkeley Symposium on Mathematical Statistics and Probability, vol. 1, pp. 281–297 (1967)
8. Mega, M.S., Cummings, J.L., Fiorello, T., Gornbein, J.: The spectrum of behavioral changes in Alzheimer's disease. Neurology **46**(1), 130–135 (1996)
9. Nordahl, C., Boeva, V., Grahn, H., Netz, M.: Organizing, visualizing and understanding households electricity consumption data through clustering analysis. In: 2nd Workshop on Aging, Rehabilitation and Independent Assisted Living, IJCAI Workshop (2018)
10. Nordahl, C., Persson, M., Grahn, H.: Detection of residents' abnormal behaviour by analysing energy consumption of individual households. In: 2017 IEEE International Conference on Data Mining Workshops (ICDMW), pp. 729–738. IEEE (2017)
11. Rousseeuw, P.J.: Silhouettes: a graphical aid to the interpretation and validation of cluster analysis. J. Comput. Appl. Math. **20**, 53–65 (1987)
12. Sakoe, H., Chiba, S.: Dynamic programming algorithm optimization for spoken word recognition. IEEE Trans. Acoust. Speech Signal Process. **26**(1), 43–49 (1978)
13. Zagler, W.L., Panek, P., Rauhala, M.: Ambient assisted living systems-the conflicts between technology, acceptance, ethics and privacy. In: Dagstuhl Seminar Proceedings. Schloss Dagstuhl-Leibniz-Zentrum fr Informatik (2008)
14. Zhang, Y., Chen, W., Black, J.: Anomaly detection in premise energy consumption data. In: 2011 IEEE Power and Energy Society General Meeting, pp. 1–8. IEEE (2011)

DNAS-STriDE Framework for Human Behavior Modeling in Dynamic Environments

Muhammad Arslan[1]([✉]) [iD], Christophe Cruz[1] [iD],
and Dominique Ginhac[2] [iD]

[1] Laboratoire d'Informatique de Bourgogne (LIB),
Univ. Bourgogne Franche-Comté (UBFC), 7534 Dijon, France
muhammad.arslan@u-bourgogne.fr,
christophe.cruz@ubfc.fr
[2] Laboratoire Imagerie et Vision Artificielle,
Univ. Bourgogne Franche-Comté (UBFC), 7535 Dijon, France
dominique.ginhac@ubfc.fr

Abstract. Numerous studies have been conducted over the past few decades on human behavior modeling and simulation by incorporating the dynamic behaviors of people for different Facility Management (FM) applications. For example; the Drivers, Needs, Actions and Systems (DNAS) framework which provides a standardized way to conceptually represent energy-related occupant behaviors in buildings and allows the exchange of occupant behavior information and integration with building simulation tools. Despite numerous studies dealing with dynamic interactions of the building occupants, there is still a gap exists in the knowledge modeling of occupant behaviors for dynamic building environments. Such environments are best observed on construction sites where the contextual information linked to the building spaces evolve often over time in terms of their location, size, properties and relationships with the site environment. The evolving contextual information of a building is required to be mapped with the occupant interactions for an improved understanding of their changing behaviors. To fill this research gap, a framework is designed for providing a 'blueprint map' to integrate DNAS framework with our Semantic Trajectories in Dynamic Environments (STriDE) data model to incorporate the dynamicity of building environments. The proposed framework extends the usability of a DNAS framework by providing a centralized knowledge base that holds the mobility data of occupants with relevant historicized contextual information of the building environment to study occupant behaviors for different FM applications.

Keywords: Behavior · Knowledge modeling · Spatio-temporal · Safety management

1 Introduction

Humans are the important factor for building FM operations as they impact the building environments actively (i.e. production of heat because of their presence) and passively (i.e. operating building appliances) [1]. For ensuring an appropriate level of quality of

© Springer Nature Switzerland AG 2019
J. M. F. Rodrigues et al. (Eds.): ICCS 2019, LNCS 11540, pp. 787–793, 2019.
https://doi.org/10.1007/978-3-030-22750-0_79

services to the building occupants, the most crucial challenge faced by the facility managers is to understand the occupant behaviors and their interactions with buildings [1, 2]. Although, this is a complex activity because the occupant behaviors and the buildings are dynamic in nature and context-dependent. Here, a context refers to any information based on the contextual factors such as space, time and environment utilized for categorizing the situation of occupants [3]. Failure in understanding occupant behaviors because of inadequate integration of all relevant contextual factors associated with the occupants and the building environments can result into serious financial and management crises such as under-utilization of the building spaces, decreased productivity due to poor environmental conditions, increased energy usage, and safety hazards [1–4]. On the contrary, if the occupant behaviors are modeled and predicted effectively by incorporating all the possible contextual factors (social-personal, economic, etc.) which may influence occupant behaviors will lead to an increased physical comfort, enhanced safety at work and improved work performance of the occupants while keeping the level of building resources to the optimum [1, 5]. Existing literature [4–9] encompasses many studies for constructing occupant behavior extraction systems which help facility managers in decision making for FM operations by modeling dynamic behaviors of building occupants which change over time according to the building environment. Despite numerous existing studies which model dynamic interactions of the building occupants, there is still a gap exists in the knowledge modeling of occupant behaviors for dynamic building environments. To fill this research gap, one of the prominent occupant behavior models i.e. DNAS is selected from the existing literature [4] based on its relevance to our case-study and popularity that is perceived from its citations. For incorporating the dynamicity of the building environments in terms geometry and contextual information, a frame-work named 'DNAS-STriDE' is proposed which aims to serve an extension of the original DNAS by integrating the conceptual modeling of DNAS with our data model named 'STriDE' (Semantic Trajectories in Dynamic Environments) [10]. The resulted framework by integration has provided a centralized knowledge base that holds the mobility data of occupants with relevant historicized contextual information of the building environment to study occupant behaviors for different FM application scenarios in buildings.

The rest of the paper is organized as follows: Sect. 2 describes the background of the study. Section 3 is based on the proposed prototype system. Section 4 presents the discussion, a conclusion and an outlook of the future work.

2   Background

Behaviors are the interactions (leaving or entering a room, visual and thermal indoor conditions adjusting using windows or blinds, doors, etc.) of building occupants which can be categorized into different movements, simple presence or actions with their environment (building, its systems and appliances) which impact on the building performance (heating/cooling, indoor air quality, energy, comfort, etc.) during a building lifecycle [1, 2, 4, 5]. In majority of the situations, the occupants' presence is the precondition for any kind of behavior understanding as building occupants can only interact with the building environment if they are present inside the building [1].

Thereby, an occupant interaction which results in changing a building state (presence or absence in case of occupancy monitoring) or no interaction leaving the present state of a building unchanged are both facets of occupant behaviors [1]. For understanding the occupant interactions, the modeling process of their behaviors conventionally initiates from the data acquisition of occupants along with the building environmental and infrastructural parameters [5]. The quality of data captured from various types of sensor data acquisition systems differs greatly in terms of the resolution of the sensors deployed [1, 5]. Spatial, temporal and occupant resolutions are combined for determining the overall resolution of the system for capturing the occupant behaviors. As the resolution of the captured sensor data increases, the building space gets more precise, the occupants become more distinct based on their identities and the information from the sensor data is accessible faster [5]. For example, a low-resolution system will only capture the binary information (presence/absence) of the occupants in a specific time where the identities of the occupants are not recognized. Whereas, a high-resolution system will be able to detect the number of occupants, their identifications, as well as their activities [5].

Once data is acquired of occupant behaviors, the studies are performed on the collected data and correlations are extracted between the occupant behaviors and the building parameters using a set of contextual factors. For modeling the occupant behaviors, conventionally there exists four types of approaches [1, 5] which are: (1) static-deterministic, (2) static-stochastic, (3) dynamic-deterministic and (4) dynamic- stochastic. After applying the most appropriate modeling approach, a process of understanding occupant behaviors is evaluated to find out the reliability and effectiveness of the employed model used for extracting insights. The whole process of modeling occupant behaviors and evaluating a model is iterative which helps in tuning the model for best results [1, 5]. As a result, occupant behavior model functions as a stand-alone system or the output of the model is linked with the Building Performance Simulation (BPS) programs or Building Information Modeling (BIM)-based tools for further experimentations or generating information-enriched visualizations [5].

3 DNAS-STriDE Framework

After an extensive review of behavior modeling, it is observed that majority of the existing systems [4–9] model human behaviors by capturing the stochastic and reactive nature of their behaviors in building environments. However, these models do not incorporate the information of complex dynamic environments where the building objects evolve over time. For example, DNAS framework of Hong et al. [4] which is based on ontological modeling and used for describing the impact of occupant behaviors on energy use of buildings. This impact is represented using four components which are; drivers, needs, actions and systems. These four components of the framework encompass the building environment and cognitive processes of the occupants. DNAS framework has provided a basic ontological method for representing the energy-related occupant behaviors. However, DNAS requires more developments so that it can be used as a stand-alone or an integrated solution for extracting insights about occupant behaviors in dynamic environments. For this, DNAS-STriDE integrated framework is proposed which offers; (1) a centralized knowledge base for mining behavioral interactions of the occupants which can be in the form of occupant-to-occupant, occupant-to-building or building-to-occupant interactions, (2) a process of data enrichment of DNAS ontology using the results of a state-of-the-art machine learning model (Hidden Markov Model (HMM) in our case) in the form of 'hidden states' which are used to find the correlations and patterns in the acquired sensor data for studying the occupant behaviors, moreover (3) incorporation of the dynamicity of the building locations in data modeling stage in terms of the evolving spatial and contextual information over time.

The proposed framework (see Fig. 1) requires sensory data to perform behavioral analysis. The acquisition of relevant sensor data is based on the application requirement. For example, the safety manager of a building requires to monitor the movements of the occupants in a building. In this case, using our DNAS-STriDE framework, 'driver': monitoring movements of occupants, 'need': safety management in a building by identifying unsafe movements, 'action': tracking movements of occupants using their spatio-temporal trajectories, 'system': Bluetooth Low Energy (BLE) beacons for sensor data acquisition and 'states': (1) static (no movement), (2) normal movement ($0 <$ steps ≤ 84 and $\pi/2 \leq$ angle $< \pi$) and (3) risky/unsafe (steps > 84 angle $\geq \pi$). More information on movement states can be found in [11]. For understanding mobility-based behaviors for a safety management application, BLE beacons are installed on different building locations. After sensor data acquisition, the acquired location data is transformed into trajectories after preprocessing (i.e. filtering). More information on trajectories and their preprocessing can be found in [3]. After preprocessing the trajectories, the STriDE model is used for the semantic enrichment [3, 10] for incorporating the contextual and application-based information in processed trajectories. The STriDE model is based on a Continuum model that can stores data as well as performs semantic enrichments of moving and changing objects' trajectories [10]. For the semantic enrichment, the STriDE model uses a set of classes and properties from the existing vocabularies. The vocabularies are used for defining the concepts and their relationships by classifying the terms. Our STriDE model uses three different vocabularies which are; (1) Simple Knowledge Organization System (SKOS), (2) Dublin Core Terms (DCT) and (3) GeoSPARQL (GEO).

For tracking the evolution of building objects (users, trajectories and rooms), the STriDE model uses the concept of timeslices (TSs) [10]. A TS includes four components which are; an identity, alpha-numeric properties, a geographical and a time component [10]. At the occurrence of a change in any component of a TS excluding the identity, a new TS is generated inheriting the components of the last known TS. To show a proof-of-concept application of our DNAS-STriDE framework to hold the building evolution with the semantically-enriched trajectories to study movement behaviors for safety management, let's suppose we have a building from where the spatio-temporal data is collected (as discussed above) and the purpose of one of the building locations is changed (see Fig. 2). The location 'office' is now a 'storage room'. The STriDE model uses 'concepts' for describing the building locations. In Fig. 2, there is a hierarchy of SKOS concepts. It has a skos:hasTopConcept connects skos:Concept room. Two skos:Concept (office and room) are defined. All these concepts form a hierarchy. In addition, there is a profile named employeeProfile which

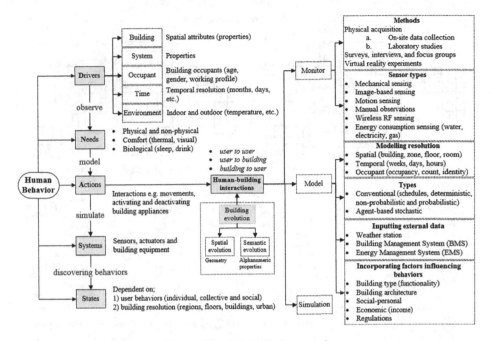

Fig. 1. DNAS-STriDE framework

gives access to all the concepts (locations). As soon as the functionality of a room (i.e. a context) is changed, a new TS is created. For instance, a user Jane is an entity of a TS ts-$jane_0$ and her position is tracked by a trajectory tr-jane. We can observe by a link between tr-$jane_0$ and room1 that Jane is in room1. The entity room1 was initially an office as suggested by the dct:subject link of tr-$room1_0$ towards the concept office. Later, this room is changed as a storage room having the same geometry as of the office represented as ts-$room1_1$.

4 Discussion and Conclusion

Once the spatio-temporal data of occupants is collected, processed and semantically-enriched using updated building information that is evolving over time, a process of knowledge modeling takes place. Knowledge modeling captures and models the knowledge into a reusable structure for the sake of historization of information, sharing as well as reapplying it for different applications [12]. Our DNAS-STriDE integrated model provides a knowledge model to study occupant behaviors using a building context with the help of stored semantic trajectories. Later, probabilistic models based on the application requirements are applied for categorizing different types of movements using the stored semantic trajectories. For our application scenario of safety management in a building, HMMs are used for categorizing the occupant movements. To categorize movements into three states which are static (stay location), normal and risky, the values of step lengths and turning angles are used for defining the hidden

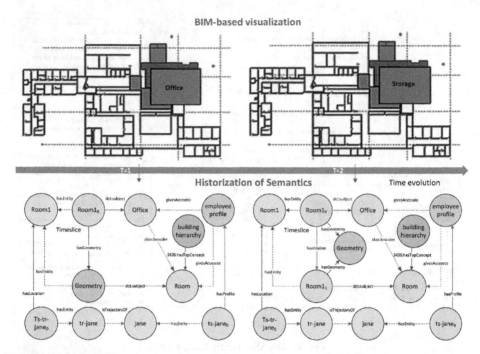

Fig. 2. DNAS-STRiDE historization of building evolution (Timeslices are in Blue, entities are in Green, SKOS concepts are in Yellow, profile is in Purple and geometry is in Red color). (Color figure online)

states. In Fig. 2, at a time instance, the most probable stay locations of occupants are shown in Green in color using the universal Architecture, Engineering & Construction (AEC) industry standard i.e. Building Information Model (BIM). Whereas, the most probable building locations at a time instance where there are the normal and risky movements of occupants are shown in Yellow and Red in color. Generated BIM-based visualizations provide insights about building occupants' movements in real-time using a building context. As a result, locations with risky movements can be easily identified, and necessary actions can be taken accordingly by the safety managers. Managing the safety in a building by acquiring spatio-temporal data, transforming them into semantic trajectories and exploring the states of the occupants to represent their movement behaviors using five components which are 'driver', 'need', 'action', 'system' and 'state' is one of the use-cases of the proposed framework. However, the proposed framework which adds a new dimension (i.e. hidden state) to study occupant behaviors in the original DNAS framework can be used in different types of other applications for managing the information of occupants and evolving building environment and infrastructure during a building lifecycle starting from a construction phase to a facility management phase.

Acknowledgments. The authors thank the Conseil Régional de Bourgogne-Franche-Comté and the French government for their funding. The authors also want to especially thank the director of the l'IUT de Dijon-Auxerre for allowing us to host the living lab and Orval Touitou for his technical assistance to this research work.

References

1. Wagner, A., O'Brien, W., Dong, B.: Exploring Occupant Behavior in Buildings: Methods and Challenges. Springer, Heidelberg (2018). https://doi.org/10.1007/978-3-319-61464-9. ISBN 978-3-319-61464-9
2. Chen, S., Yang, W., Yoshino, H., Levine, M.D., Newhouse, K., Hinge, A.: Definition of occupant behavior in residential buildings and its application to behavior analysis in case studies. Energy Build. **104**(1), 1–13 (2015)
3. Arslan, M., Cruz, C., Ginhac, D.: Semantic enrichment of spatio-temporal trajectories for worker safety on construction sites. Pers. Ubiquitous Comput. 1–16 (2019). https://doi.org/10.1007/s00779-018-01199-5
4. Hong, T., D'Oca, S., Turner, W.J., Taylor-Lange, S.C.: An ontology to represent energy-related occupant behavior in buildings. Part I: introduction to the DNAs framework. Build. Environ. **92**, 764–777 (2015)
5. Yan, D., et al.: Occupant behavior modeling for building performance simulation: current state and future challenges. Energy Build. **107**, 264–278 (2015)
6. Chen, H., Luo, X., Zheng, Z., Ke, J.: A proactive workers' safety risk evaluation framework based on position and posture data fusion. Autom. Constr. **98**, 275–288 (2019)
7. Goh, Y.M., Ubeynarayana, C.U., Wong, K.L., Guo, B.H.: Factors influencing unsafe behaviors: a supervised learning approach. Accid. Anal. Prev. **118**, 77–85 (2018)
8. Fang, W., Ding, L., Luo, H., Love, P.E.: Falls from heights: a computer vision-based approach for safety harness detection. Autom. Constr. **91**, 53–61 (2018)
9. Ding, L., Fang, W., Luo, H., Love, P.E., Zhong, B., Ouyang, X.: A deep hybrid learning model to detect unsafe behavior: integrating convolution neural networks and long short-term memory. Autom. Constr. **86**, 118–124 (2018)
10. Cruz, C.: Semantic trajectory modeling for dynamic built environments. In: IEEE International Conference on Data Science and Advanced Analytics (DSAA), pp. 468–476. IEEE (2017)
11. Arslan, M., Cruz, C., Ginhac, D.: Understanding worker mobility within the stay locations using HMMs on semantic trajectories. In: 14th International Conference on Emerging Technologies (ICET), pp. 1–6. IEEE (2018)
12. Pittet, P., Cruz, C., Nicolle, C.: An ontology change management approach for facility management. Comput. Ind. **65**(9), 1301–1315 (2014)

OPENCoastS: An Open-Access App for Sharing Coastal Prediction Information for Management and Recreation

Anabela Oliveira[1($\boxtimes$)], Marta Rodrigues[2], João Rogeiro[1],
André B. Fortunato[2], Joana Teixeira[1], Alberto Azevedo[2],
and Pedro Lopes[1]

[1] Information Technology in Water and Environment Division,
LNEC, Av. do Brasil, 1700-066 Lisbon, Portugal
aoliveira@lnec.pt

[2] Estuaries and Coastal Zones Division, LNEC, Av. do Brasil,
1700-066 Lisbon, Portugal

Abstract. Coastal forecast systems provide coastal managers with accurate and timely hydrodynamic predictions, supporting multiple uses such as navigation, water monitoring, port operations and dredging activities. They are also useful to support recreational activities. Still, the widespread use of coastal forecasts is limited by the unavailability of open forecasts for consultation, the expertise needed to build operational forecast systems and the human and computational resources required to maintain them in operation every day. A new service for the generic deployment of forecast systems at user-specified locations was developed to address these limitations. Denoted OPENCoastS, this service builds circulation forecast systems for user-selected coastal areas and maintains them in operation using the European Open Science Cloud (EOSC) computational resources. OPENCoastS can be applied to any coastal region and has been in operation since 2018, forced by several regional and global forecasts of the atmospheric and ocean dynamics. It has attracted over 150 users from around 45 institutions across the globe. However, most users come from research institutions. The only requirement needed to use this service – a computational grid of the domain of interest – has proven difficult to obtain by most coastal managers. Herein, a new way to bring coastal managers and the general public to the OPENCoastS community is proposed. By creating an open, scalable and organized repository of computational grids, shared by expert coastal modelers across the globe, the benefits from the use of OPENCoastS can now be extended to all coastal actors.

Keywords: Open data repositories · Coastal forecasts · Unstructured grids

1 Introduction

Open Science is increasingly recognized as the avenue that brings the knowledge created by the scientific community and the companies to society, promoting their recognition and the socio-economic impact of research on society. The paradigm of open science has

J. M. F. Rodrigues et al. (Eds.): ICCS 2019, LNCS 11540, pp. 794–807, 2019.
https://doi.org/10.1007/978-3-030-22750-0_80

been promoted in recent years, focused mainly on data and publications. Policies and recommendations from multiple fora (e.g. https://ec.europa.eu/research/openscience/index.cfm?pg=home, https://project-open-data.cio.gov/) have been strongly pushing towards a new way to conduct research and share knowledge [1, 2]. Funding agencies, such as the H2020 program, are promoting open science, recognizing its impact on innovation and therefore on the economy and the quality of life.

While these efforts have gradually been successful and accepted by the scientific community for publication and for access to scientific data, most of the highly specialized software development has been far away from the desired openness. In the field of ocean modelling, the current trend is for community-based, open-source codes (examples of popular open models originated in Europe include DELFT3D, MOHID and TELEMAC, and in the USA, ADCIRC, SCHISM and ROMS). However, most IT water platforms that integrate these models to provide services to the community are not freely available for usage or even for improvement and reuse for other purposes. These platforms are fundamental to end-users and stakeholders for their management chores but are also important for the general public activities.

The forecast system development area is still an example of little openness, mostly due to the difficulty to operate coastal forecast systems outside research institutes infrastructures. Handing over these systems to water management authorities is a very difficult task, as expensive hardware and specialized human resources are needed to keep them in operation. Recently, the integration and availability of computing resources provided through the European Open Science Cloud (EOSC) allowed a first opportunity to open coastal forecast systems to the coastal community.

OPENCoastS, a new Web platform developed under the concept of open science, is an open service integrated in the thematic services and marketplace of the project EOSC-Hub [3, 4]. It aims at generating on-demand coastal forecast systems with minimal user intervention, and is applicable worldwide. The platform guides the user through seven simple steps towards the generation of an operational forecast system in any coastal region. The user only has to provide an unstructured grid of the study area and information on the river flow (if needed). The platform includes: (i) the definition of boundary conditions, selected from several options available, (ii) the model parameterization of the simulations and (iii) the choice of the online stations for model validation, automatically identified by the platform. The platform also includes interfaces for visualizing the results and managing the forecasts.

The open science paradigm, which was at the foundation of the OPENCoastS service, facilitated by the availability of resources in the EOSC community, is reinforced herein with a new complementary service: a computational grid repository. The availability of computational grids remains the major limitation for the uptake of this service by the coastal stakeholder community and the general public, in spite of the training efforts by the OPENCoastS development team and the availability of free grid-generation codes. The creation of a public, open repository of computational grids proposed here, implemented through organized and easy-to-access technologies, shared by the expert numerical modelers across the globe, will strengthen the accessibility of OPENCoastS to all coastal actors.

This paper is organized as follows. Section 2 provides an overview of the OPENCoastS service, along with a description of its interconnection with global ocean and atmospheric forecast providers. The grid repository is described in Sect. 3. The use of this repository is illustrated in Sect. 4, for the Algarve coast circulation prediction. The paper closes with some final considerations.

2 Overview of the OPENCoastS Platform

2.1 Goals and Main Characteristics

The OPENCoastS service builds on-demand circulation forecast systems for user-selected regions of the coast and maintains them running operationally for the time frame defined by the user. This daily service generates forecasts of water levels and depth-averaged velocities over the spatial region of interest, based on numerical simulations of all relevant physical processes. This service takes advantage of two e-infrastructures for computational and storage resources: the National Distributed Computing Infrastructure – INCD (part of the Portuguese Roadmap for Infrastructures) and IFCA (Institute of Physics of Cantabria, Spain). OPENCoastS is supported by the EGI computational resources (European Grid Initiative), through the H2020 EOSC-Hub project, being one of its thematic services (https://www.eosc-hub.eu/catalogue/OPENCoastS). Linkage to EUDAT storage services is underway.

The architecture of the OPENCoastS service includes (Fig. 1):

- the user interface component, a web-based portal;
- the computation component, where simulation results are generated and post-processed;
- the archive component, responsible for preserving all relevant data.

The service is composed generically by a frontend (the web platform) and a backend (that deals with all computing tasks). The frontend was developed with the Django Python Web framework using libraries such as: matplotlib, shapely, netCDF4 and numpy, and is supported by a PostGIS spatial object-relational database. The frontend viewer application also uses the ncWMS2 software [5] to serve the forecasts outputs, composed of UGRID-compliant NetCDF files, as Web Map Services (WMS). The OPENCoastS service is available at https://opencoasts.ncg.ingrid.pt.

The backend is responsible for generating the forecast results, handling all tasks of the simulation chain established for each forecast deployment. The simulation chain is updated and operated daily, to produce 48 h predictions, based on the previous status. Each simulation involves a suite of files that remain unchanged throughout the deployment life span, such as the computational grid supplied by the user, and the daily files, such as the forcing conditions to be applied at the boundaries (oceanic, riverine and atmospheric). The platform is strongly anchored on the Water Information Forecast Framework, WIFF [6], which simplifies the assembly and execution of the recurring tasks needed for every simulation. WIFF is a generic forecasting platform, adaptable to any geographical location, which integrates a set of numerical models that run periodically. Forecast systems typically entail computationally demanding tasks. Usually, these tasks are offloaded to computational resources that can handle them. This approach is necessary to cope with very demanding jobs, but it also introduces additional difficulties, such as the heterogeneity of the execution environments and the underprivileged access to the resources. These constraints are overcome in OPENCoastS by employing an underprivileged container technology (udocker [7]) which offers a homogeneous environment without requiring administrative permissions. Udocker allows software to be encapsulated together with all its dependencies and to be executed independently of the Linux distribution used by the host systems. This set of processes, denoted as container, is thus placed in a fully isolated environment, with a given amount of resources, such as CPU or RAM. These features are possible through the use of control groups and namespaces isolation, advanced features of modern Linux kernels.

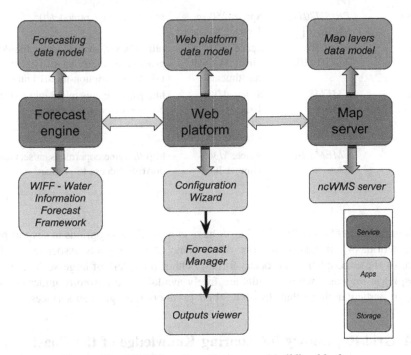

Fig. 1. The OPENCoastS architecture and building blocks.

2.2   OPENCoastS Linkage to Large-Scale Ocean and Atmosphere Forecasts and Data Sources

In order to provide local forecasts for any coastal region in the world, this service is connected to several oceanic and atmospheric forecast providers. This linkage is necessary to provide reliable and accurate predictions at local scales forced by robust systems that are integrated with global monitoring networks through data assimilation. The main providers for forcings in OPENCoastS are summarized in Table 1.

Users can select the configuration that best fits their purposes in terms of physical processes (tides, tides + storm surges, tides + storm surges), model resolution and area of coverage. Simple comparison with elevation in-situ data is provided after a user selection of data stations from EMODnet-Physics, allowing the user to download data and model results.

Table 1. OPENCoastS forcing providers.

Type of forcing	Name, provider, reference	Resolution	Web address
Atmospheric	*GFS, NOAA* [8]	Space: 0.25°, Time: 1 h	https://nomads.ncep.noaa.gov:9090/dods/gfs_0p25_1hr
Atmospheric	*ARPEGE, MétéoFrance* [9]	Space: 0.1°, Time: 1 h	https://donneespubliques.meteofrance.fr/?fond=produit&id_produit=130&id_rubrique=51
Ocean	*PRISM2017, LNEC* [6]	Space: 250 m, Time: 1 h	http://ariel.lnec.pt/node/40
Ocean	*FES2014, AVISO* [10]	Space: 1/16°, Time: 34 tidal constituents	https://www.aviso.altimetry.fr/en/data/products/auxiliary-products/global-tide-fes/description-fes2014.html
Ocean	*CMEMS Global, Copernicus* [11]	Space: 1/12°, Time: 1 h	http://marine.copernicus.eu/services-portfolio/access-to-products
Ocean	*CMEMS IBI, Copernicus* [11]	Space: 1/36°, Time: 1 h	http://marine.copernicus.eu/services-portfolio/access-to-products

Several applications of OPENCoastS have exploited these options at several places in the world. [4] discusses in more detail some of the impacts associated with the choice of the atmospheric and ocean forcing. Other providers of large-scale ocean and atmospheric forecasts whose results are freely available are currently under consideration, in particular those that distribute their products through web services.

3   A Grid Repository for Sharing Knowledge of the Coast

Over the past year, the OPENCoastS platform has been used by over 150 users, from all continents, and forecasts have been produced for over 120 different systems, with an average of 9 activated systems per month. A considerable effort was placed in training, through both hands-on courses at specialized conferences and wide-scope in-situ training in several countries, broadcasted simultaneously by web streaming and video conferencing (the course material is available at http://opencoasts.lnec.pt/index_en.php#eventos). About 90% of current users originate from the research and academic community, and only 10% of them originate from the coastal management community. The general public is not represented at all in the user community.

To further promote the uptake of the OPENCoastS service by the coastal management community as well as the general public, two strategies can be followed:

1. to provide the option of open access to specific forecast deployments
2. to provide free access to the computational grids necessary to make a deployment.

The first strategy is one of the next steps in the OPENCoastS development. It will allow users to access model results from a list of available open deployments. This option minimizes work for end-users and the general public when these shared deployments are publicized on an individual basis, but cannot be properly organized and used as part of a global repository for grid sharing. The indexing and organization of this deployment list, towards facilitating the search for a specific geographical area, is not possible as each deployment has specific metadata chosen by its owner. Also, the conditions for the forecast setup are also selected by its owner, preventing users that access these forecast's outputs to customize the outputs to their needs. It is however relevant for deployment sharing within teams.

The second strategy, introduced herein, gives all the necessary tools for a user to set up his/hers deployment based on an openly shared grid. By making a repository of computational grids available, along with the necessary metadata on the geographical reference system and vertical positioning of each grid, this solution provides full freedom while overcoming the barrier of building a finite element grid for a particular coastal area. Search mechanisms as well as geographic location are available, making the finding of the site's grid simple and fast.

In order to publish the available grids as Open Data, a grid repository was created using Github: https://github.com/LNEC-GTI/OPENCoastS-Grids (Fig. 2). Among other

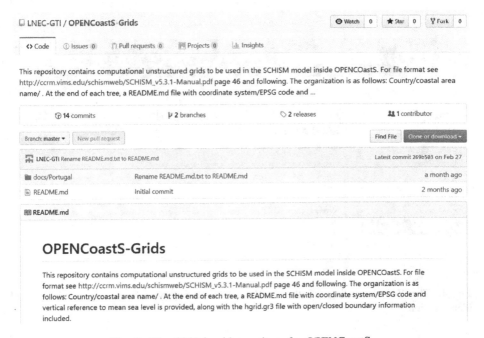

Fig. 2. The GitHub grid repository for OPENCoastS.

advantages, this approach provides a versioning system for files, allowing to keep track of the shared datasets. Within this repository a hierarchical organization, country by country, is provided, followed by the estuary/coastal area name. An example is provided in Fig. 3 for a shared grid of the Guadiana estuary in Portugal and Spain. Multiple grids for the same system can be stored in subsequent directories, as different versions or name identifiers. In order to preserve this repository, facilitate the access to the shared grids and provide them with a DOI, a link to this GitHub repository was made in Zenodo (Fig. 4). Each grid corresponds to one release within the repository, each with its own identifier (Fig. 5).

Fig. 3. The GitHub grid repository entrance for the Guadiana grid.

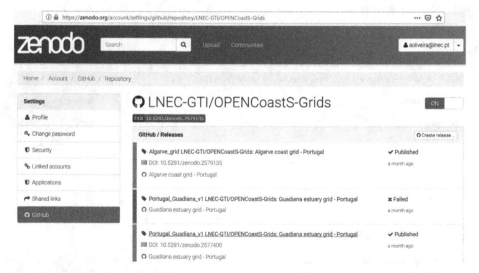

Fig. 4. The Zenodo listing of the GitHub grid repository entries, supported by grid identification.

After downloading the desired grid to their drive, the users can then set up their forecast systems using OPENCoastS at https://opencoasts.ncg.ingrid.pt. The information on the origin of the grid can be stored in OPENCoastS deployment metadata and be shared with others in the future if the user allows it. Through this solution, sharing (in a searchable, organized way) and preserving grids is accomplished in a citable way that gives credit to the authors providing the grids for public use. Moreover, this solution sets the stage for the future option within OPENCoastS to facilitate the access to predictions under a specific deployment for other people besides the deployment owner. Similarly, a bathymetry uploader can be added to the repository service to allow the update of the bathymetry over time.

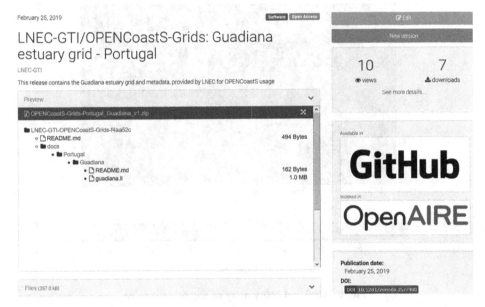

Fig. 5. Sample shared grid (for the Guadiana estuary, Portugal) available in Zenodo from the GitHub grid repository, with a unique identifier.

4   Demonstration of the OPENCoastS Sharing Service at the Algarve Coast

4.1   Grid Generation

The main input the user has to provide to the OPENCoastS service is a triangular unstructured grid. Although many grid generators are available to create this type of grids, the inexperienced user may find the generation of such grids to be a daunting task. Here, the generation of a grid is illustrated using the Algarve coast (southern Portugal) as an example. Two freeware codes are used herein to generate the grid: XMGREDIT, a semi-automatic grid generator designed for coastal applications [12]; and NICEGRID, a post-processor to automatically improve grid quality [13].

The first step in grid generation is the definition of the domain. The southern Portuguese coast is approximately 140 km long, from the mouth of the Guadiana estuary to the East to the Cape of St. Vincent to the West (Fig. 6). To the East, the domain was extended along the Spanish coast up to an area where the coastline is approximately straight; to the West, the domain was cut a few kilometers before Cape of St. Vincent. These choices aimed at minimizing potential numerical problems. To the South, the boundary was defined as a circular arch with an 80 km radius. The choice of a circular open boundary also aims at minimizing numerical problems due to discontinuities.

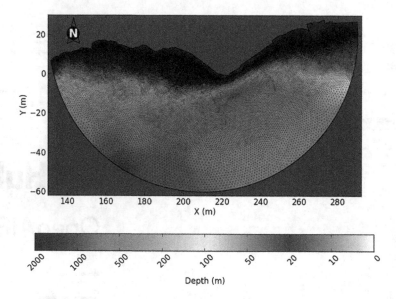

Fig. 6. Grid and bathymetry of the Algarve coast.

Nodes were placed automatically using XMGREDIT. The grid spacing was set to 250 m between the coastline and the 50 m isobath, 500 m at the 100 m isobath, 2000 m at the open boundary, and followed a smooth variation in between. Once the set of nodes were defined, they were triangulated (i.e., triangular elements were defined based on those nodes) to provide a preliminary grid. Although this preliminary grid has the desired grid spacing, many elements are highly skewed. The grid was then automatically improved using NICEGRID. This code reduces skewness by moving, adding and deleting nodes, targeting a smooth transition between element sizes and quasi-equilateral triangles. The final grid has about 50 thousand nodes (Fig. 6), and its generation took about 2–3 h.

4.2   Sharing the Grid Through the GitHub/Zenodo Repository

Using the procedure described above, this grid was made available in GitHub (Fig. 7) with the corresponding metadata in the README.md file. A link was created in Zenodo, through a release in GitHub (Fig. 8), providing a DOI (https://doi.org/10. 5281/zenodo.2579135). This grid can be downloaded to a local disk, and then uploaded in the OPENCoastS configuration assistant as illustrated below.

Fig. 7. Grid of the Algarve coast integrated in the repository.

4.3   Using OPENCoastS to Build a New Forecast Deployment from a Shared Grid

A new forecast system was then implemented for the Algarve coast using the OPENCoastS service. To deploy this system, the following steps were performed in the Configuration Assistant: (1) the model selection, (2) the input of the horizontal grid (Fig. 9); (3) the definition of the boundary conditions (Fig. 10), (4) the definition of stations; (5) the definition of input parameters (e.g. time step); and (6) the definition of spatially-varying parameters (e.g. friction coefficient). After the completion of these steps, a summary of the deployment is presented and the user can submit the forecast. The Algarve coast forecast deployed herein is forced by tides from FES2014 at the ocean boundary and by the atmospheric forecasts from ARPEGE at the surface. Two time steps were tested: the one suggested within the OPENCoastS application (360 s) and a user-defined time step (60 s). Several virtual stations were selected along the Algarve coast (Fig. 11), providing information about the water levels and velocities.

Forecast results are available from the Outputs Viewer, where the user can generate maps and time series of water levels and velocities (Fig. 11). The user can also download the model forecasts for comparison with data from other sources. A comparison between the data from the Lagos tidal gauge (ftp://ftp.dgterritorio.pt/Maregrafos/Lagos) is presented, illustrating the good quality of OPENCoastS predictions (Fig. 12). Both forecasts provided feasible results but further data (in particular velocities data) would be required to assess which setup (regarding the time step) provides more accurate results.

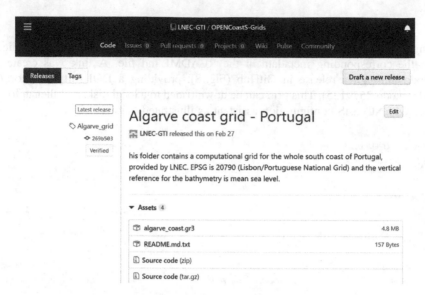

Fig. 8. Release in GitHub, allowing for the linkage with Zenodo.

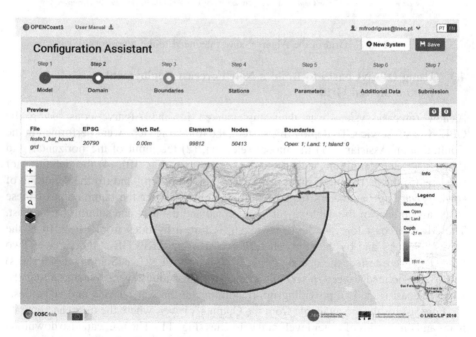

Fig. 9. Configuration Assistant - Step 2, domain definition.

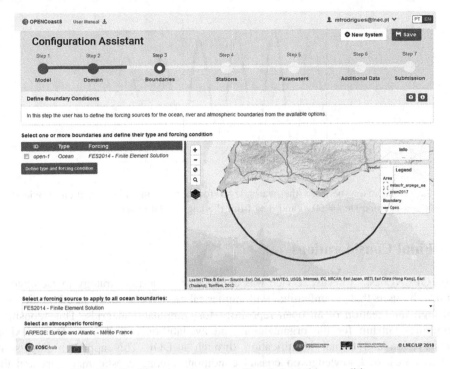

Fig. 10. Configuration Assistant - Step 3, boundary conditions.

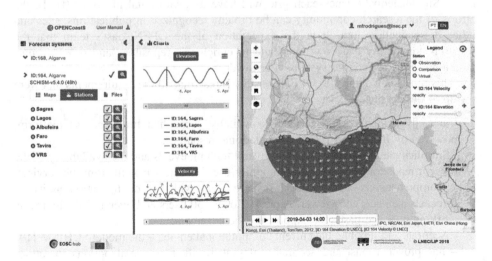

Fig. 11. Outputs Viewer, with map and time series of the selected stations views.

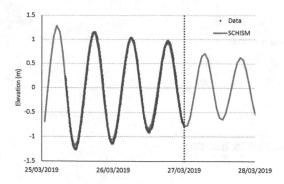

Fig. 12. Comparison between the water level forecasts at the Lagos station provided by OPENCoastS (time step = 60 s) and data from the Lagos tidal gauge.

5   Final Considerations

The OPENCoastS forecast service was enhanced with a new strategy to facilitate its uptake by users that are unfamiliar with numerical models but need coastal predictions, through the creation of an open repository for computational grids. This repository combines GitHub for file organization and availability, and Zenodo for indexing, preservation and digital identification through a DOI. This approach targets an enhancement of knowledge on coastal conditions among coastal managers and the general public, and the sharing of tools from the research community to the society at large. Simultaneously, since each grid will have its own digital identifier (DOI), the efforts of the research community can be broadly recognized in publications, sites and other dissemination fora. It is the expectation of the OPENCoastS team that the research community can now contribute to enrich this repository and provide computational grids for coastal systems worldwide.

This repository completes OPENCoastS v1, reaching all the requirements that were set up at design stage [3, 4]:

1 - broadly available: this service is available through a broad access platform, powered by INCD and IFCA computational resources;
2 - Simplicity and usability: through several training events and the availability of the present repository, all coastal community users can now benefit from the service;
3 - Comprehensive: all forecast tasks are dealt with the configuration assistant, forecasts manager and viewer. A pool of over 150 users and over 120 deployments working over the past year support this achievement;
4 - Accurate and flexible: the current computational engine – the model SCHISM [14] – has proven robust and accurate, integrated with the several options for atmospheric and ocean forcings and for input customizations;
5 - Modular: OPENCoastS was built in a modular way, to provide support to grow from the current 2DH barotropic physics to other processes.

OPENCoastS is being extended to other processes. In the near future, both 3D baroclinic and coupled wave-current simulations will be offered to the users. In the long run, extensions to water quality and morphodynamic problems are also envisioned.

Acknowledgements. This work was partially funded by EC H2020 project EOSC-hub (Grant Agreement No 777536), by Lisboa2020 Operational Program through the INCD project (LISBOA-01-0145-FEDER-022153) and by FCT, UBEST project (PTDC/AAG-MAA/6899/2014).

References

1. European Commission: Cost Benefit Analysis for FAIR research data (2018). https://publications.europa.eu/en/publication-detail/-/publication/d3766478-1a09-11e9-8d04-01aa75ed71a1/language-en. Accessed 02 Apr 2019
2. Burwell, S.M., VanRoekel, S., Park, T., Mancini, D.J.: M-13-13 — Memorandum for the Heads of Executive Departments and Agencies (2019). https://project-open-data.cio.gov/policy-memo/. Accessed 02 Apr 2019
3. Fortunato, A.B., et al.: Generation of operational forecasts on demand: the OPENCoastS platform. In: 17th IMUM Book of Abstracts (2018). http://imum2018.mpimet.mpg.de/fileadmin/user_upload/imum2018/template/img/Book_of_Abstracts_revised.pdf
4. Oliveira, A., et al.: A platform for the automatic generation of coastal forecast systems: the OPENCoastS service. J. Oper. Ocean. (in preparation)
5. Blower, J.D., Gemmell, A.L., Griffiths, G.H., Haines, K., Santokhee, A., Yang, X.: A Web Map Service implementation for the visualization of multidimensional gridded environmental data. Environ. Model Softw. **47**, 218–224 (2013)
6. Fortunato, A.B., et al.: Operational forecast framework applied to extreme sea levels at regional and local scales. J. Oper. Ocean. **10**(1), 1–15 (2017)
7. Gomes, J., et al.: Enabling rootless Linux Containers in multi-user environments: the udocker tool. Comput. Phys. Commun. **232**, 84–97 (2018)
8. NOAA data page. https://www.ncdc.noaa.gov/data-access. Accessed 02 Apr 2019
9. Arpege home page. https://donneespubliques.meteofrance.fr/?fond=produit&id_produit=130&id_rubrique=51. Accessed 02 Apr 2019
10. FES2014 description. http://www.egsiem.eu/images/static/ProjMtg1/Annex06_WP2.pdf. Accessed 02 Apr 2019
11. CMEMS home page. http://marine.copernicus.eu/services-portfolio/access-to-products/. Accessed 02 Apr 2019
12. Turner, P., Baptista, A.M.: ACE/gredit User's Manual. Software for Semi-automatic Generation of Two-Dimensional Finite Element Grids. Center for Coastal and Land-Margin Research, Oregon Graduate Institute of Science & Technology (1993)
13. Fortunato, A.B., Bruneau, N., Azevedo, A., Araújo, M.A.V.C., Oliveira, A.: Automatic improvement of unstructured grids for coastal simulations. J. Coast. Res. Spec. Issue **64**, 1028–1032 (2011)
14. Zhang, Y.J., Ye, F., Stanev, E.V., Grashorn, S.: Seamless cross-scale modeling with schism. Ocean Model. **102**, 64–81 (2016)

Author Index

Abdallah, Sawssen Ben 528
Abdildin, Yerkin G. 282
Aguilar-Rivera, Anton 223
Ajmi, Faiza 528
Alekseev, A. K. 315
Álvarez-García, Juan Antonio 15
Angello, Aaron 389
Anitescu, Mihai 251
Antão, Samuel 269
Araar, Oualid 137
Arbore, Andrea 84
Arslan, Muhammad 787
Arzoglou, Ektor 499
Avramenko, Viktor V. 293
Azevedo, Alberto 794

Baczynski, Jack 710
Belousov, Konstantin 110
Ben Othman, Sarah 528
Benkner, Siegfried 514
Benmerzoug, Djamel 573
Beydoun, Ghassan 237
Bezati, Endri 717
Bischof, Christian 535
Bloch, Aurelien 717
Böck, Carina Stefoni 688
Boeva, Veselka 779
Boldrini, Luca 445
Bolgova, Ekaterina 674
Bolten, Tobias 164
Bondarev, A. E. 315, 325
Bondarev, Egor 125
Boubeta-Puig, Juan 559
Bouhired, Saadi 137
Bücker, H. Martin 363
Burger, Michael 535

Cabaleiro, José Carlos 483
Cai, Ningning 3
Cai, Xing 681
Campion, Corey 389
Caparrini, Fernando Sancho 15
Cardoso, Pedro J. S. 55
Carreño, Amanda 702

Carrillo, C. 521
Cartwright, Eoin 771
Castelli, Mauro 656
Cellini, Francesco 445
Chien, Steven W. D. 612
Child, Chris 543
Chinnici, Marta 84
Chli, Maria 749
Chrpa, Lukáš 453
Chung, Yongmann M. 468
Chuprina, Svetlana 27, 110
Conopoima, Cesar Augusto 627
Cortés, A. 521
Costa, Nuno 70
Coti, Camille 566
Crane, Martin 771
Crum Jr., William B. 389
Cruz, Christophe 787
Cui, Zhenyu 664

da Silva Bragança, João Felipe 710
da Silva, Allan Jonathan 710
Damiani, Andrea 445
Danielak, Brian 374
de Freitas Assad, Luiz Paulo 688
de With, Peter H. N. 125
Deniz, Oscar 15
Díaz, Gregorio 559
Dokulil, Jiri 514
Dondi, Riccardo 656
Dos Santos, Rodrigo Weber 627
Dourado Loula, Abimael Fernando 627
Drira, Khalil 573

El-Amin, Mohamed F. 437
Elisseev, Vadim 269
Elo, Tommi 499
Enríquez, Fernando 15
Espinosa, A. 521

Fernandes, Hortênsio C. L. 55
Fernandez Llatas, Carlos 445
Fioriti, Vincenzo 84

Fortunato, André B.   794
Franco, Santiago   491
Fulginei, Francesco Riganti   581
Furuichi, Mikito   476

García-Barriocanal, Elena   408
García-Martínez, Arturo   303
Garg, Deepeka   749
Gatta, Roberto   445
Ghahremani, Amir   125
Gimeno, Jesús   97
Ginestar, Damian   702
Ginhac, Dominique   787
Golubchikov, Egor   461
Golubev, Kirill   674
Grabowski, Michał   764
Grahn, Håkan   779
Groen, Derek   725
Gu, Jingzi   41
Guedes, Anne Goni   688
Guermouche, Nawal   573
Guerreiro, Pedro M. M.   55
Guseva, Ekaterina   461

Hafstein, Sigurdur   421
Hamfelt, Andreas   334
Hammadi, Slim   528
Han, Jizhong   207, 507
He, Yang-Hui   543
Hernandez-Gress, Neil   346
Hervert-Escobar, Laura   346
Hosono, Natsuki   476
Hu, Songlin   207, 507
Huang, Longtao   207
Hvatov, Alexander   635

Iamsiri, Saowanee   149
Igreja, Iury   627

Jarvis, Chad   681
Jastrząb, Tomasz   741
Johnson, David   334

Kageyama, Akira   588
Kalashnikov, Viacheslav   293, 303
Kalashnykova, Nataliya   293, 303
Kalyuzhnaya, Anna   635
Kapturczak, Marta   261
Karsakov, Andrey   674

Kesarev, Sergey   674
Kiev, Vladislav   461
Kitchin, Diane   453
Kong, Yitian   125
Kou, Jisheng   437
Kovalchuk, Sergey   674
Krämer, Juliane   535
Krupiński, Robert   177
Kurzyk, Dariusz   764
Kuvshinnikov, A. E.   315
Kuzin, Alexey   461

Laggoune, Ali   137
Landau, Luiz   688
Langguth, Johannes   681
Laso, Ruben   483
Laudani, Antonino   604
Laure, Erwin   612
Leal-Coronado, Mariel A.   303
Lech, Piotr   177
Lee, Young Choon   756
Leng, Jiaxu   664
Lenkowicz, Jacopo   445
León, Arabella   733
Levada, Alexandre L. M.   191
Li, Bo   41
Li, Jiacheng   507
Li, Jingfa   437
Lin, Zheng   41
Lindsay, Alan   491
Lines, Glenn Terje   681
Liu, Tingwen   551
Liu, Xinlian   389
Liu, Ying   664
Lopes, Pedro   794
Lozito, Gabriele Maria   581, 604
Lucaferri, Valentina   581, 604

Ma, Can   3
Macià, Hermenegilda   559
Maldonado, Daniel Adrian   251
Manzoni, Sara   656
Margalef, T.   521
Markidis, Stefano   612
Martins Rocha, Bernardo   627
Marumo, Koichi   619
Masciocchi, Carlotta   445
Maslyaev, Mikhail   635
Mattavelli, Marco   717

Mauri, Giancarlo 656
Maximiano, Marisa 70
McCluskey, Thomas Lee 491
McKeever, Steve 334
Meng, Dan 3, 41
Metsker, Oleg 674
Mizuno, Norihito 468
Monniaux, David 566
Mora-Cantallops, Marçal 408
Morita, Ryuta 619
Moussiou, Sami 137

Nakajima, Kengo 681
Niemeyer, Kyle E. 401
Nikander, Pekka 499
Nordahl, Christian 779
Nsimba, Cédrick Bamba 191

O'Shea, Brian 374
Okarma, Krzysztof 177
Oliveira, Anabela 794
Olshevsky, Vyacheslav 612
Osudin, Daniil 543

Paçacı, Görkem 334
Parodi, Mauro 581
Paulo, Tiago 70
Peña, Fabian C. 237
Perera, Thilina 595
Pérez, Manolo 733
Persson Netz, Marie 779
Pessoa, Maria Eduarda 688
Phung, James 756
Pileggi, Salvatore F. 237
Pohle-Fröhlich, Regina 164
Portalés, Cristina 97, 733
Postanogov, Igor 27
Prakash, Alok 595
Przewięźlikowski, Marcin 764
Pyatlin, Andrey 461

Queiroz, Dyellen Soares 688
Quintiliano Bezerra Silva, Andre 642

Radicioni, Martina 581, 604
Ramirez-Velarde, Raul 346
Reis, Catarina I. 70
Reshetnikov, Vyacheslav 461
Rivera, Francisco F. 483

Rodrigues, Marta 794
Rogeiro, João 794
Rosa, Iolanda 70
Rostami, M. Ali 363
Ruskin, Heather J. 771
Ryabinin, Konstantin 110
Rycerz, Katarzyna 764

Salvini, Alessandro 581
Sánchez-Alonso, Salvador 408
Sanevas, Nuttha 149
Santos, Ivo 70
Sawko, Robert 269
Schanen, Michel 251
Sebastián, Jorge 97
Sevilla, Javier 97, 733
Shabrov, Nikolay 461
Shi, Jinqiao 551
Sicilia, Miguel-Angel 408
Silvia, Devin 374
Sishtla, Chaitanya Prasad 612
Skillen, Alex 269
Soria, Luis Miguel 15
Srikanthan, Thambipillai 595
Suleimenova, Diana 725
Sun, Shuyu 437

Tecław, Mateusz 177
Teixeira, Joana 794
Tönnies, Klaus D. 164
Toste, Raquel 688

Valentini, Vincenzo 445
Valero, Edelmira 559
Valero, Valentín 559
Vallati, Mauro 445, 453, 491
Vallez, Noelia 15
Vanhille Campos, Christian 725
Velasco, Francisco 15
Verdú, Gumersindo 702
Vermandel, Sébastien 528
Vidal-Ferràndiz, Antoni 702
Vieira, Amanda 70
Villamil, Maria Del Pilar 237
Vogiatzis, George 749

Wang, Weiping 3, 41
Watcharopas, Chakrit 149
Wattuya, Pakaket 149

Xie, Weihong   664

Yakovlev, Alexey   674
Yamagiwa, Shinichi   619
Yamakawa, Masashi   468
Yamamoto, Kohei   588
Yu, Hang   566

Zang, Liangjun   207
Zarour, Karim   573
Zhang, JinChao   41

Zhang, Liang   664
Zhang, Tianlin   664
Zhao, Jiapeng   551
Zhao, Lin   207
Zhou, Wei   507
Zhuravlev, Alexey   461
Zieniuk, Eugeniusz   261
Zimoń, Małgorzata J.   269
Zomaya, Albert Y.   756
Zoppis, Italo   656

Printed in the United States
By Bookmasters